Atmospheric Ozone Variability

Implications for Climate Change, Human Health and Ecosystems

Springer
London
Berlin
Heidelberg
New York
Barcelona
Hong Kong
Milan
Paris
Santa Clara
Singapore
Tokyo

Kirill Ya. Kondratyev and Costas A. Varotsos

Atmospheric Ozone Variability

Implications for Climate Change, Human Health and Ecosystems

Springer

Published in association with
Praxis Publishing
Chichester, UK

PRAXIS

Professor K. Ya. Kondratyev, Counsellor of the Russian Academy of Sciences, Scientific Research Centre for Ecological Safety, St Petersburg, Russia

Dr C. A. Varotsos, Associate Professor, Department of Physics, Division of Applied Physics, University of Athens, Greece

---

SPRINGER–PRAXIS BOOKS IN ENVIRONMENTAL SCIENCES
SUBJECT *ADVISORY EDITOR*: John Mason B.Sc., Ph.D.

ISBN 1-85233-635-8 Springer-Verlag Berlin Heidelberg New York

British Library Cataloguing in Publication Data
Kondratyev, K. Ya. (Kirill IAkovlevich)
Atmospheric ozone variability : implications for climate change, human health and ecosystems. – (Springer-Praxis books in environmental sciences)
1. Atmospheric ozone – Reduction – Physiological aspects
2. Atmospheric ozone – Reduction – Environmental aspects
3. Atmospheric ozone – Reduction – Health aspects
I. Title II. Varotsos, Costas, 1956–
363.7′3875

Library of Congress Cataloging-in-Publication Data
Kondrat′ev, K. IA. (Kirill IAkovlevich)
Atmospheric ozone variability : implications for climate change, human health, and ecosystems / Kirill Ya. Kondratyev and Costas Varotsos.
p. cm. – (Springer-Praxis series in environmental sciences)
Includes bibliographical references.
1. Atmospheric ozone – Environmental aspects. 2. Climatic changes. 3. Public health. 4. Ecosystem management.
I. Varotsos, Costas, 1956– II. Title. III. Series
QC879.7.K67 2000
363.738′75 – dc21 99-086290

Printed by MPG Books Ltd, Bodmin, Cornwall, UK

Cover design: Jim Wilkie
Typesetting: Originator, Gt. Yarmouth, Norfolk, UK

Printed on acid-free paper supplied by Precision Publishing Papers Ltd, UK

Preface

Since the discovery of the ozone layer in the stratosphere and recognition of the fact that life on the Earth's surface is possible only because of the absorption of shortwave ultraviolet radiation by stratospheric ozone, problems of atmospheric ozone have acquired outstanding significance and attention in both scientific literature and in the mass media. Such attention has enormously increased after the discovery of the 'ozone hole' (spring minimum of total column ozone – TOZ) in Antarctica, which has been explained as mainly the result of halocarbon emissions into the atmosphere. The subject of anthropogenic impact on the global ozone layer is recognized as a top-priority environmental problem of similar importance as that of global climate change (the coupled nature of these two global problems is very significant).

The principal purpose of this book is to discuss basic aspects of ozone variability in the atmosphere (including both stratospheric and tropospheric ozone) as well as implications of ozone variability for climate change, human health and the biosphere. A brief survey of the totality of these problems can be found in Chapter 1 (the recent WMO/UNEP *Scientific Assessment of Ozone Depletion: 1998* contains a detailed consideration of many ozone problems).

Chapter 2 dwells upon subjects whose significance have obviously been underestimated earlier: measurement techniques (development for ground-based, balloon, aircraft and satellite observations), calibration problems, relevant intercomparisons of instrumentation and estimates of measurement errors. Some most important results of TOZ and ozone vertical-profile observations are also discussed. The results of a number of outstanding field observational programmes are analysed.

The aim of Chapter 3 is to consider stratospheric ozone-formation processes with special emphasis on interaction between atmospheric photochemistry and dynamics. Both observations and numerical modelling results are discussed (special attention is paid to some contemporary three-dimensional transport/photochemistry global ozone-dynamics models). Chapter 3 considers problems such as polar ozone dynamics, volcanic eruptions and ozone itself, the impact of aviation on the ozone layer, solar activity influence on global ozone dynamics.

The problem of surface ultraviolet radiation, its variations in dependence on TOZ change and the relevant impact on aquatic and terrestrial ecosystems as well as human and animal health is the subject of Chapter 4.

Chapter 5 is devoted to tropospheric ozone (TrOZ) variability and its dependence on various precursors of TrOZ with an emphasis on gas components and aerosols of anthropogenic origin. Special attention has been paid to TrOZ precursors of biogenic origin produced mainly because of emissions by vegetation and oceans. Numerical modelling of TrOZ variability has also been considered in detail.

Discussion of the interaction between global climate and ozone change is the principal aim of Chapter 6. After a detailed introduction the problem of radiative forcing (RF) due to various gas and aerosol components is considered. But, on the other hand, some limitations relevant to the use of RF are analysed. Stratospheric ozone is an ultraviolet absorber of solar radiation, while tropospheric ozone functions mainly as a greenhouse gas. This results in different impacts of TOZ and TrOZ on ozone-induced climate change, which are discussed in Chapter 6.

KIRILL YA. KONDRATYEV and COSTAS A. VAROTSOS

Contents

List of abbreviations

α-T	Albedo Temperature
ABL	Atmosphere Boundary Layer
ABLE	Arctic Boundary Layer Expedition
ACE	Aerosol Characterization Experiment
ADEOS	Advanced Earth Observing Satellite
AEC	Aerosol Extinction Coefficient
AEM-2	Applications Explorer Mission-2 (satellite)
AES	Atmospheric Environment Service
AF	Adjusted Forcing
AGCM	Atmospheric General Circulation Model
ALA	Aerosol Limb Absorption
ALE	Atmospheric Lifetime Experiment
ALOMAR	Arctic Lidar Observatory for Mid-Atmospheric Research
AMAS	Advanced Millimetre-wave Atmosphere Sounder
AMF	Airmass Factor
AMS	American Meteorological Society
AMSR	Advanced Microwave Radiometer
AOD	Aerosol Optical Depth
APARE	East Asia/North Pacific Area Regional Experiment
ARC-TOC	Arctic Troposphere O_3 Chemistry
ARGOS	Advanced Remote Gaseous Oxides Sensor
ASHOE	Airborne Southern Hemisphere Ozone Experiment
ASUR	Airborne Submillimeter Receiver
AT	Albedo/Temperature Feedback
ATLAS	Atmospheric Laboratory for Applications and Science
ATMOS	Atmospheric Trace Molecule Spectroscopy
AUSPEX	Atmospheric Utility Signatures, Predictions and Experiments

AVHRR	Advanced Very High Resolution Radiometer
BAD	Biologically Accumulated Dosage
BAPoMON	Background Air Pollution Monitoring Network
BCC	Basal Cell Carcinoma
BIBEX	Biomass Burning Experiment: Impact on the Atmosphere and Biosphere
BIC	Balloon Intercomparison Campaign
BLISS	Balloonborne Laser In-Situ Sensor
BM	Brewer–Mast (electrochemical cell ozonesonde)
BMLS	Balloon Microwave Limb Sounder
BOREAS	Boreal Ecosystem–Atmosphere Study
BUV	Backscattered Ultraviolet
CACTUS	Chemistry, Aerosols and Climate Tropospheric Unified Simulation
CAO	Central Aerological Observatory
CB4	Carbon Bond 4 Mechanism
CCD	Convective Cloud Differential
CCM	Climate Community Model
CCN	Cloud Condensation Nuclei
CEOS	Committee on Earth Observation Satellites
CF	Cloud/Aerosol Attenuation Factor
CFC	Chlorofluorocarbons
chl	chlorophyll
CHMI	Czech Hydrometeorological Institute
CHS	Contact Hypersensitivity
CIMS	Chemical Ionization Mass Spectrometer
CIRRIS	Cryogenic Infrared Radiance Instrument for Shuttle
CLAES	Cryogenic Limb Array Electron Spectrometer
CNRS	Centre National de la Recherche Spatiale
COL	Cut Off Low
COS	Carbonylsulphide
CPV	Circumpolar Vortex
CRISTA	Cryogenic Infrared Spectrometer and Telescopes for the Atmosphere
CTM	Chemical Transport Model; Chemical Tracer Model
CTMK	Chemistry Transport Model KNMI
CZCS	Coastal Zone Color Scanner
DAAC	Distributed Active Center
DCC	Dobson Calibration Centre
DIAL	Differential Absorption Laser; Differential Absorption Lidar
DIPR	Daily-integrated Photodissociation Rate

DKRZ	Deutsches Klimarechenzentrum
DLR	German Air and Space Research Organization
DMOC	Daily Maximum Ozone Concentration
DMS	Dimethylsulphide
DOAS	Differential Optical Absorption Spectroscopy
doc	dissolved organics
DOM	Dissolved Organic Matter
DTH	Delayed-type Hypersensitivity
DU	Dobson Unit
EASOE	European Arctic Stratospheric Ozone Experiment
ECC	Electrochemical Concentration Cells (ozonesonde)
ECHAM	European Centre Hamburg Chemistry–General Circulation Model
ECMWF	European Centre for Medium-range Weather Forecasts
ECOC	Electrochemical Ozone Sensor
EER	Erythemally Weighted Irradiance
EI	Emission Index
EMEFS	Eulerian Model Evaluation Field Study
EMF	Electromotive Force
EN-SCI	Environmental Science Corporation
ENSO	El Niño/Southern Oscillation
EOF	Empirical Orthogonal Functions
EOS CHEM	Earth Observing System
EPOP	ESA Polar Platform
ERBE	Earth Radiation Budget Experiment
ERBS	Earth Radiation Budget Satellite
ESL	Equivalent Solar Latitude
ESMOS	European Stratospheric Monitoring Stations
ESRANGE	ESRO Sounding Rocket Launching Range
EUROTRAC	European Experiment on Transport and Transformation of Environmentally Relevant Trace Constituents
EUV	Erythemal UV
EXOS-C	(satellite)
EXPRESSO	Experiment for Regional Sources and Sinks of Oxidants
FB	Feedback
FDH	Fixed Dynamical Heating
FFT	Fast Fourier Transform
FOZAN	Fast Ozone Analyser
FP	Filter Probes
FTIR	Fourier Transform Infrared
FTS	Fourier-Transform Spectrometer

FURM	Full Retrieval Method
GAA	Greater Athens Area
GAGE	Global Atmospheric Gases Experiment
GAMETAG	Global Atmospheric Measurements Experiment on Tropospheric Aerosols and Gases
GASP	Global Air Sampling Programme
GAW	Global Atmospheric Watch
GCM	General-Circulation Models
GCMAM	Global Climate/Middle Atmosphere Model
GCR	Galactic Cosmic Rays
GCTM	Global Chemistry/Transport Model
GEIA	Inventory of Global Emissions
GHG	Greenhouse Gas
GISS	Goddard Institute for Space Studies
GKSS	German Scientific Centre
GO_3OS	WMO Global Ozone Observing System
GOME	Global Ozone Monitoring Experiment
GOMETRAN	Global Ozone Monitoring Experiment Transfer Code
GOMOS	Global Ozone Monitoring by Occultation of Stars
GOMR	Global Ozone Monitoring Radiometer
GRIPS	GCM Reality Intercomparison Project for SPARC
GROMOS	Ground-based Millimeter-wave Ozone Spectrometer
GSFC	Goddard Space Flight Center
GTE	Global Tropospheric Experiment
GTOP	Global Tropospheric Ozone Project
GUV	Ground-based Ultraviolet Radiometer
GW	Gravitational Waves
HADAM	Hadley Centre Atmospheric General-Circulation Model
HALOE	Halogen Occultation Experiment
HARW	2-D Global Model
HCFC	Hydrochlorofluorocarbons
HFC	Hydrofluorocarbons
HIRDLS	High Resolution Dynamics Limb Scanner
HIRS	High Resolution Infrared Sounder
HMS	Hydroxymethane sulphonate
HSCT	High-speed Civil Transport
IAGA	International Association of Geomagnetism and Aeronomy
IAM	Interface Adapter Model
IAMAS	International Association of Meteorology and Atmospheric Sciences
IAV	Interannual Variability

IGAC	International Global Atmospheric Chemistry
IGBP	International Geosphere–Biosphere Programme
IGS	Infrared Grating Spectrometer
ILAS	Improved Limb Atmospheric Spectrometer
IMAGES	Intermediate Model of the Annual and Global Evolution of Species
IMG	Interferometric Monitor for Greenhouse Gases
IOC	International Ozone Commission
IPCC	Intergovernmental Panel on Climate Change
IPD	Immediate Pigment Darkening
IRHS	Infrared Heterodyne Spectroscopy
IRIS	Michelson Interferometer
ISAMS	Improved Stratospheric and Mesospheric Sounder
ISCCP	International Satellite Cloud Climatology Project
ISN	Isoprene nitrates
ISO	Isoprene
ITCZ	Intertropical Convergence Zone
JGOFS	Joint Global Ocean Flux Study
JPL	Jet Propulsion Laboratory
KI	Potassium Iodide (method for ozone measurement)
KNMI	Royal Meteorological Institute (The Netherlands)
LA	Laboratoire d'Aerologie
LAS	Limb–Atmospheric Infrared Spectrometer
LCO	Light Climatic Observatory
LFB	Lightning–Ozone Feedback
LHS	Laser Heterodyne Spectrometer
LIDAR (lidar)	Light Detection and Ranging
LIMS	Limb Infrared Monitor of the Stratosphere
LOICZ	Land–Ocean Interactions in the Coastal Zone
LORE	Limb Ozone Retrieval Experiment
LRIR	Limb Radiance Inversion Radiometer
LS	Lower Stratosphere
LSC	Lead Service Center (e.g. NASA-LSC)
LTE	Local Thermodynamic Equilibrium
LW	Longwave
M-83	Filter Ozonometer
MACR	Methylacrolein
MAESA	Measurements of Atmospheric Effects of Stratospheric Aircraft
MAHRSI	Middle Atmosphere High Resolution Spectrographic Investigation
MAPS	Measurement of Air Pollution from Satellites

MAS	Millimetre-wave Atmospheric Sounder
MASP	Multiangle Aerosol Spectrometer Probe
MASTER	Mm-wave Acquisitions for Stratosphere–Troposphere Exchange Research
MBO	Methylbutenol
MCF	Methylchloroform
MCM	Master Chemical Mechanism
MCO_3	Peroxymethacrylic radical
MCT	Mesoscale Chemistry Transport
MD	Ozone-Mass Deficiency
MED	Minimum Erythema Dose
MEDCAPHOT	Mediterranean Campaign of Photochemical Tracers
MFRSR	Multi-filter Rotating Shadowland Radiometer
MFT	Middle Free Troposphere
MGC	Minor Gaseous Components
MICROTOPS	Total Ozone Portable Spectrometer
MIPAS	Michelson Interferometer for Passive Atmospheric Sounding
ML	Mixed Layer
MLOPEX	Mauna Loa Observatory Photochemistry Experiment
MLS	Microwave Limb Sounder
MOGUNTIA	Model of the Global Universal Tracer Transport in the Atmosphere
MOHp	Meteorological Observatory of Hohenpeissenburg
MOZAIC	Measurements of Ozone by Airbus In-service Aircraft
MOZART	Model for Ozone and Related Chemical Tracers
MPAN	Peroxymethacrylic nitrate
MR	Microwave Radiometer
MRI	Meteorological Research Institute (Japan)
MSD	Mean-Square Deviation
MSU	Microwave Sounding Unit
MT	Mid-troposphere
MVIRI	Meteosat Visible and IR Instrument (Scan Radiometer)
MVK	Methylvinyl ketone
NARE	North Atlantic Regional Experiment
NASDA	National Aeronautic and Space Development Administration
NAT	Nitric Acid Trihydrate
NCAR	National Center for Atmospheric Research
NCEP	National Centers for Environmental Prediction
NCSR	National Center for Scientific Research
NDSC	Network for the Detection of Stratospheric Change
NFB	No Feedback

NHGOS	Northern Hemisphere Ground Ozone Stations
NILU	Norwegian Institute for Air Research
NIRS	Near-infrared Spectrometer
NIWA	National Institute of Water and Atmospheric Research (New Zealand)
NMC	National Meteorological Center
NMHC	Non-methane Hydrocarbons
NOAA	National Oceanic and Atmospheric Administration
NPOESS	National Polar Orbiting Environmental Satellite System
NPOP	NASA's Polar Orbiting Platform
NT	Normalized Transmission
NTF	No Tropospheric Feedback
NWP	Numerical Weather Prediction
O_3s	Ozone originating in the stratosphere
O_3t	Ozone originating in the troposphere
OCS	Carbonylsulphide
ODIN	(satellite)
ODS	Ozone Depleting Substances
ODS	Ozone-depleting Substance
ODUS	Ozone Dynamics Ultraviolet Spectrometer
OGOS	Operational Ground-based Ozone Sensor
OHP	Observatoire d'Haute Provence
OLME	Ozone Layer Monitoring Experiment
OMI	Ozone Monitoring Instrument
ONIT	Organic nitrate
OOAM	Orbiting Ozone and Aerosol Monitor
OPE	Ozone Production Efficiency
ORNI	Organic Nitrates
OSN	Ozone Sounding Network
OSORT	Ozone and Solar Radiation Research Team
OVP	Ozone Vertical Concentration Profiles; Ozone Vertical Profile
OZAFS	Ozone and Aerosol Fine Structure
PAN	Peroxyacetyl nitrate
PAR	Photosynthetically Available Solar Irradiance; Photosynthetically Active Radiation
PAUR	Photochemical Activity and Solar Ultraviolet Radiation
PBL	Planetary Boundary Layer
PDB	Pee Dee Belemnite (calcite), V-PDB: Vienna ...
PEM	Pacific Exploratory Mission
PERPA	National Network for Athenian Air Pollution
PIRAMHYD	Passive Infrared Atmospheric Measurements of Hydroxyl

PMN	Peroxymethacryloyl nitrates
POAM	Polar Ozone and Aerosol Monitor; Polar Ozone and Aerosol Measurement
PPN	Peroxyacetyl nitrates from RCHO
PRIRODA	Module
PSC	Polar Stratospheric Clouds
PSS	Photochemically Stable State
PTFE	Polytetrafluoroethylene
PV	Potential Velocity; Potential Vorticity
PVU	Potential Vorticity Units
QBO	Quasi-biennial Oscillation
RADM	Regional Acid Deposition Model
RAF	Radiation Amplification Factor
RAFT	Radiation Field in the Troposphere
RAM	Radiometer for Atmospheric Measurements
RB	Robertson–Berger
RF	Radiative Forcing
RFD	Radiative Flux Divergence
RIS	Retro-reflector in Space
ROCOZ	Rocket Ozonesonde
ROSE	Rural Oxidants in the Southern Environment
RS	Rayleigh Scattering
RSAS	Rayleigh Scattering Altitude Sensor
SABE	Solar Absorption Balloon Experiment
SAFARI	Southern Africa Fire–Atmosphere Research Initiative
SAFIRE	Spectroscopy of the Atmosphere using Far Infrared Emission
SAGE	Stratospheric Aerosol and Gas Experiment
SAM	Stratospheric Aerosol Measurement
SAMS	Stratosphere and Mesosphere Sounder
SAOZ	System for Analysis of Observation at Zenith
SAQM	SARMAR Air Quality Model
SAT	Surface Air Temperature
SBUV	Solar Backscatter Ultraviolet
SC	Solar Constant
SCC	Squamous Cell Carcinoma
SCIAMACHY	Scanning Imaging Absorption Spectrometer for Atmospheric Chartography
SeaWiFS	Sea-viewing Wide Field of view Scanner
SESAME	Second Stratospheric Arctic and Mid-latitude Experiment
SFRG	Stratospheric Radiative Forcing Gradient
SIRHS	Stratospheric Sounding by Infrared Heterodyne Spectroscopy

SIRIS	Stratospheric Infrared Interferometer Spectrometer
SJVAQS	San Joaquin Valley Air Quality Study (ozone episodes)
SLIMCAT	Single Layer Isentropic Model of Chemistry and Transport
sm	suspended minerals
SME	Solar Mesosphere Explorer (satellite)
SMM	Solar Maximum Mission (programme)
SMOW	Standard Mean Ocean Water; V-SMOW: Vienna ... (isotopic signature)
SOAPEX	Southern Ocean Atmospheric Photochemistry Experiment
SOC	Surface Ozone Concentration
SOI	Southern Oscillation
SOLAR	Company name
SOLSE	Shuttle Ozone Limb Sounding Experiment
SOLSTICE	Solar–Stellar Irradiance Camparison Experiment
SOLVE	SAGE-III Ozone Loss and Validation Experiment
SOO	Solar and Ozone Observatory (CHMI)
SOPRANO	Sub-millimetric Observations of Processes in the Atmosphere Noteworthy for Ozone (submillimetre limb sounder)
SOS	Southern Oxidants Study Nashville/Middle Tennessee Ozone Study
SPARC	Stratospheric Processes and Their Role in Climate
SPAS	Shuttle Pallet Satellite (satellite)
SPE	Solar Proton Event
SPEAM	Sunphotometer Earth Atmosphere Measurement
SPF	Sun Protection Factor
SPOT	Système Pour l'Observation de la Terre (satellite)
SPS	Star Pointing Spectrometer
SRFG	Stratospheric Radiative Forcing Gradient
SSA	Stratospheric Sulphate Aerosol
SSBUV	Shuttle Solar Backscatter Ultraviolet
SSC	Scientific Steering Committee
SSF/IC	Stiga Intercomparison of Dobson Instruments
SST	Sea-surface Temperature
SSU	Stratospheric Sounding Unit
STAAARTE	Scientific Training and Access to Aircraft for Atmospheric Research Throughout Europe
STARE	Southern Atlantic Region Experiment
STARS	Study of Transport and Chemical Reactions in the Stratosphere
STE	Stratospheric–Tropospheric Exchange
STEM	Sulphur Transport Eulerian Model
STH	Higher Stratosphere

STL	Lower Stratosphere
STRATOZ	Stratospheric Ozone Experiment
SUCCESS	Subsonic Aircratft: Cloud and Contrail Effects Special Study
SUL	Sulphates
SUMAS	Submillimeter Atmospheric Sounder
SUVR	Solar Ultraviolet Radiation
SW	Shortwave
SYMOCS	System for the Monitoring of Stratospheric Compounds
SZA	Solar Zenith Angle
TDLHS	Tunable Diode Laser Heterodyne Spectrometer
TERRA	(satellite)
TES	Tropospheric Emission Spectrometer
TFA	Trifluoroacetic acid
TG	Trace Gases
THESEO	Third European Stratospheric Experiment on Ozone
THIR	Temperature Humidity Infrared Radiometer
TIROS	Television Infrared and Observational Satellite
TM	Tracer Model
TOC	Troposphere Ozone Chemistry
TOGA	Tropical Ocean Global Atmosphere Data
TOMS	Total Ozone Mapping Spectrometer
TOMS/EP	Earth Pole Total Ozone Mapping Spectrometer
TOR	Tropospheric Ozone Research
TOTEX, TOTE	Tropical Ozone Transport Experiment
TOVS	TIROS (Television Infrared and Observational Satellite) Operational Vertical Sounder
TOZ	Total Column Ozone
TRACE	Transport and Chemical Evolution
TRACE-A	Transport and Atmospheric Chemistry near the Equator – Atlantic
TROPOZ	Tropospheric Ozone Experiment
TROZ, TrOZ	Tropospheric Ozone
UARS	Upper Atmosphere Research Satellite
UFT	Upper Free Troposphere
UGAMP	Universities Global Atmospheric Modelling Programme
UGCM	UGAMP General Circulation Model
UKMO	UK Meteorological Office
UNEP	United Nations Environment Programme
UT	Upper Troposphere
UVI	Ultraviolet Index
UVR	Ultraviolet Radiation

UVS	Ultraviolet Spectrometer
UVSP	Ultraviolet Spectrometer Polarimeter
VMR	Volume Mixing Ratio
VOC	Volatile Organic Compounds
VOTE	Vortex Ozone Transport Experiment
WAOWE	Weybourne Atmospheric Observatory Winter Experiments
WCRP	World Climate Research Programme
WMO	World Meteorological Organization
WODC	World Ozone Data Centre
WOUDC	World Ozone and Ultraviolet Data Centre

1

Introduction

It is well known that the ozone layer in the stratosphere (about 90% of ozone resides in a layer of the stratosphere between approximately 10 and 50 km above the Earth's surface, called the 'ozone layer') protects life on earth against harmful ultraviolet (UV) solar radiation (Bojkov, 1999; *Scientific* ..., 1999). The on-going depletion of the ozone layer because of anthropogenic impact results in the enhancement of surface UV radiation, which can lead to dangerous consequences for man and the biosphere.

In 1970 Professor P. Crutzen pioneered the discussion of the adverse impact on the ozone layer of nitrogen oxides from fertilizers and from supersonic aircraft. In 1974 Professors S. F. Rowland and M. J. Molina identified CFCs (chlorofluorocarbons) as ozone destroyers, which have been confirmed by subsequent studies. In recognition of their pioneering work, these scientists were awarded the Nobel Prize for Chemistry in 1995.

Tropospheric ozone (which amounts to about 10% of total ozone content – TOZ) is a significant greenhouse gas (GHG). Changes in stratospheric and tropospheric ozone contribute differently to climate change and have different effects on humans and ecosystems. An important specific phenomenon relevant to tropospheric ozone change is photochemical smog formation. Specific functions of stratospheric and tropospheric ozone have caused stratospheric ozone to be considered as *good* ozone and tropospheric ozone as *bad* (Low, 1998).

It is thus obvious that atmospheric ozone is a component of the atmosphere that deserves special attention. This is why the Montreal Protocol on Substances that Deplete the Ozone Layer (1987) to the Vienna Convention for the Protection of the Ozone Layer (1985) and subsequent Amendments and Adjustments to the Protocol have been signed (Bojkov, 1999; Bojkov *et al.*, 1994, 1995a, b, 1998; Koehler and Hajost, 1990; Sarma, 1998). In accordance with the Helsinki Declaration on the Protection of the Ozone Layer (2 May 1989) it has been agreed to phase out the production and consumption of CFCs controlled by the Montreal Protocol as soon as possible but not later than the year 2000.

Total phase out was recommended by the London Amendment to the Montreal

Protocol on the basis of the 1989 assessment. The discovery of the Antarctic 'ozone hole' in 1985 and conclusions of the 1991 Assessment led to the recommendations of the Copenhagen Amendment (1992) to broaden the list of ozone-depleting substances (ODS) by adding HCFCs (hydrochlorofluorocarbons) and methyl bromide. The subsequent 1997 adjustments and amendment marked 10 years of the Protocol (Sarma, 1998) and required phasing out the following categories of substances (at present there are 95 chemicals controlled by the Protocol):

Chlorofluorocarbons:
- CFCs, halons, hydrobromofluorocarbons
- HBFCs, other fully halogenated CFCs, carbon tetrachloride, 1,1,1-trichloroethane
- Methyl chloroform

Hydrochlorofluorocarbons:
- HCFCs

Hydrobromofluorocarbons:
- HBFCs

Methyl bromide.

Control measures for chemicals include (in the case of developed countries) phase out of halons by 1994, phase out of CFCs, carbon tetrachloride, methyl chloroform and HBFCs by 1996, phase out of methyl bromide by 2005 and phase out of HCFCs by 2030.

It has been recommended that developing countries phase out HBFCs by 1996, phase out CFCs, halons and carbon tetrachloride by 2010, methyl chloroform and methyl bromide by 2015 and HCFCs by 2040.

It must be pointed out, however, that the situation with regard to methyl bromide (CH_3Br, bromomethane, MeBr) is still rather controversial. Yates *et al.* (1998) have shown that new technology has been developed that could almost eliminate MeBr emissions from soil fumigaation, calling into question the need for a phase out.

It was mentioned (*UNEP* ..., 1994) that the data of recent years clearly show that growth rates of CFC-11, CFC-12, halon-1301 and halon-1211 are slowing down. According to *Scientific* ... (1999), the total combined abundance of ozone- depleting compounds in the lower atmosphere peaked about 1994 and then started to slowly decline and the combined abundance of stratospheric chlorine and bromine is expected to peak before the year 2000. The increase in total tropospheric organic chlorine in 1992 was only about 60 ppt/yr (1.6%), compared with 110 ppt/yr (2.9%) in 1989. The abundance of carbon tetrachloride is also decreasing. On the other hand, the atmospheric abundance of several CFC substitutes (HCFCs, HFCs) is increasing, as anticipated.

The result of the recommendations was a slowing down of the growth of halocarbon concentrations in the stratosphere (the first example of a noticeable positive influence by man on the atmosphere). In accordance with observations from the *UARS* satellite discussed by Russel (Russel and Lastovicka, 1998) global average (for the 70°N–70°S latitude band) total Cl and F ratio trends at an altitude of 55 km decreased between 1991 and 1997 from 124 to 116 ppt/yr (Cl) and from 92 to 82 ppt/yr (F).

Bojkov (1999) emphasized that the long period of increasing total chlorine abundance in the troposphere – primarily from CFCs, carbon tetrachloride and methyl chloroform – has ended: chlorine from major CFCs is still increasing slightly and most halons continue to increase substantially (e.g. halon-1211, 5% per year up to the present time). The abundance of HCFCs and HFCs (hydrofluorocarbons) is increasing as a result of their CFC substitutes. Stratospheric halogen loading lags behind tropospheric loading by up to 6 years: the maximum concentrations of chlorine and bromine are expected to peak at ~3.7 ppb around the year 2000 and keep close to this over the next one or two decades. Even a complete and immediate global elimination of all ODSs, emissions would result in stratospheric halogen loading returning to the pre-1980 values by the year 2033. Some major scientific findings in *Scientific* ... (1998) are as follows:

- The rate of decline in stratospheric ozone at mid-latitudes has slowed; hence, the projections of ozone loss made in the 1994 UNEP Assessment are larger than has actually occurred. In the northern polar latitudes, in six out of the last nine boreal winter–spring seasons, ozone has declined in some months by 25 to 30% below the 1960s average. Bojkov *et al.* (1998) have shown that during the 1990s over the 35–50° mid-latitude belts the ozone deficiency in the Southern Hemisphere was less than over the Northern Hemisphere by about 39%. The observed TOZ losses from 1979 to 1997 were about 5.4 and 2.8% for northern mid-latitudes in winter/spring and in summer/autumn, respectively; and −5.0% in southern mid-latitudes all the year round. There are no statistical TOZ trends in the equatorial region (20°S to 20°N).
- The springtime Antarctic ozone hole continues unabated. Although the extent of ozone depletion was strongest in 1998, it has over the last 3–4 years remained generally similar to that in the early 1990s with monthly TOZ values in September and October continuing to be 40–55% below the pre-ozone-hole values (with up to 70% decrease for periods of about 1 week).
- The late winter/spring ozone values in the Arctic were unusually low in 6 out of the last 9 years, the 6 years were characterized by unusually cold and protracted stratospheric winters with minimum Arctic circumpolar vortex (CPV) temperatures near the threshold for large chlorine activation. Ozone has declined during some spring months by 25–35% below the pre-1976 average. Elevated stratospheric halogen abundance over the next decade or so would imply the Arctic's continuing vulnerability to large ozone losses.
- Over northern mid-latitudes, the downward TOZ trend is largest between 40 and 15 km (>7% per decade) and is smallest near 30 km (2% per decade). The bulk of TOZ decline is between the tropopause and 25 km.
- Stratospheric ozone losses may have caused part of the observed cooling of the lower stratosphere in the polar and upper mid-latitudes and global average negative radiative forcing of the climate system. Much of the observed downward trend in lower stratospheric temperatures (about 0.6°C per decade

from 1979 to 1994) is attributed to the ozone loss in the stratosphere. The stratospheric ozone losses since 1980 may offset about 30% of the positive greenhouse gas effect over the same time period.

- The understanding of the reaction between increasing surface UVB radiation and decreasing column ozone has been further strengthened by ground-based observations and newly developed satellite methods show promise for establishing global trends in UV radiation. The largest UV increases occur during spring at high latitudes in both hemispheres.
- Based on past emissions of ODSs and projection of the maximum allowances under the Montreal Protocol into the future, the maximum ozone depletion is estimated to lie within the current decade or the next two decades, but its identification and the evidence for the recovery of the ozone layer lie still further ahead. It has been pointed out in *Scientific* ... (1998) that 'a full recovery of the Earth's protective ozone shield could occur by the middle of the next century, but it would require that the Protocol is fully implemented'.
- The increase of ozone in the troposphere since pre-industrial times is estimated to have contributed 10 to 20% of the warming due to the increase in long-lived greenhouse gases during the same period.

Sarma (1998) emphasized the role of the Montreal Protocol, as well as subsequent Amendments and Adjustments, in avoiding negative environmental consequences such as: (1) at least 50% ozone depletion at mid-latitudes in the Northern Hemisphere and 70% depletion at mid-latitudes in the Southern Hemisphere, about 10 times larger than today; (2) at least doubling surface UVB radiation at mid-latitudes in the Northern Hemisphere and quadrupling at mid-latitudes in the Southern Hemisphere compared with an unperturbed atmosphere. This compares with current increases of about 5 and 8% in the Northern and Southern Hemispheres, respectively, since 1980.

The implication of ozone depletion could be horrendous with nearly 19 million additional cases of non-melanoma skin cancer up to the year 2060 and 3 million more cases up to 2030. The number of eye cataracts could increase by about 130 million cases by the year 2060, about 50% of this in developing countries. There is a number of unquantifiable effects such as damage to the immune system, adverse impact on animals, lower productivity of crops, damage to aquatic ecosystems and so on. It must be pointed out, however, that a number of scientists (Ellsaesser, 1992, 1994; Maduro and Schauerhammer, 1992) have warned against overemphasizing such dangers, because of ozone depletion.

Ozone problems have become essential parts of such global-scale international programmes as the World Climate Research Programme (WCRP) with its SPARC (Stratospheric Processes and their Role in Climate) Project (Chanin and Geller, 1998) and the International Geosphere–Biosphere Programme (IGBP) including the IGAC (International Global Atmospheric Chemistry) Core Project, as well as many other ozone-related projects (Kondratyev, 1998).

Marked progress has been achieved in the development of satellite remote sensing aimed at obtaining information on global-scale spatio-temporal ozone and ODS variability and trends (Fortuin and Kelder, 1999; Herman *et al.*, 1999a, b). A number of comprehensive observational programmes have been carried out in the high latitudes of the Northern Hemisphere and in the Antarctic to study the stratosphere and ozone dynamics. As a result, not only has the well-known 'ozone hole' in the Antarctic been discovered, but also ozone 'mini-holes' and substantial depletion of ozone over vast territories of Siberia (*Geophysical* . . . , 1986; Harris *et al.*, 1998; Pyle and Harris, 1991; Tuck *et al.*, 1989). Additional reliable assessments of the atmospheric effects of aircraft emissions on both stratospheric and tropospheric ozone have been obtained (Brasseur *et al.*, 1998; Dameris *et al.*, 1998; Fuglestvedt *et al.*, 1999; Grewe *et al.*, 1999; Penner *et al.*, 1999; Stuhler and Frohn, 1998). Important progress has been made using interactive modelling of general atmospheric circulation and ozone dynamics (Hólm *et al.*, 1999; Takigawa *et al.*, 1999), in assessing the impact of volcanic aerosols on stratospheric processes (Timmreck *et al.*, 1999). A new analysis of ozonesonde data for the troposphere and the lower stratosphere has been carried out (Logan, 1999a, b) as well as further studies about the nature of the Antarctic ozone hole (MacKenzie *et al.*, 1999; Rex *et al.*, 1998). New studies have been conducted to assess the influence of ozone depletion on surface ultraviolet radiation (UVR) and the subsequent impact on human health and the biosphere (Zerefos *et al.*, 1998; Ziemke *et al.*, 1999).

Problems associated with tropospheric ozone change have recently attracted much more attention than before, especially in the field of numerical modelling (observations still remain inadequate). Important progress has been made in interactive simulation of the troposphere dynamics and ozone change (Høv, 1997; Isaksen, 1988). The impact of both stratospheric and tropospheric ozone on climate has been studied (*Ozone* . . . , 1989; Kondratyev and Varotsos, 1997). Special attention has been paid to heterogeneous chemical reactions on surfaces of aerosol particles in the stratosphere (especially in the case of polar stratospheric clouds) and troposphere (Kondratyev, 1989, 1998, 1999; Kulmälä and Wagner, 1996; Seinfeld and Pandis, 1998).

As has been pointed out in the *Scientific Assessment of Ozone Depletion* (1999):

- Enhancing anthropogenic emissions of ODS (nitrogen oxides, carbon monoxide and hydrocarbons) lead to large-scale production of ozone, which, through long- range transport, influences the ozone concentration in large regions of the troposphere in both hemispheres. Such changes are characterized by strong spatio- temporal variability.
- The increase of ozone in the troposphere since pre-industrial times is estimated to have augmented the average radiative forcing (RF) by $0.35 \pm 0.15\,W/m^2$, which is thought to have contributed 10–20% of the warming owing to the increase of long-lived GHGs during the same period.

Undoubtedly, marked progress has been achieved in studying atmospheric ozone and anticipating the consequences of its changes; however, quite a number of related problems remain unsolved thus far. Many aspects of ozone research have been discussed in a number of recent publications, especially in the *Scientific Assessment of Ozone Depletion: 1994* (1998), and also in NATO ASI monographs *Atmospheric*

Ozone Dynamics (Varotsos, 1997) and *Atmospheric Ozone* (Bojkov and Visconti, 1998). Nevertheless, there is a necessity for a more comprehensive and integrative consideration of all ozone problems in their full complexity, and this precisely is the principal purpose of this book, which is addressed to broad circles of experts and non-experts (including students), who may be interested in the problem of atmospheric ozone change and the subsequent impact on man and the biosphere.

1.1 STRATOSPHERIC OZONE

1.1.1 Observations and interpretation of observation data

Although TOZ observations started about 150 years ago, with the discovery of the existence of the ozone layer in the stratosphere, the global archive of TOZ data has been accumulated mainly during the past few decades because of satellite observations. Various remote-sensing techniques have been used for this purpose using satellite measurements of backscattered solar UV radiation, attenuation of UV radiation by the atmosphere under conditions of 'occultation geometry', and thermal emission in the 9.6 μm ozone absorption band (Kondratyev, 1989; *UNEP* ..., 1994; *Scientific* ..., 1999).

The most important results from observations are detection of the mean global TOZ decrease and the appearance of the 'ozone hole' in the Antarctic as well as a short-term but sharp TOZ decrease in Northern Hemisphere high latitudes.

Over the past few years substantial data on ozone depletion in the Arctic have been obtained. Hansen and Chipperfield (1999) studied, for instance, ozone depletion at the edge of the Arctic polar vortex in winter 1996/1997 when the Arctic polar stratospheric vortex was extremely long lived. The ozone lidar at the Arctic Lidar Observatory for Middle Atmosphere Research (ALOMAR) operated from mid- December 1996 until mid-May 1997 when the vortex ceased. Observational data indicate that for most of its lifetime the vortex was centred at the pole, and its edge was almost permanently located over Northern Scandinavia. Ozone depletion of up to 40% was observed at the 475 and 550 K levels. Maximum depletion occurred around 5 May at levels up to 550 K and around 20 April at 675 K. Model analysis performed by Hansen and Chipperfield (1999) showed that while much of the early spring ozone depletion was due to halogen chemistry, associated with chlorine activation on polar stratospheric clouds, the ongoing depletion in late April and early May was due to 'summertime' NO_x ($NO + NO_2$) chemistry. The unusual persistence of the vortex, with the isolation of high-latitude air masses until early May, allowed this depletion to occur.

Rinsland *et al.* (1999) discussed observations made inside the November 1994 Antarctic stratospheric vortex and inside the April 1993 remnant Arctic stratospheric vortex using the Atmospheric Trace Molecule Spectroscopy (ATMOS) Fourier transform spectrometer. By comparing vortex and extravortex observations of NO_y (total reactive nitrogen) obtained at the same N_2O volume mixing ratios,

Arctic vortex denitrification of 5 ± 2 ppbv at 470 K ($\sim$18 km) has been inferred (these results are robust for a wide range of winter conditions).

Since stratospheric ozone is an intensive absorber of solar radiation and tropospheric ozone is a greenhouse gas (Kondratyev, 1998; Low, 1998; Wang and Isaksen, 1995), there is no doubt that changes in the content of ozone in the atmosphere must affect the climate.

UNEP ... (1994) and *Scientific* ... (1999) along with other publications categorize the following as the most important aspects of the problem:

- From 1979, the rate of mean annual TOZ decline in mid-latitudes of both hemispheres was within 2.6–3.7% over a period of 10 years. Analysing ozone decline in the northern polar and mid-latitudes during winter and spring Bojkov *et al.* (1998) concluded that the ozone-mass deficiency O_3 (MSD) (defined from the pre-1976 base average and areal extent with negative deviations greater than $\sim$2 and $\sim 3\sigma$), integrated for the first 105 days of each year, has increased dramatically from $\sim$2,800 Mt in the early 1980s to $\sim$7,800 Mt in the 1990s, exceeded 12,000 Mt in the winter–spring periods of 1993 and 1995. Bojkov (1999) pointed out that total ozone levels in the latitudes 60°N–60°S were at their lowest in 1993 because of the large increase in stratospheric aerosol caused by the eruption of Mount Pinatubo in 1991 (Kondratyev and Galindo, 1997; Kondratyev, 1999). Since 1992/1993, TOZ values over this part of the globe have been variable around a fairly constant level. Table 1.1 illustrates the TOZ trends (Bojkov, 1999).
- The amplitude of the annual cycle of ozone at mid to high latitudes has decreased by 15% in the last two decades, because larger declines of maximum ozone values have occurred during winter and spring.
- Ozone depletion increases with latitude, particularly in the Southern Hemisphere. Little or no downward trends are observed in the tropics (20°N–20°S). Analysis of global TOZ data through early 1998 shows substantial decreases in ozone in all seasons at mid-latitudes (35–60°) of both hemispheres. There were downward trends in the Northern Hemisphere of about 6% per decade between 1979 and 1998, observed in winter and spring and about 1.5–3% per decade were observed in summer and fall. In the Southern Hemisphere

Table 1.1. Total ozone trends (percent per decade since January 1979) calculated as averages of individual station trends in the indicated regions and groups of months (ground-based observations). All values are negative and statistically significant at 2σ.

Region	DJF	MAM	JJA	SON	Year
Arctic	7.9	7.7	2.5	3.6	5.7
35–60°N	4.1	5.7	2.9	1.6	3.7
35–60°S	2.9	2.2	3.4	2.0	2.6
Antarctic	6.3	2.4	6.5	20.0	8.9

the difference between seasons was smaller and the mid-latitude trends averaged within 2.0–3.4% per decade. Satellite and ozonesonde data show that much of the downward trend in ozone levels occurs below 25 km.

- According to *Scientific* ... (1999) TOZ decreased significantly at mid- latitudes (25–60°) between 1979 and 1991, with estimated linear downward trends of 4.0%, 1.8% and 3.8% per decade, respectively, for northern mid-latitudes in winter/spring, northern mid-latitudes in summer/fall and southern mid-latitudes year round. However, since 1991 the linear trend observed during the 1980s has not continued, but rather TOZ was almost constant at all mid-latitudes in both hemispheres after recovery from the 1991 Mount Pinatubo eruption. Present-day understanding of how changes in halogen and aerosol loading affect ozone explains why linear extrapolation of the pre-1991 ozone trend to the present is not suitable.
- A complex 3D-spatial structure of ozone distribution was recently discovered taking the form of lamination (Varotsos, 1997) and formation of low-ozone air pockets during the northern winter in the middle stratosphere outside the polar vortex.
- Nair *et al.* (1998) pointed out that the rapid ozone loss localized in the pockets is due to the isolation of air at high latitudes (and high solar zenith angles). Thus low ozone levels are due to a decrease in the odd oxygen-production rates and not an increase in the loss rate by reaction with halogen species, as in the classical ozone hole. For an explanation of low-ozone pockets see Morris *et al.* (1998) who developed a simulation model that takes account of photochemical processes and stratospheric dynamics.
- Russel and Lastovicka (1998) concluded, on the basis of ozonesonde data, that the number of thin ozone layers per one ozone vertical profile decreased from 1970 to 1993 from about 1.9 to 1.0 (with depletion of the ozone content in such layers of about 50%).
- The lowest values of TOZ were obtained in 1992–93 at the ozone hole in the Antarctic and the lowest TOZ values in the whole observational period were obtained over the densely populated regions of the Northern Hemisphere. The conclusion that anthropogenic chlorine and bromine compounds, coupled with surface chemistry on natural polar stratospheric particles, are the cause of polar ozone depletion has been further strengthened. The links to halogen chemistry have also been established with regard to ozone losses detected in the Arctic winter stratosphere.
- According to *Scientific* ... (1999) the large ozone losses during the spring in the Southern Hemisphere continued unabated with approximately the same magnitude and areal extent as in the early 1990s. In Antarctica, the monthly total ozone in September and October has continued to be 40 to 55% below the pre-ozone-hole values of approximately 320 m-atm cm (Dobson Units, DU), with up to a 70% decrease for periods of a week or so. This depletion occurs primarily over the 12 to 20-km-altitude range, with most of the ozone in this layer disappearing during early October.
- Observations made after the eruption of Mount Pinatubo in 1991 resulted in the

discovery of the substantial impact such a major volcanic eruption has on stratospheric ozone (Kondratyev and Varotsos, 1995b; Timmreck *et al.*, 1999).

- Impacts of subsonic and supersonic aircraft on upper tropospheric and lower stratospheric ozone have been confirmed (Andronache and Chameides, 1998; Baumgardner *et al.*, 1998; Brasseur *et al.*, 1996; Campos *et al.*, 1998; Penner *et al.*, 1999; Vay *et al.*, 1998; Weisenstein *et al.*, 1998).
- The earlier conclusion about the substantial role of methyl bromide as an ozone-destroying component has been verified, the main sources of this are crop fumigation, biomass burning and cars. The most recent data (*Scientific* ..., 1999) indicate, however, that the role of methyl bromide as an ozone-depleting compound is now considered to be less than was estimated in *UNEP* ... (1994), although significant uncertainties remain (the science of atmospheric methyl bromide is complex and still not well understood).
- The planned countermeasures to control emissions of ozone-depleting compounds will not greatly change the maximum level of TOZ decline forecast to be reached during the next 10 years.
- According to model simulations much of the observed downward trend in lower stratospheric temperatures (about 0.6°C per decade over 1979–1994) is attributed to the ozone loss in the lower stratosphere (*Scientific* ..., 1999). The decrease in stratospheric ozone since 1980 may have offset about 30% of greenhouse forcing, which favoured global climate warming. However, the increase in tropospheric ozone in the same period has intensified the atmospheric greenhouse effect in the Northern Hemisphere by ~20%. The problem of ozone trends with opposite signs in the stratosphere and in the troposphere deserves special attention with respect to TOZ variation on UV radiation.
- From some published data (Michaels *et al.*, 1994), the TOZ decline in 1992–93 was not followed by any increase in UVB radiation. The necessity of a thorough analysis of data on the trends of TOZ, UVB and related biological consequences was expressed by Ellsaesser (1992, 1994). However, Feister and Grewe (1995) observed higher UV radiation from low ozone values at northern mid-latitudes in 1992 and 1993 and pointed out that this could lead to adverse biological effects. Numerous further studies have confirmed this conclusion.
- *Scientific* ... (1999) concludes that the understanding of the relation between increasing surface UVB radiation and decreasing TOZ has been further strengthened by ground-based observations, and newly developed satellite methods show promise for establishing global trends in UV radiation. The satellite estimates for 1979–1992 indicate that the largest UV increases occur during spring at high latitudes in both hemispheres.

Logan (1999b) presented an analysis of ozonesonde data with the aim of using them to evaluate models of transport and chemistry. These data show that between 10 and 30% of the ozone column is located between 100 hPa and the tropopause at mid and high latitudes, and this region drives much of the seasonal TOZ variation. The sonde data quantify the build-up of ozone in the lowermost stratosphere of the Northern Hemisphere in winter and its loss in late spring and summer. The amount of ozone between the tropopause and 100 hPa in the Northern Hemisphere decreases from 175 to 75 Tg from March to September, with the maximum rate of decrease in May to June; about half of the decrease is caused by the increase in the height of the tropopause. There is a lag of about a month when ozone starts to accumulate in the lowermost stratosphere from September near 40°N to October or November near 80°N, and a similar lag in the time when ozone starts to decrease from February or March near 40°N to March or April near 80°N.

Despite caution against overemphasis of the ozone impact on the environment and ecosystems, this problem undoubtedly deserves very serious attention.

Prather *et al.* (1996) posed a number of important questions in this context:

(1) What would have happened if chlorofluorocarbon (CFC) use had followed a free-market growth into all sectors and countries before the observation of stratospheric ozone depletion?
(2) What if the ozone hole had been discovered in 1985 (as indeed it was) without the prior suspicion about chlorine from CFCs?
(3) When should we expect to see recovery of the ozone layer as provisions of the Montreal Protocol begin to show their effect on global atmosphere?
(4) How much stratospheric chlorine and associated ozone depletion could the global atmosphere have been committed to in a worst-case scenario?

Answering these questions, Prather *et al.* (1996) emphasized that:

(1) whether we had ignored the scientific evidence or not, ozone depletion would be dramatically worse than that we are experiencing today;
(2) if CFCs had followed free- market growth until 2002, the Antarctic ozone hole would be a permanent fixture through the twenty-first century, instead of disappearing by 2050 as predicted in the Copenhagen '92 scenario;
(3) ozone depletion is expected to reverse and recover measurably in the first decades of the next century; unambiguous detection of the recovery will take a decade or more.

Kaye (1998) and McCormick (1998) have summarized discussions of satellite TOZ observations using TOMS (Total Ozone Mapping Spectrometer) and SAGE (Stratospheric Aerosol and Gas Experiment) instrumentation, respectively. Up to the present time a total of four TOMS instruments have flown in space: Nimbus-7 TOMS (1978–1993), Meteor-3 TOMS (1991–1994), ADEOS TOMS (1996–1997) and Earth Probe TOMS (1996–present). TOMS instruments have provided excellent data for TOZ, although there are some limitations that must be considered in certain applications, especially for studies of tropospheric ozone (currently no data are released for zenith angles greater than 84°, although data are in principle available for up to 88°).

Assimilation of TOZ satellite measurements in a three-dimensional tracer transport model has been discussed by Jeuken *et al.* (1999). Fioletov *et al.* (1999) assessed the world ground-based TOZ network performance by comparison with satellite data (TOMS and SBUV instrumentation on board the Nimbus-7 and Meteor-3 satellites). The results indicate that about 80% of all Dobson, Brewer and filter ozonometer stations have standard deviations of monthly mean differences with TOMS that are less than 2.5%. Typically, results of the comparisons between ground-based stations and SBUV are similar to those for ground-based and TOMS comparisons.

Bhatt *et al.* (1999) evaluated the quality of satellite Halogen Occultation Experiment (HALOE) ozone profiles in the lower stratosphere by comparison with coincident ozonesonde profiles. It was found that the HALOE ozone agrees, on average, to within 10% of their coincident ozonesonde measurements down to 100 hPa at tropical/subtropical latitudes and to 200 hPa at extratropical latitudes. Bacmeister *et al.* (1999) conducted an intercomparison of satellite and aircraft observations of ozone, CFC-11 and NO_y using trajectory mapping. Correlation coefficients for coincident pairs of satellite and aircraft data are between 0.68 and 0.80.

Apart from TOZ data, processing of TOMS information allows retrieval of tropospheric ozone content and surface flux of ultraviolet radiation, especially if ratios such as UVA/UVB (UV radiation longward and shortward of 320 nm) are considered. TOMS instrumentation made it possible to accumulate a data set covering the period from late 1978 till February 1998 for zonal mean TOZ. One of the applications of the data obtained was the study of TOZ long-term trends and interannual variability in mid-latitudes (particular emphasis was placed on dynamically driven contributions). The results suggest that changes in lower stratospheric temperature and/or tropopause height could contribute significantly to mid-latitude decreases observed by TOMS (Schubert and Munteanu, 1998). It was seen that short-term meteorological events can significantly affect the zonal mean distribution of TOZ, especially in February, when there can be enormous variation from one winter to the next.

To test retrieval algorithms, observational data are available from Satellite Ozone Limb Sounding Experiment (SOLSE) and Limb Ozone Retrieval Experiment (LORE) instruments that flew on the space shuttle in the fall of 1997 are available. Measurements of a number of species, with UV absorption in the ultraviolet, using the Global Ozone Monitoring Experiment (GOME) instrument aboard the ERS-2 satellite have been conducted (column distributions of BrO, ClO and SO_2 have been studied).

In continuation of earlier SAGE observations three new ozone-measuring instruments for SAGE-III have been developed (McCormick, 1998). The first flight took place in 1999 (Russian Meteor-3M satellite in a sun-synchronous orbit) while the second flight will be an opportunity to fly into a mid to high-inclination orbit. The third SAGE-III instrument will be flown aboard the International Space Station (ISS), scheduled for mid to late 2002. Using solar occultation, SAGE-III data will allow retrieval of 1-km vertically resolved mid-atmosphere profiles of ozone, aerosols and a number of aerosol properties, water vapour, temperature, nitrogen dioxide and cloud presence. In addition, using lunar occultation, it will be possible to retrieve nitrogen trioxide and chlorine dioxide vertical profiles. For validation purposes the SAGE-III Ozone Loss and Validation Experiment (SOLVE) has been planned, whose aim is to use a joint balloon and airborne measurement campaign in the Arctic designed to examine the processes that control polar to mid-latitude stratospheric ozone levels and also aid in SAGE-III validation. An important role in solving validation problems belongs to the ozonesonde network. Guirlet *et al.* (1998) described the Third European Stratospheric Experiment on Ozone (THESEO) conducted in 1998-1999 as a continuation of the earlier European Arctic Stratospheric Ozone Experiment (EASOE, 1991–1992) and the Second Stratospheric Arctic and Mid-latitude Experiment (SESAME) with the principal purpose of better understanding those processes that determine stratospheric ozone trends in mid-latitudes on the basis of studying interactions between relevant chemical and dynamical processes. The aim is to investigate ozone formation both within and outside of the circumpolar vortex. The observations covered a broad range of latitudes from the tropics to the Arctic using ozonesondes, lidars, small balloons and aircraft. On 19 December 1999 the NASA Terra satellite was launched, opening new perspective of ozone observation.

Discussing ozone problems, it is appropriate to quote the Policy Statement on Atmospheric Ozone accepted by the American Meteorological Society Council, 28 January 1996 (Hales, 1996):

1. The AMS recognizes ozone as an atmospheric constituent that has a number of important beneficial as well as detrimental effects on the atmosphere, on surface ecosystems and on humankind. These include effects on both ultra-violet and longwave radiation, resulting effects on atmospheric wind systems and direct impacts on plants and animals. Moreover, ozone is recognized as the dominant progenitor of much of the trace-gas chemistry that occurs in both the troposphere and the stratosphere. The AMS also notes that extensive couplings exist among these chemical, radiative and dynamical components of ozone's behavior and add substantial uncertainty to many of the currently available assessments of ozone's impact.
2. Despite many uncertainties, ample evidence exists to substantiate that atmospheric ozone has been affected in important and even critical ways by human activity. In case of the stratosphere, the existence of the Antarctic ozone hole is unquestionable and our evidence that it results from human-produced halocarbons is overwhelming. Ozone depletion in the mid-latitude stratosphere is less dramatic; however, the general agreement of the large numbers of available column and profile measurements make it reasonable to speculate that such depletion is indeed occurring and that it is largely human-induced. Anthropogenic activities also significantly influence tropospheric ozone. Humankind is directly responsible for the excessive ozone levels that often occur near the surface in and down-wind from populated areas. These effects are also felt throughout significant regions of the free troposphere. The effects in both the stratosphere and troposphere are sufficiently profound to mandate substantial concern, on both a local and a global basis.
3. While the above manifestations of human impact on ozone are clear, there remain important gaps in our understanding of ozone's complex behavior. For example, currently we cannot quantitatively explain the mid-latitude observa-

tions of insignificant surface UV increases. Impacts of tropospheric ozone build-up on the free troposphere's chemistry are also in a highly speculative state. These issues and others must be resolved in order to forecast potential future manifestations satisfactorily and provide a firm basis for policy analysis and associated policy actions.

4. Many of the noted uncertainties arise because of the strong couplings between chemistry, radiation and atmospheric dynamics. Because of this, their resolution will require substantial combined effort by scientists having chemical and meteorological backgrounds. The AMS intends to actively provide one component of a continuing forum for this scientific interaction and act in conjunction with other like-minded organizations for this purpose.
5. Current ozone-control legislation (such as the US Clean Air Act, with 1990 amendments) and international agreement (such as the Montreal Protocol) tend to reflect thc uncertainties noted above. The State Implementation Plan approach of the current US Clean Air Act is burdened by older and largely invalid concepts of chemistry and atmospheric transport, which make ozone-standard attainment difficult if not impossible at many locations. Future amended versions of this act should incorporate more manageable and scientifically valid approaches to ozone control and substantial current meteorological and chemical research should be directed to determining effective strategies for this purpose. With regard to the Montreal Protocol, it is encouraging to note that halocarbon limitations under this agreement appear to have resulted, recently, in decreases of some of the shorter-lived halogen-containing species. Owing to the noted uncertainties and complexities associated with stratospheric ozone depletion, however, the effects of this and other international agreements must be monitored continuously and carefully to ensure their effectiveness and to establish the basic understanding required for more effective maintenance during future years.

1.1.2 Theory of ozone-layer formation and changes

As pointed out by Crutzen (1998), the theory suggested by Sydney Chapman in 1930 to explain ozone-layer formation was considered adequate until about the mid-1960s (we shall follow Professor Crutzen's summary of the history of ozone theory). In accordance with this theory, the formation of 'odd oxygen' – O and O_3 – is due to photolysis of molecular oxygen O_2 by solar radiation at wavelengths shorter than 240 nm:

$$O_2 + h\nu \longrightarrow 2O \qquad (\lambda \leq 240\,\text{nm}) \tag{1.1}$$

This atomic oxygen reacts quickly with molecular oxygen O_2 to form ozone O_3, which, however, will rapidly decompose again unless some energy-absorbing molecule M (N_2 and O_2) is present. Thus:

$$O + O_2 + M \longrightarrow O_3 + M \tag{1.2}$$

The loss of ozone is due to photolysis:

$$O_3 + h\nu \longrightarrow O + O_2 \tag{1.3}$$

Reactions (1.2) and (1.3) lead to the rapid establishment of a steady state for the concentration of O and O_3 without affecting the concentration of odd oxygen whose destruction (counteracting its production through photolysis (reaction (1.1)) occurs by reaction:

$$O + O_3 \longrightarrow 2O_2 \tag{1.4}$$

Later (in 1950) David Bates and Marcel Nicolet with Sydney Chapman proposed that catalytic reactions involving hydroxyl (OH) and HO_2 radicals could counter-balance the production of odd oxygen in the meso and thermosphere which was necessary because reaction (1.4) was too slow. Taking this proposal into account John Hampson and Garry Hunt suggested the reactions:

$$OH + O_3 \longrightarrow HO_2 + O_2 \tag{1.5}$$

$$HO_2 + O_3 \longrightarrow OH + 2O_2 \tag{1.6}$$

$$2O_3 \longrightarrow 3O_2$$

The primary source for OH radicals was photolysis of O_3 by solar UV radiation of wavelengths shorter than about 320 nm, which leads to electronically excited $O(^1D)$ atoms. Most of $O(^1D)$ reacts with O_2 and N_2 to reproduce O_3, rapidly producing a null cycle with no effect on concentrations of ozone or odd oxygen. Therefore:

$$O_3 + h\nu \longrightarrow O(^1D) + O_2 \tag{1.7}$$

$$O(^1D) + H_2O \longrightarrow 2OH \tag{1.8}$$

$$O(^1D) + M \longrightarrow O + M \tag{1.9}$$

Professor Crutzen demonstrated, however, that reactions (1.5) and (1.6) were too slow and therefore discarded the theory of Hampson and Hunt. The influence of nitrogen-oxide compounds (potentially NO and NO_2 act as catalysts for O_3 destruction) on the photochemistry of the ozone layer was considered instead. The presence of active nitrogen oxides ($NO_x = NO + NO_2$) in the atmosphere results from the following reactions:

$$OH + NO_2(+M) \longrightarrow HNO_3(+M) \tag{1.10}$$

$$HNO_3 + h\nu \longrightarrow OH + NO_2 \tag{1.11}$$

In accordance with the views mentioned the following catalytic reactions were suggested:

$$NO + O_3 \longrightarrow NO_2 + O_2 \tag{1.12}$$

$$O + NO_2 \longrightarrow NO + O_2 \tag{1.13}$$

Combining reactions (1.12) and (1.13):

$$O + O_3 \longrightarrow 2O_2$$

As can be seen the net result of reactions (1.12) and (1.13) is equivalent to direct reaction (1.4). It later became clear that enough NO is produced in the following reaction (1.14):

$$N_2O + O(^1D) \longrightarrow 2NO \qquad (1.14)$$

to make reactions (1.12) and (1.13) the most important ozone-loss reactions in the stratosphere between altitudes of 25 and 45 km.

Nitrous oxide is a natural byproduct of microbial processes in soil and water but, on the other hand, there are many anthropogenic sources of N_2O emissions (e.g. application of nitrogen fertilizers). At the present time it is known that the rate of increase in atmospheric N_2O concentration (0.2–0.3% per year) – observed during the past few decades – was due to anthropogenic activities. The discovery of the important role of NO_x in ozone-layer formation and change was very significant for assessment of the impact of subsonic and supersonic aircraft on the ozone layer (Brasseur *et al.*, 1998; Campos *et al.*, 1998; Dubey *et al.*, 1997; Penner *et al.*, 1999).

During the 1970s intensive study took place of the potential for Cl and ClO to destroy stratospheric ozone. In September 1974 Professor Crutzen published a model analysis of potential ozone depletion resulting from the continued use of CFCs, which indicated the possibility of ozone depletion by as much as 40% near the 40-km altitude as a result of continued use of these compounds at 1974 rates. Emphasis has been put on the use of chlorine chemistry to understand and quantitatively assess anthropogenic chlorine-catalysed ozone destruction. Observations confirmed that halogen chemistry plays a larger role in the chemical destruction of ozone in the mid-latitude lower stratosphere than that expected from gas-phase chemistry. Bromine compounds require special attention, because bromine is almost 100 times more dangerous to ozone than chlorine, on an atom-to-atom basis.

An important consequence of the discovery of the ozone hole was recognition of the decisive role of heterogeneous chemical reactions on aerosol particles in the atmosphere. In the case of the ozone hole, particles of polar stratospheric clouds (PSCs) formed under extremely low temperatures as a result of condensation of water vapour and nitric acid. Since formation and stability of PSCs depend strongly on meteorological conditions in the stratosphere, climate and ozone dynamics become closely connected. The chemical reactions responsible for stratospheric ozone depletion are extremely sensitive to temperature.

In this context Shindell *et al.* (1998) investigated the interplay between projected future emissions of greenhouse gases (GHGs) and levels of ozone-depleting halogen species using a global climate model that incorporates simplified ozone- depletion chemistry. Numerical modelling results demonstrate that temperature and wind change induced by increasing concentrations of GHGs alter planetary-wave propagation, reducing the frequency of sudden stratospheric warming in the Northern Hemisphere. This results in a more stable circumpolar vortex, with significantly colder temperature in the lower stratosphere and concomitantly increased ozone depletion. So, increase in concentrations of GHGs may be a cause for the very large loss of Arctic ozone – expected to reach a maximum between 2010 to 2019, roughly a decade after the abundance of stratospheric chlorine reaches a maximum, with mean losses comparable with those in the

Antarctic during the early 1990s (up to two-thirds of Arctic ozone loss is predicted as geographically localized). The result of 'greenhouse' stratospheric cooling over the coming decades should be a more severe and durable Antarctic ozone hole.

However, Salawitch (1998) pointed out that the Arctic winter of 1997–98 was fairly warm, with considerably less O_3 loss than occurred in the previous five winters. This behaviour is not necessarily inconsistent with the model of Shindell *et al.* (1998), but complicates ascribing climatic influence to temperature and O_3 change that have occurred during the previous five winters. Shindell *et al.* (1998) emphasized the need for further analysis of observational data as well as additional numerical modelling to better understand the factors that regulate the temperature of polar vortices.

Considine *et al.* (1999) conducted a comparison of observed TOZ values and two-dimensional modelling (both photochemistry and dynamics were taken into account) and also assessed (Monte Carlo analysis) the uncertainty (due to input parameters) in predicting TOZ trends between 1979 and 1995. The comparison of observed and calculated TOZ values indicates no significant difference. Large differences between models and measurements of input-parameter uncertainty are seen, however, in the meridional gradient of trends and peak-to-peak variation in trends over an annual cycle.

Further important progress was made by Takigawa *et al.* (1999). They have developed a fully coupled mid-atmosphere general-circulation model that included the photochemistry for ozone and other chemical species (19 photolysis and 52 chemical reactions). Horizontal resolution (T21) was about 600 km with 30 layers in the vertical. The results obtained indicate that seasonal TOZ evolution is quite realistic, although the model slightly underestimates total tropical ozone. The model also underestimates ozone amounts around the equatorial tropopause. In the case of temperature calculations the model has a cold bias of about 25 K in the lower stratospheric Northern Hemisphere winter and 5 K in the Southern Hemisphere winter. The model temperature around the equatorial tropopause is cooler than observed. A future model version is supposed to include the following processes:

(1) the heterogeneous chemistry on aerosols and PSCs;
(2) new transport scheme;
(3) modified formulation of the chemical processes, especially for more detailed ClO_x and BrO_x chemistry.

Prather (1998) investigated timescales in atmospheric chemistry connected to perturbations to N_2O (which is the third most important greenhouse gas), NO_y (odd nitrogen usually including NO, NO_2, NO_3, N_2O_5, $ClONO_2$, HNO_4, HNO_3) and O_3. This group of chemical compounds is fully coupled through stratospheric photochemistry: N_2O emitted at the surface enters the stratosphere and releases NO_y, which catalytically destroys O_3 in the mid-stratosphere. As a result, more ultraviolet sunlight is available for photolytic destruction of N_2O. Prather (1998) simulated such a coupling using a representative one-dimensional model to show how photolytic coupling leads to a more rapid (10–15%) decay of N_2O perturbations than anticipated from the N_2O lifetime based on steady-state budgets. This pattern will be excited by emissions of N_2O or by almost any chemical perturbation (e.g.

those affecting stratospheric ozone). Such coupled patterns might be discernible in the atmosphere.

Jobson *et al.* (1999) reviewed several archived datasets to examine the relationship between mixing-ratio variability and lifetime for hydrocarbon and halocarbon species in the troposphere and stratosphere. In the case of the Harvard forest ground site, winter and summer data displayed the same lifetime dependence, $\tau^{-0.18}$ (τ is the residence time in years), which was significantly weaker than the $\tau^{-0.5}$ dependence of remote tropospheric data, indicating that source terms dominated regional variability at Harvard. Stratospheric data displayed a stronger lifetime dependence than tropospheric trends, indicating the more dominant role for sink terms in describing spatial variability in this region of the atmosphere.

Summarizing achievements in understanding TOZ change, Bojkov (1999) pointed out that contemporary simulation models, based on the observed build-up of chlorine/bromine in the stratosphere, predict ozone depletion that is in good quantitative agreement with the altitude and latitude of the measured ozone decline during the past few decades, especially over the polar regions. This clearly confirms the predictions made by Rowland and Molina in 1974 concerning a substantial reduction in upper stratospheric ozone due to release of CFCs into the atmosphere. Observations and models have also confirmed the importance of stratospheric sulphate aerosol (SSA) and polar stratospheric clouds (PSCs) in ozone-loss chemistry through heterogeneous reactions that activate halogen species and deactivate nitrogen species. Recently it was shown that cold liquid aerosol can maintain elevated ClO in non-denitrified air. Since chlorine activation in liquid particles in the lower stratosphere (both on SSA and PSCs) decreases strongly with decreasing temperature, this process is controlled mainly by temperature and water-vapour pressure and only secondarily by particle composition. A decisive factor of ozone change is, undoubtedly, interaction between the chemistry and dynamics of the stratosphere.

1.1.3 Biological implications of total ozone depletion

The biological consequences of the decreasing trend of TOZ are connected mainly with enhancement of biologically active UV solar radiation at the earth's surface. It is appropriate in this context to quote the basic conclusions made in *UNEP Scientific Assessment* (1994):

- Large increases in ultraviolet radiation have been observed in association with the ozone hole at high southern latitudes. Measured UV enhancements agree well with model calculations (*Abstracts* ..., 1998; Zerefos and Bais, 1997).
- Clear sky UV measurements at mid-latitude locations in the Southern Hemisphere are significantly larger than at a corresponding site in the Northern Hemisphere, in agreement with expected differences due to TOZ and Sun–Earth separation.
- Local increases in UVB were measured in 1992/93 at mid and high latitudes in the Northern Hemisphere. The spectral signatures of the enhancement clearly implicate the anomalously low ozone observed in those years, rather than

variability of cloud cover or tropospheric pollution. Such correlations add confidence to the ability to link ozone changes to UVB changes over relatively long timescales.

- Increases in clear-sky UV over the period 1979 to 1993 due to observed ozone changes are calculated to be greatest at short wavelengths and at high latitudes. Poleward of 45°, the increases are greatest in the Southern Hemisphere.
- Uncertainties in calibration, influence of tropospheric pollution, and difficulties of interpreting data from broadband instruments continue to preclude the unequivocal identification of long-term UV trends.
- Scattering of UV radiation by stratospheric aerosols from the Mount Pinatubo eruption did not alter total surface-UV levels appreciably (Fioletov *et al.*, 1998).

Recent information on biological implications of total ozone depletion has been discussed at the European Conference on Atmospheric UV Radiation in Helsinki (*Abstracts* ..., 1998) as well as in the monograph by Martens (1998) and in a number of journal publications (Arnone *et al.*, 1998; Bigelow *et al.*, 1998; Katsambas *et al.*, 1997; Lubin *et al.*, 1998; Madronich, 1998; Ziemke *et al.*, 1998, 1999, and others). As pointed out in *Scientific* ... (1999) increases in surface UVB irradiance (e.g. between 1989 and 1997 15% per decade at 300 nm, 8% per decade at 305 nm) have been detected with a few ground-based spectroradiometers at mid-latitudes (near 40°) that are consistent with expected changes from decreasing amounts of ozone. New satellite estimates of global (±65°) UV irradiance have been obtained from TOMS data which made it possible to produce climatological maps of UV irradiance on a daily basis. Satellite data between 1979 and 1992 indicate the presence of annual erythemal UV-irradiance increases per decade equal to $3.7 \pm 3\%$ at 60°N and $3 \pm 2.8\%$ at 40°N (while $3.6 \pm 2\%$ at 40°N and ±6% at 60°S). No statistically significant trends were observed within ±30° latitude.

Ziemke *et al.* (1999) examined the distribution of long-term trends in ground-level erythemally weighted UVB (wavelengths 290–320 nm) exposures in the northern latitudes from 1979 to 1991 using measurements from the *Nimbus 7* TOMS instrument. They emphasized that although the loss of stratospheric ozone alone produces increases in surface exposures at all UVB wavelengths, other factors – the presence of clouds and UVB absorbing aerosol in the troposphere – make the problem of detecting change in surface erythemal UV more difficult. The results obtained indicate that zonal mean erythemal UV data show increases of ~3–7% per decade in mid to high latitudes. In summertime most of the regional increases (exceeding 6% per decade) in the Northern Hemisphere in the latitude range 30–40°N originate from the Pacific and Atlantic Ocean regions.

An important observation by Ziemke *et al.* (1999) is that positive trends in subtropical latitudes originate mostly over oceanic regions whereas positive trends at higher latitudes originate mostly over landmasses. A portion of these increases in erythemal UV over Central Europe and the east Asian continent in summer months may be attributed to decreases per decade in cloudiness. Statistically significant positive trends over ocean in the 30–40°N latitude band suggests a less polluted environment and a more direct influence from stratospheric ozone loss.

As far as impacts of variable UV radiation on humans and ecosystems are concerned the most important aspects of the problem include human health (one of the most dangerous consequences is skin cancer (Katsambas *et al.*, 1997)), health of terrestrial and marine ecosystems (Arnone *et al.*, 1998; Kondratyev, 1998).

1.2 TROPOSPHERIC OZONE

Increasing pollution of the troposphere has stimulated studies of anthropogenic impacts on tropospheric chemical composition (Barrie and Platt, 1997; Bojkov and Visconti, 1998; Davies *et al.*, 1998; Derwent *et al.*, 1998; Ebel *et al.*, 1997, Fiore *et al.*, 1998; Newell *et al.*, 1997; Spänkuch *et al.*, 1998; Wessel *et al.*, 1998; and others). In this context tropospheric ozone occupies a special place in view of its importance for human health and ecosystem functioning as well as its role as a greenhouse gas (Houghton *et al.*, 1996). Pszenny and Brasseur (1997) pointed out that surface ozone is a human respiratory irritant. Relatively small ozone amounts can cause chest pain, coughing, nausea, throat irritation and congestion in healthy people. It may also worsen bronchitis, heart decease, emphysema and asthma. Surface ozone is also phytotoxic. It can produce foliar injuries, reductions in crop yield and biomass production, and shifts in competitive advantages of vegetation species in mixed populations. Photolysis of tropospheric ozone by UV radiation in the presence of water vapour, the primary source for hydroxyl radicals (OH), is very important. Hydroxyl radicals are responsible for the removal of many trace gases (such as CH_4, HFCs, HCFCs) through oxidation.

A number of fundamental publications (*UNEP* ..., 1994; Høv, 1997; Isaksen, 1988; *Scientific* ..., 1999) contain basic information on various features of tropospheric ozone dynamics and relevant problems.

Basic achievements up to 1994 are shown in *UNEP Scientific Assessment* (1994):

- Recent measurements of the NO_y/O_3 ratio have basically confirmed earlier estimates of the flux of ozone from the stratosphere to be in the range of 240–280 Tg (O_3) yr^{-1}, which is in reasonable agreement with results from general circulation models.
- The observed correlation between ozone and alkyl nitrates suggests a natural concentration of 20–30 ppb in the upper planetary boundary layer (at about 1-km altitude), which agrees well with the estimate from the few reliable historic data.
- Measurements of the gross ozone-production rate yielded values as high as several tens of ppb per hour in the polluted troposphere over populated regions, in good agreement with theoretical predictions. Likewise, the efficiency of NO_x in ozone formation in moderately polluted air masses was found to be in reasonable agreement with theory.
- Direct measurements of hydroxyl and peroxy radicals have become available. While they do not survey to establish a global climatology of OH, they do provide a test of our understanding of the fast photochemistry. To date,

theoretical predictions of OH concentrations (from measured trace-gas concentrations and photolysis rates) tend to be higher than the measurements by up to a factor of two.

- Measurements of peroxy radical concentrations in the remote free troposphere are in reasonable agreement with theory; however, significant misunderstanding exists with regard to the partitioning of odd nitrogen and the budget of formaldehyde.
- Measurements have shown that export of ozone produced from anthropogenic precursors over North America is a significant source for ozone in the tropics during the dry season. These findings show the influence of human activities on the global tropospheric ozone balance.
- Photochemical net ozone destruction in the remote atmosphere has been identified in several experiments. It is likely to occur over large parts of the troposphere with rates up to several ppb per day. Consequently, an increase in UVB radiation (e.g. from stratospheric ozone loss) is expected to decrease tropospheric ozone in the remote atmosphere but in some cases will increase production of ozone in and transport from the more polluted regions. The integrated effect on hydroxyl concentrations and climate is uncertain.

It was emphasized in *UNEP* ... (1994) that uncertainties in the global tropospheric ozone budget, particularly in the free troposphere, are mainly associated with uncertainties in the global distribution of ozone itself and its photochemical precursors, especially CO and NO_x. The role of heterogeneous processes including multiphase chemistry in the troposphere is not well characterized, and the catalytic efficiency of NO_x in catalysing ozone formation in the free troposphere has not been confirmed by measurements.

Recent results and problems of tropospheric ozone research have been discussed by Høv *et al.* (1997) in connection with the completion of the first phase of the Tropospheric Ozone Research (TOR) Project which is a part of the EUROTRAC Programme as well as by Pszenny and Brasseur (1997) in the context of the development of tropospheric-ozone studies within IGAC (International Global Atmospheric Chemistry) Core Project of the IGBP (Bates *et al.*, 1998; Kondratyev, 1998).

Pszenny and Brasseur (1997) pointed out that the most important tropospheric ozone precursors are nitrogen oxides ($NO_x = NO + NO_2$), methane and other 'nonmethane' hydrocarbons (NMHCs) and carbon monoxide. All of these precursors are products of fossil fuel and biomass burning, but each also has significant sources from the biosphere. The only ozone sink are depositions to surfaces (vegetation, soil, oceans), chemical destruction *in situ*, and some export back to the stratosphere in the tropics. It is believed that the increase of ozone observed in near-surface air, which may also be occurring in the free troposphere, is the result of increasing NO_x emission.

Cox (1999) developed a one-dimensional box model to describe the ozone budget and HO_x chemistry in the marine boundary layer. Results using small prescribed NO_x concentrations gave compensation points, where ozone loss by photolysis and physical removal is balanced by its production, via NO_x chemistry, of ~30 ppt to ~15 ppt at the two sites considered (Mace Head, Ireland and Cape Grim, Tasmania), in line with conclusions from observational data.

According to *Scientific* ... (1999), trends in tropospheric ozone since 1970 in the Northern Hemisphere show large regional differences, with increases in Europe and Japan, decreases in Canada and only small changes in the US. The trend in Europe since the mid-1980s has reduced to virtually zero (at two recording stations). In the Southern Hemisphere, small increases have been observed in surface ozone.

A number of field-observational programmes carried out recently obtained new important information on ozone dynamics (Ariya *et al.*, 1998; Fisher *et al.*, 1998; Godzik, 1998, Lin *et al.*, 1998; Mailhot *et al.*, 1998; Oltmans *et al.*, 1998; and others). One of the programmes was the Pacific Exploratory Mission – West A (PEM – West A) which is a major component of IGAC's East Asia/North Pacific Regional Experiment (APARE) Activity (Pszenny and Brasseur, 1997). The principal aim of PEM–West A has been investigations of the tropospheric ozone distribution over the North Pacific and of the growing emissions of ozone precursors from eastern Asia, which will be a major contributor to expected ozone increase on a hemispheric scale in the coming decades. Observed ozone trends were best described in terms of two geographical domains: the western North Pacific rim and the western tropical North Pacific. For both regions, ozone photochemical destruction decreased more rapidly with altitude than did photochemical formation. The ozone tendency was typically found to be negative below 6 km and positive from 6 to 8 km. On the basis of lidar sounding data from a DC-8 aircraft, Newell *et al.* (1997) studied the influence of Asian continental pollution on the western Pacific in September–October 1991 and February–March 1994.

Schultz *et al.* (1999) studied the budget of ozone and nitrogen oxides ($NO_x = NO + NO_2$) in the tropical South Pacific troposphere by photochemical point modelling of aircraft observations at 0–12-km altitude from PEM–Tropics A flights in September–October 1996. It was found, in particular, that chemical production of ozone equals one-half of chemical loss in the tropospheric column over the tropical South Pacific. The net loss is 1.8×10^{11} molecules $cm^{-2}\ s^{-1}$. The missing source of ozone is matched by westerly transport of continental pollution into the region. Application of a global three-dimensional model corroborates the results from the point model and reveals the importance of biomass burning emissions in South America and Africa for the ozone budget over the tropical South Pacific (biomass burning increases average ozone concentrations by 7–8 ppbv throughout the troposphere). The NO_x responsible for ozone production within the South Pacific troposphere below 4 km can largely be explained by decomposition of peroxyacetylnitrate (PAN) transported into the region with biomass burning pollution at higher altitudes.

Fenn *et al.* (1999) discussed *in situ* and laser remote measurements of gases and aerosols, made with airborne instrumentation to establish a baseline chemical signature of the atmosphere above the South Pacific Ocean during PEM–Tropics A observations. Between 8 and 52°S, biomass-burning plumes containing elevated levels of O_3, over 100 ppbv, were frequently encountered by the aircraft at altitudes ranging from 2 to 9 km. Air with elevated O_3 was also observed remotely up to the tropopause, and these air masses were observed to have no enhanced aerosol loading. Frequently, these air masses had some enhanced potential vorticity (PV) associated with them, but not enough to explain the observed O_3 levels. A relationship between O_3 and PV was developed from cases of clearly defined O_3 from stratospheric origin, and this relationship was used to estimate the stratospheric contribution to the air masses containing elevated O_3 in the troposphere.

The aim of the summer 1993 intensive of the North Atlantic Regional Experiment (NARE) was to investigate how major sources of gases in the surrounding industrial regions affect ozone distribution over the north Atlantic Ocean (Lin *et al.*, 1998). The results strongly suggest that NO_x emissions from human activities have substantially increased ozone concentrations in much of the troposphere over the north Atlantic. In the boundary layer and lower free troposphere the largest increases are seen in the industrial regions of the United States and Europe.

Lee *et al.* (1998) discussed 8-yr ozone sounding data obtained between 1989 and 1997 at Naha (Okinawa Island). These data demonstrate the continuous increase in tropospheric ozone in north-east Asia in the 1990s. It was suggested that this tropospheric ozone trend is due to the increasing emission of NO_x from north-east Asia (China, Japan, South Korea, Taiwan) during this period with a rate of 3.9% per year.

Logan (1999a, b) conducted an analysis of ozonesonde data, synthesizing what is known about the distribution of tropospheric ozone (it was pointed out that a minimum of 20 soundings are required for 95% confidence intervals of ozone monthly mean levels to be less than $\pm$30% near the extratropic tropopause; 20 soundings also ensures mean levels reliable to better than $\pm$15% for 500–500 hPa for the extratropics and 800–1,000 hPa in the tropics). The results indicate that ozone variability is higher in the upper troposphere for subtropical locations than for tropical locations, and 35 soundings are required for 400–1,000 hPa mean levels to be defined to better than $\pm$15%.

For the northern mid and high latitudes, the broad summer maximum in ozone in the mid-troposphere extends all the way up to the tropopause. Median concentrations at the tropopause are highest in June and July, typically 120–200 ppb, and are a factor of 2 smaller in winter. Highest values of ozone are in spring 2 km above the tropopause. Mean ozone values in the mid-troposphere are approximately constant from 35 to 50°N in summer. In the northern subtropics, there is a summer minimum in mid-tropospheric ozone over the Pacific and a summer maximum over the Atlantic which appear to be related to differences in circulation. Ozone is most similar in May and June over the Atlantic and Pacific, the months with minimum burning in the tropics. On the basis of ozonesonde data in combination with surface and satellite data Logan (1999a) derived a climatological grid for tropospheric ozone.

Using a large ozone-profile dataset obtained at Réunion Island (21°S, 55°E) since 1992, Taupin *et al.* (1999) showed that tropospheric ozone above the south-western Indian Ocean is strongly linked to dynamical conditions prevailing in the tropics. The average seasonal variation of TrOZ reaches 25 DU and peaks at 55 DU during the month of October. A large ozone enhancement was observed that occurs in parallel with the biomass-burning season in south-eastern Africa and Madagascar (this season extends from August to December and peaks in October). The ozone maximum concomitantly results from the transport of ozone and its precursors westward from the biomass-burning zones. High ozone values recorded in the upper troposphere during the austral winter are likely to originate from additional stratosphere–troposphere exchanges. Winter anticyclonic conditions further favour the photochemical production of ozone. The low tropospheric ozone-mixing ratios observed across the entire troposphere during the austral summer could result from the heterogeneous destruction of ozone and from less favourable conditions for its photochemical production, because of the development of deep convection.

Since the production of tropospheric ozone is due to oxidation of CO and hydrocarbons in the presence of nitrogen oxide radicals ($NO_x = NO + NO_2$), which may originate from lightning, downwelling of stratospheric air, aircraft emissions and surface emissions (including fossil-fuel combustion) transported convectively in the upper stratosphere, as well as hydrogen oxide radicals ($HO_x = OH +$ peroxy radicals), Jaeglé *et al.* (1998a) examined the origin of NO_x in the upper troposphere over the central United States using aircraft observations during the SUCCESS (Subsonic aircraft: Cloud and Contrail Effects Special Study) campaign in April–May of 1996. Simultaneous measurements of NO, NO_y, N_2O, O_3, OH, HO_2, H_2O, CO, CO_2 and CH_4 concentrations were made up to 12.5 km altitude together with aerosol, cloud and radiative observations. Observational data on correlation between NO_y (sum of NO_x and its oxidation products) and CO at 8–12-km altitude indicate that NO_x originates primarily from convective transport of polluted boundary-layer air. Lightning and aircraft emissions appear to be only minor sources of NO_x. Simulation modelling led to the conclusion that the NO_x/NO_y ratio is maintained above the chemical steady state by frequent convective injections of fresh NO_x from the polluted boundary layer and by the long lifetime of NO_x in the upper troposphere (5–10 days) (see also Molemaker and Arellano, 1998). In contrast with earlier studies, Jaeglé *et al.* (1998b) found no evidence for fast heterogeneous recycling from HNO_3 to NO_x in the upper troposphere.

The sources of HO_x (OH + peroxy radicals) and the associated production of ozone at 8–12 km over the United States were examined by Jaeglé *et al.* (1998a) by modelling observations of OH, HO_2, NO and other species during SUCCESS (the 0–D photochemical model was used, developed at Harvard University). Comparison with observations indicates that HO_x concentrations measured in SUCCESS are a factor of 3 higher than can be calculated from oxidation of water-vapour photolysis of acetone. The highest discrepancy was seen in the outflow of a convective storm. Jaeglé *et al.* (1998b) showed that convective injection of peroxides (CH_3OOH and H_2O_2) and formaldehyde (CH_2O) from the boundary layer to the upper troposphere could resolve this discrepancy. More generally, the data collected over the central United States during SUCCESS suggest that the local convection was a major source of HO_x and NO_x to the upper troposphere. An average net ozone production of 2 ppbv day^{-1} was found between 8 and 12 km over the continental United States in spring. Ozone production was NO_x limited under essentially all conditions encountered in SUCCESS. The high levels of HO_x present in the upper troposphere stimulate ozone production and increase the sensitivity of ozone to NO_x emissions from aircraft and other sources.

An important component of IGAC is the Biomass Burning Experiment: Impact on the Atmosphere and Biosphere (BIBEX). This was set up to monitor the prominent role of biomass burning as a source of atmospheric pollutants and of the important ecological impact of vegetation fires in the tropics. The Southern Africa Fire–Atmosphere Research Initiative (SAFARI-92) field experiment was conducted in the 1992 dry season in Southern Africa under the auspices of BIBEX, together with the joint Brazilian and US project Transport and Atmospheric Chemistry near the Equator – Atlantic (TRACE-A), field studies of which were also conducted in the 1992 BIBEX IGAC's Southern Atlantic Region Experiment (STARE) campaign.

The results of SAFARI confirmed that it is justified to consider biomass burning as a significant contributor to the overall increase in GHGs that has occurred over the last 150 years, accounting for some 10–25% of current emissions. A major finding of TRACE-A was that widespread biomass burning in both America and Southern Africa is the dominant source of precursor gases necessary for the formation of the huge amounts of ozone observed over the South Atlantic Ocean. The generation of ozone is usually enhanced in the upper troposphere, where relatively high concentrations of NO_x prevail.

It has been found that ozone in the tropical troposphere plays a key role in determining the global oxidizing power of the atmosphere. Most of the oxidation of long-lived gases by hydroxyl radicals takes place in the tropics, where intense sunlight and high humidity promote the formation of OH from photolysis of ozone.

The release into the atmosphere of biogenic gases produced by the African rain forest and savanna, as well as from biomass burning, and their role in the formation of tropospheric ozone and aerosol is the focus of the IGAC Experiment for Regional Sources and Sinks of Oxidants (EXPRESSO). Under IGAC, the Global Tropospheric Ozone Project (GTOP) is now being formulated so that all forthcoming data can be used intelligently to provide an understanding of the global tropospheric ozone cycle (Pszenny and Brasseur, 1997). A major aim of IGAC is to understand the role of biological processes in producing and consuming atmospheric trace gases.

An important contribution to studying tropospheric ozone dynamics was made by the TOR subproject of EUROTRAC (Borrell *et al.*, 1998). As pointed out by Høv *et al.* (1997), the main achievement of TOR was the development of suitable instrumentation for high-quality measurements of ozone precursors (NO_x, NO_y, VOC), intermediates (carbonyl compounds, RO_2) and photolysis rates as well as the implementation of quality assurance procedures for these measurements, the establishment of a large high-quality network and accumulation of a significant database. The high-quality datasets provide detailed mechanistic information on the chemical and physical processes that control the budget of ozone and its precursors in the polluted boundary layer and in the free troposphere. The vertical soundings were analysed for distribution and seasonal variation in ozone in the free troposphere over Europe and provide insight into processes – such as the exchange between the boundary layer and the free troposphere and between the troposphere and the stratosphere.

Examples of the results obtained include the concept of O_X for studying chemical ozone formation, the O_X/NO_X relation for studying the catalytic efficiency of NO_X, the use of alkyl nitrates to measure peroxy radicals, the role of individual hydrocarbons in ozone formation and the use of the H_2O/O_3 correlation for the ozone budget in the upper troposphere.

A hierarchy of models was developed and/or applied in TOR for data analysis. Three-dimensional transport models with coupled chemistry modules were developed in TOR and have been used to study important transport processes – such as convection – and for budget studies.

Of course, after the completion of the first phase of TOR many problems still remain unresolved because of the complexity of the processes that govern tropospheric ozone dynamics. A different aspect of the problem is the unseparable influence of transport and chemistry on the concentrations that are observed at a given location. It is also important that the net chemical balance of O_3 depends in non-linear fashion on the concentrations of NO_X, VOCs, H_2O and O_3 itself and on the UV radiation flux and is thus closely coupled with the atmospheric lifecycles of other trace gases. At continental surface sites, dry deposition plays an important role in the net balance of ozone, in addition to advection and vertical exchange. It is obvious that future success may be achieved only on the basis of combined analysis of sufficiently complete observational data and results of simulation modelling with the use of fully coupled models of chemistry and transport.

In this context, very important progress was recently made in simulations of tropospheric ozone distribution using a chemistry–climate model. An illustration of this success is the analysis of tropospheric O_3 over the Indian Ocean carried out by De Laat *et al.* (1999) using the European Centre Hamburg (ECHAM) chemistry–general circulation model. Such an approach allowed interactive dynamics of the atmospheric circulation and ozone to be studied. It was shown, in particular, that large-scale upper tropospheric O_3 minima were caused by convective transport of O_3-depleted boundary-layer air in the intertropical convergence zone (ITCZ). Similarly, an upper tropospheric O_3 minimum was caused by Cyclone Marlene south of the ITCZ. The mid-tropospheric O_3 maxima were caused by transport of polluted African air.

As pointed out earlier, satellite remote sensing data are important sources of information on tropospheric-ozone distribution are (Borrell *et al.*, 1998; *Scientific ...*, 1999). There are many techniques capable of retrieving tropospheric ozone from TOMS data (Kaye, 1998). One of them is the use of TOMS-SBUV residuals (in which tropospheric ozone is taken as the difference between TOMS total ozone and integrated stratospheric ozone from one of the Solar Backscatter Ultraviolet (SBUV) instruments) or the difference in ozone columns over mountains and nearby sea-level areas (as have been carried out over the Andes Mountains and nearby Eastern Pacific). Another residual technique suggested by Sushil Chandra is based on the difference between TOMS total ozone and stratospheric ozone determined by combining the Microwave Limb Sounder (MLS) and HALOE instruments aboard the *UARS* satellite. Another similar method, known as the Convective Cloud Differential (CCD) technique, derives tropospheric ozone from the average difference between TOZ over clear regions and over regions of high clouds (in the latter case reliable cloud-top information becomes very important). Initial studies used TOMS data and TOMS/SAGE residuals only, recently work has been based on a residual approach using TOMS and SBUV data (in this case precise determination of the tropopause height is very important).

The Tropospheric Emission Spectrometer (TES) planned for EOS CHEM spacecraft (Kaye and Rind, 1998) is a new method of obtaining tropospheric ozone data. TES is a Fourier Transform Spectrometer, measuring in the 650–3000 cm^{-1} wavelength region using both nadir and limb-viewing geometry; spectral resolution is 0.025 cm^{-1} in the limb mode and 0.1 cm^{-1} in the nadir mode. The primary focus of the TES instrument is measurement of ozone and its precursors in the troposphere, although TES has excellent capability for detecting a broad range of species because of its high resolution.

Munro *et al.* (1998) discussed retrieval of vertical profiles of ozone in the troposphere from the data of the European Space Agency's GOME on board the *ERS-2* satellite launched in April 1995. GOME is an ultraviolet/visible spectrometer which measures solar radiation backscattered from the Earth's atmosphere in four contiguous wavelength bands between 137 and 790 nm at moderate resolution (0.2–0.4 nm). An important advantage of GOME in comparison with the SBUV instrument is the availability of data for the ozone Huggins bands (310–340 nm). With knowledge of the temperature profile, the temperature-dependent spectral structures in the Huggins bands yield additional ozone-profile information below the ozone peak (in the troposphere and lower stratosphere). The comparison between retrieval results obtained and ozonesonde data indicates that, in cloud-free scenarios and above low clouds, ozone profiles extending down through the troposphere can be retrieved from GOME measurements. Munro *et al.* (1998) demonstrated GOME's ability to detect tropospheric ozone produced by biomass burning over Africa.

Future tropospheric ozone research faces a number of still unresolved problems, including:

(1) Accumulation of adequate global information on observed variability of tropospheric ozone for sufficiently long time periods (the lack of data is especially severe in the tropics and subtropics); observations of atmospheric-ozone precursors are equally important.
(2) Investigations about the chemistry and photochemistry of tropospheric ozone under cloudy conditions.

As emphasized by Crutzen (1998), the role of clouds as transporters of chemical constituents (such as reactive hydrocarbons, CO and NO and their oxidation products) from the boundary layer to the mid and upper troposphere (and possibly into the lower stratosphere) needs to be better understood and quantified, so that they can be parametrized for inclusion in large-scale photochemical models of the atmosphere. Similarly, the production of NO by lightning and its vertical redistribution by convective storms needs also to be much better quantified.

The interaction of chemical constituents emanating from the boundary layer with liquid and solid hydrometeors in the clouds are of special importance. The influence of clouds on the photochemically active UV radiation field is a potentially important research topic. An issue of key significance are interactions between gases and atmospheric aerosols. Since the continental biosphere is a large source of hydrocarbons, quantification of relevant sources in terms of physical (e.g. temperature, humidity, light levels) and biogeochemical (solid physical and chemical properties, land use) parameters are urgently needed for inclusion in atmospheric models. The formation of ozone, carbon monoxide, partially oxidized gaseous hydrocarbons and organic aerosol may be better quantified and parametrized for inclusion in a chemical transport model, if the hydrocarbon oxidation mechanisms in the atmosphere were better understood.

Monod and Carlier (1999) have conducted a box-model daytime study of the multiphase photochemistry (both gas and aqueous phases) of C1 organic compounds within a non-precipitating cloud on a local scale. According to their results, when ozone accumulates in clear-sky conditions, as soon as the cloud is formed, the tropospheric ozone changes drastically: the net production decreases by a factor of 2 or more and, depending on the NO_x concentrations and the pH values, can actually lead to a net chemical destruction. Monod and Carlier (1999) pointed out that both indirect and direct impacts, caused by the presence of the liquid phase, explain this. The indirect impact is the result of the much higher solubility of HO_2 and RO_2 radicals than NO and NO_2, resulting in lower gas-phase efficiency of the NO_x cycles in producing ozone. The direct impact is caused by the very fast reaction of ozone towards O_2^- radicals within the liquid phase. The relative importance of these two impacts has been determined for different NO_x concentrations and pH values. When $NO_x = 1$ ppbv, the direct impact is of negligible importance at pH 3 (for a total decrease in ozone production of the factor of 2), but it accounts for 6% at pH 4.16 (for a total decrease in ozone production of a factor of 6.6), 16% at pH 5.2 (for a total net chemical destruction of ozone) and 28% at pH 6 (for a total net chemical destruction of ozone). At this NO_x level and at pH 5.2 and 6, despite its small contribution, the effect of the direct impact is the net chemical destruction of ozone. When $NO_x = 0.1$ ppbv, both the direct and the indirect impact contribute to net chemical destruction at all pH values. However, the direct impact now contributes to a larger extent, accounting for 15% at pH 3, and up to 77% at pH 6. The direct impact involves the $O_3 + O_2^-$ reaction which leads to the production of aqueous-phase OH radicals. The latter species in turn react with dissolved organic compounds to produce additional HO_2 and O_2^- radicals. Therefore, the aqueous phase sink of ozone is auto-catalytic in the presence of soluble organic compounds.

Warnings about potential surprises in the photochemistry of tropospheric ozone were made by Ravishankara *et al.* (1998), who discussed the role of the hydroxyl radical OH as a natural atmospheric detergent responsible for cleansing the atmosphere of pollutants by means of their oxidation. Ravishankara *et al.* (1998) pointed out that, in general, the reactions of electronically excited species are of negligible importance in the chemistry of the lower atmosphere, but the case of $O(^1D)$ is a notable exception. Its role is pivotal. Even though most of the $O(^1D)$ is deactivated to the ground state, $O(^3P)$ – the small fraction that survives to react with H_2O and CH_4 – turns out to be the major source of OH. Knowledge of how $O(^1D)$ is formed in the atmosphere is therefore critical in understanding the creation of OH. Recent surprising findings are beginning to reveal the importance of the longer-wavelength 'tail' in the chemistry of $O(^1D)$ formation. The longer wavelengths are important because stratospheric ozone screens most of the shortwave ultraviolet from the lower atmosphere.

Bojkov (1999) emphasized the significant uncertainties that remain in the budget of tropospheric ozone.

1.3 CONCLUSIONS

Investigations of change in total ozone (especially in high latitudes) highlight the TOZ variability problem as a very important aim for the study of global-scale environmental dynamics, having great significance for humans and ecosystems. The recognition of such an environmental danger led to the signing of the Montreal Protocol and further Amendments to the Protocol to avoid dangerous environmental consequences of ozone depletion. There are many lessons to be learned from the Montreal Protocol that can be applied to solving other global environmental issues (*Scientific* ..., 1999).

The primary lesson is application of the 'precautionary principle': taking necessary action in time to prevent damage rather than waiting for proof of the damage by which time the damage could have been great and irreversible. Another important lesson of the Protocol is how to react to issues when there is no scientific certainty (in 1982 when the Protocol was signed many questions remained open). The solution has been to undertake successive steps to phase-out ozone-depleting substances and getting the scientific community to advise governments periodically on the further steps needed to protect the ozone layer and recommend alternative technologies. Over the last 10 years governments changed the Protocol four times in accordance with relevant scientific advice. Another lesson of the Protocol is in promoting universal participation, including developing countries in the Protocol by recognizing 'common and differential responsibility'. Such a position resulted in almost all countries committing themselves to protection of the ozone layer. One more important lesson is the integration of science, economics and technology both in developing control measures and implementing them.

Sarma (1998) emphasized that one measure of success of the Montreal Protocol and its subsequent Amendments and Adjustments was the forecast of 'the world that was avoided' by the Protocol:

- The abundance of ozone-depleting substances in 2050, the approximate time when the ozone layer is now projected to recover to pre-1980 levels, would be at least 17 ppb of equivalent effective chlorine (this is based on the conservative assumption of a 3% annual growth in ozone-depleting gases) which is about five times larger than today's value;

- Ozone depletion would be at least 50% at mid-latitudes in the Northern Hemisphere and 70% at mid-latitudes in the Southern Hemisphere, about 10 times larger than today; and
- Surface UVB radiation would at least double at mid-latitudes in the Northern Hemisphere and quadruple at mid-latitudes in the Southern Hemisphere compared with an unperturbed atmosphere. This compares with current increases of 5% and 8% in the Northern and Southern Hemispheres, respectively, since 1980.

One of the main causes of the current focus on tropospheric ozone is its role as a greenhouse gas and toxic component. While reviewing the sources of change in radiative forcing and their influence on climate, Jim Hansen (1997) emphasized the need for information about the exact nature of change in the vertical distribution of ozone (especially to see if increases in tropospheric ozone may have countered decreases in stratospheric ozone) and the nature of aerosol particles (especially single-scattering albedo) as well as their height. In this context further development of numerical climate modelling, taking into account various minor gaseous components, is urgently needed. Brasseur *et al.* (1998) described simulations of ozone and related chemical tracers obtained by the Model for Ozone and Related Chemical Tracers (MOZART) which was developed at the National Center for Atmospheric Research (NCAR). One of the results was the simulation of changes in surface ozone since pre-industrial time, which shows the changes in ozone amounts of some 30–40 ppb in July over the US and much of Europe along with those of 20–30 ppb over most of Eurasia. The average radiative forcing coming from these increases should be approximately 0.45 W/m^2 (higher in the Northern Hemisphere, lower in the Southern Hemisphere). Predictions for the future (2050) show the largest changes of surface ozone are likely to come in the tropics.

Jacob *et al.* (1995) described the Chemistry, Aerosols and Climate Tropospheric Unified Simulation (CACTUS) effort at Harvard University. The present targets of this effort include the climatic effects of changes in emissions of precursors on tropospheric ozone formation, the feedback of climate change on tropospheric ozone, the climatic effects of changes in emissions of sulphate precursors and the response of sulphate aerosols to climate change.

1.4 REFERENCES

Abstracts of the European Conference on Atmospheric UV Radiation, 29 June–2 July 1998, Helsinki (1998) Finnish Meteorological Institute, Helsinki, 127 pp.

Andronache, C., Chameides, W. L. (1998) Interactions between sulphur and soot emissions from aircraft and their role in contrail formation. 2. Development. *J. Geophys. Res.*, **103**, ND9, 10,787–10,802.

Ariya, P. A., Jobson, B. T., Sander, R., Niki, H., Harris, G. W., Hopper, J. F., Anlauf, K. G. (1998). Measurements of C_2–C_7 hydrocarbons during the Polar Sunrise Experiment 1994: Further evidence for halogen chemistry in the troposphere. *J. Geophys. Res.*, **103**, ND11, 13,169–13,180.

Arnone, R. A., Ladner, Sh., La Violette, P. E., Brock, J. C., Rochford, P. A. (1998) Seasonal and interannual variability of surface photosynthetically available radiation in the Arabian Sea. *J. Geophys. Res.*, **103**, NC4, 7,735–7748.

Atmospheric Ozone, Proceedings XVIII Quadrennial Ozone Symposium, 12–21 September 1996, L'Aquila, Italy (1998) Parco Scientifico e Tecnologico, Abruzzo, Italy, 450 pp.

Bacmeister, J. T., Kuell, V., Offenmann, D., Riese, M., Elkins, J. W. (1999) Intercomparison of satellite and aircraft observations of ozone, CFC-11, and NO_y using trajectory mapping. *J. Geophys. Res.*, **104**, ND13, 16,375–16,390.

Barrie, L. A., Platt, U. (1997) Arctic tropospheric chemistry: Overview to *Tellus* special issue. *Tellus*, **49B**, 450–454.

Barrie, L. A., Platt, U., Shepson, P. (1998). Surface ozone depletion at the polar sunrise fueled by sea-salt halogens. *IGACtivities Newsletter*, N14, 4–7.

Bates, T. S., Huebert, B. S., Gras, J. L., Griffith, F. R., Durkee, P. A. (1998) International Global Atmospheric Chemistry (IGAC) Project's First Aerosol Characterization Experiment (ACE-1) Overview. *J. Geophys. Res.*, **103**, ND13, 16,297–16,318.

Baumgardner, D., Miake-Lye, R. C., Anderson, M. R., Brown, R. C. (1998) An evaluation of the temperature, water vapor, and vertical velocity structure of aircraft contrails. *J. Geophys. Res.*, **103**, ND8, 8,727–8,736.

Bhatt, P. P., Remsberg, E. E., Gordley, L. L., McInerney, J. M., Brackett, V. G., Russell, J. M., III (1999) An evaluation of the quality of Halogen Occultation Experiment ozone profiles in the lower stratosphere. *J. Geophys. Res.*, **104**, ND8, 9,261–9,276.

Bigelow, D. S., Slusser, J. R., Beaubien, A. E., Gibson, J. H. (1998). The USDA ultraviolet radiation monitoring program. *Bull. Amer. Meteorol. Soc.*, **79**, N4, 601–615.

Biggs, R. H., Joyner, M. E. B. (1994) *Stratospheric Ozone Depletion/UV-B Radiation in the Biosphere*, NATO ASI Series, Global Environmental Change, Vol. 18, Springer-Verlag, Berlin, 358 pp.

Bojkov, R. D., Reinsel, G. C. (1985). Trends in tropospheric ozone concentrations. *Atmospheric Ozone*, Reidel, Dordrecht, pp. 775–781.

Bojkov, R. D. (1986a). Surface ozone during the second half of nineteenth century. *J. Clim. Appl. Meteorol.*, **25**, 343–352.

Bojkov, R. D. (1986b) The 1979–1985 ozone decline in the Antarctic as reflected in ground based observations. *Geophys. Res. Lett.*, **13**, N12, 1,236–1,239.

Bojkov, R. D., Mateer, C. L., Hansson, A. L. (1988) Comparison of ground-based and Total Ozone Mapping Spectrometer using in assessing the performance of the Global Ozone Observing System. *J. Geophys. Res.*, **93**, ND8, 9,525–9,533.

Bojkov, R. D. (1988) Ozone changes at the surface and in the free troposphere. *Tropospheric Ozone*, I. S. A. Isaksen (ed.), D. Reidel, Amsterdam, pp. 83–96.

Bojkov, R. D., Zerefos, C. S., Balis, D. S., Ziomas, I. C., Bais, A. F. (1993). Record low total ozone during northern winter of 1992 and 1993. *Geophys. Res. Lett.*, **20**, 1,351–1,354.

Bojkov, R. D., Fioletov, V. E., Shalamjansky, A. M. (1994) Total ozone changes over Eurasia since 1973 based on re-evaluated filter ozonometer data. *J. Geophys. Res.*, **99**, 22,985–22,999.

Bojkov, R. D., Fioletov, V. E. (1995a) Estimating the global ozone characteristics during the last 30 years. *J. Geophys. Res.*, **100**, 16,537–16,551.

Bojkov, R. D., Bishop, L., Fioletov, V. E. (1995b) Total ozone trends from quality controlled ground-based data (1964–1994). *J. Geophys. Res.*, **100**, 25,867–25,876.

Bojkov, R. D., Fioletov, V. E. (1996). Total ozone variations in the tropical belt: an application for quality of ground based measurements. *Meteorol. Atmos. Phys.*, **58**, N1–4, 223–240.

Bojkov, R. D. (1998) Ozone variations in the Northern Polar region. *Meteorol. Atmos. Phys.*, **38**, 117–130.

Bojkov, R. D., Visconti, G. (1998) *Atmospheric Ozone. Proceedings XVIII Quadrennial Ozone Symposium, 12–21 September 1996, L'Aquila, Italy*, Parco Scientifico e Tecnologico, Abruzzo, Italy, Vol. 1, 450 pp.

Bojkov, R. D., Balis, D. S., Zerefos, C. S. (1998) Characteristics of the ozone decline in the Northern polar and middle latitudes during the winter–spring. *Meteorol. Atmos. Phys.*, **2**, 462 pp.

Bojkov, R. D. (1999). International Assessment of Ozone Depletion: 1998. *WMO Bull.*, **48**, N1, 35–44.

Borrell, P., Varotsos, C. A., Kondratyev, K. Ya. (1998). Key problems of chemistry and photochemistry of the troposphere: to the completion of the first phase of EUROTRAC. Project and future perspectives. *Ecological Chemistry*, **7**, N1, 1–12.

Brasseur, G. P., Müller, J.-F., Granier, C. (1996). Atmospheric impact of NO_x emissions by subsonic aircraft: A three-dimensional model study. *J. Geophys. Res.*, **101**, ND1, 1,423–1,428.

Brasseur, G., Amanatidis, G. T., Angeletti, G. (eds.) (1998) European Scientific Assessment of the Atmospheric Effects of Aircraft Emissions. *Atmos. Environ.*, **32**, N13, 2,327–2,422.

Brasseur, G. P, Cox, R. A., Hauglestaine, D., Isaksen, I., Lelieveld, J., Lister, D. H., Sausen, R., Schumann, U., Wahner, A., Wiesen, P. (1998) European satellite assessment of the atmospheric effects of aircraft emissions. *Atmos. Environ.*, **32**, 2,329–2,418.

Brasseur, G. P., Hauglustaine, D. A., Walters, S., Rasch, P. J., Muller, J.-F., Granier, C. and Tie, X. X. (1998) MOZART: a global chemical transport model for ozone and related chemical tracers, Part 1. Model description, *J. Geophys. Res.*, 1998.

Campos, T. L., Weinheimer, A. S., Zheng, J., Montzka, O. D., Walega, J. G., Grahek, F. E., Vay, S. A., Collins, J. E. Jr, Wade, L. O., Sachse, G. W., Anderson, B. E., Brune, W. H., Tan, D., Faloona, I., Baughcam, S. L., Ridley,

B. A. (1998) Measurements of NO and NO_x emission indices during SUCCESS. *Geophys. Res. Lett.*, **25**, N10, 1,713–1,716.
Chandra, S., Varotsos, C., Flynn, L. E. (1996) The mid-latitude ozone trends in the Northern Hemisphere. *Geophys. Res. Lett.*, **23**, N5, 555–558.
Chanin, M.-L., Geller, M. (1998) Report of the WCRP/JSC Meeting. *SPARC Newsletter*, N11, 1–2.
Considine, D. B., Stolarski, R. S., Hollandsworth, S. M., Jackman, C. H., Fleming, E. L. (1999) A Monte Carlo uncertainty analysis of ozone trend predictions in a two-dimensional model. *J. Geophys. Res.*, **104**, ND1, 1,749–1,766.
Cox, R. A. (1999) Ozone and peroxy radical budgets in the marine boundary layer: Modeling the effect of NO_x. *J. Geophys. Res.*, **104**, ND7, 8,047–8,056.
Crutzen, P. S., 1998. The Bulletin interview. *WMO Bull.*, **47**, N2, 111–123.
Dameris, M., Grewe, V., Khler, I., Sausen, R., Brühl, C., Grooß, J.-U., Steil, B. (1998) Impact of aircraft NO_x-emissions on tropospheric and stratospheric ozone. Part II: 3-D model results. *Atmos. Environ.*, **32**, 3,185–3,200.
Davies, W. E., Vaughan G., O'Connor F. M. (1998) Observation of the near-zero ozone concentration in the upper troposphere in midlatitudes. *Geophys. Res. Lett.*, **25**, N8, 1,173–1,176.
Derwent, R. G., Jenkin, M. E., Saunders, S. M., Pilling, M. S. (1998). Photochemical ozone creation potential for organic compounds in north-west Europe calculated with master chemical mechanism. *Atmos. Environ.*, **32**, N14–15, 2,429–2,441.
Donahue N. M., Kroll, J. H., Anderson, J. G., Demerjian, K. L. (1998). Direct observation of OH production from the ozonolysis of olefins. *Geophys. Res. Lett.*, **25**, N1, 59–62.
Dubey, M. K., Smith, G. P., Hartley, W. S., Kinnson, D. E., Cornell, P. S. (1997). Rate parameter uncertainty effects in assessing stratospheric ozone depletion by supersonic aviation. *Geophys. Res. Lett.*, **24**, N22, 2,737–2,740.
Ebel, A., Friedrich, F., Rodhe, H. (1997) *Tropospheric Modeling and Emission Estimates*, Springer-Verlag, Heidelberg, 450 pp.
Ellsaesser, H. W. (ed.) (1992) *Global 2000 Revisited. Mankind's Impact on Spaceship Earth*, Paragon House, New York, 436 pp.
Ellsaesser, H. W. (ed.) (1994) The unheard arguments: a rational view of stratospheric ozone. *21st Century Science and Technology*, **7**, N23, 37–45.
Feister, U., Grewe, R. (1995) Higher UV radiation inferred from low ozone levels at northern mid-latitudes in 1992 and 1993. *Global and Planetary Change*, **11**, N1, 25–34.
Fenn, M. A., Browell, E. V., Butler, C. F., Grant, W. B., Kooi, S. A., Clayton, M. B., Gregory, G. L., Newell, R. E., Zhu, Y., Dibb, J. E., Fuelberg, H. E., Anderson, B. E., Bandy, A. R., Blake, D. R., Bradshaw, J. D., Heikes, B. G., Sachse, G. W., Sandholm, S. T., Singh, H. B., Talbot, R. W., Thornton, D. C. (1999) Ozone and aerosol distributions and air mass characteristics over the South Pacific during the burning season. *J. Geophys. Res.*, **104**, ND13, 16,197–16,212.

Fioletov, V. E., Griffeen, E., Kerr, J. B., Wardle, D. I., Uchino, O. (1998). Influence of volcanic sulphur dioxide on spectral irradiance as measured by Brewer spectrophotometers. *Geophys. Res. Lett.*, **25**, N10, 1,655–1,658.

Fioletov, V. E., Kerr, J. B., Hare, E. W., Labow, G. J., McPeters, R. D. (1999) An assessment of the world ground-based total ozone network performance from the comparison with satellite data. *J. Geophys. Res.*, **104**, ND1, 1,737–1,748.

Fiore, A. M., Jacob, D. S., Logan, J. A., Yin, J. H. (1998). Long-term trends in ground level ozone over the contiguous United States, 1980–1995. *J. Geophys. Res.*, **103**, ND1, 1,471–1,480.

Fisher, H., Nikitas, C., Parchatka, K., Zenker T., Harris, G. W., Matuska, P., Schmitt, R., Mihelcic, D., Muesgen, D., Paetz, H. -W., Schultz, M., Volz-Thomas, A. (1998). Trace gas measurements during the Oxidizing Capacity of the Tropospheric Atmosphere campaign 1993 at Izaña. *J. Geophys. Res.*, **103**, ND11, 13,505–13,518.

Fortuin, J. P. F., Kelder, H. M. (1999) An ozone climatology based on ozonesonde and satellite measurements. *J. Geophys. Res.*, **104** (in print).

Fuglestvedt, J. S., Berntsen, T. K., Isaksen, I. S. A., Mao, H., Liang, X., Wang, W. C (1999) Climatic forcing of nitrogen oxides through changes in tropospheric ozone and methane: Global 3D model studies. *Atmos Environ.*, **33**, 961–977.

Geophysical Research Letters, November Supplement (1986), **13**, N12, 1,191–1,362.

Godzik, B. (1998) Ground level ozone concentration in the Krakow region, southern Poland. *Environ. Pollut.*, **98**, N3, 273–280.

Grewe, V., Dameris, M., Hein, R., Köhler, I., Sausen, R. (1999) Impact of future subsonic aircraft NO_x emissions on the atmospheric composition. *Geophys. Res. Lett.*, **26**, N1, 47–50.

Guirlet, M., Kibane-Dawe, I., Harris, N. (1998). Third European Stratospheric Experiment on Ozone (THESEO). *SPARC Newsletter*, N10, 24.

Hales, J. (1996). Scientific background for AMS Policy Statement on Atmospheric Ozone. *Bull. Amer. Meteorol. Soc.*, **77**, N6, 1,245–1,253.

Hansen, J., Sato, M., Ruedy, R. (1997) Radiative forcing and climate response. *J. Geophys. Res.*, **102**, 6,831–6,864.

Hansen, G., Chipperfield, M. P. (1999) Ozone depletion at the edge of the Arctic polar vortex 1996/1997. *J. Geophys. Res.*, **104**, ND1, 1,846–1,897.

Harris, N. R. P., Kibane-Dawe, I., Amanatidis, G. T. (eds.) (1998) Air pollution research report 66. *Polar Stratospheric Ozone 1997 – Proceedings of the Fourth European Symposium, 22–26 September 1997, Schliersee, Bavaria, Germany*, European Communities (EUR 18032) EN, Brussels, 772 pp.

Herman, J. R., Krotkov, N., Celarier, E., Larko, D., Labow, G. (1999a) Distribution of UV radiation at the Earth's surface from TOMS-measured UV-backscattered radiances. *J. Geophys. Res.*, **104**, ND10, 12,059–12,076.

Herman, J. R., Larko, D., Ziemke, J. (1999b) Regional changes in global UV reflectivity from clouds and aerosols. *J. Geophys. Res.*, **104** (in print).

Hollandsworth, S. M., Mc Peters, R. D., Flynn, L. E., Planet, W., Miller, A. S., Chandra, S. (1995). Ozone trends deduced from combined Nimbus-7 SBUV and NOAA-11 SBUV/2 data. *Geophys. Res. Lett.*, **22**, N8, 905–908.

Hólm, E. V., Untch, A., Simmons, A., Saunders, R., Bouttier, F., Andersson, E. (1999) *The SODA EU Project 1996-1999: Final Report from ECMWF, Abstract, 6 July 1999*, 12 pp.

Houghton, J. T., Meiro Filho, L. G., Callander, B. A., Harris, N., Kattenberg, A., Maskell, K. (eds.) (1996) *Climate Change 1995: The Science of Climate Change*, Cambridge University Press, Cambridge, 584 pp.

Høv, Ø. (ed.) (1997) *Tropospheric Ozone Research. Tropospheric Ozone in the Regional and Sub-regional Context*, Springer for Science, The Netherlands, 455 pp.

Høv, Ø., Kley, D., Volz-Thomas, A., Beck, J., Grennfelt, P., Penkett, S. A. (1997). An overview of tropospheric ozone research. In: *Tropospheric Ozone Research*, Ø. Høv (ed.), Springer-Verlag, Berlin, pp. 1–34.

Isaksen, I. S. A. (1988) *Tropospheric Ozone. Regional and Global Scale Interactions*, Reidel, Dordrecht, Holland, 425 pp.

Jacob, D. J., Leovy, C., Schoeberl, M., McNeal, R. J. and Gleason, J. (1995) Atmospheric trace gas measurements for the year 2000 and beyond, *The Earth Observer*, **7**, 31–38.

Jaeglé, L., Jacob, D. J., Brune, W. H., Tan, D., Faloona, I. C., Weinheimer, A. S., Ridley, B. A., Campos, T. L., Sachse, G. W. (1998a) Sources and chemistry of HOx and production of ozone in the upper troposphere over the United States. *Geophys. Res. Lett.*, **25**, N10, 1,705–1,708.

Jaeglé, L., Jacob, D. J., Brune, W. H., Tan, D., Faloona, I. C., Weinheimer, A. S., Ridley, B. A., Campos, T. L., Sachse, G. W. (1998b). Sources of HO_x and production of ozone in the upper troposphere over the United States. *Geophys. Res. Lett.*, **25**, N10, 1,709–1,712.

James, P. M. (1998). An interhemispheric comparison of ozone minihole climatologies. *Geophys. Res. Lett.*, **25**, N3, 301–304.

Jeuken, A. B. M., Eskes, H. J., van Velthoven, P. F. J., Kelder, H. M., Hólm, E. V. (1999) Assimilation of total ozone satellite measurements in a three-dimensional tracer transport model. *J. Geophys. Res.*, **104**, ND5, 5,551–5,563.

Jiang, Y., Yung, Y. L., Zurek, R. W. (1996). Decadal evolution of the Antarctic ozone hole. *J. Geophys. Res.*, **101**, ND4, 8,985–8,999.

Jobson, B. T., McKeen, S. A., Parrish, D. D., Fehsenfeld, F. C., Blake, D. R., Goldstein, A. H., Schauffler, S. M., Elkins, J. W. (1999) Trace gas mixing ratio variability versus lifetime in the troposphere and stratosphere: Observations. *J. Geophys. Res.*, **104**, ND13, 16,091–16,115.

Kane, R. P., Sahai, Y., Casiccia, C. (1998). Latitude dependence of the quasi-biennial oscillation and quasi-triannual oscillation characteristics of the total ozone measured by TOMS. *J. Geophys. Res.*, **103**, ND7, 8,477–8,490.

Katsambas, A., Varotsos, C. A., Vezirgianni, G., Antoniou, Ch. (1997). Surface solar ultraviolet radiation: A theoretical approach of the SUVR reaching the ground in Athens, Greece. *Environ. Sci. Pollut. Res.*, **4**, N2, 69–73.

Kaye, J. (1998). Summary of Total Ozone Mapping Spectrometer (TOMS) science team meeting. *The Earth Observer*, **10**, N3, 15–22.

Kaye J. A., Rind, D. (1998) Summary of chemistry/climate modeling meeting. *The Earth Observer*, **10**, N3, 27–36.

Kim, J. H., Newchurch, M. J. (1998) Biomass-burning influence on tropospheric ozone over New Guinea and South America. *J. Geophys. Res.*, **103**, ND1, 1,455–1,462.

Kleinman, L. I., Daum, P. H., Lee, J. H., Lee, Y-N., Weinstein-Lloyd, J., Springston, S. R. (1998) Photochemistry of O_3 and related components over southern Nova Scotia. *J. Geophys. Res.*, **103**, ND11, 13,519–13,529.

Knudsen, B. M., Larsen, N., Mikkelson, I. S., Morcrette, J.-J., Braathen, G. O., Kyr, E., Fast, H., Gernandt, H., Kanzawa, H., Nakane, H., Dorenkov, V., Yushkov, V., Hansen, G., Gil, M., Shearman, R. J. (1998) Ozone depletion in and below the Arctic vortex for 1997. *Geophys. Res. Lett.*, **25**, N5, 627–630.

Koehler, J., Hajost, S. A. (1990) The Montreal Protocol: A dynamic agreement for protecting the ozone layer. *Ambio*, **19**, N2, 82–86.

Komhyr W. D., Grass K. D., Evans, R. D., Leonard, R. K., Quincy, D. M., Hofmann, D. J., Koenig, G. L. (1994) Unprecedented 1993 ozone increase over the United States from Dobson spectrophotometer observations. *Geophys. Res. Lett.*, **21**, N3, 201–204.

Kondratyev, K. Ya. (1989). Ozone Dynamics. Progress in Science and Technology. *Geomagnetism and Upper Atmosphere*, **11**, VINITI, Moscow, 212 pp. (in Russian).

Kondratyev, K. Ya., Varotsos, C. A. (1995a) Atmospheric ozone variability in the context of global change. *Int. J. Remote Sensing*, **16**, N10, 1,851–1,881.

Kondratyev, K. Ya., Varotsos, C. A. (1995b). Volcanic eruptions and global ozone dynamics. *Int. J. Remote Sensing*, **16**, N10, 1,887–1,895.

Kondratyev, K. Ya., Varotsos, C. A. (1997). A review of greenhouse effect and ozone dynamics in Greece. In: *Atmospheric Ozone Dynamics*, C. Varotsos (ed.), Springer-Verlag, Berlin, pp. 175–228.

Kondratyev K. Ya., Galindo, I, (1997) *Volcanic Activity and Climate*, A. Deepak, Hampton, Virginia, 382 pp.

Kondratyev, K. Ya. (1998) *Multidimensional Global Change*, Wiley/Praxis, Chichester, UK, 761 pp.

Kondratyev, K. Ya. (1999) *The Climatic Effect of Aerosols and Clouds*, Sprınger/ Praxis, Chichester, UK, 272 pp.

Kulmälä, M., Wagner, P. E. (eds.) (1996) *Nucleation and Atmospheric Aerosols*, Elsevier Science Inc., 967 pp.

Kapitza, A. P. (1996) Controversies in the theory of ozone hole formation. *Optics of the Atmosphere and Ocean*, **9**, No. 5, 1,164–1,167 (in Russian).

de Laat, A. T. J., Zachariasse, M., Roelofs, G. J., van Velthoven, P., Dickerson, R. R., Rhoads, K. P., Oltmans, S. J., Lelieveld, J. (1999) Tropospheric O_3 distribution over the Indian Ocean during spring 1995 evaluated with a chemistry-climate model. *J. Geophys. Res.*, **104**, N11, 13,881–13,895.

Landgraf, J., Crutzen P. S. (1998). An efficient method for online calculations of photolysis and heating rates. *J. Atmos. Sci.*, **55**, N5, 863–878.

Lee, S. H., Akimoto, H., Nakane, H., Kurnosenko, S., Kinjo, Y. (1998). Lower tropospheric ozone trend observed in 1989–1997 at Okinawa, Japan. *Geophys. Res. Lett.*, **25**, N10, 1,637–1,640.

Liang, L., Horowitz, L. W., Jacob, D. S., Wang, Y., Fiore A. M., Logan, J. A., Gardner, G. M., Mungerd, M. (1998). Seasonal budgets of reactive nitrogen species and ozone over the United States, and export fluxes to the global atmosphere. *J. Geophys. Res.*, **103**, ND11, 13,435–13,450.

Lin, X., Trainer, M., Hsie, E.-Y. (1998). A modeling study of tropospheric species during the North Atlantic Regional Experiment (NARE). *J. Geophys. Res.*, **103**, ND11, 13,593–13,613.

Logan, J. A. (1999a) An analysis of ozonesonde data for the troposphere: Recommendations for testing three-dimensional models and development of a gridded climatology for tropospheric ozone. *J. Geophys. Res.*, **104**, ND13, 16,115–16,149.

Logan, J. A. (1999b) An analysis of ozonesonde data for the lower stratosphere: Recommendations for testing models. *J. Geophys. Res.*, **104**, ND13, 16,151–16,170.

Low, P. S. (1998) Tropospheric ozone: The missing GHG in the UNFCCC equation. *Linkages Journal*, **3**, N4, 32–34.

Lubin, D., Jensen, E. H., Gies, H. P. (1998) Global surface ultraviolet radiation climatology from TOMS and ERBE data. *J. Geophys. Res.*, **103**, ND20, 26,061–26,091.

Madronich, S. (1998). WMO ad hoc Scientific Steering Committee on UV monitoring: Activities and scientific issues. *SPARC Newsletter*, N10, 19–26.

Maduro, R. A., Schauerhammer, R. (1992) *The Holes in the Ozone Scare. The Scientific Evidence that the Sky Isn't Falling*, 21st Century Science Associates, Washington DC, Wiley, 356 pp.

Mailhot, J., Strapp J. W., MacPherson, J. I., Benoit, R., Blair, S., Donaldson, N. R., Froude, F., Benjamin, M., Zawadzki, I., Rogers R. R. (1998) The Montreal-96 Experiment on Regional Mixing and Ozone (MERMOZ): An overview and some preliminary results. *Bull. Amer. Meteorol. Soc.*, **79**, N3, 439–442.

Martens, P. (1998). *Health and Climate Change. Modelling the Impacts of Global Warming and Ozone Depletion*, Earthscan, London, 1998, 224 pp.

Mauzerall, D. L., Logan, J. A., Jacob, D. S., Anderson, B. E., Blake, D. R., Bradshaw, J. D., Heikes, B., Sachse, G. W., Singh, H., Talbot, B. (1998). Photochemistry in biomass burning plumes and implications for tropospheric ozone over the tropical Southern Atlantic. *J. Geophys. Res.*, **103**, ND7, 8,401–8,424.

McCormack, J. P., Miller, A. J., Nagatani, R., Fortuin, J. P. F. (1998). Interannual variability in the spectral distribution of extratropical total ozone. *Geophys. Res. Lett.*, **25**, N12, 2,153–2,156.

McCormick, M. P., Veiga, R. E., Chu, W. P. (1992). Stratospheric ozone profile and total ozone trends derived from the SAGE I and SAGE II data. *Geophys. Res. Lett.*, **19**, N3, 269–272.

McCormick, M. P. (1998) Summary of the SAGE III (Stratospheric Aerosol and Gas Experiment III) science team meeting. *The Earth Observer*, **10**, N3, 7–14.

MacKenzie, I. A., Harwood, R. S., Stott, P. A., Watson, G. C. (1999) Radiative-dynamic effects of the Antarctic ozone hole and chemical feedback. *Quart. J. Roy. Meteorol. Soc.*, **125**, Part B, N558, 2,171–2,203.

Michaels, P. J., Singer S. F., Knapperberger, O. P. E., Kerr, J. B., McElroy, C. T. (1994). Analysing ultraviolet-B radiation: is there a trend? *Science*, **264**, 1,341–1,343.

Molemaker, M. J., de Arellano, J. V.-G. (1998). Control of chemical reactions by convective turbulence in the boundary layer. *J. Atmos. Sci.*, **55**, N4, 568–579.

Monod, A., Carlier, P. (1999) Impact of clouds on the tropospheric ozone budget: direct effect of multiphase photochemistry of soluble organic compounds. *Atmos. Environ.*, **33**, 4,431–4,446.

Moody, J. L., Munger, J. W., Goldstein, A. H., Jacob, D. S., Wofsy, S. C. (1998). Harvard forest regional-scale air mass composition by Patterns in Atmospheric Transport History (PATH). *J. Geophys. Res.*, **103**, ND11, 13,181–13,194.

Morris, G. A., Kawa, S. R., Douglas, A. R., Schoeberg, M. R., Froidevaux, L., Waters, J. (1998). Low-ozone pockets explained. *J. Geophys. Res.*, **103**, ND3, 3,595–3,610.

Munro, R., Siddans, R., Reburn, W. S., Kerridge, B. S. (1998). Direct measurement of tropospheric ozone distributions from space. *Nature*, **392**, 168–171.

Nair, H., Allen, M., Froidevaux, L., Zurek, R. W. (1998) Localized rapid ozone loss in the northern winter stratosphere: An analysis of UARS observations. *J. Geophys. Res.*, **103**, ND1, 1,555–1,571.

Newell, R. E., Browell, E. V., Davis, D. D., Liu, Sh. E. (1997) Western Pacific tropospheric ozone and potential vorticity: Implications for Asian pollution. *Geophys. Res. Lett.*, **24**, N22, 2,733–2,736.

Oltmans, S. J., Lebone, A. S., Scheel, H. E., Harris, J. M., Levy, H.,II, Galbally, I. E., Brunke, E.-G., Meyer, C. P., Lathorp, J. A., Johnson, B. S., Shadwick, D. S., Cuevas, E., Schmidlin, F. J., Tarasick, D. W., Claude, H., Kerr, S. B., Uchino, O., Mohnen, V. (1998). Trends of ozone in the troposphere. *Geophys. Res. Lett.*, **25**, N2, 143–146.

Ozone Depletion, Greenhouse Gases, and Climate Change (1989) National Academy Press, Washington, DC, 122 pp.

Penner, J. E., Lister, D. H., Griggs, D. J., Dokken, D. J., McFarland, M. (eds.) (1999) *Aviation and the Global Atmosphere. A special report of IPCC WG I and III*, Cambridge University Press, Cambridge, UK, 373 pp.

Pierce, R. B., Fairlie, T. D., Remsberg, E. E., Russel, J. M., III, Grose, W. L. (1997). HALOE observations of the Arctic vortex during the 1997 spring: Horizontal structure in the lower stratosphere. *Geophys. Res. Lett.*, **24**, N22, 2,705–2,708.

Prather, M., Midgley, P., Rowland, S. R., Stolarski, R. (1996) The ozone layer: the road not taken. *Nature*, **381**, N6583, 551–554.

Prather, M. S. (1998). Time scales in atmospheric chemistry: Coupled perturbations to N_2O, NO_y and O_3. *Science*, **279**, N5355, pp. 1,339–1,342.

Pyle, S. A., Harris, N. R. D. (1991) *Polar Stratospheric Ozone. Air Pollution Research Report 34*, Commission of the European Communities, Brussels, Belgium, 306 pp.

Pszenny, A., Brasseur G. (1997). Tropospheric ozone: An emphasis on IGAC Research, *Global Change Newsletter*, N30, 2–10.

Ravishankara, A. R., Hancock, G., Kawasaki, M., Matsumi, Y. (1998). Photochemistry of Ozone: Surprises and Recent Lessons. *Science*, **280**, 60–61.

Rex, M., von der Gathen, P., Harris, N. R. P., Lucic, D., Knudsen, B. M., Braathen, G. O., Reid, S. J., De Backer, H., Claude, H., Fabian, R., Fast, H., Gil, M., Kyr, E., Mikkelsen, I. S., Rummukainen, M., Smit, H. G., Staehelin, J., Varotsos, C., Zaitcev, I. (1998) In situ measurements of stratospheric ozone depletion rates in the Arctic winter 1991/1992: A Lagrangian approach. *J. Geophys. Res.*, **103**, N5, 5,843–5,853.

Ridley, B. A., Walega, S. G., Lamarque, J.-F., Grahek, F. E., Trainer, M., Hübler, G., Lin, X., Fehsenfeld, F. C. (1998). Measurements of reactive nitrogen and ozone to 5-km altitude in June 1990 over the southeastern United States. *J. Geophys. Res.*, **103**, N3, 8,365–8,388.

Rinsland, C. P., Salawitch, R. J., Gunson, M. R., Solomon, S., Zander, R., Mahieu, E., Goldman, A., Newchurch, M. J., Irion, F. W., Chang, A. Y. (1999) Polar stratospheric descent of NO_y and CO and Arctic denitrification during winter 1992–1993. *J. Geophys. Res.*, **104**, ND1, 1,847–1,861.

Russel, J. M. III, Lastovicka, J. (1998). The IAGA/ICMA Symposium Session 2.18 on 'Middle Atmosphere Trends and Low Frequency Variability'. Uppsala, Sweden, August 14, 1997. *SPARC Newsletter*, N10, 25–26.

Salawitch, R. J. (1998). Ozone depletion: A greenhouse warming connection. *Nature*, **392**, N6676, 551–552.

Sarma, K. M. (1998). Protection of the ozone layer – A success story of UNEP. *Linkages Journal*, **3**, N3, 6–10.

Sasano, Y., Bodeker, G., Kreher, K. (1998) Preliminary ILAS observations of O_3, HNO_3 and N_2O in the Arctic stratosphere during the winter of 1996/97. *SPARC Newsletter*, N10, 18–19.

Schubert, S. D., Munteanu, M.-J. (1998) An analysis of tropopause pressure and total ozone correlations. *Mon. Weather Rev.*, **116**, 569–582.

Schultz, M. G., Jacob, D. J., Wang, Y., Logan, J. A., Atlas, E. L., Blake, D. R., Blake, N. J., Bradshaw, J. D., Browell, E. V., Fenn, M. A., Flocke, F., Gregory, G. L., Heikes, B. G., Sachse, G. W., Sandholm, S. T., Shetter, R. E., Singh, H. B., Talbot, R. W. (1999) On the origin of tropospheric ozone and NO_x over the tropical South Pacific. *J. Geophys. Res.*, **104**, ND5, 5,829–5,843.

Scientific Assessment of Ozone Depletion: 1998. WMO Global Ozone Research and Monitoring Project Report No. 44 (1999) WMO, Geneva, Switzerland.

Seinfeld, J. H., Pandis, S. N. (1998) *Atmospheric Chemistry and Physics*, John Wiley, New York, 1326 pp.

Shindell, D. T., Rind, D., Lonergan P. (1998) Increased polar stratospheric ozone losses and delayed eventual recovery owing to increasing greenhouse-gas concentration. *Nature*, **392**, 589–592.

Spänkuch, D., Dhler, W., Güldner, J., Schulz, E. (1998). Estimation of the amount of tropospheric ozone in a cloudy sky by ground-based Fourier-transform infrared emission spectroscopy. *Appl. Optics*, **37**, N15, 3,133–3,142.

Staehelin, J., Kegel, R., Harris, N. R. P. (1998). Trend analysis of the homegenized total ozone series at Arosa (Switzerland), 1926–1996. *J. Geophys. Res.*, **103**, ND7, 8,385–8,400.

Stratospheric Processes and Their Role in Climate (SPARC), Proceedings First SPARC General Assembly, 2–6 December 1996, Melbourne, Australia (1997) WCRP-99 (WMO No. 814), WMO, Geneva, Switzerland, Vols. I, II, 672 pp.

Stuhler, H. Frohn, A. (1998) The production of NO by hypersonic flights. *Atmos. Environ.*, **32**, N18, 3,153–3,155.

Syvorotkin, V. L. (1998). Earth's outgassing destroys the ozonosphere. *Earth and Universe*, N1, 21–27 (in Russian).

Takigawa, M., Takahashi, M., Akigoshi, H. (1999) Simulation of ozone and other chemical species using a Center for Climate System Research/National Institute for Environmental Studies atmospheric GCM with coupled stratospheric chemistry. *J. Geophys. Res.*, **104**, ND11, 14,003–14,018.

Taupin, F. G., Bessafi, M., Baldy, S., Bremaud, P. J. (1999) Tropospheric ozone above the southwestern Indian Ocean is strongly linked to dynamical conditions prevailing in the tropics. *J. Geophys. Res.*, **104**, ND7, 8,057–8,066.

Timmreck, C., Graf, H.-F., Kirchner, I. (1999) A one-and-half-year interactive MA/ ECHAM4 simulation of Mount Pinatubo aerosol. *J. Geophys. Res.*, **104**, ND8, 9,337–9,359.

Tourpali, K., Tie, X. X., Zerefos, C. S., Brasseur G. (1997) Decadal evolution of total ozone decline: Observations and model results. *J. Geophys. Res.*, **102**, ND20, 23,955–23,962.

Tuck, A., Watson, R., Toon, B. (Eds.), (1989). The Airborne Antarctic Ozone Experiment (AAOE). *J. Geophys. Res.*, **94**, ND9, 11,179–11,737, ND14, 16,434–16,858.

UNEP Scientific Assessment of Ozone Depletion, WMO Global Ozone Research and Monitoring Project Report No. 37 (1994), WMO, Geneva, Switzerland.

Varotsos, C., Helmis, C., Cartalis, C. (1992). Annual and semiannual waves in ozone as derived from SBUV vertical global ozone profiles. *Geophys. Res. Lett.*, **19**, N9, 926–928.

Varotsos, C. A., Kondratyev, K. Ya. (1995). Ozone dynamics over Greece as derived from satellite and in-situ measurements. *Int. J. Remote Sensing*, **16**, N10, 1,777–1,798.

Varotsos, C. (Ed.), (1997) *Atmospheric Ozone Dynamics Observations in the Mediterranean Region*, NATO ASI Series, Springer-Verlag, Berlin, 336 pp.

Vay, S. A., Anderson, B. E., Sache, G. W., Collins, J. E., Jr., Podolske, J. R., Twohy, C. H., Gandrud, B., Chan, K. R., Baughcum, S. L., Wallio, H. A. (1998) DC-8-

based observations of aircraft CO, CH_4, N_2O, and $H_2O(g)$ emission indices during SUCCESS. *Geophys. Res. Lett.*, **25**, N10, 1,717–1,720.

Volkovitzky, O. A., Gavrilov, A. A., Kaidalov, V. (1996). Is man really destroying the planetary ozone layer? *Proc. Russian Acad. Sci.*, **66**, N9, 783–786 (in Russian).

Wang, W.-C., Isaksen, I. S. A. (eds.) (1995) *Atmospheric Ozone as a Climate Gas: General Circulation Model Simulations*, Springer-Verlag, Berlin, 459 pp.

Wang, H. S., Cunnold, D. M., Bao, X. (1996). A critical analysis of Stratospheric Aerosol and Gas Experiment ozone trends. *J. Geophys. Res.*, **101**, ND7, 12,495– 12,514.

Wang, Ch, Prinn, R. G., Sokolov, A. (1998). A global interactive chemistry and climate model: Formation and testing. *J. Geophys. Res.*, **103**, ND3, 3,399–3,417.

Wardel, D. I., Kerr, J. B., McElroy, C. T., Francis, D. R. (eds.) (1997) *Ozone Science: A Canadian Perspective on the Changing Ozone Layer*, University of Toronto Press, Toronto, Canada, 1997, 119 pp.

Weisenstein, D. K., Ko, M. K. W., Dyominov, I. G., Pitari, G., Ricciardulli, L., Visconti G., Bekki, S. (1998) The effect of sulfur emissions from HSCT aircraft: A 2-D model intercomparison. *J. Geophys. Res.*, **103**, ND1, 1,527–1,547.

Wessel, S., Aoki, S., Winkler, P., Weller, R., Herber, A., Gernandt, H., Schrems, O. (1998) Tropospheric ozone depletion in polar regions: A comparison of observations in the Arctic and Antarctic. *Tellus*, **50B**, 34–50.

Yates, S. R., Wang, D., Gan, J., Ernst, F. F., Yury, W. A. (1998). Minimizing methyl bromide emissions from soil fumigation. *Geophys. Res. Lett.*, **25**, N10, 1,633–1,636.

Zerefos, C. S., Ghazi, A. (eds.) (1985) *Atmospheric Ozone. Proceedings of the Quadrennial Ozone Symposium, 3–7 September 1984, Halkidiki, Greece*, D. Reidel, Dordrecht, 842 pp.

Zerefos, C. S., Bais, A. F. (eds.) (1997) *Solar Ultraviolet Radiation. Modelling, Measurements and Effect*, NATO ASI Series I: Global Environmental Change, Springer, Berlin, Vol. 52, 332 pp.

Zerefos C., Meleti C., Balis D., Tourpali K., Bais A. F. (1998) Quasibiennial and longer-term changes in clear-sky UV-B solar irradiance. *Geophys. Res. Lett.*, **25**, N23, 4,345–4,348.

Ziemke, J. R., Herman, J. R., Stanford, J. L., Bhartia, P. K. (1998). Total ozone/UVB monitoring and forecasting: Impact of clouds and horizontal resolution of satellite retrievals. *J. Geophys. Res.*, **103**, ND4, 3,865–3,871.

Ziemke, J. R., Chandra, S., Herman, J., Varotsos, C. (1999) Erythemally weighted UV trends over northern latitudes derived from Nimbus-7 TOMS measurements. *J. Geophys. Res.*, **104** (in press).

2

Observed variability of total ozone column and vertical ozone concentration profiles

The ozone (an allotrope of ordinary oxygen) is produced naturally in the stratosphere and drifts downward by mixing processes, producing a maximum in its concentration near 25 km. It is also destroyed naturally by absorbing ultraviolet radiation (0.2–0.3 μm) and through chemical reactions with other components. The natural balance between production and destruction leads to the maintenance of the atmospheric ozone layer which is subject, on the other hand, to substantial anthropogenic impacts.

Ozone is so rare in the atmosphere (about three ozone molecules per 10 million air molecules) that if it was brought down to the earth's surface and compressed, it would form a layer about as thick as a dime. The thickness of this layer at standard temperature and pressure determines total ozone content or total ozone column (TOZ). Although thin, this layer of ozone (more than 90% is found in the stratosphere) prevents harmful solar energy (the biologically damaging ultraviolet radiation) to reach the earth's surface, strongly absorbs infrared thermal radiation (heating the stratosphere) and plays a major role in photochemistry (involved in formation of nitrogen, carbon, hydrogen, oxygen and halogen chemical cycles). Various human activities on the one hand decrease the amount of ozone in the stratosphere, as first pointed out in the early 1970s, and on the other hand increase tropospheric ozone, which (in case of high concentration) is toxic to living systems (*Scientific . . .*, 1999).

Observations conducted over recent years have provided deeper understanding of the anthropogenic and natural causes of change in the chemical composition of the atmosphere and their association with ozone depletion in the stratosphere. These observations and laboratory investigations indicate that the by-products of human-made chlorofluorocarbons (CFCs) and halons are the main causes for ozone decline, especially in the Antarctic. Such a decline of a global scale started in the 1970s and still continues with statistically significant rates except over the 20°N–20°S tropical belt.

The reliable assessment of ozone depletion in the stratosphere and ozone increase in the troposphere depends on accurate monitoring of long-term change in global TOZ and in ozone vertical concentration profiles (OVP). Therefore, there is a need for continuous and long-term collection of reliable ozone data. Ground-based measurements constitute a key component of the Global Ozone Network, both on their own account and by providing ground-based calibration for satellite-based instruments. However, the uneven geographical distribution of the existing ground-based network gives rise to a spatial sampling error, when attempts are made to determine the global distribution of the ozone content. The benefit of ground-based instruments is that it is easy to maintain them in good condition, while satellite-based instruments provide better temporal–spatial coverage and resolution.

This chapter describes the instrumentation and measurement techniques that have been used for the measurement of the TOZ or the OVP. Additional information in this respect was presented elsewhere (WMO, 1990) and in a number of other reports (*UNEP*, 1994; *Scientific* ..., 1999; WMO, 1986).

2.1 OBSERVATIONS OF TOTAL OZONE COLUMN

Cornu in 1879 suggested that the cause of the rather sharp limitation of the UV part of the solar spectrum (as observed at the ground) was absorption by the atmosphere. In 1880, Hartley suggested that ozone was responsible for this absorption. Fabry and Buisson, in 1912, measured the absorption coefficients of ozone and concluded that there was about 0.5 cm (500 Dobson Units – DU) of ozone in the vertical thickness of the atmosphere. The same researchers in 1920 made measurements of the amount of ozone at Marseilles by using a special spectrograph (Fabry and Buisson, 1913).

In the early 1920s the British physicist GMB. Dobson initiated the first regular measurements of atmospheric ozone using spectrographic instruments (Dobson, 1931). Today, ozone observations are carried out daily by using different kinds of instrumentation from early ground-based devices and balloons to the more advanced devices flown on aircraft, rockets and satellites.

At the time of the International Geophysical Year (in 1957/58), the World Meteorological Organization (WMO) assumed responsibility for establishing standard international procedures to ensure uniformity and the high quality of ozone observations conducted at the present time in more than 160 ground-based ozone stations. These standard procedures have been used at the WMO Global Ozone Observing System (GO_3OS) as well as the Background Air Pollution Monitoring Network (BAPoMON), which mainly constitute the WMO Global Atmospheric Watch (GAW). Since 1989 the latter also serves as an early-warning system to detect changes in atmospheric concentrations of greenhouse gases (WMO, 1990).

In connection with the International Geophysical Year the World Ozone Data Centre (WODC) was initiated, first located at the WMO Secretariat. From 1961 it has been run in Toronto for WMO, first by the Meteorological Service of Canada and, since 1972, by its successor the Atmospheric Environment Service (AES) of Environment Canada.

More than 40 GO_3OS stations (using filter instruments) making long-term observations supplemented by another 100 stations and satellites over the last 20 years (data held by the World Ozone and Ultraviolet Data Centre – WOUDC), show that TOZ decline has been observed over the mid and polar latitudes, which is approximately at a 10% level relative to the TOZ of the 1950s and 1960s. After filtering out known natural variability (inter-annual and intra-annual cycles) TOZ decline in both hemispheres was strongest during the winter and spring (higher than 6–7% per decade) and is only half that during the summer and autumn. According to *Scientific* ... (1999) TOZ declined during all months of the year in the mid and high latitudes. In the Northern Hemisphere, trends are much larger in winter and spring (about −3 to −6% per decade), than in summer and fall (about −1 to −3% per decade).

We shall now briefly describe the basic instrumentation and the basic measurement techniques that have been used for the measurement of TOZ.

2.1.1 Ground-based instrumentation for TOZ observations

Absorption is a very important tool for TOZ monitoring from the ground spectroscopy. A brief outline of the main ground-based devices for TOZ observations is given below.

2.1.1.1 Primary ground-based instrumentation for TOZ observations

TOZ was measured from the ground at a few sites in the late 1920s and early 1930s by using the Dobson spectrophotometer, the development of which was based on earlier research with the Fabry–Buisson spectrophotometer which used photographic plates (Fabry and Buisson, 1921).

Nowadays, three main types of instruments are used for routine TOZ observations from the ground: the Dobson spectrophotometer, the Brewer spectrophotometer and the M-83/124/134 filter ozonometer. All of them use the principle of differential absorption by ozone of solar radiation within the 300–460 nm band. The determination of TOZ by the three instruments becomes feasible through measuring scattered solar radiation in the Hartley or Hartley–Huggins bands in the UV spectral region. Dobson and Brewer spectrophotometers can measure TOZ with up to 99% accuracy, while the filter ozonometers are less accurate (Bojkov *et al.*, 1988). Observations carried out these instruments over longer time periods can also be used to measure OVP with approximately 5-km resolution and TOZ with fairly high accuracy by looking directly at the Sun or Moon.

In the following, more detailed information about these basic instruments is presented.

2.1.1.1.1 Dobson ozone spectrophotometer

The Dobson spectrophotometer is the oldest device for measuring atmospheric TOZ, developed in 1924 by GMB. Dobson (Figure 2.1) Today, the observations of TOZ by

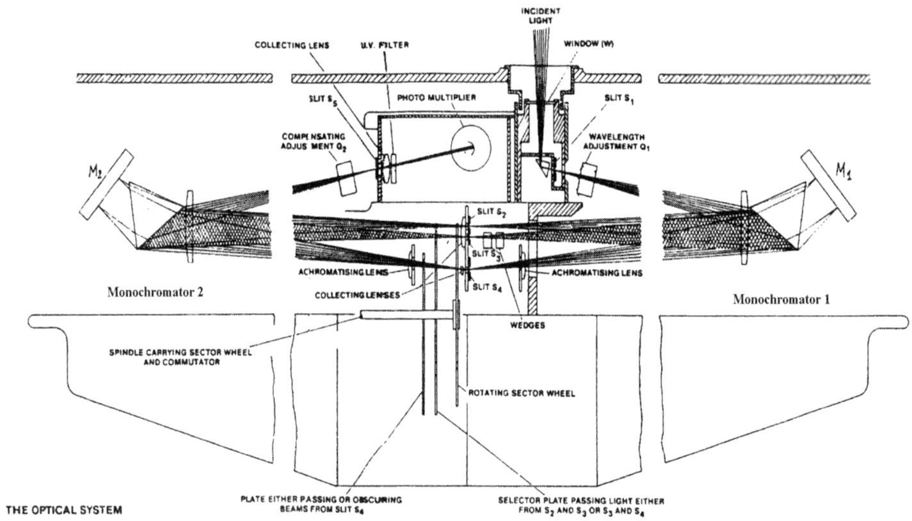

Figure 2.1. Dobson optical diagram.

this instrument constitute one of the longest geophysical-measurement series in existence (TOZ has been continuously observed over Arosa, Switzerland since 1926).

The instrument uses a quartz double monochromator and has been used to monitor TOZ ever since its development. One monochromator is used to disperse radiation and the second to reject interfering scattered radiation (Dobson, 1931; Dobson and Normand, 1957).

The principle behind operation of the instrument is taking TOZ measurements in the Huggins bands by measuring the difference between the intensity of solar light at certain wavelength pairs. These pairs are selected in such a manner that the difference of ozone absorption coefficient for each pair is as large as possible, while the difference between the corresponding wavelengths remains small. By using one pair of wavelengths a differential measurement is made. The one wavelength is significantly absorbed by ozone and the other passes through a variable optical attenuator (wedge). This measured difference combined with the extraterrestrial constant and the ozone absorption spectrum indicates the TOZ. By using two pairs of wavelengths the scattering effects can in essence be filtered out.

The pair most frequently used for most precise observations is the standard double-pair wavelengths designated AD, while the pair CC′ is used less frequently for less precise observations (A, 3055 and 3254 Å; C, 3114 and 3324 Å; D, 3176 and 3398 Å; C′, 3324 and 4536 Å). Direct Sun AD observations generally provide the most precise measurements, if the secant of the zenith angle μ is less than 3.

Dobson apparatus can be operated manually, semi-automatically or automatically (Evans *et al.*, 1994); it can also show the vertical ozone distribution by using the 'Umkehr technique' (described in Section 2.2.1.1.1). For instance, Dobson instruments D15 and D101 at the Light Climatic Observatory (LKO) in Arosa, Switzerland are partially automated and located inside a spectrodome performing Umkehr observations in a completely automated way (Hoegger *et al.*, 1992).

Nowadays (1999), over 90 Dobson spectrophotometers constitute the global primary ground-based network of stations for monitoring atmospheric concentrations of ozone. This network provides the most reliable source of ozone data and is a key component of the GO_3OS, which is now part of the GAW of the WMO.

The Dobson network also serves a valuable function in providing ground-based calibration and validation data for satellite-based measurements and various remote-sensing instruments, which are used to measure atmospheric ozone and are described in Section 2.2.3.

Since 1960, the World Dobson Spectrophotometer Central Laboratory (operated by the Climate Monitoring and Diagnostics Laboratory of the US National Oceanic and Atmospheric Administration) has organized intercomparisons of Dobson instruments, in order to maintain calibration within the network. These field intercomparisons (jointly organized by WMO and the International Ozone Commission) are usually accompanied by laboratory assessments and internal adjustments.

In 1962 the Dobson spectrophotometer No. 83 was established as the US standard for TOZ measurements. This instrument was initially calibrated relative to Dobson spectrophotometer No. 72 and additional calibrations were performed nine times between 1972 and 1987, using direct-Sun observations (Komhyr *et al.*, 1989a).

Since the early 1960s, almost all Dobson instruments have been calibrated either directly or indirectly relative to instrument No. 83.

Since 1974, most of the existing Dobson instruments have been modernized and calibrated by direct intercomparison with the WMO designated World Primary Standard Dobson Spectrophotometer No. 83 (maintained at the World Dobson Spectrophotometer Central Laboratory in Boulder, Colorado). Table 2.1. presents the results of the intercomparison of Dobson instruments held at Arosa, Switzerland in 1990. According to Basher (1995) the error in direct-Sun observations for a well calibrated and operated instrument is about 1% (1σ) and for zenith sky measurements about 3%.

It should be noted that Komhyr (1980) suggested that 'travelling' standard lamps, properly calibrated and adjusted, can also be used for basic error detection and stability testing of Dobson instruments. This kind of systematic intercomparison of Dobson instruments against travelling standard lamps have been undertaken since 1984 (Komhyr *et al.*, 1989a). In this context, satellite instruments (providing ozone observations) are often used as the 'travelling' reference instrument, in order to identify those field Dobson instruments that are systematically in error (Bojkov *et al.*, 1988). According to Basher (1995), WMO travelling standard-lamp tests showed that more than 75% of Dobson instruments differ from the standard lamp by less than $\pm 1.5\%$.

Each participant in the international intercomparison (Table 2.1) gathers together the regional instruments, as well as the World Primary Standard Dobson Instrument or Regional Secondary Standard Instrument (calibrated in Boulder), for 3 weeks in summer at a site that has relatively clear skies. In the summer of 1997 such an

Table 2.1. Preliminary results of intercomparison of Dobson instruments (22 July to 10 August 1990) at Arosa, Switzerland. Delta Xad is percentage deviation from the USA standard.

Instrument No.	Country–Location	Last Intercomparison	Initial ΔX_{AD} (%)	Final ΔX_{AD} (%)
13	Portugal – Lisbon	1987	0.28	−0.16
14	Norway – Tromso	1977	−0.99	−0.19
15	Switzerland – Arosa		−4.96	0.82
40	Belgium – Uccle	1986	−0.6	−0.15
41	United Kingdom	1985	−0.59	0.15
50	Iceland – Reykjavik	1977	−0.16	0.05
64	GDR – Postdam	1977	0.22	−0.09
74	Czech – Hradek Kralove	1986	−0.66	−0.07
84	Poland – Belsk	1986	0.35	−0.05
92	Denmark – Greenland	1986	−0.5	−0.05
101	Switzerland – Arosa	1986	0.37	0.09
104	FRG – Hohenpeinssenberg	1986	−2.4	−0.04
107	USSR – Moscow	1988	−0.95	
110	Hungary – Budapest	1988	0.93	−0.24
118	Greece – Athens	Never*	−2.27	0.03
120	Spain	1989	−1.92	
121	Rumania – Bucharest	1988	−0.06	0.04
65	USA – standard	1990	0	0

*Has never been used or calibrated before this intercomparison.

intercomparison took place in Stiga (Kalavryta), Greece. Details about this intercomparison are now discussed (Evans and Varotsos, 1999).

WMO International Intercomparison of Dobson spectrophotometers at Stiga Ski Field, Kalavryta, Greece, July–August 1997

The Stiga intercomparison (SSF/IC) of Dobson instruments was sponsored by the World Meteorological Organization (WMO), the University of Athens (Ozone and Solar Radiation Research Team – OSORT) and assisted by the NOAA Climate Monitoring and Diagnostics Laboratory's World Dobson Calibration Centre. The SSF/IC consisted of the intercomparison of Dobson Instruments with the World Secondary Standard Dobson Instrument No. 65. Instruments and personnel from the Czech Republic, Egypt, Germany, Greece, Romania, Russia and the United States attended the intercomparison. In this context, long-term quality controlled ozone data, obtained within the framework of the WMO/GAW, are essential for assessment of the state of the global ozone layer.

The main tasks of the intercomparison were:

- Technical inspection and adjustment of the instruments.
- Comparison of Dobson spectrophotometers with the World Secondary Standard Dobson Instrument No. 65 to determine the existing calibration level.

- Making repairs if necessary.
- Determination of new calibration constants for each Dobson spectrophotometer, as needed.
- To provide a forum for instruction in operating Dobson spectrophotometers at home stations, and sharing knowledge concerning the management of an ozone-observing programme.

All participating instruments left the intercomparison properly calibrated, and in correct operating order.

Discussions about station operation and instrument maintenance were held amongst the participants. From these discussions, the scientific and technical directors made the following recommendations to the WMO:

- Standardize the electronic circuits of Dobson instruments with a simple effective design using easily available components. To ensure the use of this design, WMO should provide the means of converting the various instruments to this circuit.
- Sponsor a revision of the *Operational Handbook* in light of the 15 years' experience with the instruments since the last revision.
- Sponsor a station-assessment programme to evaluate station operation and ensure data quality in the coming years.
- Continue the progress towards establishment of regional calibration centres for Dobson instruments, as a task described in the WMO/GAW Report 113.

A general recommendation was that total ozone observations and ozone vertical-profile measurements made by the University of Athens (WMO Station 293) should be continued, as these data play an important part in the understanding of both ozone climatology and ozone trends in the Eastern Mediterranean and south-eastern Europe.

Recently, the Meteorological Observatory of Hohenpeissenberg (MOHp) in Germany was designated to act as the European Ozone Centre taking on the role of a Dobson calibration centre (DCC). An initial intercomparison of three reference instruments was carried out in summer 1999 at MOHp in collaboration with Hradec Kralove (Czech Republic). The normal MOHp (DCC) support will include among others:

1. Maintenance of each neighbouring Dobson station will be guaranteed within the bounds of the usual DCC duties: regular (at least once every 4 years) check of the instrument's state and calibration as well as a kind of 'first aid', if some trouble occurs with each instrument.
2. Re-evaluation of a station's Dobson record should be done by its own staff, because they know the history of the instrument and have easy access to all necessary information, like test records and documents. DCC can give recommendations and advice on what to do and how to do it. The procedure and methods of re-evaluation are given in the NOAA Technical Report NESDIS 74 *Dobson Data Re-evaluation Handbook* and in WMO-GAW Report No. 29 *Handbook for Dobson Ozone Data Re-evaluation.*

Interference by SO_2 and NO_2 in Dobson TOZ measurements

The precision of a well-calibrated Dobson spectrophotometer is about 0.5% and its accuracy is of the order of 2–4%. The instrument is also subject to additional errors in the presence of SO_2 and NO_2 aerosols (Komhyr and Evans, 1980; Varotsos *et al.*, 1995). The sources of error in the Dobson TOZ measurement become significant in cases where the Dobson instrument is placed in an area surrounded by highly polluted atmosphere.

In the greater Athens area, where Dobson spectrophotometer No. 118 is installed, more than half of the industrial activities of the country are concentrated. Athens is situated in a relatively small basin of 450 km^2 surface area, surrounded by three mountains with heights of 100, 1,000 and 1,500 m. This creates ventilation problems, especially during meteorological conditions of low synoptic winds accompanied by high sunshine (anticyclonic conditions). Very high concentrations of urban pollutants, especially the photochemical ones (NO_2, O_3) are frequently observed under these conditions.

In order to find out the extent of error in these measurements both in winter time and summertime because of the presence of SO_2 and NO_2 in the Athenian atmosphere, the SO_2 and NO_2 concentration data of PERPA (National Service for Air Pollution Monitoring) at 13:00 LT were used. These data were obtained at the Patissia Station located in the centre of Athens, and refer to the 2-year period 1989–1990.

Because PERPA SO_2 and NO_2 concentration data are given in $\mu g\,m^{-3}$, the former must be multiplied by the factor 3.82×10^{-4} and the latter by the factor 5.32×10^{-4} so as to convert them into ppmv. The expression relating ppmv to the $\mu g\,m^{-3}$, is the following:

$$1\,\text{ppmv} = 10^{-3}[M/(M_i \rho)]\,\mu g\,m^{-3} \tag{2.1}$$

where $M = 28.97$ is the average air molecular weight, $\rho = 1.184\,kg\,m^{-3}$ is the air density for atmospheric pressure $P = 1{,}013$ hPa and $T = 298.15$ K, and M_i is the molecular weight of the specific pollutant.

Radiosonde temperature profiles as well as surface temperature (obtained from the National Meteorological Service of Greece) were also used. All these data resulted from measurements made on a daily basis at 12:00 UTC at Hellinikon Airport, Athens, over 17 years (1974–1990).

The error in TOZ measurement caused by the presence of absorbing species other than ozone is given by the equation (2.2) (Komhyr and Evans, 1980):

$$\Delta x = x_1' - x_1 = \frac{\alpha_2 m}{\alpha_1 \mu} x_2 \tag{2.2}$$

where x_1' is the apparent TOZ amount, x_1 is the true TOZ amount, without the error owing to the presence of the specific absorbing species and x_2 is the total amount of the absorbing species (all in atm-cm), α_1 is the ozone double-pair wavelength decimal absorption coefficient, α_2 is the corresponding coefficient for the absorbing species, m is the relative optical air mass and μ is the relative optical ozone mass. As an approximation we can take $\mu = m$.

Thus the percentage error in true TOZ amount x_1 because of the presence of the absorbing species will be:

$$\% \text{ error in } x_1 = \frac{100\alpha_2}{x_1\alpha_1} x_2 \tag{2.3}$$

The total amount x_2 of the absorbing species can be estimated in atm-cm by the following equation if we know its mixing ratio C in parts per million by volume (ppmv), as well as the height of the pollution mixing layer:

$$x_2 = \frac{1 - \exp[-(z/H)]}{1.251} C \tag{2.4}$$

where H is the scale height of the pollution mixing layer. It is given by the expression $H = R_d T/g_0$, where R_d is the gas constant for dry air ($= 287\,\mathrm{J\,kg^{-1}\,K^{-1}}$), T is the mean temperature of the mixing layer and $g_0 = 9.80665\,\mathrm{m\,s^{-2}}$ is the global average of the acceleration due to gravity at mean sea level. $H = 8.49$ km for a layer at 17°C, z is the mixing layer's height.

Mixing heights were calculated from radiosonde temperature profiles and surface temperature measurements by the simple thermodynamic method of Holzworth (1972). This method has been widely used in several countries all over the world and seems to be appropriate for countries such as Greece, especially for the afternoon mixing height.

The potential temperature is calculated at the surface as well as in the layers at which the radiosonde data were available, using the Poisson equation. If the potential temperature of a layer was greater than that of the surface, then the mixing height was defined at the bottom of that layer.

The mean total SO_2 amounts in the atmosphere over Athens for the periods December – March and May – August of the 2-year period 1989–1990 at local noon was estimated by using formula (2.4). Afterwards by using (2.3) the corresponding percentage errors in true TOZ amount because of the presence of SO_2 for a value of true TOZ amount of 0.300 atm-cm (300 DU) was calculated.

The results are shown in Figures 2.2 and 2.3. From these figures it is evident that the errors in true TOZ amount due to the presence of SO_2 in Athens were higher between December and March than between May and August. This is due to the fact that SO_2 concentration is greater in winter than in summer since in winter the Athenian atmosphere is supplied with SO_2 from the central-heating systems of the buildings.

In order to estimate the percentage error in true TOZ amount because of the presence of NO_2 the same method as used for SO_2 was followed. The results are presented in Figure 2.4. This figure shows that the errors in true TOZ amount because of the presence of NO_2 in Athens are higher from May to August (not shown here) than from December to March. The explanation of this is that NO_2 concentration is greater in summer, than in winter, since in summer there is greater photochemical pollution.

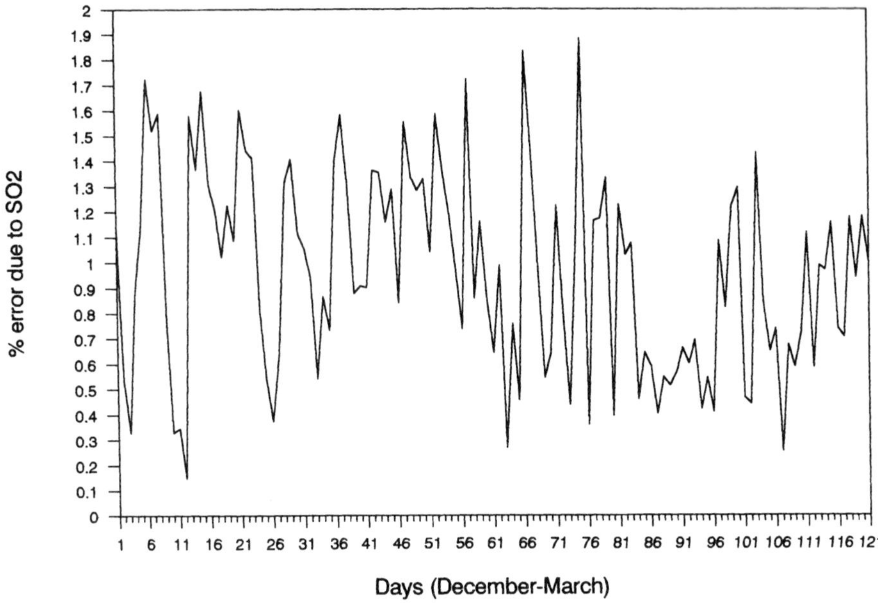

Figure 2.2. Mean daily percentage error in TOZ caused by SO_2 for December to March of the 2-year period 1989–1990.

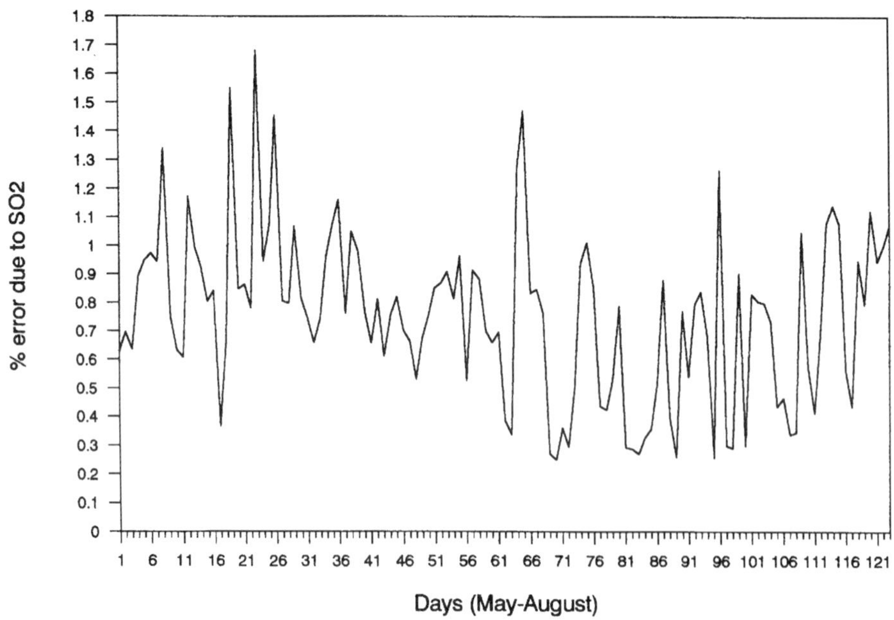

Figure 2.3. Mean daily percentage error in TOZ caused by SO_2 for May to August of the 2-year period 1989–1990.

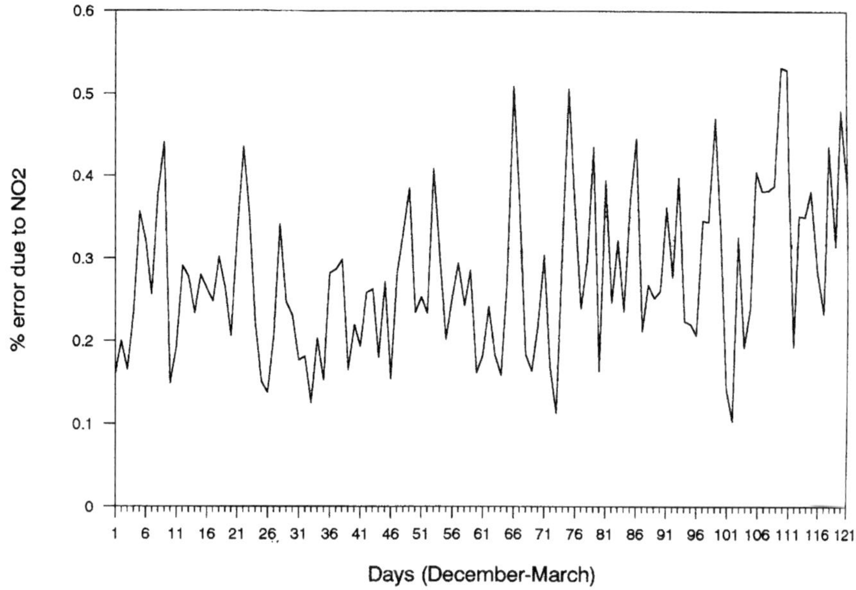

Figure 2.4. Mean daily percentage error in TOZ caused by NO_2 for December to March of the 2-year period 1989–1990.

Influence of stray light in Dobson TOZ measurements

One of the main factors affecting the accuracy of determination of TOZ measurements using a Dobson spectrophotometer arises from stray light contribution, mainly to the shorter wavelength band of the wavelength pairs measured by this instrument. This fact means that the problem is principally evident in the lower intensity band of the Dobson bandpairs. The influence of stray light on Dobson TOZ measurements is manifested by the characteristic effect of apparently declining ozone amounts with airmass above about 2.5. For direct sunlight AD measurements, errors of 1%, 3% and 10% may be present at airmasses 2.5, 3.2 and 3.8, respectively (Basher, 1982).

The term ‘stray light’ represents both unwanted radiation scattered by the atmosphere (air molecules, aerosols) into the instrument and unwanted radiation scattered within the instrument. Both kinds of stray light may be unwanted radiation of the same wavelengths as those of the measured bands as well as unwanted radiation of different wavelengths from the measured ones. Thus there exists monochromatic and heterochromatic stray light.

In 1982 Basher presented a simple monochromatic heterochromatic stray-light model for investigating non-linearity errors of the log intensity ratio versus airmass function. That model is characterized by two parameters. The first R_0 is the ratio of the energy of the monochromatic stray-light source to that of the desired band as would be measured under the imaginary conditions of zero airmass. If we consider the monochromatic and heterochromatic stray light at 370 nm then, according to Basher, the value of R_0 will be between 10^{-5} and 10^{-3}. The more able the instrument is to reject the stray light, the smaller the R_0 value is. The application of this model to total ozone X_{AD} measurements made by Basher in 1977 with the Wallops Island Dobson spectophotometer showed that the value of the R_0 parameter at that time was about $10^{-3.8}$ to $10^{-3.6}$. The application of the same stray-light model to the Belsk Dobson instrument No. 84 (Poland) showed R_0 values of 10^{-4} to $10^{-3.8}$ for all the analysed time series (Degórska and Rajewska-Wiech, 1989). The maximum value of 10^{-5} is not unattainable. This claim is verified by the airmass dependences taken with Canadian Dobson No. 77 and shown by Olafson and Asbridge (1981). The second parameter of Basher's stray-light model is the parameter a which reportedly depends on ozone amount and range from 0.7 up to 1.2.

Series of total ozone X_{AD} measurements for a large range of airmasses, extending to the value of 3.0, were taken over many days with Dobson spectrophotometer No. 118 in Athens, Greece. Among this measurement series were cases where the dependence of Dobson TOZ measurements on airmass were first observed after about the value of 3.0, while in others were observed after about the value of 2.5. The averages of the results for a range of ozone values were compared with some of the curves from Basher's stray light model. For that purpose, a reproduction of some of the stray-light model's curves is given. Figures 2.5 and 2.6 show two clusters for various a values of some of the stray-light model curves. The first corresponds to the value $R_o = 10^{-4.0}$ and the second to the value $R_0 = 10^{-3.0}$. From these figures it is obvious that the smaller the parameter R_0 the more able the specific Dobson instrument is to reject interfering stray light. Indeed, comparing any of these

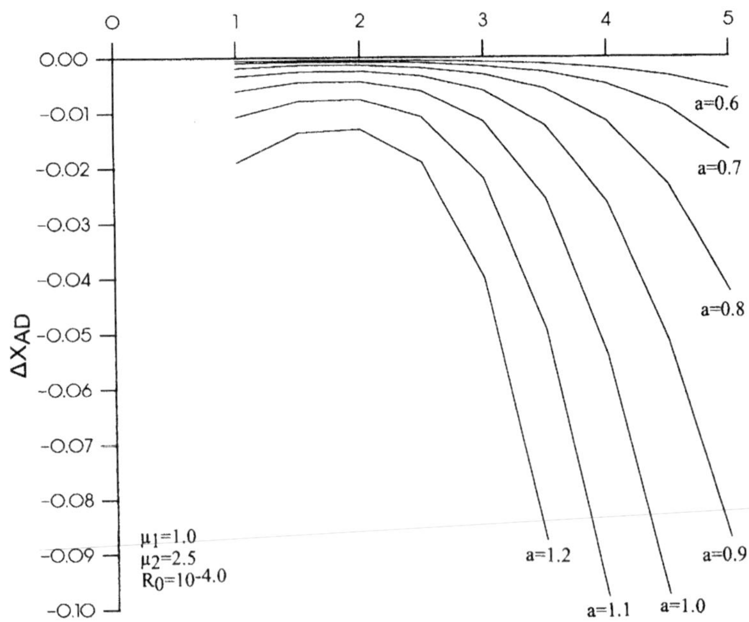

Figure 2.5. Variation in ozone error X_{AD} as a function of airmass μ for $R_0 = 10^{-4.0}$ as calculated by the Basher's stray light model.

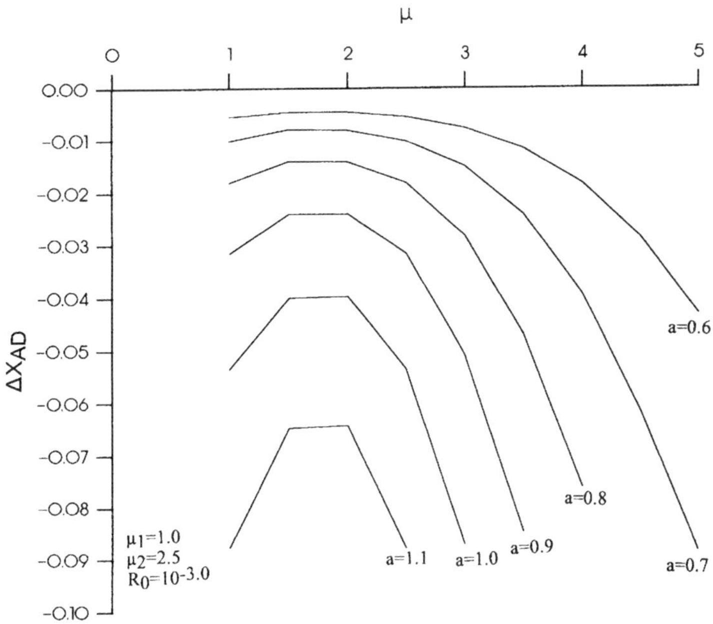

Figure 2.6. As Figure 2.5 but for $R_0 = 10^{-3.0}$.

curves corresponding to the same value of the parameter a but to different values of the parameter R_o, we can observe that in the case of $R_0 = 10^{-3}$ the curves start to decline with much smaller airmass, than for the corresponding curves with $R_0 = 10^{-4}$. Also these are more abrupt; that is, the rate of increase in error of total ozone measurement, because of the presence of stray light in the instrument, will be greater. Similar behaviour can be observed with the model curves for different values of a. The greater the value of a, the greater is the influence of stray light on Dobson total ozone measurements.

The results of the comparisons are given in Table 2.2. The first column shows the dates the measurements were made. These are dates between 1992 and 1994. The second and the third columns contain values of R_0 and a, the stray-light model parameters of the model curves fitting the data series. The fourth column contains the true total ozone amount in atm-cm as calculated by the model, while the last column shows the measured total ozone value also in atm-cm (see also Figure 2.7).

The general conclusion from the abovementioned series of TOZ measurements is that, over the time period in which these measurements were made, the value of R_0 of the Basher stray light model for the Dobson No. 118 instrument was in the range between $10^{-3.5}$ and $10^{-3.4}$. This means that this particular instrument does not suffer from stray-light interference when TOZ measurements are made with low values of airmass μ (regularly obtained observations). Another result from the same series of measurements is that there is only a weak relation between the other Basher model parameter a and the TOZ content because of the interference arising from the

Table 2.2. The results from some stray-light experiments taken by the Dobson No. 118 instrument in Athens between 1992 and 1994.

Date	R_o	a	True TOZ (atm-cm)	Measured TOZ (atm-cm)
27 November 1992	$10^{-3.4}$	0.6	0.285	0.282
	$10^{-3.5}$		0.285	
7 December 1992	$10^{-3.4}$	0.6	0.286	0.291
	$10^{-3.5}$		0.285	
23 March 1993	$10^{-3.4}$	0.6	0.303	0.306
	$10^{-3.5}$		0.303	
16 November 1992	$10^{-3.4}$	0.7	0.278	0.281
	$10^{-3.5}$		0.277	
14 January 1993	$10^{-3.4}$	0.8	0.312	0.305
	$10^{-3.5}$		0.309	
19 January 1994	$10^{-3.4}$	0.8	0.330	0.327
	$10^{-3.5}$		0.327	
24 March 1993	$10^{-3.4}$	0.8	0.307	0.307
	$10^{-3.5}$		0.305	
21 December 1992	$10^{-3.4}$	0.9	0.304	0.293
	$10^{-3.5}$		0.301	
18 March 1993	$10^{-3.4}$	0.9	0.340	0.336
	$10^{-3.5}$		0.337	

presence of SO_2 and NO_2, which absorb at the selected wavelengths of solar UVB radiation (Varotsos *et al.*, 1999).

2.1.1.1.2 Brewer spectrophotometer

Since the early 1980s, scientists around the world have been measuring TOZ, SO_2, NO_2 with Brewer spectrophotometers, which use a similar method to measure TOZ as that of Dobson instruments (Brewer, 1973).

These are grating spectrophotometers, developed in the early 1970s as an alternative to Dobson instruments in the World Ozone Network (Figure 2.8). They are a modified Ebert-type spectrometers with focal length 16 cm, width 11 cm and aperture ratio F/6. They use a single detector and are generally supplied with an automatic Sun tracker. Ground-based observations are made on both the solar and lunar disks as well as the zenith sky. They simultaneously measure the intensity of light at five wavelengths in the ultraviolet absorption spectrum of ozone with a resolution of 0.6 nm. The operational radiances at 306.3, 310.1, 313.5, 316.8 and 320 nm are monitored sequentially in 1.6 s. Measurements at these wavelengths allow correction for the effects of sulphur dioxide, a potential interferent for Dobson ozone measurements in polluted air (Section 2.1.1.1.1). Various intercomparisons of Brewer spectrophotometers, both between themselves and with Dobson instruments have been carried out mainly by the AES, Canada. Periodic intercomparison of Dobson instrument No 83 (World Primary Standard

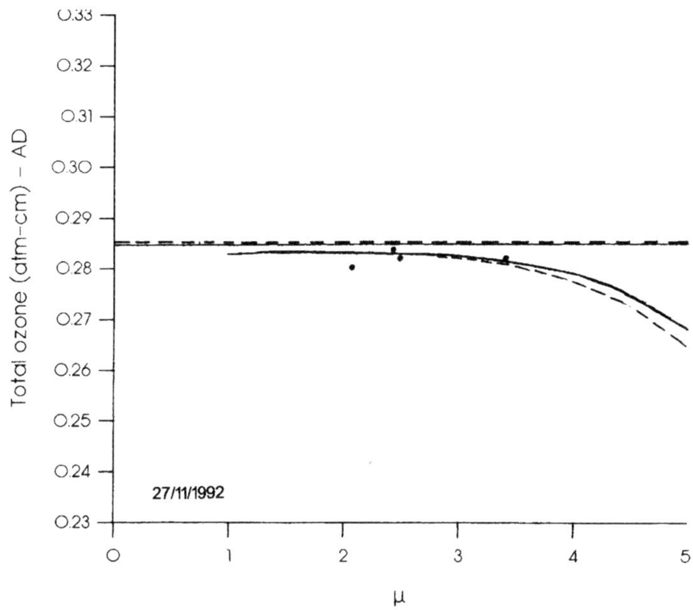

(*a*)

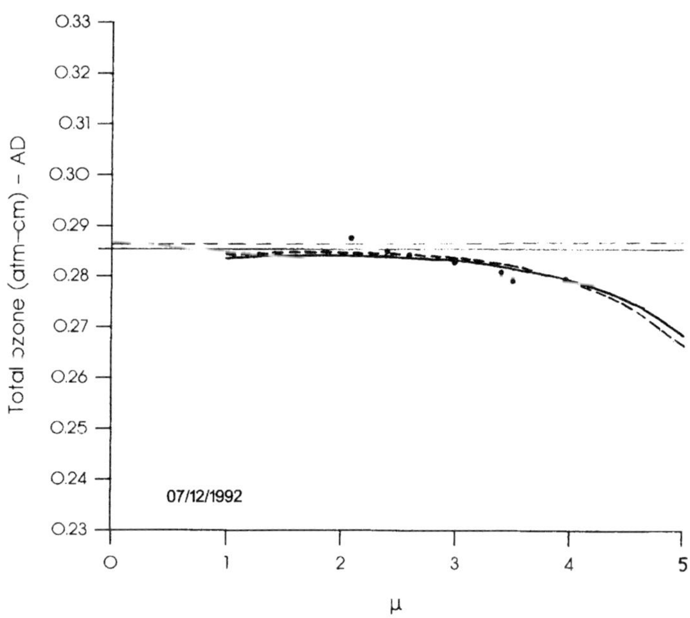

(*b*)

Figure 2.7. Comparison between experimental data and stray-light model values for (a) 27 November 1992 and (b) 7 December 1992.

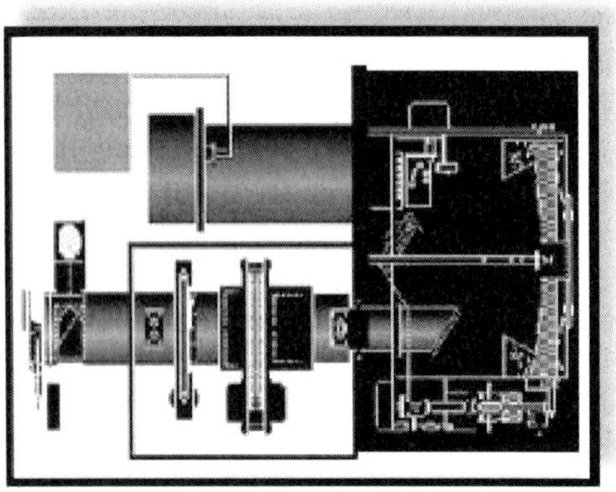

Figure 2.8. Top view of Mark II design.

Dobson Spectrophotometer) with Canada's Standard Brewer Ozone Instrument No. 17 has achieved consistent results within 1% (Kerr *et al.*, 1988).

Differences between measurements of the TOZ derived from Brewer and Dobson instruments are of the order of 3% (Köhler, 1995, 1999a; Vanicek, 1998). In 1997, over 130 Brewer spectrophotometers were in operation in 33 countries and in Antarctica.

To investigate the effects of scattered light on ozone retrieval a new double monochromator Brewer instrument was recently built by AES. Stray light in the double monochromator is several orders of magnitude less that that in a conventional single Brewer instrument. This instrument was made by linking two modified single Brewer optical frames together. The AES maintains a triad of calibrated Brewer instruments, as well as two travelling standards and one manufacturing standard. Each Brewer instrument is calibrated and maintained at WMO/AES ideal levels by means of intercomparison and travelling standards (Kerr *et al.*, 1998).

Mark II and Mark IV model Brewer instruments are fully automated to use the direct Sun, zenith sky or Moon as its light source. Instrument design and automation allows for near simultaneous observations of TOZ, SO_2 and NO_2, as well as UV spectra (290–325 nm). From a series of zenith sky observations at twilight, OVP can be determined by utilizing the 'Umkehr' inversion technique, which is described in Section 2.2.1.1.2.

2.1.1.1.3 Filter ozonometers M-83/124/134

The filter-type instrument is based upon the same principle as the Dobson spectrophotometer in using differential absorption of ultraviolet radiation in the 300–350 nm Huggins band of ozone. The M-83 filter ozonometer was developed in 1963 (Gushchin, 1963). This instrument is equipped with optical light filters, which consist of a set of optical pigmented glasses. It uses 21 nm and 15 nm bandpass channels at central wavelengths 299 and 326 nm, respectively. Improvements were made in 1972 by using the stated bandwidths and a third filter for corrections due to aerosol effects. Improved versions, instruments M-124 and M-134 developed in the 1980s,

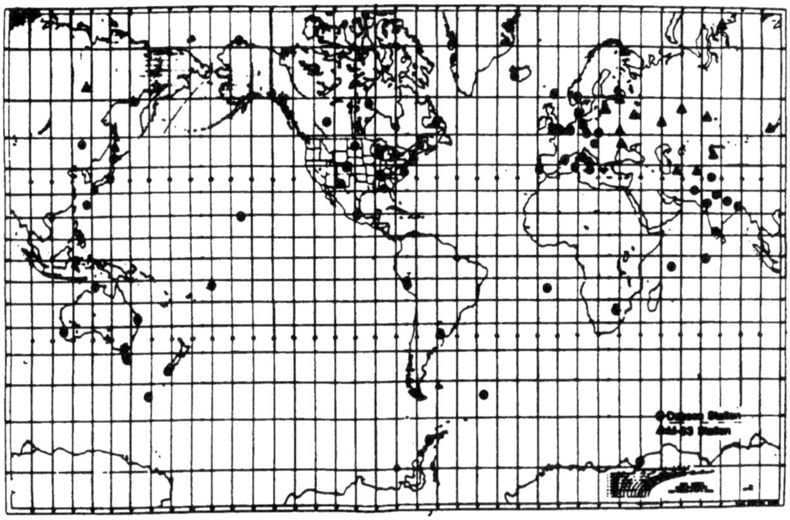

Figure 2.9. Geographical distribution of the existing network of Dobson spectrophotometers (●) and M83 type ozonometers (▲).

were introduced into the USSR's network of stations in 1985–86 (Gushchin and Sokolenko, 1984). This instrument measures ozone column content by looking directly at the Sun, using a peak spectral sensitivity at 299 nm (strong absorption by ozone) and another peak response at 326 nm (weak absorption by ozone). Like M-83, the improved instruments are intended to measure total ozone in the atmosphere, using both the direct-Sun observation method and zenith-sky one. Instead of two photoelectric receivers, the improved versions of M-83 employ only one sensing element (reducing substantially the weight and size of this device). The accuracy appears to be 2–4% and the precision of measurement is approximately 5%. A network consisting of 45 stations has been operating for 30 years in the former USSR and other countries (Figure 2.9).

The principle behind operation of M-83/124/134 ozonometers (Figure 2.10) and, in general, filter-type instruments (Gushchin ozonometers) has often been criticized mainly because of the effect of finite wavelength intervals. As glass filters are used with broadband transmission characteristics for the selection of wavelength bands, their peak transmission shifts depend on the solar zenith angle and the amount of ozone. For instance, empirical curves have to be used to determine TOZ from direct Sun or zenith-sky measurements (Feister, 1994). Some years ago, a number of researchers (Bojkov, 1969; Bojkov *et al.*, 1994) concluded that there is systematic disagreement between M-83/124/134 ozonometers and Dobson spectrophotometers, which increases with increasing atmospheric mass μ (reaching 20–30%, when $\mu > 2$). This difference becomes even larger if visibility is below 5 km (in the presence of haze). The filter-ozonometer data (covering more than one third of the northern mid and high latitudes from Eastern Europe to the Far East) were re-evaluated based on station-instrument changes and calibrations (Bojkov *et al.*, 1994). It can be assumed

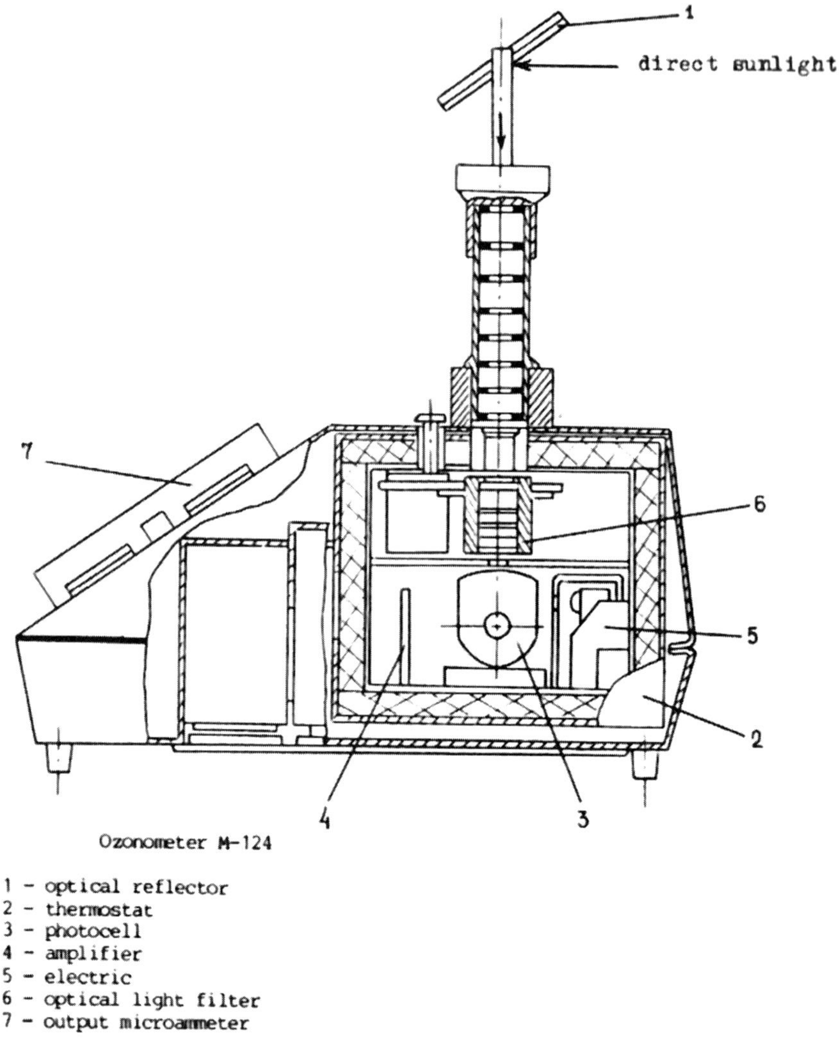

Figure 2.10. Schematic diagram of M-124 Ozonometer.

that, in order to use filter-equipped instruments, careful study of the relevant errors, like the effective transmission factors of the filters (Gushchin *et al.*, 1985) needs to be carried out in advance.

2.1.1.2 Secondary ground-based instrumentation for TOZ observations

The Global Ozone Network of primary ground-based instruments is supplemented by several similar instruments such as the Canterbury ozonometer (New Zealand) and the Sentran-ozonometer (USA), which are based on various techniques. Among these is the differential optical-absorption spectroscopy (DOAS) technique, which is one of the most powerful tools for observations of stratospheric trace gases (O_3, NO_2, NO_3, BrO and OClO). This technique was further developed (diode-array detectors and charge-couple devices) and DOAS instruments have been employed both on the ground and on aircraft during numerous measurement campaigns. In the following sections, brief information about some of these ground-based instruments is given.

2.1.1.2.1 *System for Analysis of Observation at Zenith (SAOZ)*

SAOZ measures the ozone column by application of the DOAS method (described above) to zenith-sky scattered-light spectra in the ultraviolet–visible range, taken at twilight.

In particular, it is a UV-visible diode-array spectrometer (300–600 nm) developed for regular TOZ and NO_2 monitoring at all latitudes up to the polar circle in winter. The instrument has a flat-field grating, a 512 or 1024-diode array detector and a 30° field-of-view limiter. Measurements are performed automatically from sunrise until sunset through a solar zenith angle (SZA) of approximately 94°. These are made by looking at the sunlight scattered at zenith in the visible wavelength range, twice a day, just before sunrise and just after sunset with a resolution of 0.6 nm. It should be mentioned that: (1) visible observations in the Chappuis bands at twilight (450–580 nm) have the advantages of being independent of stratospheric temperature, little contaminated by tropospheric ozone and multiple scattering and permitting observations even in winter at the polar circle; (2) measurements in the UV (by Dobson and Brewer instruments) cannot be performed at SZA larger than 80° to 82°; that is, at high latitudes in late fall and winter.

In order to retrieve the constituent vertical total column the slant column is divided by an airmass factor (AMF), which corresponds to the enhancement of the optical path compared with the vertical. The resulting differential spectrum is filtered to leave features resulting from absorption by various species in the atmosphere: O_3, NO_2, H_2O, O_4. The absorber features are then correlated with laboratory cross-sections to obtain the total amount of each present in the path.

Errors in TOZ with this method are the sum of those of the spectral analysis of the measurements and those of the AMF used in the retrieval. The SAOZ standard AMF calculated for high latitude during winter is 16.59 for O_3 and 17.77 for NO_2 at 90° SZA. On average, the standard AMF is 3% smaller than that calculated from sondes with little seasonal dependence (Barlett and Vaughan, 1998; Sarkissian *et al.*, 1997). Ozone columns are therefore overestimated by the same factor. The overall accuracy is ±10% for ozone and (+25, −45%) for NO_2 at the polar circle in winter, where the column is the smallest. In Section 2.2.2.2.9 a small balloonborne version of an SAOZ spectrometer is presented.

2.1.1.2.2 *MICROTOPS II (Total Ozone Portable Spectrometer)*

This filter instrument is a continuation of a series of hand-held ozonometers with TOPS at the beginning (Morys *et al.*, 1996). The new generation of these is MICROTOPS II, a 5-channel sun-photometer with centre wavelengths of 300,

305.5, 312.5, 940 and 1020 nm for measurements of TOZ, total water vapour and aerosol optical-thickness (Morys *et al.*, 1996). With this instrument TOZ is derived from measurements for three wavelengths in the UV region, given the site's latitude and longitude, universal time, altitude and pressure. As in Dobson instruments the measurement at an additional third wavelength enables a correction for particulate scattering and stray light. The total water vapour is determined by measurements at 940 nm and 1,020 nm. The angle of view of each optical channel is 2.5° and the resolution is 0.001 $\mu W\,cm^{-2}$. Typical agreement between various MICROTOPS II instruments (accuracy) is within 1–2% (Flynn *et al.*, 1996; Labow *et al.*, 1996). Repeatability of consecutive ozone measurements is better than 0.5%. Köhler (1999a, b) recently made a 21-months' intercomparison of the MICROTOPS II filter ozonometer with the Dobson and Brewer spectrometers. The result is that MICROTOPS can measure total ozone with an accuracy comparable with conventional spectrometers (agreement is better than ±1%), over a reasonable range of μ. Adverse conditions (clouds, haze and low sun) result in deviations of more than ±2% or even ±3%.

2.1.1.2.3 Fourier-Transform Spectrometer (FTS)

This instrument uses the Sun as a radiation source and measures absorption by IR-active molecules in the stratosphere. Solar radiation received by the instrument is directed by a beam splitter into two arms of the interferometer and is then recombined onto a detector. The length of one arm of the interferometer is usually varied and hence constructive and destructive interference effects occur for the various wavelengths. With this technique all wavelengths are measured simultaneously. That is the main advantage of interferometers compared with most grating spectrometers (Grant, 1989).

The lack of highly accurate spectral-line data and the overlaps within the instrument resolution of absorption lines of several species are the main factors that restrict measurement accuracy. Because of this a large spectral database is maintained in support of the Atmospheric Trace Molecule Spectroscopy (ATMOS) project (outlined in Section 2.2.3.1.2) that is updated continually with other field and laboratory measurements. In general, FTS measurement accuracy of ozone because of spectral parameter uncertainties is approximately 5%. The resultant accuracy from further uncertainties (temperature and pressure profiles, instrument errors, interference from other species) is from 10 to 20%.

The instrument has been used in various experimental campaigns as a ground-based, balloonborne or spaceborne device. SIRIS, for instance, is a balloonborne FTS (see also Section 2.2.2.2.5). It is an emission-mode remote-sensing instrument for measurement of the Earth limb from a balloon platform. Remote sensing of the limb allows vertical coverage from 5 to 40 km with a vertical resolution of 3 km. Another balloonborne-type instrument is the submillimeter high resolution FTS.

Ground-based FTS (with a 0.01 cm^{-1} resolution) has been used by the Jet Propulsion Laboratory (JPL) group to measure stratospheric trace gases in the Antarctic at the time of the spring ozone hole and to establish that chlorine rather than nitrogen is responsible for the rapid decrease in ozone after the spring sunrise.

A ground-based FTS instrument has also been operated in France to measure TOZ making use of two spectral regions (2084–2085 cm^{-1} and 2775–2776 cm^{-1}).

The International Scientific Station of the Jungfraujoch located at 3580 m in the Swiss Alps is one of the alpine sites that constitute the primary (Network for Detection of Stratospheric Changes) station for the northern mid-latitude and is equipped with various instruments, such as Fourier Transform Infrared (FTIR) spectrometers for TOZ measurements. The FTIR ozone data were retrieved from direct-Sun absorption spectra in various spectral regions by two FTS, a commercial Brucker IFSHR-120 spectrometer and a home-made FTS (De Maziére *et al.*, 1998).

A similar FTIR spectrometer has been used at Zugspitze/Garmisch (47.4°N, 11.0°E, 2964 m above sea level (a.s.l.)) since 1995. The main device is a Brucker IFS 120 HR spectrometer with a maximum resolution of $\Delta\nu = 0.00185\,cm^{-1}$. The optics of a 1.92-m Newtonian telescope (40-cm diameter) with a computer-controlled two-mirror solar tracker is matched to the spectrometer. For the open path or laboratory measurements the radiation of three different internal sources can be detected through the telescope and solar-tracker optics to an external retro-reflector positioned at a horizontal distance of 123 m. The reflected radiation is then directed back into the spectrometer through a beam splitter (Sussmann and Schaefer, 1997).

A ground-based Brucker 120 HR FT-spectrometer is also used at Kiruna in Sweden (67.84°N, 20.41°E, 419 m a.s.l.) to derive the isotopic ratios of ozone from solar and lunar IR absorption spectra recorded in the 10 μm spectral region (Meier *et al.*, 1998).

Finally, all thermal emission spectra have been measured also by a BOMEM, M100 Michelson interferometer at Peterborough, Ontario, since 1991 (Evans *et al.*, 1994). The resolution of this instrument was of the order of 4 cm^{-1}. The results indicate that it provides useful information on TOZ and on the tropo-stratospheric OVP (see also Section 2.2.1.2.6).

2.1.1.2.4 Light Detection and Ranging (LIDAR)

Lidar uses the difference of two laser wavelengths (308 nm, 351 nm), one absorbed by ozone (308 nm) and the other not (351 nm). The two beams are sent out into atmosphere, they are scattered and reflected and are then collected by a telescope. The intensity of the scattered light from each laser beam is compared and a measure of the concentration of ozone in the range 10 km to 50 km is received. Additional information for the lidar technique is given in Sections 2.2.1.2.1. and 2.2.2.3.2.

2.1.1.2.5 High Resolution Visible/Ultraviolet Absorption Spectroscopy

Absorption spectroscopy monitors the column abundance of various stratospheric species by measuring the absorption of light scattered by the sunlit zenith sky very precisely, typically 0.05% absorption (McKenzie and Johnston, 1982; Noxon *et al.*, 1979; Pommereau and Goutail, 1988a, 1988b). An important feature of such measurements is that they can be carried out on cloudy, as well as clear days.

Observations of stratospheric species that are achievable with ultraviolet/visible absorption spectroscopy have proven to be O_3, NO_2, NO_3 and OClO (Megie *et al.*, 1989).

An example of this type of instrument is the crossed Czerny–Turner spectrometer (Mount *et al.*, 1987), with a grating chosen so that the appropriate colour filters provide second-order light in the red region of the spectrum, approximately 605–685 nm (for NO_3 observations), and third-order light in the blue region, approximately 403 to 453 nm (for O_3, NO_2 and OClO observations). The instrument is used at approximately 0.5-nm spectral resolution. The detection system is a Reticon diode array cooled by a refrigerator to about $-70°C$; the array contains 1,024 independent silicon diodes that simultaneously measure the spectrum over the indicated wavelengths. Eleven diodes cover the full width at half maximum of the instrumental spectral profile. Thus, spectral lines are highly over-sampled. Observations are usually made using the direct light from the Sun or Moon. Direct-light observations have the advantage that weak sources (i.e. Moon) can be observed.

The 1σ standard errors of the fitted column abundances are typically 2–5% for large solar zenith angles for NO_2. Ozone columns can be determined to 10–15% accuracy using the Chappuis bands. Future development is expected to improve this considerably. The largest source of systematic error in the measurements is the evaluation of airmass factors for scattered light that are used to convert measured slant-column abundances to vertical-column abundances. These errors largely affect the accuracy, but not the precision of the data.

2.1.1.2.6 System for the Monitoring of Stratospheric Compounds (SYMOCS)

This instrument is a zenith-sky DOAS spectrometer (Mayer and Putz, 1998) developed by the Norwegian Institute for Air Research (NILU). It measures ozone and NO_2 total columns and slant columns of BrO and OClO at the spectral region 332–482 nm (using the ultraviolet/visible spectroscopy method). This instrument was installed at Spitsbergen in the fall of 1995 and participated in a NDSC (Network for the Detection of Stratospheric Change) validation campaign for ozone and NO_2 in the d'Haute Provence observatory (OHP) in June 1996. Just before winter 1997 SYMOCS moved to Andoya in Norway (ALOMAR) to participate in several satellite-validation studies (GOME and SAGE III) (Fierli *et al.*, 1998).

2.1.1.2.7 Star Pointing Spectrometer (SPS)

The instrument consists of a telescope that focuses light from stars, planets or the Moon onto a spectrometer and two CCD-array detectors. From the observations of atmospheric absorption, the atmospheric columns of several gases can be determined (Fish *et al.*, 1994). A novel version of this instrument (used for O_3, OClO, NO_2 and NO_3 retrieval) was deployed in Abisko (69°N) during EASOE. This instrument is particularly useful for polar measurements in winter, because it observes at night. By using an array detector, spectral noise from atmospheric flicker is avoided. By using a two-dimensional array as well, light from the sky adjacent to the star is measured simultaneously and subtracted. This technique is very important at twilight at the poles during aurora or in hazy conditions near city lights.

2.1.1.2.8 MDR-23 (Russian commercial device)

This is a grating scanning spectrometer with a 0.7 nm spectral resolution and a 51.2 nm min^{-1} scanning rate. It measures TOZ in the 303–333 nm spectral range at solar zenith angles lower than 70° and total NO_2 in the 435–450 nm spectral range during morning and evening twilights at solar zenith angles 84–96°. TOZ is derived using nomograms calculated using the single-scattering approach (Elokhov and Gruzdev, 1991, 1992). This kind of instrument was installed aboard the ship *Akademik Feodorov*, and TOZ was measured in the 40°S–40°N latitude band in the Atlantic Ocean in May 1988.

2.1.1.2.9 Scanning spectrometer (EVA)

This operates in the range 450–470 nm for TOZ and 430–450 for total NO_2, in zenith view made at twilight. A fully automatic version, developed later, obtained a set of 30 spectra, 200 samples each, accumulated for each measurement to reduce photomultiplier noise. Instrumental errors are estimated to be $4\text{–}6 \times 10^{15}$ mol cm^{-2} for NO_2 and $1\text{–}2 \times 10^{19}$ mol cm^{-2} for O_3 (Yela *et al.*, 1998)

2.1.1.2.10 Solar IR spectroradiometer

This instrument is basically a double monochromator on a heliostat to track the Sun. Solar radiation is focused on the entrance slit of the instrument by a Newtonian telescope, and it is converted to a square wave by the in-built chopper with frequency 163.4 Hz. Incoming radiation is dispersed by the gratings, blared at 1600 nm and passed through the filter to a detector fixed at the exit slit. The electrical signal generated by the detector is amplified and the output is fed to the computer for data processing and analysis. The maximum error in TOZ is less than 7%. The stray light is less than 0.1%, with maximum error 3% (Bose *et al.*, 1998)

2.1.1.2.11 Ground-based UV radiometer (GUV)

This is a multi-channel moderate-bandwidth UV-filter instrument measuring absolute irradiance at 305 nm and 320 nm used to derive total ozone. The optical part of the instrument consists of a Teflon diffuser, interference filters and photosensitive detectors. The instrument is temperature stabilized at 40°C and the time resolution is 1 min. Details of the instrument and data analysis are given in van Roozendael *et al.* (1998) and Dahlback *et al.* (1998). Ground-based UV/VIS (visible) measurements of O_3, NO_2, OClO and BrO were carried out over Ny-Ålesund (79°N, 12°E). The instrument operated in the near UV and visible spectrum (327–491 nm) and zenith-sky spectra were measured every day for solar zenith angles smaller than 95°. Automatic spectral calibration with a HgCd lamp and measurements of the dark current and the detector were performed daily (Wittrock *et al.*, 1998). A UV spectroradiometer (absolutely calibrated) was also used for ground-based TOZ determination by Huber *et al.* (1998). In this respect, a high-resolution double monochromator with a full bandwidth of 0.72 nm at 50% maximum was used. Spectra were taken from 290 to 500 nm. The spectroradiometer was calibrated against a 1000 W standard lamp. An automated solar UV spectrophotometer (SUVS) was presented by Dorokhov *et al.* (1989) for measurements of TOZ and

UV radiation, using a similar technique to that of the Dobson spectrophotometer (see Section 2.1.1.1.1). Spectral measurements of solar radiation were also carried out in the Arctic with a double monochromator H10D produced by Jobin-Yvon. The TOZ was derived from direct solar measurements by applying Beer's law (Henriksen and Stamnes, 1989).

2.1.2 Satellite remote sounding of TOZ

During the second half of the 1960s several attempts at satellite remote sounding of ozone and other minor gas components from Soviet manned spacecraft were first made, using a hand-held spectrograph and a complex of solar spectrometers functioning in a regime of occultation geometry (Kondratyev, 1972; Kondratyev *et al.*, 1996).

Since 1970, appropriate instrumentation launched on board various satellites, have provided a global distribution of TOZ with very good spatial and temporal coverage. For instance, TOZ monitoring by NASA's Nimbus-7 SBUV and TOMS instruments began in 1978. These measurements continue with SBUV/2 instruments on board the NOAA-9, -11 and -14 satellites, and TOMS instruments on the Russian Meteor-3, Earth Probe, and Japanese ADEOS satellites. The European Space Agency's GOME on the ERS-2 satellite, also performing BUV measurements, complements the US efforts. In addition, the Shuttle SBUV (SSBUV) experiment (conducting eight missions between October 1989 and January 1996) provided regular checks on the calibration of individual satellite instruments.

Multiple intercomparisons with ground-based instruments improved data-retrieval algorithms and therefore satellite observations of TOZ became compatible ground-based ones (McPeters and Labow, 1996).

In the US ozone-research plans include both a total ozone mapper and an ozone-profiling instrument for the National Polar Orbiting Environmental Satellite System (NPOESS). The first launch of NPOESS may not occur until 2010. An important source of minor-gas components data is the TERRA satellite launched in December 1999. In Europe the Ozone Monitoring Instrument (OMI) is being considered for the METOP satellite series, in order to measure both total ozone and ozone-vertical profile (Kaye, 1997).

A brief description of the satellite instrumentation providing TOZ observations, is given in Sections 2.1.2.1 and 2.1.2.2, grouped by type of measurement (direct absorption, indirect absorption and emission). It should be noted that several instruments provide not only TOZ, but also OVP measurements.

2.1.2.1 Direct-absorption measuring instruments

In this section brief information is given about a few instruments that are used to measure TOZ by directly measuring absorption features of ozone, either within a path defined by the instrument or by using an extraterrestrial source such as the Sun.

2.1.2.1.1 TIROS Operational Vertical Sounder (TOVS)

The Television Infrared and Observational Satellite (TIROS)-N TOVS instruments aboard NOAA's satellites provide information on the ozone mixing ratio vertical profiles as well as TOZ in the atmosphere. It is a set of three instruments, the High Resolution Infrared Sounder (HIRS/2), the Stratospheric Sounding Unit (SSU) and the Microwave Sounding Unit (MSU). All three instruments are used to provide temperature soundings by observing emission from CO_2, H_2O, N_2O or O_2. The surface temperature and the presence of clouds can be determined. TOZ is determined by the use of three HIRS filter radiometer channels (through the 9.7 micron channel).

TOVS data have been used from 1978 to the present (Neuendorffer, 1996; Kondratyev, 1997, 1998) and their standard deviation is approximately 7%. TOVS total-ozone data are only sensitive to variations in the lower stratosphere and long-term TOVS total-ozone trends are, at best, only indicative of lower stratospheric ozone trends (*Scientific* ..., 1999).

2.1.2.1.3 OZON-MIR

The PRIRODA module of the Mir Space Station was launched in mid-1996. This module has two instruments on board, known as OZON-MIR and ISTOK-1. These instruments use the occultation technique to measure concentrations of a number of trace gases and aerosols. ISTOK-1 also has a limb emission mode, although retrieval studies suggest that the errors in this mode are considerably larger than in the occultation mode (*Scientific* ..., 1999). The instruments and their specifications are shown in Table 2.3.

2.1.2.1.2 Laser Heterodyne Spectrometer (LHS) Tunable Diode LHS (TDLHS)

This spectrometer measures TOZ by performing observations of the ozone profiles in the 0.9-μm region in the infrared. A laser is used as a local oscillator to direct a

Table 2.3. Instruments on board the PRIRODA module of the Mir Space Station.

Instrument	Spectral Range	Channels/ Bandwidth	Swath	Spatial resolution	Remarks
MSU-SK	0.5–12.5 μm	5/100 nm	350 km	120 × 300 m	Conical scanner
MSU-E	0.5–0.9 μm	3/100 nm	45 km	25 m	Pushbroom scanner
MOS-A	757–767 nm	4/1.4 nm	80 km	2.7 km	Imaging spectrometer
MOS-B	408–1,010 nm	13/10 nm	80 km	650 m	Imaging spectrometer
ALISA	532 nm	–	3′	150 m (vertical)	Lidar pulse power 40 mJ
TV camera	400–750 nm	–	15°	399 m	Colour
MOMS-2P	400–750 nm	4 + 3/80 nm	44/88 km	5/16 m	Stereo/MS
OZON-MIR	0.26–1 μm	4	2′ × 25′	–	Occultation

narrow bandwidth beam onto a detector or photomixer. Usually a CO_2 laser is used as the local oscillator, but occasionally a tunable diode laser is employed. The Sun or the Moon acts as the source of radiation beyond the atmosphere. The radiation is selectively absorbed by ozone or other molecular species in the atmosphere and is detected by the spectrometer. A TDLHS system was used, as a ground-based station instrument, at the University of Denver to determine TOZ (McElroy *et al.*, 1991). A CO_2 and TDLHS can also be used to measure VOP. Such measurements can be made with a vertical resolution of 4 km, which is limited by the rate of change in the spectral weighting functions with altitude or with lower resolution, if an insufficient number of channels are used. The accuracy of the measurements is approximately 5–10% for column content and 10–30% for profile measurements. The precision is approximately 2%. The advantages of the LHS over other techniques are its ultra-high spectral resolution, high spatial resolution, high quantum detection efficiency and very good signal-to-noise ratio. The high resolution makes the system very selective as the interference problems due to overlapping lines or bands are minimized (Fogal *et al.*, 1989).

2.1.2.2 *Indirect absorption measuring instruments*

Satellite observations of TOZ have been obtained for over a decade using the backscattered ultraviolet (BUV) technique, which is briefly described here. Incident solar radiation at 317.5 and 360 nm passes through the ozone within the upper atmosphere, and is backscattered largely within the troposphere at altitudes below the peak in ozone concentration. This backscattered radiation passes once again through the ozone to the satellite. The radiation at 317.5 nm is attenuated by ozone during both passages through the atmosphere. However, measurements at the 317.5 nm BUV alone cannot be used to determine TOZ without knowledge of the varying reflectivity of the backscattering region. In order to determine this reflectivity, the 360 nm BUV (which is not attenuated by ozone) is measured with the BUV technique. The first synoptic global images of TOZ were obtained in the sunlit hemisphere, using the imaging instrumentation on board the spacecraft *Dynamics Explorer-1* (DE-1).

The instruments that are presented in this section have been developed to measure TOZ from space by measuring backscattered UV solar radiation.

2.1.2.2.1 Total Ozone Mapping Spectrometer (TOMS)

The TOMS uses a single monochromator and scans across the sub-orbital track, sampling radiation backscattered from the underlying surface and atmosphere. It uses an IR cloud-cover photometer to avoid problems from clouds (*Scientific* ..., 1999). This instrument observes in six wavelength regions from 312.5 to 380 nm (similar to the BUV and SBUV) and TOZ is inferred, by utilizing the wavelength dependence of the Earth's ultraviolet albedo in the Huggins band of the ozone absorption spectrum. The first four wavelength regions are used in pairs to provide three estimates of TOZ by the differential absorption method, while the other two (free of ozone absorption) are used to determine the effective background albedo.

Table 2.4. TOMS main characteristics.

Spectral bands	6
Wavelengths	312.5, 317.5, 332.2, 331.2, 339.8, 360, 380 nm
Bandwidth	1.0 nm
Total ozone accuracy	< 2%
Ozone trend accuracy (goal)	1% per decade
Sulphur dioxide accuracy	±25%
Scan time	6.3 seconds
Date rate	approx. 700 bps
Mass	< 33 kg
Power	< 20 W

The Temperature Humidity Infrared Radiometer (THIR) on board the satellite is used to determine cloud locations and heights.

The precision is given as 2% or better (small drift in diffuser-plate reflectance of 0.4% per year). The new absorption cross-sections of ozone absorption yielded TOZ values 6% higher than the old cross-sections. The TOMS instrument on board the Nimbus-7 satellite, since late 1978, has been used to measure global distribution of TOZ. In Table 2.4 the main characteristics of the TOMS instrument are given.

2.1.2.2.2 Meteor-3/TOMS

The measurements made using the TOMS experiment aboard the Meteor-3 satellite began on 22 August 1991 and ended on 28 December 1994. This experiment provided daily global coverage of the Earth's TOZ by measuring the backscattered Earth radiance. These data were archived at the Goddard Space Flight Center (GSFC), Distributed Active Center (DAAC) and have been processed independently by the Central Aerological Observatory (CAO) of the Russian State Committee on Hydrometeorology. The $1\frac{1}{2}$-yr overlap between Nimbus-7 and Meteor-3 permitted an intercalibration that allows the data from the two satellites to be used to form a continuous 16-year dataset, which can be used to derive ozone trends from November 1978 through December 1994.

The basic differences between the Meteor-3 TOMS (engineering model) and the Nimbus-7 TOMS instrument are the following:

- An Interface Adapter Model (IAM) was added to make the TOMS instrument compatible with Meteor-3, which was launched on 15 August 1991 in an orbit with an 82.5° inclination. The orbit proceeded relative to the Earth–Sun line with a period of 212 days, unlike Nimbus-7 which had a Sun-synchronous orbit. For two periods within this cycle the instrument was close to the terminator and made ozone retrieval difficult.
- Replacement of the single diffuser with a three-diffuser carousel allowed calibration by comparison of diffusers with different rates of exposure.

- The higher altitude of Meteor-3/TOMS (1200 km compared with 996 km for Nimbus-7/TOMS) meant greater overlap for the Meteor-3/TOMS than for Nimbus-7/TOMS.
- The Meteor-3 instrumentation was further improved when the diffraction grating was replaced, the mirrors were resurfaced, flight-qualified electronics were added, and a solid-state data recorder replaced the tape recorder, which had been used for Nimbus-7 and so almost no data were lost during the 3-year mission.

To obtain daily high-resolution global maps of atmospheric ozone, Meteor-3/TOMS measured solar irradiance and the radiance backscattered by the Earth's atmosphere in six (1-nm) selected wavelength bands in the ultraviolet. The experiment used a single monochromator and scanning mirror to sample the backscattered SUVR at 35 sample points at 3° intervals. It scanned the Earth in 3° steps to 51° on each side of the sub-satellite point in a direction perpendicular to the orbital plane.

In normal operation, the scanner measured 35 scenes, one for each scanner-view angle stepping from right to left. It then quickly returned to the first position, not performing measurements on the retrace. Eight seconds after the start of the previous scan, another scan would begin. The observations used for ozone retrieval were made during the sunlit portions of the orbit (Figure 2.11).

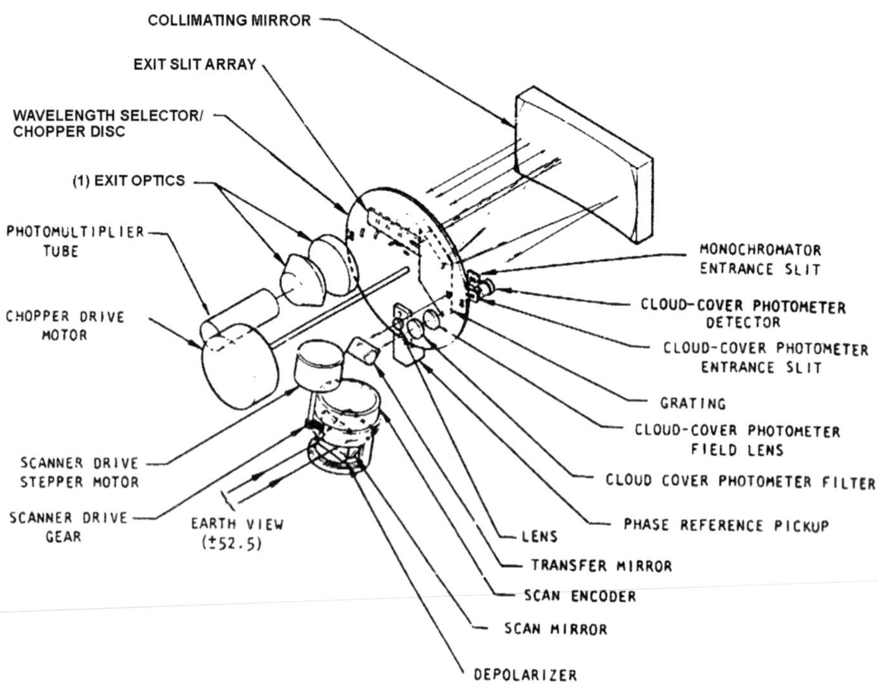

Figure 2.11. TOMS optical diagram.

A detailed description of the adaptations to TOMS for Meteor-3 is provided by Herman *et al.* (1996) and the pre-launch and post-launch calibrations are described by Jaross *et al.* (1995). Because of the post-launch calibrations severe degradation took place of an aluminium diffuser plate deployed to reflect sunlight into the instrument. Aboard Nimbus-7 severe degradation of the diffuser plate was also observed as time passed (Wellemeyer *et al.*, 1996). The three-diffuser system aboard Meteor-3 reduced the exposure and degradation of the diffuser and allowed calibration through comparison of signals reflected by diffusers with different rates of exposure. As mentioned above, diffusers were arranged at the sides of an equilateral triangle and mounted on a carousel, so that a given diffuser could be placed into view on demand. For Meteor-3 TOMS TOZ, the absolute error is ±3% and the drift for 40 months is less than ±1%. Meteor-3 TOMS agrees with both Nimbus-7 TOMS and ground-based measurements to better than 1% (McPeters *et al.*, 1996).

An additional TOMS instrument is scheduled for launch in 2000 aboard a Russian Meteor-3M satellite (Kaye, 1997).

2.1.2.2.3 Earth Probe Total Ozone Mapping Spectrometer (TOMS/EP)

TOMS launched on board an *Earth Probe* satellite (TOMS/EP) in July 1996 was developed by NASA in order to measure ozone by monitoring ultraviolet light. It observes both incoming solar energy and backscattered ultraviolet radiation at six wavelengths. TOMS EP is on a 500-km orbit, which provides for higher horizontal resolution, but does not provide full daily global coverage. It is due to this fact that 2 days are required to produce a global map (*Scientific* ..., 1999).

The instrument makes 35 observations every 8 s, each covering a 30 to 125 mile width on the ground. Almost 200,000 daily observations cover every single spot on the Earth, except for areas near one of the poles, where the Sun remains close to (or below) the horizon during the entire 24 hour period.

In addition to ozone, this instrument measures sulphur dioxide released in volcanic eruptions. These observations are of great importance in detection of volcanic ash clouds that are hazardous to commercial aviation.

2.1.2.2.4 Advanced Earth Observing Satellite (ADEOS I,II)

The Japanese ADEOS satellite carried the fourth TOMS into orbit by an H-II rocket on 17 August 1996. It remained in range at 10:45 AM throughout the ozone data record from 11 September 1996 to 29 June 1997. Its orbital inclination was 98.6° and the nominal orbit altitude was 800 km with a period of 100.9 min.

The approach taken to calibrate the three diffuser plates (by examining the differences in degradation of diffuser reflectivity resulting from different rates of exposure), was first used with Meteor-3 TOMS and proved to be successful. In addition the Japanese ADEOS/TOMS is equipped with UV lamps for monitoring the reflectivity of the solar diffusers.

For ADEOS/TOMS TOZ, the absolute error is ±3%, the random error is ±2% and the drift over the 9-month data record is less than ±0.5%.

The ADEOS/TOMS observations of TOZ are approximately 1.5% higher than a 45-station network of ground-based measurements (McPeters and Labow, 1996; Krotkov *et al.*, 1998; Seftor *et al.*, 1997; Herman *et al.*, 1997; Torres *et al.*, 1995; Torres and Bhartia, 1999).

The complementary instrumentation of ADEOS includes the Improved Limb Atmospheric Spectrometer (ILAS), the Interferometric Monitor for Greenhouse Gases (IMG) and the Retro-reflector in Space (RIS) instrument. ILAS is described in Section 2.2.3.3.10. IMG is a nadir-observing Michelson-type FTS (see Sections 2.1.1.2.3 and 2.1.2.2.11) designed to measure density profiles of CO_2 and H_2O, total ozone column and mixing ratios of CH_4, N_2O and CO in the troposphere. RIS, which is used together with laser ground stations, supports vertical profile and/or column measurements of a small number of gases.

ADEOS-II (a sun-synchronous orbit satellite) is being developed to include sensors like the Advanced Microwave Scanning Radiometer (AMSR), an advanced microwave radiometer developed by the National Space Development Agency of Japan (NASDA). ILAS-II, an improved spectrometer for measuring infrared radiation at the edge of the atmosphere, was developed by the Environmental Agency of Japan (Kondratyev *et al.*, 1998).

2.1.2.2.5 Global Ozone Monitoring Experiment (GOME)

GOME was launched in 1995 aboard the European Space Agency's Earth Resources Satellite (ERS-2) in a polar sun-synchronous orbit. It is a nadir-viewing multi-channel spectrometer measuring solar irradiance and earthshine radiance in the wavelength range 240–790 nm at moderate spectral resolution (0.2–0.4 nm). It exploits the fact that near 260 nm the penetration depth of solar radiation into the atmosphere strongly increases with increasing wavelength, because at this specific wavelength the ozone absorption cross-sections have their maximum. When the radiation has wavelength greater than about 310 nm, it penetrates into the tropopause reaching the surface. Therefore, the measurement of backscattered radiation in the UV-visible region provides information about the vertical-ozone concentration profile. In general, GOME measures TOZ, ozone vertical profile and total column of several trace constituents, including BrO, NO_2 and OClO and also obtains information on clouds, aerosols and surface spectral reflectance. It can measure TOZ at a higher horizontal resolution than TOMS, and thus complements TOMS observations. Retrieval of OVP from backscattered UV-visible spectra is based on the 'eigenvector method', which uses *a priori* information (climatological mean profile with the corresponding covariance matrix) as the original optimal estimation approach. Alternatively, the Full Retrieval Method (FURM) can be used (developed at the Institute of Remote Sensing of the University of Bremen). This is mainly based on the optimal estimation approach and contains the radiative transfer code GOMETRAN as an essential tool (Rozanov *et al.*, 1997a, b). Very recently an explicit treatment of clouds in the retrieval algorithm based on a new GOMETRAN cloud parametrization has been developed (Kurosu *et al.*, 1997).

Ozone observations by GOME on ERS-2 during the Arctic springs of 1997 and

1998 showed that extensive regions of low TOZ were observed. In particular, the major decrease is dominant in the lower and mid-stratosphere inside the polar vortex. Also the winter of 1997/98 was warmer than the year before and less ozone depletion was observed. In spring 1998 an ozone mini-hole event was observed by GOME, and ozone profiles under ozone mini-hole condition were derived for the first time.

2.1.2.2.6 Global Ozone Monitoring Radiometer (GOMR)

GOMR is an operational instrument for carrying out long-term ozone measurements that started in 1998. It supplants SBUV/2 and is the next in the series of NOAA operational satellite ozone-monitoring systems, subsequent to the current Advanced TIROS-N (*Scientific* ..., 1999). It consists of two sub-systems: a UV nadir and side-scan sounder based on Nimbus-7 TOMS technology (providing TOZ in the sunlit portion of the Earth), and an infrared limb sounder based on EOS limb sounding instruments (providing stratospheric profiles of ozone and other minor constituents) (Megie *et al.*, 1989) (Figure 2.12).

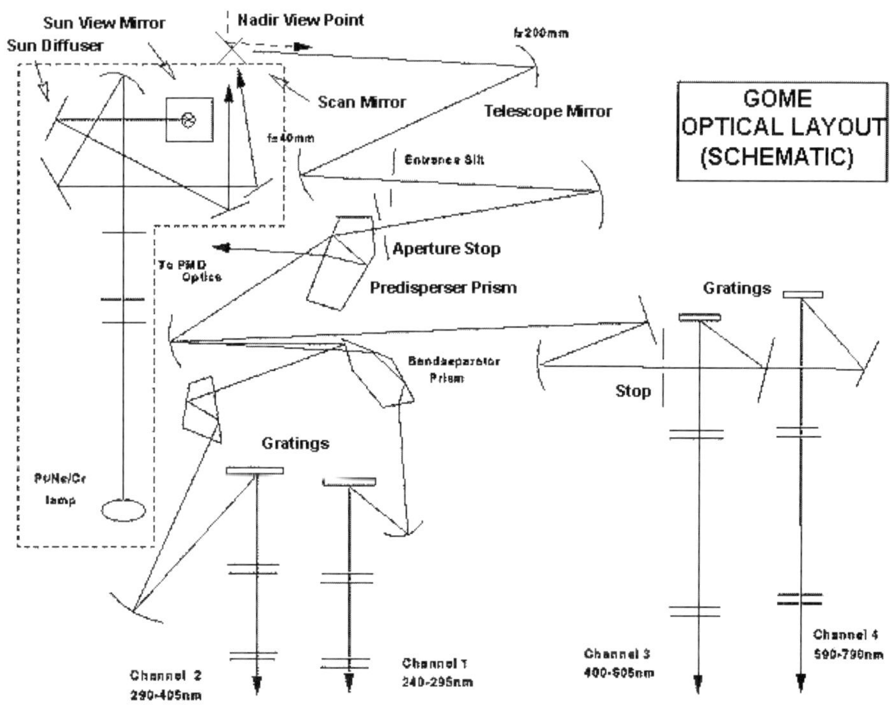

Figure 2.12. Schematics of GOME optics. The GOME instrument is a four-channel spectrometer. Adjacent to the spectrometer is a calibration unit housing a Pt/Cr/Ne hollow cathode discharge lamp and the fore optics for solar viewing.

2.1.2.2.7 ESA Polar Platform (EPOP)

The European Space Agency has selected the following three instruments for stratospheric ozone measurements on the first EPOP-1 flight (Megie *et al.*, 1989):

1. Global Ozone Monitoring by Occultation of Stars (GOMOS), that measures ozone by stellar occultation, observing the full UV/visible/near-infrared spectrum of stars, using a double-grating spectrometer.
2. Scanning Imaging Absorption Spectrometer (SCIAMACHY). Relevant information is given in Section 2.2.3.2.7.
3. Michelson Interferometer for Passive Atmospheric Sounding (MIPAS) that is a limb sounder operating in the mid-infrared region of the spectrum (see also, Section 2.2.3.3.3)

2.1.2.2.8 NASA's Polar Orbiting Platforms (NPOP-1/2)

NPOP-1 includes an infrared limb sounder and SAGE III occultation instruments providing observations of ozone with high vertical and horizontal resolution, high accuracy and precision from cloud tops to 80 km. In parallel to this, a SAGE III instrument, as an attached payload for the Space Station, has provided, since 1997 ozone profiles at low to mid-latitudes, in order to complement the NPOP-1 high-latitude coverage. NPOP-2 includes limb sounders operating in the infrared, sub-millimeter and microwave portions of the spectrum providing additional measurements of ozone profiles (Chiou *et al.*, 1998a, b).

2.1.2.2.9 Ozone Dynamics Ultraviolet Spectrometer (ODUS)

This is a Japanese ultraviolet instrument measuring ozone and several other species. ODUS is going to be flown on the EOS CHEM platform, currently scheduled for 2002. It uses techniques similar to those used by TOMS and GOME (see Sections 2.1.2.2.1, 2.1.2.2.5).

2.1.2.2.10 Ozone Layer Monitoring Experiment (OLME)

This instrument consists of UV cameras using both charge-coupled devices and UV photodiode detectors. It was flown aboard Chile's FaSAT Bravo satellite in 1997 and measured ozone in the Antarctic and sub-Antarctic regions of Chile. Another instrument OM-2 (developed in Israel) uses filters for the measurement of total ozone and the ozone-vertical profile aboard the Israeli Techsat-1 satellite (Parnes, 1992).

2.1.2.2.11 Interferometric Monitor for Greenhouse Gases (IMG)

This is a Michelson-type interferometer that was launched on 17 August, 1996 on the Japanese ADEOS platform (see Section 2.1.2.2.4). This instrument monitors IR (3.3–14 μm) from the Earth's surface and the atmosphere, observing with a nadir view. For inversion of data the atmospheric temperature profile is needed in order to retrieve the total amount of O_3, CO, CH_4 and a few points of the vertical profile. For these data an inversion algorithm based on neural networks techniques is currently used (Hadji-Lazaro *et al.*, 1998).

2.1.2.2.12 Infrared Atmospheric Sounding Interferometer

This instrument is a Michelson-type interferometer providing spectra between 3.4 and 15.5, with an apodized resolution of $0.5\,cm^{-1}$. It will be launched in 2001 on the European METOP platform (Hadji-Lazaro *et al.*, 1998).

Let us briefly discuss some results illustrating TOZ observations (some more results will be considered in Chapter 3).

2.1.3 Observed variability in total ozone column

TOZ is characterized by significant temporal and spatial variability. As far as temporal variability is concerned, it consists of large periodic and aperiodic components. Periodic variations have timescales ranging from day-to-day changes, through seasonal and annual variations to an 11-year solar cycle. Aperiodic variations include the irregular quasi-biennial oscillation (QBO) with a period of roughly 26–30 months, the irregular El Niño Southern Oscillation (ENSO) with a period of 4–7 yr and other interannual signals. Important phenomena are ozone trends which will be considered in Section 2.3.1.

As far as spatial variability is concerned, it is characterized by latitudinal and longitudinal dependence. Further information is given in Section 2.1.3.1.

2.1.3.1 Latitudinal variation of TOZ

A very important feature of global TOZ distribution is the strong latitudinal gradient with lower values over the equator and tropics and higher values over mid and high latitudes. This gradient is characterized by a well-pronounced annual cycle, reaching a maximum in spring and a minimum in fall. The amplitude of this annual cycle is a function of latitude, with maximum at about 60° north and south latitude (Varotsos *et al.*, 2000). In the tropics seasonal variations are small, with ozone maxima in summer. Such latitudinal distribution results from the relatively long lifetime (months to years) of ozone in the lower stratosphere and the Brewer–Dobson circulation that transports stratospheric ozone from the tropics toward the poles and downwards at high latitudes.

Bojkov and Fioletov (1995), based on re-evaluated TOZ observations from over 100 Dobson and filter radiometer stations from pole to pole, presented the following TOZ trend results:

- TOZ through early 1995 continues its decline (which started in the 1970s), with statistically significant year-round and seasonal trends except over the equatorial belt.
- The cumulative year-round TOZ reduction over the 35°–60° belts of both hemispheres since the early 1970s is up to 8%. While in the southern mid-latitudes it is difficult to distinguish seasonal dependence of TOZ trends, the cumulative decline in the northern mid-latitudes in winter and spring is about 9% and 4–6% for summer and fall.
- Observations from 12 Dobson polar stations demonstrated that the northern polar region shows the same ozone decline as northern mid-latitudes or even a

Table 2.5. TOZ trends from January 1979 to March 1995 for five regions of northern mid-latitudes, Arctic and Antarctica.

	Trend % per decade			
	Year	Dec.–March	May–Aug.	Sept.–Nov.
North America (36°–60°N)	-4.2 ± 1.4	-4.5 ± 2.2	-4.5 ± 1.9	-2.6 ± 1.4
Europe (40°–60°N)	-4.8 ± 1.6	-7 ± 3.2	-4.3 ± 1.9	-1.7 ± 2.0
European Russia (45°–60°N)	-4.4 ± 1.6	-6.3 ± 3.0	-2.8 ± 1.9	-2.1 ± 1.8
Western Siberia (50°–60°N)	-5.8 ± 1.7	-8.2 ± 3.1	-3.6 ± 1.9	-3.6 ± 2.1
Eastern Siberia and Far East (43°–62°N)	-6.4 ± 1.7	-7.8 ± 2.3	-5.2 ± 2.4	-5.7 ± 2.2
Arctic (60°–80°N)	-5.3 ± 1.7	-7.7 ± 3.2	-3.8 ± 1.9	-2.5 ± 2.2
Antarctica (65°–90°S)		-5.4 ± 3.0		-22 ± 8

slightly stronger one (the cumulative decline is about 7% year-round and 9% for winter and spring). In the southern polar region the December–March trend is also nearly the same as the mid-latitude trend reaching about 10%, while during September–November the decline reaches 40%.

- The eruption of Mount Pinatubo in 1991 did not significantly influence the overall declining TOZ trends.
- TOZ trends will continue possibly until the peak levels of chlorine are reached in the stratosphere, currently expected to take place at the turn of the century.

Trends of TOZ for various geographic regions from January 1979 to March 1995 are shown in the Table 2.5 (Bojkov and Fioletov, 1998)

Chandra *et al.* (1996) and Varotsos *et al.* (1997), using measurements of TOZ by Nimbus-7 TOMS version 7, suggested that the trends over the latitudes centred at 40°N of the Northern Hemisphere vary from −3 to −9% per decade during winter and within −1 to −3% per decade during summer. They also found, using 30-hPa temperature as an index of dynamical variability, that large negative trends in column ozone are reduced in magnitude by 1 to 3% per decade. In order to establish the dynamical component of interannual variability, they constructed similar time series of stratospheric temperature from the National Meteorological Center (NMC) analyses and channel 4 (50–150 hPa) of the Microwave Sounding Unit (MSU) on the TIROS series of NOAA operational satellites (Randel and Cobb, 1994).

Figure 2.13 depicts the relative changes in regional and local trends with respect to zonal mean trends. Chosen as an example, the regional and zonal trends correspond to the Mediterranean region (35–45°N, 0–30°E) and Athens, Greece (38°N, 24°E). This figure also shows the ranges (shaded area) in the seasonal trends over the longitude circle at 40°N. The Figure 2.13(a) is based on the linear regression model which includes the QBO signal (based on the 30-hPa Singapore zonal wind) and the solar cycle signal (based on F10.7 radio emission flux at the wavelength 10.7 cm). A comparison of Figure 2.13(a) with a similar figure in Chandra *et al.* (1996) shows that trends in version 7 TOMS data are found to be about 2% per decade in winter,

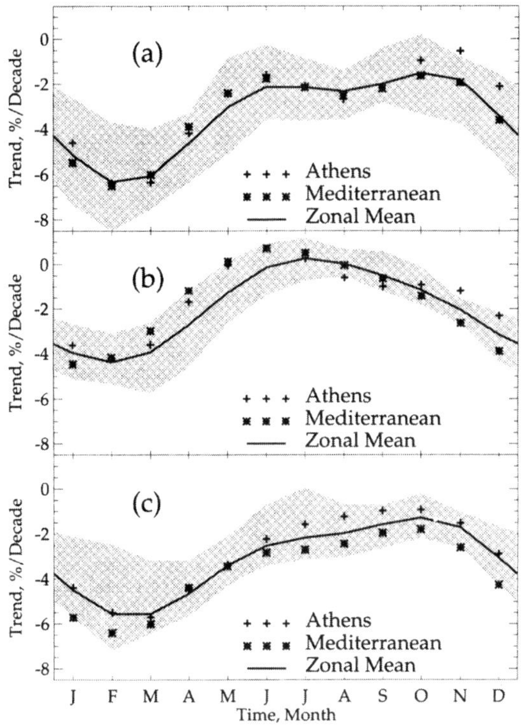

Figure 2.13. Monthly trends in total ozone in the Mediterranean region and at Athens compared with the trends derived from zonally averaged data. Also shown in this figure is the range of variability in seasonal trends over the latitude circle centred at 40°N (shaded). The regression models in the three panels contain (a) QBO and solar cycle terms, (b) lower stratospheric temperature (MSU) term, and (c) NMC 500-hPa temperature.

less negative compared to version 6 with no significant change during summer months.

Taking the effects of dynamical perturbations on TOZ trends into account, it seems plausible to use tropospheric or stratospheric temperature as an index of dynamical variability. The use of temperature has an additional advantage, since the effects of QBO and to some extent the F10.7 (or UV radiation) are implicitly included in temperature changes (Chandra *et al.*, 1996). Figures 2.13(b) and 2.13(c), are calculated by replacing QBO and F10.7 time-series terms with time series based on the MSU temperature (weighted in the 50 to 150-hPa layer) and the NMC 500-hPa temperature, as was done in Chandra *et al.* (1996). A comparison of Figures 2.13(b) and 2.13(c) with the corresponding figures in Chandra *et al.* (1996) shows that the winter trends are about 2% less negative compared with version 6 with no significant change during summer.

An inspection of Figures 2.13(a) and 2.13(b) suggests a significant change in both zonal and regional trends in all seasons even though the relative differences between the Mediterranean and the zonal mean trend are not changed significantly. In general, TOZ trends are reduced by 1 to 3% as a result of using the MSU temperature as a proxy for dynamical variability. It is mainly due to the negative trends in the lower stratospheric temperature, which vary from −1.5 to −2 K per decade during winter (Chandra and Stolarski, 1991; Randel and Cobb, 1994). Similar analysis using the NMC temperature at a smaller change in TOZ trends (1% or less) is shown in Figure 2.13(c). It is probably due to the positive trend of about 0.4 to 0.8 K during the same period in the 500-hPa temperature time series.

The afore-mentioned phase change in the trend of temperature from the troposphere to the stratosphere is consistent with the changes in stationary wave patterns suggested by Hood and Zhou (1998).

2.1.3.2 Longitudinal variation of TOZ

Re-examination of the influence of interannual variability on satellite-derived trends in total ozone during winter and spring confirms that the dynamical perturbations in the atmosphere cause large longitudinal spread in total ozone trends at mid-latitudes (Varotsos and Kondratyev, 1998a).

Figure 2.14 is a cross-section of TOZ trends (% per decade) as a function of the month of the year and longitude from 180°W to 180°E in the region covering the latitude band from 35°N to 45°N (centered at 40°N). This figure shows negative trends in all seasons over the entire latitude belt centred at 40°N and is generally consistent with the pattern discussed in Chandra *et al.* (1996a). The trends vary from −1 to −2% per decade during summer to −2 to −6% per decade during winter. The 2σ uncertainties associated with these trends are in the range of 2 to 3% in summer and 3 to 6% in winter. Chandra *et al.* (1996a), using the same analysis but with

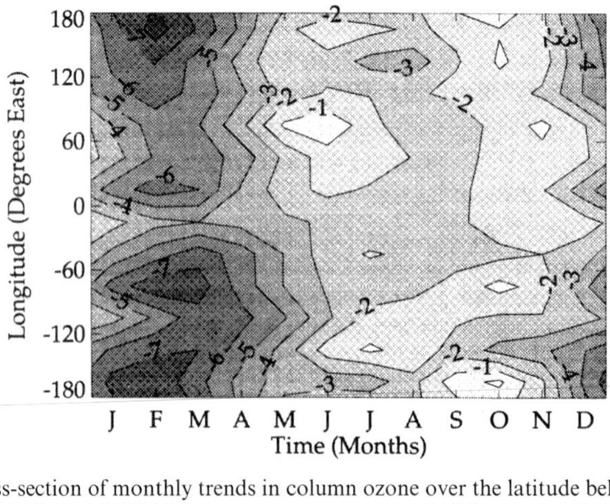

Figure 2.14. A cross-section of monthly trends in column ozone over the latitude belt 35°N–45°N derived from the 13 years of Nimbus-7 TOMS measurements (version 7). The trend values are given as % per decade. The negative and positive longitudes correspond to west and east longitudes, respectively.

version 6 TOMS data, found that the trends vary from −2 to −3% per decade during summer to −3 to −9% per decade during winter.

2.1.3.3 Total ozone measurements over Athens: intercomparison between Dobson, TOMS (version 7) and SBUV measurements

In this section we are primarily concerned with an intercomparison of total ozone data from SBUV, TOMS and Dobson measurements. This has been done for Athens, Greece where the Dobson spectrophotometer has been installed since 1989 (Varotsos and Kondratyev, 1998b). It should be noted that because of the sunny weather that characterize that location, the obtained correlation between the observations of the various instruments is much better.

2.1.3.3.1 Dobson, SBUV and TOMS comparison

TOMS data have been obtained in several versions, version 7 being the most recent. Version 7 data are corrected for drift, caused by the degradation of the instrument's diffuser plate, using the internal pair justification technique. Version 7 data are obtained after careful intercomparison with ground-based measurements. In principle, better agreement between version 7 data and ground-based measurements should be expected when compared with version 6 data.

Comparison between total ozone measurements from TOMS and ground-based measurements has shown significant differences between ground stations (Heese *et al.*, 1992). These authors presented the relative mean difference in TOMS data between 1986 and 1989. This is as high as +6.9% and +8.3% at Oslo and Longyeabyen respectively, but lower at other stations and within the accuracy of ±3.6% for this comparison. Trømso and Resolute have a mean difference of +2.9% and +1.2% respectively, while Reykjavik is the only station with a negative mean difference of −2.5% (Heese *et al.*, 1992). The Norwegian stations in higher northern latitudes have solar zenith angles often higher than 80° over a significant part of the year. The relative difference between TOMS and Dobson total-ozone results shows a higher scatter at large solar zenith angles, so that daily ozone values in particular may be in error.

The problem of high scatter at large solar zenith angles does not arise for measurements at lower latitudes. One of the aims of the present section is to make a detailed intercomparison between TOMS (version 7) and Dobson measurements over Athens (37.59°N, 23.44°E), Greece, which is treated as a representative ground-based station for the mid-latitudes, but we also want to make a comparison with SBUV data too.

So, data from a period when overlapping TOMS, SBUV and Dobson measurements were available is chosen. This period covering 1 March 1991 until 31 December 1991 is the only suitable period when all data were available simultaneously.

2.1.3.3.2 Results and discussion

In this section the plausible SO_2 and NO_2 interference to Dobson total ozone data is estimated. Estimated SO_2 and NO_2 contribution is not very large ($<1\%$), so the Dobson total ozone correction with regard to these trace gases is very slight and therefore can be neglected (Section 2.1.1.1.1).

In Figure 2.15 scatter plots for TOMS (version 7) and Dobson observations are shown and in Figure 2.16 those for SBUV and Dobson measurements. The slope of

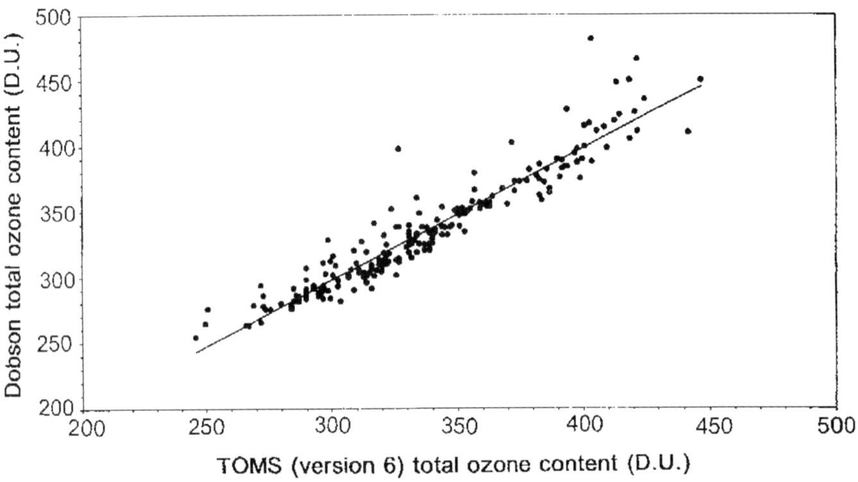

Figure 2.15. Scatter diagram for TOMS (version 7) and Dobson observations at Athens, Greece (1 March 1991–31 December 1991). All days (cloudy and cloud-free) were considered.

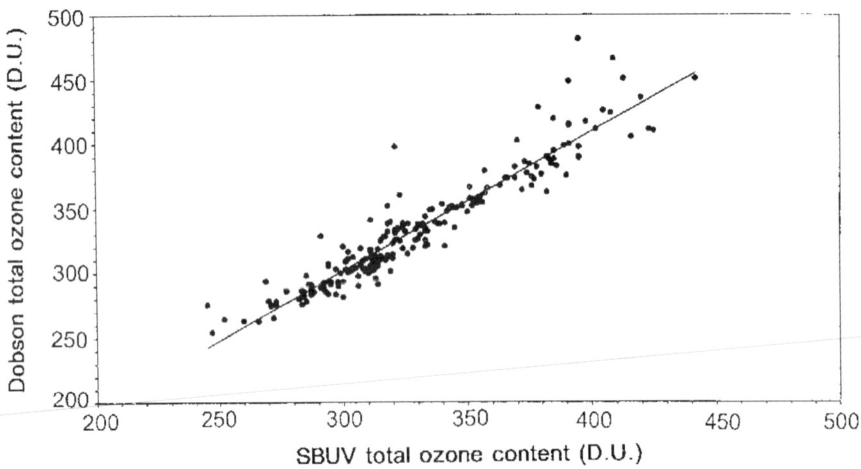

Figure 2.16. Scatter diagram for SBUV (version 7) and Dobson observations at Athens, Greece (1 March 1991–31 December 1991). All days (cloudy and cloud-free) were considered.

the regression line for Figure 2.15 (TOMS version 7) is 1.006 and that for Figure 2.16 (SBUV) is 1.076 indicating that both TOMS (version 7) and SBUV measurements, in general, broadly agree very well with Dobson measurements and that, based on the slopes of the regression line, the agreement for version 7 TOMS data and the Dobson data is slightly better than for the SBUV. A closer inspection of these figures, however, shows a considerable scatter, when the total ozone content exceeds 380 DU (see the far right of both Figures 2.15 and 2.16). In particular, the scatter is greatest when the total ozone content is between 370 and 420 DU.

In view of the above observation all cloudy days in the scatter plots shown in Figures 2.17 and 2.18 corresponding to TOMS (version 7) and SBUV data respectively are excluded. While the slope of the regression line is 0.932 in Figure 2.17 (TOMS version 7 versus Dobson), it is 0.994 in Figure 2.18 (SBUV versus Dobson), which clearly indicates that on sunny days the agreement between SBUV and Dobson measurements is much better than version 7 TOMS measurements.

Note that on cloudy days (refer to the scattered patches in Figures 2.15 and 2.16 discussed above) the Dobson measurements are higher than the TOMS (version 7) and SBUV measurements. This observation points to a possible source of error in the correction of both TOMS and SBUV measurements for tropospheric ozone, which is only partially measured, particularly so on cloudy days, because the effective reflecting surface for backscattered radiance lies in the upper troposphere. So, the amount of tropospheric ozone below this layer depends on the standard ozone profiles used in the TOMS and SBUV algorithm. These standards are fixed climatological annual means. Thus, seasonal and day-to-day variations of up to 50% in tropospheric ozone and a long-term trend of 0.5–1.0 % per year observed by

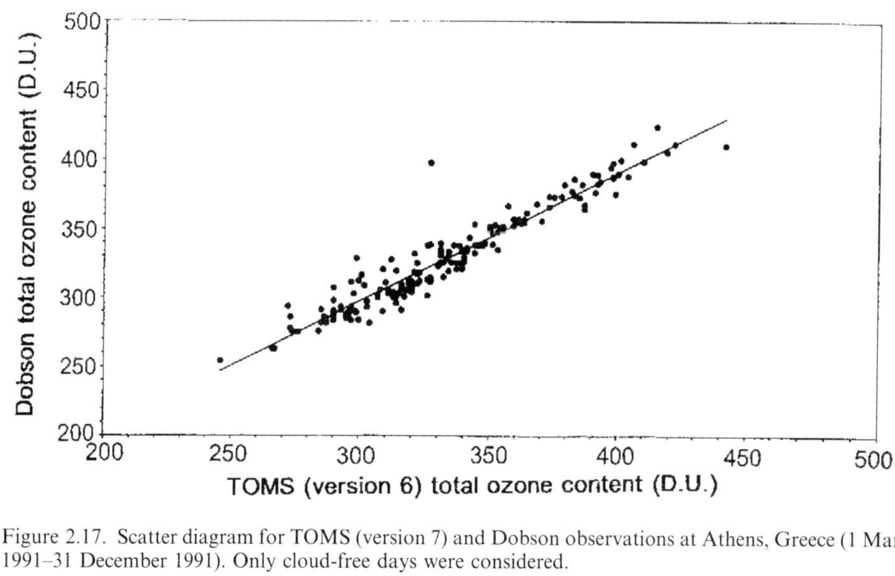

Figure 2.17. Scatter diagram for TOMS (version 7) and Dobson observations at Athens, Greece (1 March 1991–31 December 1991). Only cloud-free days were considered.

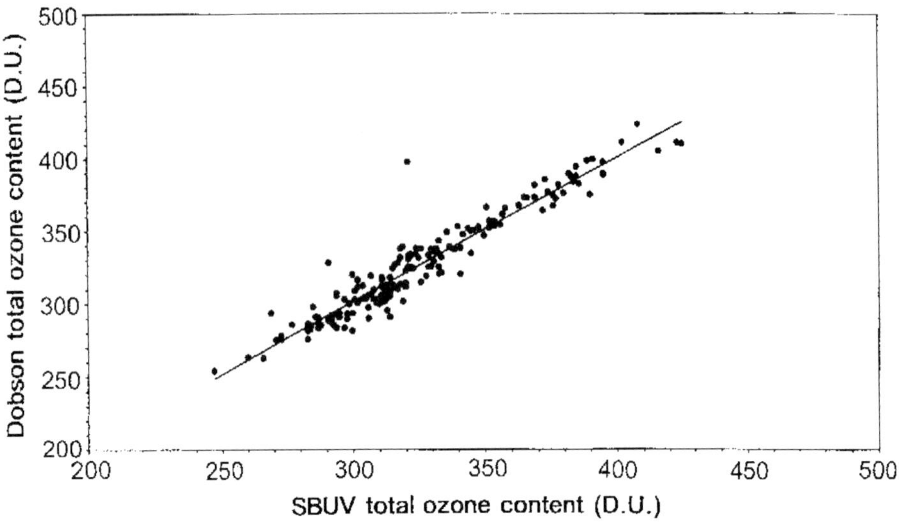

Figure 2.18. Scatter diagram for SBUV (version 7) and Dobson observations at Athens, Greece (1 March 1991–31 December 1991). Only cloud-free days were considered.

sonde measurements at several stations in the Northern Hemisphere (Bojkov, 1988), will not be recognized by TOMS and SBUV.

Varotsos and Kondratyev (1995) studied the mid-latitude total ozone trends in the Northern Hemisphere. In his analysis of the total ozone trend over Athens, Greece (Fleig *et al.* (1982)) pointed out the inadequacy of TOMS measurements when there are thick clouds). Varotsos (1995) further pointed out that, in the presence of a cut-off low system, vertical transport of air from the stratosphere to the troposphere is encountered as pointed out by Ebel *et al.* (1991). This results in the shift of significant amounts of ozone to the upper troposphere. Although TOMS measurements are corrected for the total ozone content to account for the 'hidden' ozone quantity under the cloud top, in every such case, since the ozone quantity within the mixed stratospheric air exceeds the background tropospheric value, satellite observation underestimates the total ozone under the cloud top. Thus, the air within the cut-off low vortices is isolated for at least a few days and hence any reduction in its ozone value (measured from a satellite) can be attributed to the vertical transport of ozone within the troposphere under the cloud (Varotsos, 1995).

2.1.3.3.3 Conclusions

From the preceding discussion it is clear why there is a better correlation between SBUV ozone measurements and ground-based Dobson measurements on sunny days, with a deterioration when cloudy days associated with cut-off lows over Athens are included in the analysis.

Exclusion of cloudy days from the analysis and consideration only of overlapping common periods between TOMS (version 7) and SBUV data and the ground-based

Dobson measurements leads to a more objective comparison. It was found SBUV data compare better with Dobson data on sunny days than does TOMS version 7 data. For more detail about this, see *Scientific* ... (1999).

2.2 OBSERVATIONS OF OZONE VERTICAL PROFILE

In 1881 Hartley reached the principal conclusion that ozone is a normal constituent of the higher atmosphere, being in larger proportion there than near the Earth's surface, and its quantity is quite sufficient to account for the ultraviolet limitation of the solar spectrum at ground level (Dobson, 1968). The first theoretical explanation of the ozone vertical profile (OVP) was given at a conference in Paris (May 1929) by Chapman (1930). However, in 1930, it was not yet possible to reach a general understanding because the ozone peak was misplaced at an average height of about 45 km.

The first measurements of the OVP were made by an indirect method (Umkehr) introduced by Götz in 1929 (Dobson, 1968). According to these observations the average height of the ozone layer is only of the order of 25 km and the main changes in OVP appear to be centred between 10 and 20 km. It should be stressed that these studies of OVP were made before direct soundings with the use of optical instruments (Dobson, 1968).

Since 1965 the use of chemical instruments has become the basis for the first analysis of the meridional pattern of OVP (Griggs, 1964). Today, observations of OVP are made with the utilization of ground-based, airborne and spaceborne instrumentation.

2.2.1 Instrumentation for ground-based OVP observations

A number of instruments and techniques have been used to measure OVP from the ground. However, only a few of these have produced reliable and long-term datasets. In the following sections, brief information about the main instruments and techniques for OVP measurements from the ground is presented.

2.2.1.1 Primary ground-based instrumentation for OVP observations

OVP is mainly retrieved from the ground using zenith-sky measurements in the so-called Umkehr mode, which takes its name from the German word for retrieval.

In 1929, C. Götz registered spectra of the zenith skylight while the Sun was rising or setting while he was operating two spectrographs (a Fabry type and a Fabry-Buisson type) at Spitzbergen. He found that, when the Sun was fairly high, the shorter wavelengths decreased in intensity more rapidly than longer wavelengths with increasing zenith distance of the Sun, because of the greater absorption of the shorter wavelengths. He found, however, that when the Sun was very low the reverse occurred. The explanation he gave is that when the Sun is high most of the short wavelength zenith skylight is scattered from the direct solar beam below the ozone region, but when the Sun is very low the absorption of the direct sunlight by the ozone is so great that the amount of light scattered above the ozone region and which passes down to the instrument by the short vertical path becomes predominant. Near the end of 1930 he wrote to G. M. B. Dobson in Oxford telling him of the effect and suggested that the ratio of two wavelengths should show a similar effect. He also suggested that this effect might be used as a means of estimating the OVP. G. M. B. Dobson, not really believing in Götz's suggestion, started to make measurements using his own instrument, the Dobson spectrophotometer (see Section 2.1.1.1.1), on the first clear day early in January 1931. He started before sunrise and he was surprised to find that effect proposed by Götz was real. It was the first 'Umkehr curve' (the term was suggested by Götz for 'turning back' of the curves, when the Sun is low) ever obtained. After that, in 1932 G. Götz and G. M. B. Dobson (joined later by A. R. Meetham) worked together for 6 weeks at Arosa and showed with certainty that the average height of the ozone in the atmosphere was about 22 km and not about 40–50 km as had been previously thought (Chapman, 1930).

Umkehr observations may be made at the ground or from space with Dobson or Brewer-type spectrophotometers. The ground-based instrument performs observations of the blue zenith sky while the Sun is between the horizon and an elevation of 30° during morning or afternoon hours.

Umkehr observations share most of the errors associated with measurements performed by ground-based instruments. However, these observations suffer some additional errors which do not significantly influence TOZ measurements. Apart from instrument calibration problems, the inversion algorithm, the temperature dependence of the absorption cross-section and the optical effects due to both stratospheric and tropospheric aerosols constitute major sources of uncertainty.

A new Umkehr inversion algorithm employing improved first-guess ozone profiles and improved ozone absorption coefficients was later proposed (Mateer and DeLuisi, 1992; Reinsel *et al.*, 1994), which in layer 9 displays less error sensitivity to stratospheric aerosols normally existing in the Junge layer (see Section 2.2.1.1.1).

2.2.1.1.1 Standard Dobson Umkehr

As already mentioned, the Dobson instrument (see Section 2.1.1.1.1) can be used to provide information on the OVP. Measurements of the downward scattered radiation at two wavelengths can be made for solar zenith angles (SZA) within 60–86.5°. From the ratio of these radiances as a function of SZA (which reverses for SZA near 90°) the OVP can be inferred.

To reduce the time required for an Umkehr measurement, the Short method was developed, which uses the A–C–D Dobson wavelength pairs (Mateer and DeLuisi, 1985). By using the triple pair (A–C–D), OVP information is obtained, when measurements are made during a solar zenith angle change from 80 to 89°. In contrast, the C pair, for which the Standard Umkehr method provided most of the past data, requires a SZA change from 60 to 90°.

Although retrieval provides data for Umkehr layers 1–9 (0–48 km), only data for layers 4–8 (19–43 km) are reliable. The vertical resolution of retrieval for these layers is from 11 to 14 km. The systematic uncertainty is estimated to be 15–20 and 5–12% (DeLuisi *et al.*, 1989). As mentioned above, Mateer and DeLuisi (1992) proposed a new inversion algorithm using new temperature-dependent ozone absorption coefficients (Bass and Paur, 1985). Therefore, the initial estimates have been revised, considering that the aerosol corrections needed for the new and old retrievals are the same (Reinsel *et al.*, 1994).

There are presently seven automated Dobson instruments that routinely make Standard and Short Umkehr measurements. These instruments were developed by NOAA/GMCC (Komhyr *et al.*, 1985) and are located at Perth (Australia); Lauder (New Zealand); Huancayo (Peru); Mauna Loa (Hawaii, USA); Boulder (Colorado, USA); Haute Provence (France) and Fairbanks (Alaska, USA). The frequency of Umkehr measurements significantly increased with the advent of automated Dobson instruments, because observers are not required and a shorter observing time (for the Short Umkehr) decreases the chance of cloud interference.

2.2.1.1.2 *Brewer Umkehr*

A Brewer Umkehr observation consists of zenith-sky radiance ratio measurements for three wavelength pairs over the SZA range from 60 to 90°. The profile retrieval data cover the altitude range of layers 1–10 (0–53 km). A new five-wavelength version of the Brewer Umkehr inversion algorithm has recently been suggested by McElroy *et al.* (1994). The performance of this new algorithm was assessed by comparing the OVP retrieved with those profiles generated using the 6-wavelength algorithm, showing that there exists a much-reduced sensitivity of the 5-wavelength algorithm. Until a complete error analysis has been carried out for the Brewer Umkehr system, the profiles for the standard Dobson Umkehr should be considered as only approximately applicable. Results of an intercomparison of Brewer Umkehr, Dobson Umkehr and ozonesonde data indicated that the quality of the Brewer Umkehr technique is comparable with that of the Dobson 'Short' Umkehr method (McElroy *et al.*, 1989).

A recently developed Brewer Umkehr retrieval algorithm produces ozone profiles directly from observations made by the Brewer spectrophotometer. It uses the climatological ozone natural variability as a function of TOZ, the season, and the year (Elansky *et al.*, 1998; Kosmidis *et al.*, 1998) to predict information about OVP (such as TOZ). Furthermore, the estimation of measurement errors by using the new method leads to the determination of OVP performing observations at one solar zenith angle (express-Umkehr method). Note that the express-Umkehr method allows for OVP measurements (1) between clouds and varying aerosol conditions and (2) at polar sites (the Standard and Short Umkehr are not applicable because of small changes in solar zenith angle).

2.2.1.2 *Secondary ground-based instrumentation for OVP observations*

Besides the direct *in situ* stratospheric techniques, remote-sensing observations have been made over a wide range of the electromagnetic spectrum, from microwave to ultraviolet. Of these, microwave and millimetre-wave observations for rotational transitions offer certain advantages. First, calculations of spectral-line intensities are simpler and more certain because the vibrational and electronic wave functions do not need to be known. Second, millimetre-wave radiation, unlike ultraviolet and visible radiation, is unaffected by atmospheric aerosols. Third, the high spectral resolution permits the measurement of pressure-broadened line shapes, from which information can be obtained on the ozone altitude distribution.

2.2.1.2.1 Lidar

The Differential Absorption Laser technique (DIAL) is the main basis of lidar measurements of the OVP in the troposphere and the stratosphere (from the ground up to 50 km). This technique requires measurements of emission at two laser wavelengths with different ozone-absorption cross-sections. The choice of the two wavelengths used in the experiment depends directly on the altitude range monitored. The laser radiation that is scattered back to the surface is collected by a telescope (see Sections 2.1.1.2.4. and 2.2.2.3.2). The derivative of the logarithm of each signal is computed and the ozone number density is obtained from the difference of the derivatives, divided by the differential ozone cross-section (Ancellet *et al.*, 1988; Eisele and Trickl, 1998; Godin *et al.*, 1989; McDermid *et al.*, 1990; Megie *et al.*, 1977; Pelon and Megie, 1982; Pelon *et al.*, 1986; Uchino *et al.*, 1978; Werner *et al.*, 1983; Zuev and Kabanov, 1987). Vertical profiles of ozone concentrations in the Planetary Boundary Layer (PBL) were measured at the Schauinsland in the southern Black Forest with the Advanced Remote Gaseous Oxides Sensor (ARGOS) lidar (Bisling *et al.*, 1998).

Two types of error affect ozone-lidar measurement; the random or statistical error (related to the signal-to-noise ratio) and the systematic error (related to errors in the value of ozone-absorption cross-sections and to uncertainties in the Rayleigh and Mie scattering or the absorption by other species such as NO_2, SO_2). To obtain an ozone profile in the 15–50 km-altitude range with resolution of 0.5–8 km, the corresponding uncertainty is approximately 2 to 10%.

With respect to passive systems the lidar system has several advantages: (a) control of the emitting source and (b) higher altitude resolution of OVP.

However, in the presence of Polar Stratospheric Clouds (PSCs) conventional DIAL systems are strongly affected by scattering by cloud particles. An alternative to reduction of the interference of particle scattering is to use the Raman DIAL technique (instead of conventional Rayleigh-DIAL). Such an instrument was built and has been operated by Forschungszentrum Geestacht GmbH (GKSS) at ESRANGE near Kiruna, Sweden since the winter of 1996/97. This lidar uses two laser systems as emitters (Nd:YAG and XeCl excimer). The backscattered light is collected by an afocal Cassegrain telescope. It is separated by a filter polychromator with eight different channels. Both lasers are synchronized on a mechanical chopper wheel, which protects the photomultipliers against the very intense backscattered light from the lower atmosphere. The Meteorological Research Institute (MRI) in Japan developed an ozone lidar, the MRI Mark II, which consists of a XeCl excimer laser for the on-line wavelength, an Nd : YAG laser for the reference line and an 80-cm-diameter receiving telescope (Fujimoto *et al.*, 1998).

2.2.1.2.2 Microwave radiometry

The Microwave Radiometer (MR) measures thermal emission from rotational transitions of the molecule under investigation. The frequency employed is typically below 300 GHz (1 mm wavelength). The sensor operates on the super-heterodyne principle, which provides a very high spectral resolution, allowing determination of the exact shape of the emission line. The altitude range over which high-quality ozone profiles can be determined from MR measurements is typically 15–70 km (Zommerfeld and Kunzi, 1989) with an altitude resolution of $\sim$ 10 km. By extending the bandwidth of the receiver from 200 MHz to 1200 MHz it is possible to include the lower stratosphere (Rubin and Künzi, 1989). The accuracy claimed for this sensor is 1 ppmv and the statistical error is of the order of 15%. The measuring time required for a complete profile is approximately 1 h.

Note that accuracy can be improved by using observing sites at high altitudes, thus reducing the attenuation by tropospheric water vapour. Model calculations (Kunzi and Rubin, 1988) indicate that for >90% of available observing time the sensor has to be at an altitude of 3,000 m in the tropics, whereas in polar regions the site can be at any altitude. Ozone profile data were also taken with microwave instruments in southern California between 1989 and 1995 (Parrish *et al.*, 1998).

The operation and data retrieval of an MR can be fully automated and require only little maintenance (see also the MLS instrument and RAM described later on in this section).

2.2.1.2.3 Ground-based Millimeter-wave Ozone Spectrometer (GROMOS)

This spectrometer has been performing continuous measurements of the pressure-broadened millimetre-wave line emission of ozone at 142.175 GHz at the University of Bern, Switzerland since November 1994. Strato and mesospheric ozone is measured on an operational basis at Bern as a part of ESMOS (European Strato-spheric Monitoring Stations) using the GROMOS (Jost *et al.*, 1998). A similar instrument is the Operational Ground-based Ozone Sensor (OGOS) providing ozone profiles from 15 to 75 km (Zommerfelds and Künzi, 1989). The collected ozone profiles are retrieved from microwave observations using the optimal estimation method. Retrieval of ozone profiles is carried out for the altitude range of 12–80 km, with an altitude resolution of 10–15 km and a time resolution of 1–2 hrs (Peter *et al.*, 1998). A ground-based millimetre-wave radiometer observing the pressure-broadened 204-GHz ClO transition lines is installed at the International Scientific Station Jungfraujoch in the Swiss Alps. In 1995 the radiometer was modified and equipped with an additional front-end to measure the ozone transition line at 195 GHz (Siegenthaler *et al.*, 1998).

2.2.1.2.4 Stratospheric Sounding by Infrared Heterodyne Spectroscopy (SIRHS)

Trace constituents in the stratosphere may be detected by exploiting their infrared vibrational–rotational lines. The measurements of the individual line shapes provides the density–altitude profile by finding an inverse solution of the radiative transfer equation. However, the measurements of individual line intensity profiles is very difficult with conventional spectroscopy, because of the pressure broadening wings and Doppler cores in the line profiles.

Infrared heterodyne spectroscopy (IRHS) gives high specific detection sensitivity and ultrahigh spectral resolution (~ 5 MHz), and therefore individual line shapes can be measured without significant instrumental distortion (Abbas *et al.*, 1979). Stratospheric measurements may be performed either in absorption or in self-emission. In a heterodyne receiver, the infrared signal from the source is mixed with coherent radiation from a local oscillator (such as a CO_2 laser or a tunable diode laser), and the difference frequency signal is detected by radio-frequency techniques. The vertical mixing-ratio profiles and the total abundance of some constituents may be determined with high vertical resolution (~ 4 km) and the diurnal variation may be followed by tracking the Sun or the Moon. The Rutherford Appleton Laboratory has developed a high sensitivity airborne 500 GHz heterodyne receiver designed to measure the spectral-line emission of chlorine monoxide, bromine monoxide, ozone, nitrous oxide and nitric acid (Siddans *et al.*, 1998).

2.2.1.2.5 Radiometer for Atmospheric Measurements (RAM)

RAM is a ground-based passive microwave radiometer (installed at Ny-Alesund; 78.9°N, 11.9°E) that detects mixing-ratio profiles of ozone as well as chlorine monoxide. It can be operated all year round independently of weather, performing 1–5 profiles per hour. Its vertical resolution is 10–15 km and it is dependent on altitude. Its altitude range is 12–55 km with 0.1 ppm precision at a height of 20 km.

A similar radiometer was installed at the Bordeaux Observatory, France (45°N, 05°W) in 1995. This instrument is tuned to the 110.836 GHz and ozone profiles are retrieved using the optimal estimation method in the altitude range 25–65 km.

Also, two ground-based microwave instruments, capable of measuring OVP between 20 and 70 km, are operated at the Jet Propulsion Laboratory. These measurements are considered accurate because the wavelength at which these instruments operate (2.7 mm) is very large compared with aerosol particle size (Parrish *et al.*, 1998).

Since November 1994, the RAM is part of the permanent installation at the International Network for the Detection of Stratospheric Changes (NDSC).

2.2.1.2.6 Ground-based infrared solar spectroscopy

Ground-based high-resolution infrared (IR) solar spectroscopy is a powerful tool, used to measure OVP, in the study of vertical distribution of various atmospheric trace gases (Pougatchev *et al.*, 1995). For instance, OVP in the troposphere and the lower stratosphere has been retrieved from a series of 0.0035–0.013 cm^{-1} resolution infrared solar spectra recorded by Fourier transform spectrometers at Kitt Peak in USA (32°N, 112°W) and at Lauder in New Zealand (45°S, 169.7°E). A direct comparison of OVP from IR ground-based high-resolution solar spectra with ozonesonde, microwave and Dobson showed that IR total and O_3 columns above 20 km agree with Dobson and microwave measurements better than 2% (see Section 2.1.1.2.3). For the lowest layer and 12–20 km zone the differences between IR and integrated ozonesonde columns are-7 ± 5 and 2 ± 4%, respectively (Pougatchev *et al.*, 1998).

2.2.2 Airborne instrumentation for OVP observations

Airborne measurements of the vertical profile of ozone concentration is very accurate because the instruments are brought into direct contact with the atmosphere. However, airborne measurements cannot provide a global picture of ozone distribution because of the local area they cover. The most common types of airborne measurements are from balloons, aircraft and rockets using the instrumentation briefly described in the following sections.

2.2.2.1 Ozonesondes

Ozonesondes are balloon or rocketborne instruments, flown with standard meteorological radiosondes, performing *in situ* measurements of ozone vertical profiles up to altitudes of about 35 km. Ozonesonde measurements, after correction for residual ozone at altitudes above the balloon-burst level, are normalized in order to reach agreement between the integrated ozone of the sonde and the TOZ observed by a nearby TOZ-measuring instrument (i.e. Dobson Spectrophotometer). The current ozonesonde network is geographically uneven with the large majority of stations in Europe and North America. The two most commonly used devices are the electrochemical and optical ozonesondes, which are now briefly discussed.

2.2.2.1.1 Electrochemical ozonesondes

Electrochemical ozonesondes utilize iodine–iodide redox electrodes and have been used for many years to measure the temporal and spatial variation of the vertical distributions of ozone in both the troposphere and lower stratosphere.

Various versions of them have so far been constructed. The most commonly used are the electrochemical concentration cell (ECC) ozonesonde (Komhyr, 1969, 1987; Komhyr and Harris, 1971), comprised of platinum electrodes and the Brewer–Mast (BM) electrochemical cell ozonesonde (Brewer and Milford, 1960; Mast and Saunders, 1962; Griggs, 1964) that uses one platinum and one silver electrode. Both types measure ozone concentration via its reaction with an aqueous solution of potassium iodide. A small pump is used to pass air through ozonesonde cells. When air containing ozone is passing through these cells, each ozone molecule releases an iodine (I_2) molecule from the electrolyte, which is transformed by the cell to iodide ($2I'$), resulting in current flowing through the external circuit because of the flow of two electrons. Note that in Eastern Europe the version principally used is the OSE. In Japan model RSII-KC79 ozonesondes have been used for weekly observations at four sites since 1989 (Fujiwara *et al.*, 1998). This ozonesonde is a potassium–bromide type based on a similar electrochemical reaction to that of BM. Another electrochemical ozonemeter, the ECOC instrument is identical to a balloonborne ECC ozonesonde and is described in Section 2.2.2.3.4. Recently, the EN-SCI Corporation

(Environmental Science Corporation) produced model KZ-ECC which is a portable atmospheric ozone-sounding system.

The precision of ECC for ozone measurement in the troposphere is 5–13% and the accuracy ±10%. In the stratosphere (10 hPa), the accuracy is ±5% but varies from ±5 to ±20% at pressure levels lower than 10 hPa. The Brewer–Mast sondes perform a little poorer.

More detailed information about these basic instruments is given in the following subsections.

Brewer–Mast (BM) ozonesonde. The Mast Model 730-10 Bubbler is an update version of the older 730-5 ozonesonde. The overall configuration of 730-5 has been retained and the only difference in using the 730-10 is due to changes in the radiosonde.

The Mast 730-8 and 730-10 ozonesondes use a patented electrochemical sensing cell developed by Dr. Alan W. Brewer.

The chemical solutions used are found to be relatively stable in storage. It is suggested that the following procedure be used in preparing the solution:

A. Fill a 500 ml plastic bottle approximately two-thirds full of a good grade distilled water.
B. Add the chemicals in the amounts indicated and in order. Shake and thoroughly dissolve after each addition

$$1.261\,\mathrm{g\,Na_2HPO_4\cdot 12\,H_2O}\ \text{or}\ 0.944\,\mathrm{g\,Na_2HPO_4\cdot 7\,H_2O}$$

$$0.575\,\mathrm{g\,NaH_2PO_4\cdot H_2O}$$

$$0.500\,\mathrm{g\,KI}$$

C. Add distilled water to make 500 ml total solution volume.

The output of the bubbler cell is a current proportional to the quantity of ozone sampled. The current is converted to a frequency which modulates the transmitter. The frequency is independent of the radiosonde modulator frequency. Adjustments for the radiosonde reference shift often cause a corresponding error in the ozone signal. The following relationship is used:

$$\mathrm{P_3}\ (\mu\mathrm{mb}) = 4.31 \times 10^{-3}\, ITt$$

where I is the current (μA), T is the air pump temperature (K), t is the time in seconds to pump 10 ml of sample.

The calibration curve assumes a constant air-pump temperature of 300 K (DeBacker *et al.*, 1998). Table 2.6 gives an approximate correction for the air-pump temperature. The percentage increase is to be added to the partial pressure (add) depending upon the altitude (ambient pressure).

Versions of the instrument are manufactured in the USA, Germany and India (e.g. the Wyoming digital electrochemical ozonesonde). Improvement in data quality is achieved by normalization of all ozonesonde ascents to Dobson spectrophotometer TOZ.

Table 2.6

Ambient pressure (hPa)	Add (%)
1013–250	2
250–120	1
120–70	0
70–55	−1
55–38	−2
38–25	−3
25–17	−4
17–11.5	−5
11.5–7.7	−6
7.7–5.2	−7

ECC ozonesondes. The model 6A Electrochemical Concentration Cell (ECC) ozonesonde is a balloonborne instrument that measures the vertical distribution of atmospheric ozone up to 40-km altitude. The ozone sensor used in the sonde, developed by Komhyr (1964), is an iodine–iodide redox electrode concentration cell that consists of two bright platinum electrodes immersed in potassium iodide solutions of different concentrations, contained in separate cathode and anode chambers. The chambers are fabricated from polytetrafluoroethylene (Teflon-PTFE resin) and are linked together by an ion bridge that serves as an ion pathway and retards mixing of the cathode and anode electrolytes. The electrolyte in each chamber also contains potassium bromide and a buffer. The concentrations of these chemicals are the same in each half-cell. A similar technique is currently used in the model 2Z ECC ozonesonde produced by EN-SCI Corporation.

The ECC sensor does not require application of an external electromotive force (EMF) for operation (as in the BM sensor). Power for the cell is derived from the difference in the potassium iodide concentrations present in the two half-cells (Komhyr, 1969).

As air containing ozone flows into the cathode solution, a chemical reaction occurs. The reaction is an iodide redox reaction. The platinum electrodes do not take part in the chemical reaction, since they are inert; they only carry electrons between cells of the sensor. Electrochemical reactions take place in the boundary layers of the electrodes. An electrical current is generated proportional to the rate at which ozone enters the cell. The electronic interface relays the electric current data to the radiosonde. The ozone and meteorological data are telemetered to a receiving station where, using a microcomputer and appropriate software, they are transformed into readable information. Ozone concentration in the air is determined by an equation that also takes into account the air-flow rate, air pressure and pump temperature. ECC ozonesondes yield TOZ that agree well with the Dobson spectrophotometer ozone value derived from observations on AD wavelengths (Komhyr *et al.*, 1989b).

In this respect, the EN-SCI Corporation conducted laboratory tests to check the performance of sample ECC sensors using a Thermo Environmental Instrument Incorporated Primary Standard Ozone Calibrator with calibration standards set by the US National Institute of Standards and Technology (NIST). This quality control indicates that ECC sondes, when charged with standard 1% KI cathode electrolyte, overestimate ozone by 5–7% in the region of the atmospheric ozone maximum. The problem is corrected by using a 0.5% KI concentration of ECC sensor cathode electrolyte instead of the 1% KI solution concentration currently used. Additionally, ECC ozonesondes made using the less concentrated KI cathode electrolyte yield integrated ozone amounts that are in excellent agreement with Dobson TOZ observations (Komhyr *et al.*, 1995).

2.2.2.1.2 Optical ozonesonde

This instrument consists of a multi-filter UV radiometer launched aboard an ARCAS or a Super-Loki rocket to a height of about 70 km and permitted to fall back to Earth with a parachute. Direct solar irradiance in several channels is measured as the instrument descends on a parachute and ozone concentration is determined by using the Beer–Lambert–Bouguer law. The precision and accuracy depend on the accuracy of the ozone absorption cross-sections and the knowledge of the filter transmission. The precision is given as approximately 2.5% and the accuracy is estimated at 5–7%. An example of this instrument is the optical Rocket Ozonesonde (ROCOZ), which uses four filters from 267 to 309 nm. The shortest wavelength is used to measure ozone from 39 to 54 km and the longer from 20 to 47 km. Samples are taken every 0.6 s, giving measurements every 80 m at 55 km and every 9 m at 25 km. The improved rocket ozonesonde (ROCOZ-A) includes electronics that provide essentially drift-free outputs throughout ozone soundings. It should be stressed that the fundamental ozone value measured by ROCOZ-A is the vertical ozone overburden, as a function of geometric altitude (Barnes and Simeth, 1986). Measurements by ROCOZ-A ozonesonde were initiated by NASA in 1983.

2.2.2.2 *Other balloon instrumentation*

In the previous section, the basic versions of ozonesondes that can be carried into the troposphere and stratosphere, using either balloons or rockets, were discussed. Other instruments that perform *in situ* measurements of atmospheric ozone, mostly balloonborne, are presented in this section.

2.2.2.2.1 Chemiluminescence ozonometer

Regener (1960, 1964) developed a very fast-response chemiluminescent ozonesonde that often exhibited considerable variations in sensitivity to ozone; this was later corrected. The chemiluminescence ozonesonde or rocketsonde exploits the chemiluminescence reaction between ozone and Rhodamine B in order to measure the stratospheric ozone profile (20–65 km) with a vertical resolution of 0.5–1 km as the instrument descends by parachute or ascends by balloon. Atmospheric air is brought into a reaction chamber that is shielded from sunlight (Rhodamine B fluoresces in the 500–615-nm region and is sensitive to sunlight). In the chamber, ozone reacts with the dye and the resulting luminescence is monitored (Krueger, 1986). Sometimes gas-phase chemiluminescence is also used by employing such gases as 2-methyl-2-butene and NO (see also FOZAN described later in Section 2.2.2.2.8). A novel chemiluminescence sonde was described by Speuser *et al.* (1989) which can measure ozone up to 1.5 mb.

The accuracy of the chemiluminescence rocketsonde/ozonesonde is about 12% and the precision approximately 6%. Note that three types of ozonometric apparatus were used on the meteorological rockets M-100B: heterophase demi-luminescent, gas-phase chemiluminescent and optical (Perov and Tishin, 1989).

2.2.2.2.2 Mass spectrometer

This instrument has been used to measure vertical distribution of the ozone density in the stratosphere and in the mesosphere. For instance, a liquid-helium-cooled mass spectrometer was developed and demonstrated in the 1990s for measuring ozone (and other species), when balloonborne in the stratosphere.

This instrument uses liquid helium to trap most of the gases entering it and ionizes the molecules, which travel along certain pathways. Then, using magnetic fields, various ions are separated into different paths in the measurement chamber. So, if the gas flowrate is known then the fractional composition of the gas can be identified. The measurement accuracy for ozone is about 2% (Mauersberger *et al.*, 1985).

2.2.2.2.3 Balloon Microwave Limb Sounder (BMLS)

The balloonborne version of this instrument operates in the 200-300 GHz range and it is the first use of a microwave remote-sensing technique for limb sounding. The brightness temperatures at various frequencies, within emission lines from various molecular species in the stratosphere, can be measured using microwave local oscillators and a filter bank (35 filters per local oscillator). From these measurements and by looking at various limb angles the altitude profiles of concentrations of species can be inferred (ClO, O_3, H_2O, NO_2, HNO_3, N_2O and HCN). For OVP measurements the O_3 spectral lines at 206.132 GHz were chosen because of the insensitivity of O_3 emission to stratospheric temperature variations. For instance, a temperature variation of 10°C changes the O_3 emission by only 1% (Waters *et al.*, 1993).

An important version of the MLS has been used on aircraft and in the Upper Atmospheric Research Satellite (UARS) launched in 1991. Another MLS version with additional frequencies between 63 and 640 GHz is currently used on the Earth Observing System (EOS) platform, which is in a 98° Sun-synchronous orbit (permitting measurements in the polar regions).

2.2.2.2.4 Far-infrared emission measurements

The instruments used to study molecular species in the stratosphere (emission spectral region from 7 to 200 cm^{-1}) are Fourier-transform spectrometers. To reduce detector noise cryogenically cooled detectors are used and then resolution varies from 0.5 to 0.0033 cm^{-1}. Direct comparison of the atmospheric emission with the blackbody signal at liquid-nitrogen temperature leads to determination of the emissions made by atmospheric molecular species. These instruments are pointed outward by a few degrees above the horizon. The 'onion-peeling' technique is used to separate signals into components from various altitudes.

SAFIRE (Spectroscopy of the Atmosphere using Far IR Emission) will be used for the Earth Observing System (EOS) to be launched by NASA over the coming few years. The main aim of the SAFIRE experiment will be to improve understanding of mid-atmospheric ozone distribution by performing global-scale measurements of the most important chemical, radiative and dynamical processes that influence ozone changes. The instrument will measure ozone in the 10–100 km-altitude range, with an accuracy of 10%. It will operate day and night and cover the latitude belt 86°N–86°S (Megie *et al.*, 1989).

A Fourier Transform Infra-Red spectrometer, built at JPL in 1984 is the Mk IV balloon interferometer (high resolution – 0.01 cm^{-1}) operating in solar absorption mode (Osterman, 1999).

2.2.2.2.5 *Stratospheric Infrared Interferometer Spectrometer (SIRIS)*

Infrared emission spectroscopy using Fourier-transform spectroscopy (Section 2.1.1.2.3) can be employed to measure vertical distributions of ozone and other molecular species in the stratosphere. It is necessary to measure the vertical profiles of both temperature and absolute radiance to conduct such measurements. The stratospheric IR Interferometer Spectrometer using dichroic filters divides the 700–1,920-cm^{-1} region into four channels. The region between 1125 and 1425 cm^{-1} has been used to perform observations for O_3 and other species in the 12–40-km-altitude region. The instrument which is balloonborne allows (by scanning below the horizon) measurements with 3 km of vertical resolution (Grant, 1989).

The total error associated with SIRIS, varies from 14% at 38 km to 42% at 16 km.

2.2.2.2.6 *Balloonborne Laser In-Situ Sensor (BLISS)*

This instrument is a microprocessor-controlled high-resolution absorption spectrometer operating in the mid-infrared region using tunable diode laser (TDL) sources.

It performs observations of ozone and other stratospheric trace species by measuring the absorption from balloonborne diode lasers and reflected back from a retro-reflector suspended ~500 m below the balloon. For each gas of interest in the stratosphere, a narrow-band tunable diode laser is scanned over a few selected infrared absorption lines and the mean gas concentration over the fixed pathlength is determined from the depth of the observed absorption lines by Beer's law. This instrument's calibration uncertainty is 4.8% overcoming various factors of uncertainty like temperature, pressure and laser half-width at half-maximum. As far as accuracy is concerned, this is governed by precision in combination with uncertainty in the absorption cross-section of the ozone line and its temperature dependence and broadening coefficient, which lead to a total 10% accuracy in the 28–38-km-altitude range (Grant, 1989). Another balloonborne *in situ* ozonometer was built in France

(Moreau *et al.*, 1989) and was based on the absorption of a laser beam by ozone in the Hartley band.

2.2.2.2.7 Ultraviolet Absorption Photometer

This is an *in situ* instrument that performs observations of ozone concentrations by measuring the absorption at 253.65 nm (near the UV ozone peak). It is actually a modified version of the Dasibi ozone monitor that is often used for ozone measurements at ground level of air pollution. Dasibi-like devices use a mercury lamp as the spectral source, a narrow-band filter, a silicon photodiode as the detector and a pump to pull air through the sample volume. For half of the measurement time, an ozone scrubber removes the ozone thus providing a reference atmosphere (Sections 2.2.5.3 and 2.2.2.2.3). In general UV photometers have good precision (3%) and accuracy (5%). They are typically used on large balloons with other instruments and during return to the ground observe fine structure in ozone distribution. For instance, a dual-beam UV-absorption photometer (or balloon dual-beam UV in-situ O_3 photometer) for balloonborne measurements of ozone has response time 1 second and an uncertainty at the ozone maximum of 3.6%, where 2% of this is the accuracy of the ozone cross-section. This photometer has two identical absorption chambers, each alternating between reference mode (ozone free) and sample mode by means of a four-port valve and ozone scrubber. The ratio of the absorption signals, along with the known lengths and ozone absorption cross section, yields the ozone concentration (Proffitt and McLaughlin, 1983). The Deutsches Zentrum für Luft- und Raumfahrt (DLR) research aircraft *Falcon* has also been used for ozone measurement using a UV-absorption photometer (Ziereis *et al.*, 1998). A balloon-borne suntracking photometer operating at six wavelength bands centred at 210, 215, 290, 750, 860 and 950 nm was developed and used to measure the OVP in the 10–35-km-altitude region (Lat *et al.*, 1989).

In the past this instrument has flown aboard gondolas carrying other instruments to provide data for the validation of the ozone measurements of several satellite instruments including VARS MLS, HALOE, CLAES and ADEOS ILAS. Recently, a suite of in-situ instruments has been assembled and flown as the Observations of the Middle Stratosphere (OMS) gondola to expand the database obtained with the ER-2 (NOAA) instruments.

2.2.2.2.8 Fast Ozone Analyser (FOZAN)

This is a chemiluminescent sensor based on the heterophase luminescence of ozone in a solid-state matrix filled with gallic-acid dye. Inside a chemical reactor, the sensor is placed along a line where the airflow to be analysed is provided by a pump. The luminescent light produced by the sensor is modulated by an optical liquid-crystal chopper, regulated by a controller unit and registered by the photomultiplier. A double-entry valve connects to the chemical reactor, either with a built-in calibrating ozone generator or the airflow to be analysed. The calibrating ozone generator supplies an airflow, with a strictly specified ozone concentration, to the chemical reactor. The generator produces ozone because of the effect of UV radiation from a mercury lamp on the airflow. The instrument can operate in two modes: (a) the measurement mode in which the airflow (after being thermostatted) is passed through the valve to the reactor to undergo the ozone-sensor chemiluminescent reaction; (b) the calibration mode in which the valve passes the airflow to the reactor through the ozone calibration cell and the calibrating ozone generator. The two modes are switched automatically. This instrument was used onboard the high-altitude aircraft M-55 *Geophysica* at the beginning of 1997 (Georgiadis *et al.*, 1998).

2.2.2.2.9 SAOZ sonde

This is a small balloonborne version of the SAOZ UV-visible spectrometer (Pundt and Pommereau, 1998), which was outlined in Section 2.1.1.2.1. It measures vertical profiles of O_3, NO_2, H_2O and OClO by solar occultation in the ultraviolet and visible range (Pommereau and Piquard, 1994a, 1994b). Measurements are performed at altitudes of 28–31 km and during the balloon ascent. One complete measurement cycle is made every 30 s for every 200 m in altitude during the balloon ascent, and every 300–1000 m during solar occultation depending on altitude and latitude.

2.2.2.3 *Aircraft instrumentation*

Two of these balloonborne instruments (the UV photometer (Section 2.2.2.2.7) and the diode laser *in situ* (Section 2.2.2.2.6)) are also used on aircraft. Among other devices providing measurements of airborne ozone-vertical profile are the Infrared Grating Spectrometer and the Differential Absorption Lidar Technique, which are discussed in more detail in the following two sections.

2.2.2.3.1 Infrared Grating Spectrometer (IGS)

This instrument performs measurements of ozone and other trace species in the stratosphere. It observes absorption in the long-optical path by the instrument, while observing the Sun at sunrise or sunset.

A modified version of this device, the grille spectrometer, is a correlation instrument with large-area geometrical patterns (grilles), instead of slits, at both the entrance and the exit of the spectrometer. The favourite arrangement has a set of hyperbolic curves in four quadrants, with alternate shaded bands adjacent to quadrants of mirror images. This specific arrangement scans rapidly ($0.2\,cm^{-1}\,s^{-1}$) over a limited spectral range (5–$10\,cm^{-1}$) (Grant, 1989).

This instrument was used to measure the latitudinal variations in abundances of ten stratospheric species including ozone, onboard aircraft between 62°N and 62°S. It can also be used on a balloon to measure H_2O, NO_2, HNO_3 and O_3 in the altitude range 23–40 km.

2.2.2.3.2 Differential Absorption Lidar Technique

The instrument consists of: (1) a telescope to capture backscattered photons; (2) a filter to reject out-of-band radiation; (3) a photomultiplier tube and (4) signal-chain electronics (see Sections 2.1.1.2.4 and 2.2.1.2.1). The conventional differential absorption lidar (DIAL) consists of a light source (e.g. a XeCl-gas laser) emitting pulses (25 Hz) at two wavelengths, one of which lies in the ozone absorption band. Backscattered radiation (Rayleigh scattering) is observed by telescope with a photon-counting detector. High resolution is achieved by measuring the time between emission and detection of the light pulses. For ozone measurements, wavelengths of 308 nm (ozone absorbent) and 353 nm (ozone almost transparent) are chosen (Krueger *et al.*, 1989; Godin *et al.*, 1989). Accuracy for ozone-mixing ratios is estimated to be 10–15% in the mid-stratosphere, with a decrease above and below this region. Airborne lidar provides OVP from aircraft cruise altitudes up to 25 km. Vertical and the horizontal resolution is at 0.5 Km and 50–100 km, respectively (Megie *et al.*, 1989).

2.2.2.3.3 UV photometers

The Measurements of Ozone by Airbus In-Service Aircraft (MOZAIC) programme co-ordinated by Laboratoire d'Aerologie (LA/CNRS, Toulouse) consists of three phases in order to assess the impact of aircraft emissions on the atmosphere. The first phase (January 1993 to September 1996) involved development and installation of instrumentation and both making and interpreting O_3 and H_2O in-flight measurements (onboard an Airbus A340). The second 2-yr phase (October 1996) continued the O_3 and H_2O measurements and their interpretation. The third phase aims to certify and install new devices for NO_y and CO measurements.

The ozone device (see Sections 2.2.2.2.7 and 2.2.5.3) is a dual-beam UV absorption analyser (Thermo-Electron). A Teflon pump is used for pulling air through the analyser. The detection limit is 2 ppbv and overall precision is ±2 ppbv ±2% (Thouret *et al.*, 1998a, b). Measurements are taken every 4 s, starting after take-off and continuing up to landing. During the MOZAIC programme, seven ozone analysers were used with the seventh one remaining at LA/CNRS and acting as the reference analyser. In a similar programme, the ozone-mixing ratio of ambient air around a Transall C-160 aircraft was detected *in situ* using a UV-absorption monitor (Horiba APOA-350E) (Zahn *et al.*, 1998). During the Arctic winters of 1991/1992 and 1994/1995, airborne UV-DOAS (Differential Optical Absorption Spectroscopy) on board a Transall aircraft measured the stratospheric trace gases NO_2, O_3, BrO and ClO (Grund *et al.*, 1998). Note that on the Solar Absorption Balloon Experiment (SABE-2) the NASA-LSC and NASA Goddard 'Dasibi' photometers measured lower ozone than SBUV instruments for altitudes above 35 km (Weinstock and Anderson, 1989).

2.2.2.3.4 Electrochemical Ozone Sensor (ECOC)

This instrument performs *in situ* ozone measurements onboard aircraft. For instance, during the APE/POLECAT campaign the M55 aircraft *Geophysica* sampled vortex filaments with sudden and intense tracer variation in *in situ* ozone data from ECOC instrument onboard the aircraft, while flying on an isentropic surface. This sensor consists of a Teflon sampler with cell, two pumps and control electronics. The cell is filled with a KI–water solution similar to standard ozonesondes, but specially designed for use on a high-altitude research aircraft. Both pumps are identical. The second pump is used to eliminate possible contamination of the electrochemical cell by the air leaving the pressure sensor during ascent.

This pump also shortens the time taken by air samples to get from the air intake to the cell. The KI ions react with the ozone of air circulated through the cell, which induces a detectable electrical current (see Section 2.2.2.1.1). The time resolution of the cell is 5 s and the accuracy of individual ozone readings better than 5%. The instrument is temperature controlled and is assembled in a metal box and placed in an unpressurized compartment (Steffanutti *et al.*, 1999).

2.2.2.3.5 Airborne Submillimeter Receiver (ASUR)

ASUR and its predecessor SUMAS (Submillimeter Atmospheric Sounder) have been used for several campaigns such as EASOE (Urban *et al.*, 1998). It is a passive heterodyne radiometer operating at submillimeter wavelengths (in the frequency range 620–650 GHz) and employing novel superconductor–insulator–superconductor detector technology. Vertical concentration profiles are derived from pressure-broadened emission lines of the species under examination. It uses the German Air and Space Research Organization's (DLR) research aircraft *Falcon* as a platform. Among the molecules measurable (which show rotational/vibrational transitions in emission) by ASUR are the chlorine species HCl and ClO, O_3 and N_2O (Kleipool *et al.*, 1998).

Another version of remote sensing instrument measuring abundance profiles of ClO, O_3, HCl, HNO_3 and N_2O is the Submillimeterwave Limb Sounder (SLS). It is a heterodyne radiometer measuring thermal emission spectra near 604 and 640 GHz designed for use on high altitude balloons and aircraft (see also Section 2.2.3.3.4).

2.2.3 Satellite instrumentation for OVP observations

Satellite-based techniques or instrumentation for OVP observations is presented in this section. Note that the satellite-based solar-occultation technique and satellite-based nadir measurements have important advantages over ground-based measurements, because they provide near-global distributions, as opposed to point measurements from ground-based *in situ* techniques. It is self-calibrating because unattenuated sunlight is measured during each event allowing data to be normalized to extra-atmospheric values. The longer absorption path in the limb geometry provides higher sensitivity than that from nadir-looking geometry. Moreover, higher vertical resolution is inherent in this technique because most of the absorption occurs at the tangent altitude. Finally, this technique simplifies data interpretation because the Sun is used as a constant background source.

Note that the first instrument used to measure ozone in the 55–85-km region was an ultraviolet spectrometer and polarimeter onboard the Solar Maximum Mission spacecraft launched in February 1980. This used the technique of solar occultation in the UV region by the Earth's limb. The first attempts of satellite occultation soundings were conducted from Soviet manned spacecraft (Kondratyev, 1972, 1998; Kondratyev *et al.*, 1996). The Russian Ozone and Aerosol Fine Structure (OZAFS) experiment obtained a few ozone profiles obtained by satellite occullation sounding aboard the Soviet Salyut 7 orbiting station in 1965.

2.2.3.1 *Direct-absorption measuring instruments*

A few instruments use the Beer–Lambert–Bouguer law to determine the ozone concentration by directly measuring absorption features of ozone either within a path defined by the instrument or by using an extraterrestrial source (e.g. Sun). The mean advantage of direct-absorption measuring instruments is that in the data-processing algorithm only a few assumptions have to be made regarding quantities that are not directly measurable. However, their mean disadvantage is the uncertainty of the absorption cross-section and also the limited signal-to-noise ratio at the extreme measurement range.

In general, direct-absorption measuring instruments are divided into two groups according to the source of the radiation used. Those instruments that rely on an external radiation source (i.e. the Sun or the Moon) are called passive instruments (i.e. optical ozonesondes, IR spectrometers and laser heterodyne spectrometers). Passive instruments usually measure an average molecular density along the atmospheric path (observing at sunrise or sunset from above the atmosphere or descending through the atmosphere).

Instruments with their own radiation source are called active instruments and include the source, the atmospheric pathlength, the collector, the aperture, the spectral analyser, the detector, the signal chain, the data-storage unit and the instrument controller (i.e. DIAL, UV photometer).

2.2.3.1.1 Stratospheric Aerosol and Gas Experiment (SAGE)

The SAGE instruments are satelliteborne multi-channel radiometers and measure gas-extinction profiles using solar occultation. SAGE I was launched aboard the dedicated Applications Explorer Mission-2 (AEM-2) in February 1979 and operated for 34 months until November 1981, when the spacecraft's electrical system failed. SAGE II was launched on the Earth Radiation Budget Satellite (ERBS) in October 1984 and globally monitored the vertical distribution of stratospheric aerosols, ozone, water vapour and nitrogen dioxide, by measuring the extinction of solar radiation throughout the atmosphere during ERBS solar occultations (Maudlin *et al.*, 1985). Both SAGE I and SAGE II are in approximately 600-km circular orbits with inclination angles of 56 and 57°, respectively. They measure the absorption of sunlight by ozone with four channels (SAGE I), at 0.385, 0.45, 0.6 and 1.0 μm at spacecraft sunrise and sunset (solar-occultation geometry). The 0.6-μm channel is used to measure stratospheric ozone by using the Chappuis band. The 0.45-μm channel is used to measure the concentration of stratospheric NO_2 and the outer two channels are used to determine the aerosol optical depth. Vertical resolution is approximately 1 km for altitudes above 10 km. SAGE II is a seven-channel instrument and, in addition to ozone, NO_2 and aerosols, it measures water-vapour concentrations. It makes measurements up to 65 km versus 55 km for SAGE I. The extra channels allow for better corrections for aerosols. SAGE III is an improved version with higher spectral resolution, greater spectral wavelength coverage and a lunar-occultation capability, which allows for a broader range of latitudes (than that of solar occultation) and for measurements of NO_3 and OClO. In SAGE III (for the

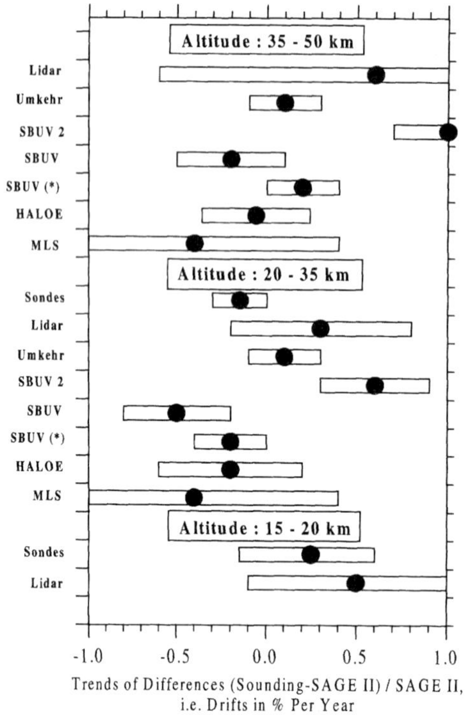

Figure 2.19. Trends of differences (i.e. drifts) between ozone measurements made by various ozone-profiling instruments and SAGE-II, in % per year [(Sounding − SAGE-II)/SAGE-II]. Trends of ozonesondes, lidar and Umkehr differences are presented as averages for eight northern mid-latitude sounding stations. Trends of satellite differences are presented as global means. The average differences are indicated by the dots and the bars represent the 99% confidence interval of the drift estimation. The entries for SBUV are for SBUV/SAGE-II comparison; entries for SBUV* are for SBUV and combined SAGE-I and SAGE-II comparison.

Earth Observing System Programme) the filters are replaced by a spectrometer, a diode array is used for detecting radiation in the various wavelength regions and the spectral range is extended to 1.55 μm for observation of aerosols with larger particles and aerosols near the Earth's surface. One SAGE-III instrument was launched aboard the Russian spacecraft Meteor-3M in 1998 (polar Sun-synchronous orbit) and another is planned to be used at the International Space Station in approximately 2001 (51.5° inclination orbit). A third SAGE-III instrument is planned for a polar Sun-synchronous orbit early next century.

Regarding precision and accuracy, the major uncertainty in ozone concentrations, measured by either SAGE-I or SAGE-II, is the Chappuis-band ozone absorption cross-sections and their temperature dependencies, which are estimated to give a 6% uncertainty. The measurement precision of SAGE-II is 4–7% in the 15–52-km altitude range, becoming 20% at 60 km and 50% at 10 km. When statistical and systematic uncertainties are combined, the accuracy is better than 10% for heights between 15 and 52 km. Figure 2.19 shows the regression slopes for time series of coincident differences between SAGE-II and the other measurements (*Scientific* ..., 1999).

2.2.3.1.2 Atmospheric Trace Molecule Spectroscopy (ATMOS)

The first chemically comprehensive set of space-based measurements was made by the ATMOS instrument, an infrared occultation interferometer. This experiment uses a Fast Fourier Transform (FFT) interferometer, which operates in the near and mid-infrared region of the spectrum to detect many molecular constituents in the upper atmosphere and to measure their vertical and horizontal distributions. It was first flown aboard the shuttle *Challenger* (as a part of *Spacelab 3*) in April–May 1985 and measured solar radiation at sunsets and sunrises (as seen from the orbiter). Twenty occultations were recorded from this experiment and these produced simultaneous concentration profiles for a large number of minor upper-atmospheric species, several of which had not previously been observed. The simultaneity of the observations is a very important aspect is the study of the highly interactive nature of photochemical processes. Subsequent flights of the ATMOS instrument (3/92, 4/93, 11/94) extended ATMOS coverage to the tropics and high latitudes, and were used for trend determination by comparison of the mid-1985 and late-1994 observations (Grant, 1989).

2.2.3.1.3 Polar Ozone and Aerosol Monitor (POAM)

The POAM II instrument is a satelliteborne multichannel radiometer similar to SAGE-II. It was carried onboard the Système Pour l'Observation de la Terre (SPOT-3) satellite in an orbit similar to that of the Stratospheric Aerosol Measurement (SAM)-II sensor on the Nimbus-7 satellite (i.e. at a Sun-synchronous 98° inclination 833-km orbit with a 10.30 LT equatorial crossing time). In normal operation, it provided for Southern Hemisphere observations much closer to the South Pole, extending to 88°S at the vernal equinox from 15 October 1993 until 14 November 1996. It is a solar occultation instrument with nine narrow-band channels between 352 and 1,060 nm that retrieve stratospheric ozone, aerosol extinction, NO_2 and water vapour (Glaccum *et al.*, 1996). Its ozone profiles were used to study the breakup of the 1993 Antarctic polar vortex (Bevilacqua *et al.*, 1995), polar vortex dynamics during the 1994 and 1995 northern winters (Randall *et al.*, 1995), and polar mesospheric clouds in the Southern Hemisphere (Debrestian *et al.*, 1997; Fromm *et al.*, 1997).

One POAM instrument called the Orbiting Ozone and Aerosol Monitor (OOAM) flew on a 45° inclination orbit in the summer of 1997. Another POAM III instrument flew recently (early 1998) on the French SPOT-4 satellite in a polar Sun-synchronous orbit.

2.2.3.1.4 Halogen Occultation Experiment (HALOE)

This instrument (Figure 2.20) is flown on the Upper Atmosphere Research Satellite (UARS) covering two longitude sweeps each day (15 profiles each): one at the latitude of sunsets and one at the latitude of sunrises. Each sunrise or sunset observation is separated by 1.6 h and 24° longitude. It has eight infrared channels

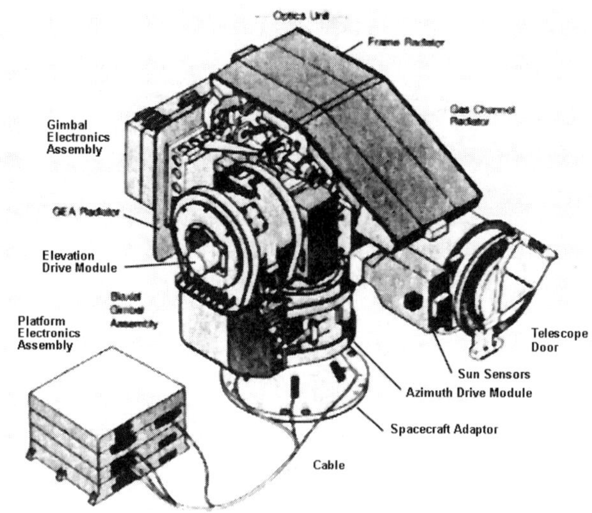

Figure 2.20. HALOE instrument.

(from 2.45 to 10.01 μm) and four of these are broadband (60–120 cm^{-1}) filter radiometer channels for CO_2, H_2O, O_3 and NO_2 and four channels are gas-filter correlation radiometer channels for HF, HCl, CH_4 and NO.

The HALOE instrument uses the principle of the satellite solar-occultation technique and the measurement principle is based on both gas-filter correlation radiometry and broadband-filter radiometry. In the first technique incoming radiation is split into different channels for each species and passed through a narrowband filter; it is then further split into two paths for each species. One path includes a cell containing the gas to be measured. As the atmosphere passes in front of the Sun, the path containing the cell will see little change in the absorption, because the appropriate portion of the Sun's spectrum is already absorbed by the gas in the cell. The conventional broadband technique is described in other sections describing other instruments (e.g. LIMS in Section 2.2.3.3.1). The vertical resolution for retrieval of aerosol, NO_2, H_2O and O_3 is 2 km. The accuracy for ozone-mixing ratios is estimated to be 10–15% in the mid-stratosphere, with a decrease in accuracy above and below this region (Grant, 1989).

Stellar occultations obtained by the Solar–Stellar Irradiance Comparison Experiment (SOLSTICE) onboard UARS are also used to study the ozone concentration in the mesosphere between 55 and 85 km (de Toma *et al.*, 1998).

2.2.3.1.5 Sunphotometer Earth Atmosphere Measurement (SPEAM)

This Canadian instrument was flown onboard the US space shuttle Challenger in October 1984 and provided observations at 315 and 324 nm at 63.34°S, 91.96°E, from which OVPs are inferred. Observation started several minutes before the occultation time. The instrument was manually pointed at the Sun through the quartz window, and this was maintained until the occultation was over. The small size of the instrument, and the fact that it can be returned to the laboratory for calibration between flights, make this technique very useful. It was also flown on the Space Shuttle in 1992 making measurements of O_3, NO_3 and NO_2. It was actually a dual-wavelength interference filter sunphotometer, similar to the commercial single-channel device formerly used in the Canadian solar-radiation network. The flight version of it (STS 41-G) differs from its conventional groundbased version in that it can observe simultaneously at two different wavelengths (McElroy *et al.*, 1991). SPEAM made measurements as a part of a Canadian payload aboard two Space Shuttle flights – STS-41G in October 1984 and STS-52 in October 1992.

2.2.3.2 Scatter-measuring instruments

A number of instruments measure ozone vertical profiles and TOZ from space by measuring backscattered UV solar radiation. The spectral distribution of backscattered radiation to space by the Earth's atmosphere depends on the presence of gases with selective absorption bands and molecular (Rayleigh) as well as aerosol scattering. The instruments can use spectrometers or filters for various spectral channels and they can observe in the nadir direction (as for solar backscattered UV) or through scanning cross-tracks (as with total ozone mapping spectrometry (Grant, 1989)). These instruments operate best above 15–20 km (where there is strong absorption by ozone in the Hartley band).

2.2.3.2.1 Solar Backscattered Ultraviolet Radiometer (SBUV)

The BUV instrument is a downward-viewing double monochromator with a 1-nm bandpass, which is stepped through 12 wavelength increments every 32 s through the 255–340-nm spectral region. In addition, there is a separate filter photometer with a 5-nm bandpass centred at 380 nm, which is used to measure backscattered radiance. This instrument is designed to determine the vertical ozone profile, while measuring backscattered solar UV radiation directed vertically upwards from the ground. Two wavelength pairs are used for measuring total ozone and eight short wavelengths are used to profile the ozone. The precision of this instrument is better than 2% for zenith angles less than about 60° and clear conditions. The accuracy is limited to approximately 6%. The BUV experiment was flown onboard the Nimbus series of satellites, and was used to measure global distribution of ozone for 7 yr. A UV spectrometer was also flown aboard the EXOS-C/Ohzora satellite (launched in February 1984) to measure backscattered UV in the wavelength region 250–320 nm with a spectral resolution of 1 nm for the period March 1984–September 1987 (Ogawa *et al.*, 1989). Complete global coverage of total ozone was obtained. However, an anomaly in the Earth's magnetic field in the South Atlantic affects the distribution of the Van Allen radiation belts and, therefore, these radiation levels send too much noise to the detectors. This was overcome in the Solar Backscattered UV instrument (SBUV) by adding a 50-Hz optical chopper. The first of the operational SBUV/Model 2 (SBUV/2) instruments was launched onboard NOAA-9 in late 1984 and overlaps with the new instrument, launched on NOAA-11 in 1998. Several additional SBUV/2 instruments are planned for NOAA satellites between now and 2007. The SBUV and SBUV/2 instruments have a spectral-scan capability, which is used periodically, and long-term solar irradiance data can be obtained from them. Because of their degradation in space, data about solar spectral irradiance measured by the SSBUV (Section 2.2.3.2.2.) instrument have been important. Note that a grating drive problem with the SBUV/2 aboard NOAA-14 led to discontinuation of its use in spectral-scan mode (Kaye, 1997).

2.2.3.2.2 Shuttle Solar Backscatter Ultraviolet Spectral Radiometer (SSBUV)

This instrument was designed and developed at the Goddard Space Flight Center to calibrate Nimbus and NOAA instruments. It carries out measurements of ozone concentrations by comparing solar ultraviolet radiation with that scattered back from the atmosphere. In late 1989, the Space Shuttle *Atlantis* carried the instrument for the first time, in an appropriate orbital flightpath to assess performance by directly comparing data from identical instruments (SBUV) aboard NOAA spacecraft and Nimbus 7.

SSBUV flew for the sixth time in early 1994 aboard STS-62, housed in a canister in the shuttle's cargo bay. Its principal purpose was comparison of observations from several ozone-measuring instruments aboard NOAA-9, NOAA-11, Nimbus-7 and UARS. The quality of the in-orbit calibration depends on the flight-to-flight calibration repeatability in SSBUV. Given accurate measurements of the backscattered radiance, it is necessary to account for differences in solar-zenith angle and effective surface reflectivity (Heath *et al.*, 1993).

2.2.3.2.3 Shuttle Ozone Limb Sounding Experiment (SOLSE)

An ozone-measuring payload flown aboard the Space Shuttle in the fall of 1997 included two UV instruments (Section 2.2.3.2.4). One was the Shuttle Ozone Limb Sounding Experiment (SOLSE). This is a spectrometer that uses UV wavelengths, consisting of a two-dimensional detector (altitude/wavelength). It measures ozone concentration in the stratosphere and upper troposphere (Hilsenrath *et al.*, 1991).

2.2.3.2.4 Limb Ozone Retrieval Experiment (LORE)

LORE and SOLSE (Section 2.2.3.2.3.) were the two UV instruments of the ozone-measuring payload flown aboard the Space Shuttle in the fall of 1997. LORE is a multiwavelength version of the Rayleigh Scattering Attitude Sensor (RSAS), a single channel (355 nm) instrument, that flew with SSBUV on STS-72 in January 1996. It uses optical filters and covers UV, visible and infrared channels. It obtains ozone profiles as low as the 10–15 km altitude range (Hilsenrath *et al.*, 1991).

2.2.3.2.5 Solar Mesospheric Explorer-Ultraviolet Spectrometer (SME-UVS)

The ultraviolet spectrometer (UVS) on the Solar Mesosphere Explorer (SME) satellite is designed to measure the ozone density of the mesosphere in the 1.0–0.1 hPa (48–70 km) region at sunlit latitudes. To achieve this, the instrument measures altitude profiles of Raylight-scattered solar photons in the 200–340-nm spectral region.

This instrument has an off-axis parabolic telescope, an Ebert–Fastie grating spectrometer and a pair of photomultiplier tubes. The spectrometer is used at a pair of fixed wavelengths separated by approximately 30 nm (265.0 and 296.4 nm) and operates with a spectral resolution of 1.5 nm. The spacecraft rotates at 5 rpm and observes the Earth's limb, leading and trailing the SME ground track. It measures the altitude profiles of Rayleigh-scattered UV. The random errors vary from 3.4% at a height of 48 km to 10.2% at 68 km. The temperature and pressure errors vary from 4.5% at 48 km to 10.1% at 68 km. The SME-UVS provided a number of ozone density datasets in the measured ozone concentration, which help to explain ozone photochemistry (Grant, 1989). The Solar Maximum Mission (SMM) spacecraft carried a high-resolution ultraviolet spectrometer polarimeter (UVSP), employed since 1980 to measure the ozone profile by the technique of solar occultation (Aikin, 1989).

2.2.3.2.6 Solar Mesosphere Explorer (SME) Near-IR Spectrometer

This instrument uses an off-axis parabolic mirror and an Ebert–Fastie spectrometer. There are two channels with a resolution of 12 nm and a separation of 0.6 μm. The 1.27-μm channel is used for emission measurements and the 1.87 μm channel is used for reference purposes (i.e. to correct data for radiances that are due to oxygen emission). Random-error sources for ozone density include temperature (3% at 50 km, 1% at 90 km); systematic errors include calibration errors (10% in the laboratory before launch and 20% after launch).

The SME Near-Infrared Spectrometer has generated a number of datasets on spatial and temporal variation in ozone concentration since its launch in October 1981. A similar instrument was developed in Japan (Infrared Air-glow-IRA) and flew onboard the EXOS-C satellite (Ogawa *et al.*, 1989).

2.2.3.2.7 Scanning Imaging Absorption Spectrometer for Atmospheric Cartography (SCIAMACHY)

This is a multichannel spectrometer measuring radiation in the spectral range 240–1750 nm (1.9–2.4 μm) aboard the European satellite ENVISAT-1, launched in 1999. This instrument was designed to measure the column and profile distribution of a number of gases, including high horizontal-resolution measurements of ozone. It is a combination of two spectrometers also operating in the UV/visible/near-infrared part of the spectrum to observe transmitted, reflected and scattered light. It provides measurements from nadir (scattered and reflected solar photons), limb (scattered and reflected solar photons), occultation (transmitted solar or lunar photons) and solar reference (extraterrestrial solar photons), achieving global coverage within 3 days.

From limb measurements, stratospheric profiles of pressure, temperature and several trace gases (O_3, NO_2, H_2O, ...) can be retrieved (Rozanov *et al.*, 1997a, b).

2.2.3.3 Emission-Measuring instruments

A few instruments measure ozone in emission, observing features in the near-IR to millimetre wavelengths. These instruments observe the Earth's limb from a spaceborne platform, and some are used in nadir-viewing directions (Grant, 1989). The signals in emission are generally weaker than those in absorption and, therefore, the instruments are generally cooled, except the millimetre-wave ones (they can observe in very narrow spectral regions using heterodyne-detection techniques).

2.2.3.3.1 Limb Infrared Monitor of the Stratosphere (LIMS)

The first space-based measurements of stratospheric trace constituents (besides ozone), were made from the LIMS (Figure 2.21) and the Stratosphere and Mesosphere Sounder (SAMS) aboard Nimbus-7.

LIMS is a cryogenically cooled limb-scanning infrared filter radiometer, launched onboard Nimbus-7 in 1978 with a lifetime of 7 months. It was developed from the earlier Limb Radiance Inversion Radiometer (LRIR) with a limited number of channels. It has six channels (centred between 6.2 and 15.0 μm) that permit observation of emissions by CO_2, HNO_3, O_3, H_2O and NO_2. Two channels near 15 μm can be used to retrieve vertical temperature profiles. The other channels are used for molecular-species measurements. Ozone, for example, is observed in the

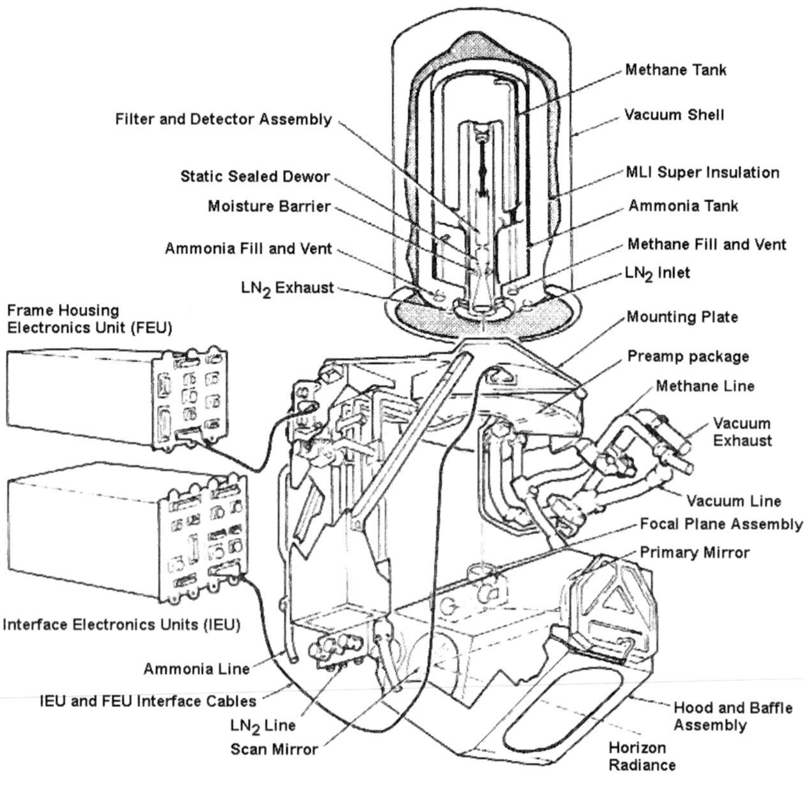

Figure 2.21. Cutaway diagram of the LIMS instrument.

926–1141 cm^{-1} spectral region, with a vertical field of view of 1.8 km and a horizontal field of view of 18 km from 15 to 64-km altitudes. Because of its geometry, this technique has an inherently high vertical resolution and can sound at high altitudes. The instrument scans from above the atmosphere to below the Earth's horizon and back every 24 s. The precision of this instrument was found to vary from 2–4% at 30 hPa to 1% at 0.4 hPa. The absolute accuracy was about 10%. Many experiments provided data about distribution and variability of temperature and several minor constituents (O_3, H_2O, HNO_3 and NO_2) over the latitude range 84°N to 64°S. Some data on ozone and aerosol profiles in the stratosphere was also obtained by the Aerosol Limb Absorption (ALA) instrument which flew onboard the Japanese Ohzora satellite, launched in February 1984. ALA was a two-channel sunphotometer (6,500, 1,000 nm) that used the solar-occultation technique. Another occultation instrument, the Limb-Atmospheric Infrared Spectrometer (LAS), also flew aboard Ohzora (Ogawa *et al.*, 1989).

2.2.3.3.2 Solar Mesosphere Explorer Near-Infrared Spectrometer (IRS/SME)

The near-infrared spectrometer (NIRS) onboard the SME satellite is designed to carry out ozone measurements between 50 and 90 km over most latitudes at 3:00 PM local time. The instrument measures emission from $O_2(^1\Delta g)$ at 1.27 μm that is primarily due to photodissociation of ozone. Vertical resolution is better than 3.5 km. Note that at the time of the El Chichón eruption (1982), there was no SAGE instrument operating and data about the eruption was obtained from SME (Grant, 1989).

2.2.3.3.3 Michelson Interferometer (IRIS, MIPAS)

This instrument (IRIS-D) is designed to record the infrared emission spectrum of the Earth and its atmosphere between 400 and 1,600 cm^{-1}, in order to provide the vertical profiles of temperature, humidity and ozone concentration. IRIS-D launched onboard Nimbus-4 is an advanced version of IRIS-B launched onboard Nimbus-3 a year earlier, on 14 April 1969. The subsequent improvement resulted in an increase in spectral resolution from 5 to 2.8 cm^{-1}.

The MIPAS (Michelson Interferometer for Passive Atmospheric Sounding) instrument is a high spectral-resolution limb sounder. It measures temperature, ozone and other chemical species (especially nitrogen and chlorine-containing species), with a vertical resolution of 3 km, from the mid-troposphere to the mesosphere (Carlotti, 1995). It also measures aerosols and PSCs and has been used on Transall C-160 aircraft (Gulde *et al.*, 1994). It has been carried onboard the ENVISAT satellite since 1999.

2.2.3.3.4 Microwave Limb Sounder (MLS)

The MLS was originally an airborne millimetre wavelength (1.0–1.5 mm) radiometer. The present version operates in the 200–300-GHz range. Using local oscillators and filter banks (35 filters per local oscillator), brightness temperatures (day and night) at various frequencies, along emission lines from various molecular species, are measured with a vertical resolution of about 5 km. MLS was found to have a precision of 4% and an accuracy of the order of 9%. MLS was used on board the Upper Atmospheric Research Satellite (UARS) and proposed for the Earth Observing System (EOS). It was also used in the Balloon Intercomparison Campaign (BIC). MLS has been acquiring millimetre-wavelength emission measurements of the stratosphere in both hemispheres since late September 1991. Measurements are made as the instrument's field of view is vertically scanned through the atmosphere limb in a plane perpendicular to the UARS trajectory. MLS pointing geometry, together with the inclination of the UARS orbit, leads to a measurement latitudinal coverage extending from 80° on one side of the equator to 34° on the other (Froidevaux *et al.*, 1996).

A new MLS instrument, a significantly enhanced version of the UARS MLS, is going to be carried by the EOS-CHEM platform, which is due to be launched in 2002 and will measure temperature, ozone and other chemical species in the lower stratosphere and the upper troposphere. Note that given the length of the MLS dataset, their use in studies of long-term trends and shorter-term phenomena such as the quasi-biennial oscillation (QBO) is becoming possible (de Toma *et al.*, 1998).

2.2.3.3.5 ODIN satellite

ODIN is a small Swedish satellite built jointly by Canada, France and Finland, launched in the second half of 1998 to a Sun-synchronous 600-km orbit. It has two instruments, a sub-millimeter radiometer using microwave wavelengths to measure ozone, chlorine monoxide, water vapour and other constituents, and an optical spectrograph and infrared imaging system using UV-visible and near-infrared wavelengths for studying ozone, NO_2, aerosols and other constituents. It is the first satellite to employ submillimeter (480–580 GHz) radiometry of atmospheric thermal emission in a limb sounding mode to measure global distributions of several species important for ozone chemistry in the stratosphere (O_3, ClO, N_2O and HNO_3). These observations are performed simultaneously, while another mode is focused on odd hydrogen chemistry, giving H_2O, O_3, HO_2 and CO. The retrieval method is the Optimal Estimation Method (von Schéele, 1997).

2.2.3.3.6 Cryogenic Limb Array Etalon Spectrometer (CLAES)

This instrument measures concentrations of members of the nitrogen and chlorine families – as well as ozone, water vapour, methane and carbon dioxide – by observing infrared thermal emissions at wavelengths from 3.5 to 12.7 μm. In normal operation it utilizes a telescope, a spectrometer and linear array of 20 detectors to perform simultaneous observations for 20 altitudes ranging from 10 to 60 km. To avoid interference from thermal emissions of the detectors and optics the CLAES cryogenic system is used, although it limits the lifetime of the instrument. This system consists of two subsystems: a block of solid neon at −260°C (cooling the detectors) and a surrounding block of solid carbon dioxide at −150°C (reducing emission from the optical system). It has been used onboard the UARS research satellite (Grant, 1989).

2.2.3.3.7 Improved Stratospheric and Mesospheric Sounder (ISAMS)

This is a filter radiometer that observes, with eight detectors, infrared molecular emissions by means of a movable off-axis reflecting telescope from 4.6 to 16.6 μm. It utilizes a Stirling-cycle refrigerator to cool its detectors to −195°C and carries samples of some of the gases to be measured in cells within the instrument. Installing these cells in front of the detectors, the atmospheric radiation collected by the telescope passes through them, allowing the spectra of the gases in the cells to be matched with the spectra observed in the atmosphere. As some gases cannot be confined in cells (because of their chemical activity) ISAMS employs broadband filters to isolate portions of the spectrum and so permit measurements of the gases. The instrument onboard UARS is an improved version of that aboard Nimbus 7 and measures the concentrations of nitrogen chemical species, as well as ozone, water vapour, methane and carbon monoxide. CLAES and ISAMS data, through their excellent spatial and comprehensive chemical information content, have provided significant information on both chemical and transport processes in the stratosphere, including issues such as tropical mid-latitude transport of tracers (Koutoulaki *et al.*, 1998).

2.2.3.3.8 High Resolution Dynamics Limb Scanner (HIRDLS)

This is a multi-channel limb scanning infrared radiometer developed jointly by US and UK. It will be carried by the EOS-CHEM platform. HIRDLS is going to measure temperature, ozone and a number of trace gases with a vertical resolution of 1 km from the tropopause to the mesopause. It should provide information on small-scale variability in the atmosphere for use in transport studies (Dials *et al.*, 1998).

2.2.3.3.9 Cryogenic Infrared Spectrometers and Telescopes for The Atmosphere (CRISTA)

This is an infrared limb sounder designed to measure densities of mid-atmospheric trace gases with high temporal and spatial resolution. It contains three telescopes for three different viewing directions (the lateral two are separated in azimuth by +18° and −18°) and cools the optics to cryogenic temperatures. The telescopes (used to sense three atmospheric volumes 600 km apart) are each followed by infrared grating spectrometers; the centre telescope is additionally equipped with a spectrometer for the far-infrared spectrum. A complete spectrum is measured in about 1.2 s, while a complete altitude scan is achieved in less than 1 min, providing trace-gas profiles on a grid of lower than 500 km × 600 km and with vertical resolution of about 2–3 km.

CRISTA was successfully used aboard the CRISTA-SPAS (Shuttle Pallet Satellite) free flier together with ATLAS-3 (Kay and Miller, 1996) during Space Shuttle mission STS-66 in November 1994. A second flight of CRISTA and the Middle Atmosphere High Resolution Spectrographic Investigation (MAHRSI) for NO and OH observations took place in August 1997, as part of Space Shuttle mission STS-85. The US Cryogenic Infrared Radiance Instrument for Shuttle (CIRRIS) instrument designed mainly for airglow studies, flew aboard the Space Shuttle in April–May 1991 (STS-39) and relayed information on OVP and other traces.

2.2.3.3.10 Improved Limb Atmospheric Spectrometer (ILAS)

ILAS is an occultation instrument, getting vertical-profile measurements at high northern and southern latitudes. It consists of two spectrometers, one using infrared for trace constituent measurements and the other visible wavelengths to derive temperature and pressure profiles.

This spectrometer was developed by the Environmental Agency of Japan and installed onboard the ADEOS spacecraft, which was launched on 17 August 1996. ILAS measures the vertical profiles of O_3, NO_2, HNO_3, N_2O, CH_4, H_2O, CFC-11, CFC-12, N_2O_5 and aerosols in the altitude range of 10–60 km with an altitude resolution of 1–2 km. It performs 14 measurements per day on an Arctic and on an Antarctic latitude circle (58–73°N, 65–90°S). ILAS O_3 profiles are in good agreement with ozonesonde profiles between 20 and 24 km. Higher ILAS ozone below 20 km, may be attributed to the presence of aerosols, while lower ILAS O_3 at 24–32 km looks like a systematic difference between ILAS and sondes, owing probably to the ILAS operational algorithm (Yushkov *et al.*, 1998). A modified version of ILAS-II is scheduled to fly onboard the Japanese ADEOS-2 satellite (see Section 2.1.2.2.4).

2.2.3.3.11 Millimetre-wave Atmospheric Sounder (MAS)

This is a Shuttleborne limb-scanning radiometer that measures molecular emission spectra of O_3, H_2O, ClO and O_2 in the Earth's atmosphere, in order to provide their geographical and vertical distribution. MAS belongs to the Atmospheric Laboratory for Applications and Science (ATLAS) Shuttle payload package. It measured OVP at altitudes between 20 and 80 km over more than 90% of the Earth's surface during the three ATLAS missions (1992, 1993, 1994).

The MAS instrument uses a single 1-m diameter off-axis parabolic antenna (pointed toward the Earth's limb) and observes radiation along paths through the Earth's atmosphere at tangent heights from 10 to 140 km. The O_3 spectrum is measured with 50 RF filters, arranged in three banks (wide, intermediate and narrow having widths of 40, 2 and 0.2 MHz, respectively).

Precision of mixing-ratio results is estimated at 2–4% at altitudes of 29–62 km, increasing at both higher and lower altitudes to about 8%. Taking measurement errors into account the total uncertainty is estimated to be about 5% near the O_3 peak (Daehler *et al.*, 1998).

2.2.3.3.12 Advanced Millimetre-wave Atmospheric Sounder (AMAS)

This instrument was designed as a limb-viewing, heterodyne spectrometer with two sub-millimeter receivers to measure ClO and BrO at 500 GHz and HCl at 625 GHz and receivers to measure O_3, N_2O and O_2 in the 300-GHz region, H_2O near 325 GHz and CO and HNO_3 close to 345 GHz. AMAS is designated for launch alongside NASA's TOMS on the Russian Meteor-3M/2 in 2000 and will monitor global distributions of important constituents in the stratosphere and upper troposphere (Reburn *et al.*, 1998). The European Space Agency is investigating several scenarios for future satelliteborne limb sounders in the millimetre and submillimetre spectral range. For instance, MASTER is planned to be an instrument in the millimetre range focusing on the upper troposphere and on exchange processes between the troposphere and the stratosphere. SOPRANO is planned to focus on stratospheric chemistry with a submillimetre receiver. PIRAMHYD is planned to measure at even smaller wavelengths than SOPRANO (Bühler *et al.*, 1996). These instruments are planned to operate in the millimetre and submillimetre spectral range, allowing the retrieval of atmospheric mixing-ratio profiles of several molecular species.

2.2.4 Observed variability in vertical ozone distribution

In the troposphere, the ozone concentration falls (on the average) with increasing altitude until the tropopause is reached. In the stratosphere, ozone concentration increases rapidly with altitude to a maximum near 5 hPa, with an often appearing secondary maximum in the lower stratosphere near 100 hPa. As mentioned in Chapter 1, a long-term decrease in stratospheric ozone has occurred at mid-latitudes in both hemispheres. In the Northern Hemisphere the decrease is larger in winter and spring (11% since 1979) than in summer or autumn (4% since 1979). The most dramatic changes have been seen at high latitudes in wintertime vortices over the Arctic and Antarctic. Analyses of measurements of ozone in the 1994/95, 1995/96 and 1996/97 winters have shown chemical losses of up to 50% at some altitudes in each winter. Current ozone datasets came from various ozone-measuring platforms using different techniques with different spatial and temporal coverage. *Scientific ...* (1999) provided the most reliable datasets with time records longer than 5 to 10 years (Table 2.7).

The Arctic vortex has been colder than usual in recent winters with record low temperatures observed during the three winters leading up to the 1997/98 winter. Noted that no major mid-winter warming of the stratosphere has occurred since 1990/91. Prior to that time, they had occurred every couple of years, and the longest gap in the 40-year record had been 4 years. In cold conditions (temperatures below

Table 2.7. Inventory of the most 'reliable' datasets with time records longer than 5 years.

Platform	Middle/Upper stratosphere ($Z > 25$ km)	Low stratosphere ($Z < 25$ km)	Troposphere
Satellites			
SAGE I	2/1979–11/1981	2/1979–11/1981	
SAGE II	10/1984–present	10/1984–present	
SBUV + SBUV/2	1978–present		
HALOE	10/1991–present	10/1991–present	
MLS	10/1991–present	10/1991–present	
Ground-based			
Microwave	1989–present	1989–present	
Umkehr/Dobson	1957–present		
Lidar	1990–present	1985–present	1990–present
Balloonborne sondes		1965–present	1965–present

approximately 195 K or −78°C), polar stratospheric clouds (PSCs) can form. As mentioned earlier, chemical reactions on PSC particles convert chlorine compounds into forms that can rapidly destroy ozone in the presence of sunlight.

We shall now briefly discuss some outstanding field campaigns.

2.2.4.1 EASOE

A major European campaign, the European Arctic Stratospheric Ozone Experiment (EASOE), was organized to study the polar regions during the winter of 1991/92. Particular goals were:

(a) to measure the change in ozone concentration with altitude throughout the winter;
(b) to measure the concentrations of other trace chemical species (especially chlorine and nitrogen species);
(c) to investigate the role of polar stratospheric clouds, and in particular to study dehydration and denitrification of the Arctic stratosphere;
(d) to study the meteorological processes that move chemically perturbed air southward.

Measurements were made using groundbased instruments, at 16 sites ranging from the Arctic Circle to southern Europe, from large stratospheric balloons and from three research aircraft. Forty-three stratospheric balloons, capable of carrying scientific payloads of up to 500 kg directly into the ozone layer, were launched from the Swedish Space Corporation rocket range ESRANGE near Kiruna, Sweden. In addition, over 1,000 ozonesondes were launched on small balloons from more than 20 sites. A wide variety of meteorological information was provided by the various European meteorological services, including some information provided especially for EASOE scientists.

Much new information was gained. Despite this activity, many intriguing questions remain: What causes the mid-latitude loss? How are the losses over the poles linked to those at mid-latitudes? (See Sections 2.3.1 and 2.4.)

2.2.4.2 SESAME

In 1994 and 1995 European scientists conducted SESAME, the Second European Stratospheric Arctic and Mid-latitude Experiment, to investigate the processes occurring at both high and mid-latitudes and to see how they are linked (Sections 2.3.2 and 2.4). At the same time, a US-led expedition was looking at similar processes in the Southern Hemisphere.

2.2.4.3 THESEO

The Third European Stratospheric Experiment on Ozone (THESEO) was organized to investigate long-term ozone loss over northern high and mid-latitudes during 1998 and 2000. The processes involved include both those that occur over mid-latitudes and those resulting from the exchange of air between the mid-latitude stratosphere and the neighbouring parts of the atmosphere. In the first few months field measurements were made mainly at mid- and high latitudes and focused on a limited number of specific processes. At the end of November until the beginning of February (1997/1998) the temperature within the vortex fell below the critical temperature for the formation of Polar Stratospheric Clouds (PSCs). However, because of the occurrence of a series of mid-stratospheric warmings the extent of the PSCs was not large. Moreover, outside the vortex in the upper stratosphere unusually high temperatures were observed. This winter was finally warmer than the previous three (exceptionally cold) and, therefore, TOZ values were higher than those in the past few winters. These higher TOZ values masked any chemical ozone loss (estimated to be 15–20% inside the vortex).

2.2.4.4 *The experience of the Athens Ozone Station*

2.2.4.4.1 *Introduction*

Special features in vertical-ozone structure, especially in the lower stratosphere, have received significant scientific attention in the past decades. For example, many researchers have reported the existence of layers with enhanced and depleted ozone amounts in vertical-ozone profiles (Hering, 1964; Hering and Borden 1964; Reid and Vaughan, 1991; Varotsos *et. al.*, 1992).

Dobson (1973) examined the occurrence of such a lamination structure. In particular, he made a systematic analysis of the laminar ozone structure in the stratosphere over a wide latitudinal and longitudinal range. In Dobson's analysis, the criterion for detection of laminae within a certain height interval was the change in ozone partial pressure to a level greater than 3 mPa. In the same study, ozone profiles were separated into three groups: group 0 with profiles containing at least two laminae, group 1 with profiles exhibiting moderate lamination and group 2 with profiles extensively laminated. One of Dobson's findings was that the features within the laminated structure vary with latitude and season. In particular, between January and April the lamination phenomenon was most frequently present. Whereas at latitudes below about 20°N during spring and below about 30°N during autumn, a laminated ozone structure was very seldom found.

Reid and Vaughan (1991) examined the ozone laminar structure in the altitude range of 9.5–21.5 km. Laminae were classified according to their vertical extent (depth), as well as the difference in ozone partial pressure (magnitude). They also used criteria to separate genuine laminar features from instrumental noise or large-scale features in the ozone profiles. The acceptable depth for laminae was chosen to be between 200 m and 2.5 km, while the ozone partial pressure anomaly was at least 2 mPa. They also showed that laminae are most frequently found below 18 km at high latitudes during winter and spring, which is in agreement with the findings of Dobson. But Reid and Vaughan (1991) disagree with Dobson about the origin of laminae. Based on the observation that their magnitudes are greatest and their depths least during winter and spring, they ruled out Dobson's suggestion that they all originate near the subtropical jetstream, although the possibility remains that a small number of laminae may be generated by the stratosphere–troposphere exchange, as Begum (1989) believes. Reid and Vaughan (1991) suggested that laminae may represent evidence for a process that can cause exchange of air in and out of the polar vortex. This argument is supported by more recent polar ozone campaigns, when particularly sharp and deep laminae near the winter polar vortex were detected (Reid, Vaughan and Kyrö, 1993). Analysis of data from 800 experimental ozonesoundes at 18 stations during EASOE (European Arctic Stratospheric Ozone Experiment) showed that laminae are absent above the 450 K level within the polar vortex, and are most abundant near its edge. Additionally, the small amplitude of inertia gravity waves could not totally support the appearance of laminae features (Reid *et al.*, 1994).

Orsolini *et al.* (1995a), using a high-resolution off-line transport model, demonstrated that large-scale isentropic advection can create the laminar structure of a vertical tracer profile. This is an additional indication that a lamination profile relates to the exchange of air in and out of the polar vortex.

Dobson (1973) also observed a characteristic ozone minimum in the vertical-ozone distribution at 14–17 km, but no explanation was given for both the incidence of minima at the preferred height region and its constancy with latitude. It was stated though that there is a strong correlation between the existence of the characteristic ozone minimum at 14–17 km and the occurrence of the double tropopause, especially at latitudes around 40°N. Note that Dobson made the assumption that the laminated structure and the characteristic ozone minimum are of the same origin, since variations in appearance frequency are very similar for both season and latitude.

Attempts were made to examine the role of atmospheric circulation in the formation of these special features on the vertical-ozone distribution at the Northern Hemisphere lower mid-latitudes. To this end, vertical-ozone profiles were classified according to their characteristics and the prevailing meteorological situation. In addition to previous classifications (Varotsos *et al.*, 1992, 1994, 1995) a new group was added to examine the ozone-depletion effect, which occurred in the region where the ozone partial pressure takes its maximum value.

2.2.4.4.2 Observation Data

The ozonesondes used for vertical soundings at the Athens Station were the Electrochemical Concentration Cells (ECC, Scientific Pump; see Section 2.2.2.1.1) as explained by Komhyr (1969). For measurements of temperature and humidity, as well as data transmission, RS-80 radiosondes (Väisalä system) were used. Interface cards for digitization of the signals and the receiving system were also from Väisalä. The data were taken during the ascent of the balloon (TOTEX 1200) with a rate of $5\,m\,s^{-1}$. Ozonesoundings were performed in most cases at 10:00 UTC.

The previous analysis, made by the same group, contains all the relevant information for ozonesoundings performed by 1995 (Varotsos *et al.*, 1992, 1994, 1995). During the last winter period (November 1996–April 1997) a total of 25 ozonesoundings were performed at the Athens Station. These ozonesoundings were classified into five groups, according to their specific features.

1. The first group consists of profiles in cyclonic flow at 200 hPa, generally on the western or southern side of an elongated trough. The main characteristics of the profiles of the first group are (a) high tropopause; (b) small secondary maximum in ozone (< 7 mPa); and (c) ozone minimum at 100 hPa.
2. The second group is similar meteorologically to the first, but consists of profiles with a rather larger secondary layer in the lower stratosphere and again an ozone minimum at 100 hPa.
3. The third group is mainly characterized by many thin layers in the lower stratosphere (laminated structure). The profiles of the third group were observed during fairly cyclonic flow or on the eastern side of the trough.
4. The fourth group contains profiles with one thick layer in the lower stratosphere and an ozone minimum at 14–17 km. The profiles of the fourth group were usually observed during cut-off low events.
5. Finally, the fifth group consists of profiles with an ozone-depletion effect in the region, where the partial pressure of ozone takes the maximum value. The meteorological situation at the latter category is characterized by the movement of the polar vortex to the mid-latitudes of the Northern Hemisphere, resulting in more intense north-western circulation above the site.

2.2.4.4.3 Frequency occurrence of special features in vertical-ozone structure

From the available ozonesounding dataset for the Athens station, the monthly distribution for the occurrence of each group is examined for the time period from November to April (Figure 2.22). The first group (ozone minima at the 100 hPa and small secondary maximum) is more likely to occur in early winter, while the second group (same characteristics as the first group but with larger secondary maximum) at the end of the winter. Both groups are characterized by the ozone minimum in the lower stratosphere. Therefore, in order to examine the occurrence of the characteristic minimum the two groups were merged into a single one (1 + 2). The characteristic ozone minimum shows maximum occurrence during the winter period. This is in agreement with Dobson's findings (1973).

The third group (laminated ozone structure) is more likely to occur at the end of the winter and early spring. This is also in agreement with Dobson (1973), who observed the maximum in the lamination during February and March.

The fourth group (thick ozone layer at the lower stratosphere and ozone minimum at 14–17 km) mostly occurs in early spring while the fifth group (ozone decline effect in the altitude range of the maximum partial pressure) is more likely to occur during late winter and early spring.

2.2.4.4.4 Plausible impact of circulation in the lower stratosphere on ozone-vertical structure

Five measured vertical-ozone profiles are presented below as the most characteristic of each group.

On 24 January 1997 (Figure 2.23a) the characteristic ozone minima in the lower stratosphere (100 hPa) was observed. On 6 April 1997 (Figure 1.23b) the characteristic ozone minima again was observed, as well as a larger secondary maximum in the

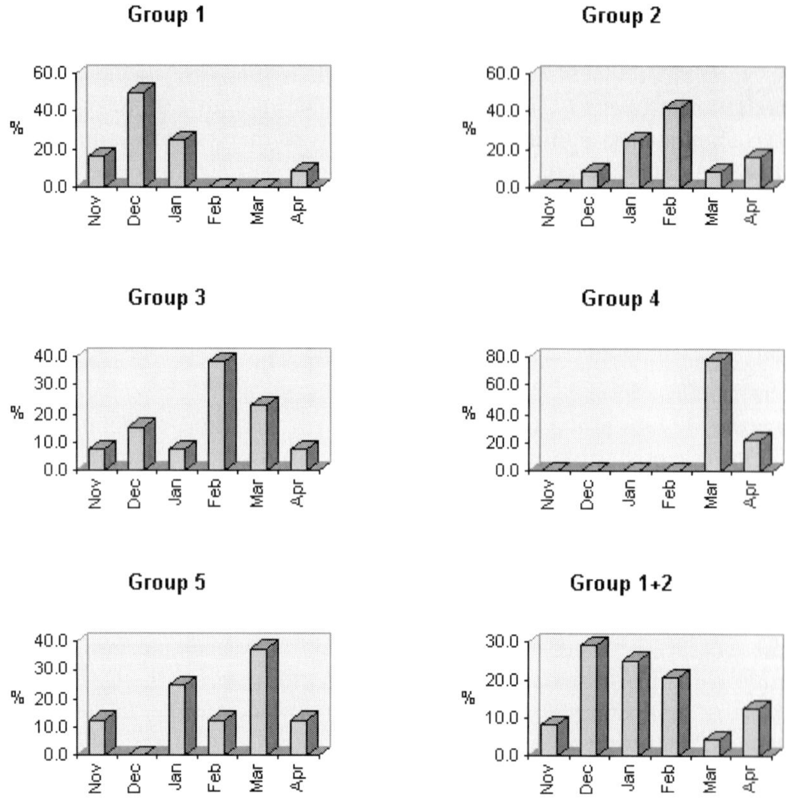

Figure 2.22. Monthly distribution for the occurrence of each group of ozonesoundings for the period from November to April over Athens, Greece

lower stratosphere. By examining the weather maps (not shown) it is evident that Athens was influenced by the subtropical jet-stream at that time. This is in agreement with observations made by Dobson (1973) that the 15-km minimum was associated with air coming from the subtropical region.

On 10 January 1997 (Figure 2.24a) the laminated ozone structure was observed. This was strongly associated with the establishment of a north-western circulation over the region, implying that Athens station was under the influence of air of polar origin.

On 1 April 1997 (Figure 1.24b) a cut-off low system was observed over the region. The ozone profile contains a thick layer in the lower stratosphere. This is in agreement with the observation that a cut-off low event seems to be combined with high ozone partial pressure at low levels. On the same time the laminated structure was observed in the region of 8–13 km.

On 6 February 1997 (Figure 2.25) the most interesting feature in the vertical-ozone structure was the ozone-depletion effect that occurred at the maximum ozone partial-pressure altitude region. This new finding of strong ozone depletion, which

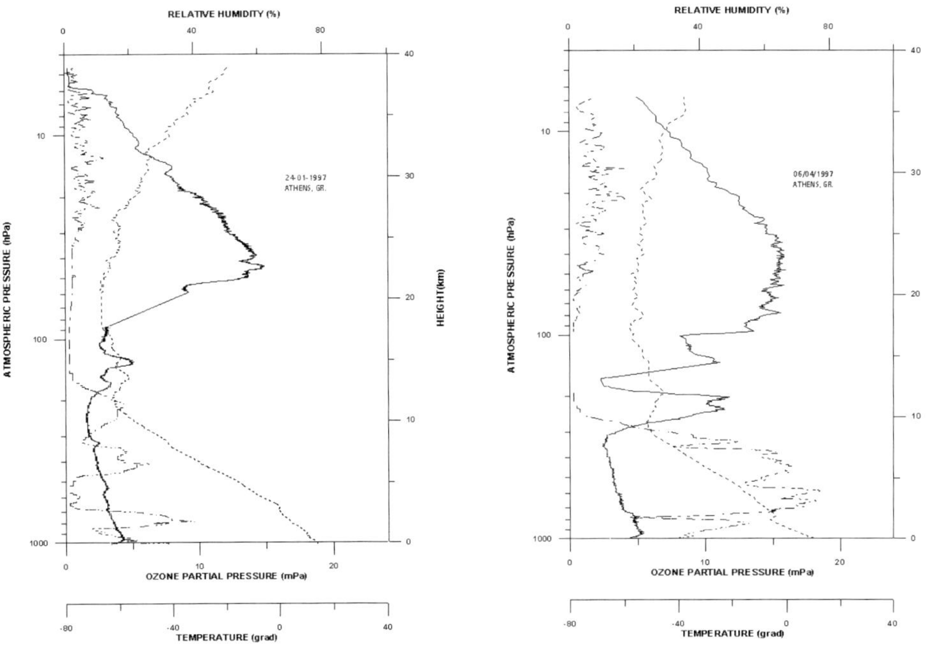

Figure 2.23. Ozonesoundings for (a) 24 January 1997 and (b) 6 April 1997.

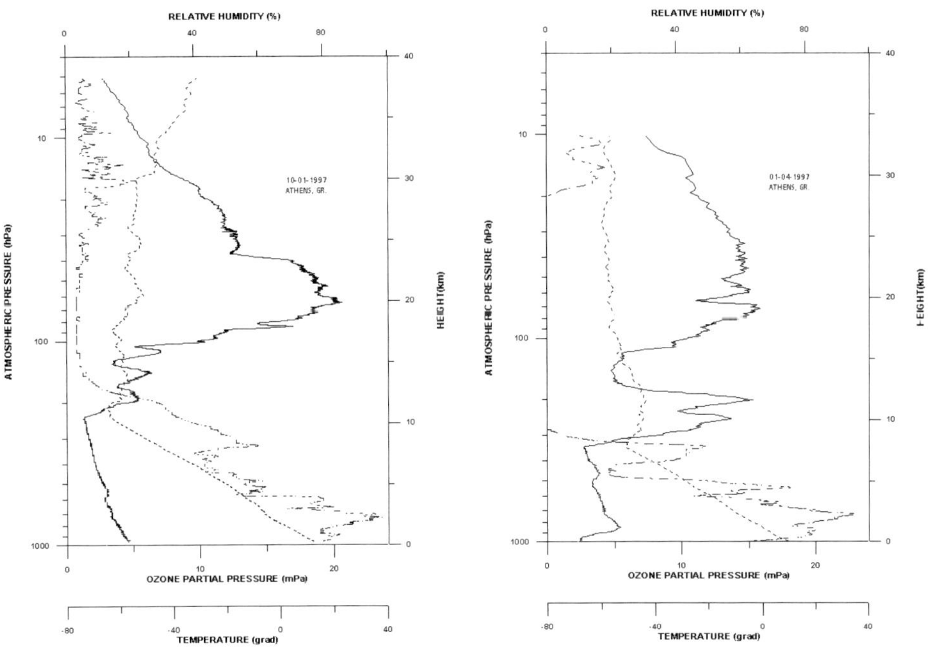

Figure 2.24. Ozonesoundings for (a) 10 January 1997 and (b) 1 April 1997.

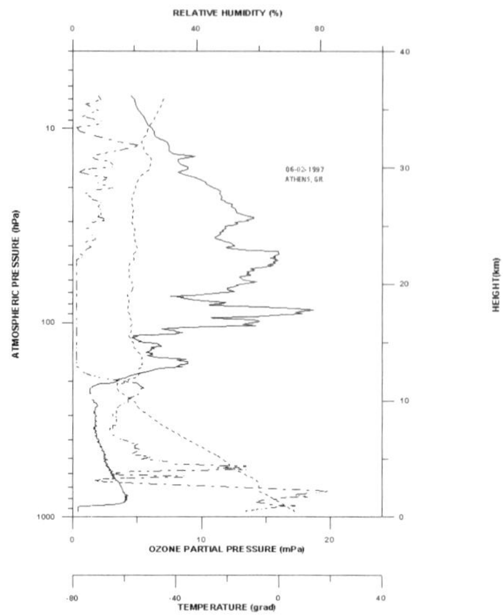

Figure 2.25. Ozonesounding for 6 February 1997.

occurred at the level of the ozone maximum concentration, should be considered in the context of conventional knowledge; that is, maximum ozone depletion occurs in the layer just below the layer of the ozone maximum partial pressure. The meteorological pattern in the latter case is characterized by the extension of the polar vortex to the mid-latitudes of the Northern Hemisphere, resulting in more intense north-western circulation above the site.

2.2.4.5 *Experience from the Northern and Southern Hemisphere*

2.2.4.5.1 Introduction

As mentioned in Chapter 1, strong ozone depletion has been recorded in the lower polar stratosphere in both hemispheres. In the Antarctic ozone depletion became evident in the 1980s, with most of the ozone loss occurring in the lower stratosphere and attributed to chemical ozone-destruction mechanisms (*Scientific* ..., 1999). Recently, in the Arctic lower stratosphere, chemical ozone losses have been observed from the beginning of the 1990s (von der Gathen *et al.*, 1995). The strongest ozone loss in the lower stratosphere occurs in the Antarctic in October and in the Arctic in March. It is commonly accepted that these ozone losses are mainly caused by Polar Stratospheric Clouds (PSC) and related chemical destruction of ozone inside the stratospheric vortex. However, negative ozone trends outside the polar regions are not well understood. In order to discuss the relation between strong negative ozone trends in the polar regions and those in lower latitudes, with respect to the mean global circulation and transport, data on altitude-dependent ozone trends are useful. Sutton (1994) found, using Lagrangian calculations of transport phenomena in mid-atmosphere, that airflow in a seasonal timescale begins in the mesosphere in early winter and is directed towards the upper polar stratosphere. On the other hand, the mean hemispheric meridional circulation in the lower stratosphere is also directed from the tropics towards polar altitudes.

In order to study potential connections between globally observed ozone trends and those in polar latitudes, altitude-dependent meridional ozone trends have been studied for spring season (i.e. October and March monthly zonal means), when the strongest ozone depletion occurs in the lower polar stratosphere. During the last decade vertical-ozone distributions in the polar stratosphere were mainly obtained by ozonesoundings. These observations give a detailed description of the variability in ozone up to altitudes of about 30 km, i.e. 10 hPa. Data from greater altitudes can be obtained by using satellite observations such as Solar Backscatter Ultra-Violet (SBUV) and Stratospheric Aerosol and Gas Experiment (SAGE) (McCormick *et al.*, 1992).

2.2.4.5.2 Data analysis

Altitude-dependent ozone trends were retrieved from satellite SBUV measurements between 1979 and 1993 (NASA/GSFC and NOAA) (Hood *et al.*, 1993; Hollandsworth *et al.*, 1995). Over this period two SBUV instruments were in use. The first instrument (SBUV) was onboard the Nimbus-7 satellite (N7SBUV) from 1979 to 1990 and the second (SBUV/2) onboard the NOAA-11 satellite from 1989. The N11-SBUV/2 data were processed using the version 6 algorithm, in almost the same way as for the SBUV data. The ozone data used here consisted of two separate time series derived from SBUV and SBUV/2 observations, taking into consideration the deviation between observations during normal operation in the 18-month overlap period (Hollandsworth *et al.*, 1995). It is well documentated that retrieval of ozone trends from SBUV data for the lower stratosphere is uncertain, since data for the four lowest layers (i.e. below 32 hPa) primarily reflect the differences between column ozone and the profile amounts above 32 hPa. So, we have compared these SBUV trends with those obtained from long-term ozonesonde observations carried out at Antarctic Station Syowa (69°S; 39°E) (Gernandt *et al.*, 1995). The coincidence of both trends, zonal mean one from SBUV data and local one from sonde data, is fairly good up to altitudes of about 20 hPa (Figure 2.25). By considering the result of this intercomparison we conclude that the trends calculated by means of SBUV data are qualitatively reliable for this southern latitude.

For this analysis, zonal means of ozone content for Umkehr layers 1 to 11, i.e. up to 0.25 hPa, were calculated for 10° latitude bands. The volume mixing ratio for the mean pressure altitude at each Umkehr layer was calculated from the zonal mean ozone content. Applying the linear regression technique, the trend of monthly zonal means for October and March for each Umkehr layer was obtained expressed in ppmv per year. The statistical significance of these trends was assessed by means of regression coefficients. By this method a rough profile of altitude-dependent trends is obtained for October and March.

2.2.4.5.3 Some results

Figure 2.26 shows the trend of ozone-mixing ratio (in ppmv per year) at Umkehr layers 1–11, as obtained from SBUV data for each October since 1979 for the 60–70°S

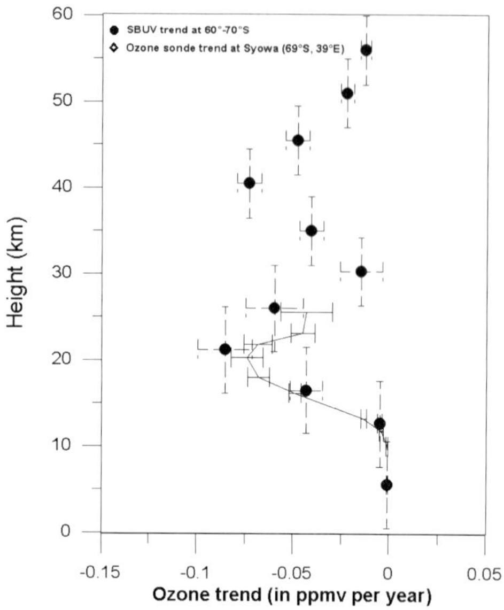

Figure 2.26. Profiles of ozone-mixing ratio trend (in ppmv/year) for Umkehr layers 1 to 11 in the Southern Hemisphere for October for 60–70°S and the ozonesounding trend from Syowa Station (69°S; 39°E) for the period 1966–1995.

latitude zone. In addition we have used the trend profile as derived from ozonesounding data collected at Antarctic Station Syowa (69°S; 39°E) from 1966 to 1995 for the same month (Gernandt *et al.*, 1995).

It is evident from Figure 2.26 that during the winter and spring, there are two significant trend maxima in the lower and upper stratosphere over high latitudes. The analysis for October, when the maximum ozone loss occurs, shows clearly that the zonal mean total ozone loss accounts in the same order of about −0.08 ppmv per year in both the upper stratosphere and the lower stratosphere.

An inspection of Figure 2.26 also denotes that both maxima of negative trends are clearly separated by an altitude region from about 10 to 20 hPa, where no significant trend is seen except in polar latitudes. Inside the polar vortex, air is being transported downward due to diabatic cooling. Therefore, this ozone-poor air comes to the lower stratosphere, where by PSC-related chemistry, ozone is further depleted in these airmasses. This downward displacement of air might explain the downward extension of air with decreased ozone from the mid-stratosphere.

The chemical ozone loss in the lower polar stratosphere amplifies the negative trend. So, it is assumed that the negative trend in the lower polar stratosphere is caused by both downward transport of air with low ozone concentration and additional chemical loss in the lower polar stratosphere.

Figures 2.27a and 2.28a depict the global distribution of ozone trends for October and March. The altitude range where these trends are significant (95%) is shaded. There is a clear separation between the negative trend in the upper stratosphere and

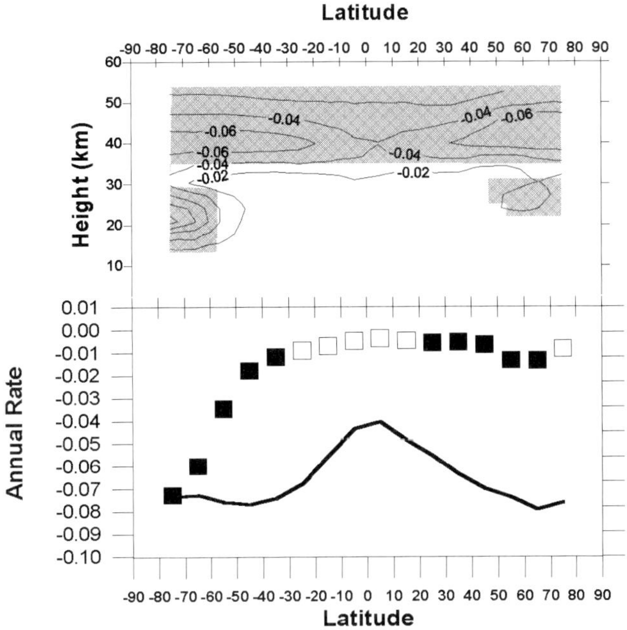

Figure 2.27. (a) Altitude–latitude cross-section for the global distribution of ozone trend for Umkehr layers for October. The shaded area is significant with 95%. (b) Latitudinal distribution of the TOZ trend (in 10^{-2} DU/yr) and the corresponding contributions by the altitude range where the maximum trend occurs (in ppmv/yr). Values with 95% significance level are indicated by full squares and dots.

in the lower stratosphere, except for the spring polar stratosphere, i.e. October in the Southern and March in the Northern Hemisphere. Inside the polar vortex the negative trend is significant (within 95%) at all stratospheric altitudes during spring. While in the upper stratosphere significant negative trends are obtained at all latitudes, in the lower stratosphere close to 30 to 40° latitude in both hemispheres for the months October and March, the trends do not continue to be significant. The trends observed at the upper stratosphere coincide very well with the results of SBUV data analysis (Hollandsworth *et al.*, 1995; Hood *et al.*, 1993), where negative ozone trends of about 6–8% per decade for the 2–4 hPa layer for October at 60°S is observed.

From the figures shown, it is also evident that the upper trend maximum displaces upwards from the equator to polar latitudes in the summer hemispheres (i.e. in spring and fall). In the lower stratosphere the trend maximum turns back to mid-latitudes. Along these trajectories the negative trend is increasing with latitude towards both polar regions in the upper stratosphere (Figures 2.27b and 2.28b). Within the polar region the negative trend is amplified downward during spring, and it decreases toward mid-latitudes in the lower stratosphere and upper troposphere.

The latitudinal distribution of total ozone content (TOZ) and contributions of maximum negative trends are shown for the same months in Figures 2.27b and

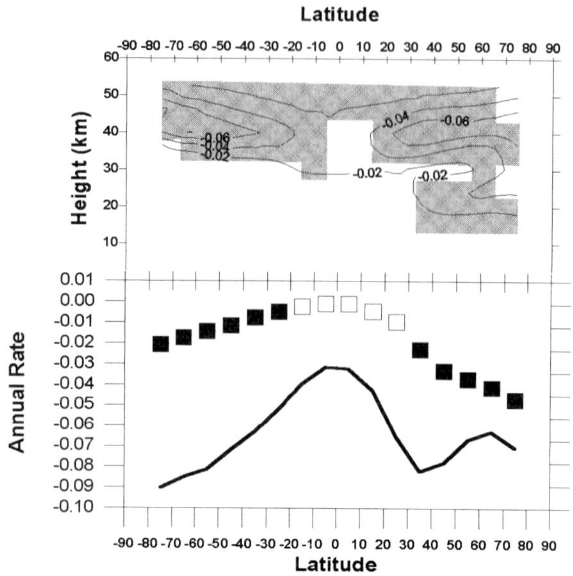

Figure 2.28. Same as Figure 2.27 but for March.

2.28b. While the TOZ trend is not significant (empty squares) in tropical latitudes the maximum trends in the upper stratosphere are significant (full squares) by the 95% level.

From these figures, it can also be concluded that trajectories of the maximum negative trend coincide very well with mean circulation patterns. From the tropical upper stratosphere and mesosphere, air is transported downward and towards polar latitudes by the meridional circulation.

The upper trend maximum follows this pattern. On the other hand, a decrease in meridional transport may also cause decrease of ozone towards high latitudes. Inside the vortex, air is transported downward to regions where chemical ozone loss occurs in spring; that is, mainly in September and October in the Antarctic and from January to March in the Arctic (von der Gathen *et al.*, 1995). During spring in the lower stratosphere the increasing chemical ozone loss amplifies the negative trend. The altitude range of significant negative ozone trends (shown in Figures 2.27a and 2.28a) is extended towards mid-latitudes in the spring hemisphere. The maximum of negative ozone trend is displaced downward into the upper troposphere at latitudes lower than 50°.

By considering these results and the global meridional circulation a connection might be found between atmospheric regions through transport from the tropical upper stratosphere via the winter polar vortex to the upper troposphere in mid-latitudes where finally some portions of the airmass arrives with decreased ozone. Along this path the trend increases from tropical to polar latitudes, which might be caused by chemical or dynamic reasons. The trend is significantly amplified by PSC-related chemical ozone loss inside the polar vortex in spring. Because of the convection of air the net ozone loss can be greater when ozone-depleted air is transported downward from the mid-stratosphere. Mesoscale exchange processes between polar and mid-latitudes may finally transport stratospheric air with depleted ozone into the upper troposphere. Therefore, some impact can appear in the upper troposphere due to stratosphere–troposphere exchange processes in mid-latitudes.

2.2.5 Surface ozone measurements

Three methods are widely used for detecting surface ozone in clean air – the chemiluminescent method, the KI method and the UV absorption method – which are briefly discussed in the following sections.

2.2.5.1 *Chemiluminescent method*

Several commercial instruments are available that in general consist of an air-intake filter, detector cell, photomultiplier, amplifier, recorder, sample pump, ethylene supply and flowmeter. Ambient air and ethylene are delivered simultaneously to a mixure zone of a detection cell, where the ozone reacts with the ethylene thus emitting light which is detected with a photomultiplier tube, after the amplification of the photocurrent. The sensitivity is approximately 2 µg O_3/m^3. These instruments are calibrated against artificially generated test atmospheres of ozone (WMO, 1978).

2.2.5.2 *Electrochemical Potassium Iodide method*

The apparatus mainly consists of the measuring cell and air pump (integrated in one unit). The principle of the method makes use of the oxidation of the neutral buffered KI solution by ozone. Two electrons per reacting ozone are produced on a platinum electrode at the cathode. The electric current from the measuring cell is amplified by a DC amplifier. The instrument has a sensitivity of 2 to 5 µg O_3/m^3 and a range from this detection limit to clean-air ozone concentrations of about 200 µg/m^3 (Feister *et al.*, 1989). The method is considered accurate to within ±5% (WMO, 1978).

2.2.5.3 *UV absorption method*

The main instruments that use the UV absorption method are Dasibi-1003-1008, ThermoElectron 49 and Environment S.A. (Sections 2.2.2.2.7 and 2.2.2.3.3). In contrast to the Dasibi types, the ThermoElectron is a dual-beam instrument and has the fastest response (Helas *et al.*, 1989). Another instrument that is based on the ultraviolet absorption technique is the Dylec Model 1006 AHJ produced under license from Dasibi Inc. The cycle time of measurements is about 12 s (Ogawa and Komala, 1989).

2.3 GLOBAL OZONE DYNAMICS

As mentioned in Chapter 1, the most characteristic feature of global total ozone dynamics is the presence of a strong spatio-temporal variability, which was documented and analysed in detail in recent reports UNEP-1994 and WMO/UNEP-1999. The development of various kinds of ozone observations constitutes a great step forward. However, quite a number of problems still remain unresolved. This is why the need for long-term continuity of ozone measurements was recently discussed (Kaye and Readings, 1998). Several important conclusions were made in this context by the Committee on Earth Observation Satellites (CEOS) Analysis group (Kaye and Readings, 1998):

- The need to use all relevant observational techniques for ozone observations, including spacebased measurements of TOZ and vertical-ozone profile, ozonesondes, aircraft and groundbased measurements (especially Dobson, Umkehr, microwave and lidar remote-sounding techniques). Long-term trend studies should be a principal purpose of data interpretation.
- Simultaneous monitoring of various other quantities, such as meteorological parameters, solar forcing, aerosols and other atmospheric trace constituents is a prerequisite of adequate ozone-data interpretation.
- Special attention must be paid to problems of calibration and data quality. In this context combined groundbased, balloon, aircraft and satellite observations are important for data verification.

As mentioned in the SPARC Report (Harris and Hudson, 1998; Harris *et al.*, 1998) global ozone observations display both long-term and considerable interannual variability. Sources of long-term variability include changes in solar output, temporal changes in the abundance of stratospheric aerosols and variability in the stratospheric circulation leading to changes in ozone transport on seasonal, interannual and decadal timescales. Several components of interannual variability have been identified in ozone observations, in particular quasi-biennial oscillation (QBO) and the 11-year solar cycle. Moreover, numerous anthropogenic impacts (through ozone-depleting substances) on the ozone layer have to be mentioned as well as very complex ozone–climate interactions.

To summarize the achievements in the study of global TOZ dynamics, we begin with the problem of TOZ trends. This problem was discussed in detail in two recent proceedings of scientific meetings (*Atmospheric Ozone*, 1998; *Stratospheric Processes* ..., 1997) as well as in *Scientific* ... (1999).

2.3.1 Total ozone trends

The problem of filtering out regular trends in strong spatio-temporal variability is fairly complicated and includes two principal aspects: (1) developing an adequate statistical approach; (2) analysis of instrumental biases. We briefly discuss below only the second aspect (see Section 2.1–2.3).

The problem of instrumental bias has been discussed throughout the whole history of ozone measurements, but acquired a special importance after satellite observations started. Even at the early stages of satellite TOZ observations very serious attention was paid to this problem. After comparison of Nimbus 7 TOMS data from 1978 to 1985 with surface observations, Bojkov *et al.* (1988) discovered a significant bias. According to TOMS data, an inhomogeneous trend of TOZ depletion was observed of about 3.5% compared with Dobson data, the cause of which was probably ageing of the TOMS diffuser plate. After filtering out the drift of TOMS sensitivity it became possible to use satellite information for verifying the reliability of the global net of 92 groundbased ozone stations. The result was that there was close agreement (within ±2% for about two-thirds of the stations) between satellite and Dobson data, whereas in the case of M-83 filter ozonometers much stronger deviations were observed.

Results of the most recent TOZ observations at various sites were considered in a number of papers (*Atmospheric Ozone*, 1998). For instance, Denis *et al.* (*Atmospheric Ozone*, 1998) discussed observations at Reunion Island (21°S) in the western Indian Ocean and at Tarawa (01°N) in the western Pacific which started in 1992–1993 in this intertropical area within the NDSC (Network for Detection of Stratospheric Change) Programme. Besides the signature of stratospheric transport in the annual cycle at the tropics and the QBO modulation at the equator, the data indicate the presence of a significant tropospheric component in TOZ variability. At both stations TOZ minima occur at the season of the maximum development of the ITCZ.

Dubrovsý and Kalvová (*Atmospheric Ozone*, 1998) studied the mean annual TOZ cycle and its correlation with meteorological conditions on the basis of a 29-year series (1962–1990) of Dobson measurements in Hradec Králové, Czech Republic.

Fioletov *et al.* (*Atmospheric Ozone*, 1998) analysed the great difference between TOZ trends derived from groundbased (Dobson measurements) and satellite (SBUV, SBUV/2) observations in the Southern Hemisphere (the ratio of relevant TOZ values is nearly 1 : 2 at 30–55°S). It was shown that a very strong austral winter TOZ decline, calculated from SBUV-SBUV/2 data, was mostly related to instrument and/or algorithm errors for high solar-zenith angles (>83°) as well as to specific features of the orbit of the NOAA-11 satellite. Actual ozone decline (winter and year-round) is approximately the same for both hemispheres in the mid-latitudes (except Antarctica).

Hare and Fioletov (*Atmospheric Ozone*, 1998) discussed typical problems with the use of TOZ data from the World Ozone and Ultraviolet Radiation Data Centre (WOUDC) located in Toronto, since the early 1960s (this data series is nearly 40 years long).

Hilsenrath *et al.* (*Atmospheric Ozone*, 1998) summarized the contributions of SSBUV (eight Space shuttle missions from October 1989 to January 1996) in support of the US long-term ozone-monitoring programme (first of all from the viewpoint of validation of satellite observations). A significant decrease in Northern Hemisphere TOZ from the winter of 1992 to the following winter was observed.

Similar validation problems (homogenization of ozone records) were discussed by Köhler and Claude (*Atmospheric Ozone*, 1998) in connection with the interpretation of the long TOZ time series (since 1967) of Dobson observations in Höhenpeissenberg.

Analysis of re-evaluated Dobson TOZ data, obtained at five stations on the US mainland between 1979 and 1994, made by Kohmyr *et al.* (*Atmospheric Ozone*, 1998) yielded average linear downward trends of −4.54, −5.26, −2.35 and −2.06% per decade, respectively, for winter, spring, summer and autumn seasons, and −3.68% on an annual basis. The contribution of various causes of TOZ change (QBO, Pinatubo eruption, etc.) were assessed.

Reinsel *et al.* (1989) carried out the trend analysis of volcanic aerosol-corrected Umkehr ozone-profile data from 10 stations between 1977 and 1987. The results indicated a significant overall negative trend, exclusive of trend variations associated with solar flux variations, of the order of −0.5% per year in Umkehr layers 7–9 over the period mentioned, and a significant positive solar cycle association in all layers 4–9 with the maximum solar effect estimated to occur in layer 9. A comparison was made by Reinsel *et al.* (1989) between the solar backscattered ultraviolet (SBUV) monthly average ozone-profile data near the 10 Umkehr stations and the corresponding Umkehr data, corrected for aerosol errors, to investigate possible drifts. The results showed that a substantial overall negative linear drift in SBUV data relative to corrected Umkehr data in layers 7–9 took place, with estimated values of the drift of the order of −1.0% per year for layers 8 and 9. The problem of volcanic aerosol correction of Umkehr observations of ozone profile for five northern mid-latitude Dobson stations was also considered by DeLuisi *et al.* (1989). Using corrected data they reached the conclusion that between 1979 and 1986 decreases in ozone concentrations in layers 8 and 9 were, respectively, 9% and 15%. According to Dobson Umkehr data, considered by Tiao *et al.* (1986), ozone trends for 3A, 3B and 4A layers (100–50 hPa) were about −0.5% per year.

Karol *et al.* (1992) studied TOZ linear trends in the Northern Hemisphere on the basis of groundbased Dobson and filter observations between 1975 and 1986. They discovered strong changes in not only amplitude but even sign of mean monthly TOZ trends dependent on season and time period. TOZ trends also varied according to phase of a 11-year solar activity cycle (rise and fall in 1975–1980 and 1981–1986, respectively) and QBO phases. Fioletov (1992) analysed long-term variations in TOZ mean zonal values in the Northern Hemisphere as well as TOZ trends for two latitudinal zones (40–52°N and 53–64°N), again using groundbased observation data. In the case of long-term variations, QBO-induced variations were very distinctly shown.

Satellite observational data series are still comparatively few. So existing long-term groundbased observations become very important (the history of groundbased observations was discussed by Farman (1989)). Bojkov *et al.* (1990) carried out a statistical seasonal trend analysis of revised Dobson TOZ data for 29 Northern Hemisphere stations located between 15 and 64°N latitude using data for 1986. For explanation of solar and quasi-biennial oscillation-induced ozone variation, the

10.7-cm solar flux series and the 50-hPa equatorial wind series were used. The following statistical model for monthly ozone series y_t was applied:

$$y_t = \sum_{i=1}^{12} \mu_i I_{i;t} + \sum_{i=1}^{12} \beta_i I_{i;t} R_t + \gamma_1 (Z_{1;t} - 120) + \gamma_2 Z_{2;t-k} + N_t \qquad (2.1)$$

where μ_i is the ozone mean in month $i (i = 1, \ldots, 12)$; $I_{i;t}$ is the indicator series for the ith month of the year: 1 if month t corresponds to month i of the year and 0 otherwise; β_i is the trend in Dobson units/yr after December 1969 in month i of the year; R_t is the linear ramp function, beginning in January 1970, $(t - t_0)/12$ if $t > t_0$ (where t_0 corresponds to December 1969) and 0 if $t \leq t_0$; $Z_{1;t}$ is the solar 10.7-cm flux series, $\tilde{a}_1$ is the associated coefficient; $Z_{2;t-k}$ is the QBO series lagged k months, $\tilde{a}_2$ is the associated coefficient; N_t is the residual noise series.

The results obtained by Bojkov *et al.* (1990) indicate significantly more negative trends during winter (December–March: −1.2%, −2.1% and −3.0% per decade for latitudes 35, 45 and 55°N, respectively) than during summer (May–August: −0.6% per decade, without clear dependence on latitude), especially at higher latitudes, with trends in winter becoming more negative with increasing latitude. Mean annual trend for all latitudes is about $-0.84 \pm 0.82\%$ per decade. The trends also display some regional variation, being considerably less negative in Japan than in North America and Europe. Bojkov *et al.* (1990) have concluded that trend results, based on unrevised published data, were on average less negative than trends from revised Dobson data for European stations, by about 1.0% per decade across all seasons, with only small average differences for stations in North America and Japan. Later on, Bojkov *et al.* (1995) considered the revised Dobson TOZ trend over the Northern Hemisphere for 1964 to 1994.

Staehelin *et al.* (1998a, b) conducted a trend analysis of the homogenized Dobson total ozone series at Arosa, Switzerland (47°N, 1,800 m amsl) between 1926 and 1996 (earlier results were considered by Dütsch and Staehelin (1989)). The analysis of this longest TOZ series in the world was based on the use of statistical model (2.1). Ozone trends for annual means are $-(2.3 \pm 6)\%$ per decade when a simple linear change from 1970 to 1996 is assumed. Strongest depletions (by about −4% per decade) were found between January and March. A different approach of applying multiple regression, when terms for solar cycle, stratospheric aerosol loading and high mountain temperature record were included, resulted in annually averaged ozone trends of the order of $-(1.9 \pm 0.6)\%$ per decade. Staehelin *et al.* (1998a) demonstrated that the recent TOZ decrease cannot be accounted for by the higher average aerosol content in the second half of the century. Assessment of the levels of total and stratospheric ozone depletion led to the conclusion that the decrease of ozone content in the stratosphere was approximately 30% larger than that of total ozone (this assessment is based on a crude estimate of the increase in tropospheric ozone). It has been assumed that the acceleration observed in TOZ trends between the 1970s and the 1980s over northern mid-latitudes might be partially attributed to the larger increase in tropospheric ozone in the 1970s.

Applying a methodology of TOZ signal processing similar to (2.1), Stolarski *et al.* (1991) deduced total ozone trends from Nimbus-7 TOMS data. The linear trend obtained from the data averaged between 65°N and 65°S latitudes was $-0.26 \pm 0.14\%$ per decade over the 11.6-year time period from November 1978 through May 1990. The trend was near zero ($0.0002 \pm 0.2\%$ per year) at the equator and increased towards both poles. At 50°N the annually averaged trend is $-0.5 \pm 0.21\%$ per year, but there were strong seasonal variations at this latitude from more than -0.8% per year in winter and early spring (February and March) to about -0.2% per year in summer (July and August). The solar component of TOZ variability (for global average data) was 3.7 ± 0.6 DU/100 units of F10.7 or 1.5% over the solar cycle, while QBO contribution was -0.4 ± 0.1 DU/10 knots of 30 hPa wind or 1% over a QBO cycle.

Special attention has been paid to QBO-induced TOZ variability (Bowman, 1989; Zerefos *et al.*, 1992; Yang and Tung, 1994, 1995). Yang and Tung (1995) have reached the following three conclusions: (1) the QBO signal was strongest in the mid and high latitudes and was present mainly in the winter–spring season; (2) the extratropical ozone QBO signal was out of phase with the equatorial ozone signal, the latter being in phase with the equatorial stratospheric zonal wind QBO signal; (3) there was no gradual phase propagation. Instead, there were distinct regions, in each of which the ozone QBO signal had fairly constant phase and large phase shifts occurred in-between the transition regions.

Kane *et al.* (1998) studied latitude dependence of quasi-biennial oscillation (QBO) and quasi-triennial oscillation (QTO) characteristics using TOMS total ozone data between 1979 and 1992 in the form of 12-month moving averages. Comparison of these data was made with the stratospheric low-latitude zonal wind and equatorial eastern Pacific sea-surface temperature (SST). The equatorial ozone dynamics was characterized by a strong QBO component with a period of about 30 months, and its maxima tallied with westerly wind maxima. At other latitudes, TOZ maxima spacing was often different over 30 months, more so in the Northern Hemisphere.

A spectral analysis carried out by Kane *et al.* (1998) showed that both hemispheres had one peak at 20 months and another peak at ~30 months, only up to ~50° latitude. At higher latitudes these peaks shifted to ~23 and ~36 months. For 0–50° latitude the ~30 month periodicity in the Northern Hemisphere showed abrupt changes, while the southern latitudes showed a roughly gradual phase shift. The Northern Hemisphere had an additional periodicity at ~4 years, roughly matching the SST. Analysis of the same data without 12-month averaging showed the above characteristics more clearly and revealed additional peaks at ~0.7 years (8–9 months) and ~3 years (16–17 months). All of these peaks showed shifts (mostly increases) at higher latitudes.

Powerful perturbations in general atmospheric circulation, connected with the El Niño/Southern Oscillation (ENSO) phenomena, had a significant impact on the ozone layer (Shiotani, 1992; UNEP, 1994; *Scientific* ..., 1999; Zerefos *et al.*, 1992). Using statistical analysis of TOMS data, Randel and Cobb (1994) studied in detail the global response of TOZ to ENSO dynamics. The southern oscillation index is usually used for quantitative description of ENSO.

Table 2.8. The characteristics of total ozone anomalies in January in the equatorial region during negative SOI anomalies.

Time period	SOI minima	Characteristics of centres of total ozone anomalies Pacific Ocean			Africa		
		Latitude S	Longitude E	ΔO_3	Latitude N	Longitude W	ΔO_3
June 1982–April 1983	−4.6	15	120	−1.3	10	10	−1.8
November 1986–September 1987	−2.0	5	160	−2.2	5	0	−2.2
September 1991–April 1992	−3.4	0	135	−1.2	10	5	−2.7
March 1997	3.2	10	130	−4.1	5	0	−2.5

As is now known, (1) El Niño events result in a positive (negative) response in the extratropics (near equator), which is characterised by a strong longitudinal asymmetry; (2) this effect is stronger pronounced in northern winter months. Chernikov *et al.* (1999) summarized recent ENSO observation results (Table 2.8). These point out that El Niño's influence on total ozone occurs by a strengthening of upward motions over oceans, because of SST increase with subsequent TOZ decrease in correspondence with the Norman–Dobson principle. Photochemical ozone destruction is possibly strengthened by an increase of cloudiness and albedo during El Niño events. The importance of these causes of ENSO-induced TOZ change is confirmed by TOZ increase in the extratropics (the closure of a meridional circulation cell here results in downward airflows and TOZ increase) and the stability of local TOZ minima positions during El Niño's development in different years (Table 2.8).

Using Nimbus-7 TOMS (1978–1993) and Earth Probe (1996–1998) total ozone data as well as relevant global meteorological information, Chernikov *et al.* (1999) carried out an analysis of interrelationships between global fields of TOZ and meteorological parameters. The following conclusions were reached by Chernikov *et al.* (1999):

- The El Niño event of 1997/98, as one of the most intensive phenomena in the history of observations, strongly affected the ozone layer and resulted in a decrease of column ozone amounts. Basic regions with negative TOZ anomalies in the equatorial zone were located over the central and eastern Pacific, as well as near the Greenwich Meridian.
- Negative total ozone anomalies were associated with regions of horizontal wind divergences in the upper troposphere (200 hPa) over the eastern equatorial Pacific and eastern Africa. Positive total ozone anomalies were observed in regions with wind convergence in the upper troposphere over the western equatorial Pacific and eastern part of the Indian ocean.

- Short-term TOZ variations (of the order of a few days) were discovered for the first time with typical scales of about several thousand kilometers and with absolute values within 200–225 DU. Such variations were observed mainly over the El Niño region. The largest values of their characteristics (amplitude, areal extent, frequency, etc.) were registered in mid-January 1998 with a time lag of 2–3 weeks after a maximum in SST in the El Niño region.
- Ozone mini-holes formed on both sides of the equator in the El Niño event over the Pacific during the second half of December 1997, the sizes of which increased with time reaching about 1 million km^2 by the end of December. TOZ values decreased to 200–225 DU.
- It was found that negative and positive components of SOI are better predictors than net SOI. Both negative and positive SOI led to a decrease in total ozone in the equatorial region, but with different intensities. This may be an indication of a limited applicability of the linear assumption in the parametrization of interrelationships between ENSO and total ozone variations.
- The unusually strong influence of the El Niño event of 1997–98 on the ozone layer could have influenced global climate because of a chain of processes: ozone decrease, ocean warming, troposphere warming and stratosphere cooling, and the intense interaction of the latter two. A similar situation existed after the strongest El Niño event in 1982/83.

Considering both satellite and groundbased observational data available for March 1991, Stolarski *et al.* (1992) analysed relevant trends. In the case of groundbased observations the average trend for the 26–64° N latitude belt was −2.7% per decade in the winter, −1.3% per decade in the summer and −1.0% per decade in the fall. Stolarski *et al.* (1992) emphasized that all of these declines are statistically significant.

Table 2.9 contains a list of satellite instrumentation that has been used for BUV ozone observations (Hilsenrath *et al.*, 1998). As pointed out earlier, an important contribution was also made by SAGE-I and II data (SAGE-III was recently

Table 2.9. Satellite instrumentation for BUV ozone observations.

Instrumentation	Satellite	Observation period
BUV	Nimbus-4	1970–1975
SBUV	Nimbus-7	1979–1990
SBUV/2	NOAA-9	1984–1997
SBUV/2	NOAA-11	1989–1994
SBUV/2	NOAA-14	1995–
SSBUV	Shuttle	Eight flights, 1989–1996
TOMS	Nimbus-7	1979–1992
TOMS	Meteor-3M	1991–1994
TOMS	Earth Probe	1996–
TOMS	ADEOS	1996–1997
GOME	ERS-2	1995–

launched). An improved version of GOME was installed on the Envisat satellite launched in 1999.

Hilsenrath *et al.* (1998) pointed out that, since 1978, there was a mean global TOZ depletion of about 2–3%, while the ozone content in the upper stratosphere decreased by 7–8%. Because these trends are higher than trends predicted by numerical modelling, further studies of TOZ variability and its causes are needed. From the viewpoint of the quality of observational data, reliable groundbased and onboard calibration became very important. Another important aspect of the problem is adequate control of the sensitivity of instrumentation (in the case of Nimbus-7 such control reduced sensitivity drift by 1.0% for each 10-yr period). One possibility to control the drift of spectral sensitivity would be comparison of UV albedo data with the results of 'fresh' Shuttle observations.

Also important are intercomparisons between observational data from various satellites. One such intercomparison, made for Nimbus-7 and NOAA-11 data over 18 months, indicated (Hilsenrath *et al.*, 1998) that TOZ value differences did not exceed 1% (except high latitudes where they reached 4%).

Rathman *et al.* (1997) considered the results of a preliminary comparison between TOZ data from TIROS Operational Vertical Sounder (TOVS) and Global Monitoring Experiment (GOME). Two comparisons were made: (a) TOVS ozone analysis maps versus GOME level 2 data; (b) TOVS data located at the Northern Hemisphere Ground Ozone Stations (NHGOS) versus GOME data. Rathman *et al.* (1997) emphasized that both analyse consistently showed an offset in TOZ values between the datasets: 35 DU and 10 DU for (a) and (b) comparisons, respectively, despite good correlation between spatial and temporal features of the datasets. However, a noticeably poor correlation was observed in latitudinal belts 10–20°N and 10–20°S. The smallest region that was statistically representative of the ozone value correlation dataset of TOVS data at NHGOS and GOME level-2 data was determined to be a region enclosed by an effective radius of 0.75 arc-degrees (83.5 km). Future plans include carrying out synchronous combined satellite, balloon and aircraft observations for validation purposes.

The *UNEP Scientific Assessment* (*UNEP* ..., 1994) contains a survey of TOZ trends from groundbased data through February 1994 and SBUV/2 satellite data through May 1994 (Tables 2.10–2.12). Figure 2.29 illustrates specific features of TOZ trend annual variations. The representative trends (annual averages, in % per decade) for north and south mid-latitudes and the tropics are given in Table 2.13 (uncertainties (±) are expressed at the 95% confidence limits (two standard errors).

The principal conclusions made in the *UNEP Assessment Report* 1994 were as follows:

- Northern Hemisphere mid-latitude trends are significantly negative in all seasons, but are much larger in winter/spring (about 6% per decade) than in summer/fall (about 3% per decade).
- Tropical (approximately 20°S–20°N) trends are slightly negative, but not statistically significant, when suspected drift in satellite data is incorporated into the uncertainty.

Table 2.10. SBUV 2 trends in % decade by seasons and latitudinal zone from January 1979 to May 1994, with 95%-uncertainty limits (two standard errors, labelled 2se).

Zone	Dec–Feb	2se	Mar–May	2se	Jun–Aug	2se	Sep–Nov	2se	Year	2se
65 N	−5.6	4.2	−6.3	2.9	−3.5	1.4	−4.3	1.6	−5.0	2.0
55 N	−6.0	3.4	−6.1	2.6	−3.0	1.6	−3.7	1.5	−4.8	1.9
45 N	−6.4	2.9	−5.7	2.3	−2.8	1.5	−3.1	1.5	−4.6	1.8
35 N	−4.9	2.4	−4.5	2.5	−3.1	1.5	−2.9	1.5	−3.9	1.8
25 N	−3.2	2.0	−2.7	2.5	−2.7	1.6	−3.0	1.3	−2.9	1.6
15 N	−2.0	1.6	−1.8	1.8	−2.0	1.9	−2.6	1.4	−2.1	1.4
5 N	−1.3	1.8	−1.7	2.2	−1.8	1.5	−1.6	1.8	−1.6	1.6
5 S	−1.5	1.3	−1.8	1.7	−2.5	1.3	−2.1	1.5	−2.0	1.2
15 S	−0.7	1.1	−0.3	1.1	−1.5	1.8	−1.0	1.5	−0.9	1.1
25 S	−3.1	0.9	−2.7	1.3	−3.6	2.6	−2.7	1.8	−3.0	1.4
35 S	−4.4	1.0	−5.3	1.8	−6.5	2.6	−3.9	2.0	−5.0	1.5
45 S	−4.4	1.4	−5.0	1.7	−6.6	2.5	−3.5	2.2	−4.9	1.5
55 S	−4.6	1.6	−6.3	1.9	−10.7	3.0	−6.3	3.3	−7.0	2.1
65 S	−5.8	1.4	−7.6	2.1	−14.3	3.8	−13.6	5.2	−10.4	2.4

Table 2.11. Short-term Dobson trends in % decade using data from January 1979 to February 1994, showing averages of individual trends within latitude zones, with 95%-uncertainty limits (two standard errors, labelled 2se).

Zone	N	Dec–Feb	2se	Mar–May	2se	Jun–Aug	2se	Sep–Nov	2se	Year	2se
50–65 N	7	−6.2	1.5	−6.9	0.5	−4.6	1.4	−2.2	1.1	−5.2	0.5
40–50 N	9	−5.4	1.5	−6.1	0.6	−3.0	0.6	−2.0	0.9	−4.3	0.5
30–40 N	8	−3.9	1.3	−3.5	1.4	−1.6	1.4	−0.9	0.8	−2.5	1.0
20–30 N	5	−1.7	0.7	−1.8	0.2	−0.9	1.7	−0.5	0.9	−1.2	0.2
0–20 N	3	0.5	1.1	−0.0	0.4	−0.7	0.6	−0.8	0.4	−0.3	0.2
30–0 S	5	−1.1	0.7	−1.1	1.4	−2.0	0.7	−1.4	0.5	−1.4	0.6
55–30 S	6	−3.6	1.9	−2.9	1.2	−3.6	1.7	−2.8	1.4	−3.2	1.3

Table 2.12. Long-term Dobson trends in %/decade using data from January 1964 to February 1994 from January 1970, showing averages of individual trends within latitude zones with 95%-uncertainty limits (two standard errors, labelled 2se).

Zone	N	Dec–Feb	2se	Mar–May	2se	Jun–Aug	2se	Sep–Nov	2se	Year	2se
50–65 N	7	−4.0	1.0	−3.4	0.5	−1.4	0.4	−1.2	0.5	−2.6	0.4
40–50 N	9	−3.7	0.8	−3.6	0.9	−1.8	0.6	−1.3	0.4	−2.7	0.5
30–40 N	8	−2.4	1.0	−1.8	0.7	−0.6	0.5	−0.7	0.5	−1.4	0.6
20–30 N	5	−1.5	0.8	−1.1	0.5	0.0	0.3	−0.4	0.7	−0.7	0.4
0–20 N	3	−0.4	0.8	−0.0	0.4	−0.8	0.7	−0.7	0.9	−0.3	0.3
30–0 S	5	−1.2	0.7	−1.2	1.5	−1.7	0.3	−1.4	0.7	−1.4	0.7
55–30 S	6	−1.8	1.0	−1.9	0.8	−2.5	0.6	−1.6	0.8	−2.0	0.7

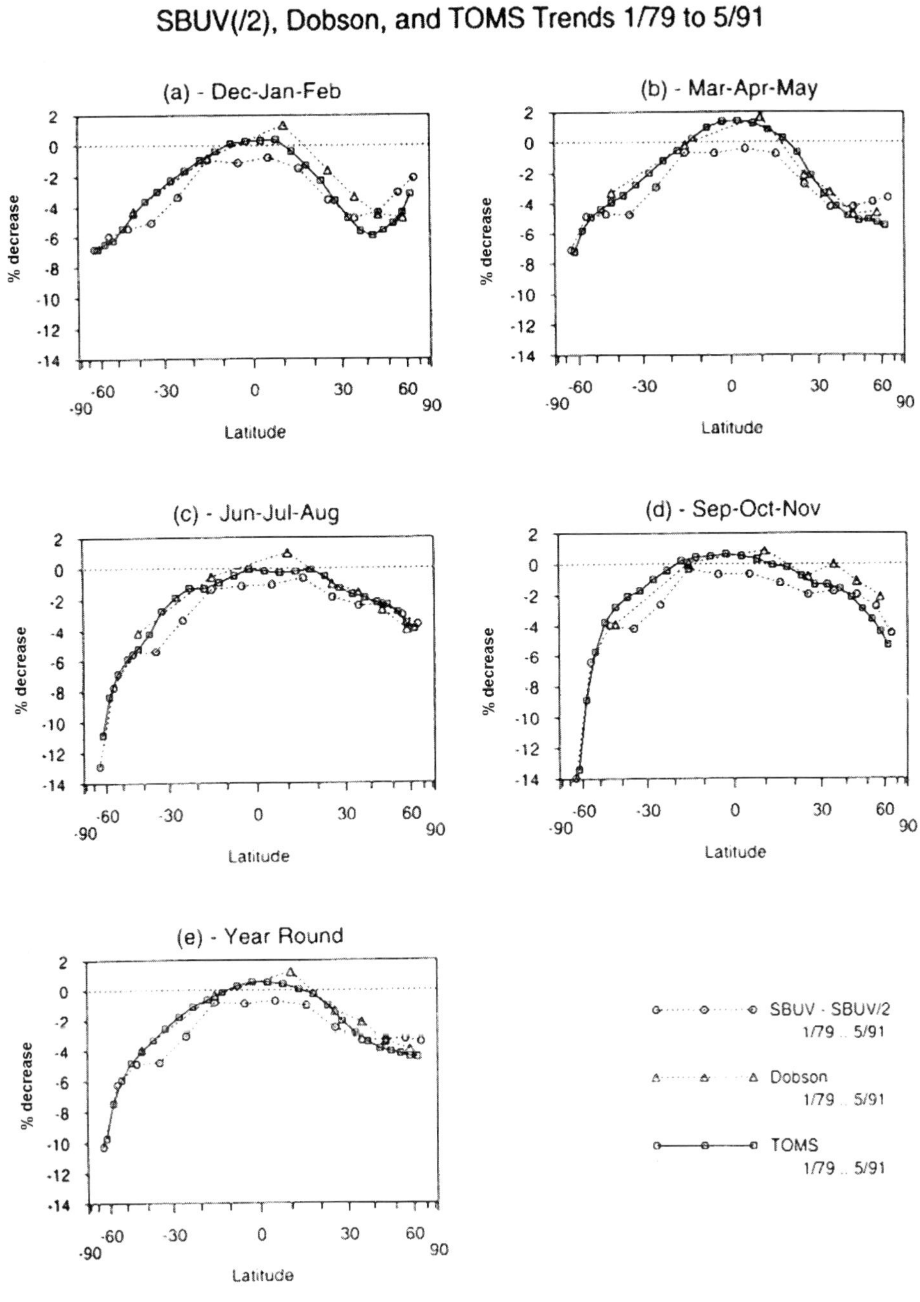

Figure 2.29. SBUV/2, Dobson and TOMS seasonal total ozone trends in %/decade by latitude up to May 1991. Open circles are SBUV/2 trends between January 1979 and May 1991; open triangles are Dobson trends from January 1979 to May 1991; open squares are TOMS trends from January 1979 to May 1991. The Dobson trends are averages within latitudinal zones of individual trends at 59 Dobson stations.

Table 2.13. Representative TOZ trends.

		Latitude		
		Mid-south	Equatorial	Mid-north
Recent				
January 1979–May 1994	SBUV + SBUV/2	−4.9 ± 1.5	−1.8 ± 1.4	−4.6 ± 1.8
January 1979–February 1994	Dobson network	−3.2 ± 1.3	−1.1 ± 0.6	−4.8 ± 0.8
January 1979–February 1994	Ozonometer (former USSR)	n.a.	n.a.	−4.9 ± 0.8
Pre Pinatubo				
January 1979–May 1991	SBUV + SBUV/2	−4.9 ± 2.3	−0.8 ± 2.1	−3.3 ± 2.4
January 1979–May 1991	TOMS	−4.5 ± 2.1	+0.4 ± 2.1	−4.0 ± 2.1
January 1979–May 1991	Dobson network	−3.8 ± 1.3	+0.2 ± 2.1	−3.9 ± 0.7
January 1979–May 1991	Ozonometer (former USSR)	n.a.	n.a.	−3.8 ± 1.0

The corresponding ozone loss (in %) accumulated over 15.3 years for trends calculated through 1994, are:

	Latitude		
	Mid-south	Equatorial	Mid-north
SBUV + SBUV/2	−7.4 ± 2.3	−2.7 ± 2.2	−7.0 ± 2.7
Dobson network	−4.8 ± 2.1	−1.7 ± 0.9	−7.3 ± 1.3
Ozonometer (former USSR)	n.a.	n.a.	−7.5 ± 1.3

- Southern mid-latitude trends are significantly negative in all seasons, and increase in magnitude for high latitudes.

Similar more recent data can be found in *Scientific* ... (1999).

2.3.2 Ozone-vertical profile variability

Reliable information on ozone-vertical profile (OVP) variability in the stratosphere is very important for solving a number of problems, such as ozone impact on climate change and ozone depletion due to anthropogenic CFC emissions. In this context, Perluski and London (1989) discussed satellite-observed long-term averaged seasonal and spatial ozone variation in the stratosphere. They emphasized the presence of various kinds of temporal changes: annual, semi-annual, quasi-biennial, daily and solar induced variability as well as long-term trends. Analysis of TOMS data between October 1978 and September 1987 made it possible to identify basic features of ozone annual and semi-annual variation in the lower, mid and upper stratosphere for the latitude interval 65°S–65°N.

Bojkov (1999) pointed out that up to 1996 in northern mid-latitudes, combined OVP trends (SAGE I/II, Umkehr, ozonesondes) were negative at all altitudes between 12 and 50 km and were statistically significant at the 2σ level. The downward trend was largest near 40 km and 15 km ($\sim$7.5% per decade) and smallest at $\sim$30 km (2% per decade). The peak ozone loss was observed in the 15–20 km layer. The largest trends were in winter/spring.

Annual ozone variation was most clearly pronounced in the subpolar stratosphere of both hemispheres at 1.5, 8.0 and 40 hPa, where contribution of annual components to total ozone variability exceeds 80%. Minimum amplitude of annual variation takes place in the mid and upper stratosphere near 10°N. Maximum amplitude of averaged semi-annual variations was observed near 2–3 hPa in subpolar latitudes, decreasing towards mid-latitudes. Semi-annual ozone variation in the tropical middle stratosphere are comparable with or even stronger than annual variations. A negative trend of the annual variation amplitude ($\sim$0.1 ppm/yr) was observed in the upper stratosphere of southern subpolar latitudes. Interannual changes in amplitudes of both annual and semi-annual variation are small as a rule (except the tropical mid-stratosphere, where the influence of the El Chichón eruption was substantial, and the subpolar upper stratosphere in the Southern Hemisphere).

Assessments of relative contributions of photochemical and dynamical processes to ozone concentration field formation confirmed earlier conclusions. Ozone distribution in the lower stratosphere (below 30 hPa) is controlled (especially in winter) mainly by large-scale atmospheric circulation, while in the upper stratosphere (above 5–10 hPa) radiative and photochemical processes dominate (except subpolar latitudes in winter when atmospheric transport is also important). In the tropics this zone is located at a lower level because of the shorter photochemical lifetime in the equatorial belt under conditions of high insolation.

London and Liu (1992) used ozonesonde observations to analyse long-term tropospheric and lower stratospheric ozone variation over 20 years for the four seasons. Results indicated the presence of minimum ozone-mixing ratios in the equatorial and tropical troposphere except in regions where net photochemical production is dominant. In the mid and upper troposphere, and lower stratosphere to 50 hPa, ozone increases from the subtropics to subpolar latitudes in both hemispheres. In mid-stratosphere, the ozone-mixing ratio is a maximum over the tropics.

London and Liu (1992) pointed out that the observed ozone vertical gradient is small in the troposphere but increases rapidly in the stratosphere. The seasonal variation at a typical mid-latitude station (Höhenpeissenberg) showed a summer maximum in the low to middle troposphere, shifting to a winter–spring maximum in the upper troposphere and lower stratosphere and spring–summer maximum at 10 hPa. The amplitude of the annual variation increased from a minimum in the tropics to a maximum in polar regions. The amplitude increased with height at all latitudes up to about 30 hPa where the phase of the annual variation changes abruptly. The annual long-term ozone trends were significantly positive at about +1.2% per year in the mid-troposphere (500 hPa) and significantly negative at about −0.6% yr^{-1} in the lower stratosphere (50 hPa). The results demonstrated the importance of ozonesonde observations and illustrated the need for combined analysis of both ozonesonde and satellite OVP data.

Spänkuch and Schulz (1998) substantiated a multiple regression approach to assess the amount of layer ozone ΔO_3 from predictors accessible from model outputs of meteorological analysis, in addition to TOZ (which is the most important predictor). This approach was verified using Höhenpeissenberg data.

Furrer *et al.* (1992) discussed the results of long-term Dobson and balloon ozone measurements in the vicinity of Berlin. The results of observations over 24 years (1964–1988) indicated an increase in ozone concentration within the mid-stratosphere (below 31 km) as well as in the troposphere but, on the other hand, ongoing ozone depletion in the lower stratosphere became evident. The large-scale vertical redistribution of atmospheric ozone in the troposphere and in the lower stratosphere is in agreement with model calculations and trend predictions, because of change in the chemical composition and the ozone photochemistry induced by anthropogenic emissions of ozone-depleting substances. Claude *et al.* (*Atmospheric Ozone*, 1998) analysed TOZ and OVP trends obtained by Dobson and ozonesonde observations in Höhenpeissenberg (both since 1967). The trend for lower stratospheric ozone (12–17 km) was −5% per decade obtained from OVP data for the Jungfraujoch site in the Swiss Alps. De Maziére *et al.* (*Atmospheric Ozone*, 1998) substantiated a climatological ozone/vertical-ozone distribution model.

SAGE I and II satellite data are important sources of information concerning ozone/vertical ozone profiles. McCormick *et al.* (1989, 1992) discussed SAGE data with the aim of revealing basic features of annual, semi-annual and quasi-biennial variation as well as long-term trends that depend on altitude and latitude. The results indicated, for instance, the existence of an ozone-depletion trend (~ few percent) in mid-latitudes near 40-km altitude between 1980 and 1986, whereas ozone changes above and below this level were much smaller. During the whole time period considered, ozone depletion at the height of 25 km was about 3%. Total ozone content above 17 km averaged for the latitude interval 65°S–65°N was characterized by a linear trend of about $-0.30 \pm 0.19\%$ per year (or −3.6% for the whole period of observations from February 1979 through April 1991). Statistically significant ozone trends were concentrated near 23-km altitude (negative trends at all latitudes) and in the 38–50-km layer (10°N–35°S), where a negative ozone trend was observed. Column ozone losses occurred below 25 km, with most of the loss coming from the region between 17 and 20 km. McCormick *et al.* (1992) emphasized that the concentration of mid-latitude ozone depletion in the lower stratosphere (17–20 km) imposes certain constraints on the choice of potential mechanisms that were responsible for this depletion. Fussen (1998) assessed biases of long-term ozone trends from SAGE-II data because of the impact of volcanic aerosol. An important improvement in the SAGE-II inversion algorithm was suggested by Rusch *et al.* (1998).

Simultaneous observations of O_3, H_2O and aerosols made it possible to analyse interrelationships between mixing ratios of ozone and aerosol within the circumpolar vortex in Antarctica during winter and spring (when the ozone hole is formed) and also the synchronous variability of ozone and water vapour (see Chapter 3).

Very important progress in studying both ozone dynamics near the tropopause and stratosphere–troposphere ozone exchange resulted from the work of the Measurement of Ozone and Water Vapour by Airbus In-Service Aircraft (MOZAIC) programme over a 2-yr period, from September 1994 till August 1996 (Thouret *et al.*, 1998a, b). This programme was the continuation of a number of dedicated aircraft campaigns: Global Atmospheric Measurements Experiment on Tropospheric Aerosols and Gases (GAMETAG), Stratospheric Ozone Experiment (STRATOZ), Tropospheric Ozone Experiment (TROPOZ), Pacific Exploratory Mission-West (PEM-West), Transport and Atmosphere Chemistry Near the Equatorial Atlantic/Southern Africa Fine-Atmosphere Research Initiative (TRACE A/SAFARI-92), Troposphrische Ozon (TROZ) project, Global Air Sampling Programme (GASP).

From data collected at cruise level by five Airbus A340s using automatic ozone-measuring devices (dual-beam UV absorption instruments) an accurate ozone climatology at 9–12-km altitude was generated over the Northern Hemisphere (0°–80°N, 130°W–140°E) and down to 30°S over South America and Africa. Between September 1994 and December 1997, 7500 flights were made over continents and the Atlantic Ocean. Measurement accuracy was estimated at $\pm(2\,\text{ppbv} + 2\%)$, but much better in-flight levels were in fact observed: average discrepancy (between different devices) ranging from 1 ppbv at tropospheric concentrations to a few ppbv at stratospheric concentrations. MOZAIC data for the whole troposphere and lower stratosphere (0–12 km) were compared with groundbased observations made at eight stations of the Ozone Sounding Network (OSN): Höhenpeissenberg, Wallops Island, Tateno, Palestine, Pretoria, Goose Bay, Biscarosse and Poona. This comparison demonstrated a reasonably high level of agreement, despite the different nature of the programmes (techniques, platforms, sampling frequencies, spatial distribution and operation periods). Mean concentrations derived from ozonesondes are about 3 to 13% higher than those obtained by the MOZAIC programme in the free troposphere, in a similar geographical location (Thouret *et al.*, 1998b). Results of the intercomparisons supported the assumption that it is not advantageous to scale the ozonesonde data to the overhead ozone column; the scaling appears to cause overestimation of tropospheric O_3 concentration by about 3–6% at Höhenpeissenberg, and to cause more scatter in the sonde–MOZAIC differences. The correspondence between OSN and MOZAIC ozone climatology confirms their usefulness for validation studies.

Thouret *et al.* (1998a) pointed out that there are still large uncertainties in the budget of atmospheric ozone and especially with regard to the relative importance of photochemical formation and exchanges between the stratosphere and troposphere, and in determining future ozone trends. In this context certain regular features of ozone spatial distribution were analysed. Thouret *et al.* (1998a) reached the conclusion that north of 35°N the distribution at 9–12-km altitude is dominated by the influence of ozone-rich air of stratospheric origin; farther south, ozone-poor air from the troposphere prevails. To identify the stratospheric and tropospheric components and to help in interpreting the data a classification based on a threshold of 100 ppbv of ozone was used.

Distinct seasonal ozone variability was observed in stratospheric air (maximum in spring, minimum in winter) and in tropospheric air (maximum in summer, minimum in autumn), related to the intensities of dynamic (tropopause variation; stratosphere/troposphere exchange) and chemical (photochemistry) processes. The tropics exhibit variation that are heterogeneous in time and space, and reflect the influences of active photochemical processes, deep convection and biomass burning emission. Ozone concentrations decrease with latitude in both the stratosphere and troposphere. As far as ozone-vertical profile is concerned, it is more strongly expressed in the stratosphere (with a strong vertical gradient) than in the troposphere, where ozone distribution is much more homogeneous. Smaller ozone concentrations over the Atlantic Ocean than over the continents, because of the zonal variation of the polar front and the position of the ridge/trough pressure systems, are important features of geographical distribution.

A substantial contribution to studies of ozone vertical profiles has been made by various groundbased and aircraft remote-sensing observations with a special role played by lidar soundings. For instance, the airborne UV differential absorption lidar (DIAL) system participated in the Tropical Ozone Transport Experiment/Vortex Ozone Transport Experiment (TOTE/VOTE) in late 1995/early 1996 (Grant *et al.*, 1998). This system allowed retrieval of the ozone-vertical profile above the aircraft from approximately 2 km up to about 16 km with vertical resolution of about 1.3 km (the horizontal resolution was 70 km). An important purpose of the ozone soundings were intercomparisons with results from a groundbased DIAL as well as from a number of satellite stratospheric instrumentation: HALOE, MLS and SAGE-II (Section 2.2.3). Grant *et al.* (1998) pointed out that ozone profiles generally agreed within random-error estimates for various instruments in the middle of the profiles in the tropics, but regions of significant systematic differences, especially near or below the tropopause or at the higher altitudes, were also found. The comparisons strongly suggested that the airborne UV DIAL system can play a valuable role as a mobile lower-stratospheric validation instrument.

Solomonov *et al.* (1992) applied microwave remote-sensing techniques for soundings of the ozone layer using 142.175 GHz O_3 line ($10_{0.10}$–$10_{1.9}$ rotational transition) located in the 2-mm transparency window. Observations were conducted near Moscow between January 1987 and April 1990. The results indicated the presence of ozone enhancement above 27 km because of the impact of stratospheric warming. Large-scale atmospheric circulation was observed to have a very substantial influence on ozone change.

Kawabata *et al.* (1997) conducted groundbased radiometry observations of strato-mesospheric ozone employing a superconductive NbSIS mixer receiver cooled by a closed-cycle 4 K refrigerator. The $J = 6_{1.5}$–$6_{0.6}$ (110.836 GHz) O_3 line was measured. Sporadic observations have been carried out on the Higashiyama Campus of Nagoya University (35°N, 137°E, 70 m a.s.l.) since January 1992. The results obtained indicate that amplitudes of seasonal variations are small at 20, 3.2, and 0.056 hPa levels and the phase or type of seasonal variation appears to change at these altitudes. Well-known seasonal variations were observed below 20 hPa. The summer O_3 maximum, caused by photochemical production, can be clearly seen at 5.6 and 10 hPa levels. The earlier-discovered temperature-dependent ozone-distribution cycle with winter maximum (Froidevaux *et al.*, 1994) was observed in the altitude range between 1.8 and 0.1 hPa. There was a similarity between observed seasonal variation at the 1.8 hPa level and UARS MLS data at 2.2 hPa in mid-latitudes (32–64°). Strong enhancement of the ozone mixing ratio between 1.0 and 0.1 hPa was observed in the winters of 1992/1993 and 1994/1995 but not in the 1993/1994 winter. It can be assumed that seasonal variation in the layer between 1.0 and 0.1 hPa are of the QBO type, but observations were not long enough for a reliable assessment.

Sandor and Clancy (1998) carried out groundbased microwave emission measurement of HO_2 (at 254.55153 and 268.56586 GHz), O_3 (249.7886 GHz) and $H_2^{18}O$ (203.40752 GHz) at the National Radio Astronomy Observatory (Kitt Peak, Arizona: 32°N, 112°W). Microwave ($\lambda \sim 1.2$ mm) emission of the upper stratosphere and mesosphere (45–80 km) was measured in April 1992, January 1993 and April and December 1996 with the purpose of studying HO_2, OH and H rapid interconversion during HO_x catalytic ozone-destruction cycles. Comparison between co-ordinated diurnal (09:00–18:00 local time) measurements of mesospheric O_3 with model calculations revealed a model unpredictability of ozone, which can be resolved by a 40% reduction in the rate coefficient of the reaction $H_2O + O_3 \rightarrow 2OH + O_2$.

Substantial contributions to ozone-vertical-profile studies were presented at the International Ozone Symposium in 1996 (*Atmospheric Ozone*, 1998).

Grainger and Atkinson (*Atmospheric Ozone*, 1998) carried out the first comprehensive global three-dimensional 6-hourly analyses of ozone mixing-ratio data from satellite observations (MLS HALOE, SAGE II, SBUV/2, TOVS), which were interpolated onto a 2.5° grid with 19 levels between 1,000 and 0.1 hPa (data for October 1994 were considered). There are still, however, some problems with the homogenization of observational data from various platforms.

Chan *et al.* (*Atmospheric Ozone*, 1998) discussed OVP observations (ozonesonde data) in Hong Kong, which demonstrates the complicated structure of ozone-vertical profiles (basically bimodal with a laminated structure when observed in the lower stratosphere in winter and spring).

Cunnold and Wang (*Atmospheric Ozone*, 1998) investigated the effects of temperature uncertainty on interpretation of ozone trends in the upper stratosphere and emphasized that estimates of long-term trends in SAGE ozone measurements and in solar-cycle response of ozone can be different dependent on whether they are based on ozone concentrations at altitude levels or ozone mixing ratios on pressure surfaces.

De Luisi *et al.* (*Atmospheric Ozone*, 1998) discussed aerosol corrections to retrieved Umkehr ozone profiles. A new Brewer Umkehr retrieval algorithm was substantiated by Elansky *et al.* (1998). A number of other papers were also devoted to improving Umkehr retrieval algorithms.

Fujimoto *et al.* (*Atmospheric Ozone*, 1998) conducted an intercomparison of lidar, ozonesonde and SAGE-II data using lidar and ozonesonde observations made in Tsukuba (36°N, 140°E). The results indicate that SAGE-II and lidar data agree well enough (within 10% in the stratosphere) as also do ozonesonde and lidar data (below 32 km).

Mich and Kalenka (*Atmospheric Ozone*, 1998) studied fine structure (lamination) of OVP using ozonesonde data. Validation and results of 9 years of lidar (DIAL) sounding at Höhenpeissenberg were considered by Steinbrecht *et al.* (*Atmospheric Ozone*, 1998). Since 1987 lidar, in agreement with SAGE, has shown significant ozone depletion by 0.5 to 1% per year between 35 and 40 km. Stübi *et al.* (*Atmospheric Ozone*, 1998) discussed homogenization of the Payerne Brewer/Mast ozone sounding series, which started in 1966.

The most representative statistical information on ozone-vertical profiles was obtained on the basis of satellite and ozonesonde observations. These results were considered in detail in the UNEP Scientific Assessment (*UNEP* ..., 1994) and *Scientific* ... (1999). Only some of the results are mentioned here.

Figure 2.30 from *UNEP* ... (1994) presents OVP trends derived from SAGE/II data for the time period 1979–1991. Figure 2.31 illustrates an intercomparison of

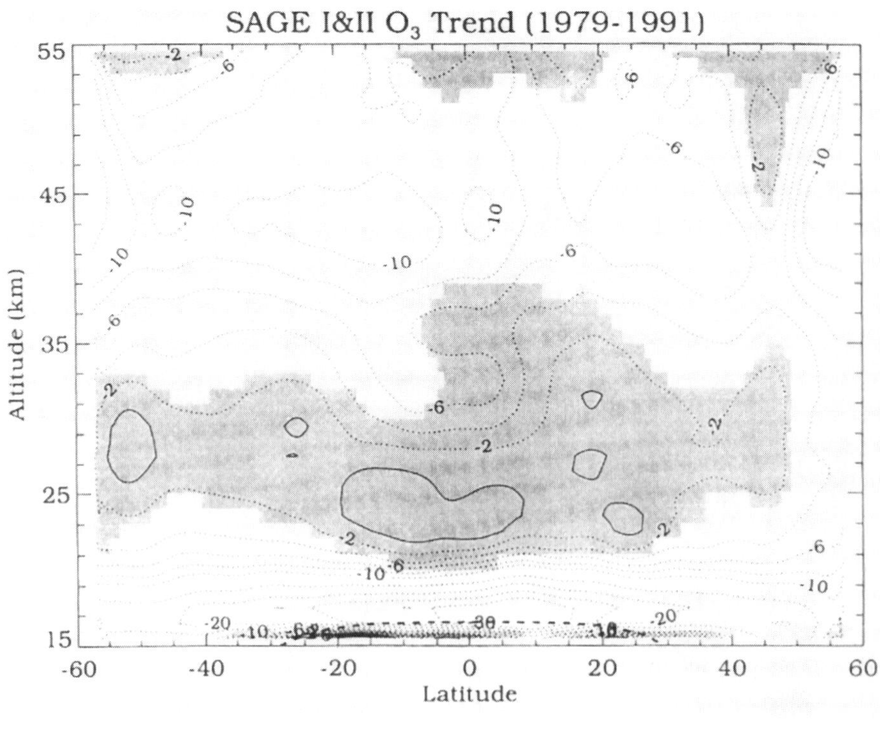

Figure 2.30. Trends calculated for SAGE I/II for 1979–91. Hatched areas indicate trends calculated to be insignificant at the 95% confidence level. The dashed line indicates the tropopause. The altitudes of SAGE-I measurements have been adjusted by 300 metres.

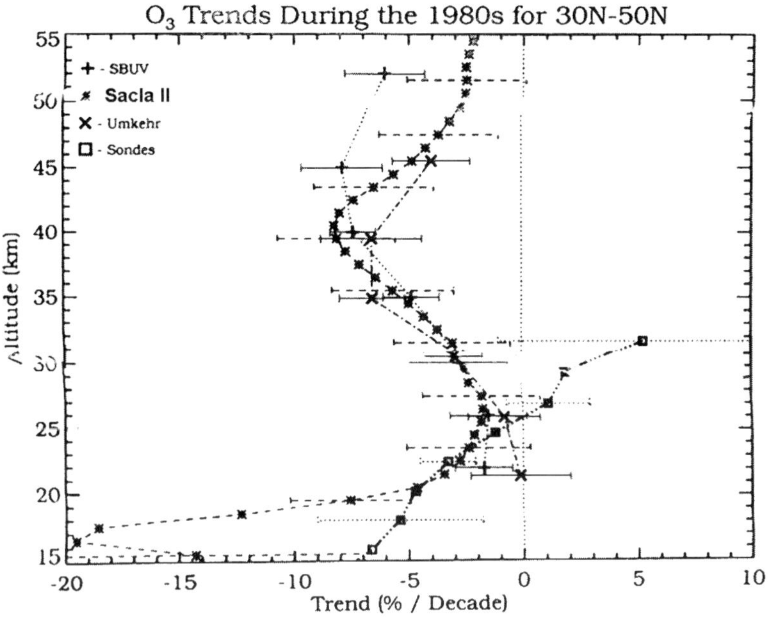

Figure 2.31. Comparison of trends in the vertical distribution of ozone during the 1980s. Ozonesonde and Umkehr trends are from Miller *et al.* (1994). 95% confidence limits are shown.

SBUV, SAGE I/II, Umkehr and ozonesondes. These data are in reasonable agreement and demonstrate that there is no significant ozone depletion in the 25–30-km layer. The agreement continues down to about 20 km, where statistically significant reductions of $7 \pm 4\%$ per decade were observed between 30 and 50°N by both ozonesonde and SAGE I/II. According to SAGE data there was a decrease of more than 20% per decade in the equatorial region just above the tropopause between about 30°N and 30°S. In absolute terms this loss is small because of a rather low ozone concentration at these altitudes. The height of the maximum ozone decrease is about 16 km and the region of decrease becomes broader away from the equator. At northern mid-latitudes (Figure 2.31) the SAGE trend at 16–17 km is $-20 \pm 8\%$ per decade, compared with an average trend from the ozonesondes of $-7 \pm 3\%$ per decade. Over the longer period (1968–1991) the ozonesonde record indicates a trend of $-4 \pm 2\%$ per decade at 20 km at northern mid-latitudes. At 35–45 km, ozone declined 5–10% per decade at 30–50°N and slightly more at southern mid-latitudes.

Harris and Hudson (1998) summarized updated results concerning ozone trends obtained from four sources of information: SAGE I/II, SBUV, SBUV/2, Umkehr and ozonesonde observations (Figure 2.32). SAGE data cover the time period from February 1979 through July 1996 with a 3-yr interval of missing data starting from November 1981. SBUV data are from 1978 up to 1996. Umkehr observational data started in 1957 and ozonesonde data from the beginning of 1960s. For comparison

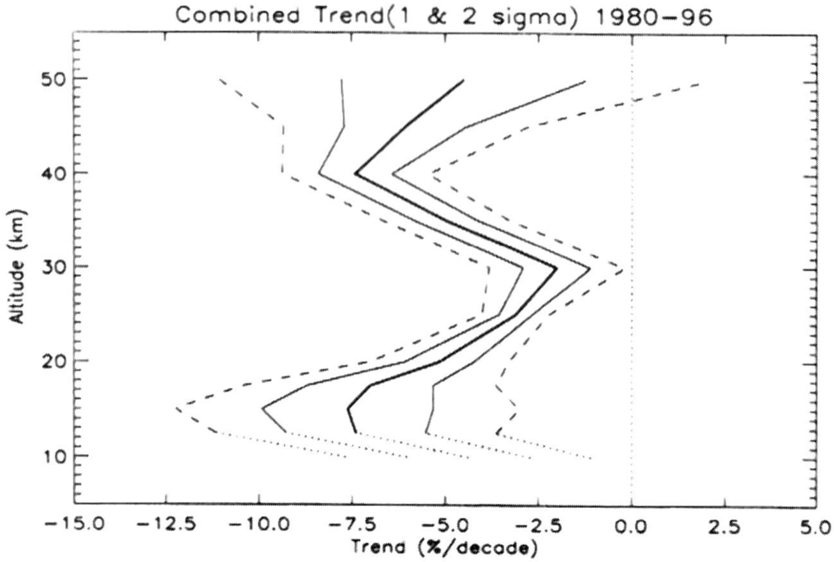

Figure 2.32. Estimate of the mean trend in the vertical distribution of ozone that occurred over northern mid-latitudes from 1980 to 1996 (heavy solid line) calculated using trends derived from SAGE I/II, ozonesondes, SBUV and Umkehr measurements. Combined uncertainties are shown as 1σ (light solid lines) and 2σ (dashed lines). Combined trends and uncertainties are extended down to 10 km as shown by the light dotted lines. Results below 15 km are a mixture of stratospheric and tropospheric trends and exact numbers should be viewed with caution. Combined trends have not been extended lower into the troposphere because there are concerns regarding the representativeness of any mean trends derived from the small sample of sonde stations.

and validation purposes, other sources of O_3 information were also used (UARS data, lidar and microwave soundings). Serious attention was paid to methodological aspects (data quality control) with special emphasis on aerosol correction (SAGE-II data for altitudes below 10 hPa after the Pinatubo eruption were excluded). It was shown that potential sensitivity drift for all four instrumentation types did not exceed 5% per 10 years (within two standard errors). Accuracy of global average SAGE-I/II trend data for the 20–40-km layer is 0.2% per year (at the 95% significance level), while for SAGE-I data for the lower stratosphere it is about 0.25% per year.

Analysis of trends revealed the presence of statistically significant trends for all altitudes within the 12–50 km layer with two distinct maxima near 15 km and 40 km. According to new observation data an ozone-concentration decrease from −6 to −8% per decade takes place at 40–45 km but is characterized by substantial annual variation with the winter maximum (noticeable hemispheric differences are absent). In the lower stratosphere (at altitudes below 30 km), the principal information comes from SAGE (20–30 km) and ozonesonde (up to 27 km) observations. At 20 km the ozone concentration trend varies (due to ozonesonde data for different points) from −3 to −11% per decade. Annual variations (which are absent above 20 km) are noticeable mainly in the 10–20-km layer, but are characterized by substantial geographic variability. According to data from European stations, a maximum ozone drop is observed in winter/spring, whereas Canadian data revealed the maximum in spring/summer time. There is agreement between satellite and ozonesonde observations with regard to annual variation in the 10–20 km layer in mid-latitudes with a distinct summer/spring maximum in the Northern Hemisphere and much weaker variation in the Southern Hemisphere.

It was emphasized in the SPARC Report (Harris *et al.*, 1998) that it is difficult to make generalizations concerning trends in tropospheric ozone. Some relevant results will be discussed in Chapter 3.

2.3.3 General features of ozone global distribution

Since basic features of TOZ global distribution were discussed in Chapter 1, we shall dwell only upon a number of recent results. Note, however, that the first fundamental attempt to substantiate global empirical models of total ozone and ozone-vertical profiles (for the 10–32-km layer) was made by Bekoryukov (1991), who used for this purpose all available (at the beginning of the 1990s) groundbased observational data.

A new stage in the study of global ozone characteristics was marked by the publication of Bojkov and Fioletov (1995), who used all available TOZ data from over 150 past and present Global Ozone Observing System (GO_3OS) stations (after careful quality re-evaluation) to analyse observed global ozone dynamics. The importance of groundbased observations is due to their long time series in comparison with all other data. Bojkov and Fioletov (1995) had the principal aim of deducing basic ozone characteristics both for pre-ozone hole and during 'ozone hole' time periods. TOMS data facilitated estimation of the longitudinal inhomogeneity of the TOZ distribution, which made it possible to use groundbased data for establishing long-term zonal as well as hemispheric and global ozone variation from 1964 to 1994. The difference between estimations of monthly zonal variation from groundbased and TOMS data for the overlapping period of 1979-1993 was less than 1% in latitudes 40°S–60°N. Ozone changes are several times larger than possible errors of the estimated values; therefore, the results are highly reliable.

The results obtained by Bojkov and Fioletov (1995) indicated that the Northern Hemisphere average TOZ value was ~312 and the southern value was ~300 DU in the pre-ozone-hole decades (1964–1980), and that the global average from 1984 to 1993 was lower by ~3% (from 306 ± 1.0 down to 297.7 ± 2.2 DU). The Southern Hemisphere contributed ~64% of the overall ozone decline. Levels of annual ozone maximum reduced by 5.8% in the Southern Hemisphere and 3.2% in the Northern Hemisphere, and the levels of ozone minimum by 2.1% and 1.2%, respectively.

As far as temporal global mean TOZ variation is concerned, three major components of natural ozone variation (annual and solar cycles and QBO) can be identified. There are two major maxima in the temporal TOZ spectrum: 26–28 months (QBO) and 100–130 months (solar cycle). A rather small ENSO signal (about 40 months) has also been identified. A major component of global ozone variation is its annual cycle. The annual TOZ course is antisymmetric between the two hemispheres, with large differences both in absolute values and time of appearance of the maximum in polar regions.

Figure 2.33 data from Bojkov and Fioletov (1995) illustrate latitude–time cross-sections of TOZ deviations from 1964–1980 'norms' (TOMS data) in percent (the calculations were made for 5° latitude zones). These data clearly confirm (middle and bottom part of Figure 2.31 relevant to overlapping groundbased and TOMS data) the agreement between ground-based and satellite observations. Clearly seen is the appearance of QBO-related change, the very dramatic decline in the southern polar region, well-known TOZ anomalies in 1983, 1985 and 1987, a very strong decline in the 1992–1993 northern winter, etc.

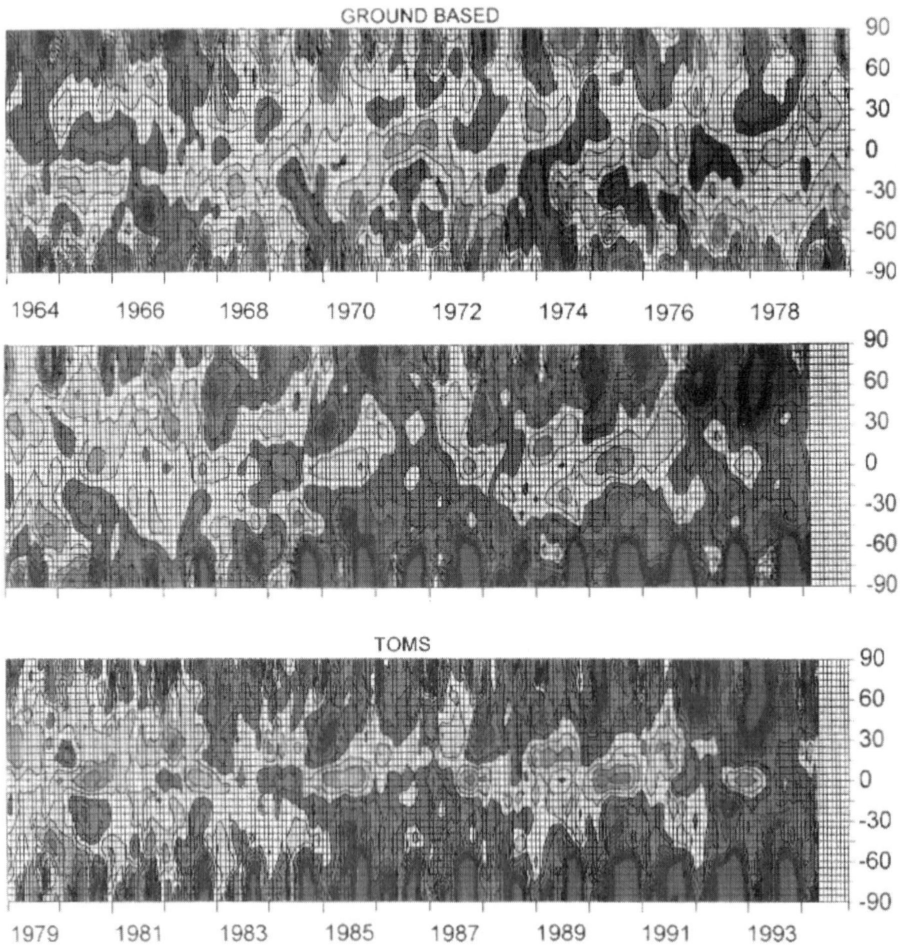

Figure 2.33. Latitude–time cross-sections of total ozone deviations from 1964–1980 'norms' in percent: (top) and (middle) groundbased data; (bottom) TOMS. Deviations below −15% are plotted by the same colour as for the interval −13 to −15%.

Analysis of long-term TOZ variation for Northern and Southern Hemispheres indicated that a weak decline which started in the 1970s became a major change in hemispheric ozone during the 1980s. This decline was interrupted in the NH between 1981 and 1982 (solar maximum years). The overall decline in the SH was so strong between 1984 and 1985 that annual TOZ maximum levels were almost always lower than annual minimum levels of the pre-1980 period.

Mean latitudinal–time cross-sections are shown on Figure 2.34 for two time intervals. Comparison of the data for these time periods demonstrate very substantial change in the southern polar region. Bojkov and Fioletov (1995) pointed out

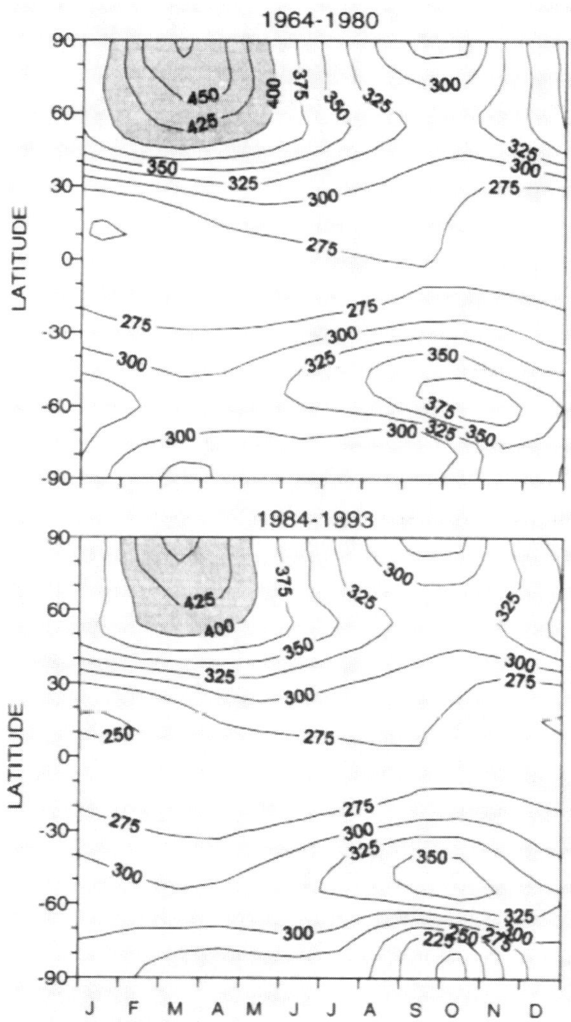

Figure 2.34. Variation in total ozone with latitude and season (in milliatmosphere centimeters) for two time intervals: (top) 1964–1980 and (bottom) 1984–1993.

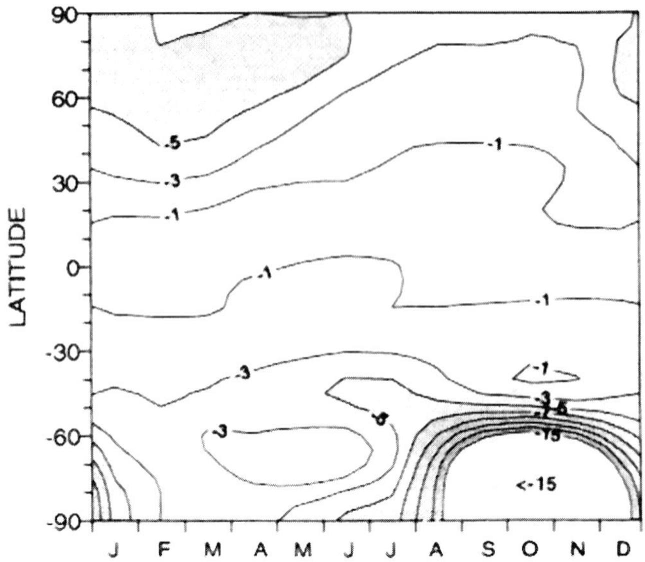

Figure 2.35. Difference (in %) between total ozone values for the two periods (1964–1980 and 1984–1993) reported in Figure 2.41. Southward of 60°S from September to November the difference is more than 15% with a maximum of 35% poleward of 80° in October.

that hemispheric differences in TOZ distribution are more pronounced in months of increasing ozone, and in the value of the month that maximum is achieved, than during the ozone-declining phase. Figure 2.35 illustrates the differences between the two periods in detail. The difference is marked in winter–spring over the northern polar and upper mid-latitudes, reaching more than 6%. The difference is lower in spring and fall. Southward of 60°S from September to November the difference is more than 15% with a maximum of 35% poleward of 80° in October. Table 2.14 shows long-term TOZ trends, which are characterised by a pronounced seasonal dependence varying by ~30 to 8% (and even more for the SH) on the cumulative decline since 1970 for mid-latitudinal bands (35–65°). The ozone decline during the 'ozone hole' period is almost twice as large in comparison with 1970. The cumulative year-round global TOZ decline is $4.8 \pm 0.6\%$, but over mid and polar latitudes it is more than 7%.

A principal conclusion reached by Bojkov and Fioletov (1995) was that natural variation, as known from the entire observing period, cannot explain the observed ozone decline, indicating the likelihood of anthropogenic influence. Another important conclusion was that the use of complimentary ground and spacebased ozone-observing facilities had great information value from the viewpoint of the analysis of global ozone variability.

A new very important step forward in understanding basic features of ozone vertical profiles and their trends was taken by scientific assessments carried out jointly by the WCRP Project on Stratospheric Processes and Their Role in Climate

Table 2.14. Long-term total ozone trends derived from groundbased data.

Region	DJFM	MJJA	SON	Year-round
Slope from 1970 to 1994				
35°N–65°N	-3.2 ± 1.0	-1.1 ± 0.9	-1.1 ± 0.6	-1.9 ± 0.8
10°N–35°N	-1.3 ± 0.8	-0.3 ± 0.7	-0.5 ± 0.7	-0.7 ± 0.7
10°N–10°S	-0.2 ± 0.9	-0.5 ± 0.9	-0.2 ± 0.9	-0.2 ± 0.7
10°S–35°S	-1.2 ± 0.5	-1.1 ± 0.7	-1.4 ± 0.7	-1.2 ± 0.4
35°S–65°S	-1.6 ± 0.8	-2.6 ± 0.8	-3.5 ± 1.3	-2.4 ± 0.7
35°N–90°N	-3.4 ± 1.1	-1.3 ± 1.0	-1.0 ± 0.7	-2.0 ± 0.8
35°S–90°S	-1.8 ± 1.0	-2.7 ± 0.8	-6.0 ± 1.6	-3.2 ± 0.8
Northern Hemisphere	-2.2 ± 0.7	-0.8 ± 0.7	-0.6 ± 0.5	-1.3 ± 0.6
Southern Hemisphere	-1.3 ± 0.7	-1.8 ± 0.5	-3.5 ± 0.8	-2.0 ± 0.5
Global	-1.8 ± 0.5	-1.3 ± 0.4	-2.1 ± 0.5	-1.7 ± 0.4
Slope from 1980 to 1994				
35°N–65°N	-5.8 ± 1.7	-2.6 ± 1.5	-2.5 ± 1.0	-3.8 ± 1.3
10°N–35°N	-2.3 ± 1.4	-0.6 ± 1.3	-1.1 ± 1.1	-1.3 ± 1.1
10°N–10°S	-0.6 ± 1.5	-1.3 ± 1.5	-0.3 ± 1.5	-0.8 ± 1.3
10°S–35°S	-1.8 ± 1.2	-2.0 ± 1.2	-2.3 ± 1.3	-2.0 ± 0.8
35°S–65°S	-3.6 ± 1.2	-4.9 ± 1.3	-7.3 ± 2.0	-5.0 ± 1.0
35°N–90°N	-6.1 ± 1.8	-3.0 ± 1.5	-2.5 ± 1.1	-4.1 ± 1.2
35°S–90°S	-4.2 ± 1.5	-5.1 ± 1.3	-11.9 ± 2.6	-6.4 ± 1.2
Northern Hemisphere	-4.0 ± 1.1	-1.9 ± 1.1	-1.6 ± 0.9	-2.6 ± 0.9
Southern Hemisphere	-2.7 ± 1.0	-3.4 ± 0.8	-6.6 ± 1.5	-3.9 ± 0.8
Global	-3.5 ± 0.5	-2.7 ± 0.5	-4.3 ± 0.7	-3.4 ± 0.4

Estimates were made for 'natural' ozone seasons (DJFM, December–March; MJJA, May–August; SON, September–November) using data from January 1964 to February 1994, with a linear trend fit over the 1970–1994 and 1980–1994 periods. Values are given in % per decade $\pm 2\sigma$.

(SPARC) and the International Ozone Commission (IOC), in close co-operation with WMO's Global Atmospheric Watch Programme (GAW). The principal aim of SPARC/IOC/GAW 'Assessment of Trends in the Vertical Distribution of Ozone' (Harris *et al.*, 1998) was the critical review of measurements and trends of the ozone vertical profile (OVP). Analysis of all available information resulted, for the first time, in a single profile of observed ozone depletion at northern mid-latitudes.

Many recent studies have been devoted to studies of various aspects of stratospheric ozone dynamics. One of the key aspects is connected with assessment of the role of stratosphere–troposphere exchange in stratospheric ozone dynamics.

2.3.3.1 *Stratosphere–troposphere exchange*

Many recent studies show that there is an increasing trend with background values of ozone in the troposphere (Bojkov, 1986). This is of serious concern because of the damage caused by high ozone concentrations to human beings, animals and plants, and also because of possible climatic effects. Stratospheric ozone leaks into the troposphere forming a natural background. In addition man-made and natural emissions of nitrogen oxides and hydrocarbons lead to the production of ozone in the sunlit troposphere.

In order to assess the anthropogenic component of free atmospheric ozone, the nature of stratosphere–troposphere exchange must be well understood. It was believed until recently that the main transfer of stratospheric ozone occurs in tropopause folding events associated with rapid cyclogenesis. Varotsos and Kondratyev (1995) summarized the level of knowledge in this field and discussed relevant observations made in Athens.

Major tropopause folding events occur at the flanks of large-scale northern flow. A threefold tropopause structure model has been proposed to simulate such events. According to this model the sounding station in Athens is situated at such a latitudinal band (36–38°N) that stratospheric air intrusions should occur by means of tropopause foldings between the polar and the mid-latitude troposphere. This fact makes the Athens region subject to the influence of the polar front and sometimes to the impact of the subtropical jetstream. Influence of the meteorological situation have been revealed by observations of the laminated structure of the vertical profile of ozone over Athens discussed above.

Observations made near a cut-off low in the upper troposphere suggest that a cut-off low may also contribute to the removal of ozone from the stratosphere. But a limited number of studies have been conducted so far, and there still remain many uncertainties about this type of stratosphere–troposphere exchange.

In the following we examine the stratosphere–troposphere ozone exchange over Athens, Greece, during the winter 1991–1992 EASOE campaign, in relation to the presence of a cut-off low in the upper troposphere or a possible tropopause folding, at a time when the station is close to the flank of the upper-level trough. Note that this was the first time such an intensive sounding campaign was performed in Athens. We examined four ozone soundings that correspond to the above meteorological situations. Note also that the nearest ozone sounding station to Athens (Höhenpeissenberg, Germany: 48°N, 11°E) is too far away to obtain information about the situation over the Athens region, which seems to be quite interesting from the atmospheric dynamics point of view, as mentioned above.

2.3.3.1.1 Cross-tropopause exchange during a cut-off low event

A cut-off low system developed over the Athens region on 18 and 20 March 1992. On the first day (18 March) the lowest height for the pressure level of 200 hPa during our campaign was observed. This resulted in a very low tropopause (approaching the 400 hPa level).

We also observed very high ozone concentration values with a maximum of 190 nb and a strong secondary maximum (about 120 nb) in the lower stratosphere (about 5-km deep). The synoptic map showed that the site was influenced by the northern circulation, which brings ozone-rich air to the region. From another point of view, we may say that a cut-off low event is combined with high ozone partial pressure at low levels. Because the stratospheric air was then at higher pressure than normal the ozone partial pressure was similarly increased. So, the remarkable fact is the minimum in the partial pressure near 15 km, which our ozone data clearly shows. The ozone profile on that day did not show any characteristics of detectable cross tropopause exchange.

On 20 March (the day of the next ozone sounding) we observed that the cut-off low system was moving eastwards, and that our sounding site was influenced by polar air. The intensity of the low was weakened and consequently the height of the tropopause increased.

On 18 March, the already observed laminated ozone structure in the region of 100–300 hPa became much more pronounced. This is in agreement with the conclusions made earlier, according to which the laminar structure of ozone profile over Athens is strongly associated with the establishment of the north-western circulation over the region. If we examine the tropospheric temperature profile of both days we find that on 20 March it is higher by 3–5°C. This agrees fairly well with findings of similar studies of cut-off low systems. This might be associated with cooling of the lower stratosphere and lifting of the tropopause, both of which were in fact observed on that day. The most important feature observed on 20 March was a significant tropopause peak of ozone (about 10 nb) at 600-hPa level with a sharp decrease in relative humidity suggesting that this air might originate from the stratosphere. One can also see, on the same day, that the ozone partial pressure increases sharply before arriving at the tropopause, which demonstrated another example of stratospheric ozone intrusion into the troposphere.

2.3.3.1.2 Troposphere–stratosphere exchange at the flank of the polar vortex

On 27 March and 8 April the sounding site was situated on the eastern edge of the upper-level trough. The strong laminae observed during those days was also an indication that we were close to the vortex boundaries. As mentioned above, in this type of situation folding occurs, which creates a potential for troposphere–stratosphere exchange. In fact, we observed tropospheric ozone peaks of 5–10 nb in the 400–600-hPa level. These peaks were less pronounced than the broad peak observed on 20 March under a cut-off low regime. Also these peaks were not associated with a sharp decrease of relative humidity. In addition, on 8 April we observed a sharp tropospheric ozone increase, before the beginning of the cross-tropopause exchange. A slight positive temperature gradient right above the tropopause, as observed on the same day, is another indication of the ongoing troposphere–stratosphere exchange.

2.3.3.2 Low ozone pockets

As pointed out earlier, ozone spatio-temporal variability is characterized by a high level of complexity. One of the features of this complexity is a recently discovered low-ozone pocket. Morris *et al.* (1995, 1998) emphasized that, although the lowest ozone mixing ratios are typically found in mid-stratosphere within the winter circumpolar vortex (CPV) (according to both chemistry and dynamics), they were also frequently observed (according to satellite remote sounding using LIMS and

LMS instrumentation) outside the CPV. Pockets of low ozone were registered on numerous occasions in anticyclones at mid to high latitudes in mid-stratosphere, typically in the 20–2-hPa layer (27–43 km or 750–1250 K) (Manney *et al.*, 1995b).

According to analysis of the LIMS data, ozone pockets are formed during mid to late winter in conjunction with stratospheric wave-breaking events. During such events, air masses with relatively high amounts of ozone move poleward over the polar cap, while a poor or low ozone forms in a strong anticyclone. Airmasses within the anticyclones at altitudes near 850 K ($\sim$ 32 km) contain $\sim$ 25% less ozone than the surrounding air. Trajectory calculations have revealed that much of the air within the pockets originated in the tropics or subtropics at higher altitudes several weeks earlier.

To achieve better understanding of the nature of low ozone pockets, Morris *et al.* (1998) examined in detail an event in February–March 1993 using a Lagrangian model with initialization based on observations from UARS. The results indicated that the model is able to realistically reproduce observational data and explain observed variation in terms of chemistry and dynamics. The primary mechanism responsible for development of these low-ozone regions is the isolation of air at high latitudes for periods of time long enough that significant ozone loss toward the local photochemical equilibrium can occur. Such behaviour is different from that the surrounding airmasses which move from mid to high latitudes and back again over the course of low days.

A case study of 9 March 1993, considered by Manney *et al.* (1995b), showed that the polar vortex is shifted off the pole during a stratospheric warming event, and a strong anticyclone (the Aleutian high) was set up over North America. Each of these dynamical features coincided with a region of low ozone. Relatively high values of ozone (more typical of mid-latitudes) extended over the polar cap, separating the two regions of low ozone. Morris *et al.* (1998) mentioned that the formation of the low-ozone region centred over Europe, inside the polar vortex, is familiar, whereas the one in the anticyclone over Canada evolved quite differently and therefore required special analysis. Comparison of ozone and water-vapour distribution showed that the low-ozone pocket does not come from the vortex. The water-vapour data strongly suggested tropical origins for the air contained within the pockets.

Manney *et al.* (1995b) pointed out a number of issues, related to the formation of low-ozone pockets, that needed to be explained by modelling efforts, including the following:

(1) the pockets appear in the anticyclone several days after the peak of a stratospheric warming;
(2) they are confined vertically to the region between 15 and 5 hPa;
(3) they appear downstream of a warm region;
(4) they originate in the tropics;
(5) they contain air parcels, which have been confined together for 1–3 weeks;
(6) they contained some amounts of HNO_3.

The purpose of the numerical modelling carried out by Morris *et al.* (1998) was to answer these questions by examination of the February–March events at the 850 K

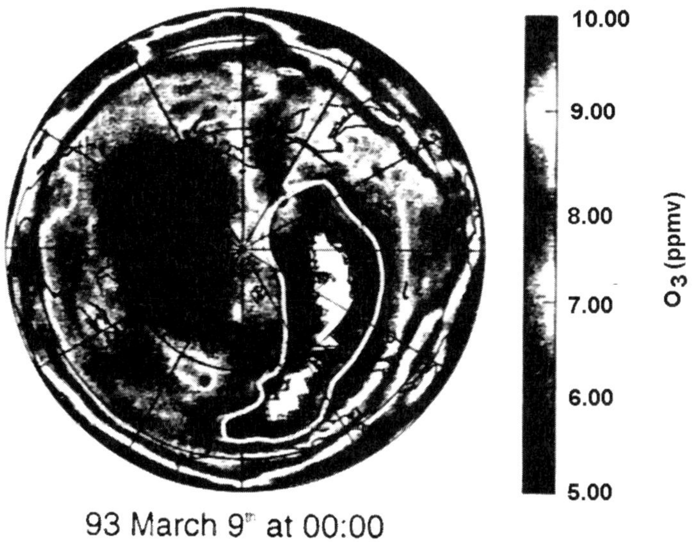

Figure 2.36. Barnes' map of gridded output ozone from the Lagrangian chemical model (Kawa *et al.*, 1993) at 0000 GMT on 9 March 1993, on the 850-K surface. The Lagrangian model reproduced the observed ozone gradients, including low ozone in both the vortex and the anticyclone. White regions indicate data values below 5 ppmv.

surface. Calculated ozone distribution at 00.00 GMT on 9 March (Figure 2.36) compared well with MLS ozone data. In both cases when low ozone (~5 ppmv) took place within the vortex boundary, high ozone values were observed in the tropics (> 10 ppm) and low values (< 6 ppmv) in the region of the anticyclone over northern Canada. Calculations indicated that local values measured by MLS and predicted by the model matched the photochemical equilibrium values for latitudes between 15 and 45°N. North of 45°, however, the observed ozone exceeded the photochemical equilibrium values.

Figure 2.37 illustrates the extent to which air within the anticyclone has been isolated at high latitudes. For this purpose for each air parcel (included in chemical calculations), calculations were made of the equivalent solar latitude (ESL), a quantity related to the dynamic history of an air parcel (Morris *et al.*, 1998). The ESL value characterizes the latitude at which an air parcel, with a given solar exposure, would be found if it had remained at a fixed location in the atmosphere. ESL relates the solar-weighted average latitude at which an air parcel has been found over the previous 2-week period: parcels with high ESL have spent most of their time at high latitudes, while those with low ESL have spent most their time at low latitudes. The ESL field is very highly correlated with the MLS ozone field. High ESL values within the vortex and the anticyclone confirm that air parcels within these regions have been isolated at high latitudes for the previous 2 weeks.

Vertical boundaries of low-ozone pockets (20–2 hPa) were successfully simulated by examination of the photochemical lifetime of ozone as a function of latitude and

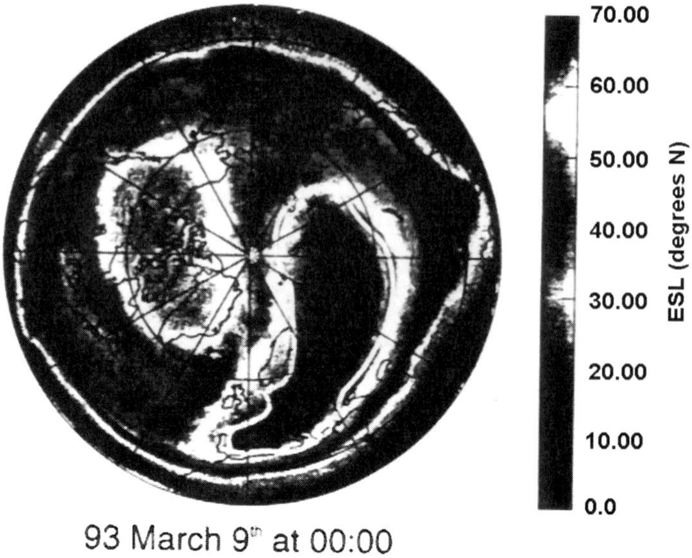

Figure 2.37. Equivalent solar latitude (see text) along 14-day diabatic back trajectories for air parcels used in the trajectory chemical model on 9 March 1993, on the 850-K surface. Parcels surrounding the anticyclone have been exposed to significantly more sunlight than those trapped inside it. The solar exposure associated with parcels outside the anticyclone is typically found at 40°–50°N, while the solar exposure inside the anticyclone is typically found near 70°N.

potential temperature obtained using the 3-D transport/chemistry model developed by Douglass *et al.* (1996). Morris *et al.* (1998) showed that the layer in which ozone pockets evolve is one in which the photochemical lifetime of ozone is similar in magnitude to the duration of strong containment, provided by dynamical features such as Aleutian anticyclones. The near balance between these photochemical and dynamical timescales makes development of low-ozone pockets possible at these altitudes.

Sensitivity calculations made by Morris *et al.* (1998) revealed that temperature history and the specific chemical initialization (including O_3, overhead column O_3, NO_y and Cl_y) are less important for development of low-ozone pockets, than is isolation at high latitudes.

Similar results were obtained by Nair *et al.* (1998) who undertook a comprehensive study of December 1992 and February/March 1993 events, using a Lagrangian photochemical model, to simulate the time-dependent ozone-loss rate within parcels of air following specific trajectories leading to observed low-ozone pockets and within parcels leading to neighbouring regions with higher ozone. These calculations were made for several different altitudes. The Caltech/JPL photochemical model was adapted to operate in a Lagrangian mode. The full model atmosphere consists of 40 levels from the surface to 0.316 hPa. Table 2.15 presents important reactions for odd-oxygen (the sum of O, $O(^1D)$ and O_3) balance. The list of species taken into account in the model is presented in Table 2.16. The model successfully simulates the observed formation of low-ozone pockets in the 6–10-hPa altitude range within measurement uncertainties but overestimates the ozone-loss rate at higher altitudes. Under the conditions considered, the low-ozone levels are due to a decrease in the odd-oxygen production rate, and not to an increase in the loss rate by reaction with halogen species, as in the 'classical' ozone hole. Trajectory calculations for both events considered indicate that when air is held at high solar-zenith angles for periods of time longer than the photochemical lifetime of ozone, a low-ozone pocket forms. Air outside these pockets and outside the polar vortex typically swings back and forth between mid and high latitudes, such that the average production rate of odd oxygen is high relative to that of air in the low-ozone pocket. Nair *et al.* (1998) underlined the fact that significant uncertainties remain in our understanding of photochemical processes controlling ozone in the upper stratosphere.

In the context of the stratosphere–troposphere exchange problem, Pan *et al.* (1997) analysed hemispheric asymmetries and seasonal variations of lowermost stratospheric water vapour and ozone, derived from SAGE-II data (the lowermost stratosphere is defined as that portion of the stratosphere between the tropopause and approximately 380-K potential temperature surface). Pan *et al.* (1997) emphasized that the lowermost stratosphere makes an interesting case for the study of transport, since the dynamics of this region is influenced by both the downward circulation from the so-called overworld (where the extratropical tropopause intersects potential temperature surfaces between about 300 to 380K) and the stratosphere–troposphere exchange across the extratropical tropopause. Trace gases of tropospheric origin can enter this region through the tropical tropopause into the overworld and then be transported into this region via downward circulation, or they can enter directly from the troposphere as a part of extratropical stratosphere–troposphere exchange. According to Pan *et al.* (1997), both water vapour and ozone data suggest that troposphere–stratosphere exchange across the extratropical tropopause has a significant influence on the lowermost stratosphere, especially for the Northern Hemisphere in the summer season.

As discussed earlier, the existence and seasonal variability of the Atlantic subtropical potential vorticity barrier, controlling stratosphere–troposphere exchanges between the lowermost extratropical stratosphere and the upper equatorial troposphere, was studied within the Measurements of Ozone by Airbus In-Service Aircraft (MOZAIC) project. In this context Cammas *et al.* (1998) investigated episodes of high-ozone mixing ratio sampled along MOZAIC flight tracks over the subtropical Atlantic with mixing ratio exceeding 100 ppbv on lengthscales larger than 200 km on flights between 11 and 12 km (a total of 154 high-ozone episodes were extracted from the MOZAIC database from August 1994 to April 1997). All these high-ozone episodes were observed north of 15°N and have lengths ranging from subsynoptic to synoptic scales.

Cammas *et al.* (1998) demonstrated that the barrier effect, located at 15°N over the central/eastern Atlantic, fits with the southernmost latitude of the subtropical jetstream during the period of interest. South of the jetstream, within the latitude band where the ITCZ oscillates, tens of ozone-rich transients (high-ozone episodes with lengthscales smaller than 80 km) were sampled within the upper equatorial Atlantic

Table 2.15. Important reactions for odd-oxygen balance.

Number	Reaction	ΔO_x	Number	Reaction	ΔO_x
(R1)	$O_2 + h\nu \longrightarrow 2O$	+2	(R147)	$N + NO \longrightarrow N_2 + O$	+1
(R2)	$O_2 + h\nu \longrightarrow O + O(^1D)$	+2	(R148)	$N + NO_2 \longrightarrow N_2O + O$	+1
(R5)	$HO_2 + h\nu \longrightarrow OH + O$	+1	(R149)	$NO + O + M \longrightarrow NO_2 + M$	−1
(R8)	$H_2O + h\nu \longrightarrow 2H + O$	+1	(R150)	$NO + O_3 \longrightarrow NO_2 + O_2$	−1
(R10)	$NO + h\nu \longrightarrow N + O$	+1	(R158)	$NO_2 + O_3 \longrightarrow NO_3 + O_2$	−1
(R11)	$NO_2 + h\nu \longrightarrow NO + O$	+1	(R166)	$NO_3 + O \longrightarrow O_2 + NO_2$	−1
(R12)	$NO_3 + h\nu \longrightarrow NO_2 + O$	+1	(R193)	$O + OClO + ClO + O_2$	−1
(R16)	$N_2O_5 + h\nu \longrightarrow NO + NO_3 + O$	(R194)	$O + OClO + M \longrightarrow ClO_3 + M$		−1
(R24)	$ClO + h\nu \longrightarrow Cl + O$	+1	(R195)	$O + Cl_2O \longrightarrow 2ClO$	−1
(R25)	$ClOO + h\nu \longrightarrow ClO + O$	+1	(R196)	$O + ClNO_3 \longrightarrow O_2 + ClONO$	−1
(R26)	$OClO + h\nu \longrightarrow ClO + O$	+1	(R197)	$O + HCl \longrightarrow OH + Cl$	−1
(R36)	$ClNO_3 + h\nu \longrightarrow O + ClONO$	+1	(R198)	$O + HOCl \longrightarrow OH + ClO$	−1
(R83)	$CO_3 + h\nu \longrightarrow CO + O$	+1	(R199)	$O(^1D) + Cl_2 \longrightarrow Cl + ClO$	−1
(R95)	$2O + M \longrightarrow O_2 + M$	−2	(R200)	$O(^1D) + HCl \longrightarrow Cl + OH$	−1
(R99)	$O + O_3 \longrightarrow 2O_2$	−2	(R201)	$O(^1D) + HCl \longrightarrow ClO + H$	−1
(R102)	$O(^1D) + O_3 \longrightarrow 2O_2$	−2	(R202)	$O(^1D) + CHFCl_2 \longrightarrow CFCl_2 + OH$	−1
(R104)	$O + H_2 \longrightarrow OH + H$	−1	(R203)	$O(^1D) + CHF_2Cl \longrightarrow CF_2Cl + OH$	−1
(R105)	$O + OH \longrightarrow O_2 + H$	−1	(R204)	$O(^1D) + CCl_4 \longrightarrow CCl_3 + ClO$	−1

(R106)	$O + HO_2 \longrightarrow OH + O_2$	−1	(R205)	$O(^1D) + CFCl_3 \longrightarrow CFCl_2 + ClO$	−1
(R107)	$O + H_2O_2 \longrightarrow OH + HO_2$	−1	(R206)	$O(^1D) + CF_2Cl_2 \longrightarrow CF_2Cl + ClO$	−1
(R108)	$O(^1D) + H_2 \longrightarrow H + HO$	−1	(R207)	$O(^1D) + CF_3Cl \longrightarrow CF_3 + ClO$	−1
(R109)	$O(^1D) + H_2O \longrightarrow 2OH$	−1	(R217)	$O(^1D) + CF_3CFCl_2 \longrightarrow Cl + 4F + ClO$	−1
(R112)	$H + O_3 \longrightarrow OH + O_2$	−1	(R219)	$O(^1D) + COCl_2 \longrightarrow 2ClO + CO$	−1
(R115)	$H + HO_2 \longrightarrow H_2O + O$	+1	(R220)	$O(^1D) + COFCl \longrightarrow ClO + FCO$	−1
(R116)	$2OH \longrightarrow H_2O + O$	+1	(R259)	$Cl + O_3 \longrightarrow ClO + O_2$	−1
(R118)	$OH + O_3 \longrightarrow HO2 + O2$	−1	(R313)	$ClO + O \longrightarrow Cl + O_2$	−1
(R122)	$HO2 + O_3 \longrightarrow OH - 2O_2$	−1	(R315)	$ClO + O_3 \longrightarrow OClO + O_2$	−1
(R125)	$O + NO_2 \longrightarrow NO + O_2$	−1	(R316)	$ClO + O_3 \longrightarrow ClOO + O_2$	−1
(R126)	$O + NO_2 + M \longrightarrow NO_3 + M$	−1	(R331)	$OClO + O_3 \longrightarrow ClO + 2O_2$	−1
(R127)	$O + HNO_3 \longrightarrow OH + NO_3$	−1	(R353)	$O + ClCO \longrightarrow Cl + CO_2$	−1
(R128)	$O + HO_2NO_2 \longrightarrow OH + NO_2 + O_2$	−1	(R358)	$O + ClCO_3 \longrightarrow Cl + O_2 + CO_2$	−1
(R129)	$O + N_2O_5 \longrightarrow 2NO_2 + O_2$	−1	(R491)	$O + CH_4 \longrightarrow CH_3 + OH$	−1
(R130)	$O(^1D) + N_2 + M \longrightarrow N_2O + M$	−1	(R494)	$O + H_2CO \longrightarrow OH + HCO$	−1
(R131)	$O(^1D) + N_2O \longrightarrow 2NO$	−1	(R496)	$O(^1D) + CH_4 \longrightarrow CH_3 + OH$	−1
(R132)	$O(^1D) + N_2O \longrightarrow N_2 + O_2$	−1	(R497)	$O(^1D) + CH_4 \longrightarrow CH_3O + H$	−1
(R133)	$O_3 + CH_3ONO \longrightarrow CH_3ONO_2 + O_2$	−1	(R498)	$O(^1D) + CH_4 \longrightarrow H_2CO + H_2$	−1
(R143)	$N + O_2 \longrightarrow NO + O$	+1	(R532)	$CH_3 + O \longrightarrow H_2CO + H$	−1
(R144)	$N + O_3 \longrightarrow NO + O_2$	−1	(R535)	$CH_3 + O_3 \longrightarrow CH_3O + O_2$	−1

Table 2.16. Species in photochemical model

O	O(^{1}D)	H	H_2	OH	HO_2
H_2O_2	N	NO	NO_2	NO_3	N_2O_3
N_2O_4	N_2O_5	HNO	HNO_2	HO_2NO_2	CH_3ONO
CH_3ONO_2	$CH_3O_2NO_2$	Cl	Cl_2	ClO	ClOO
OClO	ClO_3	Cl2	Cl2O_2	Cl_2O_3	ClNO
$ClNO_2$	ClONO	$ClNO_3$	HCl	HOCl	CH_2Cl
CCl_3	$CFCl_2$	CF_2Cl	CH_3Cl	CH_2FCl	$CHFCl_2$
CHF_2Cl	CCl_4	$CFCl_3$	CF_2Cl_2	CH_3CCl_3	CF_3CCl_3
CF_3CFCl_2	ClCO	CHClO	$COCl_2$	COFCl	CH_2ClO
CH_3OCl	CH_2ClO_2	CCl_3O_2	$CFCl_2O_2$	CH_2ClO_2H	O_2ClNO_3
CCl_3NO_4	$CFCl_2NO_4$	CH_3	CO	CO_2	HCO
H_2CO	CH_2OH	CH_3O	CH_2O_2	CH_2OOH	CH_3OH
CH_3O_2	CH_3OOH	$HOCH_2OO$			

troposphere (9–12 km). The origin of these tropical ozone-rich transients is not clear as yet. Cammas *et al.* (1998) pointed out that some of the ozone-rich transients may be interpreted as the final result of tropopause foldings and small scale mixing processes. Analysis of the seasonal variability of subtropical barriers indicates, that this variability clearly agrees with that of the position of the subtropical jetstream as derived from mean isotach data.

2.4 SOME OUTSTANDING FIELD OBSERVATIONS

Recently, a new method (Lagrangian) was proposed to assess stratospheric ozone-depletion rates in the Arctic winter, by using data collected in the frame of EASOE (von der Gathen *et al.*, 1995; Rex *et al.*, 1998), which is presented in the following.

2.4.1 A Lagrangian approach for assessment of chemically induced ozone loss

2.4.1.1 Ozone measurements

Approximately 1,200 electrochemical-type ozonesondes were launched during the 1991–1992 winter from 26 stations within EASOE and from Canadian stations, most of them electrochemical concentration cell (ECC) types 4A and 5A manufactured by the Science Pump Corporation (Komhyr and Harris, 1971). Brewer–Mast sondes were used at three stations that have been using them for over 25 years. In the lower stratosphere, Brewer–Mast sondes compared well with ECC-type sondes (Kerr *et al.*, 1994; De Backer *et al.*, 1995). Most sondes (approximately 70%) were released between 10:00 and 13:00 UT on the day of the launch. All stations were situated in mid and high northern latitudes (of the Table 2.17).

Ozone mixing ratios measured by ozonesondes have a vertical resolution of about 150 m resulting from the response time of the electrochemical cell to changes in the

Table 2.17. Ozonesonde stations during EASOE.

Station	Latitude north (deg)	Longitude east (deg)	Sonde type
Aberystwyth	52.4	355.9	ECC 5A
Alert	82.5	297.7	ECC 4A
Athens	37.9	23.8	ECC 5A
Bear Island	74.3	19.0	ECC 5A
Churchill	58.8	265.9	ECC 4A
Dikson	73.5	80.3	ECC 5A
Edmonton	53.6	245.9	ECC 4A
Egedesininde	68.7	307.3	ECC 5A
Gardermoen	60.1	11.0	ECC 5A
Goose Bay	53.3	299.6	ECC 4A
Heiss Island	80.6	58.1	ECC 5A
Hohenpeißenberg	47.5	11.0	Brewer–Mast
Iqaluit	63.8	291.5	ECC 4A
Jülich	50.8	6.1	ECC 5A
Kiruna	67.9	21.1	ECC 5A
Lerwick	60.1	358.8	ECC 5A
Lindenberg	52.2	14.1	ECC 5A
Ny-Ålesund	78.9	11.9	ECC 5A
Payerne	46.8	6.9	Brewer–Mast
Polarfront (ship)	$\sim$66 (varying)	$\sim$2 (varying)	ECC 5A
Keflavik	64.0	337.4	ECC 5A
Resolute	74.7	265.0	ECC 4A
Scoresbysund	70.5	338.0	ECC 5A
Sodankyla	67.2	26.4	ECC 5A
Thule	76.5	291.3	ECC 5A
Uccle	50.8	4.4	Brewer–Mast

ambient ozone concentration, which is of the order of 30 s (Komhyr and Harris, 1971). The sampling rate of the ozonesonde is always 10 Hz, and although data-reduction protocols vary among different stations, the vertical interval between data points reported in the ozone profile was usually 10 s. However, some stations report less dense data and gaps may occur in the profiles because of problems with the sensor or telemetry link to the sonde. So, to find ozone-mixing ratios at the required isentropic levels, we interpolated linearly over the logarithm of atmospheric pressure, avoiding cases where gaps in the profile exceeded 500 m or where the isentrope of interest was more than 200 m from the nearest ozone data point.

2.4.1.2 Trajectories

For the entire period of the EASOE campaign, 10-day isentropic back trajectories were daily calculated on six potential temperature levels, their endpoints located at the 26 ozonesonde stations (Knudsen and Carver, 1994). Afterwards, trajectories were calculated on a further five levels in the same manner. The trajectories were based on wind analysis of the European Centre for Medium-Range Weather Forecasts (ECMWF) in $15^\circ \times 15^\circ \times 6$ hr resolution. Geographic position, Ertel's potential vorticity (PV) and temperature along the trajectories were available every 2 h. Each trajectory terminated at 12:00 UT in a vertical column above each station.

Each set of back trajectories, ending on the same day as an ozonesonde launch, was examined for cases where an ozonesonde had previously made a measurement in the same air parcel. Such a case is called a 'match' (Figure 2.38). The criteria for the selection of matches are discussed below.

Only a small part of all theoretically existing matches was located at standard levels used for trajectories. The following procedure was therefore used to find additional matches at intermediate levels. Some 800 additional trajectories were calculated for situations where a match on a particular day at a particular station might occur. Indications of where a match might be expected were found by vertical interpolation between six trajectories on isentropic levels of the first set of trajectories. We did not use a straightforward interpolation scheme which would work linearly in space between simultaneous points of two of the six supporting trajectories, because this would produce large errors. We used an interpolation method based on optimal connection lines between the trajectories (Rex, 1993). The vertical interpolation between standard trajectories was used to trace the backward motion of the whole column of air probed by a single sonde. Cases where one of these interpolated columns approached the ascent path of a previous sonde at any height were identified, and for these cases additional trajectories were produced from

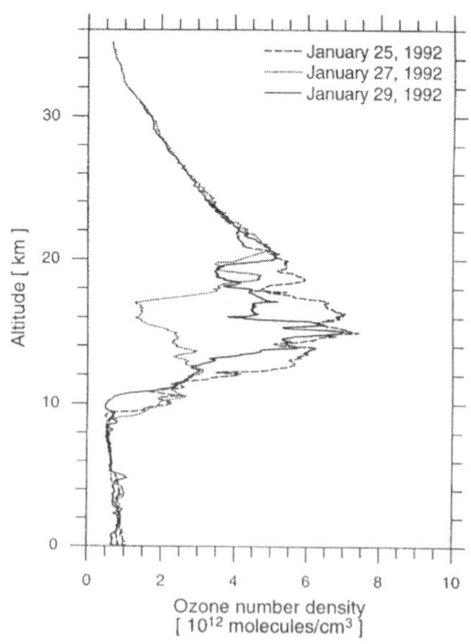

Figure 2.38. Three ozone profiles measured over 4 days by ozonesondes from Ny-Ålesund (78.9°N, 11.9°E). The large variability in the profiles was caused by dynamical effects (Rex *et al.*, 1998).

the original wind fields at the estimated 'optimal' match level. The endpoint for each of these special trajectories was the exact position (in space and time) of the second ozonesonde.

2.4.1.3 Selection of matches

In addition to the restrictions discussed above, which were chosen to limit the statistical uncertainty connected with vertical interpolation in the ozone profiles, a number of selection criteria were imposed to ensure that the data quality for the matches was as high as possible and to minimize the statistical uncertainty in calculated ozone-loss rates.

2.4.1.3.1 Match radius

There is an inevitable difference in the location of the ozonesonde and the air parcel trajectory, which can lead to uncertainties, especially in regions where the horizontal ozone gradient is large. For both ozonesonde measurements of each match the respective horizontal distance between the location of the ozonesonde in the match level and the simultaneous position of the air parcel (as defined by the trajectory) was calculated. The sum of the two distances defines the individual match radius for each match (Figure 2.38). The match radius is the most important parameter defining the quality of a match.

To locate the co-ordinates of the ozonesonde observation, the drift of the sonde during its ascent was considered. Wind data measured by the ozonesonde were used when available (in about 70% of all soundings), otherwise wind data from a nearby (< 10 km) radiosonde were used. No wind data were available for Kiruna Station, so vertical ascents were assumed. Statistical analyses of drift of ozonesondes from other stations in this latitude band showed that the true position of the Kiruna sondes might deviate by about 20–100 km from the assumed vertical ascent. The location of the air parcel at the moment of ozonesonde measurement was determined by linear interpolation (or extrapolation at the end of the trajectory in cases where the second ozonesonde of the match was launched after 12:00 UT) along a great circle of the Earth between the points of the trajectory (Figure 2.39).

Only matches where the match radius was smaller than a certain limit were used to determine ozone-loss rates. This limit was found by examining the way in which the scatter in the difference between two ozone measurements of a match varied as a function of the match radius. The scatter, defined as the root mean square of the differences of ozone for individual matches after correction for ozone loss determined for the set of matches, is plotted for the 475-K surface in Figure 2.40 as a function of the match radius. The way ozone-loss rates were determined from the set of matches is described later. Figure 2.40 shows that scatter is about 5–6% when the match radius is less than 400 km and that it gradually increases above this to 10% at 1,000 km. The number of usable matches increases with the limit for the allowed match radius and the statistical uncertainty in the determined ozone loss rates is found to reach a nimimum value for a limit of 475 km.

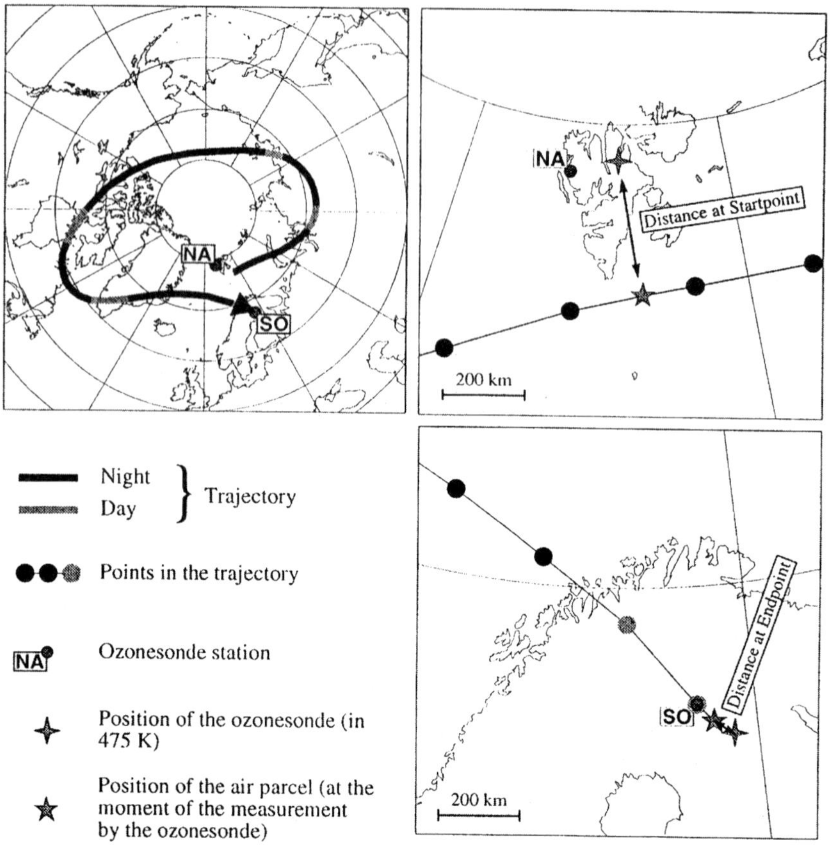

Figure 2.39. One example of a match on the 475-K potential temperature surface: the sounding at Ny-Ålesund on 24 January 1992 was linked by a trajectory with the sounding on 29 January at Sodankylä. The upper left-hand panel shows the trajectory of the air parcel. The grey sections indicate sunlit conditions. The upper right-hand panel is an enlargement of the situation at Ny-Ålesund. The sonde, launched at 10:35 UT, drifted during the ascent and reached the 475-K level at the position marked in the plot at 11:16 UT. The simultaneous position of the air parcel is marked in the plot, too. The distance between the air parcel and the ozonesonde is given by the length of the plotted black arrow. The lower right-hand panel is an enlargement of the situation at Sodankylä. Here the trajectory ended at 12:00 UT exactly above the station. In this case the sonde reached the 475-K level shortly after 12:00 UT. The extrapolated position of the air parcel at this time is marked in the plot. Again the small distance between the position of the sonde (including drift during ascent) and the air parcel is given by the length of the black arrow. The sum of the two distances defines the match radius (Rex *et al.*, 1998).

2.4.1.3.2 Gradient in ozone profile

Matches where one or both ozone profiles showed a strong vertical ozone gradient near the point of interest were ignored since results from such matches were more sensitive to diabatic uncertainty and to small-scale vertical diffusion processes. Specifically, measurements were not used when the difference between the maximum and minimum ozone mixing ratio in the 10-K interval centred around the match level was greater than 15% of the ozone value at the match level.

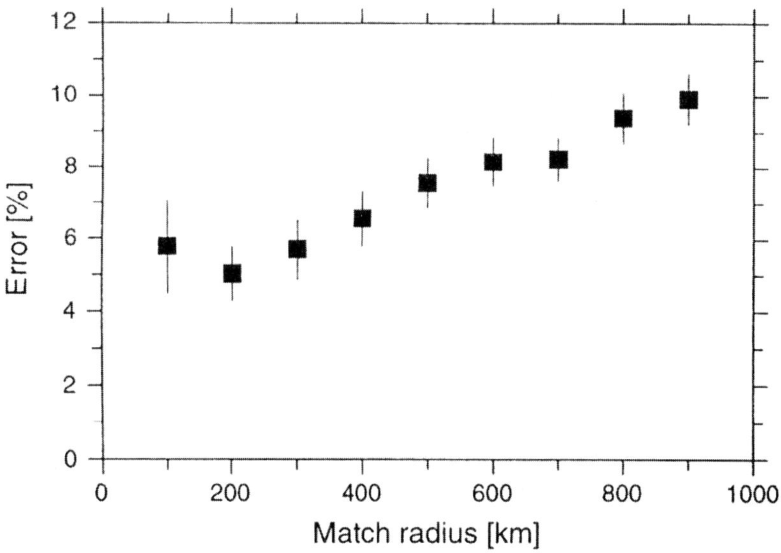

Figure 2.40. Statistical error of an individual match as a function of the match radius. Matches from 4 January to 9 February 1992, in the potential temperature range 475 ± 10 K were considered. Each marker gives the root mean square of the corrected differences in ozone-mixing ratio of matches with a match radius in a ±100-km interval centred around the position of the marker. The measured differences for each match were corrected by the expected ozone loss for the respective match. The expected ozone-loss was calculated by multiplying the sunlit time of the match with the ozone-loss rate calculated from the set of matches. See text for definition of the sunlit time of the match and calculation of the ozone-loss rate from a set of matches (Rex *et al.*, 1998).

2.4.1.4 Results

We report results from the polar vortex including the vortex-edge region. In addition, we analyse how ozone loss rates differ between the core and the edge of the vortex. Uncertainties associated with the results outside the vortex are too large for meaningful conclusions to be reached.

2.4.1.4.1 Definition and coverage of the vortex

We define the edge of the vortex in terms of potential vorticity. As height-independent numbers can be used, we have employed quantity-scaled PV (Dunkerton and DeLuisi, 1986), which is defined as $PV/(g(d\theta_0/dp))$, where PV is Ertel's potential vorticity, g is the acceleration of gravity and $(d\theta_0/dp)$ is the static stability at the isentropic surface in question taken from a fit of a mean $\theta_0(p)$ profile. The mean $\theta_0(p)$ profile was taken from data of more than 500 radiosondes launched since 1988 in December, January and February at Ny-Ålesund/Spitsbergen. Between 350 K and 700 K this mean profile is very well reproduced by the fit, given by the function:

$$p = p_0 \exp(\alpha_0 + \alpha_1\theta_0 + \alpha_2\theta_0^2 + \alpha_3\theta_0^3)$$

with $\alpha_0 = 12.48$, $\alpha_1 = -3.212 \times 10^{-2}\,\mathrm{K}^{-1}$, $\alpha_2 = 3.708 \times 10^{-5}\,\mathrm{K}^{-2}$, $\alpha_3 = -1.627 \times 10^{-8}\,\mathrm{K}^{-3}$ and $p_0 = 1\,\mathrm{hPa}$.

Finally, we multiply scaled PV by a constant factor of 2.65×10^{-5} and refer to this quantity as normed PV. The advantage of using normed PV is that Ertel's PV in PVU ($10^6\,\mathrm{km}^2\,\mathrm{s}^{-1}\,\mathrm{kg}^{-1}$) and normed PV in s^{-1} have the same values at the isentropic level 475 K.

Tables 2.18 and 2.19 show that the position of the maximum PV gradient, expressed in normed PV, does not vary very much with time and altitude during the relevant time period and above 400-K potential temperature. It generally lies near 38–39 s^{-1} normed PV. Since we want to include the region where the PV

Table 2.18. Vertical variation of the position of the maximal PV gradient in normed PV.*

Potential temperature (K)	Normed PV at the maximal PV gradient (s^{-1})
350	No well-defined maximum
380	43.6 ± 0.6
400	40.9 ± 0.8
425	38.9 ± 0.4
450	38.5 ± 0.3
475	38.8 ± 0.3
500	38.6 ± 0.5
525	38.2 ± 0.5
550	37.7 ± 0.5

* For each PV isoline the radius (r) of the circle equal in area was calcuated. The given values are mean values from 1 January to 9 February of the PV value where the quantity $d\mathrm{PV}/dr$ reaches its maximum.

Table 2.19. Temporal variation of the position of the maximal PV gradient in normed PV.*

Day of the year 1992	Normed PV at the maximal PV gradient (s^{-1})	Fraction area of the vortex enclosed by the maximal PV gradient (%)
7	39.8 ± 0.7	89
21	38.8 ± 1.2	91
35	37.0 ± 1.0	95
49	38.1 ± 1.6	87
63	38.9 ± 0.6	86
77	37.7 ± 2.0	86

* See Table 2.18 for details. In addition, the fraction of the vortex area (as defined by the 36-s^{-1} normed PV isoline) which is enclosed by the PV isoline at the maximal PV gradient is given.

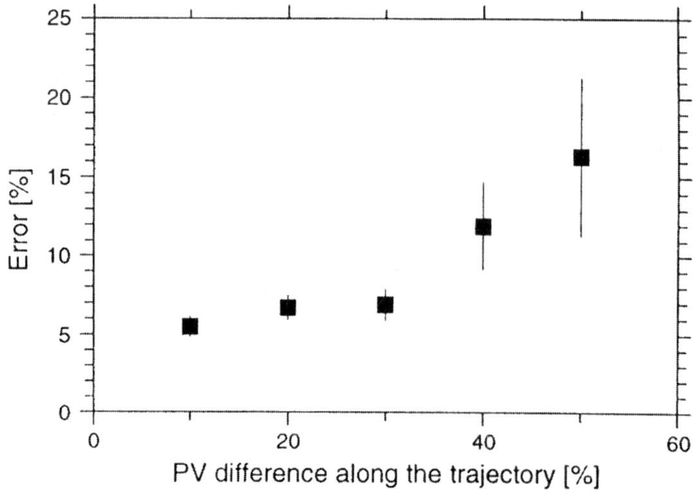

Figure 2.41. Statistical error of an individual match as a function of the percentage difference between the largest and the smallest PV value along the trajectory between both ozone measurements of the match. A 24-hr running mean was applied to the PV data before calculating the extreme values. Matches with PV changes in a ±10% interval centred around the position of the marker contributed to each point. The definition of the statistical error, time period and altitude range is as in Figure 2.40.

gradient is strong, we choose the 36-s^{-1} normed PV isoline to define the vortex edge. With this definition the vortex is about 10–15% larger than the area enclosed by the maximum PV gradient.

Figure 2.41 shows how scatter in the measured ozone varies as a function of the difference between the maximum and minimum of the 24-hr running mean of the calculated PV along the trajectory

Figure 2.42 shows how homogeneously the used matches are distributed in the polar vortex. During January the coverage is quite uniform. During February and March the vortex is still covered reasonably well, although the innermost 20% of the vortex is underrepresented in the results.

2.4.1.4.2 Sunlight dependence of the ozone loss

Figure 2.43 shows that the residual ozone of each match is strongly correlated with the time the air parcel was exposed to sunlight between the two measurements (here referred to as the sunlit time of the match). The correlation between ozone difference and sunlit time is much stronger than the correlation between the ozone difference and the total time that elapsed between the two measurements (von der Gathen *et al.*, 1995). A linear regression over 56 matches between 1 January and 9 February 1992 on the 475 ± 10-K potential temperature level inside the polar vortex yields a mean ozone loss rate of 5.5 ± 0.9 ppbv per sunlit hour. This loss corresponds to a percentage loss of 0.17 ± 0.03% per sunlit hour. Maximum loss rates were derived from 19 matches within the shorter time period between 18 January and 1 February 1992. In this period the loss rate reached 10.0 ± 1.3 ppbv per sunlit hour which

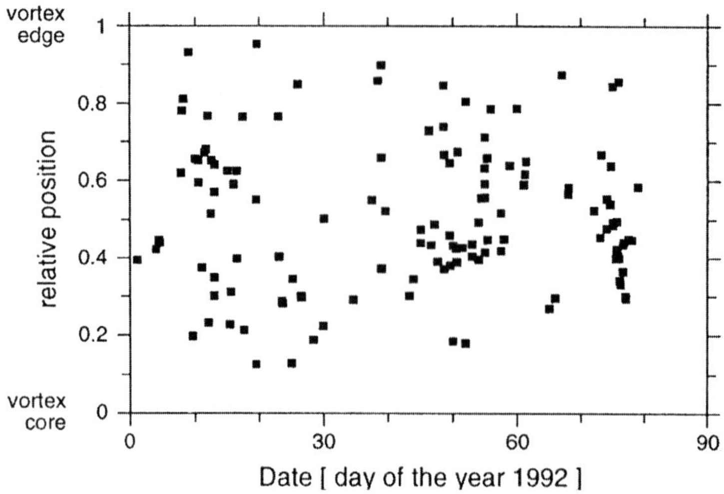

Figure 2.42. Coverage of the vortex with match events. The relative PV value of each match is plotted against date. To elucidate the homogeneity of the sampling, the PV values of the matches were mapped on a relative PV scale which runs from 0 for the maximum PV reached at the centre of the vortex on a given day to 1 at the vortex edge (36 PVU). The scaling is chosen in such a way that, for each day, equal intervals in the relative PV scale correspond to equal fraction areas of the vortex; that is, a relative PV of 0.4 means that 40% of the vortex area is enclosed by the respective isoline of relative PV (Rex *et al.*, 1998).

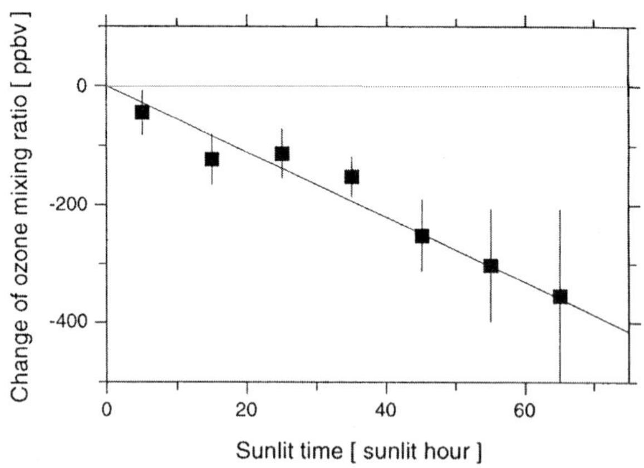

Figure 2.43. Change in ozone-mixing ratio plotted against the number of sunlit hours the respective air parcel encountered along the trajectory between both ozone measurements. Matches from 4 January to 9 February 1992 in the potential temperature range 475 ± 10 K were considered. Between 18 and 27 matches contribute within ± 10 sunlit hours to the first four data points, and between four and six matches contribute to the last three data points. Error bars denote the 1σ uncertainty (Rex *et al.*, 1998).

corresponds to $0.31 \pm 0.04\%$ per sunlit hour. All errors mentioned here denote the 1σ uncertainty.

The possibility of ozone loss occurring in the dark was examined in more detail by using a simple bivariate model which includes a term allowing for ozone loss in the dark as well as a term allowing for loss in sunlight:

$$[O_3]_B - [O_3]_A = Lt_s + Dt_d$$

where $[O_3]_A$ and $[O_3]_B$ denote ozone-mixing ratios measured by the two ozonesondes of a match, L the ozone-loss rate during sunlight, D the ozone-loss rate during darkness and t_d, t_s the dark and sunlit time the air parcel encountered between both measurements.

We define the sunlit time as the time period when the Sun is visible from the position of the air parcel in space. Because of the large airmass factor at sunrise/sunset, the flux of relevant shortwave radiation is already small for solar zenith angles smaller than at sunrise/sunset. However, with this definition we are sure that, during time periods defined as 'dark', only very small photon fluxes are present. The loss rates presented here in terms of loss per sunlit time depend on this definition of sunlit time. The values given for ozone loss per sunlit time are values averaged from sunrise to sunset over the daily course of the ozone-loss rate.

Bivariate regression gives both loss rates (1 January to 9 February 1992, on the 475 ± 10-K potential temperature level): $L = -7.0 \pm 1.5$ ppbv per sunlit hour and $D = 0.5 \pm 0.4$ ppbv per dark hour. These results clearly show that no significant ozone loss occurs during dark periods along the trajectories. In contrast to this observed sunlight dependency of the ozone loss, any dynamical influence on the ozone-mixing ratio in an air parcel (e.g. because of small-scale diffusive mixing) or any effect caused by errors in the trajectories should be independent of sunlight. The results of bivariate analysis show that such processes do not influence the result; at least their effect is smaller than the calculated value for D including its error bar.

2.4.1.4.3 Temporal development of the ozone-loss rate during early 1992

The variation of ozone-loss rates during January, February and March 1992 is shown in Figure 2.44. Each point in this figure represents the coefficient of a linear regression between the residual ozone and sunlit time of many matches in the respective time period of 14 days (as shown for a longer time period in Figure 2.43). Vortex-averaged ozone-loss rates in units ppbv per day were found by multiplying the loss rates in ppbv per sunlit hour with the mean sunlit time per day. The latter quantity was found by first calculating for each hour the fractional areas of the vortex (defined by the PV values of the respective day), which were exposed to sunlight. Daily averages were then obtained which each represent the mean sunlit time for a typical inner vortex air parcel on the respective day. This sophisticated method was used to avoid effects from possible inhomogeneous sampling of the vortex as far as possible. If, for example, the probed air parcels got more sunlight compared with the vortex average, ozone loss per day would be overestimated by any simpler method.

The area of the Northern Hemisphere where ECMWF temperatures were cold enough to form polar stratospheric clouds (PSCs), according to the nitric acid trihydrate (NAT) formation temperatures reported by Hanson and Mauersberger (1988), is also shown in Figure 2.44. A constant water-vapour mixing ratio of 4.6 ppmv and a profile of the nitric-acid mixing ratio, as measured by the limb

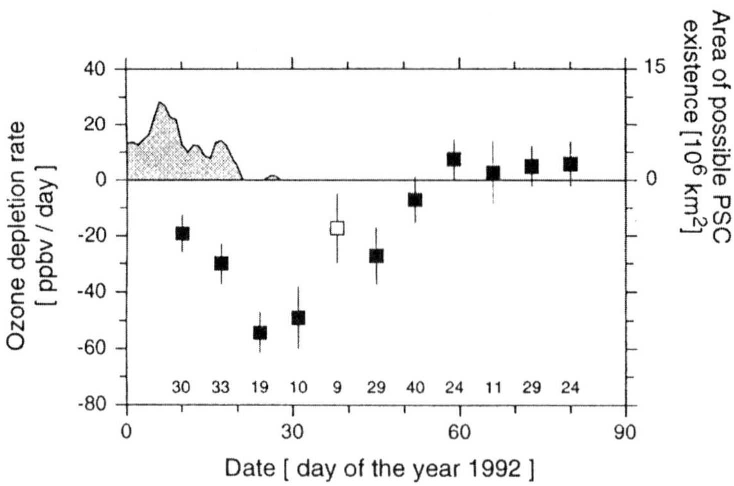

Figure 2.44. Temporal development of ozone-loss rate at 465–485 K. The shaded curve in the top panel represents the geographical area with temperatures below the NAT condensation limit. Each marker represents the result of a linear regression between ozone change and sunlit time of matches in a 17-day interval centred around the day number given by the position of the marker. The numbers at the bottom of the plot give the number of matches contributing to each point. The single point which relies on less than 10 matches is plotted light. Only matches inside the polar vortex were considered. Error bars denote the 1σ uncertainty (Rex *et al.*, 1998).

infrared monitor, of the stratosphere in January 1978 was used to calculate the condensation temperature (Braathen *et al.*, 1997). The areas shown here are only a crude indication of where PSCs might have existed as there is considerable uncertainty in the details of PSC-formation processes, ECMWF temperatures and validity of the H_2O and HNO_3 profiles used throughout the winter. The region of cold temperatures in January 1992 did not coincide with the core of the vortex but covered a sector of the vortex stretching from the vortex core to the vortex edge. In such a situation a large part of inner-vortex air can pass through areas with PSC temperatures, because of the circulation of vortex airmasses. Newman *et al.* (1993) showed that until day 15, nearly every airmass inside the vortex encountered temperatures low enough for PSCs to form at least once.

Ozone loss in the vortex initially coincided with and then slightly lagged the PSC period. During March no significant chemical ozone loss could be observed.

2.4.1.4.4 Differences between inner vortex and outer parts of the vortex

Figure 2.45 shows that, during January, ozone loss per sunlit hour is not significantly different throughout the vortex. However, because of longer sunlit times per day the loss per day is about four times greater in the outer 50% of the vortex than in the inner 50%.

During the recovery phase (7–22 February) this picture changed. Then the loss per sunlit hour (and even the loss per day) was already small in the outer part of the vortex compared with the core, indicating that the period of chemical activation

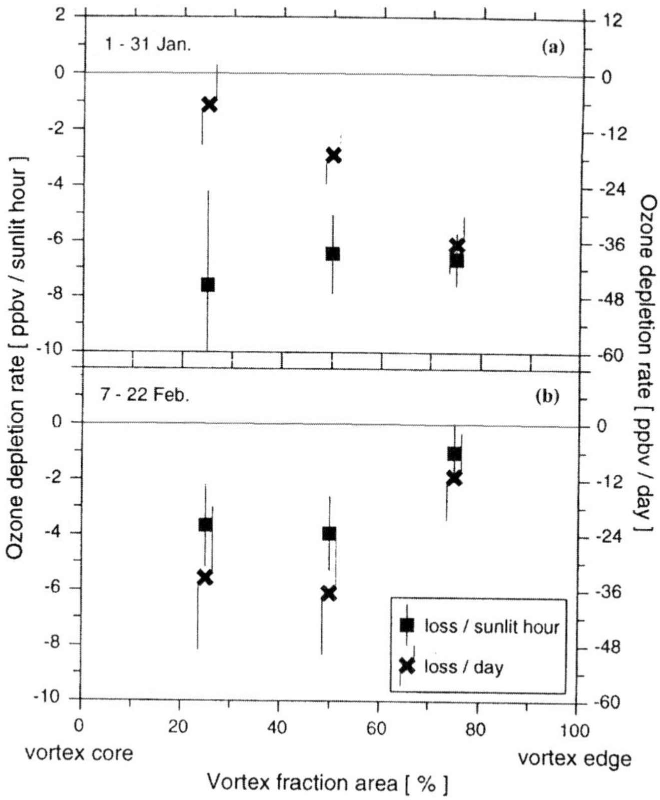

Figure 2.45. Ozone-loss rates in different parts of the vortex at the 465–485-K level. Mean ozone loss for (a) 1–31 January and (b) 7–22 February. Ozone-loss rates are plotted against the position in the vortex that is given in the fraction area of the vortex with higher PV values. Each square marker represents the results of linear regression between ozone change and sunlit time of matches in a ±25% interval centred around the position given by the marker. Cross markers give the loss in ppbv per total time. Error bars denote the 1σ uncertainty.

ceased earlier in the outer regions of the vortex. It is possible that this was caused by an earlier termination of PSC processing in the vortex edge. However, low temperatures at the end of the cold period were at the edge of the vortex, strongly suggesting that the last PSC processing happened there.

Assuming that the observed ozone loss was caused by catalytic destruction of ozone by ClO_x, a possible faster transformation of ClO_x into $ClONO_2$ in the vortex edge, compared with the inner vortex, seems to be the more likely explanation. This would be feasible because of faster production of NO_2 by faster photolysis of HNO_3 in the region of longer sunlit times (Webster *et al.*, 1993). The ClO and NO observations (13 and 22 February 1992) of Toohey *et al.* (1993) also support this mechanism.

2.4.1.4.5 Height distribution of ozone loss

Figure 2.46 shows mean ozone-loss rates between 1 January and 9 February 1992 on a number of potential temperature surfaces. Each point in this figure represents the coefficient of a linear regression between the residual ozone and the sunlit time of many matches in a 20-K interval centred around the respective level (as shown for the 475-K surface in Figure 2.42). The largest ozone loss occurred near 475 K. At 500 K, loss rates were smaller by a factor of 3 compared with the value at 475 K, while no significant ozone loss was observed at 550 K. Below 475 K, loss rates declined and reached insignificant values at 400 K and below. This ozone-loss profile correlates very well with the occurrence of temperatures low enough for PSCs to form (also shown in Figure 2.45).

2.4.1.4.6 Integrated loss of ozone

We use the vortex-averaged loss rates in ppbv per day to estimate the integrated loss of ozone over the whole winter in one subsiding airmass. Subsidence of the airmass has to be considered since the loss rate varies with altitude, as shown in Figure 2.46. Because of subsidence, airmasses passed through the region of maximum ozone-loss rates during the course of the winter. Here we use a mean diabatic cooling rate inside

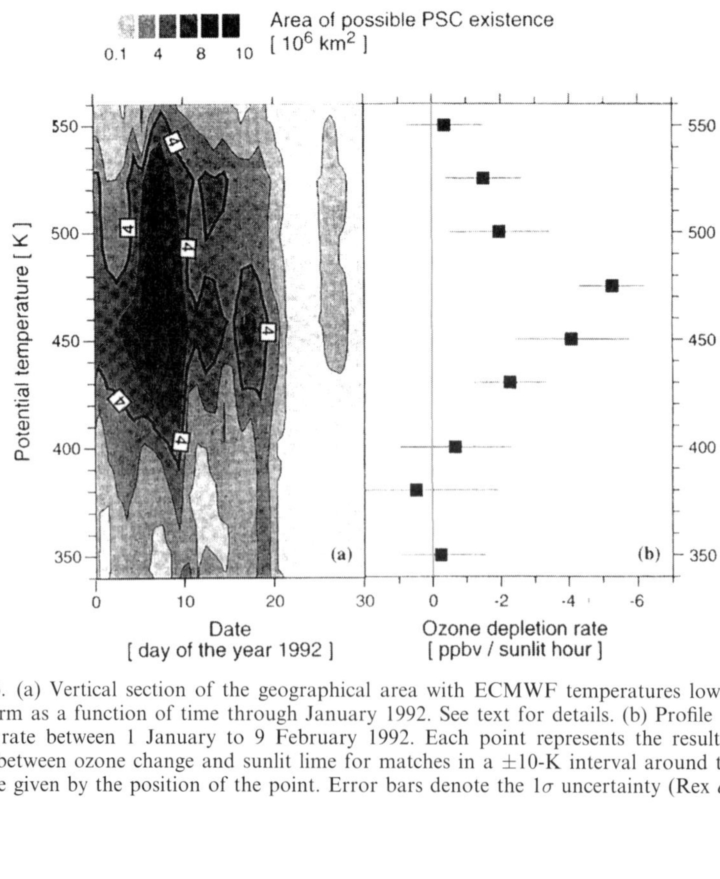

Figure 2.46. (a) Vertical section of the geographical area with ECMWF temperatures low enough for PSCs to form as a function of time through January 1992. See text for details. (b) Profile of the mean ozone-loss rate between 1 January to 9 February 1992. Each point represents the result of a linear regression between ozone change and sunlit lime for matches in a ±10-K interval around the potential temperature given by the position of the point. Error bars denote the 1σ uncertainty (Rex *et al.*, 1998).

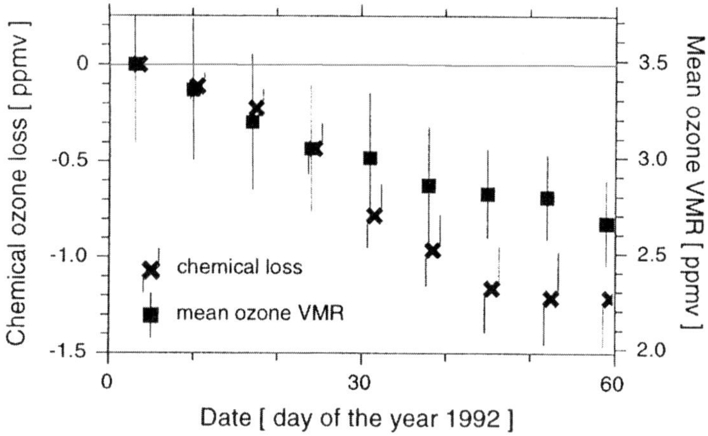

Figure 2.47. Accumulated loss of ozone in airmasses descending from 500 K (1 January) to 460 K (29 February) inside the vortex (cross markers). Error bars denote the 1σ uncertainty. Squares give the mean ozone volume mixing ratio (VMR) of all soundings inside the vortex on the same descending surface. Error bars denote scattering of the values (1σ) (Rex *et al.*, 1998).

the polar vortex of 0.7 K per day (in terms of potential temperature) as calculated by the UGAMP model (described above). We used the loss rates shown in Figures 2.44 and 2.46 to integrate ozone-loss rates for an airmass, which started at the 500-K level on 1 January and reached 460 K on 28 February 1992. The total loss is given in Figure 2.47. In this airmass the loss accumulated to -1.1 ± 0.2 ppmv ($-30 \pm 7\%$) by the end of February.

The observed mean ozone mixing ratio on the descending surface is also shown in Figure 2.47. The difference between both curves is attributed to the influence of horizontal transport of airmasses across the vortex edge. One likely effect is that relatively ozone-poor air from the edge region of the vortex was torn off the vortex and mixed with air from outside. The mean ozone-mixing ratio inside the shrinking vortex would increase this erosion at the vortex edge.

2.4.1.5 Discussion

In order to compare the outcome of different approaches measuring chemical ozone-loss rates the underlying time and space ranges have to be considered carefully. Browell *et al.* (1993) estimated an ozone loss of 23% at 460 K within the vortex (PV > 27.5 PVU) from the first set of measurements (14–23 January 1992) up to the last set of measurements (19–20 March 1992) by using airborne ozone lidar data. From the plotted mean profile of January can be read a reference ozone-mixing ratio of 3.0 to 3.25 ppmv between 460 and 500 K (considering the diabatic descent) which resulted in a loss of about 0.72 ppmv (23%). The PV at 460 K converts into about 33 s^{-1} normed PV. From 18/19 January to 29 February our analysis showed an ozone loss of 0.89 ppmv for a typical air parcel descending from 500 K (1 January) to 460 K (29 February) by diabatic cooling (compare Figure 2.47). We found no further ozone loss in March. We used 36-s^{-1} normed PV for the space limit of our measurements. Since time and space ranges are nearly identical, both ozone losses can be compared. They agree quite well within the error limits. Additionally, the height of maximum ozone loss (460 K given by Browell *et al.* (1993) and 475 K in this analysis) differs only slightly.

Proffitt *et al.* (1993) used a different definition of the vortex edge (one that relies on simultaneously measured ClO values), and they probed only the outer ring of the polar vortex. Therefore, direct comparison of the results of the present study with those from Proffitt *et al.* (1993) is not possible.

The observed time course of the ozone-loss rate inside the vortex and the vortex edge (Figure 2.45) and the sunlight dependency of the loss rate (result of bivariate regression) strongly support our current understanding of the halogen-catalysed ozone-loss mechanism. The total loss of ozone is somewhat larger than those found by the previous studies mentioned in Chapter 1. The cause of this difference is unknown, but it could be related to the separation between dynamical and chemical effects being incomplete in those works.

The magnitude of maximum ozone-loss rates found is similar to those reported from the Antarctic polar vortex (Austin *et al.*, 1989; Murphy, 1991). In our current understanding this implies that inside the polar vortex a large fraction of the inorganic stratospheric chlorine reservoir (Cl_y) was transformed to active compounds (ClO_x). This picture is confirmed by measurements of ClO made by the Microwave Limb Sounder (MLS) onboard the upper-atmosphere research satellite on 12 January 1992 (Waters *et al.*, 1993). Waters *et al.* (1993) reported up to 2 ppbv ClO in the sunlit portion of the Arctic polar vortex. Between 12 January and 15 February, no MLS measurements are available. Toohey *et al.* (1993) found on an ER-2 flight on 20 January that in a large portion of the vortex sampled by the ER-2 a large fraction of the available inorganic chlorine (greater than 70%) was converted into active forms. The time course of the ozone loss, shown in Figure 2.45a, also indicates that chlorine activation continued after 12 January. This amount of chlorine activation is similar to the wintertime chlorine activation in the Antarctic polar vortex (Waters *et al.*, 1993). The main difference between the hemispheres is how long the loss continues into spring. As shown in this work, the period with fast-ozone loss rates lasted only 2–3 weeks in the 1991/1992 Arctic winter. Therefore the integrated loss of ozone was limited to about 30%. The reasons for the earlier termination of the Arctic ozone-loss period compared with the Antarctic situation are discussed by Waters *et al.* (1993).

2.4.1.6 Match observations in the Arctic winter 1996/1997

In the 1996/1997 winter, very low ozone columns were observed above the Arctic (Newman *et al.*, 1997). These were partly caused by dynamical effects (Lefèvre *et al.*, 1998). The winter was characterized by a strong symmetric polar vortex that developed unusually late in winter and remained stable until the beginning of May. Through January the vortex was inside the polar night. Temperatures dropped at the beginning of February and stayed below PSC coexistence temperatures until the end of March at 50 hPa (Naujokat *et al.*, 1999). Here we use the Match technique (von der Gathen *et al.*, 1995a, b; Rex *et al.*, 1997; 1998; 1999) to quantify the chemical ozone inside the vortex in early 1997.

2.4.1.6.1 Measurement strategy and Analysis

A co-ordinated Match campaign, described by Rex *et al.* (1999), was carried out between 7 January and 11 April 1997, both inside and outside the polar vortex. During this time, forward trajectories were calculated from European Centre for Medium—Range Weather Forecasts (ECMWF) analyses and forecasts, using locations of ozonesonde measurements as starting points, in order to track the probed air parcels. In cases when a parcel was predicted to pass over one of the 36 ozone sounding stations, participating in the campaign (Figure 2.48), a launch request was made. A total of 746 sondes was successfully launched from these stations.

For the post-campaign analysis, a second set of trajectories was calculated from analysed ECMWF data, using cooling rates from the SLIMCAT model (Chipperfield, 1999). The criteria for accepting a match were chosen as in winter 1994/1995, with the exception of the match radius, which was reduced to 400 km (250 km), to improve the statistical error between 450 K and 500 K by about 5%. A match was assigned to the vortex when the mean normalized Potential Vorticity (PV) along the

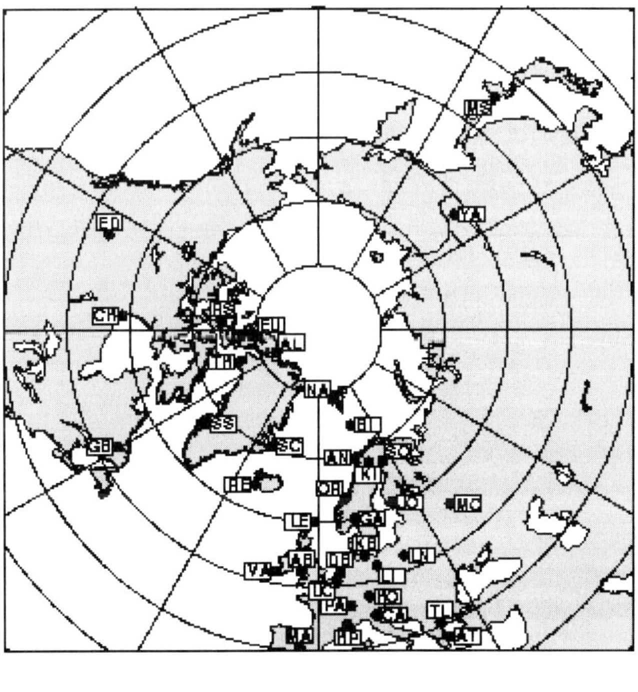

Figure 2.48. Map of all participating ozonesonde stations in the 1996/1997 Match campaign.

trajectory was higher than $36\,s^{-1}$ (Dunkerton and De Luisi, 1986), multiplied by a constant factor of 2.65×10^5 so that it had the same numerical values as Ertel's PV on the 475-K isentropic surface. In the final analysis the data of 285 sondes was used. The chosen matches sampled the vortex homogeneously in terms of PV values most of the time. Calculated loss rates can thus be interpreted as vortex averages, with some caveats discussed below. Ozone-loss rates per sunlit hour were determined by calculating linear regressions of the change in ozone-mixing ratio versus sunlit time. For these regressions matches of a 14-day period and a 20-K vertical region were used. Finally these rates were multiplied by the mean sunlit time per day inside the vortex to obtain loss rates per day. The given error bars are 1s uncertainties of the regression coefficients and do not include possible systematic effects, such as systematic errors in the cooling rates used.

2.4.1.6.2 Results and discussion

Figure 2.49 shows the ozone loss per day as a function of time and potential temperature. Results are obtained between the end of January and the end of March, covering potential temperature levels between 400 K and 520 K to 550 K. Above 425 K the depletion started in early February and continued into March. Maximum loss rates between 25 and 45 ppbv/day were reached. This period coincided with the findings of Goutail *et al.* (1998) who reported the highest chemical loss in column ozone between 1 February and 10 March. The integrated ozone loss during the period and vertical region shown in Figure 2.49 is 43 ± 9 DU. Since the results covered most of the time and vertical region where extensive PSC conditions (see

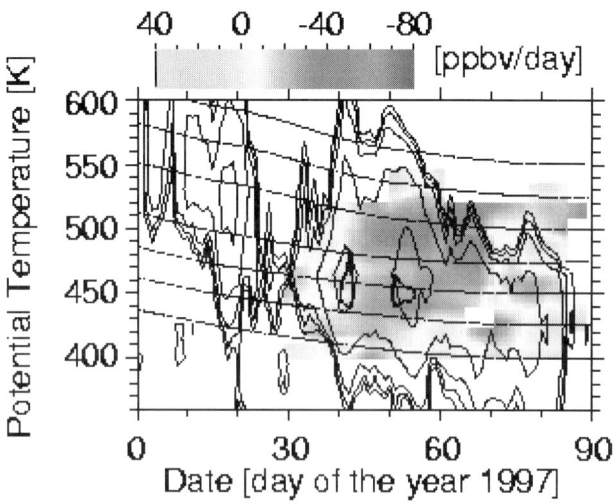

Figure 2.49. Ozone loss rates per day as a function of time and potential temperature. The fine lines indicate the 0.3, 0.7, 1.5, 4.0 and 8.0 10^6 km^2 isolines for the area of possible PSC type-I existence, the bold line is the 0.3×10^6-km^2 isoline for the area of possible PSC type-II existence as derived from ECMWF analysis. The dashed lines describe the diabatic descent (vortex average) of airmasses during the winter as determined by the SLIMCAT model.

below) existed it is likely that the integrated loss reflected a large fraction of the total accumulated column loss in 1996/1997. This is compatible with the results of Müller *et al.* (1997), who calculated 50–70 DU for the 1996/1997 column ozone loss. The derived value is much smaller than column losses in 1994/1995 and 1995/1996 with 120–160 DU (Rex *et al.*, 1999; Goutail *et al.*, 1999; Müller *et al.*, 1997), supporting the model-based finding of Lefévre *et al.* (1998) that the extreme low-ozone columns measured in 1997 (Newman *et al.*, 1997) were partly caused by dynamical effects.

Figure 2.50A shows the temporal evolution of the ozone loss rate at 475 + 10 K. The shaded curve represents the area of the Northern Hemisphere, where temperatures (derived from ECMWF analyses) might allow PSC existence, assuming a constant H_2O mixing ratio of 4.6 ppmv and the LIMS January 1979 HNO_3 profile and following the analysis of NAT formation by Hanson and Mauersberger (1988).

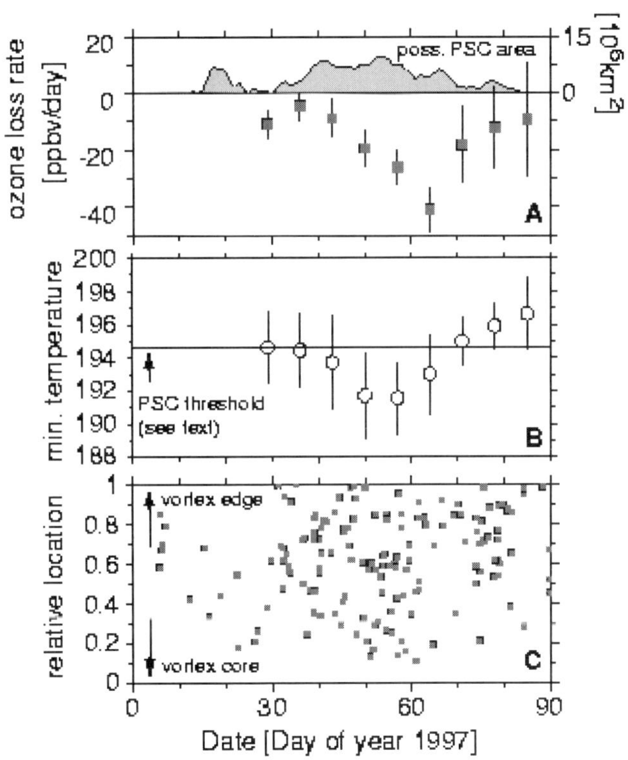

Figure 2.50. Panel A: ozone-depletion rates per day at 475 ± 10 K with 1σ error bars as a function of time. Each data point represents a linear regression between ozone change and sunlit time of matches in a 14-day period around the given date. The shaded curve is the area at 475 K with temperatures allowing PSC existence (see text). Panel B: minimum temperatures along a 10-day backward trajectory and the trajectory between the two soundings for the corresponding matches. Panel C: relative position of individual match events inside the polar vortex. The scale is chosen in a way that 0 refers to the vortex centre (highest PV value) while 1 represents the vortex edge. On a given day equal intervals on the scale correspond to equal area fractions of the vortex, while decreasing numbers represent increasing PV (e.g. 0.3 represents the PV isoline enclosing 30% of the vortex area).

While there were not enough soundings to obtain results for the first cold period in January, the loss rates started to increase some days after the beginning of February, shortly after temperatures dropped below the PSC threshold in substantial parts of the Northern Hemisphere. Ozone-loss rates increased further until the first third of March, when they dropped considerably, although temperatures inside the vortex still allowed the existence of PSCs. Figure 2.50B shows the average of the minimum temperatures in a 10-day history for the air parcels involved in the corresponding data points in panel A. For each air parcel the minimum temperature is the lowest temperature on the trajectory that links the matching pair of sondes or on a 10-day backward trajectory starting from the first sounding. A clear relation can be seen between the loss rates and the minimum temperatures in the air-parcel histories. To our knowledge this is the most direct observation of the dependence of ozone loss on low temperatures so far.

Comparing this temperature analysis with the shaded curve in panel A suggests that the smaller ozone-loss rates in the second half of March were determined by a sample of matches that did not reflect the mean vortex conditions. Further, during that period Figure 2.50C shows only few matches in the vortex core, which is where any PSCs would be expected (Figure 2.51, bottom panel). An underestimation of vortex averaged loss rates in the second half of March is therefore possible.

Figure 2.51 shows the mean ozone loss per sunlit hour between 10 February and 10 March, averaged from 450 to 500 K with respect to the relative location inside the vortex. The depletion rates vary from low values at the vortex edge to about 6 ppbv/h at the vortex core. This is in qualitative agreement with the results of Müller *et al.*, (1997) who reported larger observed ozone losses towards the vortex interior. During the same period the vortex-averaged loss rate did not exceed 3.9 ± 0.8 ppbv/h. This is considerably lower than maximum vortex-averaged loss rates per sunlit hour observed in previous years, which are 10 ± 1 ppbv/h in 1991/1992 and 1994/1995 and 10 ± 3 ppbv/h in 1995/1996 (Rex *et al.*, 1997, 1998, 1999). Corresponding to the inhomogeneous ozone loss, the minimum temperatures experienced by the air parcels in a 10-day history (Figure 2.51, open circles) were more than 5 K higher at the vortex edge than in the vortex centre. The largest areas with temperatures below the PSC threshold were situated towards the vortex centre between the first third of February and mid-March (Figure 2.51, bottom panel). Thus only a part of the vortex was affected by major ozone depletion, which kept the overall ozone loss low.

Manney *et al.* (1997) gave calculated ozone loss based on MLS data, for the end of February 1997. For the 465-K potential temperature level they calculate a loss rate of 1.3%/day between 20 and 26 February. For a comparable height and time region (465 ± 10 K and 16 February to 2 March) we determine a lower loss rate of 25 ± 6 ppbv/day, which corresponds to 0.9 ± 0.2%/day assuming an initial mixing ratio of 2.7 ppmv. However, the results agree within 2σ, and differences may also be introduced by the different periods used for the two analyses, so the results are not contradictory.

Knudsen *et al.* (1998) calculated vortex-averaged ozone depletion for different isentropic levels using ozonesonde data. The integrated ozone loss in 450 K (475 K) is

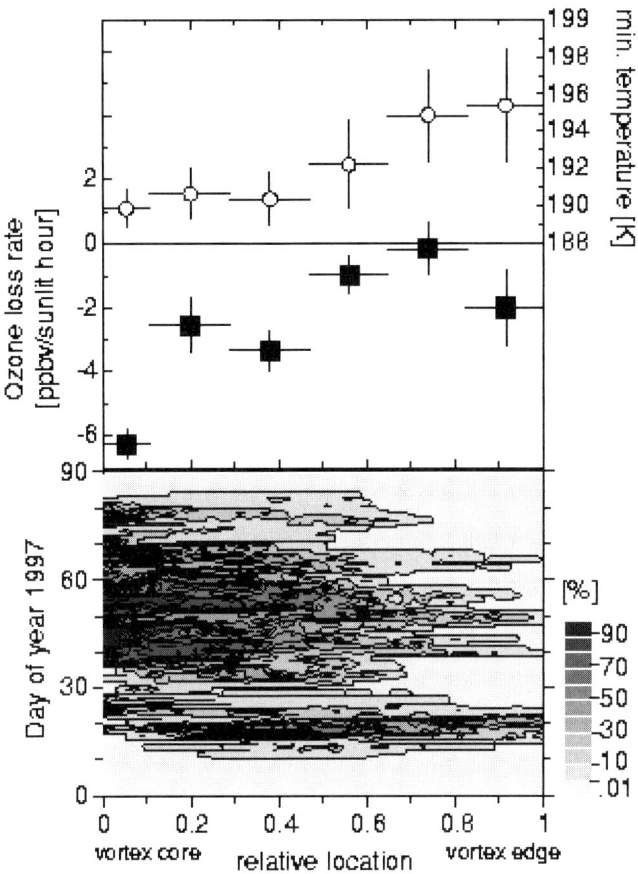

Figure 2.51. Top panel: ozone-loss rates per sunlit hour as a function of relative location (see Figure 2.50) inside the vortex. The squares symbolize linear regressions of matches between 10 February and 10 March and between 450 and 500 K. Horizontal bars indicate limits for contributing matches. Open circles represent the corresponding minimum temperatures in a 10-day history (as in Figure 2.50B). Bottom panel: (derived from ECMWF analysis) the greyscale indicates the percentage of area with temperatures allowing PSC existence (see text) in the 475-K level as a function of relative location and date.

1.1 ppmv (1.2 ppmv) compared with 0.9 ± 0.2 ppmv (0.9 ± 0.2 ppmv) as determined in this work. The slight difference between these results can be assigned to the application of different cooling rates, since values determined by them decrease to 0.9 ppmv (1.0 ppmv) when the same cooling rates are used as in this study (B. M. Knudsen, private communication). Cooling rates calculated by Knudsen *et al.* (1998) used observed ozone profiles that are mapped into a PV/theta space and are up to 80% higher than the rates used in this study, which are based on climatological ozone profiles and apply a global-flux correction (Chipperfield, 1999). This leads to the conclusion that, depending on which of the cooling rates are more precise, the ozone loss reported in this study might be underestimated by 1σ.

2.4.1.6.3 Conclusion

Significant ozone loss in the Arctic stratosphere was observed between mid-February and mid-March 1997, but the vortex-averaged ozone-loss rates determined in 1996/1997 are lower than in 1992, 1995 and 1996. This supports the picture that the extreme low ozone values observed in 1997 can be attributed at least partly to dynamical effects. The ozone loss was mainly concentrated in the vortex core, coinciding with the lowest temperatures. A direct correlation between the determined ozone loss and the temperature history of the air parcels was observed.

2.4.2 Ozone laminae in ozone vertical profile (OVP)

Reid *et al.* (1991, 1993, 1994, 1996) performed a study of ozone laminae in OVP by using quasi-isentropic trajectories, contour advection and photochemical trajectory model simulations. A summary of this study is presented in the following.

2.4.2.1 Lamina lifetimes derived from ozonesonde observations linked by diabatic trajectories: the lamina-match experiment

During the period January–March 1995, as part of the SESAME experiment, a subset of diabatic forward trajectories was generated to link observations of ozone laminae, hereafter termed lamina matches. The horizontal and vertical divergence of air parcels containing the laminae was indicated by using clusters of trajectories initialized in a 100 km $\times$ 100 km box around, and at 5 K above and below, each lamina. The application of the match technique to laminae enabled mid-latitude lifetimes of these features to be estimated.

Before proceeding, two aspects of the study merit discussion: (i) the use of potential vorticity as a dynamical tracer for air parcels containing the laminae, and (ii) the influence of the match radius R, defined as the distance by which the second ozonesonde (as it passes through the appropriate isentrope) misses the match trajectory (von den Gathen *et al.*, 1995a, b).

Typical vertical scales (or depths) of ozone laminae are 1–2 km (15–30 K in isentropic co-ordinates in the lower stratosphere), comparable with the vertical separation between model levels in the ECMWF assimilation of the wind field. It was shown that the smallest horizontal and vertical scales on which a correspondence between ozone mixing ratio and PV can be expected in mid-latitudes, based on 1.125 by 1.225° resolution data, are 300 km and 15 K, respectively (DeHaan and Reid, 1996). In order for laminae to be present in the PV field, their cross-flow scales must be at least 300 km. Indee, the existence of PV laminae near the vortex boundary was clearly demonstrated by Orsolini *et al.* (1997).

The match radius varies from a few tens of kilometres up to 500 km (the upper limit permitted for a match), and it might be expected that R would contribute significantly to the observed change in ozone concentration within laminae. Indeed, such trajectories are invariably started from regions where the vertical gradient in ozone is large (a situation avoided during the main match experiment), and the second ozonesonde is unlikely to sample exactly the same airmass (in the Lagrangian sense), thereby introducing a systematic error. However, the large meridional scales of many laminae (as inferred above) compensate for the influence of *R*. Linear regression of change in ozone-mixing ratio within laminae, as a function of the match radius, resulted in a correlation coefficient of 0.11, clearly indicating a lack of dependence on *R*.

Further support for the lamina-match approach to study lamina lifetimes derives from the progressive manner in which the ozone-mixing ratio inside laminae, when sampled three or more times along the same trajectory (for example, Figure 2.52) changes progressively rather than randomly, which a strong dependence on K would produce.

There were numerous examples in the dataset where the vertical distribution of ozone was well preserved between ozonesonde soundings, even over a week (particularly at high latitudes). To optimize the comparison of ozone concentrations within laminae as a function of time, the present study is confined to cases where (a) matches occurred on at least three isentropic surfaces in the same pair of ozone profiles, and (b) the divergence of the trajectory cluster was no more than a few degrees in latitude per day.

The results of this experiment confirm earlier findings (e.g. Reid *et al.*, 1993) that few laminae are found in ozone profiles where several matches have occurred, but the air parcels remain within the polar vortex. However, on those occasions when laminae were recorded, their ozone-mixing ratios changed very little during the period between observations, ofter 1 week or more. This result is not surprising given

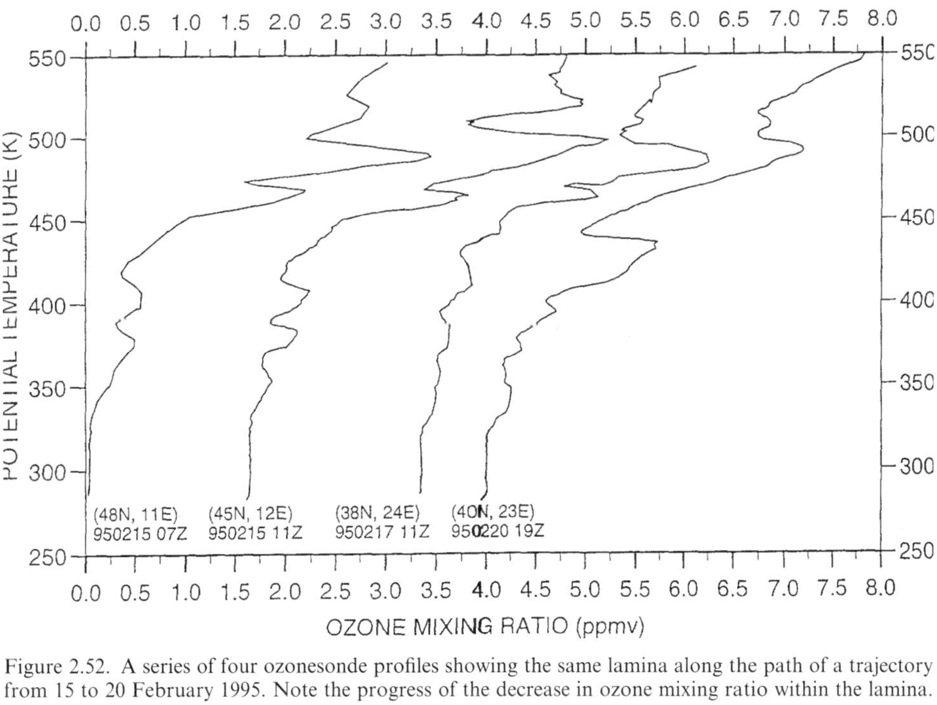

Figure 2.52. A series of four ozonesonde profiles showing the same lamina along the path of a trajectory from 15 to 20 February 1995. Note the progress of the decrease in ozone mixing ratio within the lamina.

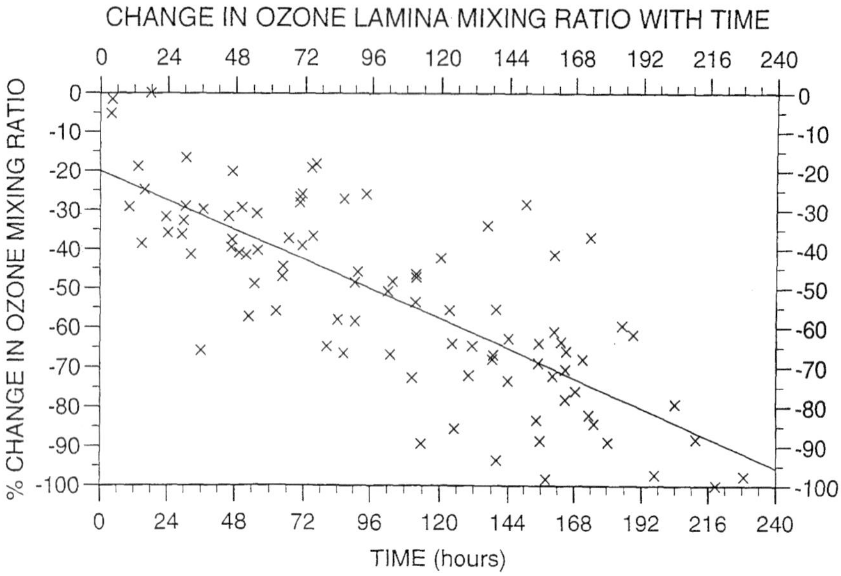

Figure 2.53. The distribution of the change in ozone-mixing ratio within ozone-enhanced laminae sampled as a function of time, for cases where laminae advected from polar to mid-latitudes. The straight line represents a linear regression performed on the data; the correlation coefficient for the fit is −0.78.

that the vortex does not generally undergo vigorous mixing. However, laminae which were first detected at extra-vortex subpolar latitudes (defined here to mean the region between 55 and 70°N, although this was variable, changing in response to the location of the vortex itself), and subsequently advected to mid-latitudes, diminished markedly in ozone concentration on a similar timescale (7–10 days). For laminae first observed by mid-latitude stations, back trajectories demonstrated that, 79% of the time, the air had been in subpolar latitudes during the preceding week.

Figure 2.53 presents the distribution of ΔO_3 in laminae moving from polar to mid-latitudes as a function of time. The straight line through the data shows the typical rate of dissipation assuming a linear relationship between ΔO_3 and Δt (correlation coefficient −0.78), leaving only 39% of the total variance unexplained by fitting the regression. However, the distribution of ΔO_3, as a function of both time and potential temperature, reveals that the rate of lamina dissipation in mid-latitudes increases with altitude, with the laminae disappearing on timescales of 7–10 days above 430 K. On lower surfaces, particularly below 400 K, some laminae prevail for longer periods.

The extent to which trajectory accuracy affects these results is difficult to evaluate, although trajectories encountering strong wind shear (e.g. in the jet around the vortex edge) are likely to be less reliable than those in mid-latitudes (Krudsen and Carver, 1994). The principle indication of trajectory robustness applied in the match experiment, and by extension the lamina-match exercise, was the degree of divergence of the trajectory cluster initiated inside and around the laminae, and cases where the cluster diverged quickly after initialization were removed from the analysis.

In view of the rate of dissipation of laminae deduced from this experiment, and given that few laminae are observed with magnitudes greater than 700 ppbv, even at high latitudes (Reid and Vaughan, 1991), the total lifespan of these features, from formation to dissipation, can rarely exceed 2 weeks. This is commensurate with the duration of filamentary structures generated by contour advection schemes.

2.4.2.2 *Case study of the ozone lamina observed on 15 February 1995 above central Europe*

The European network of regular coincident ozonesonde soundings, which was maintained during the SESAME winters of 1993/1994 and 1994/1995, resulted in a few instances where ozone laminae were observed simultaneously in a number of ozonesonde soundings made at different locations. One such example occurred on 15 February 1995, when a set of laminae were recorded at Lerwick (60.1°N, 1.2°W), Legionowo (52.4°N, 22.0°E), Lindenberg (52.1°N, 22.0°E), Uccle (50.8°N, 4.4°E), Hohenpeissenberg (47.8°N, 11.0°E), Payerne (46.8°N, 7.0°E) and San Pietro Capofiume (44.7°N, 11.6°E). A very similar lamina was present near 490 K in the ozone sounding made at Debilt (52.1°N, 5.2°E) at 13 Z (Zulu time) on the previous day which, like the event of the 15 February, was subsequently matched with another ozone sounding at Athens (37.9°N, 23.7°E) at 11 Z on 17 February, suggesting that it is part of the same feature.

Six of the ozone profiles for 15 February are shown in Figure 2.54. The cross-flow dimension of approximately 700 km, derived from these observations, supports the horizontal sheet-like nature of these phenomena suggested by Orsolini *et al.* (1995a, b). Note that, although the sounding at Hohenpeissenberg precedes the others by 4–5 h, the lamina was already present at this time; with a wind speed of approximately 13 m/s at 490 K the along-flow distance traversed during that time was approximately 210 km. However, an ozone sounding made at Valentia (51.9°N, 10.3°W) at 12 Z on 15 February (over which the trajectory cluster had passed the previous day) showed no evidence of the lamina at 490 K.

After passing over Europe, the air parcel decelerated and began to lose its integrity (Figure 2.53B); by 25 February the trajectories meandered in the vicinity of the subtropical jetstream, and transport from polar latitudes became clearly irreversible.

In addition to the event at 490 K, the two smaller ozone-enhanced laminae at 375 and 400 K, which are also present in the ozone profiles in Figure 2.54B–F, are embedded in air near latitude 35° on 11 February. In this region, the European Meteorological Bulletin Chart for 100 hPa showed the presence of a blocking anticyclone which extended over the United States and Canada; whilst some of the trajectories followed the flow around the edge of the anticyclone, several interacted with a small cyclonic feature at its edge. Since laminae on both large and small magnitudes are present in observations spanning around 7° in latitude, it is clear that lamina magnitude is not indicative of horizontal scale.

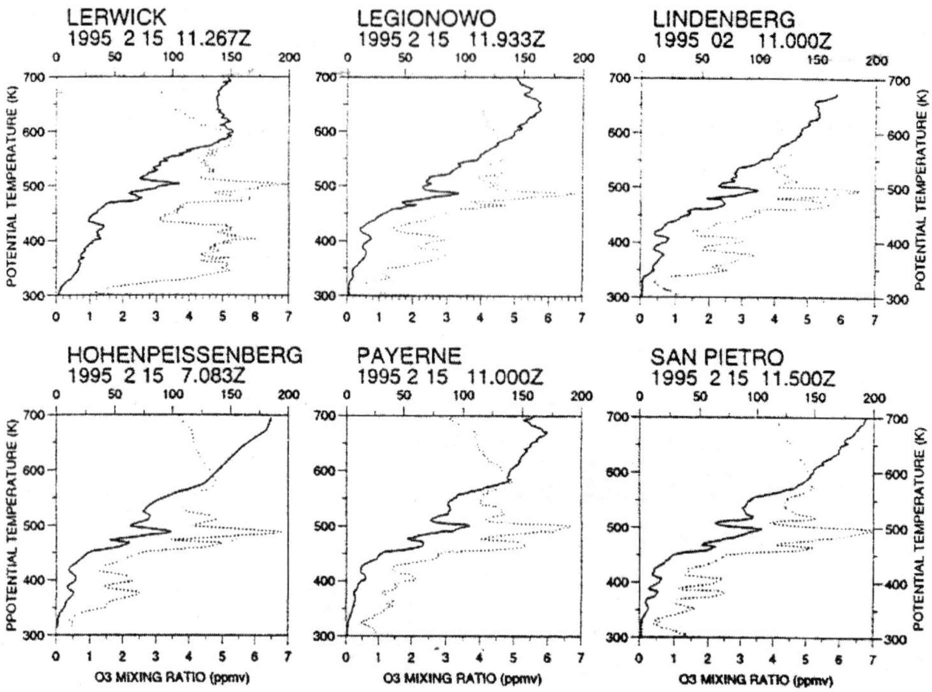

Figure 2.54. Six ozonesonde soundings over central Europe between 7 Z and 12 Z on 15 February 1995. The solid heavy line shows ozone partial pressure (in nanobars) and the lighter dotted line ozone-mixing ratio (ppbv). The same ozone-enhanced laminae between 460 and 520 K appear in all six soundings, whilst those between 360 and 410 K appear in the soundings for Legionowo, Lindenberg, Hohenpeissenberg and San Pietro Capofiume.

Both back (Figure 2.55A) and forward (Figure 2.55B) trajectories were initialized from each of the sites shown in Figure 2.54 on adjacent isentropes and in a box 100×100 km around the lamina near the 490 K observation. Figure 2.55A shows the back trajectories all converging 4 days earlier on 11 February and hitherto ran parallel to the dynamically defined edge of the polar vortex for several days. Trajectories initialized around the lamina above Debilt on 14 February followed a similar path and also converged in the same region. The shear in each cluster of trajectories was greatest on 10 February between 75–80°N and 70–100°W, where the blocking anticyclone, mentioned previously, extended into the polar regions, abutting the polar vortex and producing considerable acceleration in the flow at this point. A longitude–height cross-section of potential vorticity at 80°N showed an extra-vortex fold-like structure of elevated PV centred on 80°W, extending from above 500 K down to 450 K. This corresponded to a region in the PV analysis at 475 K where the PV gradient at the vortex edge was slack and encompassed air with the PV signature of extra-vortex air. It is possible that exchange across the vortex boundary occurred at this point, and that this resulted in the production of an ozone lamina.

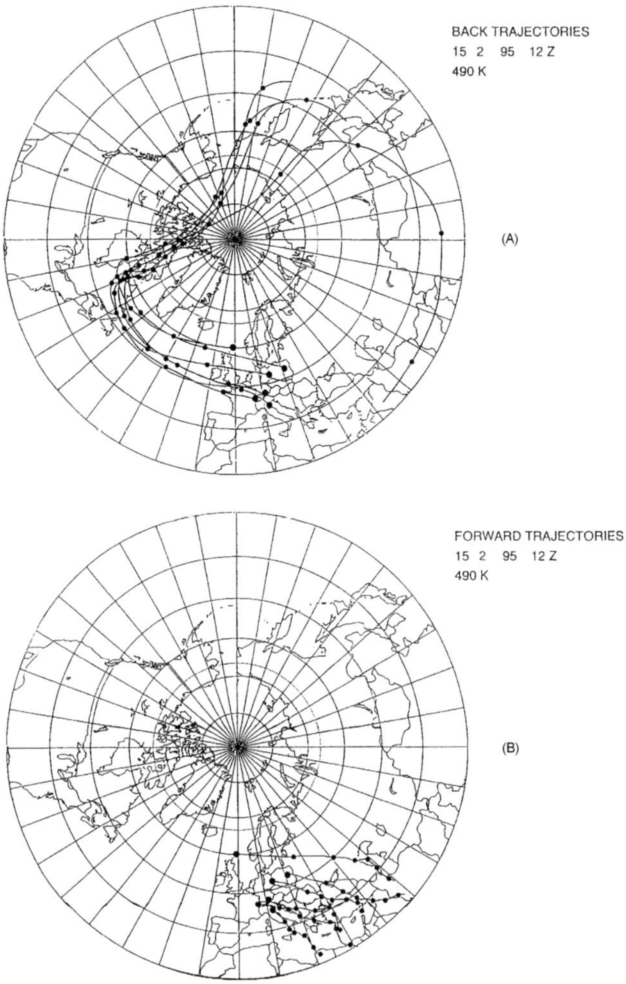

Figure 2.55. Diabatic back (A) and forward (B) trajectories from the same sites as in Figure 2.54. The back trajectory converged around 10 February between 75°N, 100°W and 80°N, 130 W in the vicinity of strong wind shear created by the presence of a blocking anticyclone abutting the cyclonic flow at the vortex boundary.

The concept of laminae produced by air shearing off the vortex boundary (e.g. Tuck *et al.*, 1992; Proffitt *et al.*, 1993) or by exchange across the boundary (e.g. Reid *et al.*, 1993) is well established in the literature, and many researchers have shown that mid-latitude air parcels bearing the chemical signature of the polar vortex can be matched up with tracer filaments generated by contour advection schemes (e.g. Pyle *et al.*, 1995). In addition, Orsolini *et al.* (1997) demonstrated the presence of a contour advection filament in the same vicinity as an ozone-enhanced lamina and concurrent aerosol-poor lamina at 450 K over northern Norway, also on 15 February 1995.

2.4.2.3 Photochemical trajectory model calculations of chemical ozone-loss rates within ozone-enhanced laminae of polar origin

The sharp increase in ClO a few degrees inside the polar vortex boundary (e.g. Webster *et al.*, 1993) means that reservoir air extruded in the form of laminae is likely to have been chemically activated at times, when temperatures were low enough to form polar stratospheric clouds. Moreover, ozone laminae are extremely numerous during winter and spring in the subpolar lower stratosphere (Reid and Timmis, 1996).

A stratospheric trajectory model, including gas phase and heterogeneous chemistry coupled with microphysical calculations, was used to study the impact of chemically activated ozone laminae, formed by the exchange of air across the boundary of the polar vortex, as they advect into mid-latitudes. Gas-phase chemistry and photodissociation calculations performed by the model are the same as those in the Oslo 2-D model (Stordal *et al.*, 1985); it includes a total of 63 gas-phase components in the O, H, N, Cl and Br families with a reaction scheme consisting of 150 chemical reactions. Radiative transfer is calculated using the two-stream approximation and includes absorption and multiple scattering for calculation of solar fluxes driving the photodissociation (Isaksen *et al.*, 1977). Heterogeneous reactions represented by the model are assumed to take place on PSC particles and sulphate aerosols (Fløisand *et al.*, 1996), while a microphysical module describes the formation and evolution of PSCs (Larsen, 1991), which depends on temperature and the gas-phase mixing ratio of H_2O and HNO_3. The total surface area of sulphate particles is based on measurements by SAGE-II (Thomason, 1997; Trepte *et al.* 1993).

Air parcels were initialized with chemical-species concentrations for the appropriate location and time from the Oslo 2-D model. Partitioning of the chlorine species was altered to fit different scenarios and the model was allowed 1 day to stabilize the partitioning of all the chemical constituents. It was then used to run photochemistry along the diabatic trajectories containing the laminae observed on 15 February 1995 (Figure 2.54), starting from 12 Z on 10 February, when the back trajectories from the 15 February converge and the shear across the trajectory clusters is greatest. The first run used chlorine-species concentrations derived from the Oslo 2-D model which provide climatological chemical data. It is followed by a scenario more representative of a situation with highly activated chlorine (high ClO concentrations), like those observed by the NASA ER-2 on 19 January 1989. This situation is assumed to describe the stratospheric conditions present through a substantial part of the 1994/1995 winter.

This simulation suggests that chemically activated winter-time laminae exported from the vortex would result in a 10% loss of ozone within the laminae: 55% of the loss due to the ClO dimmer cycle and 25% to the reactions of ClO with BrO.

2.4.2.4 Discussion

The lamina-match results presented above indicate that polar laminae often advect irreversibily to mid-latitudes and mix with the ambient air on timescales of 1 to 2 weeks. Lamina lifetimes appear to increase on going from high (500 K) to low (350 K) isentropes; in the upper part of this range the laminae dissipate in mid-latitudes on timescales of 7–10 days, whereas below 400 K their lifetimes may exceed several weeks. The small number of laminae observed within the polar vortex also changed very little on timescales of a week. It was not possible to determine whether these features were dynamic or chemical in origin: in the former case their persistence may be due to the poor mixing characteristic of the vortex interior, in the latter, they may have formed when the vortex was displaced into sunlight and subsequently remained largely unperturbed in the polar night.

The presence of the laminae over Europe on 15 February, successfully captured by contour advection analysis, strengthens the argument that laminae and filaments are different representations of the same phenomenon. Thus, the widespread occurrence of ozone-enhanced laminae throughout the surf zone may be indicative of the amount of vortex air, which moves irreversibly into mid-latitudes each winter. The large interannual variability in lamina abundance, first noted by Reid and Vaughan (1991), corresponds with the occurrence of mid-winter stratospheric warmings (Reid and Timmis, 1996), and may be linked to fluctuations in observed annual ozone-loss rates over mid-latitudes, if at least some laminae contain activated air. This view is supported by model calculations of Norton and Chipperfield (1996), which indicate that extrusions of PSC-activated air from the vortex are more frequent in some years than in others.

Accepting that ozone laminae are indeed a 1-D view of the filamentary structures seen in contour advection simulations, and combining this with typical lamina depths of 1.5 km and magnitudes of 300–700 ppbv, oxone-enhanced laminae may initially contain of the order of 10^{23} ozone molecules. According to the NILU photochemical trajectory model, around 10% (or 10^{31}) of these molecules could be destroyed in each lamina by photochemistry over a 10-day period, comparable with the lifetimes of the laminae themselves. Whilst the magnitude of ozone loss by the above mechanism cannot account for more than a fraction of the observed mid-latitude trend in ozone during winter and spring, the sheer abundance of ozone laminae observed at these times suggests that they could nevertheless make a significant contribution.

2.5 REFERENCES

Abbas, M. M., Kunde, V. G., Mumma, M. J., Kostink, T., Buhl, D., Frerking, M. A. (1979) *J. Geophys. Res.*, **84**, 2,681–2,690.

Aikin, A. C. (1989) Mesospheric ozone behaviour determined from satellite data. In: *Proceedings of the Quadrennial Ozone Symposium 1988, Gottingen, Germany*, R. D. Bojkov and P. Fabian (eds.), Deepak Publications, Hampton, VA., pp. 388–390.

Aliwell, S. R., Jones, R. L., Fish, D. J. (1997) Mid-latitude observations of the seasonal variation of BrO. 1. Zenith-sky measurements. *Geophys. Res. Lett.*, **24**, N10, 1,195–1,198.

Ancellet, G., Papayannis, A., Pelon, J., Megie, G. (1988) Tropospheric ozone lidar measurements. In: *NATO Workshop on Tropospheric Ozone 1987, Lillehammer, Norway*, I. S. Isakson (ed.), D. Reidel, Dordrecht.

Annales Geophysicae Supplement (1998) *23rd General Assembly of the European Geophysical Society. Part III. Space & Planetary Sciences*, Supplement III to Volume 16, pp. 817–1,051.

Aranda, A., Le Bras, G., La Verdet, G., Poulet, G. (1997) The BrO + CH_30_2 reaction: Kinetics and role in the atmospheric ozone budget. *Geophys. Res. Lett.*, **24**, N22, 2,745–2,748.

Arctic ripe for ozone depletion (1989) *EOS*, **70**, N9, p. III.

Arnold, F. (1990) Ozonzerstörung in der arctischen Stratosphäre. *Phys. Unserer Zeit*, **24**, N4, 175–177.

Austin, J. *et al.* (1989) Langrangian photochemical modelling studies of the 1987 Antarctic spring vortex. 2. Seasonal trends in ozone, *J. Geophys. Res.*, **94**, 16,717–16,735.

Atmospheric Ozone, Proceedings XVIII Quadrennial Ozone Symposium, 12–21 September 1996, L'Aquila, Italy. (1998) R. D. Bojkov and G. Visconti (eds.), Edigrafital S.p.A, Italy, Vol. 1, 480 pp.

De Backer, H., DeMuer, D., Schoubs, E., Allaart, M. (eds.) (1998) A new pump correction profile for Brewer–Mast ozonesondes. *Proceeding of the XVII Quadrennial Ozone Symposium*, R. D. Bojkov and G. Visconti (eds.), **2**, 891–894.

De Backer, H., Schoubs, E., Allart, M. (1995) Comparison of Brewer–Mast and ECC ozonesonde profiles at Uccle and De Bilt. In: *Polar Stratospheric Ozone, Proceedings of the Third European Symposium on Stratospheric Ozone Research, Schliersee, Bavaria, Germany*, J. A. Pyle, N. R. P. Harris, G. T. Amanatidis (eds.), Office for Official European Commission, Luxembourg, pp. 512–515.

Barlett, L. M., Vaughan, G. (1998) Ozone and NO_2 measurements by ground-based spectrometer at Aberystwyth since 1991. *Proceedings of the XVIII Quadrennial Ozone Symposium*, R. D. Bojkov and G. Visconti (eds.), Parco Scientifico e Tecnologico d'Abruzzo, L'Aquila, Italy, **1**, 9–12.

Barnes, R. A., Simeth, P. G. (1986) Design of a rocket-borne radiometer for stratospheric ozone measurements. *Rev. Sci. Instrum.*, **57**, 544.

Basher, R. E. (1982) Units for column amounts of ozone and other atmospheric gases. *Quart. J. Roy. Meteorol. Soc.*, **108**, 460.

Basher, R. E. (1995) *Survey of WMO-Sponsored Dobson Spectrophotometer Intercomparisons, WMO Global Ozone and Monitoring Project Report No. 19.* (1995) WMO, Geneva, Switzerland.

Bass, A. M., Paur, R. J. (1985) The ultraviolet cross section of ozone. 1. The measurements. *Atmospheric Ozone. Proceedings of the Quadrennial Ozone Symposium, Halkidiki, Greece*, C. S. Zerefos, A. Ghazi (eds.), D. Reidel, Hingham, MA, p. 606.

Begum, D. A. (1989) Studies of the total ozone field around the subtropical jet stream. PhD thesis, University of Wales.

Bekoryukov, V. I. (1991) Spatio-temporal total ozone field structure and its connection with stratospheric dynamics. Summary of the PhD dissertation, Institute of Applied Geophysics, Moscow, 33 pp. (in Russian).

Bekoryukov, V. I., Borisov, Ya. A., Zvyagintzev, A. M., Kruchenitsky, G. M., Perov, S. P., Rudakov, V. V. (1994) Negative anomalies in the ozone layer over Europe in the beginning of winter 1992–1993. Proceedings of the RAS. *Physics of Atmos. and Ocean*, **30**, N6, 807–811 (in Russian).

Bevilacqua, R. M., Hoppell, K., Hornstein, J., Lucke, R., Shettle, E. P., Ainsworth, T. L., Debrestian, D., Fromm, M., Lumpe, J., Krigman, S., Glaccum, W., Olivero, J., Clancy, R. T., Rusch, D. W., Randall, C., Dalaudier, F., Deniel, C., Chessefiere, E., Brogniez, C., Lenoble, J. (1995) First results frm POAM II: The dissipation of the 1993 Antarcic ozone hole. *Geophys. Res. Lett.*, **22**, N8, 909–912.

Bisling, P. G. F., Glauer, J., Köhler, S., Weitkamp, C., Knaps, A., Möllmann-Coers, M., Geiss, H., Voltz-Thomas, A., Kley, D. (1998) Dispersion, stratification and transport of tropospheric ozone observed by lidar and tracer experiments. In: *Stratospheric Ozone, Proceedings of the XVIII Quadrennial Ozone Symposium*, R. D. Bojkov, G. Visconti (eds.), Parco Scientifico e Tecnologico d'Abruzzo, L'Aquila, Italy, **2**, 319–322.

Bojkov, R. D. (1969) Differences in Dobson spectrometer and filter ozonometer measurements on total ozone. *J. Appl. Meteoral.*, **8**, 362.

Bojkov, R. D. (1986) Surface ozone during the second half of the nineteenth century. *J. Climat. and Appl. Meteorol.*, **25**, 343–351.

Bojkov, R. D., Mateer, C. L., Hansson, A. L. (1988) Comparison of ground-based and total ozone mapping spectrometer measurements used in assessing the performance of the global ozone observing system. *J. Geophys. Res.*, **93**, 9,525.

Bojkov, R. D., Bishop, L., Hill, W. J., Reinsel, G. C., Tiao, G. C. (1990) A statistical trend analysis of revised Dobson total ozone data over the Norther Hemisphere. *J. Geophys. Res.*, **95**, ND7, 9,785–9,807.

Bojkov, R. D., Fioletov, V. E. (1995) Estimating the global ozone characteristics during the last 30 years. *J. Geophys. Res.*, **100**, ND8, 16, 537–16,551.

Bojkov, R. D., Fioletov, V. E. (1998) Latitudinal and regional total ozone from quality controlled ground-based data. In: *Stratospheric Ozone. Proceedings of the XVIII Quadrennial Ozone Symposium*, R. D. Bojkov, G. Visconti (eds.), Parco Scientifico e Tecnologico d'Abruzzo, L'Aquila, Italy, **2**, 13–16.

Bojkov, R. D. (1999) International assessment of ozone depletion 1998. *WMO Bull.*, **48**, NI, 35–44.

Bose, S., Ghosh, A. B., Lal, M., Maitza, A., Sharma, R. C. (1998) Ground based observations of ozone and nitrogen dioxide during the solar eclipse 1995. *Proceeding of the XVIII Quadrennial Ozone Symposium*, R. D. Bojkov, G. Visconti (eds.) **2**, 891–894.

Bowman, K. P. (1989) Global patterns of the quasi-biannual oscillation in total ozone. *J. Atmos. Sci.*, **46**, N21, 3,328–3,343.

Bouwman, A. F., Lee, D. S., Asman, A. H., Dentener, F. J., van der Hoek, K. W., Olivier, J. G. J. (1997) A global high resolution emission inventory for ammonia. *Global Biogeochemical Cycles*, **11**, 561–587.

Braathen, G. O., M. Rummukainen, Kyrö, E., Gernandt, H., Mikkelsen, I. S., Gil, M. (1997) Ozone trends and PSC incidence in the Arctic vortex during the seven winters from 1988–89 to 1994–95. *J. Atmos. Chem.*, in press.

Braathen, G. O., Rummukainen, M., Kyrö, E., Schmidt, U., Dalilbaek, A., Jørgensen, T. S., Fabian, R., Rudakov, V. V., Gil, M., Borehers, R. (1994) Temporal development of ozone within the Arctic vortex during the winter of 1991/92. *Geophys. Res. Lett*, **21**, 1,407–1,410.

Bregman, A., van Veithoven, P. F. J., Wienhold, F. G., Fischer, H., Zenker, T., Waibel, A., Frenzel, A., Arnold, F., Harris, G. W., Bolder, M. J. A., Lelieveld, J. (1995) Aircraft measurements of O, HNO and NO in the winter Arctic lower stratosphere during the Stratosphere–Troposphere Experiment by Aircraft Measurements (STREAM) 1. *J. Geophys Res.*, **100**, 11,245–11,260.

Brewer, A. W., Milford, J. R. (1960) The Oxford–Kew ozonesonde. *Proc. R. Soc. London Ser. A*, **256**, 470.

Brewer, A. W. (1973) A replacement for the Dobson spectrophotometer?. *Pure Appl. Geophys.*, **106–108**, 919.

Browell, E. V., Butler, C. F., Fenn, M. A., Grant, W. B., Ismail, S., Schoeberl, M. R., Toon, O. B., Loewenstein, M., Podolske, J. R. (1993) Ozone and aerosol changes during the 1991–1992 Airborne Arctic Stratospheric Expedition. *Science*, **261**, 1,155–1,158.

Bühler, S., Eyring, V., Wehr, T., Kunzi, K. (1996) Limits imposed on millimeter and submillimeter wave limb sounders by continuum emissions. *Proceedings of the XVIII Quadrennial Ozone Symposium*, R. D. Bojkov, G. Visconti (eds.), Parco Scientifico e Tecnologico d'Abruzzo, L'Aquila, Italy, **2**, 883–886.

Burris, J., Heaps, W., Gery, B., Hoegy, W., Lait, L., McGee, T., Gross, M., Singh, U. (1998) Lidar temperature measurements during the Tropical Ozone Transport Experiment (TOTE)/Vortex Ozone Transport Experiment (VOTE) mission. *J. Geophys. Res.*, **103**, ND3, 3,505–3,510.

Butler, J. H., Montzka, S. A., Clarke, A. D., Lobert, J. M. (1988) Growth and distribution of halons in the atmosphere. *J. Geophys. Res.*, **103**, ND1 , 1,503–1,511.

Cammas, J.-P., Jacoby-Koaly, S., Suhre, K., Rosset, R., Marenco, A. (1998) Atlantic subtropical potential vorticity barrier as seen by measurements of ozone by airbus in-service aircraft (MOZAIC) flights. *J. Geophys. Res.*, **103**, ND18, 25,681–25,693.

Cao, M., Marshall, S., Gregson, K. (1996) Global carbon exchange and methane emissions from natural wetlands: Application of a process-based model. *J. Geophys. Res.*, **101**, ND9, 14,399–14,414.

Carlotti, M., Barbis, A., Carli, B. (1989) Stratospheric ozone vertical distribution from far-infrared balloon spectra and statistical analysis of the errors. *J. Geophys. Res.*, **94**, ND13, 16,365–16,372.

Carlotti, M., Paruolo, G., Alboni, F., Carli, B., Mencaraglia, F., Ridolfi, M. (1995) Determination of atmospheric pressure and temperature distributions from MIPAS limb emission line spectra. *ESA Contract Report No. 142956*.

Carver, G. D., Norton, W. A., Pyle, J. A. (1994) A case study in forecasting the stratospheric vortex during EASOE. *Geophys. Res. Lett.*, **21**, 1,451–1,454.

Chameides, W. L., Fehsenfeld, F., Rodgers, M. O., Cardelino, C., Martinez, J., Parrish, D., Lonneman, W., Lawson, D. R., Rasmussen, R. A., Zimmermann, P., Greenberg, J., Middleton, P., Wang, T. (1992) Ozone precursor relationship in the ambient atmosphere. *J. Geophys. Res.*, **97**, ND5, 6,037–6,056.

Chandra, S., Stolarski, R. S. (1991) Recent trends in stratospheric total ozone – Implications of dynamical and El Chichon perturbations. *Geophys. Res. Lett.*, **18**, 2,277–2,280.

Chandra, S., Varotsos, C., Flynn, L. E. (1996) The mid-latitude total ozone trends in the Northern Hemisphere. *Geophys. Res. Lett.*, **23**, N5, 555–558.

Chandra, S., Jackman, C. H., Fleming, E. L., Russell, J. M., III (1997) The seasonal and long-term changes in mesospheric water vapor. *Geophys. Res. Lett.*, **24**, N6, 639–642.

Chang, A. Y., Salawitch, R. J., Michelsen, H. A., Gunson, M. R., Abrams, M. C., Zander, R., Rinsland, C. P., Elkins, J. W., Dutton, G. S., Volk, C. M., Webster, C. R., May, R. D., Fahey, D. W., Gao, R.-S., Loewenstein, M., Podolske, J. R., Stimpfle, R. M., Kohn, D. W., Proffitt, M. H., Margitan, J. J., Chan, K. R., Abbas, M. M., Goldman, A., Irion, F. W., Manney, G. L., Newchurch, M. J., Stiller, G. P. (1996) A comparison of measurements from ATMOS and instruments aboard the ER 2 aircraft: Halogenated gases. *Geophys. Res. Lett.*, **23**, N17, 2,393–2,396.

Chapman, S. (1930) A theory of upper-atmosphere ozone. *Mem. Roy. Meteorol. Soc.*

Chernikov, A. A., Borisov, Ya., Zvjaginzev, A. M., Kruchenitsky, G. M., Perov, S. P., Sidorenkov, N. S., Stasjuk, O. V. (1999) Influence of El Niño event 1997/1998 on ozone layer of the Earth. *Geophys. Res. Lett.* (in print).

Chiou, E. W., Chu, W. P., Thomason, L. W. (1998a) Optimization of SAGE III spectral channels for water vapor retrieval. *9th Conference on Satellite Meteorology and Oceanography, 25–29 May 1998, Paris, France.*

Chiou, E. W., Chu, W. P., Thomason, L. W. (1998b), Simulation studies for the retrieval of water vapor profiles from the SAGE III experiment. *10th Symposium on Meteorological Observations and Instrumentation, 11–16 January 1998, Phoenix, Arizona.*

Chipperfield, M. P. (1999) Multiannual simulations with a three-dimensional chemical transport model, *J. Geophys. Res.*, **104**, 1,781–1,805.

Cornu, A. (1879) *Compt. Rend.*, **88**, 1,101.

Coy, L., Nash, E. R., Newman, P. A. (1997) Meteorology of the polar vortex: Spring 1997. *Geophys. Res. Lett.*, **24**, N22, 2,693–2,696.

Daehler, M., Bevilacqua, R. M., Kriebel-Hamilton, D. L., Goldizen, D., Olivero, J. J., Hartmann, G. K., Degenhardt, W., Richards, M. L., Connor, B. J., Aellig, C. P. (1998) Validation of Millimeter-wave Atmospheric Sounder (MAS) ozone measurements. *J. Geophys. Res.*, **103**, ND16, 19,615–19,639.

Dahlback, A., Hansen, G., Svenøe, T., Opedal, L., Toennessen, F. (1998) Monitoring of total ozone by means of multi-channel UV filter instruments. *Proceedings of the XVIII Quadrennial Ozone Symposium*, R. D. Bojkov, G. Visconti (eds.), Parco Scientifico e Tecnologico d'Abruzzo, L'Aquila, Italy, **2**, 887–890.

Dalyuk, I., Pokrovski, O. (1997) Statistical analysis of total ozone spatial variability in the Mediterranean. In: *Atmospheric Ozone Dynamics. Observations in the Mediterranean Region*, C. Varotsos (ed.), Springer-Verlag, Berlin, pp. 23–29.

Debrestian, D., Lumpe, J., Bevilacqua, R. M., Shettle, E. P., Hornstein, J. S., Olivero, J. J. (1997) POAM II Observations of polar mesospheric clouds in the Southern Hemisphere. *Adv. Space Res.*, **19**, 587–590.

Degórska, M., Rajewska-Wiech, B. (1989) The effect of stray light on total ozone measurements at Belsk, Poland. In: *Proceedings of the Quadrennial Ozone Symposium, 1988, Gottingen, Germany*, R. D. Bojkov, P. Fabian (eds.), Deepak Publications, Hampton, VA., p. 759.

Dials, M., Gille, J. C., Barnett, J. J., Whitney, J. G. (1998) Description of the High Resolution Dynamics Limb Sounder (HIRDLS) instrument. *Proc. SPIE*, 3,437.

Dobson, G. M. B. (1931) A photoelectric spectrophotometer for measuring the amount of atmospheric ozone. *Proc. R. Soc.*, **43**, 324.

Dobson, G. M. B., Normand, C. W. B. (1957) Observers handbook for the ozone spectrophotometer. *Ann. Int. Geophys.*, Year 5, 46.

Dobson, G. M. B. (1968) Forty years research on atmospheric ozone at Oxford: a history. *Appl. Opt.*, **7**, 387.

Dobson, G. M. B. (1973) The laminated structure of the ozone in the atmosphere. *Quart. J. Roy. Meteorol. Soc.*, **99**, 599–607.

Donovan, D. P., Fast, H., Makino, Y., Bird, J. C., Carswell, A. I., Davies, J., Duck, T. J., Kaminski, J. W., McElroy, C. T., Mittermeier, R. L., Pal, S. R., Savastiouk, V., Velkov, D., Whiteway, J. A. (1997) Ozone, column ClO, and PSC measurements made at the NDSC Eureka observatory (80°N, 86°W) during the spring of 1997. *Geophys. Res. Lett.*, **24**, N22, 2,709–2,712.

Dorokhov, V. M., Khattatov, V. U., Klimova, T. N., Pankratov, A. V., Torgovichev, V. A. (1989) An automated spectral complex for measurements of atmospheric ozone and UV radiation. In: *Proceedings of the Quadrennial Ozone Symposium, 1988, Gottingen, Germany*, R. D. Bojkov, P. Fabian (eds.), Deepak Publications, Hampton, VA., pp. 766–771.

Dorokhov, V. M., Potapova, T. E., Gutile, F., Pommerau, J.-P. (1996) Investigation of total ozone negative anomaly in the region of east-Siberian ozone mazimum during 1995 winter-spring period. *Meteorol. and Hydrol.*, N6, 53–62 (in Russian).

Douglass, A. R., Weaver, C. J., Rood, R. B., Coy, L. (1996) A three dimensional simulation of the ozone annual cycle using winds from a data assimilation system. *J. Geophys. Res.*, **101**, 1,463–1,474.

Dunkerton, T. J., De Luisi, D. P. (1986) Evolution of potential vorticity in the winter stratosphere of January–February 1979. *J. Geophys. Res.*, **91**, 1,199–1,208.

Dütsch, H. U., Staehelin, J. (1989) Discussion of the 60 year total ozone record at Arosa based on measurements of the vertical distribution and meteorological parameters. *Planet. Space Sci.*, **37**, N12, 1,587–1,594.

Dyakonova, V. D., Kramynin, A. P. (1997) On spectra of total ozone variations in the earth's atmosphere. *Proc. Ussury Astrophys. Obs.*, **2** 21–29 (in Russian).

Ebel, A., Hass, H., Jakobs, H. J., Laube, M., Memmesheimer, M., Oberreuter, A., Geiss, H., Kuo, Y.-H. (1991) Simulation of ozone intrusion caused by a tropopause fold and cut-off low. *Atmos. Environ.*, **25A**, 2,131–2,144.

Eichmann, K., Bramstedt, K., Weber, M., Rozanov, V. V., Hoogen, R., Burrows, J. P. Spring 1997 and 1998. Submitted to *Phys. and Chem. of the Earth.*

Eisele, H., Trickl, T. (1998) Lidar sounding of tropospheric ozone at Garmisch-Partenkirchen. In: *Atmospheric Ozone. Proceedings of the XVIII Quadrennial Ozone Symposium*, R. D. Bojkov, G. Visconti (eds.), Parco Scientifico e Tecnologico d'Abruzzo, L'Aquila, Italy, **1**, 351–354.

Elansky, N. F., Postylyakov, O. V., Mitin, I. V., Bais, A. F., Zerefos, C. S. (1998) The extended Umkehr method for retrieving ozone profiles from Brewer observations. *Proceedings of the XVIII Quadrennial Ozone Symposium*, R. D. Bojkov, G. Visconti (eds.), Parco Scientifico e Tecnologico d'Abruzzo, L'Aquila, Italy, **1**, 119–122.

Elokhov, A. S., Gruzdev, A. N. (1991) *Optics of the Atmosphere and Ocean*, **4**, 1,006–1,009.

Elokhov, A. S., Gruzdev, A. N. (1992) *Proceedings of the Quadrennial Ozone Symposium, Charlottesville, A*, R. D. Hudson (ed.), NASA Conference Publication, pp. 695–698.

Evans, W. F. J. (1990) Ozone depletion in the Arctic vortex at Alert during February 1989. *Geophys. Res. Lett.*, **17**, 167–170.

Evans, R. D., Komhyr, W. D., Grass, R. D. (1994) *Proceedings of the Quadrennial Ozone Symposium, Charlottesville*, NASA Conference Publication, pp. 778–781, NASA, Goddard Space Flight Center, Greenbelt, MD..

Evans, W. F. J., Puckrin, E. (1994) *Proceedings of the Fourth European Symposium*, NASA Conference Publication, pp. 749–752, NASA, Goddard Space Flight Center, Greenbelt, MD.

Evans, S. J., Toumi, R., Harries, J. E., Chipperfield, M. P., Russell, J. M., III (1998) Trends in stratospheric humidity and the sensitivity of ozone to these trends. *J. Geophys. Res.*, **103**, ND8, 8,715–8,726.

Evans, R., Varotsos, C. (1999) *WMO International Intercomparison of the Dobson Spectrophotometers*, WMO, Geneva, Switzerland (in press).

Fabry, C., Buisson, H. (1913) L'Absorption de l'ultraviolet par l'ozone et la limite du spectre solaire. *J. Phys. (Paris)*, **3**, 196.

Fabry, C., Buisson, H. (1921) Etude de l'extremité ultraviolette du spectre solaire. *J. Phys. Rad.*, N2, 197.

Farman, I. C., Gardiner, B. G., Shanklin, J. D. (1985) Large losses of total ozone in Antarctica reveal seasonal ClO_x/NO_x interaction. *Nature*, **315**, 207–210.

Farman, J. C. (1989) Measurements of total ozone using Dobson spectrometers: Some comments on their history. *Planet and Space Sci.*, **37**, N12, 1,601–1,604.

Feister, U, Grasnick, K.-H., Jakobi, G. (1989) Surface ozone and solar radiation. In: *Proceedings of the Quadrennial Ozone Symposium, 1988, Gottingen, Germany*, R. D. Bojkov, P. Fabian (eds.), Deepak Publications, Hampton, VA, pp. 441–446.

Feister, V. (1994) *Proceedings of the Quadrennial Ozone Symposium, Charlottesville, VA*, NASA Conference Publication, pp. 770–773.

Fierli, F., Hauchecorne, A., Nedeljkovic, D. (1998) Relationship between PCS events and temperature measured by ALOMAR RMR Lidar at Andøya (69°N, 16°E). In: *Atmospheric Ozone. Proceedings of the XVIII Quadrennial Ozone Symposium, 12–21 September 1996, L'Aquila, Italy*, R. D. Bojkov, G. Visconti (eds.), Parco Scientifico e Tecnologico d'Abruzzo, L'Aquila, Italy, **1**, pp. 503–506.

Fiocco, G., Cacciani, M., Di Girolamo, P., Fuà, D., De Luisi, J. (1992) Stratospheric clouds at South Pole during 1988. 1. Results of lidar observations and their relationship to temperature. *J. Geophys. Res.*, **97**, ND5, 5,939–5,946.

Fioletov, V. E. (1992) About the trends of total ozone variations during 1957–1986. In: *Investigations of Atmospheric Ozone*. Gidrometeoizdat, Moscow, pp. 16–20 (in Russian).

Fioletov, V. E., Kerr, J. B., Wardle, D. I., Davies, J., Hare, E. W., McElroy, C. T., Tarasick, D. W. (1997) Long-term ozone decline over the Canadian Arctic to early 1997 from ground-based and balloon observations. *Geophys. Res. Lett.*, **24**, N22, 2,705–2,708.

Fish, D. J., Jones, R. L., Freshwater, R. A., Roscol, H. K., Oldham, D. J. (1994) *Proceedings of the Quadrennial Ozone Symposium, Charlottesville, VA*, R. D. Hudson (ed.), NASA Conference Publication 3266, pp. 671–674.

Fish, D. J., Aliwell, S. R., Jones, R. L. (1997) Midlatitude observations of the seasonal variation of BrO. 2. Interpretation and modelling study. *Geophys. Res. Lett.*, **24**, N10, 1,199–1,202.

Fleig, A. J., Klenk, K. F., Bhartia, P. K., Gordon, D. (1982) User's guide for the Total Ozone Mapping Spectrometer (TOMS) instrument first year ozone-T dataset. *NASA RP-1096*, 50 pp.

Fløisand, I., Stordal, F., Larsen, N., Reid, S. J., Braathen, G. O., Gernandt, H., Dorokhov, V., Knudsen, B., Rognerud, B., Isaksen, I. S. A. (1996) Photochemical trajectory calculations of ozone depletion, compared with ozonesonde data from 1994 and 1995. Submitted to the SESAME edition of the *Journal of Atmospheric Chemistry*, February 1996.

Flynn, L. E., Labow, G. J., Beach, R. A., Rawlins, M. A., Flitner, D. E. (1996) Estimation of ozone with total ozone portable spectroradiometer instruments. I. Theoretical model and error analysis. *Appl. Optics*, **35**, N30, 6,076–6,083.

Fogal, P. F., McElroy, C. T., Goldman, A., Murray, D. G. (1989) Tunable Diode Laser Heterodyne spectrophotometry of ozone. In: *Proceedings of the Quadrennial Ozone Symposium, 1988, Gottingen, Germany*, R. D. Bojkov, P. Fabian (eds.), Deepak Publications, Hampton, VA., pp. 732–735.

Froidevaux, L., Waters, J. W., Read, W. G., Elson, L. S., Floweer, D. A., Jarnot, R. F. (1994) Global ozone observations from the UARS MLS: An overview of zonal-mean results. *J. Atmos. Sci.*, **51**, 2,846–2,866.

Froidevaux, L., Read, W. G., Lungu, T. A., Cofield, R. E., Fishbein, E. F., Flower, D. A., Jarnot, R. F., Ridenoure, B. P., Shippony, Z., Waters, J. W., Margitan, J. J., McDermid, I. S., Stachnik, R. A., Peckham, G. E., Braathen, G., Deshler, T., Fishman, J., Hofmann, D. J., Oltmans, S. J. (1996) Validation of UARS Microwave Limb Sounder ozone measurements. *J. Geophys. Res.*, **101**, ND6, 10,017–10,060.

Fromm, M. D., Lumpe, J. D., Bevilacqua, R. M., Shettle, E. P., Hornestein, J., Massie, S. T., Fricke, K. H. (1997) Observations of Antarctic polar stratospheric clouds by POAM II: 1994–1996, *J. Geophys. Res.*, **102**, 23,659–23,672.

Fuà, D., Cacciani, M., Di Girolamo, P., Fiocco, G., di Sarra, A. (1992) Stratospheric clouds at South Pole during 1988. 2. Their evolution in relation to atmospheric structure and composition. *J. Geophys. Res.*, **97**, ND5, 5,947–5,952.

Fujimoto, T., Uchino, O., Nagai, T. (1998) Four years intercomparison of MRI ozone lidar, ozonesonde and SAGE II data. In: *Atmospheric Ozone. Proceedings of the XVIII Quadrennial Ozone Symposium, 12–21 September 1996, L'Aquila, Italy.* R. D. Bojkov, G. Visconti (eds.), Parco Scientifico e Tecnologico d'Abruzzo, L'Aquila, Italy, **1**, 127–130.

Fujiwara, M., Kita, K., Ogawa, T., Komala, N., Saraspriya, S., Suripto, A., Sano, T. (1998) Total ozonc enhancement in September and October 1994 in Indonesia. In: *Atmospheric Ozone. Proceedings of the XVIII Quadrennial Ozone Symposium, 12–21 September 1996, L'Aquila, Italy*, R. D. Bojkov, G. Visconti (eds.), Parco Scientifico e Tecnologico d'Abruzzo, L'Aquila, Italy, **1**, 363–366.

Furrer, R, Döhler, W., Kirsch, H.-J., Plessing, P., Görsdorf, U. (1992) Evidence for vertical ozone redistribution since 1967. *J. Atmos. Terr. Phys.*, **54**, N11/12, 1,423–1,445.

Fussen, D. (1998) A critical analysis of the Stratospheric Aerosol and Gas Experiment II spectral inversion algorithm. *J. Geophys. Res.*, **103**, ND7, 8,455–8,464.

von der Gathen P., Rex, M., Harris, N. R. P., Lucic, D., Knudsen, B. M., Braathen, G. O., De Backer, H., Fabian, R., Fast, H., Gil, M., Kyrö, E., Mikkelsen, I. S., Rummukainen, M., Staehelin, J., Varotsos, C. (1995a) Observational evidence for chemical ozone depletion over the Arctic in winter 1991–92. *Nature*, **375**, 131–134.

Geleyn, J. F., Hollingsworth, A. (1979) An economical analytical method for the computation of the interaction between scattering and line absorption of radiation. *Beitr. Phys. Atmos.*, **52**, 1–16.

Geller, M. A., Wu, M. F., Nash, E. (1989) Satellite data analysis of ozone differences in the Northern and Southern Hemispheres. *Pure and Appl. Geophys.*, **130**, N213, 263–275.

Georgiadis, T., Orciari, R., Bonasoni, P., Fortezza, F., Alberti, L., Leoncini, G., Venieri, L., Gnami, V., Montanari, T., Rambelli, E. (1998) Airborne measurements for the study of vertical layering of photochemical ozone over a coastal area. In: *Atomospheric Ozone. Proceedings of the XVIII Quadrennial Ozone Symposium, 12–21 September 1996, L'Aquila, Italy*, R. D. Bojkov, G. Visconti (eds.), **1**, 367–370.

Gernandt, H., Goersdorf, U., Claude, H., Varotsos, C. A. (1995) Possible impact of polar stratospheric processes on mid-latitude vertical ozone distributions. *Int. J. Rem. Sensing*, **16**, N10, 1,839–1,850.

Glaccum, W., Lucke, R., Bevilacqua, R. M., Shettle, E. P., Hornstein, J. S., Chen, D. T., Lumpe, J. D., Krigman, S. S., Debrestian, D. J., Fromm, M. D., Dalaudier, F., Chassefiere, E., Deniel, C., Randall, C. E., Rusch, D. W., Olivero, J. J., Brogniez, C., Lenoble, J., Kremer, R. (1996) The Polar Ozone and Aerosol Measurement (POAM II) Instrument. *J. Geophys. Res.,* **101**, D9, 14,479–14,487.

Glatthor, N., Blom, C. E., von Clarmann, T., Fischer, H., Gulde, T., Piesch, C., Erle, F., Pfeilsticker, K., Chipperfield, M. P., Lee, A. M., Pyle, J. A. (1998) Airborne remote sensing of NO_2 in the Arctic winter of 1994–1995 and comparison with a three-dimensional chemical transport model. *J. Geophys. Res.*, **103**, ND11, 13,315–13,326.

Glazkov, V. N., Ivanovsky, A. K., Fedopov, V. V. (1997) Analysis of the statistical structure of long-term changes of total ozone field changes on the basis of surface measurements. *Proc. Russian Acad. Sci. Physics of the Atmosphere and Ocean*, **33**, N3, 348–359 (in Russian).

Godin, S., Megie, G., Pelon, J. (1989) Systematic lidar measurements of the stratospheric ozone vertical distribution. *Geophys. Res. Lett.*, **16**, 547.

Gogoshev, M., Petkov, B., Gogosheva, T. S. (1997) Simulation of an artificial impact on the stratosphere with the aim to restore ozone deficit. *Studying the Earth from Space*, N3, 19–24 (in Russian).

Goutail, F. *et al.* (1998) Total ozone reduction in the Arctic vortex during the winters of 1995–96 and 1996–97. *Proceedings of the Fourth European Workshop on Polar Stratospheric Ozone, 22–26 September 1997, Schliersee, Germany*, pp. 277–280.

Goutail, F. *et al.* (1999) Depletion of column ozone in the Arctic during the winters of 1993–94 and 1994–95. *J. Atm. Chem.*, **24**, 1–34.

Grant, W. B. (ed.) (1989) *Ozone Measuring Instruments for the Stratosphere*, Vol. 1, *Collected Works in Optics*, Optical Society of America, Washington, D.C.

Grant, W. B., Fishman, J., Browell, E. V., Brackett, V. G., Nganga, D., Minga, A., Cros, B., Veiga, R. E., Butler, C. F., Fenn, M. A., Nowicki, G. D. (1992) Observations of reduced ozone concentrations in the tropical stratosphere after the eruption of Mt. Pinatubo. *Geophys. Res. Lett.*, **19**, N11, 1,109–1,112.

Grant, W. B., Fenn, M. A., Browell, E. V., McGee, T. J., Singh, U. N., Gross, M. R., McDermid, I. S., Froidevaux, L., Wang, P.-H. (1998) Correlative stratospheric ozone measurements with the airborne UV DIAL system during TOTE/VOTE. *Geophys. Res. Lett.*, **25**, N5, 623–626.

Griggs, M. (1964) Design of the bubbler ozone detector. *J. Aircraft*, **1**, 306.

Grund, A., Klüpfel, T., Grendel, A., Krug, B., Perner, D. (1998) Ozone depletion potential from stratospheric OClO and BrO observations during Arctic winters 1992–96 and the comparison with observed ozone columns. In: *Atmospheric Ozone. Proceedings of the XVIII Quadrennial Ozone Symposium, 12–21 September 1996, L'Aquila, Italy*, R. D. Bojkov, G. Visconti (eds.), Parco Scientifico e Tecnologico d'Abruzzo, L'Aquila, Italy, **1**, 209–212.

Gulde, T., Piesch, C., Blom, C. E., Fisher, H., Fergg, F., Wildgruber, G. (1994) The airborne MIPAS infrared emission experiment. *Proceedings of the First International Airborne Remote Sensing Conference and Exhibition*, Environmental Research Institute of Michigan, Ann Arbor, ML), **11**, 301–311.

Gumbel, J., Witt, G. (1998) In situ measurements of the vertical structure of a noctilucent cloud. *Geophys. Res. Lett.*, **25**, N4, 493–496.

Gurevich, G. S., Krueger, A. J. (1997) Optimization of TOMS wavelength channels for ozone and sulfur dioxide retrievals. *Geophys. Res. Lett.*, **24**, N17, 2,187–2,190.

Gushchin, G. P. (1963) Universal ozonometer. *Proc. Main Geophys. Obs. Leningrad,* **141**, 83–98 (in Russian).

Gushchin, G. P., Sokolenko, S. A. (1984) Test model of the new instrument for total ozone measurements. *Proc. Main Geophys. Obs. Leningrad*, **472** (in Russian).

Gushchin, G. P., Sokolenko, S. A., Dudko, B. G., Lagutina, V. V. (1985) Ozoneometer M-124. *Proc. Main Geophys. Obs. Leningrad*, **499**, 60–67 (in Russian).

DeHaan, L. L., Reid, S. J. (1996) On the correlation between potential vorticity and ozone mixing ratio. In: *Proceedings of the Third European Symposium on Polar Stratospheric Ozone*, Office for Official Publications of the EC, Belgium, p. 67.

Hadji-Lazaro, J., Clerbaux, C., Megie, G. (1998) Simultaneous measurements of the concentrations of ozone, carbon monoxide and methane in the tropopshere at the global scale from a satellite. *Proceedings of the Advanced Study Course on Tropospheric Chemistry and Space Observation*, N. Chaumerliac, G. Megie, N. Papineau, G. Angeletti (eds.), EU Report No. 65.

Hanchecorne, A. (1998) Meridional transport of ozone in the lower stratosphere (METRO), an EC project. *SPARC Newsletter*, N11, 29–31.

Hannegan, B., Olsen, S., Prather, M., Zhu, X., Rind, D., Lerner, J. (1998) The dry stratosphere: A limit on commentary water influx. *Geophys. Res. Lett.*, **25**, N10, 1,649–1,652.

Hanson, D. R., Mauersberger, K. (1988) Laboratory studies of the nitric acid trihydrate: Implications for the south polar stratosphere, *Geophys. Res. Lett.*, **15**, 855–858.

Harries, J. E., Russell, J. M., III, Tuck, A. F., Gordley, L. L., Purcell, P., Stone, K., Bevilacqua, R. M., Gunson, M., Nedoluha, G., Traub, W. A. (1996) Validation of measurements of water vapor from the Halogen Occultation Experiment (HALOE). *J. Geophys. Res.*, **101**, ND6, 10,205–10,216.

Harris, N., Hudson, B. (1998) Summary of the SPARC-IOC assessment of trends in the vertical distribution of ozone. *SPARC Newsletter*, N10, 11–14.

Harris, N., Hudson, R., Phillips, C. (eds.) (1998) SPARC/IOC/GAW assessment of trends in the vertical Distribution of Ozone. *SPARC Report No. 1, WMO Ozone Research and Monitoring Project Report No. 43*, WMO, Geneva, Switzerland, 289 pp.

Hartley, W. N. (1880) *Chem. News.*, **42**, 268

Heath, D. F., Wei., Fowler, W. K., Nelson, V. W. (1993) Comparison of spectral radiance calibrations of SSBUV-2 satellite ozone monitoring instruments using integrating sphere and flat-plate difusser technique. *Metrologia*, **30**, 259–264.

Heese, B., Barthel, K., Hov, Ø. (1992) A comparison of total ozone data from satellite and ground-based observations at northern latitudes. *J. Geophys. Res.*, **97**, ND4, 3,825–3,830.

Helas, G., Andreae, M. O., Fontan, J., Cros, B., Delmas, R. (1989) Ozone measurements in equatorial Africa during DECAFE 88. In: *Proceedings of the Quadrennial Ozone Symposium, 1988, Gottingen, Germany*, R. D. Bojkov, P. Fabian (eds.), Deepak Publications, Hampton, VA., pp. 430–432.

Henriksen, K., Stannes, K. (1989) Solar irradiance and ozone column density in polar regions derived with an UV sensitive double monochromator. In: *Proceedings of the Quadrennial Ozone Symposium, 1988, Gottingen, Germany*, R. D. Bojkov, P. Fabian (eds.), Deepak Publications, Hampton, VA., pp. 407–410.

Hering, W. S. (1964) *Ozone Observations over North America*, Vol. 1, A.F.C.R.F.

Hering, W.S., Borden, T. R. (1964) *Ozone Observations over North America*, Vol. 2, A.F.C.R.F.

Herman, J. R., Bhartia, P. K., Krueger, A. J., McPeters, R. D., Wellemeyer, C. G., Seftor, C. J., Barry, G. J., Schlesinger, M., Torres, O., Labow, G., Byerly, W., Taylor, S. L., Swissler, T., Cebula, R. P., Gu, X.-Y. (1996) *Meteor-3 Total Ozone Mapping Spectrometer (TOMS) Data Products User's Guide*, NASA Reference Publication 1996.

Herman, J. R., Bhartia, P. K., Torres, O., Hsu, C., Seftor, C., Celarier, E. (1997) Nimbus TOMS/absorbing aerosols. *J. Geophys. Res.*, **102**, 16,911–16,922.

Hilsenrath, E., Cebula, R. P., Hynes, S. J., Caffrey, R. T. (1991) Implications of Space Shuttle flight on the calibration of instruments observing atmospheric ozone and the solar irradiance. *Metrologia*, **28**, 301–308.

Hilsenrath, E., Bhartia, P. K., Cebula, R. P., Mohnen, V. A., Wellmeyer, C. G. (1998) BUV ozone data: An example of calibration and validation needed for detecting trends from satellite observations. *SPARC Newsletter*, N10, 15–18.

Hintsa, E. J., Boering, K. A., Weinstock, E. M., Anderson, J. G., Gary, B. L., Pfister, L., Daube, B. C., Wofsy, S., C., Loewenstein, M., Podolske, J. R., Margitan, J. J., Bui, T. P. (1998) Troposphere-to-stratosphere transport in the lowermost stratosphere from measurements of H_2O, CO_2, N_2O and O_3. *Geophys. Res. Lett.*, **25**, N14, 2,655–2,658.

Hoegger, B., Levrat, G., Staehelin, J., Schill, H., Ribordy, P. (1992) Recent developments of the Light Climatic Observatory ozone measuring station of the Swiss Meteorological Institute (LKO) at Arosa. *J. Atmos. Terr. Phys.*, **54**, N5, 497-505

Hofmann, D. J., Deshler, T., Aimedieu, P., Matthews, Johnston, P. V., Kondo, Y., Sheldon, W. R., Byrne, G. J., Benbrook, J. R. (1989) Stratospheric clouds and ozone depletion in the Arctic during January 1989. *Nature*, **340**, 117–121.

Hofmann, D. J., Deshler, T. (1991) Evidence from balloon measurements for chemical depletion of stratospheric ozone in the Arctic winter of 1989–90. *Nature*, **349**, 300–305.

Hofmann, D. J., Oltmans, S. J. (1992) The effect of stratospheric water vapour on the heterogeneous reaction rate of $ClONO_2$ and H_2O for salfuric acid aerosol. *Geophys. Res. Lett.*, **19**, N22, 2,211–2,214.

Hofmann, D. J., Oltmans, S. J., Harris, J. M., Johnson, B. J., Lathrop, J. A. (1997) Ten years of ozonesonde measurements at the south pole: Implications for recovery of springtime Antarctic ozone. *J. Geophys. Res.*, **102**, ND7, 8,931–8,944.

Hollandsworth, S. M., McPeters, R., Flynn, L., Planet, W., Miller, A. J., Chandra, S. (1995) Ozone trends deduced from combined Nimbus 7 SBUV and NOAA 11 SBUV/2 data. *Geophys. Res. Lett.*, **22**, 905–908.

Holzworth, G. C. (1972) *Mixing Heights, Wind Speeds and Potential for Urban Air Pollution Throughout the Contiguous United States*, PB-207-102, Office of Air Programs, Research Triangle Park, NC.

Hood, L. L., McCormack, J. P. (1992) Components of interannual ozone change based on Nimbus and TOMS data. *Geophys. Res. Lett.*, **19**, N23, 2,309–2,312.

Hood, L. L., McPeters, R. D., McCormack, J. P., Flynn, L. E., Hollandsworth, S. M., Gleason, J. F. (1993) Altitude dependence of stratospheric ozone trends based on Nimbus 7 SBUV data. *Geophys. Res. Lett.*, **20**, 2,667–2,670.

Hood, L. L., Zhou, S. (1998) Stratospheric effects of 27-day solar ultraviolet variations: An analysis of UARS MLS ozone and temperature data. *J. Geophys. Res.*, **103**, ND3, 3,629–3,637.

Huber, M., Blunthaler, M., Ambach, W., Staehelin, J. (1995) Total atmospheric ozone determined from spectral measurements of direct solar UV irradiance. *Geophys. Res. Lett.*, **22**, N1, 53–56.

Huber, M., Blumthaler, M., Ambach, W. (1998) Ground based total atmospheric ozone determination with an absolutely calibrated UV spectrometer. In: *Atmospheric Ozone. Proceedings of the XVIII Quadrennial Ozone Symposium, 12–21 September 1996, L'Aquila, Italy*, R. D. Bojkov, G. Visconti (eds.), Parco Scientifico e Tecnologico d'Abruzzo, L'Aquila, Italy, **2**, 911–914.

Huffman, R. E. (1992) *Atmospheric Ultraviolet Remote Sensing*, Academic Press, New York, 320 pp.

Investigations of Atmospheric Ozone (1992) N. A. Zaitzeva and V. K. Khattatov (eds.), Gidrometeoizdat: Moscow, 172 pp. (in Russian).

Iraci, L. T., Fortin, T. J., Tolbert, M. A. (1998) Dissolution of sulfuric acid tetrahydrate at low temperatures and subsequent growth of nitric acid trihydrate. *J. Geophys. Res.*, **103**, ND7, 8,491–8,498.

Isaksen, I. S. A., Rognerud, B., Stordal, F., Coffey, M. T., Mankin, W. G. (1990) Studies of Arctic stratospheric ozone in a 2-D model including some effects of zonal asymmetries. *Geophys. Res. Lett.*, **17**, 557–560.

Isaksen, J. S. A., Midtborn, K. H., Sunde, J., Crutzen, P. J. (1977) Simplified method to include molecular scattering and reflection in calculation of photon fluxes and photodissociation rates, *Geophysica Norvegica*, **31**, 11–26.

Jaeglé, L., Jacob, D. J., Wang, Y., Weinheimer, A. J., Ridley, B. A., Campos, T. L., Sachse, G. W., Hagen, D. E. (1998) Sources and chemistry of NO_x in the upper troposphere over the United States. *Geophys. Res. Lett.*, **25**, N10, 1,705–1,708.

James, P. M., Peters, D., Greisiger, K. M. (1997) A study of ozone mini-hole formation using a tracer advection model driven by barotropic dynamics. *Meteorol. and Atmos. Phys.*, **64**, N1–2, 107–120.

Jaross, G., Krueger, A. J., Cebula, R. P., Seftor, C., Hartman, U., Haring, R., Burchfield, D. (1995) Calibration and postlaunch performance of the Meteor-3/TOMS instrument. *J. Geophys. Res.*, **100**, 2,985–2995.

Jiang, Y., Yung, Y. L., Douglass, A. R., Tung, K. K. (1998) The standard deviation of column ozone from the zonal mean. *Geophys. Res. Lett.*, **25**, N6, 911–914.

Johnson, B. J., Deshler, T., Zhao, R. (1995) Ozone profiles at McMurdo Station Antarctica, during the spring of 1993: Record low ozone season. *Geophys. Res. Lett.*, **22**, N3, 183–186.

Jones, A. E., Shanklin, J. D. (1995) Continuous decline of total ozone over Halley, Antarctica, since 1985, *Nature*, **376**, N6539, 409–411.

Jones, A. E., Bowden, T., Turner, J. (1998) Predicting total ozone based on GTS data: Applications for South American high-latitude populations. *J. Appl. Meteorol.*, **37**, N5, 477–485.

Jost, H., Peter, R., Kämpher, N. (1998) Comparison of ground-based millimetre wave ozone profiles with Dobson Umkehr. In: *Atmospheric Ozone. Proceedings of the XVIII Quadrennial Ozone Symposium, 12–21 September 1996, L'Aquila, Italy*, R. D Bojkov, G. Visconti (eds.), Parco Scientifico e Tecnologico d'Abruzzo, L'Aquila, Italy, **1**, 131–134.

Kadyshevitch, E. A., Elansky, N. F. (1993) Measurements of surface ozone and nitrogen oxide concentrations in Moscow. *Proc. Russian Acad. Sci. Physics of the Atmos. and Ocean*, **29**, N3, 346–352 (in Russian).

Kämpher, N. (1995) Microwave remote sensing of the atmosphere in Switzerland. *Optical Engineering*, **34**, N8, 2,413–2,424.

Kane, R. P. (1988) Long-term variation of total ozone. *Pure and Appl. Geophys.*, **127**, N4, 143–154.

Kane, R. P., Sahai, Y., Casiccia, C. (1998) Latitude dependence of the quasi-biennial oscillation and quasi-triennial oscillation characteristics of total ozone measured by TOMS. *J. Geophys. Res.*, **103**, ND7, 8,477–8,490.

Karol, I. L., Klyagina, L. P., Shalamyansky, A. M. (1992) Trend estimation of total ozone in the principal airmasses of the Northern Hemisphere in 1975–1986 (21st cycle of solar activity). *Investigations of Atmospheric Ozone*, Gidrometeoizdat, Moscow, pp. 11–15 (in Russian).

Karol, I. L., Klyagina, L. P., Romaskkina, K. I., Shalamyansky, A. M. (1995) Extremely low total ozone over Russia during the winter of 1995. *Meteorol. and Hydrol.*, N6, 115–116 (in Russian).

Kashkin, V. B., Romasko, V. Yu., Salnikova, O. E., Sukhinin, A. I. (1997) Ozone layer in Siberia due to NOAA satellite data in the IR region in the summer of 1995. *Studying the Earth from Space*, N5, 93–96 (in Russian).

Kawa, S. R., *et al.* (1993) Interpretation of NO_x/NO_y observation from AASE II using a model chemistry along trajectories. *Geophys. Res. Lett.*, **20**, 2,507–2,510

Kawabata, K., Ogawa, H., Yonekura, Y., Suzuki, H., Suzuki, M., Iwasaka, Y., Shibata, T., Sakai, T. (1997) Ground-based radiometry of stratomesospheric ozone employing a superconductive receiver. *J. Geophys. Res.*, **102**, ND1, 1,371–1,377.

Kaye, J. (1997) Summary of space-based observations of atmospheric chemistry. *SPARC Newsletter*, N8, 21–27.

Kaye, J. A. Miller, T. L. (1996) The ATLAS series of shuttle missions. *Geophys. Res. Lett.*, **23**, N17, 2,285–2,288.

Kaye, J. A., Readings, C. (1998) Long-term continuity of ozone measurements. *CEOS Newsletter*, N10, 7.

Kaye, J. A., McPeters, R. D., Herman, J. R., Bhartia, P. K., Kruegar, A. J. (1998) New TOMS instrument measures ozone and aerosols. *EOS*, **79**, N5, 57–63.

Kerr, J. B., Asbridge, I. A., Evans, W. F. J. (1988) Intercomparison of total ozone measured by the Brewer and Dobson spectrophotometers at Toronto. *J. Geophys. Res.*, **93**, 11,129–11,140.

Kerr, J. B. *et al.* (1994) The international 1991 WMO ozonesonde intercomparison at Vanscoy, Canada, *Atmos. Ocean*, **32**, N4, 685–716.

Kerr, J. B., McElroy, C. T., Wardle, D. W. (1998) The Brewer instrument calibration centre 1984–1996. In: *Atmospheric Ozone. Proceedings of the XVIII Quadrennial Ozone Symposium, 12–21 September 1996, L'Aquila, Italy*. R.D. Bojkov, G. Visconti (eds.), **2**, 915–918.

Kilbane-Dawe, I., Harris, N. (1998) Report on the Fourth European Symposium on stratospheric ozone research. *SPARC Newsletter*, N10, 25–26.

Kinnersley, J. S., Tung, K.-K. (1998) Modeling the global interannual variability of ozone due to the equatorial QBO and extratropical planetary wave variability. *J. Atmos. Sci.*, **55**, 1,417–1,428.

Kiselev, A. A., Frolkis, V. A. (1995) Role of various anthropogenic sources in expected tropospheric and stratospheric ozone changes in the 21st century. *Meteorol. and Hydrol.*, N11, 52–62 (in Russian).

Kleipool, Q. L., de Jonge, A. R. W., Whyborn, N. D., van den Broek, M., Schrijver, H., Goede, A. P. H., Urban, J., Wohlgemuth, J., Eyring, V., Küllmann, H., Schwaab, G., Hübers, H. W. (1998) Chlorine measured by ASUR over the Arctic during the SESAME campaigns. *Proceedings of the XVIII Quadrennial Ozone Symposium*, R. D. Bojkov and G. Visconti (eds.), Parco Scientifico e Tecnologico d'Abruzzo, L'Aquila, Italy, **2**, 535–538.

Knudsen, B. M., Carver, G. D. (1994) Accuracy of the isentropic trajectories calculated for the EASOE campaign. *Geophys. Res. Lett.*, **21**, 1,199–1,202.

Knudsen, B. M. *et al.* (1998) Ozone depletion in and below the Arctic vortex for 1997. *Geophys. Res. Lett.*, **25**, 627–630.

Ko, M. K. W., Sze, N.-D., Rodriguez, J. M., Weistenstein, D. K., Heisey, C. W., Wayne, R. P., Biggs, P., Canosa-Mas, C. E., Sidebottom, H. W., Treacy, J. (1994) CF_3 chemistry: Potential implications for stratospheric ozone. *Geophys. Res. Lett.*, **21**, N2, 101–104.

Köhler, U. (1999a) A comparison of the new filter ozonometer MICROTOPS II with Dobson and Brewer spectrometers at Hohenpeissenberg. *Geophys. Res. Lett.*, **26**, N10, 1,385–1,388.

Köhler, U. (1999b) Total ozone record at Hohenpeissenberg derived from Dobson and Dobson–Normalized Brewer spectrometer data. Submitted to *Geophys. Res. Lett.*

Komhyr, W. D. (1969) Electrochemical concentration cells for gas analysis. *Ann. Geophys.*, **25**, 203–210.

Komhyr, W. D., Harris, T. B. (1971) Development of the ECC ozonesonde. *NOAA Technical Report ERL 200-APCL 18*, U.S. Department of Communications, NOAA Environmental Research Laboratory, Boulder, CO.

Komhyr, W. D. (1980) Operations handbook – ozone observations with a Dobson spectrophotometer. *WMO Global Ozone Research and Monitoring Project Report No. 6*.

Komhyr, W. D., Evans, R. D. (1980) Dobson spectrophotometer total ozone measurement errors caused by interfering absorbing species such as SO_2, NO_2 and photochemically produced O_3 in polluted air. *Geophys. Res. Lett.*, **7**, 157.

Komhyr, W. D. (1987) Ozonemessungen bis 40 km Hohe mit ECC-Ozonsonden. *Promet. Meteorol. Forbildung*, **17**, 29.

Komhyr, W. D., Gammon, R. H., Harris, T. B., Waterman, L. S., Conway, T. J., Taylor, W. R., Thoning, K. W. (1985) Global atmospheric CO distribution and variations from 1968–1982 NOAA/GMCC CO flask sample data. *J. Geophys. Res.*, **90**, ND3, 5,567–5,596.

Komhyr, W. D., Grass, R. D., Leonard, R. K. (1989a) Dobson spectrophotometer 83: A standard for total ozone measurements, 1962–1987. *J. Geophys. Res.*, **94**, 9,847.

Komhyr, W. D., Oltmans, S. J., Franchois, P. R., Evans, W. F. J., Matthews, W. A. (1989b) The latitudinal distribution of ozone to 35 km altitude from ECC ozonesonde observations, 1985–1987. In: *Proceedings of the Quadrennial Ozone Symposium 1988, Gottingen, Germany*, R. D. Bojkov, P. Fabian, Deepak Publications, Hampton, VA., pp. 147–150.

Komhyr, W. D., Lathrop, J. A., Opperman, D. P., Barnes, R. A., Brothers, G. B. (1995) ECC ozonesonde performance evaluation during STOIC 1989. *J. Geophys. Res.*, **100**, ND5, 9,231–9,244.

Kondratyev, K. Ya. (ed.) (1972). *Explorations of the Environment from Manned Spacecraft*. Gidrometeoizdat, Leningrad, 297 pp. (in Russian).

Kondratyev, K. Ya., Varotsos, C., Cracknell, A. P. (1994) Total ozone trend at St. Petersburg as deduced from Nimbus 7 TOMS observations. *Int. J. Remote Sensing*, **15**, N13, 2,669–2,678.

Kondratyev, K. Ya., Buznikov, A. A., Pokrovsky, O. M. (1996) *Global Change and Remote Sensing*, Wiley/Praxis, Chichester, UK, 396 pp.

Kondratyev, K. Ya. (1997) A new stage of middle atmosphere satellite remote sensing. *Studying the Earth from Space*, N2, 127–144 (in Russian).

Kondratyev, K. Ya. (1998) *Multidimensional Global Change*, Wiley/Praxis, Chichester, UK, 761 pp.

Kondratyev, K. Ya., Nakajima, T., Sumi, A., Tanaka, T. (1998) Review article: Priorities of global change and the development of remote sensing in Japan. *Int. J. Rem. Sensing*, **19**, N7, 1,259–1,282.

König-Langlo, G., King, J. C., Pettré, P. (1998) Climatology of the three coastal Antarctic stations Dumont d'Urville, Neumayer and Halley. *J. Geophys. Res.*, **103**, ND9, 10,935–10,946.

Kosmidis, E., Zerefos, C. S., Balis, D., Bais, A. F. (1998) Testing of the New Brewer Umkehr algorithm. *Proceedings of the XVIII Quadrennial Ozone Symposium*, R. D. Bojkov and G. Visconti (eds.), Parco Scientifico e Tecnologico d'Abruzzo, L'Aquila, Italy, **1,** 139-142

Kottmeier, C., Fay, B. (1998) Trajectories in the Antarctic lower troposphere. *J. Geophys. Res.*, **103**, ND9, 10,947–10,959.

Kourtidis, K., Borchers, R., Fabian, P. (1998) Vertical distribution of methyl bromide in the stratosphere. *Geophys. Res. Lett.*, **25**, N4, 505–508.

Koutoulaki, K., Rodgers, C. D., Taylor, F. W. (1998) Non-LTE effects in the ISAMS ozone data. *Proceedings of the XVIII Quadrennial Ozone Symposium*, R. D.

Bojkov, G. Visconti (eds.), Parco Scientifico e Tecnologico d'Abruzzo, L'Aquila, Italy, **2**, 919–922.

Krotkov, N. A., Bhartia, P. K., Herman, J. R., Fioletov, V., Kerr, J. (1998) Satellite estimation of spectral surface UV irradiance in the presence of tropospheric Aerosols. 1. Cloud-free case. *J. Geophys. Res.*, in press.

Krueger, A. J. (1986) Rocket techniques for measurements of ozone and related neutral minor constituents. In: *Handbook for MAP*, R. A. Goldberg (ed.), **19**, pp. 29–55.

Krueger, A. J. (1989) The global distribution of total ozone: TOMS satellite measurements. *Planet. Space Sci.*, **37**, N12, 1,555–1,565.

Krueger, B. C., Neuber, R., El Naggar, S., Walther, H. (1989) Measurements of ozone profiles by a lidar method during the Arctic winter and spring 1988. In: *Proceedings of the Quadrennial Ozone Symposium, 1988, Gottingen, Germany*, R. D. Bojkov, P. Fabian (eds.), Deepak Publications, Hampton, VA., pp. 160–163.

Kunzi, K. F., Rubin, G. (1988) The use of microwave sensors in a global ozone monitoring network. *International Radiation Symposium, 18–24 August 1988, Lille, France.*

Kurosu, T., Rozanov, V. V., Burrows, J. P. (1997) Parameterization schemes for terrestrial water clouds in the radiative transfer model GOMETRAN. *J. Geophys. Res.*, **102** , ND18 , 21,809–21,823

Labow, G. J., Flynn, L. E., Rawlins, M. A., Beach, R. A., Simmons, C. A. (1996) Estimation of ozone with total ozone portable spectroradiometer instruments. II. Practical and comparisons. *Appl. Optics*, **35**, N30, 6,084–6,089.

Larsen, N. (1991) *Polar Stratospheric Clouds: A Microphysical Simulation Model.* Danish Meteorological Institute, Copenhagen (DMI Scientific Report 91–2).

Larsen, N., Knudsen, B., Steen Mikkelsen, I., Jargensen, T. S., Eriksen, P. (1994) Ozone depletion in the Arctic stratosphere in early 1993. *Geophys. Res. Lett.*, **21**, 1,611–1,614.

Lat, S., Subbaraya, B. H., Acharya, Y. B., Ramani, S. V., Borchers, R., Fabian, P. (1989) Vertical distribution of ozone over Hyderabad using a balloon-borne suntracking photometer. In: *Proceedings of the Quadrennial Ozone Symposium, 1988, Gottingen, Germany*, R. D. Bojkov, P. Fabian (eds.), Deepak Publications, Hampton, VA., pp. 136–139.

Laurila, T. (1995) Ozone in Finland. Formation, concentrations and effects. *Finish Air Pollution Prevention News*, N3, 9–11.

Laurila, T., Plathan, P., Hakola, H., Koskinen, T., Lätlilä, H. (1995) Pallas, a site for studies of photochemically-active trace species in the European Arctic. *EUROTRAC Newsletter*, N15, 2–4.

Law, K. S., Nisbet, E. G. (1996) Sensitivity of the CH_4 growth rate to changes in CH_4 emissions from natural gas and coal. *J. Geophys. Res.*, **101**, ND9, 14,387–14,397.

Lefèvre, F. *et al.* (1998) The 1997 Arctic ozone depletion quantified from three-dimensional model simulations. *Geophys. Res. Lett.*, **25**, 2,425–2,428.

Legrand, M., Léopold, A., Dominé, F. (1996) Acid gases (HCl, HF, HNO_3, NCOOH, and CH_3COOH): A review of ice core data and some preliminary discussions on their air–snow relationships. In: *Chemical Exchange Between the Atmosphere and*

Polar Snow, E. W. Wolff, R. C. Bales (eds.), NATO ASI Series Vol. 143, Springer-Verlag, Berlin, Heidelberg, pp. 19–43.

Legrand, M., Pasteur, E. C. (1998) Methane sulfonic acid to non-sea-salt sulfate ratio in coastal Antarctic aerosol and surface snow. *J. Geophys. Res.*, **103**, ND9, 10,991–11,006.

Legrand, M., Ducroz, F., Wagenbach, D., Mulvaney, R., Hall, J. (1998) Ammonium in coastal Antarctic aerosol and snow: Role of polar ocean and penguin emissions. *J. Geophys. Res.*, **103**, ND9, 11,043–11,056.

Liptay, K., Chanton, J., Czepiel, P., Mosher, B. (1998) Use of stable isotopes to determine methane oxidation in landfill cover soils. *J. Geophys. Res.*, **103**, ND7, 8,243–8,250.

Liu, X., Blatherwick, R. D., Murcray, F. J., Keys, J. G., Solomon, S. (1992) Measurements and model calculations of HCl column amounts and related parameters over McMurdo during the Austral spring in 1989. *J. Geophys. Res.*, **97**, ND18, 20,795–20,804.

Lobert, J. M., Butler, J. H., Montzka, S. A., Geller, L. S., Myers, R. C., Elkins, W. (1995) A net sink for atmospheric CH_3Br in the East Pacific Ocean. *Science*, **267**, 1,002–1,005.

Logan, J. (1994) Trends in the vertical distribution of ozone: An analysis of ozonesonde data. *J. Geophys. Res.*, **99**, 25,553–25,585.

London, J., Liu, S. C. (1992) Long-term tropospheric and lower stratospheric ozone variations from ozonesonde observations. *J. Terr. Atmos. Phys.*, **54**, N5, 599–625.

DeLuisi, J. J., Longenecker, D. U., Mateer, C. L., Wuebbles, D. J. (1989) An analysis of Northern mid-latitude Umkehr measurements corrected for stratospheric aerosol for 1979–1986. *J. Geophys. Res.*, **94**, ND7, 9,837–9,846.

Manney, G. L. *et al.* (1994) Chemical depletion of ozone in the Arctic lower stratosphere during winter 1992–93. *Nature*, **370**, 429–434.

Manney, G. L., Zurek, R. W., Froidevaux, L., Waters, J. W. (1995a) Evidence for Arctic ozone depletion in late February and early March 1994. *Geophys. Res. Lett.*, **22**, 2,941–2,944.

Manney, G. L., Froidevaux, L., Waters, J. W., Zurek, R. W., Gille, J. C., Kumer, J. B., Mergenthaler, J. L., Roche, A. E., O'Neill, A., Swinbank, R. (1995b) Formation of low ozone pockets in the middle stratosphere anticyclone during winter. *J. Geophys. Res.*, **100**, 13,939–13,950.

Manney, G. L., Santee, M. L., Froidevaux, L., Waters, J. W., Zurek, R. W. (1996a) Polar vortex conditions during the 1995–96 Arctic winter: meteorology and MLS ozone. *Geophys. Res. Lett.*, **23**, 3,203–3,206.

Manney, G. L., Froidevaux, L., Waters, J. W., Santee, M. L., Read, W. C., Flower, D. A., Jarnot, R. F., Zurek, R. W. (1996b) Arctic ozone depletion observed by UARS MLS during the 1994–95 winter. *Geophys. Res. Lett.*, **23**, 85–88.

Manney, G. L., Froidevaux, L., Santee, M. L., Zurek, R. W., Waters, J. W. (1997) MLS observations of Arctic ozone loss in 1996–1997. *Geophys. Res. Lett.*, **24**, N22, 2,697–2700.

Mayer, M., Putz, E. (1998) Stratospheric trace gas measurements by means of DOAS at an Alpine monitoring station in Austria. In: *Atmospheric Ozone. Proceedings of the XVIII Quadrennial Ozone Symposium, 12–21 September 1996, L'Aquila, Italy*, R. D. Bojkov, G. Visconti (eds.), Parco Scientifico e Tecnologico d'Abruzzo, L'Aquila, Italy, **1**, 65–68.

De Maziére, M., Van Roozendael, M., Hermans, C., Simon, P. C., Demoulin, P., Roland, G., Zander, R. (1998) Quantitative evaluation of the post-Mount Pinatubo NO_2 reduction and recovery, based on 10 years of Fourier transform infrared and UV-visible spectroscopic measurements at Jungfraujoch. *J. Geophys. Res.*, **103**, ND9, 10,849–10,858.

Mast, G. M., Saunders, H. E. (1962) Research and development of the instrumentation of ozone sensing. *Inst. Soc. Am. Trans.*, **1**, 325.

Matafonov, G. K., Kazimirovsky, E. S. (1998) Continental scale and orographic 'structures' in the global distribution of total ozone content. *J. Atmos. Solar-Terr. Phys.*, **60**, N10, 993–995.

Mateer, C. L., DeLuisi, J. J. (1985) A comparison of ozone profiles derived from Standard Umkehr and Short Umkehr measurements from 15 stations, In: *Atmospheric Ozone. Proceedings of the Quadrennial Ozone Symposium, Halkidiki, Greece*, C. S. Zerefos, A. Ghazi (eds.), pp. 290–294, D. Reidel, Dordrecht.

Mateer, C. L., DeLuisi, J. J. (1992) A new Umkehr inversion algorithm. *J. Atmos. Terr. Phys.*, **54**, N5, 537–556.

Mauersberger, K., Anderson, S., Murphy, D., Morton, J. (1985) Precise ozone measurements using a mass spectrometer beam system. In: *Atmospheric Ozone. Proceedings of the Quadrennial Ozone Symposium, Halkidiki, Greece*, C. S. Zerefos, A. Ghazi, D. Reidel, Hingham, MA, p. 493.

Maudin, L. E., III, McCormick, Jr., M. P., McMaster, L. R., Vaughan, W. R. (1985) The stratospheric aerosol and gas experiment II (SAGE II) design and in-orbit performance. *Proc. Soc. Photo-Opt. Instrum. Eng.*, **589**, 104.

McConnell, J. R., Bales, R. C., Stewart, R. W., Thompson, A. M., Albert, M. R., Ramos, R. (1998) Physically based modeling of atmosphere-to-snow-to-firn transfer of H_2O_2 at South Pole. *J. Geophys. Res.*, **103**, ND9, 10,561–10,570.

McCormack, J. P., Hood, L. L. (1996) Apparent solar cycle variations of upper stratospheric ozone and temperature: Latitude and seasonal dependencies. *J. Geophys. Res.*, **101**, ND15, 20,933–20,944.

McCormack, J. P., Hood, L. L., Nagatani, R., Miller, A. J., Planet, W. G., McPeters, R. D. (1997) Approximate separation of volcanic and 11-year signals in the SBUV-SBUV/2 total ozone record over the 1979–1995 period. *Geophys. Res. Lett.*, **24**, N22, 2,729–2,732.

McCormack, J. P., Miller, A. J., Nagatani, R., Fortuin, J. P. F. (1998) Interannual variability in the spatial distribution of extratropical total ozone. *Geophys. Res. Lett.*, **24**, N12, 2,153–2,156.

McCormick, M. P., Zavodny, J. M., Veiga, R. E., Larsen, J. C., Wang, P. H. (1989) An overview of SAGE I and II ozone measurements. *Planet. and Space Sci.*, **37**, N12, 1,567–1,586.

McCormick, M. P., Veiga, R. E., Chu, W. P. (1992) Stratospheric ozone profile and total ozone trends derived from SAGE I and SAGE II data. *Geophys. Res. Lett.*, **19**, N3, 269–272.

McCormick, M. P., Thomason, L. W., Trepte, Ch. R. (1995) Atmospheric effects of the Mt. Pinatubo eruption. *Nature*, **373**, 399–404.

McDermid, I. S., Godin, S., Lindquist, L. O. (1990) Ground based lidar DIAL system for long term measurements of stratospheric ozone. *Appl. Opt.*, **29**, N25, 3,603–3,612.

McElroy, C. T., Mateer, C. L., Kerr, J. B., Wardle, D. I. (1989) Umkehr observations made with the Brewer ozone spectrophotometer. *Proceedings of the Quadrennial Ozone Symposium, 1988, Gottingen, Germany*, R. D. Bojkov, P. Fabian (eds.), Deepak Publications, Hampton, VA.

McElroy, C. T. *et al.* (1991) SPEAM-I (SunPhotometer Earth Atmosphere Measurement) observations of high-altitude ozone from STS41-G. *Can. J. Phys.*, **69**, 1,123–1,127.

McElroy, C. T., Hare, E. W., Kerr, J. B. (1994) Ozone trends estimated from Umkehr measurements made at Edmonton Alberta, Canada. In: *Ozone in the Troposphere and Stratosphere. Proceedings of the Quadrennial Ozone Symposium, Charlottesville, VA*, R. D. Hudson (ed.), NASA Conference Publication, pp. 199–202.

McKenna, D. S., Jones, R. L., Austin, J., Browell, E. V., McCormick, M. P., Kruger, A. J., Tuck, A. F. (1989) Diagnostic studies of the Antarctic vortex during the 1987 Airborne Antarctic Ozone Experiment: Ozone miniholes. *J. Geophys. Res*, **94**, 11,641–11,669.

McKenzie, R. L., Johnston, P. V. (1982) Seasonal variations in stratospheric NO_2 at 45°S. *Geophys. Res. Lett.*, **9**, 1,255.

McPeters, R. D., Labow, G. J. (1996) An assessment of the accuracy of 14.5 years of Nimbus 7 TOMS version 7 ozone data by comparison with the Dobson network. *Geophys. Res. Lett.*, **23**, 3,695–3,698.

McPeters, R. D. *et al.* (1996) *Nimbus-7 Total Ozone Mapping Spectrometer (TOMS) Data Product's User's Guide*. NASA Ref. Publ. Panel, National Aeronautics and Space Administration, Washington, DC.

McPeters, R. D. Labow, G. J. (1996) An assessment of the accuracy of 14.5 years of Nimbus 7 TOMS Version 7 ozone data by comparison with the Dobson network. *Geophys. Res. Lett.*, **23**, 3,695–3,698.

Megie, G., Allain, J. Y., Chanin, M. L., Blamont, J. E. (1977) Vertical profile of stratospheric ozone by lidar sounding from the ground. *Nature*, **270**, 329.

Megie, G., Chanin, M. L., Ehhalt, D., Fraser, P., Frederic, J. F., Gille, J. C., McCormick, M. P., Schoeberl, M. (1989) Global trends. In: *Scientific Assessment of Stratospheric Ozone Depletion: 1998. WMO Global Ozone Research and Monitoring Project, Report No. 20*, WMO, Geneva: Switzerland.

Meier, A., Blumenstock, T., Nakajima, H. (1998) Regular isotopic observation of stratospheric ozone and its implications for the ozone formation theory. *Proceedings of the 4th European Symposium on Stratospheric Ozone, Air Pollution Research Report 66*, European Commission, pp. 216–219.

Michelson, H. A., Salawitch, R. J., Gunson, M. R. *et al.* (1996) Stratospheric chlorine partitioning: Constraints from shuttleborne measurements of [HCl], [$ClNO_3$], and [ClO]. *Geophys. Res. Lett.*, **23**, N17, 2,361–2,364.

Miller, A. J., Nagatani, R. M., Tiao, G. C., Niu, X. F., Reinsel, G. C., Wuebbles, D., Grant, K. (1992) Comparisons of observed ozone and temperature trends in the lower stratosphere. *Geophys. Res. Lett.*, **19**, N9, 929–932.

Minkin, A., Legrand, M., Hall, J., Wagenbach, J., Kleefeld, C., Wolff, E., Pasteur, E. C., Ducroz, F. (1998) Sulfur-containing species (sulfate and methanesulfonate) in coastal Antarctic aerosol and precipitation. *J. Geophys. Res.*, **103**, ND9, 10,975–10,990.

Mo, R., Bühler, O., McIntyre, M. E. (1998) Permeability of the stratospheric vortex edge: On the mean mass flux due to thermally dissipating, steady, non-breaking Rossby waves. *Quart. J. Roy. Meteorol. Soc.*, **124**, Part A, N550, 2,129–2,148.

Morcrette, J.-J. (1990) Impact of changes to the radiation transfer parametrizations plus cloud optical properties in the ECMWF model. *Mon. Weather Rev.*, **118**, 847–873.

Morcrette, J.-J. (1991) Radiation and cloud radiative properties in the European Centre for Medium Range Weather Forecasts forecasting system. *J. Geophys. Res.*, **96**, 9,121–9,132.

Moreau, G., Robert, C., Rigaud, P. (1989) A new multipass ozonometer. In: *Proceedings of the Quadrennial Ozone Symposium, 1988, Gottingen, Germany*, R. D. Bojkov, P. Fabian, Deepak Publications, Hampton, VA., pp. 755–758.

Morrey, M. W., Harwood, R. S. (1998) Interhemispheric differences in stratospheric water vapor during late winter, in version 4 MLS measurements. *Geophys. Res. Lett.*, **25**, N2, 147–150.

Morris, G. A. *et al.* (1995) Trajectory mapping and applications to data from the Upper Atmosphere Research Satellite. *J. Geophys. Res.*, **100**, 16,491–16,505.

Morris, G. A., Kawa, S. R., Douglass, A. R., Schoeberl, M. R., Froidevaux, L., Waters, J. (1998) Low-ozone pockets explained. *J. Geophys. Res.*, **103**, ND3, 3,599–3,610.

Morys, M., Mims, F. M., III, Anderson, S. E. (1996) Design, calibration and performance of MICROTOPS II hand-held ozonometer. *12th International Symposium on Photobiology, Solar Light Co., 1996, Vienna.*

Mosier, A. R., Duxbury, J. M., Freney, J. R., Heinemeyer, O., Minami, K. (1998a) Assessing and mitigating N_2O emissions from agricultural soils. *Clim. Change*, **40**, N1, 7–38.

Mosier, A. R., Duxbury, J. M., Freney, J. R., Heinemeyer, O., Minami, K., Johnson, D. E. (1998b) Mitigating agricultural emissions of methane. *Clim. Change*, **40**, N1, 39–80.

Mount, G. H., Sanders, R. W., Schmeltekopf, A. L., Solomon, S. (1987) Visible spectroscopy at McMurdo Station, Antarctica. 1. Overview and daily variations of NO_2 and O_3, austral spring. *J. Geophys. Res.*, **92**, 8,320.

De Muer, De Backer, H. (1992) Revision of 20 years of Dobson total ozone data at Uccle (Belgium): fictitious Dobson total ozone trends induced by sulfur dioxide trends. *J. Geophys. Res.*, **97**, ND5, 5,921–5,938.

Müller, R., Crutzen, P. J., Groob, J. U., Brühl, C., Russel, J. M., Tuck, A. F. (1996) Chlorine activation and ozone depletion in the Arctic vortex: Observations by the Halogen Occultation Experiment on the Upper Atmosphere Research Satellite. *J. Geophys. Res.*, **101**, 12,531–12,554.

Müller, R., Groob, J.-U., McKenna, D. S., Crutzen, P. J., Brühl, C., Russel, J. M., III, Tuck, A. F. (1997) HALOE observations of the vertical structure of chemical ozone depletion in the Arctic vortex during winter and early spring 1996–1997. *Geophys. Res. Lett.*, **24**, N22, 2,717–2,720.

Mulvaney, R., Wagenbach, D., Wolff, E. W. (1998) Postdepositional change in the snowpack nitrate from observation of year-round near-surface snow in coastal Antarctica. *J. Geophys. Res.*, **103**, ND9, 11,021–11,031.

Murphy, D., (1991) Ozone loss rates calculated along ER-2 flight tracks. *J. Geophys. Res.*, **96**, 5,045–5,053.

Nair, H., Allen, M., Froidevaux, L., Zurek, R. W. (1998) Localized rapid ozone loss in the northern winter stratosphere: An analysis of UARS observations. *J. Geophys. Res.*, **103**, ND1, 1,573–1,584.

Nakajima, T. Y., Nakajima, T., Nakajima, M., Fukushima, H., Kuji, M., Uchiyama, A., Kishino, M. (1998) Optimization of the Advanced Earth Observing Satellite II Global Imager channels by use of radiative transfer calculations. *Appl. Optics*, **37**, N15, 3,149–3,163.

Naturajan, M., Callis, L. B. (1997) Ozone variability in the high-latitude summer stratosphere. *Geophys. Res. Lett.*, **24**, N10, 1,191–1,194.

Naujokat, B., *et al.* (in press) The stratospheric winter 1996/97: extending later into spring than ever before, *Beilage zur Berliner Wetterkarte*, in press.

Nedoluha, G. E., Bevilacqua, R. M., Gomez, R. M., Siskind, D. E., Hicks, B. C., Russel, J. M., III, Conner, B. J. (1998) Increases in middle atmospheric water vapor as observed by Halogen Occultation Experiment and ground-based Water-Vapor Millimeter-wave Spectrometer from 1991–1997. *J. Geophys. Res.*, **103**, ND3, 3,531–3,543.

Neuendorffer, A. C. (1996) Ozone monitoring with TIROS-N operational vertical sounders. *J. Geophys. Res.*, **101**, 18,807–18,828.

Newell, R. E., Shipley, S. T., Connors, V. S., Reichle, H. G., Jr. (1988) Regional studies of potential carbon monoxide sources based on space shuttle and aircraft measurements. *J. Atmos. Chemistry*, **6**, 61–81.

Newman, P. L., Lait, R., Schoeberl, M., Nash, E. R., Kelly, K., Fahey, D. W., Nagatani, R., Toohey, D., Avallone, L., Anderson, J. (1993) Stratospheric meteorological conditions in the Arctic polar vortex, 1991 to 1992. *Science*, **261**, 1,143–1,145.

Newman, A. (1994) CFC phase-out moving quickly. *Environ. Sci. Technol.*, **28**, N1, 35A–37A.

Newman, P. A., Gleason, J. F., McPeters, R. D., Stolarski, R. S. (1997) Anomalously low ozone over the Arctic. *Geophys. Res. Lett.*, **24**, N22, 2,689–2,692.

Norton, W. A., Carver, G. D. (1994) Visualizing the evolution of the polar vortex in January 1992. *Geophys. Res. Lett.*, **21**, 1,455–1,458.

Norton, W. A. (1994) Breaking Rossby waves in a model stratosphere diagnosed by a vortex following coordinate system and a technique for advecting material contours. *J. Atmos. Sci.*, **51**, N4, 654–673.

Norton, W. A., Chipperfield, M. P. (1996) Quantification of the transport of chemically activated air from the northern hemisphere polar vortex. *J. Geophys. Res.*, **100**, 25,817–25,840.

Noxon, J. F., Whipple, E. C., Jr., Hyde, R. S. (1979) Stratospheric NO_2. 1. Observation method and behavior at mid-latitude. *J. Geophys. Res.*, **84**, 5,047.

Ogawa, T., Koike, M., Suzuki, K. (1989) Observation of vertical ozone profiles with the EXOS-C backscattered UV spectrometer. In: *Proceedings of the Quadrennial Ozone Symposium, 1988, Gottingen, Germany*, R. D. Bojkov, P. Fabian, Deepak Publications, Hampton, VA., pp. 214–217.

Ogawa, T., Ninong-Komala (1989) Diurnal and seasonal variations of the tropospheric ozone in tropical Asia. In: *Proceedings of the Quadrennial Ozone Symposium, 1988, Gottingen, Germany*, R. D. Bojkov, P. Fabian, Deepak Publications, Hampton, VA, pp. 437–440.

Olafson, R. A., Asbridge, I. A. (1981) Stray light in the Dobson spectrophotometer and its effect on ozone measurements. *Proceedings of the Quadrennial Ozone Symposium, Boulder, CO*, J. London (ed.), **1**, 46–47, NOAA, Boulder.

Orsolini, Y. J., Cariolle, D., Deque, M. (1995a) Filamentation and layering of an idealised tracer by observed winds in the lower stratosphere. *Geophys. Res. Lett.*, **22**, 839–842.

Orsolini, Y. J., Simon, P., Cariolle, D. (1995b) On the formation of ozone laminae at the edge of the polar vortex. *Quart. J. Roy. Meteorol. Soc.*, **121**, 1,923–1,941.

Orsolini, Y.J., Hansen, G., Hoppe, U. P., Manney, G., Fricke, K. (1997) Dynamical modelling of wintertime lidar observations in the Arctic: Ozone laminae and ozone depletion. *Quart. J. Roy. Meteorol. Soc.*, **123**.

Ostermann, G. B. (1999) The budget and partitioning of reactive nitrogen species in the Arctic stratosphere, *Geophys. Res. Lett.*, **26**, 1,157–1,160.

Pan, L., Solomon, S., Randel, W., Lamarque, J.-F., Hess, P., Gille, J., Chiou, E.-W., McCormick, M. P. (1997) Hemispheric asymmetries and seasonal variations of the lowermost stratospheric water vapor and ozone derived from SAGE II data. *J. Geophys. Res.*, **102**, ND23, 28,177–28,184.

Parnes, S. (1992) Israeli students begin work on Techsat 1 satellite. *Space News*, 34.

Parrish, D. D., Trainer, M., Holloway, J. S., Yee, J. E., Warshawsky, M. S., Fehsenfeld, F. C., Forbes, G. L., Moody, J. L. (1998) Relationships between ozone and carbon monoxide at surface sites in the North Atlantic region. *J. Geophys. Res.*, **103**, ND11, 1,357–1,376.

Parrish, A., Connor, B. J., Tsou, J. J. (1998) Ground-based microwave measurements of ozone profiles following the Pinatubo eruption. In: *Atmospheric Ozone. Proceedings of the XVIII Quadrennial Ozone Symposium, 12–21 September 1996, L'Aquila, Italy*, R. D. Bojkov, G. Visconti (eds.), Parco Scientifico e Tecnologico d'Abruzzo, L'Aquila, Italy, **1**, pp. 155–158.

Payan, S., Camy-Peyret, C., Jeseck, P., Hawat, T., Durry, G., Lefèvre, F. (1998) First direct simultaneous HCl and $ClONO_2$ profile measurements in the Arctic vortex. *Geophys. Res. Lett.*, **25**, N14, 2,663–2,666.

Pelon, J., Megie, G. (1982) Ozone monitoring in the troposphere and lower stratosphere: Evaluation and operation of a ground-based lidar station. *J. Geophys. Res.*, **87**, 4,947.

Pelon, J., Godin, S., Megie, G. (1986) Upper stratospheric (30–50 km) lidar observations of the ozone vertical distribution. *J. Geophys. Res.*, **91**, 8,667.

Perluski, L. M., London, J. (1989) Satellite observed long-term averaged seasonal and spatial ozone variations in the stratosphere. *Planet. Space Sci.*, **37**, N12, 1,509–1,525.

Perluski, L. M., Solomon, S., London, J. (1989) On the interpretation of seasonal variations of stratospheric ozone. *Planet. Space Sci.*, **37**, N12, 1,527–1,538.

Perov, S. P., Tishin, S. V. (1989) Diurnal variations of ozone in the mesosphere. In: *Proceedings of the Quadrennial Ozone Symposium, 1988, Gottingen, Germany*, R. D. Bojkov, P. Fabian, Deepak Publications, Hampton, VA, pp. 399–402.

Peter, T. (1999) Physico-chemistry of polar stratospheric clouds. In: *Ice Physics and the Natural Environment*, J. S. Wettlaufer, J. G. Dash, N. Understeiner (eds.), NASA ASI Series 1. *Global Environmental Change*, **56**, 143–168.

Peter, R., Caliseri, Y., Kämpfer, N. (1998) Variability of middle atmospheric ozone abundances derived from continuous ground-based millimeter wave measurements. *Proceedings of the XVIII Quadrennial Ozone Symposium*, R. D. Bojkov, G. Visconti (eds.), Parco Scientifico e Tecnologico d'Abruzzo, L'Aquila, Italy, **2**, 559–562.

Pilinis, C., King, D. B., Saltzman, E. S. (1996) The oceans: A source or a sink of methyl bromide? *Geophys. Res. Lett.*, **23**, N8, 817–820.

Planet, W. G., Lienesch, J. H., Miller, A. J., Nagatani, R., McPeters, R. D., Hilsenrath, E., Cebula, R. P., DeLand, M. T., Wellemeyer, C. G., Horvath, K. (1994) Northern hemisphere total ozone values from 1989–1993 determined with the NOAA-II Solar Backscatter Ultraviolet (SBUV/2) instrument. *Geophys. Res. Lett.*, **21**, N3, 205–208.

Pommereau, J. P., Goutail, F. (1988a) O_3 and NO_2 ground based measurements by visible spectrometry during Arctic winter and spring 1988. *Geophys. Res. Lett.*, **15**, 891–895.

Pommereau, J. P., Goutail, F. (1988b) O_3 and NO_2 observations at the Southern Polar Circle in summer and fall 1988. *Geophys. Res. Lett.*, **15**, 895.

Pommereau, J.-P., Piquard, J. (1994a) Observations of the vertical distribution of stratospheric OClO. *Geophys. Res. Lett.*, **21**, 1,231–1,234.

Pommereau, J.-P., Piquard, J. 1994b) Ozone and nitrogen dioxide vertical distributions by UV-visible solar occultation from balloons. *Geophys. Res. Lett.*, **21**, 1,227–1,230.

Pougatchev, N. S., Connor, B. J., Rinsland, C. P. (1995) Infrared measurements of the ozone vertical distribution above Kitt Peak. *J. Geophys. Res.*, **100**, 16,689–16, 698.

Pougatchev, N. S., Connor, B. J., Jones, N. B., Rinsland, C. P. (1998) Ground-based infrared measurements of the ozone vertical distribution in the troposphere and lower stratosphere. *Proceedings of the XVIII Quadrennial Ozone Symposium*, R. D. Bojkov, G. Visconti, **1**, pp. 167–170.

Prasad, S. S. (1997) Potential atmospheric sources and sinks of nitrous oxide. 2. Possibilities from excited O_2, 'embryonic' O_3, and optically pumped excited O_3. *J. Geophys. Res.*, **102**, ND17, 21,527–21,536.

Prasad, S.S. (1998) Potential new atmospheric sources and sinks of odd nitrogen: Sources involving the excited O_2 and the N_2O O_3 species. *Geophys. Res. Lett.*, **25**, N12, 2,173–2,176.

Proffitt, M. H., McLaughlin, R. J. (1983) Fast response dual-beam UV-absorption ozone photometer for use on stratospheric balloons. *Rev. Sci. Instrum.*, **54**, 1,719.

Proffitt, M. H., Margitan, J. J., Kelly, K. K., Loewenstein, M., Podolske, J. R., Chan, K. R. (1990) Ozone loss in the Arctic polar vortex inferred from high-altitude aircraft measurements. *Nature*, **347**, 31–36.

Proffitt, M. H., Aiken, K., Margitan, J. J., Loewenstein, M., Podolske, J. R., Weaver, A., Chan, K. R., Fast, H., Elkins, J. W. (1993) Ozone loss inside the northern polar vortex during the 1991–1992 winter. *Science*, **261**, 1,150–1,154.

Pundt, I., Pommereau, J.-P. (1998) Investigation of stratospheric bromine and iodine oxides using the SAOZ balloon sonde. *Proceedings of the XVIII Quadrennial Ozone Symposium*, R. D. Bojkov, G. Visconti (eds.), Parco Scientifico e Tecnologico d'Abruzzo, L'Aquila, Italy, **2**, 575–578.

Pyle, J. A., Chipperfield, M. P., Kilbane-Dawe, I., Lee, A. M., Stimpfle, R. M., Kohn, D., Renger, W., Waters, J. W. (1995) Early modelling results from the SESAME and ASHOE campaigns. *Faraday Discussions*, **100**.

Radionov, V. E., Sveshnikov, A. M. (1992) Dynamics of total ozone in the Antarctic. *Investigations of Atmospheric Ozone*, Gidrometeoizdat, Moscow, pp. 20–24 (in Russian).

Randall, D. W., Rusch, C. E., Clancy, R. T., Bevilacqua, R. M., Shettle, E. P., Hornstein, J. H., Lumpe, J., Krigman, S. S., Fromm, M., Debrestian, D., Olivero, J. J. (1995) Preliminary results from POAM II: Stratospheric ozone densities at northern latitudes. *Geophys. Res. Lett.*, **21**, 2,733–2,336.

Randel, W. J. (1993) Global variations of zonal mean ozone during stratospheric warming events. *J. Atmos. Sci.*, **50**, N19, 3,308–3,321.

Randel, W. J., Cobb, J. B. (1994) Coherent variations of monthly mean total ozone at low stratospheric temperature. *J. Geophys. Res.*, **99**, 5,433–5,447.

Rathman, W., Monks, P. S., Llwellyn-Jones, D., Burrows, J. P. (1997) A preliminary comparison between TOVS and GOME level 2 ozone data. *Geophys. Res. Lett.*, **24**, N17, 2,191–2,194.

Reburn, W. J., Siddans, R., Munro, R., Kerridge, B. J. (1998) Upper troposphere/lower stratosphere constituent retrieval simulations for AMAS. *Proceedings of the XVIII Quadrennial Ozone Symposium*, R. D. Bojkov, G. Visconti (eds.), Parco Scientifico e Tecnologico d'Abruzzo, L'Aquila, Italy, **2**, 583–586.

Regener, V. H. (1960) On a sensitive method for the recording of atmosphere ozone. *J. Geophys. Res.*, **65**, 3.975–3,977.

Regener, V. H. (1964) Measurement of atmospheric ozone with the chemiluminescent method. *J. Geophys. Res.*, **69**, 3,795.

Reid, S. J., Timmis, C. (1996) Source regions for ozone laminae. Submitted to *Quart. J. Roy. Meteorol. Soc.*

Reid, S. J., Vaughan, G. (1991) Lamination in ozone profiles in the lower stratosphere. *Quart. J. Roy. Meteorol. Soc.*, **117**.

Reid, S. J., Vaughan, G., Kyrö, E. (1993). Occurrence of ozone laminae near the boundary of the stratospheric polar vortex. *J. Geophys. Res.*, **98**.

Reid, S. J. (1993) Stratification in the lower stratosphere. PhD thesis, University of Wales.

Reid, S. J. , Rex, M., von der Gathen, P., Fløisand, I., Stordal, F., Carver, G. D., De Haan, L. L., Reimer, E., Krüger-Carstensen, R., Kyrö, E., O'Connor, F. M., Braathen, G. O., Murphy, G., Varotsos, C., Wenger, J., Zerefos, C. (1998) A study of ozone laminae using quasi-isentropic trajectories, contour advection and photochemical model simulations. *J. Atmos. Chemistry*, **30**, 187–207

Reid, S. J., Vaughan, G., Mitchell, N., Prichard, I. T., Smit, H. G. J., Jorgensen, T. S., Varotsos, C., de Backer, H. (1994) Distribution of ozone laminae during EASOE and the possible influence of inertia-gravity waves. *Geophys. Res. Lett.*, **21**.

Reid, S. J., von der Gathen, P., Rex, M., Reimer, E., Krüger-Carstensen, R., De Haan, L. L., Braathen, G. O., Burphy, G., Wenger, J., Zerefos, C., Varotsos, C. (1996) The evolution of ozone laminae using the match approach. In: *Proceedings of the Third European Symposium on Polar Stratospheric Ozone*, Office for Official Publications of the EC, Belgium, p. 590.

Reinsel, G. C., Tiao, G. C., Miller, A. J., Wuebbles, D. J., Connell, P. S., Mateer, C. L., DeLuisi, J. J. (1987) Statistical analysis of total ozone and stratospheric Umkehr data for trends and solar cycle relationships. *J. Geophys Res.*, **92**, ND2, 22,101–22,109.

Reinsel, G. C., Tiao, G. C., DeLuisi, J. J., Basu, S., Carriere, K. (1989) Trend analysis of aerosol-corrected Umkehr ozone profile data through 1987. *J. Geophys. Res.*, **94**, ND13, 16,373–16,386.

Reinsel, G. C., Tiao, G. C., Wuebbles, D. J., Kerr, J. B., Miller, A. J., Nagatani, R. M., Bishop, L., Ying, L. H. (1994) Seasonal trend analysis of published ground-based and TOMS total ozone data through 1991. *J. Geophys. Res.*, **99**, 5,449–5,464.

Reinsel, G. C., Tam, W.-K., Ying, L. H. (1994) Comparison of trend analyses for Umkehr data using new and previous inversion algorithm. *Geophys. Res. Lett.*, **21**, 1,007–1,010.

Rex, M. (1993) Stratosphärische Ozonabbauraten aus den Ozonsondendaten der EASOE-Kampagne. PhD thesis, Georg August University, Göttingen, Germany.

Rex, M., von der Gathen, P., Harris, N. R. P., Reimer, E., Beck, A., Alfier, R., Krüger-Carstensen, R., Knudsen, B. M., Mikkelsen, I. S., Chipperfield, M., Lucic, D., Allaart, M., De Backer, H., Braathen, G. O., Reid, S., Claude, H., O'Connor, F., Dier, H., Fast, H., Gamma, A., Gil, M., Godin, S., Guirlet, M., Kyrö, E., Rummukainen, M., Litynska, Z., Kois, B., Murphy, G., Ravegnani, F., Varotsos, C., Wenger, J., Yushkov, V., Dorokhov, V., Zerefos, C., Balis, D., Ziomas, I. (1996) Chemical ozone loss in the Arctic winters 1991/92 and 1994/95 (match). In: *Proceedings of the Third European Symposium on Polar Stratospheric Ozone*, Office for Official Publications of the EC, Belgium, p. 586.

Rex, M. *et al.* (1997) Prolonged stratospheric ozone loss in the 1995/96 Arctic winter. *Nature*, **389**, 835–838.

Rex, M., von der Gathen, P., Harris, N. R. P., Lucic, D., Knudsen, B. M., Braathen, G. O., Reid, S. J., De Backer, H., Claude, H., Fabian, R., Fast, H., Gil, M., Kyrö, E., Mikkelsen, I. S., Rummukainen, M., Smit, H. G., Stähelin, J., Varotsos, C., Zaitcev, I. (1998) In situ measurements of stratospheric ozone depletion rates in the Arctic winter 1991/1992: A Lagrangian approach. *J. Geophys. Res.*, **103**, ND5, 5,843–5,853.

Rex, M. *et al.* (1999) Chemical ozone loss in the Arctic winter 1994/95 as determined by the match technique. *J. Atm. Chem.*, **24**, 35–59.

Ricaud, P., Monnier, E., Goutail, F., Pommereau, J.-P., David, C., Godin, S., Froidevaux, L., Waters, J. W., Mergenthaler, J., Roche, A. E., Pumphrey, H., Chipperfield, M. P. (1998) Stratosphere over Dumont d'Urville, Antarctica, in winter 1992. *J. Geophys. Res.*, **103**, ND11, 13,267–13,284.

Rinsland, C. P., Boughner, R. E., Larsen, J. C., Stokes, G. M., Brault, J. W. (1984) Diurnal variations of atmospheric nitric oxide: ground-based infrared spectroscopic measurements and their interpretation with time-dependent photochemical model calculations. *J. Geophys. Res.*, **89**, ND6, 9,613–9,622.

Van Roozendael, M., Lambert, J. C., Simon, P. C., Hansen, G., DeMuer, D., Schoubs, E., Koopman, R., Vanderwoerd, H., Barbe, A., Claude, H., deLa Noe, J., Merienne, M. F., Staehelin, J. (1998) Ground-based validation of GOME total ozone measurements by means of Dobson, Brewer and GUV instruments. *Proceedings of the XVIII Quadrennial Ozone Symposium*, R. D. Bojkov, G. Visconti (eds.), Parco Scientifico e Tecnologico d'Abruzzo, L'Aquila, Italy, **2**, 665–668.

Roscoe, H. K., Kerridge, B. J., Pollitt, S., Louisnard, N., Flaud, L. M., Camy-Peyret, C., Alamichel, C., Pommerau, J.-P., Ogawa, T., Iwagami, N., Coffey, M. T., Mankin, W., Evans, W.F., McElroy, C. T., Kerr, J. (1990) Intercomparison of remote measurements of stratospheric NO and NO_2. *J. Atmos. Chem.*, **10**, N2, 111–114.

Roscoe, H. K. (1991) Review and revision of measurements of stratospheric N_2O_5. *J. Geophys. Res.*, **96**, ND6, 10,879–10,884.

Rosenfield, J. E., Newman, P. A., Schoeberl, M. R. J. (1994) Computations of diabatic descent in the stratospheric polar vortex. *J. Geophys. Res.*, **99**, 16,677–16,689.

Rozanov, V. *et al.* (1997) GOMETRAN: A radiative transfer model for the satellite project GOME – The plane-parallel version. *J. Geophys. Res.*, **102**, 16,683–16,695.

Rozanov, V., Kurosu, T., Burrows, J. P. (1998) Retrieval of atmospheric constituents in the UV-visible: A new quasi-analytical approach for the calculation of weighting functions. *JQSRT*, **60**, N2, 277–299.

Rubin, G., Künzi, K. F. (1989) Ground-based microwave sensor to retrieve stratospheric ozone profiles. In: *Proceedings of the Quadrennial Ozone Symposium, 1988, Gottingen, Germany*, R. D. Bojkov and P. Fabian (eds.), Deepak Publications, Hampton, VA, pp. 751–754.

Rusch, D. W., Randall, C. E., Callan, M. T., Horanyi, M., Clancy, R. T., Solomon, S. C., Oltmans, S. J., Johnson, B. J., Koehler, U., Claude, H., De Muer, D. (1998)

A new inversion for stratospheric aerosol and gas experiment II data. *J. Geophys. Res.*, **103**, ND7, 8,465–8,376.

Salby, M. L., Callaghan, P. F. (1993) Fluctuations of total ozone and their relationship to stratospheric air motions, *J. Geophys. Res.*, **98**, 2,715–2,727.

Sandor, B. J., Clancy, R. T. (1998) Mesospheric HO_x chemistry from diurnal microwave observations of HO_2, O_3, and H_2O. *J. Geophys. Res.*, **103**, ND11, 13,337–13,351.

Santee, M. L., Manney, G. L., Froidevaux, L., Zurek, R. W., Waters, J. W. (1997) MLS observations of ClO and HNO_3 in the 1996–1997 Arctic polar vortex. *Geophys. Res. Lett.*, **24**, N22, 2,713–2,716.

Santee, M. L., Tabazadeh, A., Manney, G. L., Salawitch, R. J., Froidevaux, L., Read, W. G., Waters, J. W. (1998) UARS Microwave Limb Sounder HNO_3 observations: Implications for Antarctic polar stratospheric clouds. *J. Geophys. Res.*, **103**, ND11, 13,285–13,313.

Sarkissian, A., Vaughan, G., Roscoe, H. K., Bartlett, L. M., O'Connor, F. M., Drew, D. G., Hughes, P. A., Moore, D. M. (1997) Accuracy of measurements of total ozone by a SAOZ ground-based zenith sky visible spectrometer. *J. Geophys. Res.*, **102**, 1,379–1,390.

Sasano, Y., Bodeker, G., Kreher, K. (1998) Preliminary ILAS observations of O_3, HNO_3, and N_2O in the Arctic stratosphere during the winter of 1996/97. *SPARC Newsletter*, N10, 18–19.

von Schéele, F. (1997) *Odin Programme Status and First Test Results*, IAF-97-IAA.11.1.08, Swedish Space Corporation, Solna, Sweden.

Schoeberl, M. R., Stolarski, R. S., Krueger, A. J. (1989) The 1988 Antarctic ozone depletion: Comparison with previous year depletion. *Geophys. Res. Lett.*, **16**, N5, 377–380.

Schoeberl, M. R., Proffitt, M. H., Kelly, K. K., Lait, L. R., Newman, P. A., Rosenfield, J. E., Loewenstein, M., Podolske, J. R., Strahan, S. E., Chan, K. R. (1990) Stratospheric constituent trends from ER-2 profile data. *Geophys. Res. Lett.*, **17**, 469–472.

Schwarzbach, S. E. (1995) Ozone depletion: CFC alternatives under a cloud. *Nature*, **376**, N6538, 297–298.

Scientific Assessment of Ozone Depletion Executive Summary: 1994 (1995) NOAA, NASA, UNEP and the WMO (jointly published).

Scientific Assessment of Ozone Depletion: 1998. WMO Global Ozone Research and Monitoring Project Report No. 44 (1999) WMO, Geneva, Switzerland.

Seckmeyer, G., Erb, R., Albold, A. (1996) Transmittance of a cloud is wavelength dependent in the UV range. *Geophys. Res. Lett.*, **23**, N20, 2,753–2,756.

Seftor, C. J., Hsu, N. C., Herman, J. R., Bhartia, P. K., Torres, O., Rose, W. I., Schneider, D. J., Krotkov, N. (1997) Detection of volcanic ash clouds from Nimbus-7/TOMS reflectivity data. *J. Geophys. Res.*, **102**, 16,749–16,759.

Sherlock, V. J., Jones, N. B., Matthews, W. A., Mucray, F. J., Blatherwick, R. D., Murcray, D. G., Goldman, A., Rinsland, C. P., Bernardo, C., Griffith, D. W. T. (1997) Increase in the vertical column abundance of HCFC-22 ($CHClF_2$) above Lauder, New Zealand, between 1985 and 1994. *J. Geophys. Res.*, **102**, ND7, 8,861–8,865.

Shiotani, M. (1992) Annual, quasibiannual and El Niño–Southern Oscillation (ENSO) time variations in equatorial total ozone. *J. Geophys. Res.*, **97**, 7,625–7,633.

Siddans, R., Munro, R., Reburn, W. J., Kerridge, B. J., Ellison, B., Moyna, B., Urban, J., Küllman, H., Kunzi, K. (1998) Stratospheric constituent retrievals from airborne 500 GHz measurements in SESAME Phase III. In: *Atmospheric Ozone. Proceedings of the XVIII Quadrennial Ozone Symposium, 12–21 September 1996, L'Aquila, Italy*, R. D. Bojkov, G. Visconti (eds.), Parco Scientifico e Tecnologico d'Abruzzo, L'Aquila, Italy, **2**, pp. 599–602.

Siegenthaler, A., Peter, R., Kämpfer, N. (1998) Ground-based millimeter wave observations of stratospheric chlorine monoxide and ozone above the Jungfraujoch in the winter 1996. In: *Atmospheric Ozone. Proceedings of the XVIII Quadrennial Ozone Symposium, 12–21 September 1996, L'Aquila, Italy*, R. D. Bojkov, G. Visconti (eds.), Parco Scientifico e Tecnologico d'Abruzzo, L'Aquila, Italy, **2**, pp. 603–606.

Simmons, A. J. (1992) Development of a high resolution, semi-Lagrangian version of the ECMWF forecast model. In: *Proceedings of the 1991 ECMWF Seminar on Numerical Methods in Atmospheric Models*, Vol. 2, pp. 281–324, European Centre for Medium Range Weather Forecast, Reading, England.

Simmons, A. J. (1993) Validation and development of the high resolution, semi-Lagrangian version of the ECMWF forecast model. In: *Proceedings of the 1991 ECMWF Seminar on Validation of Models Over Europe*, Vol. 1, pp. 153–182, European Centre for Medium Range Weather Forecast, Reading, England.

Solomonov, S. V., Kropotkina, E. P., Lujin, A. N., Ponomarenko, N. I., Rosanov, S. B. (1992) The remote sensing of the ozone layer at the millimeter waves. In: *Studies of Atmospheric Ozone*, N. A. Zaitseva, V. U. Khattatov (eds.), Gidrometeoizdat, Moscow, pp. 137–142 (in Russian).

Spänkuch, D., Schulz, E. (1998) Statistical estimate of the vertical ozone structure. *Idöjárás*, **102**, N4, 201–218.

Speuser, W., Sahand, S., Schurath, U. (1989) A novel fast response chemiluminescence sonde for routine soundings of stratospheric ozone up to 1.5 mb. In: *Proceedings of the Quadrennial Ozone Symposium, 1988, Gottingen, Germany*, R. D. Bojkov and P. Fabian (eds.), Deepak Publications, Hampton, VA, pp. 747–750.

Staehelin, J., Renaud, A., Bader, J., McPeters, R., Viatte, P., Hoegger, B., Bugnion, V., Giroud, M., Schill, H. (1998a) Total ozone series at Arosa (Switzerland): Homogenization and data comparison. *J. Geophys. Res.*, **103**, ND5, 5,827–5,842.

Staehelin, J., Kegel, R., Harris, N. R. P. (1998b) Trend analysis of the homogenized total ozone series of Arosa (Switzerland), 1926–1996. *J. Geophys. Res.*, **103**, ND7, 8,389–8,400.

Steffanutti, L., Sokolov, L., McKenzie, A. R., Peter, Th., Khattatov, V. H., Balestri, S. (1999) The MSS Geophysica as a platform for the Airborne Polar Experiment. *J. Atmos. Ocean. Technol.*, in press.

Stolarski, R. S., Bloomfield, P., McPeters, R. D., Herman, J. R. (1991) Total ozone trends deduced from Nimbus 7 TOMS data. *Geophys. Res. Lett.*, **18**, N6, 1,015–1,018.

Stolarski, R., Bojkov, R., Bishop, L., Zerefos, C., Staehelin, J., Zawodny, J. (1992) Measured trends in stratospheric ozone. *Science*, **256**, 342–344.

Stolarski, R. S., Labow, G. J., McPeters, R. D. (1997) Springtime Antarctic total ozone measurements in the early 1970s from the BUV instrument on Nimbus 4. *Geophys. Res. Lett.*, **24**, N5, 591–594.

Stordal, F., Isaksen, I. S. A., Horntvedt, K. (1985) A diabatic circulation two dimensional model with photochemistry: Simulations of ozone and long-lived tracers with surface sources. *J. Geophys. Res.*, **90**, 5,757–5,776.

Strahan, S. E., Rosenfield, J. E., Loewenstein, M., Podolske, J. R., Weaver, A. J. (1994) *J. Geophys. Res.*, **99**, 20,713–20,723.

Stratospheric Processes and Their Role in Climate (SPARC), Proceedings First SPARC General Assembly, Melbourne, Australia, 2–6 December 1996 (1997) WCRP-99 (WMO No. 814), WMO, Geneva, Switzerland, Vols. I, II, 672 pp.

Sussmann, R., Schaefer, K. (1997) Infrared spectroscopy of tropospheric trace gases. Combined analysis of horizontal and vertical column abundances. *Applied Optics*, **36**, N3, 735–741.

Sutton, R. (1994) Lagrangian flow in the middle atmosphere. *Quart J. Roy. Meteorol. Soc.*, **120**, 1,299–1,321.

Teitlbaum, H., Moustaoui, M., Van Velthoven, P. F. J., Kelder, H. (1998) Decrease of total ozone at low latitudes in the southern hemisphere by a combination of linear and nonlinear processes. *Quart. J. Roy. Meteorol. Soc.*, **125**, N552, Part B, 2,625–2,644.

The European Commission. *Second SESAME Press Release*, 30 March 1995.

Thomason, L. W., Poole, L. R., Deshler, T. R. (1997) A global climatology of stratospheric aerosol surface area density as deduced from SAGE II: 1984–1994. *J. Geophys. Res.*, **102**, 8,967–8,976

Thouret, V., Marenco, A., Nédélec, P., Grouhel, C. (1998a) Ozone climatologies at 9–12 km altitude as seen by the MOZAIC airborne program between September 1994 and August 1996. *J. Geophys. Res.*, **103**, ND19, 25,653–25,679.

Thouret, V., Marenco, A., Logan, J. A., Nédélec, P., Grouhel, C. (1998b) Comparisons of ozone measurements from the MOZAIC airborne program and the ozone sounding network at eight locations. *J. Geophys. Res.*, **103**, ND19, 25,965–25,720.

Tiao, G. C., Reinsel, G. C., Pedrick, J. H., Allenby, G. M., Mateer, C. L., Miller, A. J., DeLuisi, J. J. (1986) A statistical trend analysis of ozonesonde data. *J. Geophys. Res.*, **91**, ND12, 13,121–13,136.

Tibaldi, S., Palmer, T. N., Brankovic, C., Cuasch, U. (1990) Extended range predictions with ECMWF models 11: Influence of horizontal resolution on systematic errors and forecast skill. *Quart. J. Roy. Meteorol. Soc.*, **116**, 835–866.

De Toma, G., London, J., Chen, L., Rottman, G. J., Woods, T. N., Froidevaux, L. (1998) Solar cycle changes in ultraviolet irradiance and corresponding changes in mesospheric (45 to 65 km) ozone and temperature. In: *Atmospheric Ozone. Proceedings of the XVIII Quadrennial Ozone Symposium, 12–21 September 1996, L'Aquila, Italy*, R. D. Bojkov and G. Visconti (eds.), **1**, pp. 259–262.

Toohey, D. W., Avallone, L. M., Lait, L. R., Newman, P. A., Schoeberl, M. R., Fahey, D. W., Woodbridge, E. L., Anderson, J. C. (1993) The seasonal evolution of reactive chlorine in the northern hemisphere stratosphere. *Science*, **261**, 1,134–1,136.

Torres, O., Herman, J. R., Bhartia, P. K., Ahmad, Z. (1995) Properties of Mount Pinatubo aerosols as derived from Nimbus-7 total ozone mapping spectrometer measurements. *J. Geophys. Res.*, **100**, 14,043–14,055.

Torres, O., Bhartia, P. K. (1999) Impact of aerosol absorption on total ozone retrieval from satellite measurements of backscattered ultraviolet radiation. *J. Geophys. Res*, **104**, ND17, 21,569–21,578.

Trepte, C. R., Veiga, R. E., McCormick, M. P. (1993) The poleward dispersal of Mount Pinatubo volcanic aerosol. *J. Geophys. Res*, **98**, 18,563–18,573.

Tromp, T. K., Ko, M. K. W., Rodrigues, J. M., Sze, N. D. (1995) Potential accumulation of a CFC-replacement degradation product in seasonal wetlands. *Nature*, **376**, N6538, 327–330.

Tsvétkova, N. D., Yushkov, V. A., Dorokhov, V. M., Zaitsev, I. G. (1997) Some results of vertical ozone distribution measurements over Yakutsk in winter–spring periods of 1995 and 1996. *Meteorol. and Hydrol.*, N9 45–51 (in Russian).

Tuck, A. F., Davies, T., Hovde, S. J., Noguer-Alba, M., Fahey, D. W., Kawa, S. R., Kelly, K. K., Murphy, D. M., Proffitt, M. H., Margitan, J. J., Loewenstein, M., Podolske, J. R., Strahan, S. E., Chan, K. R. (1992) PSC-processed air and potential vorticity in the northern hemisphere lower stratosphere at mid-latitudes during winter. *J. Geophys. Res.*, **97**, 7,883–7904.

Uchino, O., Maeda, M., Yamamura, H., Hirono, M. (1978) Observation of stratospheric vertical ozone distribution by XeCl Lidar. *Appl. Phys. Lett.*, **33**, 807.

UNEP Scientific Assessment of Ozone Depletion, WMO Global Ozone Research and Monitoring Project Report No. 37 (1994) WMO, Geneva, Switzerland.

Urban, J., Eyreing, V., Küllmann, H., Kunzi, K., Mellmann, G., Wohlgemuth, J., Goede, A., de Jonge, A., Kleipool, Q., Mees, J., de Valk, J., Whyborn, N., Hetzjheim, H., Hubers, H. W., Schwaab. G., Chipperfield, M. P. (1998) Observations of stratospheric HlO, HCl, O_3, N_2O, and HO_2 at high latitudes during the winters of 1995 and 1996 with the Airborne-Submillimeter-SIS-Radiometer. In: *Atmospheric Ozone. Proceedings of the XVIII Quadrennial Ozone Symposium, 12–21 September 1996, L'Aquila, Italy*, R. D. Bojkov, G. Visconti (eds.), Parco Scientifico e Tecnologico d'Abruzzo, L'Aquila, Italy, **2**, pp. 615–618.

Valero, F. P. J., Platnick, S., Kinne, S., Pielewskie, P., Bucholtz, A. (1993) Airborne brightness temperature measurements of the polar winter troposphere as part of the Airborne Arctic Stratospheric Experiment II and the effect of brightness temperature variations on the diabetic heating in the lower stratosphere. *Geophys. Res. Lett.*, **20**, 2,575–2,578.

Vanicek, K. (1998) Differences between Dobson and Brewer observations of total ozone at Hradec, Kralove. In: *Atmospheric Ozone. Proceedings of the XVIII Quadrennial Ozone Symposium, 12–21 September 1996, L'Aquila, Italy*, R. D. Bojkov, G. Visconti (eds.), Parco Scientifico e Tecnologico d'Abruzzo, L'Aquila, Italy, **1**, pp. 81–84.

Varotsos, C., Kalabokas, P., Chronopoulos, G. (1992) On the laminated structure of ozone in the subtropical atmosphere. *Proceedings of the Quadrennial Ozone*

Symposium, Charlottesville, VA, NASA, Goddard Space Flight Center, Greenbelt, MD.

Varotsos, C., Kalabokas, P., Chronopoulos, G. (1994) Association of the laminated vertical ozone structure with the lower-stratospheric circulation. *J. Appl. Meteorol.*, **33**, N4, 473–476.

Varotsos, C. A., Kondratyev, K. Ya. (1995) Ozone dynamics over Greece as derived from satellite and in situ measurements. *Int. J. Remote Sensing*, **16**, N10, 1,777–1,798.

Varotsos, C., Kondratyev, K. Ya., Cracknell, A. P. (2000) New evidence for ozone depletion over Athens, Greece, *International Journal of Remote Sensing* (in press).

Varotsos, C., Cracknell, A. P., Sakellariou, N. K., Katsikis, S., Chronopoulos, G., Kassomenos, P. (1995) On the SO_2, NO_2 interferences in total ozone measurements made with the Dobson spectrophotometer No. 118 in Athens. *Int. J. Rem. Sensing*, **16**, N10, 1,805–1,814.

Varotsos, C., Ghosh, S., Katsikis, S., Kondratyev, K. Ya., Cracknell, A. P. (1996) A comparison of total ozone observation results from satellite and ground-based data (a case study of Athens). *Studying the Earth from Space*, N4, 18–23.

Varotsos, C., Kondratyev, K. Ya., Chronopoulos, G. (1997) On the total ozone depletion in middle latitudes of the Northern hemisphere due to satellite observations over Athens, Greece. *Doklady Russian Acad. Sci.*, **355**, N1, 104–105 (in Russian).

Varotsos, C., Kondratyev, K. Ya. (1998a) Total ozone dynamics in mid-latitudes of the Northern Hemisphere. *Doklady Akad. NAUK*, **359**, N6 (in Russian).

Varotsos, C., Kondratyev, K. Ya. (1998b) On potential interconnection between vertical ozone concentration profile in mid-latitudes and processes in the polar stratosphere. *Doklady Akad. NAUK*, **360**, N1 (in Russian).

Varotsos, C., Kasikis, S., Cracknell, A. P. (1999) On the influence of stray light on the total ozone measurements made with Dobson spectrophotometer No. 118 in Athens, Greece. *Int. J. Rem. Sensing*, N17, 3,307–3,315.

Volkovitsky, O. A., Gavrilov, A. A., Kaidalov, O. V. (1996) Is it true that man is destroying the planetary ozone layer? *Vestnik Russian Acad. Sci.*, **66**, N9, 783–786 (in Russian).

Wagenbach, D., Legrand, M., Fischer, H., Pichlmayer, F., Wolff, E. W. (1998) Atmospheric near-surface nitrate at coastal Antarctic sites. *J. Geophys. Res.*, **103**, ND9, 11,007–11,020.

Waters, J. W., Froidevaux, L., Read, W. G., Manney, G. L., Elson, L. S., Flower, D. A., Jarnot, R. F., Harwood, R. S. (1993) Stratospheric ClO and ozone from the Microwave Limb Sounder on the Upper Atmosphere Research Satellite. *Nature*, **362**, 597–602.

Watson, R. T. and Ozone Trends Panel, Prather, M. J. and Ad Hoc Theory Panel, Kurylo, M. J. and NASA Panel for Data Evaluation (1988) Present state of knowledge of the upper atmosphere 1988: An assessment report. *NASA Ref. Publ. 1208.*

Wauben, W. M. F., Fortuin, J. P. F., van Velthoven, P. F. J., Kelder, H. M. (1998) Comparison of modeled ozone distributions with sonde and satellite observations. *J. Geophys. Res.*, **103**, ND3, 3,511–3,530.

Wayne, R. P., Poulet, G., Biggs, P., Burrows, J. P., Cox, R. A., Crutzen, P. J., Hayman, G. D., Jenkin, M. E., Le Bras, G., Mooregat, G. K., Platt, U., Schindler, R. N. (1995) Halogen oxides: Radicals, sources and reservoirs in the laboratory and in the atmosphere. *Atmos. Environ.*, **29**, N20, 2,677–2,884.

Webster, C. R., May, R. D., Toohey, D. W., Avallone, L. M., Anderson, J. C., Newman, P., Lait, L., Schoeberl, M. R., Elkins, J. W., Chan, K. R. (1993) Chlorine chemistry on polar stratospheric cloud particles in the Arctic winter. *Science*, **261**, 1,130–1,134.

Webster, C. R., May, R. D., Allen, M., Jaegle, L., McCormick, M. P. (1994) Balloon profiles of stratospheric NO_2 and HNO_3 for testing the heterogeneous hydrolysis of N_2O_5 on sulfate aerosols. *Geophys. Res. Lett.*, **21**, N1, 53–56.

Weinstock, E. M., Anderson, J. G. (1989) On the continuing disagreement between in situ and satellite stratospheric ozone measurements. In: *Proceedings of the Quadrennial Ozone Symposium, 1988, Gottingen, Germany*, R. D. Bojkov, P. Fabian (eds.), Deepak Publications, Hampton, VA, pp. 238–242.

Weisenstein, D. K., Ko, M. K. W., Sze, N.-D. (1992) The chlorine budget of the present-day atmosphere: A modeling study. *J. Geophys. Res.*, **97**, ND2, 2,547–2,560.

Wellemeyer, C. G., Taylor, S. L., Jaross, G., Deland, M. T., Seftor, C. J., Labow, G., Swissler, T. J., Cebula, R. P. (1996) *Final Report on Nimbus-7 TOMS Version 7 Calibration*, NASA Contactor Report 4717, National Aeronautics and Space Adminstration, Washington, DC.

Wen, G., Frederick, J. E. (1995) The effects of horizontally extended clouds on backscattered ultraviolet sunlight. *J. Geophys. Res.*, **100**, ND8, 16,387–16,394.

Werner, J., Rothe, K. W., Walther, H. (1983) Monitoring of the ozone stratospheric layer by laser radar. *Appl. Phys. B.*, **32**, 113–118.

Wittrock, F., Eisinger, M., Ladstätter-Weibenmayer, A., Richter, A., Burrows, J. P. (1998) Ground based UV-VIS measurements of O_3, NO_2, OClO, and BrO over Ny-Ålesund (79°N, 12°E) In: *Proceedings of the XVIII Quadrennial Ozone Symposium*, R. D. Bojkov, G. Visconti (eds.), Parco Scientifico e Tecnologico d'Abruzzo, L'Aquila, Italy, **2**, pp. 623–626.

Wolff, E. W., Bales, R. C. (eds.)(1996) *Chemical Exchange Between the Atmosphere and Polar Snow*, NATO ASI Series 1 (Global Environmental Change), Vol. 43, Springer-Verlag, Berlin, 675 pp.

Wolff, E. W., Cachier, H. (1998) Concentrations and seasonal cycle of black carbon in aerosol at a coastal Antarctic station. *J. Geophys. Res.*, **103**, ND9, 11,033–11,041.

Wolff, E. W., Hall, J. S., Mulvaney, R., Pasteur, E. C.,, Wagenbach, D., Legrand, M. (1998) Relationship between chemistry of air, fresh snow and firm cores for aerosol species in coastal Antarctica. *J. Geophys. Res.*, **103**, ND9, 11,057–11,070.

World Meteorological Organisation (1986) *Atmospheric Ozone 1985: Assessment of Our Understanding in the Processes Controlling its Present Distribution and Change, WMO Global Ozone Research and Monitoring Project Report No. 16*, WMO, Geneva, Switzerland.

World Meteorological Organisation (1990) *Report of the International Ozone Trends Panel: 1988, WMO Global Ozone Research and Monitoring Project Report No. 18*, WMO, Geneva, Switzerland.

World Meteorological Organisation (12 March 1996) *Stronger Ozone Decline Continues, WMO Global Ozone Research and Monitoring Project Report No. 586*, WMO, Geneva, Switzerland.

Yang, H., Tung, K. K. (1994) Statistical significance and pattern of extratropical QBO in column ozone. *Geophys. Res. Lett.*, **21**, 2,235–2,238.

Yang, H., Tung, K. K. (1995) On the phase propagation of extratropical ozone quasi-biannual oscillation in observation data. *J. Geophys. Res.*, **100**, 9,091–9,100.

Yates, S. R., Wang, D., Gan, J., Ernst, F. F., Jury, W. A. (1998) Minimizing methyl bromide emissions from soil fumigation. *Geophys. Res. Lett.*, **25**, N10, 1,633–1,636.

Yela, M., Rodriguez, S., Gil, M., Caceneuve, H. (1998) NO_2 and O_3 total columns at threee different latitudes in the antarctic region from ground-based visible spectroscopy. *Proceedings of the XVIII Quadrennial Ozone Symposium*, R. D. Bojkov, G. Visconti (eds.), Parco Scientifico e Tecnologico d'Abruzzo, L'Aquila, Italy, **2**, 237–240.

Yushkov, V., Dorokhov, V., Khattatov, V., Lukyanov, A., Zaitcev, I., Zvetkova, N., Nakane, H., Akiyoshi, H., Ogawa, T. (1998) Evidence of ozone depletion over Yakutsk, Eastern Siberia, in 1995. In: *Proceedings of the XVIII Quadrennial Ozone Symposium*, R. D. Bojkov, G. Visconti (eds.), Parco Scientifico e Tecnologico d'Abruzzo, L'Aquila, Italy, **1**, pp. 241–246.

Zahn, A., Barth, V., Platt, U. (1998) Airborne High-Resolution Ozone and Potential Vorticity to study stratospheric–tropospheric exchange. In: *Atmospheric Ozone. Proceedings of the XVIII Quadrennial Ozone Symposium, 12–21 September 1996, L'Aquila, Italy*, R. D. Bojkov, G. Visconti (eds.), Parco Scientifico e Tecnologico d'Abruzzo, L'Aquila, Italy, **1**, pp. 183–185.

Zander, R., Mahieu, E., Gunson, M. R., Abrams, M. C., Chang, A. Y., Abbas, M., Aellig, C., Engel, A., Goldman, A., Irion, F. W., Kämpfer, N., Michelsen, H. A., Newchurch, M. J., Rinsland, C. P., Salawitch, R. J., Stiller, G. P., Toon, G. C. (1996) The 1994 northern midlatitude budget of stratospheric chlorine derived from ATMOS/ATLAS 3 observations. *Geophys. Res. Lett.*, **23**, N17, 2,357–2,360.

Zerefos, C. S., Bais, A. F., Ziomas, T. C., Bojkov, R. D. (1992) On the relative importance of quasi-biannual oscillation and El Niño/Southern Oscillation in the revised Dobson total ozone records. *J. Geophys. Res.*, **97**, 10,135–10,144.

Zhadin, E. A. (1995) Anomalies of the ozone layer and of stratospheric angular momentum. *Meteorol. and Hydrol.*, N7, 48–55 (in Russian).

Zhadin, E. A., Diansky, N. A. (1997) Analysis of interrelationships between inter-annual variations of total ozone and general circulation. *Meteorol. and Hydrol.*, N9, 25–34 (in Russian).

Ziemke, J. R., Stanford, J. L. (1994) Kelvin waves in total column ozone. *Geophys. Res. Lett.*, **21**, N2, 105–108.

Ziemke, J. R., Herman, J. R., Stanford, J. L., Bhartia, P. K. (1998) Total ozone/UVB monitoring and forecasting: Impact of clouds and the horizontal resolution of satellite retrievals. *J. Geophys. Res.*, **103**, ND4, 3,865–3,871.

Ziereis, H., Schlager, H., Schulte, P., Klemm, O. (1998) Aircraft observations of the NO_x abundance and variability in the upper troposphere over the North Atlantic. In: *Atmospheric Ozone. Proceedings of the XVII Quadrennial Ozone Symposium,*

12–21 September 1996, L'Aquila, Italy, R. D. Bojkov, G. Visconti (eds.), Parco Scientifico e Tecnologico d'Abruzzo, L'Aquila, Italy, **2**, pp. 647–650.

Zommerfeld, W., Künzi, K. F. (1989) Operational ground-based ozone sensor (OGOS). In: *Proceedings of the Quadrennial Ozone Symposium, 1988, Gottingen, Germany*, R. D. Bojkov, P. Fabian (eds.), Deepak Publications, Hampton, VA, pp. 784–787.

Zuev, V. E., Kabanov, M. V. (1987) *Optics of Atmospheric Aerosols*. Gidrometeoizdat, Leningrad, 259 pp. (in Russian).

Zuev, V. V., Smirnov, S. V. (1997) The results of oxone anomalies complex observations at the Siberian lidar station (56.5°N, 85°E). *Optics of the Atmos. and Ocean,* **10**, N12, 1,400–1,414.

Zvjaginzev, A. M., Kruchenizky, G. M. (1996) About relations of total ozone in mid-latitudes of the Northern Hemisphere with the North Atlantic Oscillation. *Meteorol. and Hydrol.*, N7, 72–77 (in Russian).

3

Stratospheric ozone-formation processes

The brief description of stratospheric ozone-formation processes to be found in Chaper 1 is no more than a schematic presentation of a reality. Observational data of total ozone and vertical ozone profile discussed in Chapter 2 illustrate a very complex picture of spatial and temporal ozone variability which depends not only on atmospheric internal processes but also on some external impacts like solar activity. This is why the level of scientific understanding of ozone layer dynamics is far from being adequate as yet.

3.1 GENERAL FEATURES OF STRATOSPHERIC OZONE CHANGES

There are many fundamental publications that contain reviews of basic processes which are responsible for ozone-layer formation and change. UNEP Scientific Assessment (*UNEP* ..., 1994) and WMO/UNEP Scientific Assessment (*Scientific* ..., 1999) deserve special attention in this context. It is enough therefore to briefly discuss the basic processes of stratospheric ozone formation.

We shall follow a survey published by Cicerone (1987) which discussed principal ozone formation and destruction processes. The production of atmospheric O_3 is due to the break in the O–O bond in O_2 by photodissociation at wavelengths less than 242 nm. Then oxygen atoms rapidly combine with O_2 to form O_3 by the reaction:

$$O + O_2 + M \longrightarrow O_3 + M \tag{3.1}$$

where M is any third-body molecule such as N_2 or O_2.

The global rate of solar UV energy that goes into breaking O_2 bonds by the reaction:

$$O_2 + h\nu \longrightarrow O + O \tag{3.2}$$

exceeds 2×10^{10} kW, or is at least three times the worldwide human energy-usage rate.

Additional O_3 is produced by radiation at wavelengths below 175 nm above the stratosphere and by chemical means in the lower stratosphere and the troposphere. The following intermediate reactions are of importance:

$$OH + CO \longrightarrow H + CO_2 \tag{3.3}$$

$$H + O_2 + M \longrightarrow HO_2 + M \tag{3.4}$$

$$HO_2 + NO \longrightarrow NO_2 + OH \tag{3.5}$$

$$NO_2 + h\nu \longrightarrow NO + O \tag{3.6}$$

These reactions are then followed by reaction (3.1). The presence of carbon monoxide here is due to natural and industrial sources whereas nitric oxide is present largely because of the reaction:

$$O(^1D) + N_2O \longrightarrow 2NO \tag{3.7}$$

in the stratosphere. The electronically excited $O(^1D)$ is a product of O_3 photolysis:

$$O_3 + h\nu \longrightarrow O(^1D) + O_2(^1\Delta) \tag{3.8a}$$

$$\longrightarrow O + O_2 \tag{3.8b}$$

The production of O_3 is counterbalanced by the well-known Chapman mechanism:

$$O_3 + h\nu \longrightarrow O + O_2 \tag{3.8b}$$

$$O + O_3 \longrightarrow 2O_2 \tag{3.9}$$

Net reaction:

$$2O_3 + h\nu \longrightarrow 3O_2 \tag{3.10}$$

The O3 loss due to reaction (3.9) is accompanied by two other oxygen-only reactions:

$$O + O + M \longrightarrow O_2 + M \tag{3.11}$$

and

$$O(^1D) + O_3 \longrightarrow 2O_2 \tag{3.12}$$

Further studies led to the discovery of the very important role that catalytic processes play in destroying stratospheric ozone. This made the theory of stratospheric ozone changes much more complex and resulted in the realization that biological processes at the Earth's surface influence the state of the global atmosphere.

The role of nitrogen oxides was first explored in connection with the following gas-phase catalytic cycle:

$$NO + O_3 \longrightarrow NO_2 + O_2 \tag{3.13}$$

$$O_3 + h\nu \longrightarrow O + O_2 \tag{3.8b}$$

$$NO_2 + O \longrightarrow O + O_2 \tag{3.14}$$

Net reaction:

$$2O_3 + h\nu \longrightarrow 3O_2 \tag{3.15}$$

Combination of reactions (3.13), (3.14) and (3.8b) constitute a catalytic cycle because the NO that is consumed in the first reaction is replaced in the final reaction. This, however, is a simplified scheme, because more components are actually participating in this cycle. A special role belongs, for example, to HNO_3 which may be formed and transported to the troposphere. Moreover, another NO_x catalytic cycle also destroys O_3 by reaction (3.13) followed by:

$$NO_2 + O_3 \longrightarrow NO_3 + O_2 \quad (3.16)$$

$$NO_3 + h\nu \longrightarrow NO + O_2 \quad (3.17)$$

Net reaction:

$$2O_3 + h\nu \longrightarrow 3O_2 \quad (3.15)$$

The following three reactions, chained with hydrogen oxide (HO_x) participation, also catalyse ozone destruction:

$$HO + O_3 \longrightarrow HO_2 + O_2 \quad (3.18)$$

$$HO_2 + O_3 \longrightarrow HO + O_2 + O_2 \quad (3.19)$$

Net reaction:

$$2O_3 \longrightarrow 3O_2 \quad (3.20)$$

$$HO + O_3 \longrightarrow HO_2 + O_2 \quad (3.18)$$

$$O_3 + h\nu \longrightarrow O + O_2 \quad (3.8b)$$

$$HO_2 + O \longrightarrow HO + O_2 \quad (3.21)$$

Net reaction:

$$2O_3 + h\nu \longrightarrow 3O_2 \quad (3.15)$$

$$O_3 + h\nu \longrightarrow O + O_2 \quad (3.8b)$$

$$HO + O \longrightarrow H + O_2 \quad (3.22)$$

$$H + O_3 \longrightarrow HO + O_2 \quad (3.23)$$

Net reaction:

$$2O_3 + h\nu \longrightarrow 3O_2 \quad (3.15)$$

These three HO_x cycles are most important in the upper stratosphere, although reactions (3.18) and (3.19) can also be significant in the lower stratosphere and the troposphere.

In connection with the ozone-hole problem, special attention has been paid to gaseous reactions, that involve chlorine (Molina and Rowland, 1974; Molina and Molina, 1987). Relevant key reactions are the following:

$$ClO + ClO + M \longrightarrow Cl_2O_2 + M \quad (3.24)$$

$$Cl_2O_2 + h\nu \longrightarrow C + ClO_2 \quad (3.25)$$

$$ClO_2 \longrightarrow Cl + O_2 \quad (3.26)$$

$$2(Cl + O_3) \longrightarrow 2(ClO + O_2) \quad (3.27)$$

Net reaction:

$$2O_3 + h\nu \longrightarrow 3O_2 \quad (3.10)$$

Since the loss by the ClO + ClO reaction is proportional to the square of the ClO concentration, the reaction becomes efficient in areas where concentrations of ClO are significantly enhanced by heterogeneous reactions (e.g. in the regions, where chemistry is strongly affected by the presence of PSCs) (Isaksen and Stordal, 1986). It is important that, apart from the reactions mentioned, other reaction sequences that involve Cl, ClO, HO_2, NO_2, HOCl and $ClNO_3$ also catalytically consume ozone (Cicerone, 1987; Kondratyev, 1989I).

Several chemical cycles involving bromine compounds have also been identified as ozone-destroying mechanisms in the stratosphere. The main catalytical bromine cycles responsible for the ozonc dcstruction in the lower stratosphere are (McElroy *et al.* 1986):

$$BrO + ClO \longrightarrow Br + Cl + O_2 \quad (3.28)$$

$$Cl + O_3 \longrightarrow ClO + O_2 \quad (3.29)$$

$$Br + O_3 \longrightarrow BrO + O_2 \quad (3.30)$$

Net reaction:

$$2O_3 + h\nu \longrightarrow 3O_2 \quad (3.10)$$

and (Solomon *et al.*, 1986):

$$BrO + HO_2 \longrightarrow HOBr + O_2 \quad (3.31)$$

$$HOBr + h\nu \longrightarrow Br + OH \quad (3.32)$$

$$Br + O_3 \longrightarrow BrO + O_2 \quad (3.33)$$

$$OH + O_3 \longrightarrow HO_2 + O_2 \quad (3.34)$$

Net reaction:

$$2O_3 + h\nu \longrightarrow 3O_2 \quad (3.10)$$

The first of these cycles also involves ClO, thereby linking stratospheric chlorine and bromine ozone-destroying chemistry together. In this way future ozone loss by the bromine will depend on the development of chlorine.

Cicerone (1987) pointed out that the efficiency of O_3-destroying processes depends on the fraction of time that the active chemicals exist as NO, NO2, H, OH, OH_2, Cl, ClO, Br and BrO compared with the time they exist in the form of inactive, reservoir species, such as HNO_3, N_2O_5, HNO_4, H_2O_2, $ClNO_3$, HCl and $BrNO_3$.

As discussed in more detail later on, heterogeneous reactions, occurring on PSCs and sulphate aerosol particles that involve HCl, $ClONO_2$, HOCl, N_2O_5 and H_2O, are believed to be the main reactions involved in converting reservoir compounds to active chlorine compounds (ClO). The following heterogeneous reactions play the most important role:

$$ClONO_2 + HCl \longrightarrow Cl_2 + HNO_3 \tag{3.35}$$

$$ClONO_2 + H_2O \longrightarrow HOCl + HNO_3 \tag{3.36}$$

$$N_2O_5 + HCl \longrightarrow ClNO_2 + HNO_3 \tag{3.37}$$

$$N_2O_5 + H_2O \longrightarrow HNO_3 + HNO_3 \tag{3.38}$$

$$OHCl + HCl \longrightarrow Cl_2 + H_2O \tag{3.39}$$

In this context Hendricks *et al.* (1999) developed a model to simulate heterogeneous reactions of nitrogen, chlorine and bromine compounds on and in sulphate aerosols under conditions encountered in the lowermost stratosphere and upper troposphere. Box-model calculations reveal that in the lowermost stratosphere the considered heterogeneous conversions effectively dilute NO_x and cause a strong enhancement of HO_x and ClO_x mixing ratios. Lowermost stratospheric NO_x/NO_y and OH/HO_2 ratios can efficiently be reduced by heterogeneous reactions.

Lary *et al.* (1999) considered the nature of carbonaceous surfaces, the means by which they are activated, the nature of some functional groups that they support and some reaction mechanisms that may be involved. It was emphasized that, because of the strong affinity of carbonaceous surfaces for organic species and because of the ease with which compounds in a high oxidation state can oxidize the carbonaceous surface, it is highly likely that carbonaceous aerosols are interacting chemically with a range of organic species in ways that have, as yet, not been fully characterized but may significantly affect the oxidizing capacity of the atmosphere. The reduction of NO_y on carbonaceous aerosols is an important mechanism of ozone-layer dynamics.

The contemporary stage of studying ozone-layer dynamics is characterized by a tendency to carry out simulation modelling, taking account of both natural and anthropogenic causes of ozone change, using interactive transport/chemistry models of the stratosphere. An important aspect of the problem is the climate/ozone interaction in the context of stratospheric temperature change, including anthropogenically induced global warming, which reveals itself in the stratosphere as climate cooling (Chapter 6).

Tie and Hess (1997) considered two effects of temperature interannual variability on stratospheric ozone: (1) the effect on the rate coefficient of heterogeneous reactions occurring at the surface of sulphate aerosols (e.g. volcanic aerosols); (2) the effect of the formation of polar stratospheric clouds (PSCs) and the spring Antarctic ozone depletion. Using a 2-D chemical/dynamical/microphysical model Tie and Hess (1997) evaluated interannual variabilities of ozone concentration affected by both processes. The results obtained show that during winter 1992–1993, stratospheric temperature at high latitudes in the Northern Hemisphere (NH) were colder than during winter 1991–1992, leading to a large rate of heterogeneous conversion from inactive chlorine to active chlorine in sulphate aerosols during winter 1992–1993. This resulted in ozone reduction because of heterogeneous reactions following the eruption of Mount Pinatubo, which enhanced in colder winters (1992–1993) and reduced in warmer winters (1991–1992). Calculated ozone interannual variation is consistent with TOMS data, indicating that the interannual variation in the formation of PSC probably plays an important role in the interannual variability of the spring Antarctic ozone hole.

Numerical modelling conducted by Danilin *et al.* (1998) indicates that stratospheric cooling could further deplete ozone (in addition to chlorine-induced depletion) via increased polar stratospheric cloud (PSC) formation and retard its expected recovery even with the projected chlorine-loading decrease. Calculations show that a 1 K cooling could provide the same total ozone depletion as an increase in chlorine by 0.4–0.7 ppbv.

Kinnersley and Tung (1998) emphasized the significance of stratospheric dynamics in driving TOZ change. By forcing an interactive chemical–dynamical model of the stratosphere with observed Singapore zonal winds and with observed daily-varying planetary wave heights just above the tropopause from March 1980 to February 1993, Kinnersley and Tung (1998) demonstrated that much of the observed interannual variability (IAV) in the monthly mean TOZ values from pole to pole could be realistically reproduced. Model results specifically indicate that the NH high-latitude ozone anomaly is influenced strongly by IAV in forcing of planetary waves. Model results show a large IAV in the TOZ during polar night in the NH that is due to IAV in forcing of planetary waves.

Tan *et al.* (1998a) used a photochemical box model, augmented by a volume exchange model of mixing, to identify situations for which effects of ClO_X deactivation through mixing with NO_X-rich air are at least as important as photochemical effects alone. A 2-D model of quasi-horizontal transport along isentropic surfaces was applied for similar assessments. In both cases the significant role of mixing-induced chlorine deactivation relative to photolysis-induced deactivation was demonstrated.

Reliability of kinetic and photochemical input data is a principal aspect of photochemical modelling. To assess errors in model predictions that result from errors in kinetic and photochemical data, Fish and Burton (1997) applied a Lagrangian photochemical model. The results indicated that Arctic ozone destruction can be modelled with an uncertainty of $\pm 25\%$ (1σ) and that mid-latitude ozone destruction can be modelled with an uncertainty of $\pm 50\%$. It was also been shown that two reactions, from a total of more than 100, are responsible for more than a third of the uncertainty in model calculations of Arctic ozone loss.

Knudsen *et al.* (1998) analysed a specific case of the ozone depletion in and below the Arctic vortex for 1997.

3.2 NUMERICAL MODELLING OF OZONE VARIABILITY IN THE TROPOSPHERE AND STRATOSPHERE IN RESPONSE TO ANTHROPOGENIC EMISSIONS[1]

A simplified simulation model has been suggested by Aloyan and Arutyunyan (1999), who considered the problem of global transport of gaseous pollutants in

* This section is by A. E. Aloyan and V. O. Arutyunyan.

the troposphere and lower stratosphere including relevant chemical transformations. The model is based on use of the spherical system of coordinates. Wind velocity and other meteorological data were prescribed from actual observational data (TOGA, etc.). The surface distribution and magnitudes of pollutant emission sources are assumed to be functions of horizontal space coordinates. Photochemical processes are considered in a rather simplified form which does not include catalytic reactions connected with Cl, I and Br-containing species.

Carbon and nitrogen cycles that can be ozone-destructive or ozone-productive depending on the height of a layer have been considered. Both gas-phase and heterogeneous (connected with the destruction of gaseous components) processes have been taken into account. The photochemical scheme includes 74 reactions and 18 species: $O(^1D)$, $O(^3P)$, O_3, CO, CH_2O, CH_4, CH_3OOH, CH_3O_2, OH, HO_2, H_2O_2, NO, NO_2, N_2O, NO_3, HNO_3, HNO_4, N_2O_5.

As is well known, ozone is of critical importance in the chemistry of the troposphere and stratosphere. Since its dynamics depends on numerous chemical and photochemical reactions, a decisive prerequisite of the reliability of simulation modelling is the availability of adequate information on reaction rates. The rate constant is an important characteristic for a reaction, depending on the activation energy, collision probability, reactivity of molecules, etc. Most reaction-rate constants depend on temperature which is described by the Arrhenius equation. The photodissociation rate (in s^{-1}) is calculated by the following relation:

$$J = \int_0^{\infty} I_0(\lambda)\tau(\lambda)\sigma(\lambda)\varphi(\lambda)\, d\lambda$$

where $I_0(\lambda)$ is extra-atmospheric solar irradiance; $\tau(\lambda)$ is the attenuation of radiation at wavelength λ on the way from the upper limit of the atmosphere to a reference level; $\sigma(\lambda)$ is the absorption cross-section; $\varphi(\lambda)$ is the quantum yield of a reaction, giving the probability of molecule dissociation in absorbing the quantum.

3.2.1 Formulation of the problem

In order to assess to what extent the influence of remote pollution sources on the atmospheric in a given region can be significant, the equations of pollutant transport in the atmosphere with due regard to the turbulent exchange and interaction of pollutants with the Earth's surface are used.

A coordinate system (λ, ψ, z) is used where λ is the longitude, ψ is the supplement of latitude and z is the altitude measured from the surface. The main equation of pollutant transport on a sphere is expressed in the following form (Marchuk and Kondratyev, 1992; Marchuk and Aloyan, 1995):

$$\frac{\partial \varphi_i}{\partial t} + \frac{u}{a \sin\psi}\frac{\partial \varphi_i}{\partial \lambda} + \frac{v}{a}\frac{\partial \varphi_i}{\partial \psi} + w\frac{\partial \varphi_i}{\partial z} = F_i - P(\varphi_i, \varphi_j) + \frac{\partial}{\partial z}\nu\frac{\partial \varphi}{\partial z} + \frac{1}{a^2 \sin^2\psi}\frac{\partial}{\partial \lambda}\lambda\frac{\partial \varphi}{\partial \lambda} + \frac{1}{a^2 \sin\psi}\frac{\partial}{\partial \psi}\mu \sin\psi\frac{\partial \varphi}{\partial \psi}$$

Here $\varphi_i = \varphi_i(\lambda, \psi, z, t)$ is the pollutant concentration; u, v, w are components of the wind velocity vector along λ, ψ, z directions, respectively; μ, ν are the horizontal and vertical turbulent exchange coefficients, respectively; $F_i = F_i(\lambda, \psi, z, t)$ is a function standing for emission source magnitudes; and a is the average radius of the Earth.

The problem is considered in the domain $D_t = G \times [(0, T]$, where $G = \{S \times [0, H]\}$; $S = (\lambda, \psi) : 0 \leq \lambda \leq 2\pi, 0 \leq \psi \leq \pi$; H is the upper boundary of the numerical grid domain.

The boundary conditions are given according to periodicity conditions for space coordinates:

$$\varphi_i(0, \psi, z, t) = \varphi_i(2\pi, \psi, z, t)$$
$$\varphi_i(\lambda, -\psi, z, t) = \varphi_i(\lambda + \pi, \psi, z, t)$$
$$\varphi_i(\lambda, \pi + \psi, z, t) = \varphi_i(\lambda + \pi, \pi - \psi, z, t)$$
$$\nu \frac{\partial \varphi_i}{\partial z} = a_\theta(\varphi_i - \varphi_{\text{surf}}) \qquad \text{for } z = h$$

where a_θ is determined from the surface-layer model.

3.2.2 Parametrization of the planetary boundary layer

The vertical resolution in global atmospheric models is rather coarse and it is therefore difficult to simulate the fields in the lower layers of the atmosphere with necessary accuracy. Therefore calculations of meteorological characteristics of the boundary layer have been made with the help of a parametrization scheme for the planetary boundary layer suggested by Yordanov *et al.* (1978). The following external parameters are determined at every point of the horizontal computational grid by known values of velocity and temperature at the first computational level:

$$Ro = \frac{|u_{g0}|}{|z_0|}, \qquad S_T = \frac{\beta\delta\tilde{\theta}}{1|u_{g0}|},$$
$$\eta_x = \frac{\kappa^2}{l}\frac{\partial U_g}{\partial z} = -\frac{\beta\kappa^2}{l^2}\frac{\partial\hat{\Theta}}{\partial y}, \qquad \eta_y = \frac{\kappa^2}{l}\frac{\partial v_g}{\partial z} = \frac{\beta\kappa^2}{l^2}\frac{\partial\hat{\Theta}}{\partial x}, \tag{3.40}$$

where Ro is the Rossby number; S_T is the stratification parameter; u_g, v_g are components of the geostrophic wind; l is the Coriolis parameter; η_x, η_y are baroclinicity parameters; Θ is the potential temperature; $\beta = g/\Theta$ is the buoyancy parameter; g is gravitational acceleration; $\partial\Theta$ is the difference between values of the potential temperature for the PBL and surface layer. Using Ro, S_T, η_x, η_y, we determine: $C_g = u_*/|u_g|$ ('geostrophic resistance' coefficient); α (angle between stress of the turbulent friction on surface layer and $\mathbf{u}_{g0}$); $\mu = h_0/L_0$ (dimensionless 'internal' parameter of stratification). Here the following notation is utilized: $h_0 = \kappa u_*/l$ is the internal scale of the boundary layer height; $L_0 = -c_p\rho U_*^3/\kappa\beta q_0$ is the Monin–Obukhov lengthscale; c_p is heat capacity of air; ρ is air density; q_0 is heat flux near the Earth's surface. With the values of C_g, α, μ known q_0 is determined, then u_* and turbulent exchange coefficients for $z \geq h$ are calculated (based on C_g and u_g values) by the formula:

$$\nu(\mu) = \frac{\kappa^2 u_*^2}{l^2} \begin{cases} \dfrac{h}{l + 10\mu h} & \text{for } \mu \geq 0, \\ h & \text{for } -2.33 \leq \mu \leq 0, \\ (-0.007/\mu)^{-1/3} h^{4/3} & \text{for } \mu \leq -2.33 \end{cases} \tag{3.41}$$

where h is the surface layer height.

Wind velocity components are determined in accordance with Marchuk and Aloyan (1995), using the so-called 'velocity defect' concept. The vertical velocity at the top of the atmospheric boundary layer can be found using the procedure suggested by Yordanov *et al.* (1978).

From the viewpoint of global transport studies, pollutants with small gravitational settling rates are of special interest. They remain in the atmosphere for long time periods and transported together with airmasses. In this case, the principles of turbulence theory which have been used in global circulation models for the description of temperature and moisture field dynamics could be applied to pollutants.

Computational algorithms that were applied to resolve the problem formulated above are based on the splitting technique (Aloyan and Makarenko, 1993). At each timestep a numerical scheme is used that describes the differential operator of pollutant transport along trajectories and a scheme that approximates the turbulent diffusion. At the first stage a second-order explicit monotonic conservative scheme is used, then, at the second stage, an implicit scheme is utilized (Penenko and Aloyan, 1985).

3.2.3 Influence of heterogeneous reactions on ozone and active radicals

Radicals OH and HO_2 play a decisive role in the formation of gaseous composition of the troposphere and stratosphere. Their concentrations depend considerably on hydrogen peroxide (H_2O_2) which is the main reservoir for the hydrogen species and provide OH radicals during photolysis.

Heterogeneous destruction of long-living components, with chemical transformation rates that are slower than the vertical velocity, needs consideration if $\tau_{\text{het}} \leq 10\tau_D$ where $\tau_{\text{het}} = k_{\text{het}}^{-1}$ and $\tau_D = H^2/\nu$ are characteristic timescales for heterogeneous destruction and vertical velocity, respectively. If we assume the height of homogeneous atmosphere (H) to be between 8 and 10 km, and the turbulent exchange coefficient $\nu \approx 10^4\ \text{cm}^2/\text{s}$, then the contribution of heterogeneous destruction is considerable for:

$$\gamma > \frac{0.4\nu}{C_T H^2 A} \approx \frac{5 \times 10^{-13}}{A} \quad \text{cm}^{-1}$$

where C_T is the mean thermal velocity of a molecule ($C_T = 3D/l, D$ is the molecular diffusion coefficient; l is the free pathlength); A is the aerosol specific surface. For stratospheric clouds and lower troposphere $A \approx 5 \times 10^{-6}\,cm^{-1}$ and heterogeneous reactions can be significant for values $\gamma > 10^{-7}$.

Under heavily polluted atmospheric conditions, $A \approx 5 \times 10^{-5}\,cm^{-1}$ with the corresponding value $\gamma > 10^{-8}$. In the troposphere, the main reaction connected with the destruction of hydroxyl radicals is

$$OH + H_2O_2 \longrightarrow H_2O + HO_2$$

and heterogeneous washout of H_2O_2 takes place. If $[OH] = 10^6\,cm^{-3}$ and $[HO2] = 10^7\,cm^{-3}$, then $\gamma_{OH} > (10^{-6}\,cm^{-1})/A$ and $\gamma_{HO_2} > (10^{-7}\,cm^{-1})/A$. For example, $A \approx 10\,cm^{-1}$ (at $z = 15$ km) for stratospheric clouds and $A \approx 1,000\,cm^{-1}$ (at $z = 1$ km) for urban atmosphere lead to the following values of γ for OH and HO_2: $\gamma_{OH} \approx 3.0, \gamma_{HO_2} \approx 30.0$ (stratospheric clouds) and $\gamma_{OH} \approx 0.03, \gamma_{HO_2} \approx 0.003$ (urban).

The impact of OH and HO_2 radicals on the content of tropospheric and stratospheric ozone is qualitatively distinct. In the troposphere the decrease of radicals results in ozone increase, while in the stratosphere the decrease of OH radicals leads to high levels of NO_x concentration, which results in destruction of ozone via the nitrogen cycle.

Since OH and HO_2 have clearly expressed diurnal variation characteristics, the actual heterogeneous destruction of these radicals should be considered using a 3-D photochemical model. This allows one to take into consideration the fact that in some regions daytime concentrations of OH and HO_2 grow while practically none of them is present in other regions during night.

The list of chemical reactions with corresponding rate constants is shown in Table 3.1.

3.2.4 Some results

Numerical experiments were performed to simulate the global distribution of anthropogenic pollutants above a spherical Earth's surface. Global average emissions of CH_4 and NO_x were prescribed in the calculations. Table 3.2 shows the input data utilized for numerical modelling.

Calculated concentrations of OH, O_3 and NO were obtained for $t = 12$ h and $t = 24$ h on 5 January, 1986. The corresponding maximum concentration levels are: $\varphi_{max} = 7.5 \times 10^7\,cm^{-3}$ for OH at $t = 12$ h; $\varphi_{max} = 7.7 \times 10^7\,cm^{-3}$ for OH at $t = 24$ h; $\varphi_{max} = 1.2 \times 10^{12}\,cm^{-3}$ for O_3 at $t = 12$ h; $\varphi_{max} = 1.2 \times 10^{12}\,cm^{-3}$ for O_3 at t = 24 h; $\varphi_{max} 4.5 \times 10^{11}\,cm^{-3}$ for O_3 at $t = 24$ h. The results clearly express diurnal variations which are important to take into account when dealing with heterogeneous destruction. Of similar behaviour is the radical HO_2.

3.3 3-D TRANSPORT/PHOTOCHEMISTRY GLOBAL OZONE DYNAMICS MODELS (CONTEMPORARY)

Recent years have been marked by very substantial progress in numerical modelling of global ozone dynamics (Bojkov, 1999; *Scientific* ..., 1999). Subjects of special

Table 3.1. Methane oxidation chemical scheme (with O_3, NO_x and HC) has also been taken into account.

Photodissosiation reactions		
1.	$O_2 + h\nu \longrightarrow O(^3P) + O(^3P)$	
2.	$N_2O_5 + h\nu \longrightarrow$ products	
3.	$N_2O_5 + h\nu \longrightarrow NO_2 + NO_3$	
4.	$N_2O + h\nu \longrightarrow N_2 + O(^1D)$	
5.	$CH_3OOH + h\nu \longrightarrow OH + CH_3O$	
6.	$HNO_3 + h\nu \longrightarrow OH + NO_2$	
7.	$H_2O_2 + h\nu \longrightarrow OH + OH$	
8.	$H_2O + h\nu \longrightarrow OH + H$	
9.	$O_3 + h\nu \longrightarrow O_2 + O(^1D)$	
10.	$O_3 + h\nu \longrightarrow O_2 + O(^3P)$	
11.	$NO_2 + h\nu \longrightarrow NO + O(^3P)$	
12.	$NO_3 + h\nu \longrightarrow NO_2 + O(^3P)$	
13.	$NO_3 + h\nu \longrightarrow NO + O_2$	
14.	$H_2CO + h\nu \longrightarrow H_2 + CO$	
15.	$H_2CO + h\nu \longrightarrow H + HCO$	
O_x, HO_x, NO_x, and HC reactions		
1.	$O(^3P) + O_3 \longrightarrow 2O_2$	$k_1 = 8.0 \times 10^{-12} \exp\{-2060/T\}$
2.	$O(^3P) + O_2 + M \longrightarrow O_3 + M$	$k_2 = 6.0 \times 10^{-34} N_2 \exp\{(T/300)^{-2/3}\}$
3.	$O(^1D) + N_2 \longrightarrow O(^3P) + N_2$	$k_3 = 1.8 \times 10^{-11} \exp\{(110/T\}$
4.	$O(^1D) + O_2 \longrightarrow O(^3P) + O_2$	$k_4 = 3.2 \times 10^{-11} \exp\{70/T\}$
5.	$O(^1D) + H_2O \longrightarrow 2OH$	$k_5 = 2.2 \times 10^{-10}$
6.	$O(^1D) + CH_4 \longrightarrow H_2 + CH_2O$	$k_6 = 1.4 \times 10^{-11}$
7.	$O(^1D) + CH_4(O_2) \longrightarrow CH_3O_2 + OH$	$k_7 = 1.4 \times 10^{-10}$
8.	$CH_4 + OH(O_2) \longrightarrow CH_3O_2 + H_2O$	$k_8 = 2.3 \times 10^{-12} \exp\{-1700/T\}$
9.	$OH + H_2(O_2) \longrightarrow H_2O + HO_2$	$k_9 = 3.3 \times 10^{-13} \exp\{800/T\}$
10.	$CH_3O_2 + HO_2 \longrightarrow CH_3OOH + O_2$	$k_{10} = 3.3 \times 10^{-13} \exp\{800/T\}$
11.	$CH_3OOH + OH \longrightarrow CH_3O_2 + H_2O$	$k_{11} = 3.8 \times 10^{-12} \exp\{200/T\}$
12.	$CH_3O_2 + NO(O_2) \longrightarrow NO_2 + CH_2O + HO_2$	$k_{12} = 4.2 \times 10^{-12} \exp\{180/T\}$
13.	$CH_3O_2 + CH_3O_2(O_2) \longrightarrow 2H_2CO + 2HO_2$	$k_{13} = 2.2 \times 10^{-13}$
14.	$CH_2O + OH(O_2) \longrightarrow H_2O + CO + HO_2$	$k_{14} = 1.0 \times 10^{-11}$
15.	$CH_3O_2 + CH_3O_2 \longrightarrow H_2CO + CH_3OH + O_2$	$k_{15} = 1.9 \times 10^{-13}$
16.	$CH_3O_2 + NO_2(M) \longrightarrow CH_3O_2 + NO_2$	$k_{16} = \dfrac{k_0[M]}{1 + \dfrac{k_0[M]}{k_\infty}} \cdot 0.6\{1 + [1gk[M]/k_\infty]^2\}^{-1}$, $k_0 = 1.5 \times 10^{-30}(T/300)^{-4}$, $k_\infty = 6.5 \times 10^{-12}$
17.	$CO + OH(O_2) \longrightarrow HO_2 + CO_2$	$k_{17} = 1.5 \times 10^{-13}(1 + 0.6P_{atm})$, $P_{atm} = P_{mbar} 10^{-3}$
18.	$OH + HO_2 \longrightarrow H_2O + O_2$	$k_{18} = 4.8 \times 10^{-11} \exp\{250/T\}$
19.	$O_3 + OH \longrightarrow HO_2 + O_2$	$k_{19} = 1.6 \times 10^{-12} \exp\{-940/T\}$
20.	$HO_2 + O(^3P) \longrightarrow OH + O_2$	$k_{20} = 3.3 \times 10^{-11} \exp\{200/T\}$
21.	$HO_2 + O_3 \longrightarrow OH + 2O_2$	$k_{21} = 1.4 \times 10^{-14} \exp\{-50/T\}$
22.	$HO_2 + HO_2 \longrightarrow H_2O_2 + O_2$	$k_{22} = 2.3 \times 10^{-13} \exp\{600/T\}$
23.	$HO_2 + HO_2 + M \longrightarrow H_2O_2 + O_2 + M$	$k_{23} = 1.7 \times 10^{-33}[M] \exp\{1000/T\}$
24.	$H_2O_2 + OH \longrightarrow H_2O + HO_2$	$k_{24} = 2.9 \times 10^{-12} \exp\{-160/T\}$

25.	$N_2O + O(^1D) \longrightarrow 2NO$	$k_{25} = 6.7 \times 10^{-11}$
26.	$N_2O + O(^1D) \longrightarrow N_2 + O_2$	$k_{26} = 4.9 \times 10^{-11}$
27.	$NO + HO_2 \longrightarrow NO_2 + OH$	$k_{27} = 3.7 \times 10^{-12} \exp\{240/T\}$
28.	$NO + O_3 \longrightarrow NO_2 + O_2$	$k_{28} = 2.2 \times 10^{-12} \exp\{-1400/T\}$
29.	$NO_2 + O(^3P) \longrightarrow NO + O_2$	$k_{29} = 6.5 \times 10^{-12} \exp\{120/T\}$
30.	$NO_2 + O_3 \longrightarrow NO_3 + O_2$	$k_{30} = 1.2 \times 10^{-13} \exp\{120/T\}$
31.	$OH + NO_2 + M \longrightarrow HNO_3 + M$	$k_{31} = 2.6 \times 10^{-30} \exp\{300/T\}^{3/2}$
32.	$OH + HNO_3 \longrightarrow NO_3 + H_2O$	$k_{32} = 7.2 \times 10^{-15} > \exp\{785/T\}$
33.	$HNO_4 + OH \longrightarrow H_2O + NO_2 + O_2$	$k_{33} = \dfrac{k_0[M]}{1 + \dfrac{k_0[M]}{k_\infty}} \cdot 0.6\{1 + [1gk[M]/k_\infty]^2\}^{-1}$ $k_0 = 1.8 \times 10^{-31}(T/300)^{-3/2}$, $k_\infty = 4.7 \times 10^{-12}(T/300)^{-1.4}$
34.	$HO_2 + NO_2 + M \longrightarrow HNO_4 + M$	$k_{34} = 1.3 \times 10^{-12} \exp\{380/T\}$
35.	$NO_3 + NO_2 + M \longrightarrow N_2O_5 + M$	$k_{35} = \dfrac{k_0[M]}{1 + \dfrac{k_0[M]}{k_\infty}} \cdot 0.6\{1 + [1gk[M]/k_\infty]^2\}^{-1}$ $k_0 = 2.2 \times 10^{-30}(T/300)^{-4.3}$ $k_\infty = 1.5 \times 10^{-12}(T/300)^{-0.5}$
36.	$N_2O_5 + M \longrightarrow NO_3NO_2 + M$	$k_{36} = k_p/k_{35}$, $k_p = 4.0 \times 10^{-27} \exp\{10930/T\}$

Table 3.2. Data used for numerical modelling.

Parameter	Value
Grid resolution	36 × 36
Step size by longitude	10°
Step size by latitude	5°
Number of vertical layers	15
Z_1 (first vertical layer height)	300 m
Z_2	600 m
Z_3	900 m
Z_4	1,200 m
Z_5	1,500 m
Z_6	3,040 m
Z_7	4,370 m
Z_8	7,220 m
Z_9	9,220 m
Z_{10}	11,830 m
Z_{11}	13,660 m
Z_{12}	16,250 m
Z_{13}	21,374 m
Z_{14}	25,290 m
Z_{15}	29,206 m
Time step	30 min
Meteorological data	TOGA dataset (ECMWF)
Modelling period	1 to 10 January, 1986

interest were the Antarctic ozone hole and mini-holes in the Northern Hemisphere. This problem will be discussed in detail in Section 3.4. Here discussion is mainly limited to a very important study by Grewe *et al.* (1998) who investigated the importance of dynamics for stratospheric ozone distribution in the Northern Hemisphere using multiannual simulations of the coupled dynamic–chemical general circulation model ECHAM3/CHEM (some other recent studies were mentioned in Chapter 1). As Grewe *et al.* (1998) emphasized, their purpose was to estimate not only the strength of chemical ozone depletion inside the Northern Hemisphere polar vortex but also the reduction of ozone concentration at mid-latitudes of the Northern Hemisphere because of polar vortex erosion (air exchange between the vortex and mid-latitudes).

It has been shown (Steil *et al.*, 1998) that the coupled model ECHAM3/CHEM reproduces the main climatological features regarding ozone, as well as the intra-annual and interannual variability of total ozone. However, ECHAM3 like other GCMs is characterized by a cold temperature bias in the winter stratosphere. In polar Northern Hemisphere winter regions at 50 hPa, zonal mean temperature deviations of 4–6 K occur in comparison with ECMWF analyses. Some discrepancies also take place between model PSC probabilities and those calculated on the basis of satellite data. The chemistry module CHEM is based on a family concept. It includes 13 families and species that are transported: H_2O, CH_4, N_2O, HCl, H_2O_2, CO, CH_3O_2H, $ClONO_2$, HNO_3 + type-I PSC (= HNO_3, $3H_2O$), type-II PSC (= water ice), ClO_x (= $Cl + ClO + HOCl + 2Cl_2O_2 + 2Cl_2$), NO_x (= $N + NO_x + NO_3 + NO_3 + 2N_2O_5 + HNO_4$), and O_x (= $O_3 + O(^3P) + O(^1D)$). Additionally, 17 species are treated (quasi-) diagnostically. A total of 107 photochemical reactions are considered in the model, with four heterogeneous reactions (3.42)–(3.45) and two on sulphate aerosols (equations (3.44)–(3.45)):

$$ClONO_2(g) + HCl(s) \longrightarrow Cl_2(g) + HNO_3(s) \tag{3.42}$$

$$HOCl(g) + HCl(s) \longrightarrow Cl_2(g) + H_2O(s) \tag{3.43}$$

$$ClONO_2(g) + H_2O(s) \longrightarrow HOCl(g) + HNO_3(s) \tag{3.44}$$

$$N_2O_5(g) + H_2O(s) \longrightarrow 2HNO_3(s) \tag{3.45}$$

Here (g) denotes the gas phase and (s) the particle surface.

Grewe *et al.* (1998) examined both a warm and a cold stratospheric winter to estimate the range of chemical ozone loss because of heterogeneous reactions on PSCs. They concluded that ozone depletion in the model mainly occurs inside the polar vortex, but outside the vortex an additional ozone reduction because of heterogeneous reactions on PSCs was found. Mean ozone reductions inside the polar vortex at 50-hPa level because of heterogeneous reactions on PSCs were 3.5 and 8% in a warm and in a cold model stratosphere, respectively. But the total chemical ozone depletion due to all factors should be twice as high (Bregman *et al.*, 1997). It was discovered that secondary vortex formation and vortex contraction after an elongation led to transport of airmasses with chemically reduced ozone values out of vortex. However, such events at the edge of the modelled polar vortex act as a barrier to transport. Grewe *et al.* (1998) found that during the formation of secondary vortices no additional heterogeneous reactions occur therein. Other dynamic events, such as the elongation of the polar vortex and its displacement to lower latitudes, led to an intense ozone depletion. In contrast to observations, no indications for transport into the polar vortex was found in the model results. An important conclusion is that even a minor stratospheric warming in the model caused total deactivation of chlorine compounds and further ozone depletion. In mid-latitudes, the amplitude of short-term variations are mainly caused by the displacement of the polar vortex. Grewe *et al.* (1998) emphasized that stratospheric dynamics causes high temporal variability in total ozone in Northern Hemisphere mid-latitudes.

In spite of substantial progress in modelling global ozone dynamics, there is still a number of unsolved problems which are first of all relevant to simulation of ozone photochemistry. The studies by Ravishankara *et al.* (1994, 1998) may serve as an illustration of such problems. Ravishankara *et al.* (1998) pointed out, for example, that the weakness of the chemical bond in ozone and the existence of low-lying electronically excited states of both atomic and molecular oxygen led to a number of energetically allowed dissociation channels. The most important channel is the reaction:

$$O_3 + h\nu \longrightarrow O(^1D) + O_2(^1A) \tag{3.46}$$

which has a proton energy threshold at 310 nm and forms two excited products. The other known process is the reaction:

$$O_3 + h\nu \longrightarrow O(^3P) + O_2(^3O) \tag{3.47}$$

which results in ground-state products.

Ravishankara *et al.* (1998) emphasized that, if dissociation to form $O(^1D)$ is to take place at longer wavelengths (where solar irradiance is much higher), it can only occur by the formation of an unexcited oxygen molecular partner, as in the reaction:

$$O_3 + h\nu \longrightarrow O(^1D) + O_2(^3O) \tag{3.48}$$

or by photolysis of internally excited ozone molecules. Both such processes were in the past considered to be of negligible importance. The yield of $O(^1D)$ at longer wavelengths than the 310-nm threshold was neglected. Photodissociation experiments made in the 1990s demonstrated, however, the non-negligible yield at longer wavelengths. So, above 310 nm there are important contributions to $O(^1D)$ production from internally excited ozone as well as a spin-forbidden pathway (reaction (3.48)). In addition, even the following reaction was observed:

$$O_3 + h\nu \longrightarrow O(^3P) + O_2(^1A) \tag{3.49}$$

Ravishankara *et al.* (1998) underlined the need to remember the results mentioned as lessons for future studies in atmospheric chemistry. Potential possibilities are the photolysis of chlorine and bromine nitrates, the NO3 radical, formaldehyde. About a decade ago, Yamshanov (1989) carried out a laboratory experiment to show that photolysis of nitrocompounds is accompanied by ozone destruction because of nitrogen oxide which forms as a product of photolysis. The nitrocompounds themselves may also be formed via photolysis.

Apart from unsolved problems of atmospheric photochemistry, which concentrated on studying anthropogenic impact on the ozone layer, there are serious unsolved problems about assessments of various natural mechanisms of TOZ dynamics which have not been considered in the numerical modelling of global dynamics. In this context Kapitsa (1996) emphasized the need for more careful analysis of contributions of natural versus atmospheric causes of ozone change, even in the case of Antarctic ozone hole dynamics. Gavrilov and Kaidalov (1990, 1996) showed, for example, that there is interaction between ozone variability in the Antarctic (including the ozone hole) and atmospheric semi-diurnal tides. Syvorotkin (1992, 1998) put forward the hypothesis of outgassing of the Earth as a cause of TOZ anomalies and focused attention on the fact of co-location of outgassing geological structures and TOZ anomalies. Note that all 'non-orthodox' results mentioned are of qualitative nature but they undoubtedly deserve a careful quantitative analysis.

3.4 POLAR OZONE DYNAMICS, SPRINGTIME OZONE MINIMUM IN THE ANTARCTIC

3.4.1 Introduction

A brief survey of the development of the theory of ozone-layer formation was made in Chapter 1 and the beginning of this chapter. Development of ozone-layer theory was necessitated by the discovery of important anthropogenic impacts on the ozone layer, resulting in the formation of the 'ozone hole' in the Antarctic (Farman, 1987; Hofmann, 1989; Molina and Rowland, 1974; Rycroft, 1990; Solomon *et al.*, 1986). We shall now consider this very important phenomenon. Before we begin it is worthwhile to emphasize once more the very complex and interactive nature of those processes that are responsible for stratospheric ozone formation and changes. Such changes depend on numerous interactions between heterogeneous chemical, photochemical and transport processes, including close links between ozone and climate changes as well as impacts of such external factors as volcanic eruptions, extraterrestrial irradiance variations (solar activity), cosmic rays, etc.

At the very beginning of ozone-hole studies (springtime total ozone minimum in the Antarctic), Solomon *et al.*, (1986) emphasised the importance of a unique feature of the Antarctic lower stratosphere – the high frequency of polar stratospheric clouds (PSCs) – providing a reaction site for heterogeneous chemical reactions on surfaces of cloud particles. Heterogeneous reactions between HCl and $ClONO_2$ ($\rightarrow HNO_3+Cl_2$) as well as $H_2O+ClONO_2$ ($\rightarrow HNO_3+HOCl$) were explored as possible mechanisms to explain the ozone observations. It became clear from observations and simulation modelling that heterogeneous chemistry (hydrolysis of N_2O_5 and $ClONO_2$) is operating on surfaces of aerosol particles in the stratospheric sulphate layer (*UNEP* . . . , 1994). It was pointed out in the *UNEP Scientific Assessment* (1994) that both 3-D and 2-D models were used in simulating PSC chemistry in the circumpolar vortex and how the equatorward transport of chemically perturbed polar air may affect ozone at mid-latitudes. However, lack of understanding of the detailed mechanisms for denitrification, dehydration and transport processes reduces confidence in model predictions.

It is worthwhile to illustrate the difficulties of studying ozone changes by referencing a few recent results. Danilin *et al.*, (1998) conducted simulation modelling of the sensitivity of the lower stratosphere at 70°N during winter/spring to varying stratospheric coolings and chlorine loadings. The calculations indicate that stratospheric cooling could further deplete ozone via increased PSC formation and retard its extended recovery even with the projected chlorine-loading decrease (because of the Montreal Protocol and its Amendments). It follows from the calculations that a 1-K cooling could provide the same local ozone depletion as an increase in chlorine by 0.4–0.7 ppbv for the scenarios considered. So, it became clear that climate–ozone interaction plays a very important role. Similar results were obtained earlier by Tie *et al.*, (1997) who assessed the effect of temperature interannual variability on stratospheric ozone.

Although Shindell *et al.*, (1998) showed (using the GISS 3-D climate model) that a chemical mechanism, rather than dynamical feedbacks, plays the dominant role in polar ozone recovery for the projected chlorine and greenhouse-gas loadings, an interaction between chemistry and dynamics may be important. McCormack *et al.*, (1998) demonstrated, for instance, that quasi-stationary planetary waves induce zonal asymmetries in wintertime extratropical TOZ fields with typical magnitudes of 30 to 60 DU. Model estimates of the wave-induced variability in TOZ fields agree well, both qualitatively and quantitatively, with the variability observed in lower-resolution SBUV(2) and higher-resolution TOMS satellite data, with the exception of certain years at high southern latitudes in October. McCormack *et al.*, (1998) pointed out that this disagreement is related to the displacement of the springtime ozone minimum over the pole by enhanced planetary wave activity.

One important aspect of the realistic reproduction of stratospheric chemical dynamics is the optimal choice of a set of reactions and a technique of numerical simulation. Austin (1991, 1992) discussed in this context the family approach when important chemical constituents are divided into related groups of 'families' (models containing 26, 20, 16, 14, 12, 10 and 8 families for a 12-day integration were considered).

Anderson *et al.* (1991) emphasized that three lines of evidence indicate a link between springtime TOZ minimum formation in the Antarctic and CFCs: (1) observed containment in the circumpolar vortex (CPV) of ClO concentration which is two orders of magnitude greater than normal levels; (2) *in situ* observations within the CPV during ozone-hole formation showed a decrease in O_3 concentration as ClO concentrations increased; (3) an agreement between observed ozone-loss rates and those predicted with the use of absolute concentrations of ClO and BrO, the rate-limiting radicals in an array of proposed catalytic cycles. So, there is convincing evidence that the dramatic reduction in TOZ over the Antarctic continent would not have occurred had CFCs not been added to the atmosphere.

Schoeberl and Hartmann (1991) pointed out, on the other hand, the need to take into account specific features of Antarctic stratospheric circulation in spring which results in a high degree of dynamic isolation and extremely low temperatures necessary for the formation of polar stratospheric clouds.

The decisive step (at an earlier stage of investigation) to verify the key hypothesis concerning the role of heterogeneous reactions in ozone-hole formation was made by analysis of results of aircraft observations in the Antarctic in August – September 1987 (Tuck *et al.*, 1989). Apart from successful verification of the hypothesis, analysis of observational data contributed to answering the following questions: (1) is there enough ClO in the stratosphere to explain TOZ depletion; (2) what is the role of Br and BrO; (3) characteristic features of TOZ depletion spatial distribution; (4) spatio-temporal correlations between O_3 and CCl_x, NO_y and polar stratospheric clouds (PSC); (5) the NO_y role in TOZ depletion; (6) water-vapour mixing-ratio distribution; (7) spatial distribution of PSC properties (number concentration, size distribution and chemical composition of particles); (8) dependence of PSC formation on temperature; (9) 'age' of airmasses in the reservoir of cold air below the circumpolar vortex (what is the contribution of fresh air from the troposphere); (10) possibility to separate the contribution of chemical processes and atmospheric dynamics and so avoid formation of ozone hole.

An important result of aircraft observations was the discovery of ozone depletion not only inside but also outside the polar vortex. In this context, recent observations of the ozone over Punta Arenas, Chile are of interest (Kirchhoff *et al.*, 1997).

New satellite data on odd-nitrogen partitioning (the partitioning of nitrogen species in the stratosphere into those that actively destroy ozone and those that are relatively benign) from instruments onboard Upper Atmosphere Research Satellite (UARS) were discussed by Morris *et al.* (1997). Thomason *et al.* (1997) substantiated stratospheric aerosol climatology on the basis of satellite observations, which is very important from the viewpoint of considering heterogeneous chemical reactions. Hofmann *et al.* (1997) summarized the results of 10 years of ozonesonde data at the South Pole, which may also be used to detect the future recovery of the ozone hole in the Antarctic. Springtime ozone depletion has been very severe since about 1992, with near-total loss of ozone in the 14–18-km region, but lessened somewhat in 1994 and 1995, probably because of the decay of the sulphate aerosol after the Mount Pinatubo eruption which was present at 10–16 km.

Hofmann *et al.* (1997) pointed out that a number of indicators for recovery of the ozone hole have been identified. They include an end to springtime ozone depletion at 22–24 km, a 12–20-km mid-September column ozone loss rate of less than about 3 DU per day, and a 12–20-km ozone column value of more than about 70 DU on 15 September. If recommendations of the Montreal Protocol and its Amendments are met, then it should be possible to conclusively detect the recovery of the ozone hole as early as 2008.

So, it is obvious that a sufficiently complete understanding of the nature of the ozone hole is possible only on the basis of adequate data about interactions between chemistry, radiation and dynamics, which was already pointed out at the initial stage of investigations (Kondratyev, 1989; McIntyre, 1989).

The unexpected detection of the springtime decrease of total ozone content (TOZ) in the Antarctic (happening regularly in the 1980s) should be considered, for some reasons, symptomatic. It is a paradox that with so much concern for global environmental problems, their actual manifestation in the form of 'ozone hole' has turned out to be quite unexpected and its nature is still not fully known. It is interesting that the springtime depletion of TOZ was first discovered by groundbased observations, though, of course, only satellite information made it possible to carry out a detailed analysis of the observed regularities (this again testified to the necessity of combining conventional and satellite observational means). On the whole, this situation can serve as an illustration of the preserved imperfection of the global-scale system of environmental observations and of serious gaps in theoretical studies of processes taking place in the atmosphere. The history of ozone-hole study has been marked by numerous controversial judgements. But now the situation has become much more clear.

Isaksen and Stordal (1986) suggested a model of the formation of the spring-time TOZ minimum in the Antarctic by chemical processes, based on the fact that the compounds of active chlorine result from heterogeneous reactions on the surface of ice particles which, as a rule, exist in the very cold Antarctic lower stratosphere both in winter and in early spring. This model is a modified version of the model used earlier to assess the effect of emissions of chlorine compounds on the ozone layer. The modification consists in an additional account of reactions with the participation of bromine:

$$\mathrm{BrO + O \longrightarrow Br + O_2} \tag{3.50}$$

$$\mathrm{Br + O_3 \longrightarrow BrO + O_2} \tag{3.51}$$

$$\mathrm{ClO + BrO \longrightarrow Cl + Br + O_2} \tag{3.52}$$

(the result is: $\mathrm{O + O_3 \longrightarrow 2O_2}$)

$$\mathrm{Cl + O_3 \longrightarrow ClO + O_2} \tag{3.53}$$

$$\mathrm{Br + O_3 \longrightarrow BrO + O_2} \tag{3.54}$$

(the result is: $\mathrm{2O_3 \longrightarrow 3O_2}$).

An enhanced level of ClO at the South Pole determines a particularly high efficiency of the second cycle of reactions. The following reactions are supposed to be intensive on the surface of ice particles. Two simulation calculations were carried out (Isaksen and Stordal, 1986) using: (i) low mixing ratios for chlorine and bromine (respectively, 10^{-9} and 10^{-12}) in the upper stratosphere, which simulate conditions prior to atmospheric perturbations; (ii) mixing ratios, 2.7×10^{-2} and 30×10^{-12}, from observational data for 1985. With respect to Brx compounds, their concentration for the last 10–15 years is supposed to have doubled (from 15×10^{-12} to 30×10^{-12}).

Analysis of numerical modelling results has shown that the maximum TOZ decrease in the lower stratosphere (the 15–20-km layer) because of anthropogenic emission of compounds of chlorine and bromine could have reached 14–20% in

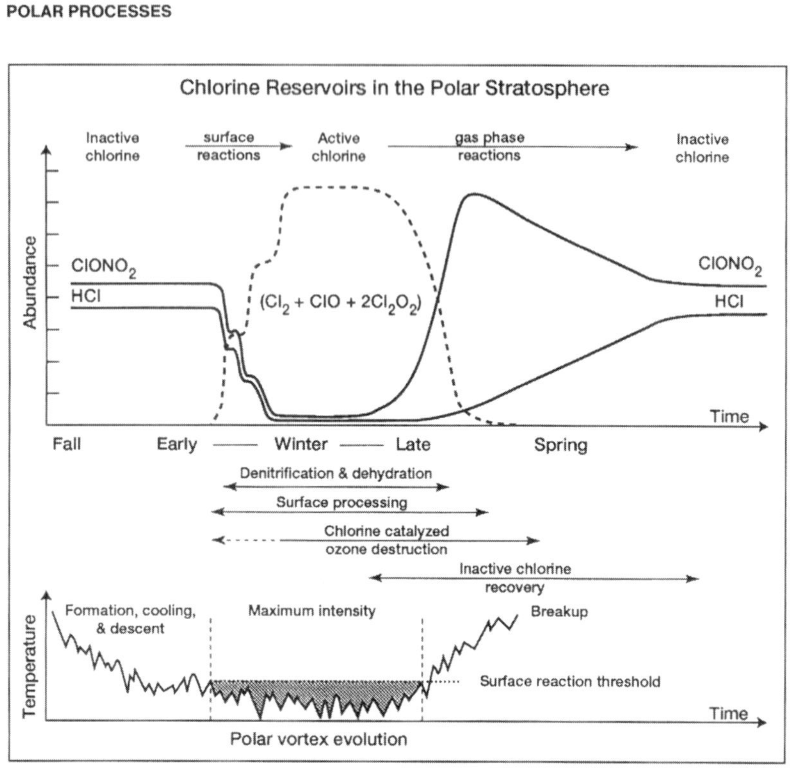

Figure 3.1. Schematic of the photochemical and dynamical features of the polar regions related to ozone depletion. The upper panel represents the conversion of chlorine from inactive to active forms in winter in the lower stratosphere and the re-formation of inactive forms in spring. The partitioning between the active chlorine species Cl_2, ClO, and Cl_2O_2 depends on exposure to sunlight after polar stratospheric cloud (PSC) processing. The corresponding stages of the polar vortex are indicated in the lower panel, where the temperature scale represents changes in minimum polar temperatures in the lower stratosphere.

October at latitude 80°S. Estimates obtained agree with the observational dataset for the previous 8 years.

Figure 3.1 is a schematic of the photochemical and dynamical features of the polar regions related to ozone depletion, as suggested in *UNEP Scientific Assessment* ... (1994). As pointed out in *UNEP Scientific Assessment*, the principal cause of ozone loss in the polar regions is photochemistry involving the halogen species, chlorine and bromine. Long-lived halogen species, primarily chlorofluorocarbons, are released into the troposphere by human activity. Photochemical degradation of these organic source molecules in the stratosphere leads to the formation of inorganic halogen species, of which chlorine monoxide (ClO), chlorine nitrate ($ClONO_2$), hydrochloric acid (HCl), bromine monoxide (BrO) and bromine nitrate ($BrONO_2$) are most important. The release of chlorine from more stable reservoirs occurs in high-latitude winter by reactions on surfaces of stratospheric aerosol particles. The formation and reactivity of these particles are enhanced at the low temperatures characteristic of the interior of the polar vortices. This reactive processing maintains high levels of active chlorine species that, along with BrO, catalytically destroy ozone as this air encounters sunlight. With sufficient insolation and higher temperatures, chlorine is returned to its reservoir forms during a photochemical recovery period and ozone destruction slows. The removal of active nitrogen by aerosol particle sedimentation in the vortex, a process defined as denitrification, strongly regulates the rate of recovery by controlling the availability of active chlorine.

The following reactions are responsible for the photochemical loss of ozone because of catalytic reaction cycles involving ClO (*UNEP* ... , 1994):

$$ClO + ClO + M \longrightarrow (ClO)_2 + M \tag{3.55}$$

$$(ClO)_2 + h\nu \longrightarrow ClOO + Cl \tag{3.56}$$

$$ClOO \longrightarrow Cl + O_2$$

$$2(Cl + O_3 \longrightarrow ClO + O_2)$$

Net reaction: $2O_3 \rightarrow 3O_2$

$$ClO + BrO \longrightarrow Cl + Br + O_2 \tag{3.57}$$

$$\longrightarrow Br + OClO \tag{3.58}$$

$$\longrightarrow BrClO_2$$

$$BrCl + h\nu \longrightarrow Br + Cl$$

$$Br + O_3 \longrightarrow BrO + O_2$$

$$Cl + O_3 \longrightarrow ClO + O_2$$

Net reaction: $2O_3 \rightarrow 3O_2$

$$ClO + O \longrightarrow Cl + O_2$$

$$Cl + O_3 \longrightarrow ClO + O_2 \tag{3.59}$$

Net reaction: $O + O_3 \rightarrow 2O_2$

where reaction steps (3.56) and (3.58) do not result in ozone destruction and where the cycles are listed in order of importance for ozone destruction inside the vortex.

A decisive condition for the beginning of ozone depletion by chlorine-induced destruction of ozone molecules is the circumpolar vortex (CPV) formation in each winter hemisphere in the polar region, schematically presented by Figure 3.2 (*UNEP* ... , 1994), as well as strong air-temperature decrease within CPV (*UNEP* ... , 1994). Moreover, the dynamical features of CPV determine the distribution of air from the vortex to lower latitudes and the incorporation of lower latitude air into CPV. As soon as solar illumination begins the huge 'photochemical machine' of ozone destruction starts functioning.

In accordance with the current understanding of the polar stratosphere (Jaeglé *et al.*, 1997), three principal heterogeneous reactions are believed to convert inorganic chlorine reservoirs to photochemically reactive chlorine forms $Cl_x = ClO + Cl +$

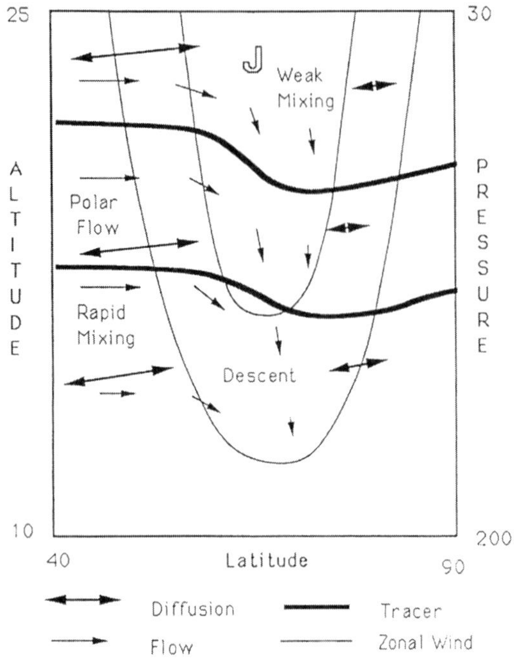

Figure 3.2. Schematic of the circulation and mixing associated with the polar vortex in the Arctic midwinter or Antarctic early spring. The vertical scale is shown in altitude (km) and pressure (mb) units. The horizontal scale is latitude in degrees. Arrows indicate mixing (double) and flow (single), with longer arrows representing larger rates. Other features are zonal wind contours (thin lines), jet core (J), and long-lived tracer isopleths (thick lines).

$2Cl_2O_2 + 2Cl_2 + OClO$ as a result of heterogeneous conversion on the surfaces of polar stratospheric cloud particles (consisting of supercooled ternary solution $HNO_3/H_2SO_4/H_2O$ (Del Negro *et al.*, 1997):

$$HCl + ClONO_2 \longrightarrow Cl_2 + HNO_3 \tag{3.60}$$

$$ClONO_2 + H_2O \longrightarrow HOCl + HNO_3 \tag{3.61}$$

$$HCl + HOCl \longrightarrow Cl_2 + H_2O \tag{3.62}$$

The gaseous products of these reactions (Cl_2 and HOCl) are rapidly converted to ClO and Cl_2O_2, the two dominant forms of Cl_x in the polar winter stratosphere. Jaeglé *et al.* (1997) emphasized that understanding the factors controlling chlorine activation and deactivation is central to predicting ozone loss in the polar regions. This is why Jaeglé et al. (1997) made for the first time simultaneous *in situ* measurements of HCl and ClO in the Southern Hemisphere with the aim of studying the processes governing chlorine activation between 15 and 20 km in the 1994 Antarctic winter. These measurements were made as part of the Airborne Southern Hemisphere Ozone Experiment/Measurements of Atmospheric Effects of Stratospheric Aircraft (ASHOE/MAESA) campaign. Data analysis indicates chemical processing of the chlorine reservoir HCl to active chlorine via the heterogeneous reaction

$$HCl + ClONO_2 \longrightarrow Cl_2 + HNO_3$$

The amount of chlorine activation is not strongly dependent on the composition of PSCs. However, HOx levels exibit different signatures on the type of heterogeneous surfaces that affect chlorine activation.

At the present time, many numerical modelling efforts have been made to assess various features of the 'photochemical machine dynamics'. For example, Pitari and Visconti (1991) applied a 2-D model to calculate the dependence of the polar ozone-loss rate on the value of the sticking coefficient γ for the reaction

$$N_2O_5 + H_2O \longrightarrow 2HNO_3 \tag{3.63}$$

of converting odd nitrogen into nitric acid on sulphate aerosols. The results indicate that if background aerosols alone are included, neglgible changes in TOZ are observed (up to about 1%) when γ is increased from 0.02 to 0.14. If PSCs are added a significant reduction in ozone-hole depth in Antarctica takes place. So, the amount of NO_x converted in HNO_3 on background aerosols seems to be critical in determining the intensity of the loss rate of ozone. Wenneberg *et al.* (1997) demonstrated the insignificance of iodine chemistry for ozone depletion in the lower stratosphere over the last few decades.

A new 3-D simulation of the Antarctic ozone hole with the aim of assessing the impact of anthropogenic chlorine on the lower stratosphere and the upper troposphere was undertaken by Brasseur *et al.* (1997). They developed a global transport-chemical model of the stratosphere (called the Study of Transport and Chemical Reactions in the Stratosphere (STARS)) which includes representation of the formation of PSCs and heterogeneous reactions on the surfaces of PSCs and sulphate aerosols. Table 3.3 illustrates gas-phase reactions included in the model and relevant reaction rates. Tables 3.4–3.7 characterize gas-phase chemical and photolysis reactions, as well as heterogeneous reactions on the surfaces of PSCs (Table 3.6) and sulphate aerosols (Table 3.7).

Brasseur *et al.*, (1997) pointed out that the formation of the observed springtime ozone hole is well reproduced by the model. A maximum of 40% TOZ depletion occurs in October. After the breakdown of the polar vortex in December, air with depleted ozone is transported to mid-latitudes in the Southern Hemisphere, resulting in a 2–4% ozone decrease at 50°S in December and 1% decrease in the subtropics. Ozone-poor airmasses are also transported to the troposphere and produce a significant decrease in upper tropospheric ozone. The model results indicate that the much weaker ozone minimum observed in Antarctica several decades ago (pre-industrial chlorine levels) was produced by natural dynamical processes.

3.4.2 Observational data and the role of stratospheric dynamics

Recent satellite observations have provided much information about the global distribution of ozone, various meteorological parameters and their spatial and

Table 3.3. Gas-phase chemical reactions included in the model corresponding reaction rates.

Reactions	Rate coefficient
$O(^1D) + O3 \longrightarrow 2O_2$	1.2e-10
$O + O + m \longrightarrow O_2 + m$	4.23e-28*am/(T*T)
$O + O_2 + m \longrightarrow O_3 + m$	6.0e-34*((300/T)**2.3)*am
$O + O_3 \longrightarrow O_2 + O_2$	8e-12* exp(-2060/T)
$O(^1D) + N_2 \longrightarrow O + N_2$	1.8e-11* exp(110/T)
$O(^1D) + O_2 \longrightarrow O + O_2$	3.2e-11* exp(70/T)
$O(^1D) + N_2 \longrightarrow N_2O$	3.5e-37*((300/T)**0.6)*am
$O(^1D) + H_2O \longrightarrow 2OH$	2.2e-10
$O(^1D) + CH_4 \longrightarrow CH_3 + OH$	1.4e-10
$O(^1D) + CH_4 \longrightarrow H_2 + CH_2O$	1.4e-11
$O(^1D) + H \longrightarrow OH + H$	1.0e-10
$H + HO_2 \longrightarrow OH + OH$	8.1e-11*0.9
$H + HO_2 \longrightarrow H_2 + O_2$	8.1e-11*0.08
$H + HO_2 \longrightarrow H_2O + O$	8.1e-11*0.02
$H + O_2 + m \longrightarrow HO_2 + m$	5.7e-32*am(300/T)**1.6
$H + O_3 \longrightarrow OH + O_2$	1.4e-10* exp(-470/T)
$O + OH \longrightarrow O_2 + H$	2.2e-11* exp(120/T)
$OH + O_3 \longrightarrow HO_2 + O_2$	1.6e-12* exp(-940/T)
$HO_2 + O_3 \longrightarrow OH + 2O_2$	1.1e-14* exp(-500/T)
$O + HO_2 \longrightarrow OH + O_2$	3.0e-11* exp(200/T)
$OH + HO_2 \longrightarrow H_2O + O_2$	4.8e-11* exp(250/T)
$OH + H_2 \longrightarrow H_2O + H$	5.5e-12* exp(-2,000/T)
$H_2 + O \longrightarrow OH + H$	8.8e-12* exp(-4,200/T)
$NO + HO_2 \longrightarrow NO_2 + OH$	3.7e-12* exp(250/T)
$HO_2 + HO_2 \longrightarrow H_2O_2 + O_2$	(2.3e-13* exp(600/T) + a25)* b25 a25 = 1.7e-33* exp(1,000/T)*am b25 = (1. + 1.4e - 21*H2O* exp(2,200/T)
$OH + H_2O_2 \longrightarrow H_2O + HO_2$	2.9e-12* exp(-160/T)
$OH + CO \longrightarrow CO_2 + H$	1.5e-13*(1 + 6*am/1,013)
$O + H_2O_2 \longrightarrow HO_2 + OH$	1.4e-12* exp(-2,000/T)
$OH + OH \longrightarrow H_2O + O$	4.2e-12* exp(-240/T)
$N + NO \longrightarrow N_2 + O$	3.4e-11
$N_2O + O(^1D) \longrightarrow N_2 + O_2$	4.9e-11
$N_2O + O(^1D) \longrightarrow NO + NO$	6.7e-11
$NO_3 + O \longrightarrow NO_2 + O_2$	1.0e-11
$NO_3 + OH \longrightarrow NO_2 + HO_2$	2.3e-11
$NO_3 + HO_2 \longrightarrow HNO_3 + O_2$	4.1e-12
$HNO_4 + m \longrightarrow HO_2 + NO_2$	k_{100}/(2.1e-27* exp(10,900/T))*am)
$N_2O_5 + m \longrightarrow NO_2 + NO_3$	k_{98}/(4.0e-27* exp(10,930/T))*am)
$O + NO_2 \longrightarrow NO + O_2$	6.5e-12* exp(120/T)
$O_3 + NO \longrightarrow NO_2 + O_2$	2e-12* exp(-1,400/T)
$N + O_2 \longrightarrow NO + O$	1.5e-11* exp(-3,600/T)
$O_3 + NO_2 \longrightarrow NO_3 + O_2$	1.2e-13* exp(-2,450/T)
$HNO_3 + OH \longrightarrow H_2O + NO_3$	7.2e-15* exp(785/T) + aux1/aux2

	aux1 = 1.9e-33* exp(725/T)*am
	aux2 = 1 + aux1/(4.1e-16* exp(1,440/T))
$OH + HO_2NO_2 \longrightarrow$ products	1.3e-12* exp(380/T)
$HNO_4 + OH \longrightarrow NO_2 + H_2O + O_2$	1.3e-12* exp(380/T)
$NO_3 + NO \longrightarrow NO_2 + NO_2$	1.5e-11* exp(170/T)
$OH + CH_4 \longrightarrow CH_3 + H_2O$	2.65e-12* exp(-1,880/T)
$O(^1D) + CF_3Cl \longrightarrow Cl +$ prod	2.3e-10
$O(^1D) + CF_2Cl_2 \longrightarrow 2Cl +$ prod	1.4e-10
$O(^1D) + Cl_2 \longrightarrow ClO + Cl$	2.8e-10
$Cl + Cl_2O_2 \longrightarrow Cl_2 + Cl + O_2$	1.0e-10
$Cl + NO_3 \longrightarrow ClO + NO_2$	2.6e-11
$ClO + NO_3 \longrightarrow Cl + NO_2 + O_2$	4.7e-13
$O(^1D) + HCl \longrightarrow OH + Cl$	1.5e-10
$OH + CH_3Cl \longrightarrow CH_2Cl + H_2O$	4.0e-12* exp(-1,400/T)
$Cl + O_3 \longrightarrow ClO + O_2$	2.9e-11* exp(-260/T)
$ClO + O \longrightarrow Cl + O_2$	3e-11* exp(70/T)
$ClO + NO \longrightarrow NO_2 + Cl$	6.4e-12* exp(290/T)
$Cl + CH_4 \longrightarrow HCl + CH_3$	1.1e-11* exp(-1,400/T)
$Cl + H_2 \longrightarrow HCl + H$	3.7e-11* exp(-2,300/T)
$Cl + HO_2 \longrightarrow HCl + O_2$	1.8e-11* exp(170/T)
$ClO + OH \longrightarrow Cl + HO_2$	1.1e-11* exp(120/T)
$OH + HCl \longrightarrow H_2O + Cl$	2.6e-12* exp(-350/T)
$O + ClONO_2 \longrightarrow$ products	2.9e-12* exp(-800/T)
$ClO + HO_2 \longrightarrow HOCl + O_2$	4.8e-13* exp(700/T)
$OH + HOCl \longrightarrow H_2O + ClO$	3e-12* exp(-500/T)
$Cl + HOCl \longrightarrow OH + Cl_2$	3e-12* exp(-130/T)
$OClO + OH \longrightarrow HOCl + O_2$	4.5e-13* exp(800/T)
$Cl + OClO \longrightarrow ClO + ClO$	3.4e-11* exp(160/T)
$OClO + O \longrightarrow ClO + O_2$	2.5e-12* exp(-950/T)
$OClO + NO \longrightarrow NO_2 + ClO$	2.5e-12* exp(-600/T)
$CH_3CCl + OH \longrightarrow 3Cl +$ products	5.0e-12* exp(-1,800/T)
$Cl_2O_2 + m \longrightarrow ClO + O_2$	3.0e-27* exp(8,450/T)
	deg $\longrightarrow$ 3.0e-27* exp(8,450/T)
$Cl_2 + OH \longrightarrow HOCl + Cl$	1.4e-12* exp(-900/T)
$Cl + ClONO_2 \longrightarrow Cl_2 + NO_3$	6.8e-12* exp(160/T)
$Cl + HO_2 \longrightarrow OH + ClO$	4.1e-11* exp(450/T)
$Cl + H_2O_2 \longrightarrow HCl + HO_2$	1.1e-11* exp(-980/T)
$O + HCl \longrightarrow OH + Cl$	1.0e-11* exp(-3,300/T)
$O + ClONO_2 \longrightarrow NO_3 + ClO$	2.9e-12* exp(-800/T)
$OH + ClONO_2 \longrightarrow NO_3 + HOCl$	1.2-12 exp(-330/T)
$BrO + ClO \longrightarrow OClO + Br$	1.6-12* exp(-430/T)
$BrO + O \longrightarrow Br + O_2$	3.e-11
$OH + HBr \longrightarrow H_2O + Br$	1.1e-11
$O(^1D) + HBr \longrightarrow OH + Br$	1.5e-10
$OH + BrO \longrightarrow HO_2 + Br$	1.0e-11
$Br + O3 \longrightarrow BrO + O_2$	1.7e-11* exp(-800/T)
$BrO + NO \longrightarrow NO_2 + Br$	8.8e-12* exp(260/T)
$BrO + ClO \longrightarrow Br + Cl + O_2$	2.9e-12* exp(220/T)
$BrO + ClO \longrightarrow BrCl + O_2$	5.8e-13* exp(170/T)

Table 3.3 (*cont.*)

Reactions	Rate coefficient
$BrO + BrO \longrightarrow Br + Br + O_2$	4.0e-12* exp(-190/T)
$Br + HO_2 \longrightarrow HBr + O_2$	1.5e-11* exp(-600/T)
$Br + OClO \longrightarrow BrO + ClO$	2.6e-11* exp(-1,300/T)
$BrO + HO_2 \longrightarrow HOBr + O_2$	6.2e-12* exp(500/T)
$CH_3Br + OH \longrightarrow CH_2Br + H_2O$	4.0e-12* exp(-1,470/T)
$O + HBr \longrightarrow OH + Br$	5.8e-12* exp(-1,500/T)

Read 1.2e-10 as 1.2×10^{-10}; am is air density (number/cm^3); rate coefficients are expressed in cm^3/s; aux is defined at the top of p. 239; a and b are a25 and b25 defined in the middle of p. 238.

Table 3.4. Gas-phase chemical reactions included in the model and corresponding reaction rates.

Reaction	k_a^{300}	n	k_b^{300}	m
$H + O_2 + m \longrightarrow HO_2 + m$	5.7e-32	1.6	7.5e-11	0
$OH + OH + m \longrightarrow H_2O + m$	6.9e-31	0.8	1.5e-11	0.0
$NO_3 + NO_2 + M \longrightarrow N_2O_5$	2.2e-30	3.9	1.5e-12	0.7
$NO_2 + OH + m \longrightarrow HNO_3 + m$	2.6e-30	3.2	2.4e-11	1.3
$NO_2 + HO_2 + m \longrightarrow HNO_4 + m$	1.8e-31	3.2	4.7e-12	0.14
$O + NO_2 + m \longrightarrow NO_3 + m$	9.0e-32	2.0	2.2e-11	0.0
$O + NO + m \longrightarrow NO_2 + m$	9.0e-32	1.5	3.0e-11	0.0
$ClO + NO_2 + m \longrightarrow ClONO_2 + m$	1.8e-31	3.4	1.5e-11	1.9
$Cl + NO_2 + m \longrightarrow ClNO_2 + m$	1.8e-32	3.6	6e-12	0.0
$ClO + ClO + m \longrightarrow Cl_2O_2 + m$	2.2e-32	3.1	3.5e-12	1.0
$BrO + NO_2 + m \longrightarrow BrONO_2 + m$	5.2e-31	3.2	6.9e-12	2.9

Table 3.5. Photolyses included in the model.

Reaction	Reaction
$O_2 + h\nu \longrightarrow O + O$	$NO + h\nu \longrightarrow O + N$
$O_3 + h\nu \longrightarrow O + O_2$	$CO_2 + h\nu \longrightarrow CO + O$
$O_3 + h\nu \longrightarrow O(^1D) + O_2$	$CH_3Cl + h\nu \longrightarrow Cl + \text{products}$
$H_2O + h\nu \longrightarrow OH + H$	$CH_3CCl_3 + h\nu \longrightarrow 3Cl + \text{products}$
$H_2O_2 + h\nu \longrightarrow 2OH$	$CCl_4 + h\nu \longrightarrow 4Cl + C$
$N_2O + h\nu \longrightarrow N_2 + O(^1D)$	$CFCl_3 + h\nu \longrightarrow 3Cl + \text{products}$
$CH_4 + h\nu \longrightarrow CH_3 + H$	$CF_2Cl_2 + h\nu \longrightarrow 2Cl + \text{products}$
$NO_2 + h\nu \longrightarrow NO + O$	$NO_3 + h\nu \longrightarrow NO_2 + O$
$HNO_3 + h\nu \longrightarrow OH + NO_2$	$Cl_2 + h\nu \longrightarrow 2Cl$
$HOCl + h\nu \longrightarrow Cl + OH$	$ClNO_2 + h\nu \longrightarrow Cl + NO_2$
$HNO_4 + h\nu \longrightarrow NO_2 + HO_2$	$BrONO_2 + h\nu \longrightarrow BrO + NO_2$
$ClONO_2 + h\nu \longrightarrow ClO + NO_2$	$BrCl + h\nu \longrightarrow BrO + Cl$
$N_2O_5 + h\nu \longrightarrow NO_2 + NO_3$	$HOBr + h\nu \longrightarrow Br + OH$
$OClO + h\nu \longrightarrow O + ClO$	$CH_3Br + h\nu \longrightarrow Br + \text{products}$
$Cl_2O_2 + h\nu \longrightarrow Cl + CLOO$	$HNO_2 + h\nu \longrightarrow NO + OH$
$HCl + h\nu \longrightarrow H + Cl$	

Table 3.6. Heterogeneous reactions on the surface of PSCs.

Reaction	Accommodation coefficient	References from Brasseur *et al.* (1997)
Type-I PSCs		
$ClONO_2 + H_2O(s) \longrightarrow HOCl + HNO_3$	0.02	Moore *et al.* (1990)
$ClONO_2 + HCl(s) \longrightarrow Cl_2 + HNO_3$	0.3	Hanson and Ravishankara (1991)
$N_2O_5 + H_2O(s) \longrightarrow 2HNO_3$	0.0006	Hanson and Ravishankara (1991)
$N_2O_5 + HCl \longrightarrow ClNO_2 + HNO_3$	0.003	Hanson and Ravishankara (1991)
$HOCl + HCl(s) \longrightarrow Cl_2 + H_2O$	0.1	DeMore *et al.* (1994)
Type-II PSCs		
$ClONO_2 + H_2O \longrightarrow HOCl + HNO_3$	0.3	Hanson and Ravishankara (1991)
$ClONO_2 + HCl \longrightarrow Cl_2 + HNO_3$	0.2	DeMore *et al.* (1994)
$N_2O_5 + H_2O(s) \longrightarrow 2HNO_3$	0.03	Quinlan *et al.* (1990)
$N_2O_5 + HCl(s) \longrightarrow ClNO_2 + HNO_3$	0.03	DeMore *et al.* (1994)
$HOCl + HCl(s) \longrightarrow Cl_2 + H_2O$	0.3	DeMore *et al.* (1994)

Table 3.7. Heterogeneous reactions on the surface of Sulfate Aerosols.

Reaction	Accommodation coefficient	References from Brasseur *et al.* (1997)
$ClONO_2 + H_2O(s) \longrightarrow HOCl + HNO_3$		Hanson *et al.* (1994)
$ClONO_2 + HCl(s) \longrightarrow Cl_2 + HNO_3$		Hanson *et al.* (1994)
$N_2O_5 + H_2O(s) \longrightarrow 2HNO_3$	0.1	DeMore *et al.* (1994)
$HOCl + HCl(s) \longrightarrow Cl_2 + H_2O$		Hanson *et al.* (1994)
$ClONO_2 + HCl(s) \longrightarrow Cl_2 + HNO_3$	0.4	Danilin and McConell (1995)
$HOBr + HCl(s) \longrightarrow BrCl + H_2O$	0.2 (60% weight of sulphate)	Hanson and Ravishankara (1995)

temporal variability. We shall first discuss some earlier results. As Brasseur (1987) notes, the distribution of ozone in the Northern and Southern Hemispheres is substantially different: the springtime TOZ maximum is located in the Northern Hemisphere near the pole, whereas in the Southern Hemisphere it is in the latitudinal band 50–60°S (Figure 3.3), which contradicts the conclusion of pure photochemical theory that specific dynamics of the stratosphere in the different hemispheres are left out of account. Possible anthropogenic impacts on ozone both at present and in the future have been broadly discussed in scientific literature (Kondratyev, 1989, 1998; *UNEP* ..., 1994; *Scientific* ..., 1999). Figures 3.4 through 3.7, taken from a review by Brasseur (1987), illustrate the results obtained.

Douglass and Rood (1986) undertook an analysis of observational data to reveal the photochemically induced variation in ozone in March at pressure levels of 2.1

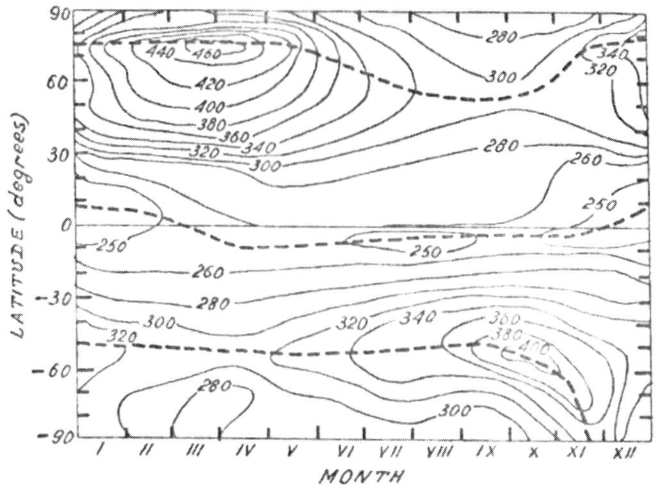

Figure 3.3. Annual variation of mean zonal TOZ values.

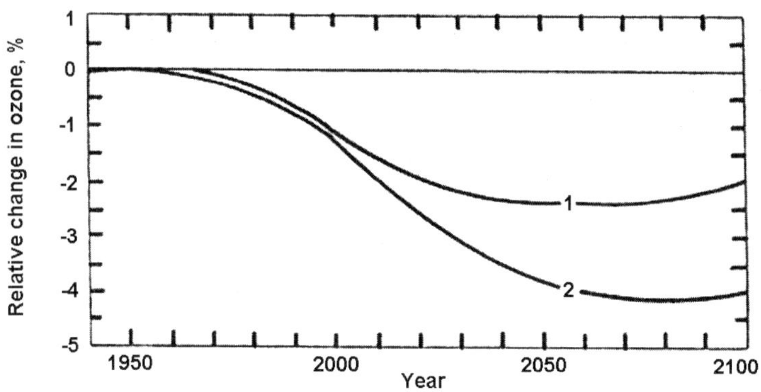

Figure 3.4. Relative temporal change in ozone (two versions).

and 0.7 hPa, using a dynamical model of ozone transport (the field of geostrophic wind is calculated from the field of temperature).

Photochemically induced variations in ozone are determined by the fact that in the case of photochemical equilibrium an inverse correlation between ozone and temperature takes place. Since the disturbances in the temperature field are determined by dynamic factors, large temperature variations are connected with an intensive transport of ozone. It follows that the negative correlation between ozone and temperature is not necessarily an indicator of photochemical equilibrium but may result from dynamic disturbances of the stratosphere.

A comparison of the calculated and observed Θ/Γ ratio, as an indicator of the agreement between photochemical theory and observations (Θ is the parameter characterizing the response of ozone to temperature variation, Γ is the parameter of

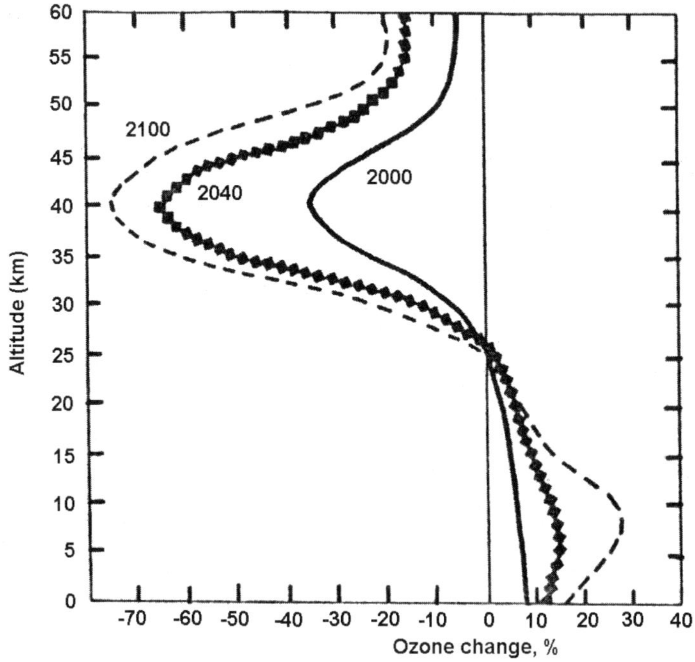

Figure 3.5. Ozone change vertical profiles.

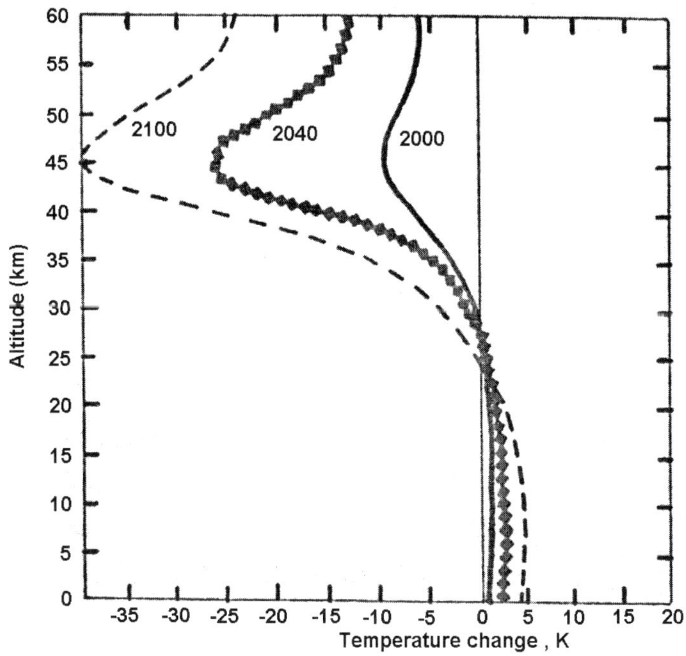

Figure 3.6. Temperature change vertical profiles.

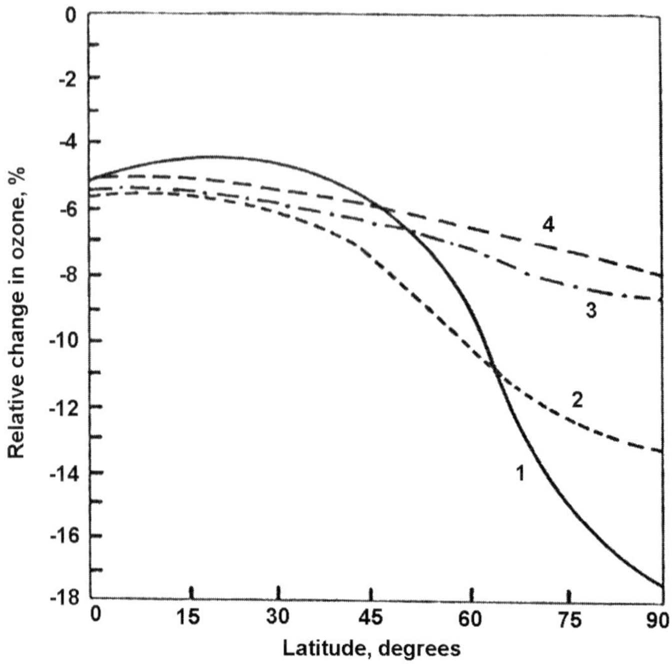

Figure 3.7. Latitudinal variation of relative TOZ change.

relaxation for odd oxygen), from the viewpoint of dependence on latitude, height and time revealed agreement within 10% (provided the effect of stratospheric dynamics is taken into account), except for the 1-hPa level near 60°N. The latter is apparently a long-since known feature of photochemical models: they happen to be unsatisfactory near the 1 hPa level. According to theoretical calculations, the ozone–temperature relationship at the 1–2 hPa level may be affected by Cl_x emissions. Therefore, a subsequent similar analysis must shed light onto the role of odd chlorine in the stratosphere from the viewpoint of its effect on photochemical processes.

As mentioned above, the most complete TOZ data came from satellite observations. For instance, Yung *et al.*, (1990) processed Nimbus-7 TOMS data. Results indicate the existence of a relatively stable collar region enclosing an interior (the vortex) that underwent large variation during the austral spring of 1987. There was evidence for quasi-periodic (15–20 days) O3 fluctuation in the collar and for upwelling of tropospheric air in late spring. The 1990s experienced an increasing trend towards stronger and long-lasting ozone holes. The most recent data for the ozone hole revealed that, in September 1998, the hole grew to a record size – two and a half times the area of Europe. The hole was also the deepest since regular measurements were started two decades ago.

Bevilacqua *et al.* (1997) presented an overview of Polar Ozone and Aerosol Measurement (POAM) II satellite-based observations of ozone in the Antarctic ozone hole in 1994, 1995 and 1996. Ozone concentrations are observed to decrease by nearly a factor of 10 near 20 km during the ozone-hole-formation period and a reduction of ozone was observed up to 20 km. The timing of ozone loss and recovery was similar each year (the period of largest depletion occurs between early September and early October). However, ozone concentrations were consistently lower (by about 10%) in 1996 than in the other two years. Maximum monthly average ozone chemical-loss rates occurred in September near 500 K each year (0.1 ± 0.004 ppmv/day in 1994 and 0.08 ± 0.004 ppmv/day in 1995 and 1996). Bevilacqua *et al.* (1997) pointed out that in late spring, after ozone-hole chemical processing is complete, ozone-mixing ratios are lower inside the Antarctic vortex (relative to the outside of the vortex) at all levels between at least 500 K and 1,500 K, presumably resulting from a combination of dynamical and chemical effects.

Much attention has been given to analysis of data of conventional (surface, balloon) observations. So, for instance, based on vertical-ozone profiles obtained at 13 stations equipped with Dobson spectrophotometers (the profiles were retrieved using the Umkehr technique), Titao *et al.* (1986) performed a statistical analysis of the ozone trend in the troposphere, lower and mid-stratosphere. On the whole, the analysis revealed a decreasing trend of ozone in the Dobson layers 3A, 3B and 4A (100–50 hPa), constituting about −0.5%/year, as well as some indications of an increasing trend in layers 1A and 1B (800–500 hPa), reaching 0.8%/year.

Assessment of the sensitivity of the results relating to special features of the data-processing technique suggested that for layers 3A and 4A this sensitivity is not substantial. According to calculations using 1-D photochemical models, during the period 1970–1982 a positive ozone trend had to be observed in layer 1A because of increased concentrations of methane and nitrogen oxides and in layer 3A there could not have been a negative trend.

In comparison with numerical modelling results and observations it should be borne in mind that: (i) the geographical distribution of ozonesonde stations is confined mainly to mid-latitudes of the Northern Hemisphere, whereas calculational data characterize hemispherical means; (ii) photochemical models are still being improved (it follows from 2-D simulations, for instance, that lower stratospheric ozone varies strongly depending on latitude, height and season). Despite all this, however, available results of calculations show a negative trend in the lower stratosphere, zero or slight positive trend in the mid-stratosphere, and strong negative trend in the upper stratosphere, agreeing qualitatively with observational data.

3.4.3 First groundbased observations of the 'ozone hole'

TOZ decrease in the Antarctic manifested by both the multi-year trend of TOZ decrease and formation of springtime ozone minimum (ozone hole), was first noticed by groundbased observations with the Dobson spectrophotometer at the British Antarctic station in Halley Bay (76°S, 27°W), and later on was confirmed by data from other stations (Figure 3.8) and also as by Nimbus-7 data (Schoeberl and Krueger, 1986).

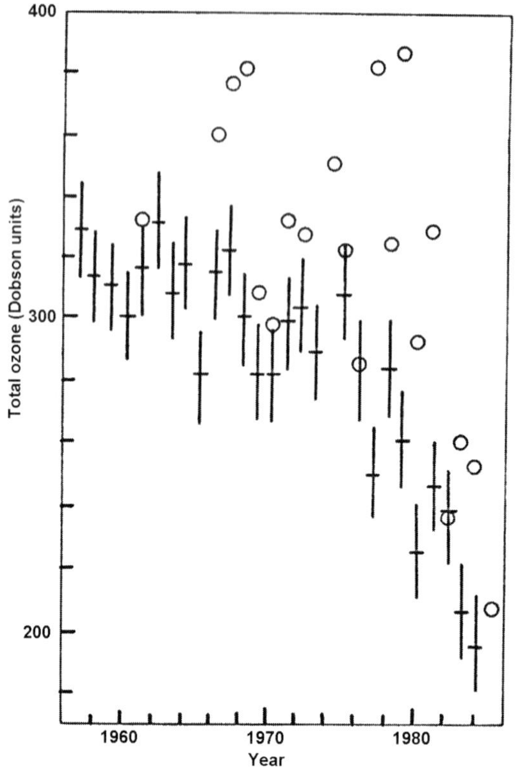

Figure 3.8. Ozone-hole dynamics.

Analysis of groundbased ozone observations at 13 Antarctic stations carried out by Bojkov (1986) showed that between 1979 and 1985 a TOZ decrease took place between late September and early November (30% of the 1957–1978 value), but during the remaining part of the year there was no further decrease of TOZ. Also, there was no substantial change of TOZ in the belt of mid-latitudes around the Antarctic, except for that caused by quasi-biennial oscillations. Data on vertical profiles of ozone concentration show that 85% of the TOZ decrease was on the 10–24 km layer. The subsequent rapid increase in ozone in spring, connected with the destruction of the wintertime circumpolar stratospheric vortex, usually starts in mid-October and exhibits a 2-month shift with respect to similar TOZ variability in the Arctic.

From a 6-year observational dataset (in the time interval 1975–1985), destruction of the circumpolar vortex occurred in mid-November, extending circumpolar circulation by 3–4 weeks. Weak (about 0.1 cm/s) upward motions in the zone of polar vortex during September and October are assumed to be one of the reasons of the TOZ decrease. To reveal the conditions existing at the time of destruction of the polar vortex and the dynamics in the TOZ field, dynamic and radiative factors governing the vortex's evolution must be studied.

Analysis of ground-based TOZ observations, made by Rood (1986), suggested that between 1958 and 1962 there was a global minimum of TOZ, the depth and structure of which were similar to those of the minimum observed in the period 1979–1983, though with substantial differences. Acccording to the data of the groundbased ozone network, the following features of TOZ dynamics were most typical: (i) a global decrease in the period 1979-1983 while a partial restoration of the TOZ level was observed between 1958 and 1962; (ii) a considerable decrease of TOZ in the South Pole zone, especially in 1984 and 1985; (iii) a substantial decrease of TOZ in the North Pole zone in the period of 1958–1962.

Though during both these periods there were similar correlations between TOZ and solar activity, the circulation factors connected with variation in wave activity should be considered a more probable reason for TOZ depletion in both the 1960s and 1980s. The minimum of TOZ in the 1960s points to the fact that global TOZ decrease, like that in the 1980s, cannot be explained by the effect of increased chlorine in the stratosphere due to CFC emission. Apparently, both minima resulted from natural variability of the global TOZ field.

Komhyr *et al.* (1986) considered an observational dataset obtained using the Dobson spectrophotometer at the South Pole station giving TOZ monthly means for the period 1964-1985. Data processing shows that, starting from the mid-1960s, a 20% decrease in TOZ has taken place. The most significant TOZ decrease was observed in the spring (October–November) but an important drop took place during autumn and winter months as well. In February the TOZ slightly increased. Starting from 1965, episodic drops with a 4–6 year periodicity were observed in the long-term decreasing trend of TOZ. These drops correlate with delays of warming of the Antarctic stratosphere as well as with the transport of ozone to the Antarctic observed every spring. Long-term variation in stratosphere warming and intensity of ozone transport have been important factors for TOZ decrease observed in the Antarctic, together with the gas-to-particle conversion and heterogeneous reactions made with the participation of ClO_x and SO_x as well as other processes.

Hofmann *et al.* (1987, 1997) (see also Hofmann, 1989) performed ozone-dust-sonde observations of the vertical ozone and aerosol profiles at McMurdo Station and the South Pole, which revealed strong variation in ozone and aerosols, correlating with the dynamics of the circumpolar vortex. Low ozone-mixing ratios in the zone of the vortex were accompanied by lack of optically active aerosol particles and increased concentrations of condensation nuclei.

Though causes of variations in ozone and aerosols are not directly interrelated, these variations reflect the complicated physicochemical nature of the processes of formation of stratospheric aerosol, as well as interesting chemical and dynamic processes as factors determining the springtime aerosol content in the Antarctic stratosphere. Analysis of observational data on the continuously decreasing aerosol content in the stratosphere (after the El Chichón eruption maximum) and on the intensified decrease of TOZ in the Antarctic raised doubts about the cause-and-effect relationship between these two phenomena.

These observations also demonstrated that the McMurdo Station may be an appropriate place to study various phenomena and processes in the zone of polar vortex and beyond it. Observations showed that on 9 November, 1985 a TOZ drop was observed at altitudes up to 70 km and higher.

3.4.4 Satellite observations of the springtime ozone depletion

According to satellite observations (Schoeberl and Krueger, 1986) the zone of minimum TOZ covers almost all the Antarctic continent with maximum decrease in the spring (September, October), when TOZ values decreased to 150 DU (Dobson Units). From the observational dataset, which started in the late 1950s when anthropogenic pollution of the stratosphere was not substantial, it can be seen that the central part of this zone, near the South Pole, is surrounded in the spring by a mid-latitudinal belt of high TOZ values, the spatial distribution of TOZ strongly varying in time. The zone of minimum TOZ meanders around the Pole, changing its shape depending on the meteorological conditions in the upper troposphere and lower stratosphere. Minimum monthly means (for October) for TOZ decreased from 260 DU in 1979 to 150 DU in 1986. A decrease of about 70 DU was also recorded in the region of maximum TOZ values.

Analysis of satellite and balloon observations shows that the atmospheric layers below 30 km contribute most to TOZ decrease. However, there is evidence for a springtime ozone decrease in the 40–50-km layer. It was first assumed that TOZ decrease was caused by increased concentrations of chlorine in the lower stratosphere (measurements of the mixing ratio of CFCs at the South Pole station revealed a continuous increase in mixing ratio of more than 5%/year).

It must be borne in mind, however, that there are no favourable conditions in the lower stratosphere for a chlorine effect on ozone, since chlorine is in a bound state (as ClONO2 forms in the presence of NO_2 and HCl). The chemical conversions needed for free chlorine to be released take place in this twilight zone too slowly to provide a rapid TOZ decrease in spring.

There are alternative possibilities of chlorine release from heterogeneous reactions on surfaces of polar strsatospheric cloud particles, as well as from chlorine oxide dimerization and reactions with participation of bromine oxides. The hypotheses, based on consideration of specific dynamic and thermal structures of the Antarctic stratosphere, proceed from accounts of the effect of upwelling and TOZ temperature dependence. The possible effect of solar activity also attracted attention.

As Bowman (1986) noted, analysis of satellite information enables relation of minimum TOZ monthly means to specific dynamics of the Antarctic stratosphere, which manifests itself through the effect of the breakdown of the Antarctic circumpolar vortex on transport and hence spatial distribution of ozone in October and November, when this breakdown takes place. Early breakdown of the circumpolar vortex (as in 1978–1979) is accompanied by large TOZ values, whereas during weak and late breakdowns (1980–1982) the TOZ values are extremely low.

Probably, then, the interannual variability of ozone is partly determined by variation in the intensity of planetary waves and circulation in the stratosphere. The observed variation in TOZ result apparently from superposing of such interannual variability and variations because of photochemical processes. There can also be a third factor: photochemically induced reactions lead to variations in radiative flux divergence, sufficiently large to cause variations in zonal mean circulation and make planetary waves penetrate the lower stratosphere, which delays the breakdown of the circumpolar vortex. The observed warming of the Antarctic stratosphere takes place 15–40 days before the breakdown of the circumpolar vortex.

Continuous ground and satellite observations in the Antarctic are needed to determine the repeatability (stability) of TOZ depletion. The restoration of the earlier TOZ level (if it happens) will testify to the failure of the hypothesis about the photochemical origin of TOZ variability. On the contrary, the repeatability of decrease (especially if this is observed in both hemispheres) will become an indicator of the importance of photochemical processes and will make the contribution of dynamic factors less probable. It is an important fact that during TOZ depletion in the Antarctic, strong fluctuations of TOZ also take place in other regions of the globe, which shows the probable global scale of this variability. In this respect the effect of solar activity on the global TOZ field may be also possible (Varotsos, 1989).

3.4.5 'Ozone hole' and stratospheric temperature

In discussions about possible causes of the springtime TOZ minimum, Newman and Schoeberl (1986) drew attention to the fact that analysis of the observational dataset on the temperature of the lower stratosphere in polar and subpolar latitudes of the Southern Hemisphere in October between 1979 and 1985 revealed a cooling trend. For 7 years the mean temperature at the South Pole at the 24-km altitude decreased by 18 K, the contribution of 1979 being significant. This cooling was partly caused by planetary waves observed in October 1979 and October 1982, but absent in other years (meaning that cooling also occurred in dynamically quiet years). Observations also showed that cooling during dynamically quiet years (1980 and 1985) took place at the 30 hPa level largely in the zone of subpolar maximum temperature (near (60°S) but not within the area of polar minimum. A substantial correlation between the interannual variability of temperature and TOZ points to one cause of this variability – the dynamics of the south-polar stratosphere.

The same conclusion was suggested by analysis of data on TOZ and vertical profile of ozone concentration from the Antarctic Syowa Station, made by Sekiguchi (1986), as well as of radiosonde data on the low stratosphere temperature profile, which revealed a very high correlation, testifying to the connection between stratosphere circulation and ozone field: the springtime TOZ minimum was always accompanied by a temperature drop in the lower stratosphere. Hence, the TOZ decreasing trend of recent years could have been caused by unusual changes in atmospheric circulation. A mutual correlation between the interannual variability of TOZ and stratospheric temperature presumably results from the effect of solar activity on the TOZ field.

Consideration of a 7-year Antarctic observational dataset (in the time interval 1975–1985) (Stolarski and Schoeberl, 1986) led to the conclusion that the annually observed September decrease in TOZ near the South Pole is compensated by a mid-latitude TOZ increase. Therefore, the stratospheric ozone content between 44°S and the South Pole remains constant during the whole period from August to November despite extreme values in high and middle latitudes. This situation should be interpreted in the sense that springtime polar ozone variations in the Southern Hemisphere are caused by a re-ditribution of ozone because of dynamic factors, whereas the role of chemical processes is apparently negligible: as a rule, theories based on consideration of chemical processes proceed from the supposition that rapid catalytic ozone losses occur in September, which is impossible to explain without taking account of chemical reactions unknown earlier. Such conclusions were not supported, however, by the more recent results mentioned above.

3.4.6 The role of long waves

Schoeberl *et al.* (1986) performed an analysis of the formation of the springtime ozone minimum in the Antarctic from TOMS data for TOZ from June to October 1981 in the Southern Hemisphere. TOZ data mapping shows that in June the TOZ field is characterized by weak horizontal gradients with maximum values near the equator. By October, a deep circumpolar TOZ minimum forms, surrounded by a band of maximum values in mid-latitudes, whereas near the equator the TOZ values remain practically constant.

The distribution of ozone in the belt of mid-latitudinal maximum is modulated by long waves (Rossby waves), whose spectrum varies over a year. So, for instance, the spring is characterized by strong modulation by waves with the wavenumber 1, with the amplitude of modulation (from minimum to maximum) reaching 100 DU, whereas the background values of the TOZ modulation field constitute about 300 DU. Since the wave minimum phase is almost fixed spatially, a region appears with an abnormally low TOZ variability in the latitudes of South America. The longitude of maximum TOZ is centred near 100°E, but strongly varies interannually.

The region of polar depletion (ozone hole) stands out on daily maps as an ellipsoidal minimum limited by a zone of strong TOZ gradient reaching about 100 DU/100 km (this is about 50% of TOZ gradient values near the polar fronts). The ozone hole is not stationary, it rotates and meanders around the Antarctic continent depending on atmospheric circulation conditions.

The spatial structure of the TOZ field is nearly similar to that of the low stratosphere temperature field (there is a positive correlation between TOZ and temperature: a TOZ decrease is accompanied by a temperature decrease), but this similarity varies from year to year, reflecting the sensitivity to external factors (these conclusions have been drawn from analysis of annual data for seven Octobers between 1979 and 1985 and of the annual change of daily TOZ fields for 1982).

Being very large, the ozone hole covers an area exceeding the Antarctic territory. So, for instance, in October 1981 the 300 DU isoline embraced the area $16.6 \times 10^6\ \text{km}^2$ and in October 1982 this area grew to $18.8 \times 10^6\ \text{km}^2$ (the area of the Antarctic is 16.1×10^6 km^2). Analysis of the dynamics of the spatial distribution of TOZ shows that a TOZ decrease in high latitudes is compensated by its increase in low latitudes (i.e. there is a redistribution of ozone).

As has been noted, analysis of ozone observations with the Nimbus-4 and 7 UV backscattered radiation spectrometer revealed a considerable decrease in TOZ after 1979, the decreasing trend manifesting itself in October south of 70°S in the longitudinal belt 0–30°W, where the activity of planetary waves is weak. Beyond this belt the TOZ trend is much weaker because of strong interannual variability of wave activity. Therefore, Chandra and McPeters (1986) performed an analysis of satellite data on TOZ and the stratospheric temperature field, taking into account the effect of vertical advection and transient long waves.

Data from 1970 to 1979 did not reveal any remarkable trend of TOZ, whereas after 1982 decrease in TOZ was the strongest trend. So, the negative correlation between the concentrations of CFCs-11 and 12 in the Antarctic troposphere and TOZ, discovered earlier, is fictitious. It follows from observations that TOZ decreases, taking place in October each year (starting in 1979), are concentrated largely in the zone of a very cold polar trough south of 70°N, where the wave activity is weak. These decreases can be explained by intensified vertical advection connected with increased net diabatic heat-flux divergence (warming) within the zone of polar trough.

An increase in aerosol content in the Antarctic stratosphere because of volcanic eruptions and/or the formation of polar stratospheric clouds (the latter is most probable during years with cold stratosphere) may be the cause of increases in diabatic warming.

The 'volcanic' hypothesis, still requiring verification by observational data, gains importance, bearing in mind that the strongest TOZ decerase took place in 1982, after the El Chichón eruption.

Using archived data of the US National Meteorological Center for 7 years (1979–1985), Newman (1986) examined the variability of the Southern Hemisphere upper stratospheric circulation from winter to summer using information about zonal winds at the 30 hPa level near 600S, horizontal temperature gradient at the level, concentration (at 30 hPa) and total ozone content (satellite information). The transition regime is classified into two extreme categories: (i) early (mid-October) final stage of stratospheric warming accompanied by intensive wave activity (1979 and 1982); (ii) late (mid November) warming stage with weak longwaves (1980 and 1981). A change in the sign of horizontal temperature gradient at the 30-hPa level is an indicator of the final stage of warming.

The onset of the regime of easterly transport near the Pole reflects the breakdown of the polar vortex. The formation of longwaves in late November–early December, which was observed in all cases considered, leads to a gradual break and finally to a disappearance of the polar vortex. Such 'wave events', together with the mean zonal mass transport, weaken zonal mean gradients of the potential vorticity and ozone content, because of intensified horizontal mixing and therefore favour breakdown of the polar vortex and filling of the ozone hole. In this connection, it was shown (Newman, 1986) that the final stage of warming and the breakdown of the polar vortex are events separated by a time interval between 15 to 40 days. The retrospective analysis of an observational dataset for the 1960s demonstrated that the time of the onset of these events has not changed radically.

Using archived data of the US National Meteorological Center for 1958–1985 for 16 Antarctic aerological stations, Angell (1986b, 1997) analysed the variability of the monthly mean temperature at the 100, 50 and 30-hPa levels (about 5,500 single values), as well as results of TOZ observations at four stations. The results obtained testify to the in-phase annual change of low stratospheric temperature and TOZ. This correlation took place not only during the springs after 1978, when in Southern Hemisphere polar latitudes strong decreases were observed in TOZ and temperature, constituting about 30% and 6–8°C, respectively, but also in the remaining seasons (in conditions of relatively small variability). The springtime decrease in stratospheric temperature (after 1978) happened to be maximum along the Antarctic coastline in the belt 30°W–60°E (which includes the stations at Halley Bay and Syowa). At the stations at latitude 10°N there was practically no temperature decrease in the lower stratosphere. The springtime temperature decrease was largely limited by the Antarctic stratosphere (it took place at levels above 300 hPa). Observational results suggest that any theory about springtime TOZ decrease must explain the subsequent temperature decrease in the lower stratosphere.

3.4.7 Estimates of the role of heterogeneous chemical reactions: polar stratospheric clouds

As mentioned above, the explanation of the springtime TOZ minimum in the Antarctic relates it to heterogeneous chemical reactions taking place on particles of polar stratospheric clouds (PSC), which are formed in the high-latitude lower stratosphere. The problem is, however, that satellite occultation soundings showed that PSCs were recorded in the Antarctic only in a rather narrow belt of latitudes (of course, in the zone of polar night the PSC observations were impossible). More representative early information on PSCs was obtained in the Northern Hemisphere from data about outgoing thermal emission near the Earth's limb (limb sounding) in 1978–1979. In this case, PSCs were recognized as corresponding to abnormally high intensities of the limb emission in the ozone channel of the spectrometer.

Comparison of the thus-obtained information about PSCs with satellite data for stratospheric humidity showed that the regions with PSCs coincided with the zones of increased humidity (Austin *et al.*, 1986). Though there were cases of saturating humidity, the mean relative humidity for 39 PSC cases constituted 58%. Probably, such a low level of relative humidity results from considerable spatial smoothing (the instrumental resolution was low), with local areas of saturation. PSCs may also form outside the regions where they have been observed.

The use of a similar approach to study the conditions of Antarctic PSC formation led to the conclusion that, after 10 September, 1979 at the 20-km altitude in the region of the circumpolar polar vortex with a maximum TOZ decrease, there were practically no PSCs. According to initial assessments by Austin *et al.* (1986), it is unlikely that the formation of the ozone hole can be connected with PSCs and with respective heterogeneous chemical reactions.

Different results were obtained by McCormick and Trepte (1986). Their analysis of PSC occultation measurements by Nimbus-7 SAM II shows that PSCs are a typical component of the wintertime Antarctic stratosphere, both in periods of volcanic eruptions and in conditions of the background stratosphere. The PSC altitude exceeding 20 km in June, July and August, decreases to 15 km by mid-September.

Usually PSCs are observed at altitudes below 15 km up to the end of September. October is characterized by a relative (in the annual course) minimum of total aerosol content in the stratosphere and, respectively, of an aerosol-induced extinction. There is practically no information at that time on the concentration and size distribution of PSC particles, or on their chemical composition, which is necessary to assess the possible effect of PSCs on the formation of the springtime ozone minimum. Apparently aerosols of volcanic origin, whose concentration in the Antarctic reached its maximum in early 1983, did not markedly affect the ozone decrease, since this decrease continued in 1984–1985. Note, however, that aerosol content in the Antarctic stratosphere in 1984 and 1985 was much lower than in 1982.

Within the Airborne Southern Hemisphere Ozone Experiment/Measurements for Assessing the Effects of Stratospheric Aircraft (ASHOE/MAESA) Project, Dye *et al.* (1996) measured particle volumes in a type-I PSC near the Antarctic polar vortex. The observations from onboard the ER-2 aircraft were made with the help of the Multiangle Aerosol Spectrometer Probe (MASP), which counts and sizes individual particles. The particle refraction index can also be estimated over a size range of 0.4 to 0.7 μm by measuring the ratio of light scattered in the forward and backward directions.

Results of observations of PSCs in the outer region of the Antarctic vortex, obtained by Dye *et al.* (1996), show similarity with previous observations in the Arctic. The existence of PSCs composed of ternary droplets was discovered in the Antarctic, thus unifying the understanding of PSC-formation processes in the two polar regions. The onset of the cloud occurred near 193 K, 3 degrees colder than nitric acid trihydrate (NAT) saturation (the same conclusion was reached by Larsen *et al.* (1996)). Dye *et al.* (1996) pointed out that the onset temperature, smooth increase of volume and enhanced NOy (HNO3 in the particles) with temperature and comparisons of observations with an equilibrium model of ternary droplet growth (Koop and Carslaw, 1996) all support the notion that much of this type-I PSC was ternary solution (H_2SO_4-HNO_3-H_2O) droplets (this provides confirmation of previous findings in the Northern Hemisphere). The ternary model does not, however, fit the observations in all regions. This may be due to the presence of some solid-phase growth in agreement with impactor observations. Hamill *et al.* (1996) showed that the mechanism of ternary solution droplets is heteromolecular condensation in which the particle is in equilibrium with environmental water vapour.

De Rudder *et al.* (1996) and Tie *et al.* (1995) carried out a model study of PSCs and their effect on stratospheric ozone. Describing the model, they emphasized that detailed microphysical processes accounting for PSC formation (in terms of heterogeneous nucleation, condensation and sedimentation) had been included in a global chemical/dynamical 2-D model to study the effect of heterogeneous reactions occurring on the surfaces of PSCs on stratospheric ozone. Three types of particles are considered in the model: (1) diluted sulphuric-acid particles, which serve as cloud condensation nuclei (CCN) for the PSC particles; (2) type-I PSC particles (PSCI) composed of a NAT shell surrounding a small sulphate core, which act (as temperature decreases further) as condensation nuclei for type-II PSC particles, (3) type-II PSC particles (PSCII) composed of an ice shell coating a PSCI core. Calculations show that PSC particle distribution in the Arctic and Antarctica is quite different. Over Antarctica, nitric acid trihydrate particles (type-I PSCs) are formed from early June to late September with a maximum surface area of 15 cm2/cm3, while in the Arctic type-I PSCs are formed only in January with a maximum surface area of 5 cm2/cm3. Ice-crystal clouds (type-II PSCs) are present over Antarctica in August, but are not seen in the Arctic. The results of simulation modelling considered by Tie *et al.* (1995) indicate that heterogeneous reactions (conversion of ClONO2 into ClO under sunlight conditions) occurring on the surface of PSCs are the likely cause of the ozone decrease from 1980 to 1990 in both the Antarctic and the Arctic. The calculations successfully simulate the development of the springtime ozone hole over Antarctica. The largest ozone depletion starts in September and reaches a maximum in October (polar sunrise). The maximum ozone depletion is located at the top of the PSC layer (20 km), where the Clx concentration is highest. In the Arctic, ozone depletion because of PSCs is much smaller owing to the shorter occurrence period and the smaller surface area available for chemical conversion.

The formation rate of type-I PSCs in both the Antarctic and the Arctic enhanced after the eruption of Mount Pinatubo in 1991 because of the dense eruptive sulphate aerosol cloud. However, the effect on ozone depletion in the Antarctic was limited because conversion from $ClNO_2$ to ClO on PSC particles was almost 'saturated' under non-volcanic conditions. Tie *et al.* (1995) therefore concluded that, for potential ozone depletion, the enlargement in the area covered by PSCs may be more important than the increase in PSC density.

Analysis of results of simultaneous retrievals of vertical profiles of ozone concentration, nitrogen dioxide and aerosol extinction coefficient (AEC), from the data of occultation soundings in the zone of the southern circumpolar vortex made on 13 October, 1981 and 13 October, 1985 by SAM-II and SAGE-II, revealed considerable variation in concentrations at all the altitudes (McCormick and Larsen, 1986). Both gas components considered are characterized by low AEC values in the region of polar vortex within the sounding altitudes from the tropopause up to 60 km (ozone) and 20–40 km (nitrogen dioxide). AEC values and total aerosol content in this region above 15–18 km were also minimum. There is a high correlation between total contents of O_3 and NO_2 and the location of respective points of observation within the polar vortex (minimum O_3 and NO_2 were recorded near the centre of the vortex). To characterize the variability of the profiles of these components within the circumpolar vortex, the vertical profiles were retrieved near 70 and 68°S (from data for 1982 and 1985, respectively), from which it follows that near the centre of the vortex the ozone content above 25 km in 1985 was greater than in 1981, but below 25 km the situation reversed.

Water solutions of nitric and hydrochloric acids, whose vapours condense in the stratosphere, are supposed to be one of the components of PSCs. This condensation may serve as a sink for nitrogen oxides and therefore affect the chemical processes responsible for the formation of the ozone hole. The following heterogeneous reactions on the surface of PSC particles can serve as an intensifying factor of the process of conversion of gaseous NO_x to HNO_3 and subsequent HNO_3 condensation on the surface of particles (Toon *et al.*, 1986):

$$N_2O_5 + H_2O \longrightarrow HNO_3$$

$$ClNO_3 + H_2O \longrightarrow HNO_3 + HOCl$$

In this case, the PSC particles will represent a concentrated HNO3 solution with an admixture of HCl and H_2SO_4. According to Solomon *et al.* (1986), the heterogeneous reaction

$$ClNO_3 + HCl \longrightarrow HNO_3 + Cl_2$$

is a probable (and substantial) source of Cl_2. As shown by theoretical calculations, the process of catalytic destruction of ozone by halogen radicals depends critically on a concentration of the oxides of nitrogen determined by the following condition:

$$[ClO] > [NO_2]$$

The rate of photolysis for HNO_3 near 75°S in late September constitutes $10^7\,s^{-1}$. If the mixing ratio for HNO_3 vapours in the stratosphere is 5 ppb, this condition will be broken in about 10 days. However, if PSCs govern the condensation of HNO_3 vapour on the surface of particles, this condition will remain. After PSC dissipation in late September the process of deepening of the springtime ozone minimum ceases. The sedimentation of PSC particles must lead to the removal of nitric-acid water-solution droplets from the stratosphere.

Hamill *et al.* (1986) analysed the observed characteristics of PSCs and the possible effect of PSCs on the formation of the springtime Antarctic ozone minimum determined by a relationship between PSC evaporation and TOZ decrease (PSC dissipation coincides in both time and space with the TOZ decrease).

Analysis of data of the occultation sounding from satellites (mainly, data of Nimbus-7 SAM II) revealed the existence of PSCs in both polar regions, and results from aircraft lidar soundings led to the conclusion that PSCs have extended cloudiness and form where the temperature drops below −85°C.

Apparently, because of the greater stability of the circumpolar vortex and lower temperature of the southern stratosphere, PSCs were observed there more often than in the northern polar region. As a rule, the coefficient of PSC extinction varies within 10^{-3}–$10^{-2}\,km^{-1}$. As in the case of pure water clouds, their extinction (in the conditions considered) must be greater, so clearly PSC particles are not pure water. Toon *et al.* (1986) made just such an assumption that PSCs consist of particles of concentrated solutions of nitric acid with a particle radius about 0.5 μm (in this case the calculated extinction agrees with that observed).

Almost all nitric acid contained in the Antarctic stratosphere can be bound in PSC droplets. Most likely, the springtime dissipation of PSCs is caused by their evaporation.

If we procced from the fact that the springtime TOZ minimum is caused by heterogeneous chemical reactions with participation of halogen radicals, then in this case the following reactions ought to play the basic role:

$$Cl + O_3 \longrightarrow ClO + O_2$$

$$ClO + O \longrightarrow Cl + O_2$$

whose efficiency depends on the concentration of nitrogen oxides.

In this context, it is supposed that during the polar night NO_2 and NO_3 produce gaseous N_2O_5, which can be absorbed by PSC particles and transform through the reaction:

$$N_2O_5 + H_2O \longrightarrow 2HNO_3$$

which provides the needed decrease in NO_x concentration. The rate of this reaction either on overcooled droplets or on the surface of solid particles can be sufficiently high. Other possible reactions producing nitric acid could be:

$$ClNO_3 + H_2O \longrightarrow HNO_3 + HOCl$$

$$ClNO_3 + HCl \longrightarrow HNO_3 + Cl_2$$

Chipperfield *et al.* (1996) developed a 3-D chemical-transport model to investigate the potential for chemical ozone depletion in the Arctic polar vortex for the past four winters. Isolating the effect of meteorological variability on polar ozone loss (prescribing the same chemical initial conditions for each winter) they showed that during 1994/1995 the conditions in the Arctic polar vortex were most conducive to ozone depletion in recent years. Calculations indicate that at 475 K (around 18 km) in the lower stratosphere the chemical ozone destruction between late December and mid-March was greater than 30% over a large area of the polar vortex. Larger percentage losses, reaching 50%, occur at slightly lower altitudes. The modelled average depletion in the entire region poleward of 60oN over the same period was around 22% at 475 K. The large depletion for 1994/1995 is a consequence of the extreme meteorological conditions during this winter.

To study the effects of heterogeneous chemistry on anthropogenic ozone depletion at northern mid-latitudes, Solomon *et al.*(1996) used a 2-D model with input data on stratospheric aerosol surface-area distributions inferred from satellite 1-μm extinction measurements. The principal conclusions made by Solomon *et al.* (1996) are as follows:

(1) the short-term (inter-annual) and longer-term (decadal) changes in aerosols very probably played a substantial role, along with trends in anthropogenic chlorine and bromine, in both triggering the ozone losses observed at northern mid-latitudes in the early 1980s and increasing the averaged long-term ozone depletion of the past decade or so;

(2) the use of observed aerosol distributions enhances the calculated ozone depletion because of halogen chemistry below about 25 km over much of the past decade, including many periods not generally thought of as affected by volcanic activity;
(3) direct observations confirm the key aspects of the model chemistry that is responsible for this behaviour and demonstrate that aerosol changes alone are not a mechanism for ozone loss in the absence of anthropogenic halogen input to the stratosphere;
(4) aerosol-included ozone changes could be confused with 11-year solar cycle effects in some statistical analyses, resulting in an overestimation of the trends ascribed to solar activity;
(5) the timing and magnitude of future ozone losses at mid-latitudes in the Northern Hemisphere are likely to be strongly dependent upon volcanic aerosol variation as well as on future chlorine and bromine loading;
(6) there is still a substantial quantitative disagreement between model predictions and observations: observed ozone changes are about 50% larger than calculated (this may be due to shortcomings in modelled photochemistry or dynamics or both).

Zhao *et al.* (1996) applied a general circulation model, coupled interactively to a 3-D stratospheric photochemistry model, to study the dynamical response of the Arctic polar vortex to ozone depletion catalysed by volcanic aerosols. It was found that temperatures may be depressed by as much as 3–7 K in late March in the lower stratosphere at northern mid and high latitudes owing to ozone-radiation-dynamics feedback effects. In the model, the reduced solar heating in late winter and early spring associated with aerosol ozone depletion acts as a trigger for the net cooling that is observed. However, a major part of the cooling is related to anomalies induced in the circulation of the lower stratosphere. Zhao *et al.* (1996) emphasized the complexity of the coupling between the dynamics and chemistry of the lower stratosphere as well as of relevant assessments of cause and effect related to perturbations in this region.

Randeniya *et al.* (1996b) included hydrolysis of $BrONO_2$ on sulphate aerosols ($BrONO_2 + H_2O_{aerosol} \longrightarrow HNO_3 + HOBr$) in 2-D model calculations of ozone change over the period 1980–1995. It was found that the inclusion of this reaction has only a marginal impact on the calculated TOZ trends between 1980 and 1995. The addition of this reaction had, however, a major impact on the calculated transient ozone loss under enhanced aerosol conditions (after the Pinatubo eruption) and approximately doubled the calculated ozone loss between 60°S and 60°N in 1992 (when maximum loss of ozone occurred). In contrast to the effects of hydrolysis of N_2O_5, the effects of hydrolysis of $BrONO_2$ are most marked at high latitudes in summer.

A new heterogeneous reaction, namely the sulphate-regulated conversion of acetone to 4-methyl-3-penten-2-one and trimethylbenzene, was studied under laboratory conditions by Duncan *et al.* (1998).

3.4.8 Nitrogen dioxide

Observations of the springtime TOZ decrease in the Antarctic revealed the possible effect on this decrease of interaction between photochemical and dynamic processes, the participation of nitrogen dioxide in chemical reactions (responsible for the TOZ decrease) being substantial. In this connection, Keys and Johnston (1986) undertook a study on variation of total NO_2 content in the atmosphere from data of groundbased observations at Arrival Heights Station (77.8°S, 166.7°E), where in the spring large-amplitude variations take place because of varying intensity of long waves. These data are compared with results of satellite sounding of ozone as well as aerological sounding.

Analysis of observational data shows that: (i) the height of the NO_2 layer rises in autumn from 20–25 to 30–35 km; (ii) in spring, nitrogen dioxide is concentrated apparently below 25 km; (iii) in autumn, at night the NO_2 content decreases; (iv) the quantities of nitrogen dioxide $(0.4–0.7) \times 10^{15}\ cm^{-2}$ observed in the early spring are apparently higher than in the late autumn.

The explanation of the springtime TOZ drop by the effect of heterogeneous chemical reactions, producing high amounts of ClO_x near the 50-hPa level, suggested in Solomon *et al.* (1986, 1987) and McElroy *et al.* (1986), requires extremely low or low values of NO_2 content in the zone of decrease. Although the indirect surface data do not permit the unique conclusion that NO_2 content is too small to provide needed high values of the $Cl_x/ClONO_2$ concentration ratio, preliminary estimates of the height of the layer do point to the possibility of such a situation.

The observational data considered show that in spring, under conditions of low atmospheric pressure, the station can be considered representative in reflecting conditions of a cold nucleus of the circumpolar vortex, which was well isolated from the atmosphere of lower latitudes (because of weak mixing) from August to October. Inside this nucleus photochemical processes intensively funtion as important factors determining total ozone and nitrogen dioxide concentrations (in the absence of the effect of transport processes). Thus, the situation discussed is characterized by the possibility of separating the effect that dynamic and photochemical processes have on the formation of the springtime ozone minimum in the Antarctic stratosphere.

Tourpali *et al.* (1997) used a 2-D chemical/transport model, which takes into account microphysical formation of PSCs and heterogeneous mechanisms under different stratospheric chlorine-loading conditions, to examine the rate of change in TOZ decline observed between 1964 and 1991 both in the Northern and the Southern Hemisphere. The results indicate that ozone depletion at high latitudes in both hemispheres is mainly caused by heterogeneous reactions of the sulphate of PSCs. At northern mid-latitudes, though, the ozone losses are caused by heterogeneous reactions on the surface of sulphate aerosols, especially after large volcanic eruptions.

Sensitivity studies carried out by Tourpali *et al.* (1997) indicate that a 2° decrease in temperature significantly affects the ozone field, especially in the Northern Hemisphere, by facilitating heterogeneous conversion processes on PSC surfaces by enhancement of particle-number concentration and by accelerating the reactions on the surface of sulphate aerosols.

Brühl *et al.* (1998) paid attention to a continuous decrease in ozone concentration, observed between May and September in the lower stratosphere at high latitudes in the Northern Hemisphere. Such low local ozone concentrations were correlated with high concentrations of NO, NO_2 and HCl. Numerical modelling results (using a detailed photochemical box model and a 2-D chemical model initialized by HALOE data) shows that the observed summertime low-ozone concentrations and much of the seasonal course of ozone is due to catalytic ozone destruction by NO and NO_2, which become the main odd-nitrogen compounds under those conditions. The observed anticorrelation between NO_2 and O_3 is mostly due to the catalytic ozone destruction cycle (Crutzen 1970, 1971):

$$NO + O_3 \longrightarrow NO_2 + O_2$$

$$O_3 + h\nu \longrightarrow O + O_2$$

$$NO_2 + O \longrightarrow NO + O_2$$

The appearance of high levels of HCl in regions with high NO_x is a consequence of the reactions:

$$ClO + NO \longrightarrow Cl + NO_2$$

$$Cl + CH_4 \longrightarrow HCl + CH_3$$

Brühl *et al.* (1998) emphasized that this effect is further enhanced as the reaction $Cl + O_3 \longrightarrow ClO + O_2$ is less effective in the case of low ozone, likewise allowing for more Cl and enhanced HCl production by the last reaction.

In connection with the important roles that NO and NO_2 radicals play in atmospheric chemistry, Prasad (1998) discussed potentially new atmospheric sources and sinks of odd nitrogen pointing out that: (1) the $O_2(B^3\Sigma) + N_2 \longrightarrow 2NO$ reaction and perhaps the photoexcitation of $N_2 . O_2$ dimers (or collision complexes) by Herzberg-band photons as well, may be potentially important atmospheric sources of odd nitrogen; (2) there is a tantalizing potential for a substantive NO_x production of the order of 1,200 NO_x/cm^3s in the upper troposphere subject to certain conditions; and (3) photodissociation of the possible $N_2O \cdot O_3$ species in the 320 to 430 nm wavelength interval may be a non-trivial source of NO_x in the upper troposphere. Prasad (1998) underlined the need for relevant experimental studies.

3.4.9 Chlorine oxides

One of the hypotheses suggested to explain the springtime ozone minimum in the Antarctic relates to the effect of photochemical reactions with participation of chlorine oxides in heterogeneous reactions, under conditions of low NO_2 content in the atmosphere. In this connection, results of groundbased spectroscopic observations (by measurements of absorption in the 130–450 nm wavelength interval) of total NO_2 in the stratosphere at Syowa Station (62.0°S, 39.6°E) between March 1983 and January 1984 were discussed (Shibasaki *et al*., 1986) as well as data of balloon measurements of the vertical profile of NO_2 concentration in November 1982 and 1983 (this was the first time balloon measurements were made in the Antarctic).

Analysis of the results obtained revealed the wintertime minimum of total NO_2 content (about $1.10^{15}\ cm^{-2}$) and the springtime maximum ($7.10^{15}\ cm^{-2}$), which agrees with the data for high northern latitudes, where both minimum and maximum values happened to be larger. The observed night-time decrease in total NO_2 was weak from the autumn till the early spring. In late September (before the TOZ minimum in mid-October) the total NO_2 content rapidly increased. The latter shows that the springtime ozone decrease in the Antarctic can be caused by both dynamic and photochemical processes.

According to balloon measurement data, vertical profiles of NO_2 concentration in 1982 and 1983 during the early summer (November) were practically the same above 25 km, like those observed in high and mid-latitudes of the Northern Hemisphere. In the lower stratosphere (below 25 km), NO_2 concentration varied widely apparently because of the effect of varying atmospheric dynamics. On the whole, the results obtained do not favour the hypothesis about photochemical mechanism for the formation of the springtime TOZ minimum.

In the spring of 1986 (1–22 September), DeZafra *et al.* (1987) performed observations at McMurdo Station of chlorine monoxide at low altitudes in the Antarctic atmosphere (12–22 km) using the spectrometer for millimetre wavelengths. The microwave emission of the atmosphere was measured at 1.1 mm (in the rotational line 278.63 GHz). The possibility of retrieving the vertical profile of the ClO mixing ratio was enhanced by the presence of a strong collision broadening, deforming the line's profile.

In usual conditions the ClO mixing ratio is maximum at an altitude of about 38 km and rapidly decreases with height over this level (at a rate of 30% per 1 km). The normal ClO mixing ratio in the tropical and mid-latitudes at a height of 20 km constitutes about 10^{-12}. From the data of a single balloon sounding at a height of 22 km the mixing ratio 10-11 was recorded. Ozonesonde data in the spring of 1986 revealed a decrease in ozone in some tropospheric layers exceeding 90%.

The results obtained (De Zafra, 1987) point to an abnormally high concentration of ClO in the lower stratosphere, which is about two orders of magnitude higher than that at a height of 20 km, forecast by theory for mid-latitudes. In this case the mixing ratio at 20 km ranges within $(0.5–2) \times 10^{-9}$ and implies that ClO is not concentrated in a narrow layer but exists at altitudes from above 20 km down to at least 15 km. Chlorine monoxide is characterized by a clear diurnal cycle and its annual course has a minimum in September. The results discussed suggest that chlorine plays an important role in the formation of the springtime ozone minimum. During an expedition to the Antarctic in the spring of 1986, OClO was also found in the stratosphere.

To reveal the long-term variation in ClO content in the lower stratosphere from observational data at McMurdo Station between 1 and 30 September 1986 using the 256-channel millimeter spectrometer (the width of the frequency band is 256 MHz), Solomon *et al.* (1987) analysed the variability of the difference of emissions measured in the daytime and at night. The results obtained point to a decrease of ClO concentration in the second half of September and a decrease to zero in early October. The dynamics of ClO reflects the important role of chlorine monoxide as a factor in the formation of the ozone minimum in the early spring.

Observational data agree with theoretical conclusions about the possible formation of a large amount of ClO by heterogeneous chemical reactions on the surfaces of PSC particles. These reactions produce chlorine because of such components as HCl and $ClONO_2$ and suppress the formation of nitrogen oxides, which leads to an increased content of chlorine monoxide. Such chemical mechanism is supported by the fact that the maximum mixing ratio for ClO is observed during the period of minimum stratospheric temperature needed for stratospheric clouds to form. Cloud particles evaporate as the stratosphere gets heated.

The reaction between gaseous BrO and ClO is assumed to be fast and to be a synergetic mechanism for the destruction of stratospheric ozone. In particular, this reaction can supposedly be one of the factors in springtime ozone decrease (up to 40%) in the Antarctic atmosphere observed between 1960 and 1985. In this connection, Hills *et al.* (1987) made laboratory measurements to estimate the rate of the reaction under discussion and to analyse its products. The results obtained confirm the supposed large contribution of this reaction to the formation of ozone minimum in the spring-time Antarctic and show that it can be a very important source of chlorine dioxide, enough to explain the ClO concentration observed in the Antarctic atmosphere, provided the stratosphere contains a sufficient amount of bromide (about 20×10^{-12}). The catalytic cycle of ozone destruction:

$$BrO + ClO \longrightarrow Br + Cl + O_2$$

$$Br + O_3 \longrightarrow BrO + O_2$$

$$Cl + O_3 \longrightarrow ClO + O_2 \qquad (\text{giving } 2O_3 \longrightarrow 3O_2)$$

is particularly substantial in conditions of the Antarctic stratosphere, since this cycle does not require solar radiation or atomic oxygen to destroy ozone. A major difficulty for all the previous theories about the ozone hole, that were based on the consideration of chlorine-catalyzed destruction of ozone, consists in the supposed low concentration of ClO. According to recent theoretical ideas, chlorine nitrate and HCl are prevalent forms of chlorine in the stratosphere.

In some publications a mechanism was proposed for conversion of nitrogen oxides to nitric acid by heterogeneous reactions on stratospheric cloud particles. Other ideas proceed from the fact that clouds form by direct condensation of nitric acid.

In connection with this mechanism, it is important to note that measurements made in 1986 revealed an abnormally high HCl concentration in the Antarctic lower stratosphere, as well as a 20–50-fold higher OClO content than was earlier believed.

This points to the possibility of the reaction:

$$BrO + ClO \longrightarrow Br + OClO$$

but the following reactions are also possible:

$$ClONO_2 + O_3 \longrightarrow OClO + NO_2 + O_2$$

$$ClO + O_3 \longrightarrow OClO + O_2$$

These reactions are slow in the gas phase but can accelerate on the surface of PSC particles. Laboratory studies of these heterogeneous reactions would be very important.

Crutzen *et al.* (1988) undertook simulation calculations of photochemical processes and vertical mixing in the belt 60–90°S from April to late October using a 1-D non-stationary radiative–convective model. Since only the vertical transport (because of turbulent diffusion with the mixing coefficient $2 \times 10^3\,cm^2/s$) was considered, the horizontal transport of trace gases to the 60–90°S zone (or from this zone) was not taken into account. As a result, calculated concentrations of trace gases at altitudes above 50 km and below 75 km were overestimated and in the 35–50-km layer they were underestimated.

In the simulation of photochemical processes the following gas families were taken into account: odd oxygen $(O + O_3)$; $NO_x(N + NO + NO_2 + NO_3 + 2N_2O_5 + HNO_4 + ClNO_3)$; HNO_3; odd hydrogen $(H + OH + HO_2)$; H_2O_2; ClOx $(Cl + ClO + OClO + 2Cl_2O_2 + ClOH + ClNO_3)$, CH_4, H_2O, H_2 and N_2O. The distribution of trace gases within individual families is determined only by photochemical processes.

Earlier calculations led to the conclusion that in daytime the ozone content gets frozen. This leads to a rapid increase in OH concentration, transformation in HCl into ClO_x and to an intensive catalytic destruction of ozone through reactions:

$$(Cl + O_3 \longrightarrow ClO + O_2) + (ClO + HO \longrightarrow ClOH + O)$$

$$+ (ClOH + h\nu \longrightarrow Cl + OH)(2OH + O_3 \longrightarrow H_2O + 2O_2)$$

$$2(Cl + O_3 \longrightarrow H_2O + O_2) + (ClO + ClO + M \longrightarrow Cl_2O_2 + M)$$

$$+ (Cl_2O_2 + h\nu \longrightarrow 2Cl + O_2)(ClO + O_3 \longrightarrow ClO + O_2)$$

$$+ (Br + O_3 \longrightarrow BrO + O_2) + (BrO + ClO \longrightarrow Br + Cl + O_2)$$

and as a result:

$$2O_3 \longrightarrow 3O_2$$

It is most likely that solid particles form in the Antarctic lower stratosphere. The temperature here often drops to 190 K, causing water to freeze on particles even when they are droplets of the supercooled H_2SO_4/H_2O solution, the equilibrium pressure of HNO_3 vapours over solid $HNO_3\ 3H_2O$ particles depending strongly on temperature. Apparently, in conditions of a very cold Antarctic night a large part of NO_x is transformed into HNO_3 by heterogeneous reactions of N_2O_5 on PSC particles catalysed by ions.

Calculations of vertical profiles of various trace gases for early May and October show that above 25 km a strong decrease in N_2O and CH_4 takes place by October because of their downward transport. Because of CH_4 decrease, the concentration of ClO_x (mainly down to HCl) substantially increases, which leads to a considerable decrease in ozone. Prognostic estimates obtained with different scenarios of CFC emissions led to the conclusion that a decrease in ozone at an altitude about 20 km may exceed 90% by the year 2025 (compared with 1960), provided the Cl_x mixing ratio increases in the upper stratosphere from 0.77 ppb (in 1960) to 5.48 ppb (in 2025).

The temperature dependence of HNO_3 vapour pressure highlights the importance of considering the effect of possible climate change. Strictly speaking, an interactive consideration of variations in ozone and temperature is needed. The results are confirmed by observations showing an increase of $ClONO_2$ concentration during the spring lower stratosphere warming, predicted theoretically. The reaction ClO + OH is critically important for substantiation of the estimates obtained. If the reaction gives HCl ($ClO + OH \longrightarrow HCl + O_2 + 233$ kJ/mol), then it seriously hinders the explanation of ozone decrease in the Antarctic stratosphere by chemical processes connected with the effect of chlorofluorocarbons (CFCs).

Since one of the suppositions about the nature of the springtime ozone minimum in the Antarctic relates to an intensified release of CFCs and other minor gaseous components, Cronn *et al.* (1986) analysed the results of chromatographic observations of concentrations of $CFCl_3$, CF_2Cl_2, CH_3CCl_3, CCl_4 and N_2O at Palmer Station (64°46′S; 64°04′E) between 4 February 1982 and 30 November 1985. During the first week of 1984 average mixing ratios for various gases were (in brackets): $CFCl_3$ (199.6×10^{-12}); CF_2Cl_2 (329.0×10^{-12}); CH_3CCl_3 (124.2×10^{-12}); CCl_4 (150.2×10^{-12}); N_2O (301.3×10^{-9}). Annual mean rates of mixing ratios increase reached: 5.87% ($CFCl_3$); 5.45% (CF_2Cl_2); 5.31% (CH_3CCl_3); 1.29% (CCl_4) and 0.24% (N_2O).

In connection with the possible photochemical nature of the springtime ozone minimum in the Antarctic (catalytic destruction of ozone through reactions with participation of bromine and chlorine), Rodriquez *et al.* (1986) noted that, for respective catalytic reactions to take place efficiently, a considerable part of chlorine in the lower stratosphere must be in the form of ClO. Earlier it was assumed that the high level of ClO concentration can be maintained because of fast heterogeneous reactions, whose rate is so far unknown.

Maximum ClO content requires other ClO_x components (Cl, Cl_2, HOCl, $ClNO_3$, OClO, Cl_2O_2) to be minimum. The latter can be favoured by the following heterogeneous reactions:

$$ClNO_3 + H_2O \longrightarrow HNO_3 + HOCl$$

$$N_2O_5 + H_2O \longrightarrow 2HNO_3$$

$$HCl + ClNO_3 \longrightarrow HNO_3 + Cl_2$$

Model calculations show that considerable amounts of OClO and Cl_2O_2 must form (at night, in particular). In this connection, calculations were made (Rodriguez *et al.*,

1986) of the diurnal change of ClO, OClO and Cl_2O_2 in various conditions of the spring Antarctic stratosphere. Analysis of the results of these calculations points to the importance of simultaneous measurements of the diurnal change in total content of ClO and OClO as a source of information about the efficiency and the limits of applicability of this chemical mechanism for the formation of the springtime ozone minimum. The destruction of ozone through catalytic reactions with participation of Cl_2O_2 may be as important as that due to BrO (with the observed level of chlorine concentration in the stratosphere), provided rapid photolysis of Cl_2O_2 takes place giving Cl and ClO_2. There is no synergetic relationship between the cycles of chlorine and bromine, since both of them depend on the level of ClO concentration.

3.4.10 Heterogeneous chemical reactions

As McElroy *et al.* (1986) noted, the observed phenomena affected by the springtime ozone minimum in the Antarctic include: (i) a decrease in total ozone content at an average rate of about 1.6 DU/day in September, starting from 1982 (between 1979 and 1982 this rate was only 1.9 DU/day); (ii) the TOZ decrease in October by about 59% starting from 1978; (iii) a TOZ–air temperature correlation at the 50-hPa level (minimum TOZ values are accompanied by minimum temperature).

An intensified TOZ decrease in September was supposed to be due to the El Chichón eruption: an increase in concentration of stratospheric aerosol particles, that serve as condensation nuclei for formation of clouds in the stratosphere, favours an increase of cloudiness and hence the growth of the number of cloud particles (of crystalline hydrates of nitric acid), on the surface of which heterogeneous reactions take place with participation of chlorine radicals, leading to a TOZ decrease. Also, clouds intensify stratospheric cooling.

The efficiency of heterogeneous reactions in ozone destruction increases considerably if bromine radicals take part in them. These reactions are very sensitive to temperature drops below 190 K. Thus, the springtime TOZ depletion in recent years can be connected with both the input to the Antarctic stratosphere of anthropogenic halogens and the temperature decrease. Upward air motions over the South Pole, because of radiative cooling of the stratosphere, and a decrease in the ozone input from lower latitudes, because of the weakened intensity of longwaves, can also be important factors in TOZ decrease.

Aikin and McPeters (1986) analysed the results of retrieving the vertical profiles of ozone concentration in the spring Antarctic from measurements with the Nimbus-7 spectrometer of backscattered UV radiation between 1979 and 1984. Analysis showed that in the zone of the circumpolar vortex there was a decrease in ozone concentration within 15–30% at all the levels from 30 to 0.7 hPa. In the spring of 1983 at latitudes 75 and 80°S the mixing ratio near the 1-hPa level continuously decreased from September till the moment of the summer solstice, after which (in late summer and in autumn) the concentration of ozone increased. The concentration at the 1 hPa level in September and October was much lower than in similar conditions in the autumn.

This asymmetry of the spring and autumn-time concentration could have resulted from sudden releases to the stratosphere of chlorine present in the bound state in the meteor substance during the polar night. Between atoms of meteors from the mesosphere and the atmosphere reactions take place giving XOH compounds, which react with HCl giving in their turn XCl salts:

$$XOH + Cl \longrightarrow XCl + H_2O$$

Since these processes can take place without sunlight, chlorine is accumulated during the polar winter. A sudden release of accumulated chlorine, because of photodissociation of XCl in spring, can lead to a partial destruction of ozone in the upper stratosphere. As a result, a substantial decrease in ozone occurs at the 30-hPa level in mid-September, remaining till mid-October, when ozone starts increasing gradually. Various mechanisms are possible here, that lower the ozone content in the upper and lower stratosphere. Verification of this hypothesis is possible by measuring the decrease in concentration of X metals resulting from the formation of HCl. The meteor substance can also participate in the formation of PSCs: polymerizing salts XCl can serve as condensation nuclei for formation of clouds. Reactions between molecules of salts and H_2SO_3 will give hydrochloric acid.

3.4.11 Application of atmospheric general-circulation models for numerical simulation of the springtime TOZ decrease

As Mahlman and Fels (1986) noted, first attempts to explain the springtime ozone minimum in the Antarctic were largely connected with the possible effects of chemical reactions determining the formation of ozone. Later on, it became clear that such an explanation faces some difficulties. A more adequate explanation consists in considering effects of dynamic factors as well. This approach is quite natural, since a large part of ozone in high latitudes is concentrated in the lower stratosphere, where the time constant for chemical reactions exceeds that for transport because of stratospheric dynamics.

The fact that total ozone content, every year since 1980, always decreases rapidly in winter suggests that in the winter troposphere of the Southern Hemisphere forcing because of planetary-scale disturbances weakens. This assumption agrees with the observed temperature decreases in the zone of the circumpolar vortex in October as well as the vertical component of the Eliassen–Palm flux divergence in September. This weakening in forcing must affect ozone transport: the meridional slope of isolines of the ozone mixing ratio must decrease with respect to isentropic surfaces (with lower values at the Poles) and the duration of winter must lengthen (and so also the lifetime of the circumpolar vortex).

This systematic decreasing trend for ozone can be partly explained by the fact that, for an equilibrium state of the polar lower stratosphere (in conditions of weakened forcing from the troposphere) to be established, several years are required. Also, a decrease in the hemispheric mean TOZ is expected because efficient removal of ozone from the mid-stratosphere is approximately proportional to the square of the velocity of diabatic meridional circulation generated by large-scale turbulence. About 10% of the TOZ decrease for 2 weeks of September can be explained by the effect of time-dependent upward motions caused by radiative (after the vernal resumption of insolation) and dynamic factors. Since this mechanism requires an additional radiation absorber in the stratosphere, PSCs are supposed to be able to play such a role. However, none of this exclude a substantial effect by chemical processes.

Cariolle *et al.* (1986) performed simulation modelling of the annual change of ozone in high southern latitudes using a 28-level (including 16 levels in the stratosphere and four in the mesosphere) spectral atmospheric general circulation model (AGCM), developed at the National Centre of Meteorological Studies in France. AGCM takes into account wavenumbers up to 10 and 13, respectively, in zonal and meridional directions. Its major feature is a detailed parametrization of radiative processes (instead of the Newton cooling scheme used earlier). The integration is made over 1 year, starting from the last day that corresponds to calculations for fixed January conditions.

Analysis of the results obtained shows that the model correctly simulates major observed features of AGC and ozone distribution. In particular, calculations simulate the formation of a very cold and mature circumpolar vortex from mid winter till spring and continuing till early November, the TOZ being minimum in the zone of vortex, constituting in the spring less than 300 DU. This value is much higher than those observed at several Dobson stations in the Antarctic. Apparently, the imperfection of photochemical schemes of AGCM is the possible reason of this difference.

The model reproduces well the phase of rapid increase in TOZ following stratospheric warming, which occurs in early November and is connected with the growth in amplitude of planetary waves (with wavenumber 1) north of 60°S. In accordance with observational data, the polar vortex moves westward along the trajectory, which starts near the Pole, crosses the Argentinian islands before the vortex dissipates in mid-latitudes under the influence of warming because of solar UV radiation absorption by ozone.

Using a 2-D (zonally averaged) AGCM, Pyle (1986) calculated the effect of variation in radiative flux divergence (RFD) on meridional circulation and annual change in TOZ in the zone of the circumpolar vortex and outside it (in mid-latitudes). With this aim in view, various vertical RFD profiles were considered, in which values of both increased radiative warming within 0.1–0.4 K/day and decreased warming were prescribed.

Calculation results showed that reasonable RFD values do not enable reproduction of the observed springtime TOZ decrease. With an assumed increase in radiative warming in late polar night in the Southern Hemisphere, time constant values turn out to be so great (much in excess of those observed) that for at least 3 months no marked changes in TOZ take place. Results obtained show that RFD changes cannot cause the springtime TOZ decrease.

Using a 1-D model of photochemical and dynamic processes in the atmosphere, Evans *et al.* (1986) performed simulation numerical experiments to reproduce the annual change in TOZ in the Antarctic, especially at the time of springtime (October) decrease in TOZ. Processes of vertical and horizontal transport are parametrized through the upward motions prescribed so that an increase in vertical velocity during the polar night and early spring provides adequate reproduction of low TOZ values in October (the alternative possibility is the delay of winter-to-summer temperatures). Calculations show that the effect of vertical motions on the annual change of TOZ are apparent throughout the whole year. In this connection, the assumption has been made that variation in the dynamics of the circumpolar vortex, caused apparently by stratospheric radiative cooling, can be more substantial in formation of the springtime TOZ minimum than photochemical processes.

3.4.12 Conclusion

Even at early stages of ozone-hole study, it was clear that manifestation of the circumpolar vortex, preventing ozone from being transported from low latitudes, was the first stage in formation of the springtime ozone minimum in the Antarctic (Shi *et al.*, 1986). However, the problem of mechanisms responsible for spatial and temporal variability of ozone within the vortex (both photochemical and dynamic processes being among them) still remains incompletely solved. Without heat transport to the zone of the circumpolar vortex, the air temperature during the polar night is close to the temperature of radiative equilibrium. As insolation renews the stratosphere gets warmed, which must stimulate upward motions, transporting poor-in-ozone tropospheric air to the stratosphere and forcing rich-in-ozone stratospheric air out of the region of the vortex. The rise of airmasses may continue till either the radiative warming is compensated by radiative cooling or the circumpolar vortex dissipates. In this case, the depth of the springtime TOZ minimum, which is determined by the intensity and duration of the upward motions, must depend on the evolution of the temperature field and on RFD in the stratosphere.

In this context, RFD calculations were made by Shi *et al.* (1986) from data of aerological soundings at the Syowa Station (69°S, 39°E) for 1982. Results of calculations show that the prevailing contribution of radiative cooling, because of CO_2 and O_3, determines the net cooling of the lower stratosphere over the whole year, provided the effect of aerosols is not taken into account. The results are very sensitive to air temperature. The contribution of aerosol to solar radiation absorption (and hence radiative warming) turned out to be less than the ozone contribution. The effect of aerosols on longwave radiation transfer, causing warming of the lower stratosphere between March and September, is more substantial. During the polar winter, aerosol-induced longwave warming is almost completely compensated by cooling, because of CO_2 and O_3, which governs formation of conditions close to radiative equilibrium.

In spring, a stage of net radiative warming sets in, which stimulates upward motions and subsequent formation of the TOZ minimum. Interannual variability in the TOZ minimum is probably connected with annual variation in temperature and aerosol content. This determines the need for more reliable information on spatial and temporal variability and properties of stratospheric aerosol.

In connection with studies about the relative role of chemical and dynamic factors in formation of the ozone hole in the Antarctic, Tung (1986) drew attention to the existence of a larger and more stable 'hole' over the tropics, which was commonly explained by the effect of upward motions due to low temperatures of the tropical lower stratosphere (from the viewpoint of the theory of photochemical equilibrium, the equatorial ozone maximum ought to have been observed). Although photochemical processes affect the formation of the vertical profile of ozone concentration a lot, observed latitudinal and annual change of ozone cannot be explained without taking stratospheric dynamics into account.

Dynamic and thermal factors determine the nature of the observed contrasts in TOZ distribution in Northern and Southern Hemispheres. A major contribution to the formation of these contrasts is determined by the Eliassen–Palm flux divergence, which characterizes irreversible transport because of large-scale turbulent mixing. The Southern Hemisphere is characterized, as a rule, by less intensive longwaves, which is explained apparently by stronger horizontal homogeneity of the surface. No doubt, this dynamic contrast must show itself in the field of zonal mean temperature. This is confirmed by observational data, which show that the zonal mean temperature of the lower stratosphere of the Southern Hemisphere is 20–30-K below that of the Northern Hemisphere.

Thermal contrasts bring forth the interhemispherical differences between radiative flux divergence and atmospheric circulation. From the point of view of the formation of the spatial distribution of ozone, upward motions in the lower stratosphere are very important: they lead to TOZ decrease, when airmasses with low concentrations of ozone come from the troposphere (this mechanism is responsible for the equatorial ozone minimum). On the other hand, when atmospheric circulation transports ozone northward and downward, this creates an ozone maximum in the Arctic. As at the equator, the effect of upward motions (in conditions of a very cold lower stratosphere) can be considered a partial cause of springtime ozone decrease in the Antarctic.

Lary (1997) reviewed the main ozone-destroying catalytic cycles at work in the stratosphere with special emphasis on the kinetic aspects of these cycles, such as the rate-limiting step and chain length. Lary (1997) pointed out that the chain length N is usually defined as the rate of propagation p divided by the rate of termination. The rate of propagation is the rate of the rate-limiting step. The rate of termination is the rate of production or destruction δ of the chain centre:

$$N = p/\delta$$

The notion of chain effectiveness E was also introduced:

$$E = Np$$

to characterize the situation when a particular radical is involved in a catalytic cycle, which has a very long chain length but is present in only small concentrations and so the effectiveness of the cycle is limited.

Table 3.6 illustrates the significance of various catalytic cycles (specified by Lary (1997) in dependence on altitude). The following conclusions were made by Lary (1997):

Table 3.8. Summary table showing the approximate altitude in kilometres at which various catalytic cycles are fastest for mid-latitudes at equinox. For this latitude the approximate chain length N is also given.

Z (km)	ρ_{max}	N		Cycle
65	8×10^6	10^3	$H/OH/HO_2$	HO_x
40	2×10^6	10^3	Cl/ClO	ClO_x
	1×10^6	10^6	O/O_3	
	2×10^4	10^4	Br/BrO	BrO_x
35	4×10^6	10^5	NO/NO_2	NO_x
	3×10^6	10^3	Cl/NO_2	ClO_x
	8×10^5	10	OH/HO_2	HO_x
30	8×10^4	80	OH/HO_2	HO_x
	3×10^4	10	HO_2/ClO	HO_x
25	9×10^4	10	ClO/NO_2	ClO_x
22	1×10^4	10^3	BrO/ClO	BrO_x
	6×10^3	10^3	BrO/NO_2	BrO_x
	2×10^3	2	ClO/ClO	ClO_x
20	5×10^3	10^3	HO_2/BrO	HO_x
	10	0.1	CF_3O/CF_3OO	CF_33O_x
	1×10^2	0.4	F/FO	FO_x
	1×10^2	0.4	$F/FO/FO_2$	FO_x

1. In the upper stratosphere, where the abundance of ozone is relatively low, the most effective catalytic cycles are those whose net reaction is $2O(^3P) \longrightarrow O_2$.
2. In mid-stratosphere, the most effective catalytic cycles are those whose net reaction is $O(^3P) + O \longrightarrow 2O_2$.
3. In the lower stratosphere, where the abundance of $O(^3P)$ is relatively low, the most effective catalytic cycles are those whose net reaction is $2O_3 \longrightarrow 3O_2$.
4. Examining the rate-limiting step alone does not give any idea as to how many times a cycle can proceed. Consequently, calculating the cycle's chain length is valuable. This is particularly true when comparing cycles which involve chain centres of very different abundance, such as chlorine and bromine cycles. In general, although bromine is less abundant than chlorine, it is still very effective in destroying ozone as the bromine cycles tend to have very long chain lengths. In this context, chain effectiveness is useful as it is the product of the cycle's rate and chain length. A cycle with a chain effectiveness greater than 10^8 mol cm^{-3} s^{-3} is likely to be an effective ozone-loss cycle.
5. In the low stratosphere, cycles involving HO_2 are important. A relatively recent revelation is that cycles involving HO_2 and halogens are particularly important. With increasing levels of atmospheric bromine the HO_2/BrO will assume a greater role in lower stratospheric and upper tropospheric ozone loss.
6. The most important ozone-loss cycles in the polar lower stratosphere, in their approximate order of importance, are BrO/ClO, HO_2/BrO and OH/HO_2, with

chain effectiveness reaching 2×10^8, 9×10^6, and 4×10^6 mol cm^{-3} s^{-1}, respectively, with the ClO dimer cycle having a much lower chain effectiveness of around 3×10^4 mol cm^{-3}s^{-1}.

7. Currently known fluorine cycles are barely catalytic and experience rapid termination. They are therefore not effective at destroying ozone.

Intercomparison between models and observations has a decisive role to play in the development of adequate models of ozone variability. An important study in this context was carried out by Wauben *et al.* (1998) who calculated the global ozone distribution in the troposphere and lower stratosphere using a 3-D chemistry/transport model of the Royal Meteorological Institute (KNMI) driven by European Centre for Medium-Range Weather Forecasts (ECMWF)-analysed meteorological fields. Results were compared with ozonesonde and TOMS data. Comparison indicates reasonably good agreement between modelled ozone profiles with climatological ozonesonde data for various stations all over the world. However, ozone-mixing ratios in the upper troposphere and lower stratosphere at mid-latitudes are generally overestimated by the model. TOMS data compare reasonably well with TOZ values obtained by combining calculated ozone distributions up to 50 hPa with climatological zonal mean data for ozone about 50 hPa. Modelled TOZ is generally underestimated in the tropics and overestimated elsewhere compared with TOZ observations. Wauben *et al.* (1998) discussed some causes of the disagreement between model and observation results. It was pointed out that, among many other unsolved problems, there was the need to improve the treatment of ozone precursors, such as NO_x emissions by lighting and biomass burning. A principal conclusion is that, in view of serious uncertainty in modelling naturally induced ozone change, the calculated atmospheric impact of anthropogenic emissions needs to be interpreted with care.

3.5 VOLCANIC ERUPTIONS AND OZONE-LAYER DYNAMICS

Powerful explosive volcanic eruptions are a unique type of natural calamity with consequences which are of great importance, not only on a regional scale but also on a global scale. Of special interest are impacts of volcanic eruptions on global climate change and the ozone layer (Kondratyev, 1984; 1988; Kondratyev and Galindo, 1997). Every explosive volcanic eruption can be regarded as a large-scale experiment in nature which opens up the possibility to study the sensitivity of nature (e.g. the climatic system, the ozone layer) to external disturbances.

Various observational means had been developed by 1991 when Mount Pinatubo erupted on Luzon Island (Philippines). As a result of this eruption, there was about a tenfold increase in the amount of aerosol in the stratosphere compared with amounts reported in recent years, both in mid and high latitudes. By that time, a very broad complex of satellite, aircraft, balloon and groundbased instrumentation for monitoring environmental dynamics had been developed and applied for various purposes. For instance, a major European campaign had been developed to study the Northern Hemisphere ozone layer in the winter of 1991/1992 under the title of the European Arctic Stratospheric Ozone Experiment (EASOE). It has become possible not only to monitor propagation of volcanic products over the globe but also to observe (with the use of both *in situ* and remote-sensing techniques) various properties of eruptive aerosols, as well as their impact on the ozone layer, the Earth's radiation budget and the climate.

Since the discovery of the Antarctic total ozone spring minimum (ozone hole), the problem of global ozone dynamics has become a key issue of environmental studies, especially in the light of serious total ozone depletions observed in highly populated mid-latitude regions of Northern Hemisphere. It is now well known that total ozone content in Northern Hemisphere mid and polar latitudes has been declining over the past two decades, that is, after chlorine loading of the stratosphere exceeded certain levels (Bojkov 1999).

Observations made after the El Chichón and Mount Pinatubo eruptions show that, apart from causes of total ozone anomalies because of varying synoptic situations and photochemical processes, there is another potential mechanism connected with the impacts of volcanic effluent. We shall now discuss this aspect of the problem.

To analyse the impact of the Mount Pinatubo eruption on atmospheric ozone total content and vertical ozone distribution, Hofmann *et al.* (1993a, b) considered ozonesonde observations at Hilo (Hawaii, 20°N) before (1985–1990) and after (1991-1992) the eruption. Ozone vertical profiles obtained since 1985 have been averaged over 2-km intervals from 14 to 30 km from which monthly averages of total ozone content have been determined as well as average ozone vertical profiles. The general nature of ozone anomalies in 1991–1992 can be summarized as lower than normal ozone below about 25 km (the largest deficits occur at 16–18 km in spring; i.e. just above the tropopause where ozone content is up to 50% below normal) and higher than normal ozone above. The net result was that total ozone content was somewhat lower than average and, during late 1992, was also as low as recorded in 1982, following the eruption of El Chichón.

Hofmann *et al.* (1993a) pointed out that elevated temperatures in the volcanic aerosol layer (heating of 2–3°C in excess of the average in the 20–25-km layer appeared in the autumn of 1991) and upward motion of the aerosol layer were observed at Hilo following the eruption. Although the nature of the perturbed ozone profile may be the result of enhanced upward motion associated with volcanic aerosol particle heating and the coupling of quasi-biennial oscillation effects, the persistent nature of the perturbation, which was present more than a year after the eruption, is not easily explained. If a portion of the anomalous ozone in 1991–1992 was indeed associated with the eruption of Pinatubo and its aftermath, then it is likely that these were predominantly ozone redistribution by dynamical effects and enhanced ozone destruction by homogeneous chemistry related to this dynamic redistribution.

Unprecedentedly low total-ozone-content values observed during the winter of 1992/1993 at all mid-latitude stations in the USA, which were considerably lower than those following the El Chichón eruption, stimulated the development of a new viewpoint on post-eruption stratospheric changes. The principal ozone losses in the lower stratosphere (below about 24 km) took place in the same region occupied by the Mount Pinatubo aerosol cloud. It was pointed out, however, that although low ozone values were quite likely related to the Pinatubo eruption, explaining the long delay in this process and the fact that ozone content was actually higher than normal values above 24 km is currently a topic of considerable debate. An important circumstance is that trajectory analyses suggest that, except at low latitudes in spring, air parcels on the days of ozone measurements generally arrived at Boulder (one of the points of surface observations) from higher latitude, although seldom higher than 60°N, and hence may have been subjected to heterogeneous chemical processing on the surface of Pinatubo aerosols, a process which is believed to be more effective under low winter sunlight levels of higher latitudes.

Dutton *et al.* (1993) emphasized that the loss of heating in the stratosphere after the Mount Pinatubo eruption, also revealed by MSU (Microwave Sounding Unit) satellite data during 1992, could be due to volcanically induced ozone reduction (a similar but weaker phenomenon was observed after the 1982 El Chichón eruption). Stratospheric cooling further contributes to tropospheric cooling. Thus a quite different picture of volcanic impact on stratospheric temperature arises, namely not just heating because of absorption by volcanic aerosols of solar radiation and upwelling longwave radiation, but also cooling because of ozone reduction in the lower stratosphere with subsequent tropospheric cooling.

Chandra (1993) undertook an analysis of changes in stratospheric ozone and temperature because of the eruption of Mount Pinatubo on the basis of satellite total-ozone content data from the Nimbus-7 TOMS and NOAA-II SBUV/2. Ozone anomalies inferred from the two instruments agree within 1–2% in the presence of large volcanic clouds produced by Mount Pinatubo. As Chandra (1993) pointed out, within a few months after the eruptions total-ozone-content values decreased by 5–6% in the tropics, 3–4% in mid-latitudes and 6–9% at high latitudes in the Northern Hemisphere.

Such total-ozone-content drops cannot be entirely explained, however, by volcanic impact only because significant contributions to total-ozone-content change were also made by quasi-biennial oscillation and as such present a different dynamical scenario for change in stratospheric ozone. After the effects of quasi-biennial oscillation and interannual variability are taken into account, the decrease in total ozone content attributed to volcanic eruption at these latitudes may not be more than 2–4% (this conclusion is in general agreement with a similar study of the El Chichón effects on stratospheric ozone).

In fact, the most noticeable effect of the Mount Pinatubo eruption, pointed out by Chandra (1993), was the breakdown of the phase relation between ozone and temperature (a similar situation was observed during the El Chichón period). This is attributed to additional heating in the lower stratosphere caused by volcanic aerosols.

Observational data show that there is no evidence of additional total-ozone-content decrease in the Southern Hemisphere where the aerosol cloud spread rapidly after the eruption. The empirical regression model, based on the 30-hPa zonal wind at Singapore, accounts for most of the observed characteristics of ozone change both at the equator and at southern latitudes. Negative total ozone content anomalies at the equator and at mid-latitudes in the Southern Hemisphere, to a large extent, are manifestations of quasi-biennial oscillation, as inferred by the 30-hPa zonal wind at Singapore. This model is not adequate, however, for the Northern Hemisphere possibly because of relatively large interference from planetary wave activity. The principal conclusion is that the study considered does not support model predictions of massive ozone loss because of heterogeneous chemical processes, which were discussed by Hofmann and Solomon (1989), and others.

The role of sulphur photochemistry in tropical ozone change after the Mount Pinatubo eruption was discussed by Bekki *et al.* (1993). Although stratospheric ozone losses following volcanic eruptions are generally attributed to the presence of sulphate aerosols, Bekki *et al.* (1993) presented model calculations which demonstrate that gas-phase sulphur chemistry may also have played an important part in tropical ozone perturbations that followed the Mount Pinatubo eruption.

This general conclusion is in accord with observational data for August and September 1991 (ozonesonde measurements) which show that the maximum ozone reduction (18–20%) was observed at an altitude around 24–25 km but, on the other hand, ozone concentrations increased by 5–10% above 28 km. TOMS data indicate local increases in the tropical total ozone content of about 2% in July, but decreases of 1% in August and 3–4% in September (zonally averaged TOMS ozone changes do not exceed 2–4%). Microwave Limb Sounder (MLS) data show in September total-ozone-content reductions in the tropical lower stratosphere, but no significant total-ozone-content increases around 30 km (since the quasi-biennial oscillation signal has not been smoothed out, the observed total-ozone-content change cannot, however, be entirely attributed to the Mount Pinatubo eruption).

The calculation made by Bekki *et al.* (1993) showed that, in the first month or so after the eruption, the large amount of SO_2 injected into the tropical atmosphere catalyses mid-stratospheric ozone production.

Ozone production catalysed by SO_2 is possible through the following sequence of reactions:

$$SO_2 + h\nu \longrightarrow SO + O \qquad (\lambda < 220\,\text{nm}),$$

$$SO + O_2 \longrightarrow SO_2 + O$$

$$2(O + O_2 + M) \longrightarrow 2(O_3 + M)$$

Net reaction:

$$3O_2 \longrightarrow 2O_3$$

SO_2 also has the potential to reduce transmission of solar radiation since it absorbs strongly between 180 and 235 nm, weakly in the 260–340-nm interval and very weakly between 340 and 390 nm. Such reduction leads to the decrease in photolysis rate for key species such as O_2 and therefore results in reducing the rate of ozone production. The simulation modelling made by Bekki *et al.* (1993) using a zonally averaged chemical–radiative-dynamical model (model intergrations cover the period from June to September 1991) demonstrated that the two effects mentioned cancel each other out at an altitude of about 25 km. After 1 or 2 months most of the SO_2 has oxidized to sulphate; the efficiency of these two mechanisms then becomes negligible (although ozone remains perturbed in the lower stratosphere because of its long photochemical lifetime in this region).

Changes in ozone concentrations in July are predominantly confined to the latitude band of volcanic injections. Ozone concentration increases in the mid-stratosphere (the top region of the volcanic cloud) by up to 7%. By September, the initial SO_2 cloud has spread throughout the tropics. Ozone increase (2%) in the mid-stratosphere remains low with reduction of up to 8%.

The rates of the competing processes mentioned depend on the amount of SO_2 present in the stratosphere. As most of the SO_2 cloud is converted to sulphate by September, the perturbations of the photochemical balance of ozone are substantially reduced at this time compared with July. Ozone production in the mid-stratosphere and oxygen photolysis is reduced by only about 10% in the lower stratosphere. Model features show good agreement with initial ozone measurements following the eruption, including both the mid-latitude switch from ozone loss to ozone gain and the increase and subsequent decrease in total ozone content.

There are, however, some discrepancies between calculations and observations; for example, underestimated model total ozone content values (in comparison with TOMS data), and the estimate of the altitude of crossing over from production to destruction of ozone ($\sim$25 km model value and observed 28 km).

An important aspect of the problem is that volcanically induced ozone changes may lead to a substantial impact on radiative forcing for the surface–troposphere system and subsequently on climate.

It has to be emphasized that, in spite of very important progress in interpretation of the Mount Pinatubo experiment results, further intensive studies in the fields of observations and numerical modelling are needed.

3.6. IMPACT ON THE OZONE LAYER OF HIGH-FLYING AIRCRAFT

More than three decades ago there was an explosion of interest in the potential impact of high-speed civil transport aircraft on the stratosphere in general and the ozone layer in particular (Crutzen, 1971; Johnston 1971). Recently a number of new studies have been carried out that are based on much better models of the stratosphere (see, in particular, Penner *et al.*, 1999). In accord with earlier results, it was shown that the principal components influencing stratospheric ozone are nitrogen oxides NO_x (NO, NO_2, NO_3, $2N_2O_5$, HNO_4, $ClONO_2$, $BrONO_2$). These compounds enhance ozone removal in the stratosphere, but contribute to ozone production through methane–NO_x smog reactions in the troposphere.

To study large-scale distribution of NO and NO_x and obtain reliable monthly mean values of these components for different seasons in a region of the atmosphere that is strongly affected by aircraft emissions, Ziereis *et al.* (1999) conducted aircraft *in situ* measurements (over 32 flights) of NO, NO_2 and O_3. The flights were made

Table 3.9. Chemical reactions with sulphur species.

No.	Reaction
(R1)	$OCS + h\nu \longrightarrow CO + S\ (SO_2)$
(R2)	$OCS + OH \longrightarrow$ products (SO_2)
(R3)	$OCS + O \longrightarrow CO + SO\ (SO_2)$
(R4)	$CS_2 + h\nu \longrightarrow$ products $(OCS + SO_2)$
(R5)	$CS_2 + OH \longrightarrow$ products $(OCS + SO_2)$
(R6)	$CS_2 + O \longrightarrow CS + SO\ (OCS + SO_2)$
(R7)	$DMS + OH \longrightarrow$ products (SO_2)
(R8)	$DMS + O \longrightarrow$ products (SO_2)
(R9)	$H_2S + OH \longrightarrow HS + H_2O\ (SO_2)$
(R10)	$H_2S + O \longrightarrow HS + OH\ (SO_2)$
(R11)	$SO_2 + OH + m \longrightarrow HSO_3 + m\ (H_2SO_4)$
(R12)	$SO_2 \longrightarrow$ washout
(R13)	$H_2SO_4 \longrightarrow$ washout

mainly in the region of the North Atlantic flight corridor in summer and late autumn mostly at altitudes between 10.5 and 11.5 km. NO mean values ranged between about 0.1 and 0.14 ppbv in late autumn. NO and NO_x did not show a significant gradient across the tropopause. Only, the correlation between NO_x and O_3 in the upper troposphere and lowermost stratosphere was very weak.

Weisenstein *et al.* (1993, 1996, 1998) developed a two-dimensional chemical transport model which contains 60 chemical species, including a complete diurnal chemistry for NO_y (NO_x, HNO_3), Cl_x (Cl, ClO, HOCl, $ClONO_2$), Br_x (Br, BrO, $BrONO_2$, HBr), O_x (O, $O(^1D)$, O_3), methyl and ethyl families along with their source gases. Model resolution is approximately 9.5° of latitude and 3.5 km in the vertical. The principal aim of these calculations was to predict changes in the stratosphere by 2015 based on prescribed changes in atmospheric pollution (Table 3.9), also taking account of atmospheric aerosol impact for two levels of stratospheric aerosol loading: background concentrations and a factor of 4 increase in aerosol loading (the latter corresponds to average global conditions after strong volcanic eruptions).

Weisenstein *et al.* (1993) pointed out that with aerosols fixed at background levels, calculated ozone changes, because of high-speed civil transport (HSCT) aircraft emissions, range from negligible up to 4–6% depletion in total ozone content at northern high latitudes. Fig. 3.9 illustrates the changes in total ozone because of a factor of 4 increase in aerosol surface area in 2015 relative to background aerosol levels for the same year in the absence of aircraft emissions. The total ozone decreases by 0.5 to 1.5% when aerosols are increased. Similar data for ozone mixing ratios are shown in Figure. 3.9(b); the maximum decrease in the 20–40-km layer reaches 4%.

Calculations of ozone changes because of aircraft exhaust products were made under the assumption that by 2015 a high-speed civil transport fleet of about 500 aircraft will be operating. Various combinations of Mach number and NO_x emission

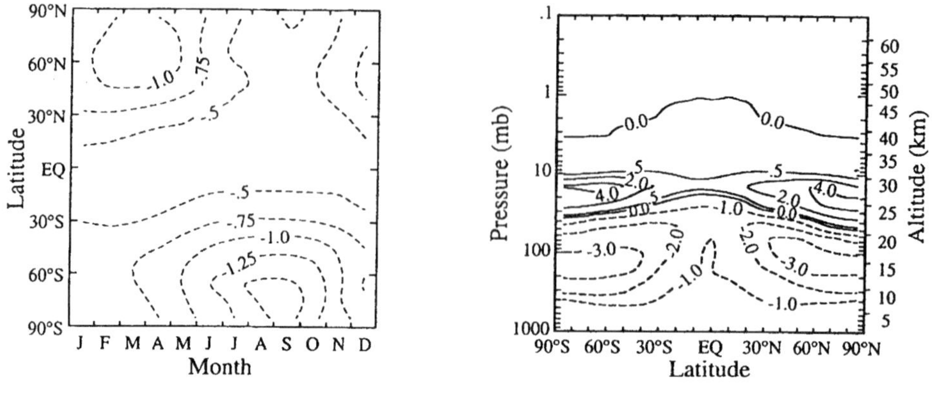

Figure 3.9. Percentage change in (a) column ozone and (b) local ozone for June because of a factor of 4 enhancement in aerosol surface area relative to background aerosol levels for 2015. Solid contour lines represent ozone increases; dashed lines represent ozone decreases.

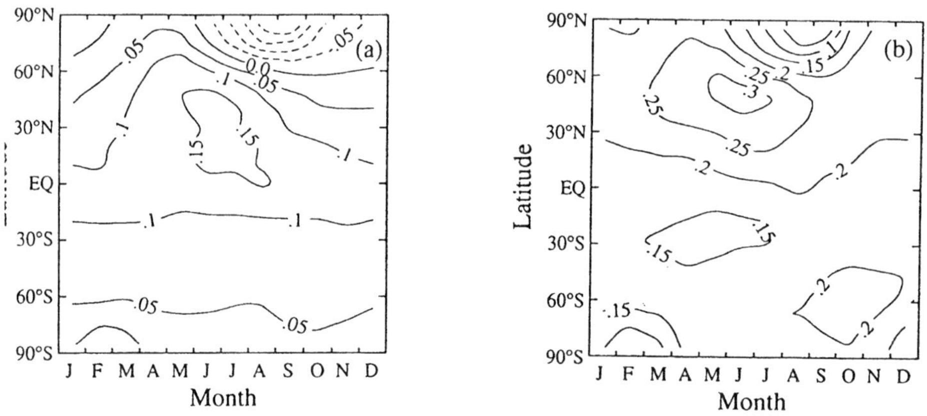

Figure 3.10. Model-calculated change in column ozone (%) because of emissions from a likely fleet of HSCT aircraft operating in 2015 at Mach 1.6 with NO_x EI = 15 and with aerosols: (a) held fixed at the background level or (b) increased by a factor of 4. Comparisons are relative to the 2015 atmosphere without HSCT aircraft but with the same aerosol loading. Solid contour lines represent ozone increases, dashed lines represent ozone decreases.

index (a parameter characterising emissions as functions of the fuel burned) for five scenarios were prescribed. Figure. 3.10 illustrates some results which indicate that Mach 1.6 aircraft with an NO_x emission index of 15 should have minimal negative effect on stratospheric ozone, though increases in tropospheric ozone could be significant. Figure 3.11 demonstrates variability in ozone concentration vertical profiles.

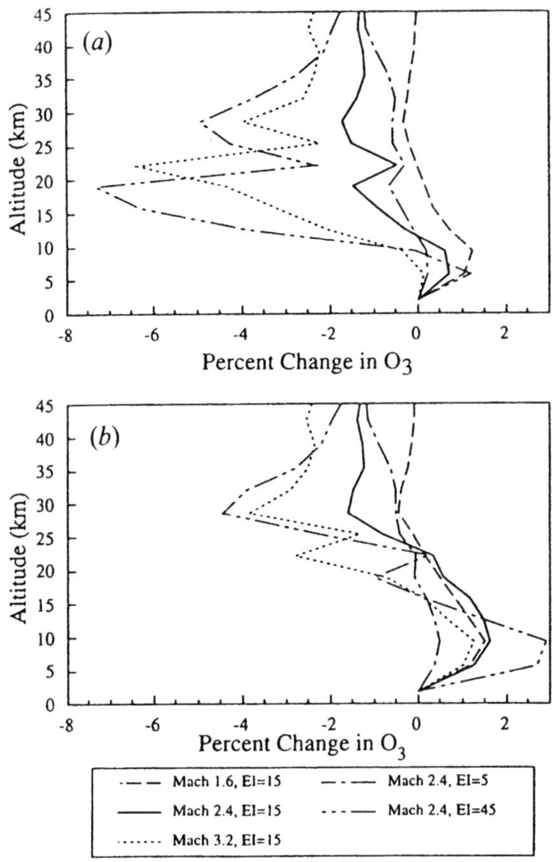

Figure 3.11. Percentage change in annual average ozone mixing ratio at 47°N because of emissions from projected fleets of HSCT aircraft: (a) background aerosols; (b) aerosols increased by a factor of 4.

Since heterogeneous chemistry could have a significant role in model-predicted ozone changes because of gas injection from high-speed civil transport aircraft, Pitari *et al.* (1993) developed a 2-D model covering the stratosphere and troposphere which includes photochemical reactions for the sulphur cycle and microphysical code for sulphuric acid aerosols. Such a model enables simulation of the impact of interactive aerosols with variable properties on ozone changes. The microphysical code can also predict the size distribution for nitric acid trihydrate and ice aerosols, globally covering a particle-radius range between 0.01 μm and about 160 μm.

Numerical modelling carried out by Pitari *et al.* (1993) for various aerosol scenarios indicates that ozone sensitivity to high-speed civil-transport-aircraft-emissions largely decreases when heterogeneous chemistry is included with respect to the pure chemistry case. Ozone sensitivity to high-speed civil-transport-aircraft emissions decreases even more when nitric acid trihydrate and ice aerosols are present. This is a consequence of aerosol-induced stratospheric denitrification which

Table 3.10. Surface boundary conditions for sulphur species.

	Species	Boundary condition
Surface mixing ratio	OCS	500 pptv
Surface emissions	SO_2	100 TgS yr^{-1}
Deposition velocity		1.0 cm s^{-1}
Surface emissions	CS_2	2.0 TgS yr^{-1}
Surface emissions	DMS	40 TgS yr^{-1}
Surface mixing ratio	H_2S	30 pptv
Deposition velocity	H_2SO_4	1.0 cm s^{-1}

Table 3.11. Heterogeneous chemical reactions.

No	Reaction
(R1)	$N_2O_5 + H_2O(\alpha) \longrightarrow HNO_3 + HNO_3$
(R2)	$ClONO_2 + H_2O(\alpha) \longrightarrow HOCl + HNO_3$
(R3)	$N_2O_5 + HCl(\alpha) \longrightarrow ClONO + HNO_3$
(R4)	$ClONO_2 + HCl(\alpha) \longrightarrow Cl_2 + HNO_3$
(R5)	$HOCl + HCl(\alpha) \longrightarrow Cl_2 + H_2O$

makes the residence time of injected odd nitrogen shorter and the relative weight of the NO_x catalytic cycle smaller. The high-speed civil-transport-aircraft-induced increase in sulphate aerosols and odd nitrogen concentration may also affect the probability for polar stratospheric cloud formation in accord with the assumption made by Hofmann and Solomon (1989) that the reaction $N_2O_5+H_2O \longrightarrow 2HNO_3$ may proceed as efficiently on sulphate aerosol surfaces as on polar stratospheric cloud particles.

Table 3.9, from Pitari *et al.* (1993), shows chemical reactions with sulphur. The scheme adopted for sulphur chemistry starts from ground fluxes of carbonyl sulphide (COS), carbon disulphide (CS_2), sulphur dioxide (SO_2), dimethylsulphide ($DMS = CH_3SCH_3$) and hydrogen sulphide (H_2S). Tropospheric oxidation of CS_2, DMS and H_2S produces sulphur radicals that are readily converted into SO_2. Then reactions of gas-to-particle conversion to sulphur-acid-aerosol droplets follows. Table 3.10 gives surface boundary conditions for sulphur species. Heterogeneous reactions considered are summarized in Table 3.11. The 2-D model has a latitudinal resolution of 10° from pole to pole and a vertical resolution of about 2.8 km from the ground to about 71-km altitude. Similar to Weisenstein *et al.* (1993), chemistry with a total of about 150 photochemical reactions was considered.

Figure 3.12 illustrates calculated ozone change from baseline in experiment *G* (gas-phase chemistry). Pitari *et al.* (1993) emphasized that in this case the impact of high-speed civil transport aircraft on ozone, which was the largest, was a direct consequence of the enhanced NOx catalytic cycle. In accord with Weisenstein *et al.*

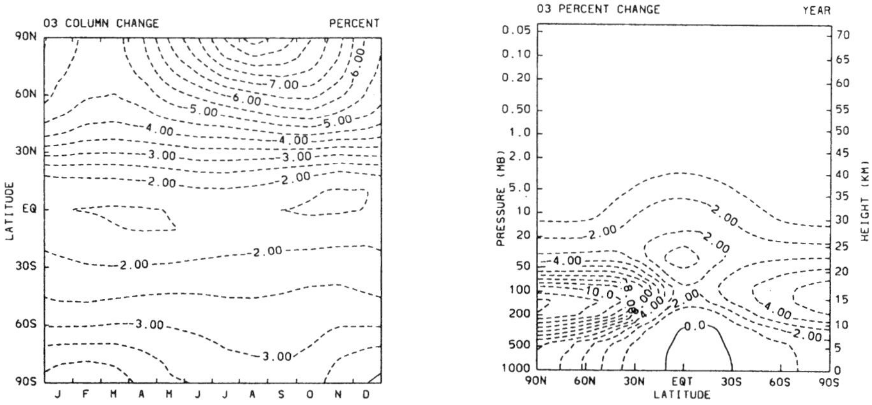

(*a*) (*b*)

Figure 3.12. Calculated ozone change from baseline in experiment G (i.e. $(G_A - G)/G$,%): (a) column and (b) yearly averaged profile. The contour step is (a) 0.5% and (b) 1%. The dashed curves are for negative values.

(1993), results of the inclusion of heterogeneous reactions on sulphate aerosols led to a large reduction in ozone depletion. Effects of heterogeneous chemistry on total ozone in atmospheric background conditions are summarized in Figure 3.13. Pitari *et al.* (1993) pointed out that the model shows a large ozone decrease over Antarctica during springtime when nitric acid trihydrate/ice aerosols are added to the sulphate layer and an ozone increase, following the inclusion of chemical reactions over the sulphate layer, with respect to the pure gas-phase chemistry case.

In a recent study, Weisenstein *et al.*, (1998) carried out numerical modelling using four independently formulated 2-D chemical transport models with sulphate aerosol microphysics to assess the possible effects of sulphur emissions from high-speed civil transport (HSCT) aircraft operating in the stratosphere in 2015. The models include the following heterogeneous reactions on sulphate aerosols

$$N_2O_5 + H_2O \longrightarrow 2HNO_3 \tag{3.64}$$

$$ClONO_2 + H_2O \longrightarrow HNO_3 + HOCl \tag{3.65}$$

$$ClONO_2 + HCl \longrightarrow HNO_3 + Cl_2 \tag{3.66}$$

$$HOCl + HCl \longrightarrow H_2O + Cl_2 \tag{3.67}$$

Reaction (3.64) is the most important on a global basis, whereas reactions (3.65)–(3.67) are important in the cold regions at high latitudes in winter and under high aerosol conditions. The aircraft emission scenarios represent 500 aircraft operating in 2015 and burning 82×10^9 kg of fuel annually. HSCT aircraft cruise at Mach 2.4 at an altitude of 18–21 km. Model-predicted changes in aerosol surface area vary by a factor of 3 among the models for identical sulphur emissions (the main cause of model differences appear to be transport, which directly affects the residence time of the emitted sulphur and thus the magnitude of the surface area perturbation).

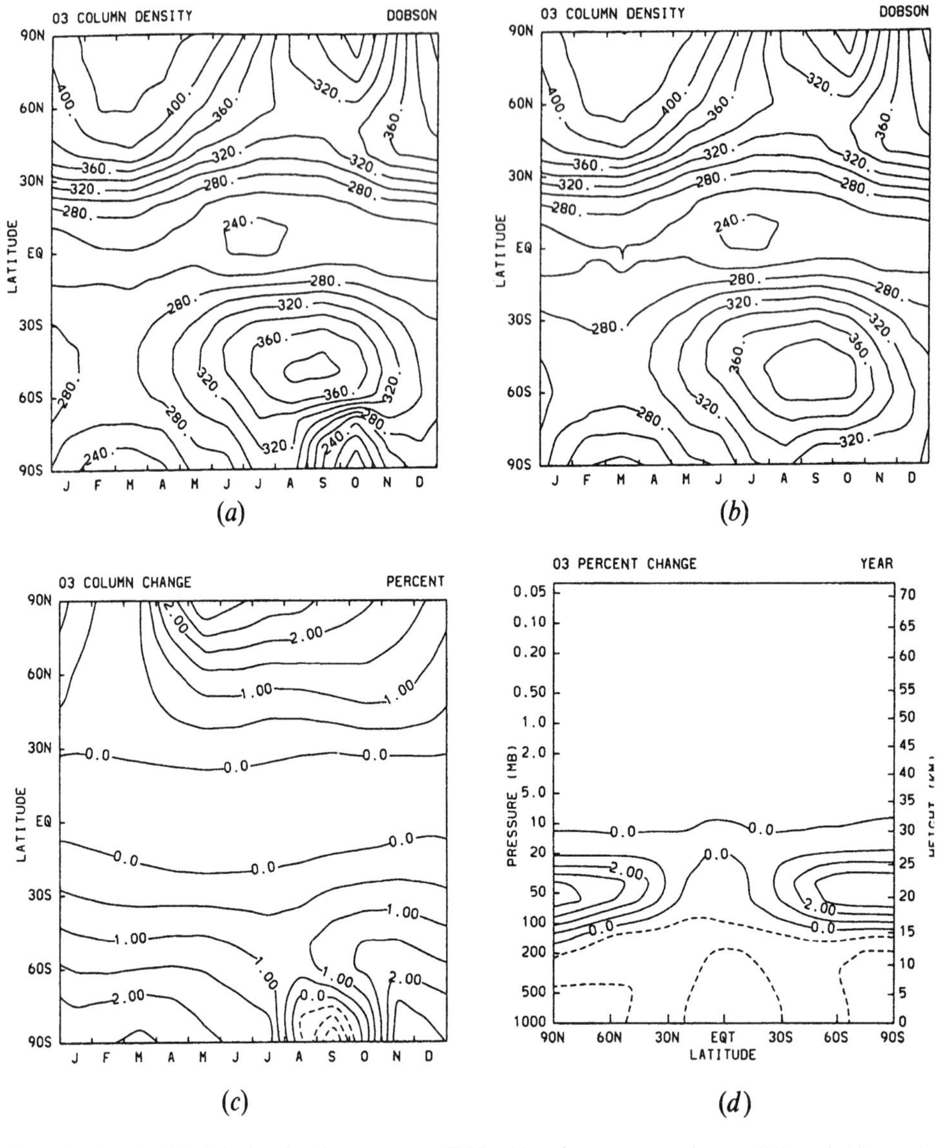

Figure 3.13. (a), (b) Calculated column ozone (DU), (c) column ozone change (%) and (d) yearly averaged change of ozone vertical profile (%). (a), (b) Experiments P and S, respectively, and (c), (d) $(S - G)/G$. (a), (b) The contour step is 20 DU, (c) 0.5% and (d) 1%.

The principal result is that all models show much larger increases in aerosol surface area when aircraft sulphur is assumed to be emitted as particles of 10-nm radius rather than as gas-phase SO_2. In the case of an emission index (EI) for SO_2 of 0.4 gm/(kg fuel burned) in 2015, maximum increases in stratospheric-sulphate-aerosol surface area range from 0.1 to 0.5 $\mu m^2/cm^3$ with sulphur emitted as SO_2

Table 3.12. Calculated change in annual average ozone column because of emissions from a Mach 2.4 HSCT fleet in 2015. All values are in %.

	AER		NSU		Italy		Cambridge	
Sulphur emission*	45°N	Global	45°N	Global	45°N	Global	45°N	Global
$EI(NO_x) = 5$								
None	−0.35	−0.21	−0.32	−0.16	−0.02	−0.02	−0.22	−0.14
0% particles	−0.57	−0.33	−0.50	−0.27	−0.11	−0.09	−0.26	−0.17
10% particles	−0.72	−0.40	−0.59	−0.32	−0.44	−0.23	−0.27	−0.21
100% particles	−1.14	−0.63	−0.83	−0.44	−1.45	−0.77	−0.36	−0.30
$EI(NO_x) = 15$								
None	−0.65	−0.30	−0.94	−0.35	−0.14	−0.12		
0% particles	−0.71	−0.36	−0.57	−0.25	−0.17	−0.15		
10% particles	−0.75	−0.39	−0.62	−0.39	−0.20	−0.15		
100% particles	−1.95	−0.52	−0.71	−0.36	−0.74	−0.46		

Background inorganic chlorine amount for 2015 was 3 ppbv.
* Sulphur emissions with $EI(SO_2) = 0.4$ are treated as 100% gas, 90% gas and 10% aerosol particles of 10-nm radius, or 100% aerosol particles of 10-nm radius. Cases with no sulphur emissions are also shown.

gas and from 1.0 to 2.5 $\mu m^2/cm^3$ with sulphur emitted as particles. This increase is due to increased heterogeneous processing on the enhanced aerosol layer, increasing ozone loss due to the HO_x and ClO_x catalytic cycles.

According to Weisenstein *et al.* (1998), calculated annual average ozone perturbations range from −0.1 to −0.6% at 45°N for sulphur emissions as SO_2 gas and from −0.4 to −1.5% with sulphur emission as 100% particles, if the following emissions have been prescribed: EI $(NO_x) = 5$, $EI(H_2O) = 1230$, $EI(SO_2) = 0.4$.

Table 3.12 illustrate TOZ change (%) at 45°N and on a global average basis in accord with the four models considered (AER, NSU, Italy and Cambridge) because of an HSCT fleet with NO_x emission indices of 5 and 15. Weisenstein *et al.*, (1998) pointed out that global ozone change is about one-half to two-thirds of the change at 45° for most scenarios. Comparison of $EI(NO_x) = 5$ and $EI(NO_x) = 15$ scenarios reveals much longer ozone depletion for $EI(NO_x) = 15$ without sulphur emissions from aircraft; but with sulphur emissions as 100% particles, ozone depletion is larger for $EI(NO_x) = 5$. This means that attempts to control NO_x emissions from aircraft without concomitant reduction in fuel sulphur may not yield desired controls for ozone depletion.

One important aspect of modelling is the sensitivity of the HSCT ozone response to Cly (a Cl_y concentration of about 2.0 ppbv is forecast for 2050). Table 3.12 shows TOZ change at 45° and global average ozone column change with $EI(NO_x) = 5$ and 12 and 2 ppbv of Cl_y.

Sensitivity analysis of the effect that includes the heterogeneous reaction:

$$BrONO_2 + H_2O \longrightarrow HNO_3 + HOBr \qquad (3.68)$$

Table 2.11. Calculated change in annual average ozone column because of emissions from a Mach 2.4 HSCT fleet in 2050. All values are in %.

	AER		NSU	
Sulphur emission*	45°N	Global	45°N	Global
$EI(NO_x) = 5$				
None	−0.43	−0.24	−0.42	−0.20
0% particles	−0.55	−0.31	−0.47	−0.23
10% particles	−0.63	−0.35	−0.52	−0.27
100% particles	−0.85	−0.46	−0.62	−0.33
$EI(NO_x) = 15$				
None	−1.03	−0.51	−1.41	−0.55
0% particles	−0.98	−0.51	−1.01	−0.44
10% particles	−0.97	−0.1	−0.91	−0.41
100% particles	−0.96	−0.52	−0.78	−0.35

Background inorganic chlorine amount for 2050 was 2 ppbv. Other long-lived trace species were retained at their 2015 levels.
* Sulphur emissions with $EI(SO_2) = 0.4$ are treated as 100% gas, 90% gas and 10% aerosol particles of 10-nm radius, or 100% aerosol particles of 10-nm radius. Cases with no sulphur emissions are also shown.

or sulphate aerosols has shown that this reaction has the effect of increasing active HO_x,ClO_x and BrO_x, mainly as a result of HOBr photolysis (Randeniya *et al*, 1996a), leading to enhanced ozone depletion.

Weisenstein *et al.* (1998) summarized their results by stating that, for future modelling studies of HSCT impact on ozone, it will be necessary to include sulphur emissions in assessment calculations, and also that knowledge of chemical and microphysical processes within the plume and wake are crucial for accurate modeling. As shown in a number of studies, fractional conversion of emissions to aerosol particles depends on processes within the engine, nozzle and near-field plume, and the resulting particle-size distribution depends on dilution within the wake.

In this context, Kärcher *et al.* (1996) considered aircraft plume chemistry in the jet regime; that is, during the first few kilometres behind the aircraft, thereby providing necessary data for follow-on modellers. The principal aim was to study the relaxation of strongly perturbed plume chemistry to background conditions. The evolution of hydrogen, oxygen, nitrogen and sulphur species was considered in detail and compared with observations that placed special emphasis on the coupling between mixing and chemistry. The case study considered relates to the impact a B-747 airliner has cruising at around 10-km altitude. An important conclusion is that gas-to-particle conversion depletes high levels of exhaust-produced gaseous H_2SO_4 and results in the generation of large numbers of new binary aerosols. Vay *et al.* (1998) carried out the first measurements of CO, CH_4, N_2O, CO_2 and H_2O in the exhaust trails of T-39, B-757 and DC-8 aircraft at cruise conditions.

Kärcher (1996) carried out model calculations, dealing with formation of aerosols and the interaction both between themselves and with exhaust gases in plumes of cruising airliners, to assess the role of the current aircraft fleet as a source of new particles within the upper troposphere and lower stratosphere. Such particles could contribute to the observed increase in stratospheric background aerosol surface area in the past decade, and thus be partly responsible for measured ozone loss at northern mid-latitudes between 10 and 25 altitudes.

In a more recent study, Kärcher (1998) developed a new coupled microphysical – chemical – dynamical trajectory box model simulating emission, formation and evolution of particles in dispersing aircraft wakes and contrails. Kärcher *et al.* (1998) discussed applications of this model and compared model results with *in situ* observations of particles generated in the wake of aircraft (gas-composition measurements were made by Tremmel *et al.* (1998)). It was shown that sulphur emissions mainly cause the formation of ultra-fine volatile particles in young aircraft plumes (mean radius $\sim$5 μm). Volatile particle numbers range between 10^{16} and 10^{17} per kg fuel burnt for average to high levels, exceeding typical soot emission indices by a factor of 10–100. Model results come into closer agreement with observations when chemical ions from fuel combustion are included in aerosol dynamics (Yu and Turko, 1998). A number of important results, discussed by Kärcher *et al.* (1998), are relevant to soot particle transformation, contrail formation temperatures, larger ice crystals. The most important unsolved problems were also emphasized.

Andronache and Chameides (1998) developed a 1-D box model simulating the temporal evolution (during the first 100–1,100 s) of a contrail of gas and aerosol components from a subsonic aircraft. The model can successfully simulate formation and growth of new H_2SO_4 and H_2O particles as well as activated soot particles by taking account of the interaction between particles and gas components. The results indicate a high concentration of $H_2SO_4 \cdot H_2O$ particles of 1–10 nm, which coagulate with soot particles leading to their activation. Particle number density and rate of growth depend on the level of sulphur emission.

The problem of contrails was also been recently discussed by Baumgardner *et al.* (1998) who evaluated temperature, water vapour and vertical velocity structure of subsonic aircraft contrails in the area behind the engine of a Lear 35 jet by measurements from another aircraft (Sabreliner), crossing the Lear 35 jet's trajectory at distances of 150 m to 5 km.

Jensen *et al.* (1998) conducted a study of environmental conditions required for contrail formation and persistence. For this purpose, *in situ* measurements of ambient temperatures and humidities were made during the Subsonic Aircraft: Contrail and Cloud Effects Special Study (SUCCESS) experiment in 1996. Chlong (1998) conducted 3-D numerical simulations of contrails to investigate the role of various external parameters and physical processes in the lifecycle of contrails.

Weisenstein *et al.* (1996) pointed out the importance of including aerosol formation within aircraft wakes because of small – scale chemical processing of emitted SO_2 in the estimates of the possible ozone impact of sulphur emissions from high-flying aircraft. In this context, Kärcher (1996) calculated the dynamics of liquid and solid particles in the plumes of B-747 and a stratospheric aircraft, cruising at pressures 216 hPa and 73 hPa, respectively, and at ambient temperatures less than 223 K. It was shown that a subset of *in situ* nucleated H_2SO_4/H_2O aerosols that spray liquid can transform into an almost pure HNO_3/H_2O mode because of enhanced uptake of HNO3. Soot embedded in a liquid coating containing H_2SO_4 and HNO_3 triggers heterogeneous freezing of water ice and leads to visible contrails. Ice particles may acquire an $HNO_3 \cdot 3H_2O$ (NAT) coating upon evaporation of H_2O, which can potentially lead to a lengthening of their lifetime. If NAT coating develops at the surface of contrail ice, the resulting persistence of such submicrometre particles could result in efficient heterogeneous chemical processing which may affect ozone. In view of current model uncertainties, it is very important to undertake an experimental verification of the results.

Hanning-Lee *et al.* (1996) undertook laboratory experiments to assess the rates of stratospheric ozone decomposition on aluminium oxide (alumina) as components of launch vehicles' solid rocket motor exhaust. Results indicate that the calculated ozone holes, formed by ozone decomposition, are rather narrow. In the worst case, ozone levels are within 5×10^{-15} of ambient values in the centre of the plume. An approximate estimate of the global impact of alumina particles on ozone decomposition indicates a potential steady-state daytime depletion of less than 206×10^{-8} at present launch rates.

Assessing the polar stratospheric clouds effect on stratospheric ozone, Tie *et al.* (1994, 1995) showed that, in the future, the density of PSCs in the Arctic could be enhanced owing to the potential emissions of water vapour and nitrogen species by projected high-speed civil transport (HSCT). The increase in PSC density could lead to a maximum ozone depletion of 10% at the northern high latitudes in winter.

Randeniya *et al.* (1996a, b) estimated (using a 2-D model) the impact of heterogeneous hydrolysis of $BrONO_2$ via sulphate aerosols in the case of the projected fleets of supersonic aircraft. The results show that addition of this reaction has a major effect, especially when heterogeneous hydrolysis of N_2O_5 has reached saturation in the regions where night length is short.

Schoeberl *et al.* (1998) obtained a Lagrangian estimate of aircraft effluent lifetime. Results indicate that effluent lifetime is larger in the stratosphere of the Southern Hemisphere than in the Northern Hemisphere, being of the order of 1 year. Emissions at 13 and 11-km levels rapidly leave the stratosphere. It is reasonable to assume that subsonic aviation makes only a small impact on the stratosphere. Estimates of effluent lifetime are very sensitive to the lower stratosphere-warming rate.

Vohralik *et al.* (1998) proved the possibility of using correlations between long-lived atmospheric species in assessment of ozone change because of the proposed fleet of supersonic aircraft or changing halocarbon concentrations. Large savings in computational time can be achieved by reducing the number of transported species using relevant correlations. A new computationally highly efficient method for on-line calculation of photolysis and heating rates, especially suited for coupled transport–chemistry models, was suggested by Landgraf and Crutzen (1998).

To assess the Space Shuttle's impact on the stratosphere, Jackman *et al.* (1996a) simulated a launch scenario of nine shuttles and three Titan rockets per year, using a

2-D photochemistry and transport model which includes heterogeneous reactions on a stratospheric sulphate aerosol (SSA) layer and polar stratospheric clouds (PSCs). As is well known, the launch of NASA's Space Shuttle and similar rockets results in injection of hydrogen chlorine, carbon monoxide, water vapour, molecular nitrogen and aluminium oxide directly into the stratosphere. Calculations indicate that maximum total inorganic chlorine enhancements are about 12 pptv (which is about 0.4% of a 3 ppbv background value) in the mid to upper stratosphere at northern mid to high latitudes. Maximum ozone decreases associated with these chlorine increases are about 0.14% in the mid to upper stratosphere at northern mid to high latitudes in the early spring. Simulations made under various assumptions gave the following mean annual global TOZ decreases: (1) 0.0056% (consideration of gas reactions only); (2) 0.010% (taking account of gas-phase reactions and heterogeneous reactions on the SSA layer); (3) 0.014% (the same as (2) plus heterogeneous reactions on PSCs). Jackman *et al.* (1996a) pointed out that stratospheric effects from heterogeneous reactions, promoted by alumina emitted from the rockets, could be larger than those predicted from chlorine emissions but need to be investigated further.

Volk *et al.* (1996) emphasized that ozone at mid-latitudes appears to be more sensitive to elevated levels of industrial chlorine than is currently predicted. Nevertheless, about 45% of the air in the tropical ascent region at 21 km is of mid-latitude origin, implying that emissions from supersonic aircraft could reach the mid-stratosphere.

Calculations made by Flatøy and Høv (1996), who applied a 3-D model to study the effect of NO_x emissions from subsonic aircraft on ozone in the upper troposphere over Europe and the North Atlantic, show that between 1 and 10 July 1991 over North Europe chemical ozone formation up to 1.8 ppbv/hr could occur, which is about a factor of 50 higher than the average for 10.5 km over continental Europe during the 10-day period. The high production rates are not sustained, however, for more than a few hours. Calculations show that 10–25 molecules of O_3 (0.03–0.05 ppt/h) are generated, on average, per NO_x molecule emitted from aircraft (2–3 ppt/h).

Using a 3-D model, Brasseur *et al.* (1996) predicted that in 2050 (with further growth in aviation) NO_x concentration at 10 km could increase by 30–60% at mid-latitudes, and the concentration of ozone could be enhanced by 7 and 2% in summer and winter, respectively.

So, one can make a general conclusion that, according to existing estimates, neither subsonic nor supersonic aviation or rocket launches are a serious danger for the stratospheric ozone layer.

3.7 SOLAR ACTIVITY IMPACT ON THE OZONE LAYER

The impact of solar activity on the atmosphere and climate is still a subject of controversy (Kondratyev, 1998). Zerefos *et al.* (1998) justly emphasized that although solar activity – ozone relations has been the subject of many recent papers, solar-induced variations are still not completely clarified, either qualitatively or quantitatively. There is, however, no doubt that solar irradiance variability influences the ozone layer.

The principal aspect of this problem is how solar irradiance changes in dependence on solar activity (SA) variation. Recent survey papers by Sklyarov (1994) and Kondratyev (1997) contain summaries of what has been achieved as a result of satellite solar 'constant' (SC) measurements. Crommelynck *et al.* (1994, 1996) discussed SC observations from the ATLAS mission. Kondratyev and Nikolsky (1995a, b) have considered spectral solar radiation observations made at the high mountain station in the Northern Caucasus. Lastovicka (1998) published a brief summary of the Second IAGA/IAMAS Workshop: Solar activity forcing of the middle atmosphere.

Satellite measurements, started in 1979, discovered mean annual SC relative variations during the 11-year solar cycle (data for cycles No. 21 and 22) within 0.1%, while 27-day variations (due to solar radiation) did not exceed 0.3%. Results of high mountain observations demonstrate the existence of energetically significant bursts of solar radiation in the UV and visible spectral regions, whose total contribution may reach $100\,Wm^{-2}$ (Kondratyev and Nikolsky, 1995b). It is important that impulse-like variations in solar radiation in the cm-wavelength range correlates with the water content in the atmosphere and presumably produces effects of molecule clustering, as well as stimulation of upper-level cloud formation.

So, some new possibilities have been revealed about the potential direct impact of extraterrestrial solar-spectrum distribution variations on the climate by the transformation of optical and radiative properties of the atmosphere that results from enhancing solar flares and radioemission bursts. Observational evidence reveals, in particular, the presence of intra-atmospheric processes of the state of water-vapour transformation directed and/or accelerated by solar activity impact.

On the basis of analysis of Nimbus-7 TOMS data, Chandra (1991) showed that solar UV-related change in total ozone content over a solar cycle is about 1.5%, which may be attributed to about 6% change in the solar UV flux near 200 nm. This estimate is also consistent with solar UV change in TOZ over a time scale of a 27-day rotation. In this later paper, Chandra and McPeters (1994) analysed the combined Nimbus-7 solar backscatterd UV and NOAA-II SBUV/2 data, covering a period of more than a solar cycle (about 11 years), to assess the response of ozone in the stratosphere to UV radiation. The principal conclusion of this study was that about 2 % change in TOZ and about 5–7% change in ozone-mixing ratio in the upper stratosphere (0.7 to 2 hPa) may be attributed to the change in solar UV flux over a solar cycle. A very important result of the analysis was that in the upper stratosphere, where photochemical processes are expected to play a major role, the observed solar-cycle variation in ozone is significantly larger than is inferred either by photochemical models or by the ozone response to the 27-day solar UV modulation. The relatively low observed sensitivity at the 10-hPa level could be caused by measurement error, since ozone measurements in this height range were seriously affected for 1–2 years by aerosol clouds injected into the atmosphere during the El Chichón and Mount Pinatubo eruptions. Causes of this disagreement are not clear as yet.

Vlachogiannis and Haigh (1998) analysed photon data from IMP-8 and GOES-7 satellites between 1989 and 1995 and studied the impact of solar proton events (SPE) on the stratosphere using a 2-D model. According to numerical modelling results, maximum O_3 depletion of more than 10% occurs over the sunlit southern polar region at altitudes between 30 and 48 km. Ozone depletion greater than 2% occurs down to latitudes ±25°. A maximum ozone decrease of 1% between 25 to 20 km, persisting for 4 years, at high latitudes was revealed. Vlachogiannis and Haigh (1998) believe that earlier estimates (e.g. Jackman *et al.*, 1996b) of larger ozone increases (~2%) in the lower stratosphere should be considered as overestimated. Since the O3 decrease occurs well above the maximum of ozone concentration, the change in TOZ does not exceed 1.7% in the spring of 1990. A general conclusion is that the impact of SPE on the structure of the lower stratosphere is small but can last more than 6 months.

The existence of large (although short-term) SC variation, discovered by Kondratyev and Nikolsky (1995a, b), demonstrates the possibility of direct impact of such variations on climate. Analysis of surface air temperature (SAT) observational data between the 12th and 15th 11-year cycles confirmed such an assumption: correlation between SAT and solar flares is very high. It must be emphasized, however, that the nature of solar activity impacts on climate is complex. A number of hypotheses concerning potential mechanisms can be suggested. Kondratyev and Nikolsky (1995a) believe that condensational and ozone hypotheses are most probable.

As far as the condensational mechanism is concerned it is probably closely related to corpuscular solar radiation which can deeply penetrate the atmosphere and even partly reaches the surface. The flux of galactic cosmic rays (GCR), which is continuously entering the atmosphere from space, loses a principal part of its energy in the lower stratosphere and upper troposphere, which results in enhancement of the ionization rate. According to balloon measurements, maximum ion concentration takes place within altitude range ~10–20 km in dependence on latitude, season and solar activity. In mid-latitudes the maximum ionization rate was observed at an altitude of about 12 km; that is, close to the tropopause. Ionization of the air in the upper troposphere stimulates the formation of cloud condensation nuclei (CCN) and thus cirrus clouds. Similar processes can also develop in the mid and lower troposphere (within altitude range 3–7 km). Since the intensity of GCR is modulated by solar activity (the magnitude of modulation may reach 30%) this leads to SA-induced variation in all relevant meteorological processes. Another potential factor about the impact on the troposphere is radioemission from active areas of the Sun, increases in which can lead to both enhancement of H_2O molecule clustering and, in the opposite direction, dissociation, depending on the frequency of radiobursts.

The ozone mechanism of solar activity impact on the atmosphere and climate is connected first of all with the influence of SA on ozone-layer dynamics and subsequent meteorological processes (Kondratyev and Nikolsky, 1995b). Various observed solar activities and other phenomena could be of importance in this respect:

- enhancement of the Sun's luminosity between 1908 and 1939 which had reached 0.3–0.4% by 1938;
- growth in magnetic activity of the Sun between 1918 and 1957–1958;
- increasing trend of the magnetic activity index K_p (from 1933 till 1957–1958) which is characterized by disturbances of both the interplanetary magnetic field (IMF) and magnetic field of the Earth;
- accumulation in the atmosphere of the Northern Hemisphere of a great amount of nitrogen oxides because of nuclear tests between 1954 and 1962;
- a number of powerful solar proton bursts (they could lead to total ozone depletion up to 30%);
- intermittent explosive volcanic eruptions (Kondratyev and Varotsos, 1995b).

All the phenomena mentioned could produce substantial impacts on the ozone layer and climate. An important attempt of numerical modelling of the SA impact on climate was undertaken by Haigh (1994). A principal conclusion of numerical modelling is that the SA impact is critically dependent on the variability of solar radiation spectral composition. Rasch *et al.* (1995) recently developed a 3-D circulation model with coupled chemistry for the mid atmosphere. Analysis of the 2-year simulation results revealed that latitude variation and seasonal evolution of the ozone column abundance is quite realistic. This model may be used for assessment of solar activity impact on the ozone layer.

Summers *et al.* (1990) developed a 1-D photochemical model and a time-dependent heat equation which incorporates non-local thermodynamic equilibrium (non-LTE) IR radiative transfer to study the response of mesosphere ozone concentration to short-term solar UV flux variation.

Substantial progress in studying solar-related ozone analysis was marked by the XVIII Quadrennial Ozone Symposium on Atmospheric Ozone (12–21 September 1996 L'Aquila, Italy,; Bojkov and Visconti, 1998). Bjarnason *et al.* (1998) examined time series of the Mg II index covering more than 14 years to correlate its variations with changes in TOZ at the equator. Confident detection of solar signals in zonally averaged TOMS data was only found at solar rotation and solar-cycle periods. On the basis of analysis of UARS data (solar ultraviolet irradiance measured by the Solar–Stellar Irradiance Comparison Experiment – SOLSTICE, as well as mesospheric ozone and temperature measured by the Microwave Limb Sounder – MLS), de Toma *et al.*, (1998) discussed relevant time series for the major portion of the declining phase of solar cycle 22. It was discovered that in the lower and mid-mesosphere the ozone mixing ratio decreased about 3% and temperature decreased about 1.5 K and this was associated with the decrease in solar radiation and the resulting changes in photolysis of O_2 and O_3 during this period. It was shown that a 2-D photochemical – radiative – dynamic model using observed solar irradiance changes during the 3-year period (1991–1994) underestimates observed ozone and temperature variations by factors of 3–5 depending on altitude.

Using the NOAA-II daily spectral scan measurements of solar UV irradiance in the 160–405-nm region from February 1989 to October 1994, De Land *et al.* (1998) produced time series of absolute solar irradiance data at photochemically important wavelengths for the period mentioned. Results indicate a TOZ decrease of 1–2% and a decrease in 2-hPa ozone of 5–7% because of the long-term decrease in solar irradiance during the declining phase of solar cycle 22.

Zerefos *et al.* (1998) studied TOZ variations during the past three solar cycles using both groundbased and satellite observational data. The principal purpose was to identify and qualify the contribution of various mechanisms of TOZ variability because of natural factors, such as seasonal changes, QBO, ENSO and trend components (including solar-activity effects). A 2-D chemical/transport model (Isaksen and Stordal, 1986) was used to simulate ozone variability. Zerefos *et al.*, (1998) emphasized the crucial role of PSCs (revealed by model runs) and the rapid drop of ozone in the early 1990s, in large part due to the synergistic effect of a westerly QBO, long lasting ENSO event, enhanced ozone destruction related to volcanic aerosol and solar activity decline. Geographical distribution of solar activity effects depends on latitude, with significant positive ozone anomalies over tropical regions during disturbed solar conditions. Towards the next solar maximum an increase in TOZ is predicted, provided that there are no complications from PSCs and future volcanic eruptions.

The fact that total-ozone-content changes are much smaller than variations in forcing between 208 and 265 nm (from 3 to 5%) is due to a significant contribution to the formation of stratospheric ozone from radiation at wavelengths shorter than 208 nm. Contrary to this situation is the case of 27-day variability, where calculated 11-year change in total ozone content is in strong disagreement with measurement results. For example, in the photochemically controlled region of the upper stratosphere the model underestimates the ozone response by more than a factor of 2. The model does not simulate the negative temperature and ozone responses observed in the lower stratosphere. Results of temperature calculations indicate that temperature change over the solar cycle reaches approximately 1.4 K at the stratopause. It is relatively constant as a function of season in the tropics but varies significantly with season at 60°N.

Brasseur (1993) mentioned that calculated (by the model) variability in the zonal wind is generally lower than 1 ms-1 and hence is not expected to affect substantially the vertical propagation of planetary and gravity waves. As far as the cause of disagreement between observation and numerical modelling is concerned, it may be understood only on the basis of high-resolution 3-D general-circulation modelling with adequate simulation of chemical, dynamical and radiative processes.

3.8 REFERENCES

Aikin, A. C., McPeters, R. D. (1986) Meteoritic material and the behavior of upper stratospheric polar ozone. *Geophys. Res. Lett.*, **13**, N12, 1,300–1,303.

Aloyan, A. E., Makarenko, S. V. (1993) Dynamics and kinetics of anthropogenic gas constituents in the atmosphere. In: *Computational Systems and Processes*. Moscow, Nauka, N10, pp. 137–163 (in Russian).

Aloyan, A. E., Arutyunyan, V. O. (1999) Numerical modelling of ozone variability in the troposphere and stratosphere in response to anthropogenic emissions (in print) (in Russian).

Ambrizzi, T., Kayano, M. T., Stefenson, C. B. (1998) A comparison of global tropospheric teleconnections using observed satellite and general circulation model total ozone column data. *Climate Dynamics*, **14**, N2, 135–150.

Anderson, J. G., Toohey, D. W., Brune, W. H. (1991) Free radicals within the Antarctic vortex: The role of CFC's in Antarctic ozone loss. *Science*, **251**, N4985, 35–46.

Anderson, B. E., Cofer, W. R., Bagwell, D. R., Barrick, J. W., Hudgins, C. H., Brunke, K. E. (1998a) Airborne observations of aircraft emissions. 1. Total nonvolatile particle emission indices. *Geophys. Res. Lett.*, **25**, N10, 1,689–1,692.

Anderson, B. E., Cofer, W. R., Barrick, J. D, Bagwell, D. R., Hudgins, C. H (1998b) Airborne observations of aircraft aerosol emissions. 2. Factors controlling volatile particle production. *Geophys. Res. Lett.*, **25**, N10, 1,693–1,696.

Andronache, C., Chameides, W. L. (1998) Interactions between sulfur and soot emissions from aircraft and their role in contrail formation. 1. Development. *J. Geophys. Res.*, **103**, ND9, 10,787–10,802

Angell, J. K. (1986a) Environmental impact of volcanism. In: *Norman D. Watson Symposium*, University of Rhode Island, pp. 9–17.

Angell, J. K. (1986b) The close relation between antarctic total-ozone depletion and cooling of the antarctic low stratosphere. *Geophys. Res. Lett.*, **13**, N12, 1,240–1,243.

Angell, J. K. (1997) Estimated impact of Agung, El Chichón, and Pinatubo, volcanic eruptions on global and regional total ozone after adjustment for the quasi-biennial oscillation. *Geophys. Res. Lett.*, **24**, N6, 647–650.

Aranda, A., LeBras, G., LaVerdet, G., Poulet, G. (1997) The $BrO + CH_3O_2$ reaction: Kinetics and role in the atmospheric ozone budget. *Geophys. Res. Lett.*, **24**, N22, 2,745–2,748.

Arnold, F. (1997) Chemical perturbation of the lowermost stratosphere through exchange with the troposphere. *Geophys. Res. Lett.*, **24**, N5, 603–606.

Arnold, F., Wohlfrom, K.-H., Klemm, M. W., Schneider, J., Gollinger, K., Schumann, K., Busen, R. (1998) First gaseous ion composition measurements in the exhaust plume of a jet aircraft in flight: Implications for gaseous sulfuric acid, aerosols, and chemions. *Geophys. Res. Lett.*, **25**, N12, 2,137–2,140.

Austin, J., Remsberg, E. E., Jones, R. L., Tuck, A. F. (1986) Polar stratospheric clouds inferred from satellite data. *Geophys. Res. Lett.*, **13**, N12, 1,256–1,259.

Austin, J. (1991) On the explicit versus family solution of the fully diurnal photochemical equations of the stratosphere. *J. Geophys. Res.*, **96**, ND7, 12,941–12,974.

Austin, J. (1992) Toward the four dimensional assimilation of stratospheric chemical constituents. *J. Geophys. Res.*, **97**, ND2, 2,569–2,588.

Austin, J. (1998) Some ideas on the further development of a stratospheric ozone climatology. *Adv. Space Res.*, **21**, N10, 1,393–1,402.

Avallone, L. M., Prather, M. J. (1996) Photochemical evolution of ozone in the lower tropical stratosphere. *J. Geophys. Res.*, **101**, ND1, 1,457–1,462.

Balachandran, N. K., Rind, D. (1995) Modelling the effects of UV variability and the QBO on the troposphere–stratosphere system. Part I. The middle atmosphere. *J. Climate*, **8**, N8, 2,058–2,075.

Batchart, N., Austin, S. (1996) On the relationship between the quasi-biennial oscillation, total chlorine and severity of Antarctic ozone hole. *Quart. J. Roy. Meteorol. Soc.*, **122A**, N529, 183–218.

Bates, T. S., Huebert, B. J., Gras, J. L., Griffiths, F. B., Durkee, P. A. (1998) International Global Atmospheric Chemistry (IGAC) Project's First Aerosol Characterization Experiment (ACE-1): Overview. *J. Geophys. Res.*, **103**, ND13, 16,297–16,318.

Bauer, E. (1978) *A Catalog of Perturbing Influences on Stratospheric Ozone*, 1955–1975, *Institute for Defence Analysis Paper P-1340*, Report No. FAA-EQ-78-20, US Department of Transportation, Washington, DC, September 1978, XV, 55 pp., Appendices.

Baumgardner, D., Miake-Lye, R. C., Anderson, M. R., Brown, R. C. (1998) An evaluation of the temperature, water vapor, and vertical velocity structure of aircraft contrails. *J. Geophys. Res.*, **103**, ND8, 8,727–8,736.

Bekki, S., Pyle, J. A. (1993) Potential impact of combined NO_x and SO_x emissions from future high speed civil transport aircraft on stratospheric aerosol and ozone. *Geophys. Res. Lett.*, **20**, 723–726.

Bekki, S., Tuomi, R., Pyle, J. A. (1993) Role of sulphur photochemistry in tropical ozone changes after the eruption of Mount Pinatubo. *Nature*, **362**, 331–333.

Bekki, S. (1997) On the possible role of aircraft-generated soot in the middle latitude ozone depletion. *J. Geophys. Res.*, **102**, ND9, 10,751–10,758.

Berntsen, T. K., Isaksen, I. S. A. (1997) A global three-dimensional chemical transport model for the troposphere. A model description and CO and ozone results. *J. Geophys. Res.*, **102**, ND17, 21,239–21,280.

Bevilacqua, R. M., Aelig, C. P., Debrestian, D. J., Fromm, M. D., Hoppel, K., Lumpe, J. D., Shettle, E. P., Hornstein, J. S., Randall, C. E., Rusch, D. W., Rosenfield, J. E. (1997) POAM II ozone observations in the Antarctic ozone hole in 1994, 1995, and 1996. *J. Geophys. Res.*, **102**, ND19, 23,643–23,657.

Bird, J. C., Pal, S. R., Carswell, A. L., Donovan, D. P., Manney, G. L., Harris, J. M., Uchino, O. (1997) Observations of ozone structures in the Arctic polar vortex. *J. Geophys. Res.*, **101**, ND9, 10,785–10,800.

Bjarnason, G. G., Rögnvaldsson, Ö. E., Erlingsson, B. (1998) Solar induced variations in equatorial total ozone. In: *Atmospheric Ozone*, R. D. Bojkov, G. Visconti (eds). Parco Scientifico e Technologico, Abruzzo, Italy, Vol. 1, 247–250.

Blavier, J.-F. (1997) Balloon-borne measurements of stratospheric radicals and their precursors: Implications for the production and loss of ozone. *Geophys. Res. Lett.*, **24**, N9, 1,107–1,110.

Bojkov, R. D., Reinsel, G. C. (1985) Trends in tropospheric ozone concentrations. In: *Atmospheric Ozone*, D. Reidel, Dordrecht, pp. 775–781.

Bojkov, R. D. (1986) The 1979-1985 ozone decline in the Antarctic as reflected in ground based observations. *Geophys. Res. Lett.*, **13**, N12, 1,236–1,239.

Bojkov, R. (1987) The 1983 and 1984 anomalies in ozone distribution in perspective. *Mon. Weather Rev.*, **115**, 2,187–2,201.

Bojkov, R. D. (1988) Ozone variations in the Northern Polar region. *Meteorol. Atmos. Phys.*, **38**, 117–130.

Bojkov, R. D., Fioletov, V. E., Shalamjansky, A. M. (1994) Total ozone changes over Eurasia since 1973 based on reevaluated filter ozonemeter data. *J. Geophys. Res.*, **99**, 22,985–22,995.

Bojkov, R. D., Fioletov, V. E. (1995a) Estimating the global ozone characteristics during the last 30 years. *J. Geophys. Res.*, **100**, 16,537–16,551.

Bojkov, R. D., Bishop, L., Fioletov, V. E. (1995b) Total ozone trends from quality controlled ground-based data (1964–1994). *J. Geophys. Res.*, **100**, 25,867–25,876.

Bojkov, R., Visconti, G. (eds.) (1998a) *Atmospheric Ozone. Proceedings of the XVIII Quadrennial Ozone Symposium.* 12–21 September 1996, L'Aquila, Italy. Parco Scientifico e Technologico, Abruzzo, Italy, Vol. 1, Vol. 2, 1,018 pp.

Bojkov, R. D., Balis, D. S., Zerefos, C. S., (1998b) Characteristics of the ozone decline in the Northern polar and middle latitudes during the winter–spring. *Meteorol. Atmos. Phys.*, **69**, 119–135.

Bojkov, R. D. (1999) International assessment of ozone depletion: 1998. *WMO Bull.*, **48**, N1, 35–44.

Bond, T. C., Charlson, R. J., Heintzenberg, J. (1998) Quantifying the emission of light-absorbing particles: Measurements tailored to climate studies. *Geophys. Res. Lett.*, **25**, N3, 337–340.

Borrmann, S., Solomon, S., Avallone, L., Toohey, D., Baumgardner, D. (1997) On the occurrence of ClO in cirrus and volcanic aerosol in the tropopause region. *Geophys. Res. Lett.*, **24**, N16, 2,011–2,014.

Bowman, K. P. (1986) Interannual variability of total ozone during the breakdown of the antarctic circumpolar vortex. *Geophys. Res. Lett.*, **13**, N12, 1,193–1,196.

Brasseur, G. (1987) The endangered ozone layer: New theories on ozone depletion. *Environment*, **29**, N1, 6–11, 39–45.

Brasseur, G. (1992a) Natural and anthropogenic perturbations of the stratospheric ozone layer. *Planet. Space Sci.*, **40**, N2/3, 403–412.

Brasseur, G. (1992b) Volcanic aerosols implicated. *Nature,* **359**, N6,393, 275–276.

Brasseur, G. (1993) The response of the middle atmosphere to long-term and short-term solar variability: A two dimensional model. *J. Geophys. Res.*, **98**, 23,079–23,090.

Brasseur, G. P., Müller, J.-F., Granier, C. (1996) Atmospheric impact of NO_x emissions by subsonic aircraft: A three-dimensional model study. *J. Geophys. Res.*, **101**, ND1, 1,423–1,428.

Brasseur, G. P., Tie, X. X., Rasch, P. S., Lefèvre, F. (1997) A three-dimensional simulation of the Antarctic ozone hole: Impact of anthropogenic chlorine on the lower stratosphere and upper troposphere. *J. Geophys. Res.*, **102**, ND7, 8,909–8,930.

Brasseur, G., Amanatidis, G. T., Angeletti, G. (eds.) (1998) European scientific assessment of the atmospheric effects of aircraft emissions. *Atmos. Environ.*, **32**, N13, 2,327–2,422.

Bregman, A., van den Broek, M., Carslaw, K. S., Müller, R., Peter, T., Scheele, M. P., Lelieveld, S. (1997) Ozone depletion in the late winter lower Arctic stratosphere: Observations and model results. *J. Geophys. Res.*, **102**, ND9, 10,815–10,828.

Brown, R. C. (1997) Aircraft sulphur emissions and the formation of visible contrails. *Geophys. Res. Lett.*, **24**, N4, 385–388.

Brühl, Ch., Crutzen, P. S., Grooß, J.-U. (1998) High-altitude, summer time NO_x activation and seasonal ozone decline in the lower stratosphere: Model calculations based on observations by HALOE on UARS. *J. Geophys. Res.*, **103**, ND3, 3,587–3,598.

Butchart, N., Austin, J. (1996) On the relationship between the quasi-biannual oscillation, total chlorine and the severity of the Antarctic ozone hole. *Quart. J. Roy. Meteorol. Soc.*, **122**, Part A, N529, 183–218.

Butler, J. H., Montzka, S. A., Clarke, A. D., Lobert, J. M., Elkins, J. W. (1998) Growth and distribution of halons in the atmosphere. *J. Geophys. Res.*, **103**, ND1, 1,503–1,511.

Cacciani, M., Fiocco, G., Colagrande, P., Di Girolamo, P., di Sarra, A., Fuà, D., (1997a) Lidar observations of polar stratospheric clouds at the South Pole. 1. Stratospheric unpertubed conditions, 1990. *J. Geophys. Res.*, **102**, ND11, 12,937–12,944.

Cacciani, M., Colagrande, P., di Sarra, A., Fuà, D., Di Girolamo, P., Fiocco, G. (1997b) Lidar observations of polar stratospheric clouds at the South Pole. 2. Stratospheric perturbed conditions, 1992 and 1993. *J. Geophys. Res.*, **102**, ND11, 12,945–12,956.

Callis, L. B. (1997) Ozone variability in the high-latitude summer stratosphere. *Geophys. Res. Lett.*, **24**, N10, 1,151–1,154.

Campos, T. L., Weinheimer, A. J., Zheng, J., Montzka, D. D., Walega, J. G., Grahek, F. E., Vay, S. A., Collins, J. E., Jr., Wade, L. O., Sachse, G. W., Anderson, B. E., Brune, W. H., Tan, D., Faloona, I., Baughcam, S. L., Ridley, B. A. (1998) Measurements of NO and NOx emissions indices during SUCCESS. *Geophys. Res. Lett.*, **25**, N10, 1,713–1,716.

Cariolle, D., Déqué, M., Morcrette, J.-J. (1986) A GCM simulation of the ozone seasonal variations at high latitudes in the Southern Hemisphere. *Geophys. Res. Lett.*, **13**, N12, 1,304–1,307.

Carslaw, K. S. (1997) Uncertainties in reactive uptake coefficients for solid stratospheric particles. 2. Effect on ozone depletion. *Geophys. Res. Lett.*, **24**, N14, 1,447–1,450.

Carslaw, K. S., Worth, M., Tsias, A., Luo, B. P., Dörnbrack, A., Lewtbacher, M., Volkert, H., Renger, W., Bacmeister, J. T., Peter, T. (1998) Particle microphysics and chemistry in remotely observed mountain polar stratospheric clouds. *J. Geophys. Res.*, **103**, ND5, 5,785–5,796.

Chandra, S. (1986) The solar and dynamically induced oscillations in the stratosphere. *J. Geophys. Res.*, **91**, 2,715–2,734.

Chandra, S., McPeters, R. D. (1986) Some observations on the role of planetary waves in determining the spring time ozone distribution in the Antarctic. *Geophys. Res. Lett.*, **13**, N12, 1,224–1,227.

Chandra, S. (1991) The solar UV related changes in total ozone from a solar rotation to a solar cycle. *Geophys. Res. Lett.*, **18**, 837–840.

Chandra, S. (1993) Changes in stratospheric ozone and temperature due to the eruption of the Mt. Pinatubo. *Geophys. Res. Lett.*, **20**, N1, 33–36.

Chandra, S., Jackman, Ch. K., Douglass, A. R., Fleming, E. L., and Contidine, D. B. (1993) Chlorine catalyzed destruction of ozone: Implications for variability in the upper stratosphere. *Geophys. Res. Lett.*, **20**, N5, 351–354.

Chandra, S., McPeters, R. D. (1994) The solar cycle variation of ozone in the stratosphere inferred from Nimbus-7 and NOAA-11 satellites. *J. Geophys. Res.*, **99**, N10, 20,665–20,671.

Chandra, S., Lean, J. L., White, O. R., Prinz, O. K., Rottman, G. S., Brueckner, G. E. (1995) Solar UV irradiance variability during the declining phase of the solar cycle 22. *Geophys. Res. Lett.*, **22**, N18, 2,481–2,484.

Chipperfield, M. P., Lee, A. M., Pyle, J. A. (1996) Model calculations of ozone depletion in the Arctic polar vortex for 1991/92 to 1994/95. *Geophys. Res. Lett.*, **23**, N5, 559–562.

Chlong, A. (1998) Large-eddy simulation of contrails. *J. Atmos. Sci.*, **55**, 796–819.

Cicerone, R. J. (1987) Changes in stratospheric ozone. *Science*, **237**, 35–42.

Considine, D. B., Dessler, A. E., Jackman, C. M., Rosenfield, J. E., Meade, P. E., Schoeberl, M. R., Roche, A. E., Waters, J. W. (1998) Interhemispheric asymmetry in the 1 mbar O_3 trend: An analysis using an interactive zonal mean model and UARS data. *J. Geophys. Res.*, **103**, ND1, 1,607–1,618.

Cordero, E. C., Nathan, T. R., Echols, R. S. (1998) An analytical study of ozone feedbacks on Kelvin and Rossby-gravity waves: Effects on the QBO. *J. Atmos. Sci.*, **55**, N6, 1,051–1,062.

Crommelynck, D., Domingo, V., Fichot, A., Lee, R. B., III (1994) Irradiance observations from the EURECA and ATLAS experiments. In: *The Sun as a Variable Star: Solar and Stellar Irradiance Variations*, J. M., Pap, C., Fröhlich, H. S., Hudson, S. R., Solanki, (eds.), Cambridge University Press.

Crommelynck, D., Fichot, A., Domingo, V., Lee, R. B., III (1996) SOLCON solar constant observations from ATLAS missions. *Geophys. Res. Lett.*, **23**, N17, 2,923–2,926.

Cronn, D. R., Ramesberger, W. L., Menzia, F. A., Waylett, S. F., Ferrara, T. W., Howard, H. M., Robinson, E. (1986) Atmospheric trace gas trends at Palmer station, Antarctica. *Geophys. Res. Lett.*, **13**, N12, 1,272–1,275.

Crutzen, P. J. (1970) The influence of nitrogen oxides on the atmospheric ozone content. *Quart. J. Roy. Meteorol. Soc.*, **96**, 320–325.

Crutzen, P. J. (1971) Ozone production rates in an oxygen, hydrogen, nitrogen oxide atmosphere. *J. Geophys. Res.*, **76**, 7,311–7,327.

Crutzen, P. J., Schmailzl, K. (1983) Chemical budgets of the stratosphere. *Planet. Space Sci.*, **31**, 1,009–1,032.

Crutzen, P. J., Brühl, C. H., Schmailzl, U., Arnold, F. (1988) Nitric acid haze formation in the lower stratosphere: a major contribution factor to the development of the Antarctic 'ozone hole'. In: *Aerosols and Climate*, M. P. McCormick and P. V. Hobbs (eds), A. Deepak Publ., Hampton, VA, pp. 287–304.

Crutzen, P. J., Muller, R., Brühl, C. H., Peter, T. (1992) On the potential importance of the gas phase reaction $CH_3O_2 + ClO \longrightarrow ClOO + CH_3O$ and the heterogeneous reaction $HOCl + HCl \longrightarrow H_2O + Cl_2$ in 'Ozone Hole' chemistry. *Geophys. Res. Lett.*, **19**, N11, 1,113–1,116.

D'Altorio, A., Masci, F., Pitari, G., Visconti, G., Rizi, V., Cervino, M., Giovanelly, G. (1992) Ground-based monitoring of Pinatubo aerosols and ozone at L'Aquila, Italy. 1. Evidence for ozone depletion during 1991/1992 winter months. *Il Nuovo Cimento*, **16C**, N1, 91–95.

Dameris, M., Grewe, V., Ponater, M., Sausen, R., Brühl, C., Steil, B. (1998a) An interactively coupled comprehensive model of atmospheric dynamics and chemistry: First results of multi-decadal integrations. In: *Atmospheric Ozone*, R. D. Bojkov, G. Visconti (eds.), Springer-Verlag, Berlin, pp. 695–698.

Dameris, M., Grewe, V., Kohler, I., Sausen, R., Brühl, C., Groos, J. U., Steil, B. (1998b) Impact of aircraft NOx emissions on tropospheric and stratospheric ozone. Part II. 3-D model results. *Atmospheric Environment*, **32**, N18, 3,185–3,199.

Danilin, M. J., McConnell, J. C. (1995) Stratospheric effects of bromine activation on/in sulphate aerosos. *J. Geophys. Res.*, **100**, 11,237–11,243

Danilin, M. Y., Rodriguez, J. M., Ko, M. K. W., Weisenstein, O. K., Brown, R. C., Miake-Lye, R. C., Anderson, M. R. (1997) Aerosol particle evolution in an aircraft wake: Implications for the high-speed civil transport fleet impact on ozone. *J. Geophys. Res.*, **102**, ND17, 21,453–21,464.

Danilin, M. Y., Sze, N.-D., Ko, M. K. W., Rodriques, J. M., Tabasadeh, A. (1998) Stratospheric cooling and Arctic ozone recovery. *Geophys. Res. Lett.*, **25**, N12, 2,141–2,144.

Dave, J. V., DeLuisi, J. J., Mateer, C. L. (1980) Results of a comprehensive theoretical examination of the optical effects of aerosols on the Umkehr measurements. *WMO Technical Conference on Regular and Global Observation of Atmospheric Pollution Relative to Climate*, Special Environmental Report., **No. 14**, pp. 15–22.

DeLand, M. T., Cebula, R. P., Hilsenrath, E. (1998) Solar UV contribution to stratospheric ozone variations 1989–1994. In: *Atmospheric Ozone*, R. D. Bojkov, G. Visconti (eds.), Parco Scientifico e Tecnologico, Abruzzo, Italy, Vol. 1, pp. 263–266.

Del Negro, L. A., Fahey, D. W., Donelly, S. G., Gao, R. S., Keim, E. R., Wamsley, R. E., Woodbridge, E. L., Dye, J. E., Baumgardner, D., Gandrud, B. W., Wilson, J. C., Jonsson, M. M., Loewenstein, M., Podolske, J. R., Webster, C. R., May, R. D., Worsnop, D. R., Tabazadeh, A., Tolbert, M. A, Kelly, K. K., Chan, K. R. (1997) Evaluating the role of NAT, NAD, and liquid $H_2SO_4/H_2O/HNO_3$ solutions in Antarctic polar stratospheric cloud aerosol: Observations and implications *J. Geophys. Res.*, **102**, ND11, 13,255–13,282.

DeLuisi, J. J. (1969) A study of the effect of haze upon Umkehr measurements. *Quart. J. Roy. Meteorol. Soc.*, **95**, 181–187.

DeLuisi, J. J. (1979) Umkehr ozone profile errors caused by the presence of stratospheric aerosols. *J. Geophys. Res.*, **84**, 1,766–1,770.

DeLuisi, J. J., Herman, B. M., Browning, S. R., Sato, R. K. (1975) Theoretically determined multi-scattering effects of dust on Umkehr observations. *Quart. J. Roy. Meteorol. Soc.*, **101**, 325–331.

DeLuisi, J. J., Mateer, C. L., Komhyr, W.D., (1984) Effects of El Chichón stratospheric aerosol cloud on Umkehr measurements at Mauna Loa, Hawaii. *Proceedings of the Quadrennial Ozone Symposium, 3–7 September 1984, Halkidiki, Greece*, D. Reidel, Dordrecht, 842pp.

De Mott, P. J., Rogers, D. C., Kreidenweis, S. M., Chen, Y., Twohy, C. M., Baumgardner, D., Heimsfield, A. S., Chan, K. R. (1998) The role of heterogeneous freezing nucleation in upper tropospheric clouds: Inferences from SUCCESS. *Geophys. Res. Lett.*, **25**, N9, 1,387–1,390.

De Rudder, A., Larsen, N., Tie, X., Brasseur, G. P., Granier, C. (1996) Model study of polar stratospheric clouds and their effect on stratospheric ozone. 1. Model description. *J. Geophys. Res.*, **103**, ND7, 12,567–12,574.

Deniel, C., Bevilacqua, R. M., Pommereau, J. P., Lefèvre, F. (1998) Arctic chemical ozone depletion during the 1994–1995 winter deduced from POAM II satellite observations and the REPROBUS three-dimensional model. *J. Geophys. Res.*, **103**, ND15, 19,231–19,244.

Deshler, T., Johnson, B. J., Rozier, W. R. (1993) Balloon-borne measurements of Pinatubo aerosol during 1991 and 1992 at 41°N: Vertical profiles, size distribution, and volatility. *Geophys. Res. Lett.*, **20**, N14, 1,435–1,438.

Dessler, A. E., Kawa, S. R., Douglass, A. R., Considine, D. B., Kumer, J. B., Roche, A. E., Mergenthaler, J. L., Waters, J. M., Russel, J. M., III, Gille, J. E. (1996) A test of the partitioning between ClO and $ClONO_2$ using simultaneous UARS measurements of ClO, NO_2, and $ClONO_2$. *J. Geophys. Res.*, **101**, ND7, 12,515–12,521.

De Zafra, R. L., Jarmillo, M., Parrish, A., Solomon, P., Connor, B., Berrett, J. (1987) High concentrations of chlorine monoxide at low altitudes in the Antarctic spring stratosphere: Diurnal variation. *Nature*, **328**, N6,129, 408–411.

Dibb, J. E., Talbot, R. W., Loomis, M. B. (1998) Tropospheric sulphate distribution during SUCCESS: Contributions from jet exhaust and surface sources. *Geophys. Res. Lett.*, **25**, N9, 1,375-1,378.

Douglass, A. R., Rood, R. B. (1986) Derivation of photochemical information near 1 mbar from ozone and temperature data. *J. Geophys. Res.*, D**91**, ND12, 13,153–13,166.

Douglass, A. R., Root, R. B., Kawa, S. R., Allen, D. J. (1997) A three-dimensional simulation of the evolution of the middle latitude winter ozone in the middle stratosphere. *J. Geophys. Res.*, **102**, ND15, 19,217–19,232.

Dubey, M. K., Smith, G. P., Harley, W. S., Kinnison, D. E., Connell, P. S. (1997) Rate parameter uncertainty effects in assessing stratospheric ozone depletion by supersonic aviation. *Geophys. Res. Lett.*, **24**, N22, 2,737–2,740.

Duncan, J. L., Schindler, L. R., Roberts, J. T. (1998) A new sulfate-mediated reaction: Conversion of acetone to trimethylbenzene in the presence of liquid sulfuric acid. *Geophys. Res. Lett.*, **25**, N5, 631–634.

Dutton, E. G., Hofmann, D. J., Oltman, S. J., Christy, J. R. (1993) Radiative transfer calculations of observed decreases in stratospheric temperatures and ozone beginning 1.5 years after the Eruption of Mt. Pinatubo. *American Geophysical Union Fall Meeting*, 6–10 December, *San Francisco*, Paper No. A12F-06, Abstract published in 26 October, supplement to *EOS*.

Dye, J. E., Baumgardner, D., Gandrud, B. W., Drdla, K., Barr, K., Fahey, D. W., Delnegro, L. A., Tabazadeh, A., Jonsson, H. M., Wilson, S. E., Loewestein, M., Podolske, J. R., Chan, K. R. (1996) In-situ observations of the Antarctic polar stratospheric cloud: Similarities with Arctic observations. *Geophys. Res. Lett.*, **23**, N15, 1,913–1,916.

Ellsaesser, H. W. (1994) The unheard arguments: A rational view of stratospheric ozone. *21st Century Science and Technology*, **7**, N3, 37–45.

Evans, W. F. L., Boville, B. W., McConnell, J. C., Henderson, G. S. (1986) The simulation of antarctic ozone with chemical and dynamical effects. *Geophys. Res. Lett.*, **13**, N12, 1,323–1,326.

Evans, S. J., Toumi, R., Harries, J. E., Chipperfield, M. P., Russel, J. M., III. (1998) Trends in stratospheric humidity and the sensitivity of ozone to the trends. *J. Geophys. Res.*, **103**, ND8, 8,715–8,726.

Farman, J. C. (1987) Recent measurements of total ozone at British Antarctic Survey station. *Phil. Trans. Roy. Soc. London*, **323**, 625–644.

Farman, J. C., Gardiner, B. G., Shanklin, J. D. (1985) Large losses of total ozone in Antarctica reveal seasonal ClO/NO_x interaction. *Nature*, **315**, 207–210.

Fish, D. J., Burton, M. R. (1997) The effect of uncertainties in kinetic and photochemical data on model predictions of stratospheric ozone depletion. *J. Geophys. Res.*, **102**, ND21, 25,537–25,542.

Flatøy, F., Høv, Ø. (1996) Three dimensional model studies of the effects of NO_x emissions from aircraft on ozone in the upper troposphere over Europe and North Atlantic. *J. Geophys. Res.*, **101**, ND1, 1,401–1,422.

Flatøy, F., Høv, Ø., Smit, H. (1996) Three-dimensional model studies of exchange processes of ozone in the troposphere over Europe. *J. Geophys. Res.*, **101**, ND6, 11,465–11,481.

Fleming, E. L., Chandra, S., Jackman, C. H., Considine, D. B., Douglass, A. R. (1995) The middle atmospheric response to short and long term solar UV variations: Analysis of observations and 2-D model results. *J. Atmos. Terrest. Phys.*, **57**, 333–365.

Forster, P. M. de F., Shine, K. P. (1997) Radiative forcing and temperature trends from stratospheric ozone changes. *J. Geophys. Res.*, **102**, 10,841–10,855.

Frölich, C. (1993) Changes of total solar irradiance. *Geophys. Monogr.*, **75**, *IUGG*, **15**, Washington, D.C., pp. 123–129.

Gavrilov, A. A., Kaidalov, O. V. (1990) A numerical model of atmospheric tides at 70–100 km altitude and its comparison with observation data. *Geomagnetism and Aeronomy*, **30**, N3, 474–481.

Gavrilov, A. A., Kaidalov, O. V. (1996) Investigation of the interaction between variations in atmospheric thermal tides and anomalous ozone concentration. *Adv. Space Res.*, **17**, N11, 157–160 (in Russian).

Geller, M. A. (1998) Presentation of SPARC at the WCRP Conference. SPARC future prospects. *SPARC Newsletter*, N10, 6–10.

Gettelman, A. (1998) The evolution of aircraft emission in the stratosphere. *Geophys. Res. Lett.*, **25**, N12, 2,129–2,132.

Gille, G. C., Smythe, C. M., Heath, D. F. (1984) Observed ozone response to variation in solar ultraviolet radiation. *Science*, **225**, 315–317.

Glatthor, N., Blom, C. E., von Clarmann, T., Fischer, H., Gulde, T., Piesch, C., Erle, F., Pfeilsticker, K., Chipperfield, M. P., Lee,. A. M., Pyle, J. A. (1998) Airborne remote sensing of NO_2 in the Arctic winter of 1994-1995 and comparison with three-dimensional chemical transport model. *J. Geophys. Res.*, **103**, ND11, 13,315–13,326.

Goodman, J. (1997) New evidence of size and composition of polar stratospheric cloud particles. *Geophys. Res. Lett.*, **24**, N5, 615–618.

Grant, W. B., Fishman, J., Browell, E. V., Brackett, V. G., Nganga, D., Minga, A., Cros, B., Veiga, R. E., Butler, C. F., Fenn, M. A., Nowicki, G. D (1992) Observations of reduced ozone concentrations in the tropical stratosphere after the eruption of Mt. Pinatubo. *Geophys. Res. Lett.*, **19**, N11, 1,109–1,112.

Grewe, D. *et al.* (1997) Heterogeneous PSC ozone loss during an ozone minihole. *Geophys. Res. Lett.*, **24**, N20, 2,503–2,506.

Grewe, V., Dameris, M., Sausen, R., Steil, B. (1998) Impact of stratospheric dynamics and chemistry on northern hemisphere midlatitude ozone loss. *J. Geophys. Res.*, **103**, ND19, 25,417–25,432.

Grooss, J.-M., Pierce, R. B., Crutzen, P. J., Grose, W. L., Russel, J. M., III. (1997) Re-formation of chlorine reservoirs in southern hemisphere polar spring. *J. Geophys. Res.*, **102**, ND11, 13,141–13,152.

Groos, J. U., Brühl, C., Peter, T. (1998) Impact of aircraft emissions on tropospheric and stratospheric ozone. Part I. Chemistry and 2-D model results. *Atmospheric Environment*, **32**, N18, 3,173–3,184.

Groves, K. S., Mattingly, S. R., Tuck, A. F. (1978) Increased atmospheric carbon dioxide and stratospheric ozone. *Nature*, **273**, 711–715.

Haigh, J. D. (1994) The role of stratospheric ozone in modulating the solar radiative forcing of climate. *Nature*, **370**, N6,490, 544–546.

Hamill, P., Toon, O. B., Turco, R. P. (1986) Characteristics of polar stratospheric clouds during the formation of the antarctic ozone hole. *Geophys. Res. Lett.*, **13**, N12, 1,288–1,291.

Hamill, P., Tabazadel, A., Kinne, S., Toon, O. B., Turco, R. B. (1996) On the growth of ternary system $HNO_3/H_2SO_4/H_2O$ aerosol particles in the stratosphere. *Geophys. Res. Lett.*, **23**, N7, 753–756.

Hanning-Lee, M. A., Brady, B. B., Martin, L. B., Siage, J. A. (1996) Ozone decomposition on alumina: Implications for solid rocket motor exhaust. *Geophys. Res. Lett.*, **23**, N15, 1,961–1,964.

Hartman, D. L., Barkstrom, B. R., Crommelynck, D., Foukal, P., Hansen, J. E., Lean, J., Lee, R. B., III, Schoeberg, M. R., Wilson, R. C. (1993) The solar total irradiance monitoring. *The Earth Observer*, **5**, N6, 23–27.

Hendricks, J., Lippert, E., Petry, H., Ebel, A. (1999) Heterogeneous reactions on and in sulfate aerosols: Implications for the chemistry of the midlatitude tropopause region. *J. Geophys. Res.*, **104**, ND5, 5,531–5,550.

Hervig, M. E., Carslaw, K. S., Peter, T., Deshler, T., Gordley, L. L., Redaelli, G., Bierman, U., Russel, J. M., III. (1997) Polar stratospheric clouds due to vapor enhancement: HALOE observations of the Antarctic vortex in 1993. *J. Geophys. Res.*, **102**, ND23, 28,185–28,194.

Hervig, M. E., Deshler, T. (1998) Stratospheric aerosol surface area and volume inferred from HALOE, CLAES, and ILAS measurements. *J. Geophys. Res.*, **103**, ND19, 25,345–25,351.

Heymsfield, A. J., Miloshevich, L. M., Twohy, C., Sache, G., Oltmans, S. (1998) Upper tropospheric relative humidity observations and implications for cirrus ice nucleation. *Geophys. Res. Lett.*, **25**, N9, 1,343–1,346.

Hills, A. J., Cicerone, R. J., Calvert, J. G., Birks, J. W. (1987) Kinetics of the BrO + ClO reaction and implications for stratospheric ozone. *Nature*, **328**, N6,129, 405–408.

Hintsa, E. J., Newman, P.A., Jonsson, H. H., Webster, C. R., May, R. D., Herman, R. L., Schoeberl, M. R., Elkins, J. W., Wamsley, P. R., Dutton, G. S., Bui, D. P., Kohn, O. W., Anderson, J. G. (1998) Dehydration and denitrification in the Arctic polar vortex during the 1995–1996 winter. *Geophys. Res. Lett.*, **25**, N4, 501–504.

Hoell, J. M., Davis, D. D., Lin, S. C., Newell, R. E., Akimoto, H, McNeal, R. S., Bendura, R. J. (1997) The Pacific Exploratory Mission – West Phase B: February–March, 1994. *J. Geophys. Res.*, **102**, ND23, 28,223–28,239.

Hofmann, D. J., Solomon, S. (1985) Ozone destruction through heterogeneous chemistry. *Geophys. Res. Lett.*, **94**, 5,029–5,041.

Hofmann, D. J., Harder, J. W., Rolf, S. R., Rosen, J. M. (1987) Balloon-borne observations of the development and vertical structure of the Antarctic hole in 1986. *Nature*, **326**, N6,108, 59–62.

Hofmann, D. J. (1989) Direct ozone depletion in springtime Antarctic lower stratospheric clouds. *Nature*, **337**, N6,206, 447–449

Hofmann, D. J., Solomon, S. (1989) Ozone destruction through heterogeneous chemistry following the eruption of El Chichón. *J. Geophys. Res.*, **94**, N4, 5,029–5,041.

Hofmann, D. J. (1993) Comparing the volcanic signal in ozone depletion after El Chichón and Pinatubo eruptions. American Geophysical Union, AGU, 1993, Annual Meeting. *EOS Suppl.*, 114.

Hofmann, D. J., Oltman, S. J., Harwis, J. M. *et al.* (1993a) Ozonesonde measurements at Hilo, Hawaii following the eruption of Pinatubo. *Geophys. Res. Lett.*, **20**, N15, 1,555–1,557.

Hofmann, D. J., Oltman, S. J., Latrop, J. A., *et al.* (1993b) Unprecendence low ozone over United States during 1992 and 1993: Dobson and Ozonesonde data. American Geophysical Union, AGU 1993, Annual Meeting. *EOS Suppl.*, 138.

Hofmann, D. J., Oltmans, S. J., Komhyr, W. D., Harris, J. M., Lathrop, J. A., Langford, A. Q. Matthews, W. A. (1994) Ozone loss in the lower stratosphere over the United States in 1992–1993: Evidence for heterogeneous chemistry on the Pinatubo aerosol. *Geophys. Res. Lett.*, **21**, N1, 65–68.

Hofmann, D. J., Oltmans, S. J., Harris, J. M., Johson, B. J., Lathrop, J. A. (1997) Ten years of ozonesonde measurements of the South Pole: Implications for recovery of springtime Antarctic ozone hole. *J. Geophys. Res.*, **102**, ND7, 8,931–8,944.

Hofmann, D. J., Stone, R., Wood, M., Deshler, T., Harris, J. (1998) An analysis of 25 years of balloonborne aerosol data in search of a signature of the subsonic commercial aircraft fleet. *Geophys. Res. Lett.*, **25**, N13, 2,433–2,436.

Hood, L. L. (1987) Solar ultraviolet radiation – indirect variations in the stratosphere and mesosphere. *J. Geophys. Res.*, **92**, 876–888.

Hood, L. L., Jirikowic, J. L., McCormack, J. P. (1993) Quasi-decadal variability of the stratosphere: Influence of long-term solar ultraviolet variations. *J. Atmos. Scien.*, **50**, N24, 3,941–3,958.

Hood, L. L., Zhou, S. (1998) Stratospheric effect of 27-day solar ultraviolet variations: An analysis of UARS MCS ozone and temperature data. *J. Geophys. Res.*, **103**, ND3, 3,629–3,637.

Horel, J., Geisler, J. (1997) *Global Environmental Change: An Atmospheric Perspective*, John Wiley and Sons, New York, pp. 152.

Hoyt, O. V., Schatten, K. M. (1993) A discussion of plausible solar irradiance variations, 1700–1992. *J. Geophys. Res.*, **98**, NA11, 18,895–18,906.

Hunt, B. G. (1977) A simulation of the possible consequences of a volcanic eruption on the general circulation of the atmosphere. *Mon. Weather Rev.*, **103**, N3, 247–260.

Hurst, D. F., Bakwin, P. S., Elkins, J. W. (1998) Recent trends in the variability of halogenated trace gases over the United States. *J. Geophys. Res.*, **103**, ND19, 25,299–25,305.

Imre, D. G. (1997) Phase transformations in sulfuric acid aerosols: Implications for stratospheric ozone depletion. *Geophys. Res. Lett.*, **24**, N1, 69–72.

Isaksen, I. S. A., Stordal, F. (1986) Antarctic ozone depletion: 2-D model studies. *Geophys. Res. Lett.*, **13**, N12, 1,327–1,330.

Isaksen, I. S. A., Rognerud, B., Stordal, F., Coffey, M. T., Mankin, W. G. (1990) Studies of Arctic stratospheric ozone in a 2-D model including some effects of zonal asymmetries. *Geophys. Res. Lett.*, **17**, N4, 557–560.

Jackman, C. H., Nielsen, J. E., Allen, D. S., Cerniglia, M. C., McPeters, R. D., Douglass, A. R., Rood, R. B. (1993) The effects of the October 1989 solar proton events on the stratosphere as computed using a three-dimensional model. *Geophys. Res. Lett.*, **20**, 459–462.

Jackman, Ch. H., Considine, D. B., Fleming, E. L. (1996a) Space shuttle's impact on the stratosphere: An update. *J. Geophys. Res.*, **101**, ND7, 12,523–12,530.

Jackman, Ch. H., Fleming, E. L., Chandra, S., Considine, D. B., Rosenfield, J. E. (1996b) Past present and future modeled ozone trends with comparisons to observed trends. *J. Geophys. Res.*, **101**, ND22, 28,751–28,756.

Jaeglé, L., Webster, C. R., May, R. D., Scott, D. E., Stimpfle, R. M., Kohn, D. W., Wennberg, P. O., Hanisco, T. F., Cohen, R. C., Profitt, M. H., Kelley, K. K., Elkins, J., Baumgardner, D., Dye, J. E., Wilson, J. C., Pueschel, R. F., Chan, K. R., Salawitch, R. J., Tuck, A. F., Hovde, S. J., Yung, Y. L. (1997) Evolution and stoichiometry of heterogeneous processing in the Antarctic stratosphere. *J. Geophys. Res.*, **102**, ND11, 13,213–13,243.

Jaffe, D. A., Berntsen, T. K., Isaksen, I. S. A (1997) A global three-dimensional chemical transport model. 2. Nitrogen oxides and nonmethane hydrocarbon results. *J. Geophys. Res.*, **102**, ND17, 21,281–21,296.

Jäger, H., Wege, K. (1990) Stratospheric ozone depletion at Northern midlatitudes after major volcanic eruptions. *J. Atmos Chem.*, **10**, N3, 273–287.

James, P. M. (1998) An interhemispheric comparison of ozone minihole climatologies. *Geophys. Res. Lett.*, **25**, N3, 301–304.

Jensen, E. J. (1997) The potential impact of soot particles from aircraft exhaust on cirrus clouds. *Geophys. Res. Lett.*, **24**, N3, 245–252.

Jensen, E. J., Toon, O. B., Kinne, S., Sachse, G. W., Anderson, B. E., Chan, K. R., Twohy, C. H., Gandrud, B., Heymsfield, A., Miake-Lye, R. C. (1998) Environmental conditions required for contrail formation and persistence. *J. Geophys. Res.*, **103** , ND4, 3,929–3,936.

Jiang, Y., Yung, Y. L. (1996) Concentrations of tropospheric ozone from 1975 to 1992 over tropical Pacific South America from TOMS data. *Science*, **272**, N5262, 714–716.

Johnston, H. S. (1971) Reduction of stratospheric ozone by nitrogen oxide catalysts from supersonic transport exhaust. *Science*, **73**, 517–522.

Kane, R. P., Sahai, Y., Casiccia. C. (1998) Latitude dependence of the quasi-biennial oscillation and quasi-triennial oscillation observations of total ozone measured by TOMS. *J. Geophys. Res.*, **103**, ND7, 8,477–8,490.

Kao, C.-Y., Glatzmaier, G. A., Malone, R. C. (1990) Global three-dimensional simulations of ozone depletion under postwar conditions. *J. Geophys. Res.*, **95**, ND13, 22,495–22,512.

Kapitsa, A. P. (1996) Controversies in the theory of ozone hole formation. *Optics in the Atmos. and Ocean*, **9**, N9, 1,164–1,167 (in Russian).

Kärcher, B. (1996) Aircraft-generated aerosols and visible contrails. *Geophys. Res. Lett.*, **23**, N15, 1,933–1,936.

Kärcher, B., Hirschberg, M. M., Fabian, P. (1996) Small-scale chemical evolution of aircraft exhaust species at cruising altitudes. *J. Geophys. Res.*, **101**, ND10, 15,169–15,190.

Kärcher, B. (1998) Physicochemistry of aircraft-generated liquid aerosols, soot, and ice particles. 1. Models description. *J. Geophys. Res.*, **103**, ND14, 17,111–17,128.

Kärcher, B., Busen, R., Petzold, A., Schröder, F. P., Schumann, U., Jensen, E. J. (1998) Physicochemistry of aircraft-generated liquid aerosols, soot, and ice particles. 2. Comparison with observations and sensitivity studies. *J. Geophys. Res.*, **103**, ND14, 17,129–17,147.

Kerr, J. G., Wardle, D. I., Tarasick, D. W. (1993) Record low ozone values over Canada in early 1993. *Geophys. Res. Lett.*, **20**, N13, 1,975–1,982.

Keys, J. G., Johnston, P. V. (1986) Stratospheric NO_2 and O_3 in Antarctica: Dynamic and chemically controlled variations. *Geophys. Res. Lett.*, **13**, N12, 1,260–1,263.

Khosravi, R., Brasseur, G. P., Smith, A. K., Rusch, D. W., Waters, J. W., Russell, J. M., III (1998) Significant reduction in the stratospheric ozone deficit using a three-dimensional model constrained with UARS data. *J. Geophys. Res.*, **103**, ND13, 16,203–16,220.

Khvorostyanov, V., Sassen, K. (1998) Cloud model simulation of a contrail case study: Surface cooling against upper tropospheric warming. *Geophys. Res. Lett.*, **25**, N13, 2,145–2,148.

Kilbane-Dawe, I., Harris, N. (1998) Report on the Fourth European Symposium on Stratospheric Ozone Research. *SPARC Newsletter*, N10, 25–26.

Kinne, S., Toon, O. B., Prather, M. J. (1992) Buffering of stratospheric circulation by changing amounts of tropical ozone: A Pinatubo case study. *Geophys. Res. Lett.*, **19**, N19, 1,927–1,930.

Kinnersley, J. S. (1996) The climatology of the stratospheric 'THIN AIR' model. *Quart. J. Roy. Meteorol. Soc.*, **122**, Part A, N529, 215–253.

Kinnersley, J. S., Tung, K.-K. (1998) Modeling the global interannual variability of ozone due to equatorial QBO and to extratropical planetary wave variability. *J. Atmos. Sci.*, **55**, N7, 1,417–1,428.

Kirchhoff, V. W. J. H., Casiccia, C. A. R. S., Zamorano, F. B. (1997) The ozone hole over Punta Arenas, Chile. *J. Geophys. Res.*, **102**, ND7, 8,945–8,954.

Klonecki, A., Levy, H., II (1997) Tropospheric ozone tendencies in CO-CH_4-NO_y-H_2O system: Their sensitivity to variations in environmental parameters and their application to a global chemistry transport model study *J. Geophys. Res.*, **102**, ND17, 12,221–12,237.

Knight, J. R., Austin, J., Grainger, R. G., Lambert, A. (1998) A three-dimensional model simulation of the impact of Mt. Pinatubo aerosol on the Antarctic ozone hole. *Quart. J. Roy. Meteorol. Soc.*, **124**, Part A, N549, 1,527–1,558.

Knudsen, B. M., Larsen, N., Mikkelsen, I. S., Morcrette, J.-J., Braathen, G. O., Kyrö, E., Fast, H., Gernandt, H., Kanzawa, H., Nakane, H., Dorokhov, V., Yushkov, V., Hansen, G., Gil, M., Shearman, R. J. (1998) Ozone depletion in and below the Arctic vortex for 1997. *Geophys. Res. Lett.*, **25**, N5, 627–630.

Komhyr, W. D., Oltman, S. J., Chopra, A. N., Leonard, R. K., Garcia, T. E., McFee, C. (1984) Results of Umkehr, ozonesonde, total ozone and SO_2 observations in Hawaii following the eruption of El Chichón in 1982. *Proeedings of the Quadrennial Ozone Symposium, 3–7 September 1984, Halkidiki, Greece*, D. Reidel, Dordrecht, 842 pp.

Komhyr, W. D., Grass, R. D., Leonard, R. K. (1986) Total ozone decrease at South Pole, Antarctica 1964–1985. *Geophys. Res. Lett.*, **13**, N12, 1,248–1,251.

Komhyr, W. D., Grass, R. D., Evans, R. D. *et al.* (1993) Ozone trends over the contiguous U.S.A., Hawaii, and Samoa, 1979–1993. American Geophysical Union, AGU, 1993, Annual Meeting. *EOS Suppl.*, 113.

Komhyr, W. D., Reinsel, G. C., Evans, R. D., Quincy, D. M., Grass, R. D., Leonars, R. K. (1997) Total ozone trends at 16 NOAA/CMDL and cooperative

Dobson spectrophotometer observatories during 1979–1996. *Geophys. Res. Lett.*, **24**, N24, 3,225–3,228.

Kondratyev, K. Ya. (1973) *The Complete Atmospheric Energetics Experiment*, GARP Publ. Series., No. 12, 38 pp.

Kondratyev, K. Ya., Hunt, G. E. (1982) *Weather and Climate on Planets*. Pergamon Press, Oxford., 756 pp.

Kondratyev, K. Ya. (1984) *Volcanoes and Climate*, World Climate Paper No. 54, (Geneva WMO), Geneva, Switzerland, 101 pp.

Kondratyev, K. Ya. (1985) Climate: Physics and chemistry of the atmosphere. *Abh. des Meteorol. Dienstes der DDR*, **43**, N134, 5–12.

Kondratyev, K. Ya. (1988) *Climate Shocks: Natural and Anthropogenic*, John Wiley and Sons: New York, 296 pp.

Kondratyev, K. Ya. (1989) The International Geosphere–Biosphere Programme: The role and place of sun–atmospheric interrelationships. *Geofís. Int.*, **28**, 453–456.

Kondratyev, K. Ya., Fedchenko, P. P. (1992) *Fine Structure of the Solar Spectrum and its Role in the Revolution of the Biosphere*, POPO, St. Petersbourg, 40pp (in Russian).

Kondratyev, K. Ya., Nikolsky, G. A. (1995a) Solar activity and climate. 1. Observational data, condensation and ozone hypotheses. *Studying the Earth from Space*, N5, 3–17 (in Russian).

Kondratyev, K. Ya., Nikolsky, G. A. (1995b) Solar activity and climate. 2. Direct impact of changes in the extra-atmospheric spectral distribution of solar radiation. *Studying the Earth from Space*, N6, 4–18 (in Russian).

Kondratyev, K. Ya., Varotsos, C. A. (1995a) Atmospheric ozone variability in the context of global change. *Int. J. Remote Sensing*, **16**, N10, 1,851–1,881.

Kondratyev, K. Ya., Varotsos, C. A. (1995b) Volcanic eruptions and global ozone dynamics. *Int. J. Remote Sensing*, **16**, N10, 1,887–1,895.

Kondratyev, K. Ya. (1997) Remote sensing of middle atmosphere from satellites: The new stage. *Studying the Earth from Space*, N2, 127–144 (in Russian).

Kondratyev, K. Ya., Galindo, I. (1997) *Volcanic Activity and Climate*, A. Deepak Publ., Hampton, VA, 382 pp.

Kondratyev, K. Ya. (1998) *Multidimensional Global Change*, Wiley/Praxis, Chichester, UK, 761 pp.

Koop, T., Carslaw, K. S. (1996) Melting of $H_2SO_4 \cdot 4H_2O$ particles upon cooling: Implications for polar stratospheric clouds. *Science*, **272**, 1,638–1,641.

Koop, T., Carslaw, K. S., Peter, T. (1997) Thermodynamic stability and phase transitions of PSC particles. *Geophys. Res. Lett.*, **24**, N17, 2,199–2,002.

Landgraf, J., Crutzen, P. J. (1998) An efficient method for online calculations of photolysis and heating rates. *J. Atmos. Sci.*, **55**, 863–878.

Larsen, N., Knudsen, B. M., Rosen, J. M., Kjome, N. T., Kyrö, E. (1996) Balloon-borne backscatter observations of type 1 PSC formation: Inference about physical state from trajectory analysis. *Geophys. Res. Lett.*, **23**, N10, 1,091–1,094.

Larsen, N., Knudsen, B. M., Rosen, J. M., Kjome, N. T., Neuber, R. N., Kyrö, E. (1997) Temperature histories in liquid and solid polar stratospheric cloud formation. *J. Geophys. Res.*, **102**, ND19, 23,505–23,517.

Lary, D. J. (1997) Catalytic destruction of stratospheric ozone. *J. Geophys. Res.*, **102**, ND17, 21,515–21,526.

Lary, D. J., Shallcross, D. E., Toumi, R. (1999) Carbonaceous aerosols and their potential role in atmospheric chemistry. *J. Geophys. Res.*, **104**, ND13, 15,929–15,940.

Lastovicka, J. (1998) Second IAGA/IAMAS Workshop: Solar activity forcing of the middle atmosphere. *SPARC Newsletter*, N10, 26–27.

Law, K. S., Plantevin, P. H., Shallcross, D. E., Rogers, H. L., Pyle, J. A., Grouhel, C., Thouret, V., Marenco, A. (1998) Evaluation of modeled O_3 using Measurement of Ozone by Airbus In-Service Aircraft (MOZAIC) data. *J. Geophys. Res.*, **103**, ND19, 25,721–25,737.

Lawrence, J. K., Ruzmaikin, A. A., (1998) Transient solar influence on terrestrial temperature fluctuations. *Geophys. Res. Lett.*, **25**, N2, 159–162.

Lawson, R. P., Heymsfield, A. J., Aulernback, S. M., Jensen, T. L. (1998) Shapes, sizes, and light scattering properties of ice crystals in cirrus and a persistent contrail during SUCCESS. *Geophys. Res. Lett.*, **25**, N9, 1,331–1,334.

Lean, J. (1991) Variations in the sun's radiative output. *Rev. Geophys.*, **29**, 505–535.

Lindesay, J. A., Andreae, M. O., Goldammer, J. G., Harris, G., Annegarn, H. J., Garstang, M., Scholes, R. J., van Wilgen, B. W. (1998) International Geosphere-Biosphere Programme/SAFARI-92 field experiment: Background and overview. *J. Geophys. Res.*, **101**, ND19, 23,521–23,530.

Loisnard, N. (1997) Impact of aircraft emissions upon the atmosphere. *SPARC Newsletter*, N8, 20.

Mackay, D., Wania, F. (1995) Transport of contaminants to the Arctic: Partitioning, processes and models. *The Science of the Total Environ.*, **160/161**, 25–38.

Mahlman, J. D., Fels, S. B. (1986) Antarctic ozone decrease: A dynamical cause? *Geophys. Res. Lett.*, **13**, N12, 1,316–1,390.

Manney, G. L., Bird, J. C., Donovan, D. P., Duck, T. S., Whiteway, J. A., Pal, S. R., Carswell, A. I. (1998) Modeling ozone laminae in ground-based Arctic wintertime observations using trajectory calculations and satellite data. *J. Geophys. Res.*, **103**, ND5, 5,797–5,814.

Marchuk, G. I., Kondratyev, K. Ya. (1992) *Problems of Global Ecology*. Moscow, Nauka, 217 pp. (in Russian).

Marchuk, G. I., Aloyan, A. E. (1995) Global transport of pollutants in the atmosphere. *Proc. Russian Acad. Sci.: Physics of the Atmosphere and Ocean*, **31**, N5, 597–606 (in Russian).

Marston, G., McGill, C. D., Richard, A. D. (1998) Hydroxyl-radical formation in the gas-phase ozonolysis of 2-methylbut-2-ene. *Geophys. Res. Lett.*, **25**, N12, 2,177–2,180.

Massie, S. T., Baumgardner, D., Dye, J. E. (1998) Estimation of polar stratospheric cloud volume and area densities from UARS, stratospheric aerosol measurement II, and polar ozone and aerosol measurement II extinction data. *J. Geophys. Res.*, **103**, ND5, 5,773–5,784.

McCormack, J. P., Hood, L. L. (1996) Apparent solar cycle variations of upper stratospheric ozone and temperature: Latitude and seasonal dependencies. *J. Geophys. Res.*, **101**, ND15, 20,933–20,944.

McCormack, J. P., Hood, L. L., Nagatani, R., Miller, A. J., Planet, W. G., McPeters, R. D. (1997) Approximate separation of volcanic and 11-year signals in the SBUV-SBUV-2 total ozone record over the 1979–1995 period. *Geophys. Res. Lett.*, **24**, N22, 2,729–2,732.

McCormack, J. P., Miller, A. J., Nagatani, R., Fortuin, J. P. F. (1998) Interannual variability in the special distribution of extratropical total ozone. *Geophys. Res. Lett.*, **25**, N12, 2,153–2,156.

McCormick, M. P., Larsen, J. C. (1986) Antarctic springtime measurements of ozone, nitrogen dioxide, and aerosol extinction by SAM II, SAGE, and SAGE II. *Geophys. Res. Lett.*, **13**, N12, 1,280–1,283.

McCormick, M. P., Trepte, C. R. (1986) SAM II measurements of antarctic PSC's and aerosols. *Geophys. Res. Lett.*, **13**, N12, 1,276–1,279.

McCormick, M. P., Thomason, L. W., Trepte, Ch. R. (1995) Atmospheric effects of the Mt. Pinatubo eruption. *Nature*, **373**, 395–404.

McElroy, M. B., Salawitch, R. J., Wofsy, S. C. (1986) Antarctic O_3: chemical mechanisms for the spring decrease. *Geophys. Res. Lett.*, **13**, N12, 1,296–1,299.

McIntyre, M. E. (1989) On the atmospheric ozone hole. *J. Atmos. and Terr. Phys.*, **51**, N1, 29–43.

McPeters, R. D., Labow, G. J., Johnson, R. J. (1997) A satellite-derived ozone climatology for balloonsonde estimation of total column ozone. *J. Geophys. Res.*, **102**, ND7, 8,875–8,886.

Meijer, E. W., van Velthoven, P. F. J., Wauben, W. M. F., Beck, J. P., Velders, G. J. M. (1997) The effects of the conversion of nitrogen oxides in aircraft exhaust plumes in global models. *Geophys. Res. Lett.*, **25**, N23, 3,013–3,016.

Mergenthaler, J. L. (1985) An estimate of the effect of El Chichón aerosol on SBUV profiling for summer 1982. *J. Geophys. Res.*, **90**, ND1, 2,355–2,359.

Mészáros, E. (1991) The history of atmospheric ozone research. *Idöjárás*, **95**, N2–3, 69–76.

Michelangeli, D. V., Allen, M., Yung, Y. L. (1989) El Chichón volcanic aerosols: Impact of radiative, thermal, and chemical perturbations. *J. Geophys. Res.*, **94**, ND15, 18,429–18,443.

Mickley, L. J., Abbatt, J. P. D., Frederick, J. E., Russel, J. M., III. (1997) Response of summertime odd nitrogen and ozone at 17 mbar to Mount Pinatubo aerosol over the southern midlatitudes: Observations from the Halogen Occultation Experiment. *J. Geophys. Res.*, **102**, ND19, 23,573–23,582.

Miller, A. J., Flynn, L. E., Hollandsworth, S. M., De Luisi, J. J., Petropavlovskikh, I. V., Tiao, G. C., Reinsel, G. C., Wuebbles, D. S., Kerr, J., Nagatani, R. M., Bishop, L., Jackman, C. H. (1997) Information content of Umkehr and solar backscattered ultraviolet (SBUV) 2 satellite data for ozone trends and solar responses in the stratosphere. *J. Geophys. Res.*, **102**, ND15, 19,257–19,263.

Minnis, P., Harrison, E. F., Stowe, L. L., Gibson, G. G., Denn, F. M., Doelling, D. R., Smith, W. L., Jr. (1993) Radiative climate forcing by the Mount Pinatubo eruption. *Science*, **259**, 1,411–1,415.

Molina, L. T., Molina, M. J. (1987) Production of Cl_2O_2 from self-reaction of the ClO radical. *J. Phys. Chem.*, **92**, 433–436.

Molina, M. J., Rowland, F. S. (1974) Stratospheric sink for chlorofluoromethanes: Chlorine atom-catalysed destruction of ozone. *Nature*, **249**, 810–812.

Morris, G. A., Considina, D. B., Dessler, A. E., Kawa, S. R., Kumer, J., Mergenthaler, J., Roche, A., Russel, J. M., III (1997) Nitrogen partitioning in the middle stratosphere as observed by the Upper Atmospheric Research Satellite. *J. Geophys. Res.*, **102**, ND7, 8,955–8,965.

Mössinger, J. C., Shallcross, D. E., Cox, R. A. (1998) UV-VIS absorption cross-section and atmospheric lifetimes of CH_2Br_2, CH_2I_2, and CH_2BrI. *J. Chem. Soc. Faraday Trans.*, **94**, N10, 1,391–1,396.

Nagai, T., Uchino, O., Itabe, T., Shibata, T., Mizutani, K., Fujimoto, T. (1997) Polar stratospheric clouds observed at EUREKA (80°N, 86°W) in the Canadian Arctic during the 1994/1995 winter. *Geophys. Res. Lett.*, **24**, N17, 2,243–2,246.

Nair, H., Allen, M., Froidevaux, L., Zurek, R. W. (1998) Localized rapid ozone loss in the northern winter stratosphere: An analysis of UARS observations. *J. Geophys. Res.*, **103**, ND1, 1,573–1,584.

Newman, P. A., Schoeberl, M. R. (1986) October antarctic temperature and total ozone trends from 1979–1985. *Geophys. Res. Lett.*, **13**, N12, 1,206–1,209.

Newman, P. A. (1986) The final warming and polar vortex disappearance during the Southern Hemisphere spring. *Geophys. Res. Lett.*, **13**, N12, 1,228–1,231.

Newman, P. A. (1997) Anomalously low ozone over the Arctic. *Geophys. Res. Lett.*, **24**, N22, 2,685–2,692.

Orsolini, Y. J., Manney, G. L., Engel, A., Ovarlez, J., Claud, C., Coy, L. (1998) Layering in stratospheric profiles of long-lived species: Balloon-borne observations and modeling. *J. Geophys. Res.*, **103**, ND5, 5,815–5,826.

Pan, L., Solomon, S., Randel, W., Lamarque, J.-E., Hess, P., Gille, J., Chion, E.-W., McCormick, M. P. (1997) Hemispheric asymmetries and seasonal variations of the lowermost stratospheric water vapor and ozone derived from SAGE II data. *J. Geophys. Res.*, **103**, ND23, 28,177–28,184.

Paparella, F., Babiano, A., Basdevant, C., Provenzale, A., Tanga, P. (1997) A Lagrangian study of the Antarctic polar vortex. *J. Geophys. Res.*, **102**, ND6, 6,765–6,774.

Pawson, S., Lamgematz, U., Radek, U., Schlese, K., Strauch, P. (1998) The Berlin troposphere-stratosphere-mesoscale GCM: Sensitivity to physical parametrizations. *Quart. J. Roy. Meteorol. Soc.*, **124**, Part B, N548, 1,343–1,372.

Penenko, V. V., Aloyan, A. E. (1985) *Models and Methods for Solving Environmental Protection Problems*, Nauka, Novosibirsk, 256 pp. (in Russian).

Penkett, S. A., Monks, L. J., Carpenter, K. C., Clemitshaw, G. P., Ayers, G. P., Gilett, R. W., Galbally, I. E., Meyer, P. (1997) Relationships between ozone photolysis rates peroxy radical concentrations in clean marine air over the Southern Ocean. *J. Geophys. Res.*, **102**, 12,805–12,817.

Penner, J. E., Lister, D. H., Griggs, D. J., Dokken, D. J., McFarland, M. (eds.) (1999) *Aviation and the Global Atmosphere. A special report of IPCC WG I and III*, Cambridge University Press, Cambridge, UK, 373 pp.

Pierce, R. B., Grooss, J.-U., Grose, W. L., Russel, J. M., III., Crutzen, P. J., Fairlie, T. D., Lingenfelsen, G. (1997) Photochemical calculations along air mass trajectories during ASHOE/MAESA. *J. Geophys. Res.*, **102**, ND11, 13,153–13,168.

Pio, C. A., Lopes, D. A. (1998) Chlorine loss from marine aerosol in a coastal atmosphere. *J. Geophys. Res.*, **103**, D19, 25,263–25,271.

Pitari, G., Visconti, G. (1991) Odd nitrogen removal on background sulphate aerosols: Implications for the ozone hole. *Geophys. Res. Lett.*, **18**, N10, 1,853–1,856.

Pitari, G., Rizi, V. (1993) An estimate of the chemical and radiative perturbation of stratospheric ozone following the eruption of Mt. Pinatubo. *J. Atmos. Sci.*, **50**, N19, 3,260–3,276.

Pitari, G., Rizi, V., Ricciardully, L., Visconti, G. (1993) High speed civil transport impact: role of sulphate, nitric acid trihydrate, and ice aerosols studied with a two dimensional model including aerosol physics. *J. Geophys. Res.*, **98**, 23,141–23,164.

Prasad, S. S. (1997) Potential atmospheric sources and sinks of nitrous oxide. 2. Possibilities from excited O_2, 'embryonic' O_3, and optically pumped excited O_3. *J. Geophys. Res.*, **102**, ND17, 21,527–21,536.

Prasad, S. S. (1998) Potential new atmospheric sources and sinks of odd nitrogen: Sources involving the excited O_2 and the $N_2O.O_3$. *Geophys. Res. Lett.*, **25**, N12, 2,173–2,176.

Prather, M. (1992) Catastrophic loss of anthropogenic ozone in dense volcanic clouds. *J. Geophys. Res.*, **97**, 10,187–10,191.

Prather, M. J. (1997) Time series in atmospheric chemistry: CH_3Br, the ocean, and ozone depletion potentials. *Global Biogeochem. Cycles*, **11**, 393–400.

Prather, M. J. (1998) Time scales in atmospheric chemistry: Coupled perturbations to N_2O, NO_y, and O_3. *Science*, **275**, 1,339–1,341.

Profitt, M. H., Fahey, D. W., Kelly, K. K., Tuck, A. F. (1989) High-latitude ozone loss outside the Antarctic ozone hole. *Nature*, **342**, N6,247, 233–237

Proffitt, M. H., Solomon, S., Loewenstein, M. (1992) Comparison of 2-D model simulations of ozone and nitrous oxide at high latitudes with stratospheric measurements. *J. Geophys. Res.*, **97**, ND1, 939.

Pueschel, R. F., Boering, K. A., Verma, S., Howard, S. D., Ferry, G. V., Goodman, J., Allen, D. A., Hamill, P. (1997) Soot aerosol in the lower stratosphere: Pole-to-pole variability and contributions by aircraft. *J. Geophys. Res.*, **102**, ND11, 13,115–13,118.

Pyle, J. A. (1986) Large ozone losses in Antarctica: The role of heating perturbations. *Geophys. Res. Lett.*, **13**, N12, 1,320–1,322.

Randel, W. S., Wu, F., Russel, J. M., III, Waters, J. W., Froidevaux, L. (1995) Ozone and temperature changes in the stratosphere following the eruption of Mount Pinatubo. *J. Geophys. Res.*, **100**, ND8, 16,753–16,746.

Randeniya, L. K., Vohralik, P. F., Plumb, I. C., Rayan, K. R., Baughcum, S. L. (1996a) Impact of the heterogeneous hydrolysis of $BrONO_2$ on calculated ozone changes due to HSCT aircraft and increased sulphate aerosol levels. *Geophys. Res. Lett.*, **23**, N4, 343–346.

Randeniya, L. K., Vohralik, P. F., Plumb, I. C., Rayan, K. R., Baughcum, S. L. (1996b) Impact of heterogeneous $BrONO_2$ hydrolysis on ozone trends and transient ozone loss during volcanic periods. *Geophys. Res. Lett.*, **23**, N13, 1,633–1,636.

Rasch, P. J., Boville, B. A., Brasseur, G. P. (1995) A three-dimensional general circulation model with coupled chemistry for the middle atmosphere. *J. Geophys. Res.*, **100**, ND5, 9,041–9,071.

Ravishankara, A. R., Turnipseed, A. A., Jensen, N. R., Barone, S., Mills, M., Howard, C. J., Solomon, S. (1994) Do hydroflourocarbons destroy stratospheric ozone? *Science*, **263**, N5,143, 71–74.

Ravishankara, A. R., Hancock, G., Kawasaki, M., Matsumi, Y. (1998) Photochemistry of ozone: Surprises and recent lessons. *Science*, **280**, 60–61.

Reinsel, G. C., Tiao, G. C., Miller, A. J., Wuebbles, D. J., Connell, P. S., Mateer, C. L., DeLuisi, J. J. (1987) Statistical analysis of total ozone and stratospheric Umkehr data for trends and solar cycle relationships. *J. Geophys. Res.*, **92**, ND2, 201–209.

Rex, M., von den Gathen, P., Harris, N. R. P., Lucic, D., Knudsen, B. M., Braathen, G. O., Reid, S. J., De Backer, H., Claude, H., Fabian, R., Fast, H., Gil, M., Kyrö, E., Mikkelsen, I. S., Rummikainen, M., Smit, H. G., Stähelin, J., Varotsos, C., Zaitcev, I. (1998) In situ measurements of stratospheric ozone depletion rates in the Arctic winter 1991/1992: A Lagrangian approach. *J. Geophys. Res.*, **103**, ND5, 5,843–5,854.

Roche, A. E., Nightingale, R. W., Kumer, J. B., Mergenthaler, J. L., Jackman, C. H., Fleming, E. L. (1998) Distribution and seasonal variation of CFCs in the stratosphere: Comparison of satellite global data and 2-D model. *Adv. Space Res.*, **21**, N10, 1,383–1,391.

Rodgers, D. C., De Mott, D. J., Kreidenweis, S. M., Chen, Y. (1998) Measurements of ice nucleating aerosols during SUCCESS. *Geophys. Res. Lett.*, **25**, N9, 1,383–1,386.

Rodriquez, J. M., Ko, M. K. W., Sze, N. D. (1986) Chlorine chemistry in the antarctic stratosphere: Impact of OClO and Cl_2O_2 and implications for observations. *Geophys. Res. Lett.*, **13**, N12, 1,292–1,295.

Rodriquez, J. M., Ko, M. K. W., Sze, N. D., Heisey, C. W., Yue, G. K., McCormick, M. P. (1994a) Ozone response to enhanced heterogeneous processing after the eruption of Mt. Pinatubo. *Geophys. Res. Lett.*, **21**, N3, 209–212.

Rodriquez, J. M., Shia, R. L., Ko, M. K. W., Heisey, C. W., Weisenstein, D. K., Miake-Lye, R. C., Kolb, C. E. (1994a) Subsidence of aircraft engine exhaust in the stratosphere: Implications for calculated ozone depletion. *Geophys. Res. Lett.*, **21**, N1, 69–72.

Rood, R. B. (1986) Global ozone minima in the historical record. *Geophys. Res. Lett.*, **13**, N12, 1,244–1,247.

Rood, R. B., Douglass, A. R., Cerniglia, M. C., Read, W. G. (1997) Synoptic-scale mass exchange from the troposphere to the stratosphere. *J. Geophys. Res.*, **103**, ND19, 23,476–23,485.

Rosen, J. M., Kjome, N. D., Larsen, N., Knudsen, B. M., Kyrö, E., Kivi, R., Karhu, J., Neuber, R., Beningra, I. (1997) Polar stratospheric cloud threshold temperature in the 1995–1996 arctic vortex. *J. Geophys. Res.*, **102**, ND23, 28,195–28,202.

Ryecroft, M. S. (1990) The Antarctic atmosphere: A hot topic in a cold cauldron. *The Geografical J.*, **156**, Part 1, 1–11.

Sausen, R., Feneberg, B., Ponater, M. (1997) Climatic impact of aircraft induced ozone changes. *Geophys. Res. Lett.*, **24**, N10, 1,203–1,206.

Schaffen, K., Myers, D. J., Sofia, S. (1996) Solar activity forecast for solar cycle 23. *Geophys. Res. Lett.*, **21**, N6, 605–608.

Schneider, J., Arnold, F., Bürger, V., Droste-Franke, B., Grimm, F., Kirchner, G., Klemm, M., Stilp, T., Wohlfrom, K.-H., Siegmund, P., van Velthoven, P. F. J. (1998) Nitric acid (HNO_3) in the upper troposphere and lower stratosphere at midlatitudes: New results from aircraft-based mass spectrometric measurements. *J. Geophys. Res.,* **103**, ND19, 25,337–25,343.

Schoeberl, M. R., Krueger, A. J. (1986) Overview of the antarctic ozone depletion issue. *Geophys. Res. Lett.*, **13**, N12, 1,191–1,192.

Schoeberl, M. R., Krueger, A. J., Newman, P. A. (1986) The morphology of antarctic total ozone as seen by TOMS. *Geophys. Res. Lett.*, **13**, N12, 1,217–1,220.

Schoeberl, M. R., Hartmann, D. L. (1991) The dynamics of the stratospheric polar vortex and its relation to springtime ozone depletions. *Science*, **251**, N4,989, 46–52.

Schoeberl, M. R., Bhartia, P. K., Hilsenrath, E., Torres, O. (1993) Tropical ozone loss following the eruption of Mt. Pinatubo. *Geophys. Res. Lett.*, **20**, N1, 29–32.

Schoeberl, M. R., Jackman, C. H., Rosenfield, J. E. (1998) A Lagrangian estimate of aircraft effluent lifetime. *J. Geophys. Res.*, **103**, ND9, 10,817–10,825.

Schulz, J. (1998) On the effect of cloud inhomogeneity on area-averaged radiative properties of contrails. *Geophys. Res. Lett.*, **25**, N9, 1,427–1,430.

Schumann, U. (1998) Editorial – Research on the effects of aircraft and spacecraft upon the atmosphere. *Atmospheric Environment*, **32**, N18, 3,065–3,066.

Schwartzkopf, M. D., Ramaswamy, V. (1993) Radiative forcing due to ozone in the 1980's: Dependence on altitude of ozone change. *Geophys. Res. Lett.*, **20**, N2, 205–208.

Scientific Assessment of Ozone Depletion: 1998. WMO Global Ozone Research and Monitoring Project Report No. 44 (1999) WMO, Geneva, Switzerland.

Searle, K. R., Chipperfield, M. P., Bekki, S., Pyle, J. A. (1998a) The impact of spatial averaging on calculated polar ozone loss. 1. Model experiments. *J. Geophys. Res.*, **103**, ND19, 25,397–25,408.

Searle, K. R., Chipperfield, M. P., Bekki, S., Pyle, J. A. (1998b) The impact of spatial averaging on calculated polar ozone loss. 2. Theoretical analysis. *J. Geophys. Res.*, **103**, ND19, 25,409–25,416.

Sekiguchi, Y. (1986) Antarctic ozone change correlated to the stratospheric temperature field. *Geophys. Res. Lett.*, **13**, N12, 1,202–1,205.

Shi, G.-Y., Wang, W.-C., Ko, M. K. W., Tanaka, M. (1986) Radiative heating due to stratospheric aerosols over Antarctica. *Geophys. Res. Lett.*, **13**, N12, 1,335–1,338.

Shia, R.-L., Ko, M. K. W., Weisenstein, D. K., Scott, C., Rodriguez, J. (1998) Transport between the tropical and midlatitude lower stratosphere: Implications for ozone response to high-speed civil transport emissions. *J. Geophys. Res.*, **103**, ND19, 25,435–25,445.

Shibasaki, K., Iwagami, N., Ogawa, T. (1986) Stratospheric nitrogen dioxide observed by ground-based and balloon-borne techniques at Syowa station (69.0°S, 39.6°E). *Geophys. Res. Lett.*, **13**, N12, 1,260–1,271.

Shindell, D. T. (1997) The potential influence of $ClO \cdot O_2$ on stratospheric ozone depletion chemistry. *J. Atmos. Chem.*, **26**, N3, 323–335.

Shindell, D. T., Wong, S., Rind, D. (1997) Interannual variability of the Atlantic ozone hole in a GCM. Part 1. The influence of tropospheric wave. *Bull. Amer. Meteorol. Soc.*, **54**, N18, 2,308–2,319.

Shindell, D. T., Rind, D., Lonergan, P. (1998) Increased polar stratospheric ozone losses and delayed eventual recovery owing to increasing greenhouse-gas concentrations. *Nature*, **392**, N6,676, 585–592.

Sillman, S. (1995) The use of NO_y, H_2O_2 and HNO_3 as indicators for ozone-NO_x-hydrocarbon sensitivity in urban locations *J. Geophys. Res.*, **100**, ND7, 14,175–14,188.

Siskind, D. E., Summers, M. E. (1998) Implications of enhanced mesospheric water vapor observed by HALOE. *Geophys. Res. Lett.*, **25**, N12, 2,133–2,136.

Sizova, I. M. (1995) Photodissociation and recombination of ozone molecules. *Optics of the Atmos. and Ocean.*, **8**, N9, 1,289–1,313 (in Russian).

Sklyarov, Ya. A. (1994) Active cavity radiometer on board Meteor-3 satellite. *Studying the Earth from Space*, N3, 20–29 (in Russian).

Solanki, S. K., Fligge, M. (1998) Solar irradiance since 1874 revisited. *Geophys. Res. Lett.*, **25**, N3, 341–344.

Solomon, S., Garcia, R., Roland, D. S., Wuebbles, D.J. (1986) On the depletion of antarctic ozone. *Nature*, **322**, N6,082.

Solomon, P. M., Connor, B., de Zafra, R. L., Parrish, A., Jaramillo, M. (1987) High concentrations of chlorine monoxide at low altitudes in the Antarctic spring stratosphere: secular variations. *Nature*, **328**, N6,129, 411–413.

Solomon, S., Mills, M., Heidt, L. E., Pollock, W. H., Tuck, A. F. (1992) On the evaluation of ozone depletion potentials. *J. Geophys. Res.*, **97**, ND1, 825–542.

Solomon, S., Portmann, R. W., Garcia, R. R., Thomason, L. W., Poole, L. R., McCormick, M. D. (1996) The role of aerosol variations in anthropogenic ozone depletion at northern midlatitudes. *J. Geophys. Res.*, **101**, ND3, 6,713–6,727.

Solomon, S., Borrmann, S., Garcia, R. R., Portmann, R. W., Thomason, L. W., Poole, L. R., Winker, D., McCormick, M. D. (1997) Heterogeneous chlorine chemistry in the tropopause region. *J. Geophys. Res.*, **102**, ND17, 21,411–21,429.

Sparling, L. C., Douglass, A. R., Schoeberl, M. R. (1998) An estimate of the effect of unresolved structure on modeled ozone loss from aircraft observations of ClO. *Geophys. Res. Lett.*, **25**, N3, 305–308.

Staehelin, J., Kegel, R., Harris, N. R. P. (1998a) Trend analysis of the homogenized total ozone series of Arosa (Switzerland), 1926–1996. *J. Geophys. Res.*, **103**, ND7, 8,385–8,400.

Staehelin, J., Renaud, A., Bader, J., McPeters, R., Viatte, P., Hoegger, B., Bugnion, V., Giroud, M., Schill, H. (1998b) Total ozone series at Arosa (Switzerland): Homogenization and data comparison. *J. Geophys. Res.*, **103**, ND5, 5,827–5,842.

Steblova, R. S. (1990) Ozone holes: A result of interaction between the Sun and the outer space with geomagnetic fields in the terrestrial atmosphere. *Doklady Russian Akad. Sci.*, **315**, N5, 1,097–1,099 (in Russian).

Stefanutti, L., Morandi, M., Del Guasta, M., Godin, S., David, C. (1995) Unusual PSC's observed by lidar in Antarctica. *Geophys. Res. Lett.*, **22**, N17, 2,377–2,380.

Steil, B., Dameris, M., Brühl, C., Crutzen, P. J., Grewe, V., Ponater, M., Sausen, R. (1998) Development of a chemistry module for GCMs: First results of multi-annual integration. *Ann. Geophys.*, **16**, 205–228.

Stephenson, J. A. E., Scourfield, M. W. J. (1992) Ozone depletion over the Polar Caps caused by solar protons. *Geophys. Res. Lett.*, **19**, N24, 2,425–2,428.

Stolarski, R. S., Butler, D. M. (1979) Possible effects of volcanic eruptions on stratospheric minor constituents chemistry. *PAGEOPH*, **117**, 486–497.

Stolarski, R. S., Cicerone, R. J. (1974) Stratospheric chlorine: A possible sink for ozone. *Can. J. Chem.*, **52**, 1610–1,615.

Stolarski, R. S. *et al.* (1997) Springtime Antarctic total ozone measurements in the early 1970s from the BUV instrument on Nimbus 4. *Geophys. Res. Lett.*, **24**, N5, 591–594.

Stolarski, R. S., Schoeberl, M. R. (1986) Further interpretation of satellite measurements of antarctic total ozone. *Geophys. Res. Lett.*, **13**, N12, 1,210–1,212.

Sukharev, B. E. (1997) On interrelationships between total ozone anomalies over north-western Russia and quasi-biannual cycle and solar activity. *Meteorol. and Hydrol.*, N9, 35–44 (in Russian).

Summers, M. E., Strobel, D. F., Bevilacqua, R. M., Zhu, X., DeLand, M. T., Allen, M., Keating, G. M. (1990) A model study of the response of mesospheric ozone to short-term solar ultraviolet flux variations. *J. Geophys. Res.*, **95**, ND13, 22,523–22,538.

Syvorotkin, V. L., Sadovsky, N. A. (1992) Riftogenesis, ozone layer and World Ocean level. *Doklady Russian Akad. Sci.*, **323**, N4, 731–733 (in Russian).

Syvorotkin, V. L. (1998) Outgasing of the Earth destroys the ozonosphere. *Earth and Universe*. N1, 21–27 (in Russian).

Tabazadeh, A. *et al.* (1997) A new parameterization of H_2SO_4/H_2O aerosol composition: Atmospheric implications. *Geophys. Res. Lett.*, **24**, N15, 1,931–1,934.

Talukdar, R. K., Mellouki, A., Schmoltner, A. M., Watson, T., Montzka, S., Ravishankara, A. R. (1992) Kinetics of the OH reaction with methyl chloroform and its atmospheric implications. *Science*, **257**, 227–229.

Talukdar, R. K., Longfellow, Ch. A., Gilles, M. K., Ravishankara, A. R. (1998) Quantum yields of O(1D) in the photolysis of ozone between 289 and 325 nm as a function of temperature. *J. Geophys. Res.*, **103**, N2, 147–150.

Tan, D. G. H., Haynes, P. H., MacKenzie, A. R., Pyle, J. A. (1998a) Effects of fluid-dynamical stirring and mixing on the deactivation of stratospheric chlorine. *J. Geophys. Res.*, **103**, ND1, 1,585–1,605.

Tan, D., Faloona, I., Brune, W. H., Weinheimer, A., Campos, T., Ridley, B., Vay, S., Collins, J., Sachse, G. (1998b) In situ measurements of HO_x in aircraft exhaust plumes and contrails during SUCCESS. *Geophys. Res. Lett.*, **25**, N10, 1,721–1,724.

Teitelbaum, H., Moustaoui, M., Van Veethoven, P. F. I., Kelder, H. (1998) Decrease of total ozone at low latitudes in the southern hemisphere by a combination of linear and nonlinear processes. *Quart. J. Roy. Meteorol. Soc.*, **124**, Part B, 2625.

Thomason, L. W., Poole, L. R., Deshler, T. (1997) A global climatology of stratospheric aerosol surface area density deduced from Stratospheric Aerosol and Gas Experiment II measurements: 1984–1994. *J. Geophys. Res.*, **102**, ND7, 8,967–8,976.

Thouret, V., Marenco, A., Nédélec, P., Grouhel, C. (1998a) Ozone climatologies at 9-12 km altitude as seen by the MOZAIC airborne program between September 1994 and August 1996. *J. Geophys. Res.*, **103**, ND19, 25,653–25,680.

Thouret, V., Marenco, A., Logan, J. A., Nédélec, P., Grouhel, C. (1998b) Comparisons of ozone measurements from the MOZAIC airborne program and the ozone sounding network at eight locations. *J. Geophys. Res.*, **103**, ND19, 25,695–25,719.

Tie, X. X., Brasseur, G., Lin, X., Friedlingstein, P., Granier, C., Rasch, P. (1994) The impact of high altitude aircraft on the ozone layer in the stratosphere. *J. Atmos. Chem.*, **18**, 103–128.

Tie, X. X., Brasseur, G. (1995) The response of stratospheric ozone to volcanic eruptions: Sensitivity to atmospheric chlorine loading. *Geophys. Res. Lett.*, **22**, N22, 3,035–3,038.

Tie, X., Brasseur, G. P., Granier, C., de Rudder, A., Larsen, N. (1995) Model study of polar stratospheric clouds and their effect on stratospheric ozone. 2. Model results. *J. Geophys. Res.*, **101**, ND7, 12,570–12,584.

Tie, X. X., Hess, P. (1997) Ozone mass exchange between the stratosphere and troposphere for background and volcanic sulfate aerosol conditions. *J. Geophys. Res.*, **102**, ND21, 25,487–25,501.

Tie, X., Granier, C., Randel, W., Brasseur, G. P. (1997) Effects of interannual variation of temperature on heterogeneous reactions and stratospheric ozone. *J. Geophys. Res.*, **102**, ND19, 23,519–23,527.

Titao, G. C., Reinsel, G. C., Pedrick, J. H., Allenby, G. M., Mateer, C. L., Miller, A. J., DeLuisi, J. J. (1986) A statistical trend analysis of ozonesonde data. *J. Geophys. Res.*, D**91**, N12, 13,121–13,136.

de Toma, G., London, J., Chen, L., Rottman, G. J., Woods, T. N., Froidevaux, L. (1998) Solar cycle changes in ultraviolet irradiance and corresponding changes in mesosphere (45 to 65 km) ozone and temperature. In: *Atmospheric Ozone*, R. D. Bojkov, G. Visconti (eds.), Parco Scientifico e Technologico, Abruzzo, Italy, Vol. 1, pp. 259–262.

Toon, O. B., Hamill, P., Turco, R. P., Pinto, J. (1986) Condensation of HNO_3 and HCl in winter polar stratospheres. *Geophys. Res. Lett.*, **13**, N12, 1,284–1,287.

Tourpali, K., Tie, X. X., Zerefos, C. S., Brasseur, G. (1997) Decadal evolution of total ozone decline: Observations and model results. *J. Geophys. Res.*, **102**, ND20, 23,955–23,962.

Tremmel, H. G., Schlager, H., Konopka, P., Schulte, P., Arnold, F., Klemm, M., Droste-Franke, B. (1998) Observations and model calculations of jet aircraft exhaust products and cruise altitude and inferred initial OH emissions. *J. Geophys. Res.*, **103**, ND9, 10,803–10,816.

Tuck, A. F., Watson, R. T., Condon, E. P., Margitat, J. J., Toon, O. B. (1989) The planning and execution of ER-2 and DC-8 aircraft flights over Antarctica, August and September 1987. *J. Geophys. Res.*, **94**, ND9, 11,181–11,222.

Tung, K. K. (1986) On the relationship between the thermal structure of the stratosphere and the seasonal distribution of ozone. *Geophys. Res. Lett.*, **13**, N12, 1,308–1,311.

UNEP Scientific Assessment of Ozone Depletion, WMO Global Ozone Research and Monitoring Project Report No. 37 (1994) WMO, Geneva, Switzerland.

Varotsos, C. (1989) Comment on connection between the 11-year solar cycle, the Q.B.O. and the total ozone. *J. Atmos. Terrestrial Phys.*, **51**, N5, 367–370.

Vaughan, G., Timmis, C. (1998) Transport of near-tropopause air into the lower midlatitude stratosphere. *Quart. J. Roy. Meteorol. Soc.*, **124**, Part A, N549, 1,559–1,578.

Vay, S. A., Anderson, B. E., Sachse, G. W., Collins, J. E., Podolske, J. R., Twohy, C. H., Gandrud, B., Chan, K. R., Baughcum, S. L., Wallio, H. A. (1998) DC-8-based observations of aircraft CO, CH_4, N_2O, and $H_2O(g)$ emission indices during SUCCESS. *Geophys. Res. Lett.*, **25**, N10, 1,717–1,720.

Vlachogiannis, D., Haigh, J. D. (1998) The impact of solar proton events on lower stratospheric ozone. In: *Atmospheric Ozone*, R. D. Bojkov, G. Visconti (eds.), Parco Scientifico e Tecnologico, Abruzzo, Italy, Vol. 1, pp. 257–278.

Vohralik, P. F., Randeniya, L. K., Plumb, I. C., Ryan, K. R. (1998) Use of corrections between long-lived atmospheric species in assessment studies. *J. Geophys. Res.*, **103**, ND3, 3,611–3,627.

Volk, C. M., Elkins, J. W., Fahey, D. W., Salawitch, R. J., Dutton, G. S., Gilligan, J. M., Profitt, M. H., Loewenstein, M., Podolske, J. R., Minschwaner, K., Marginan, J. J., Chan, K. R. (1996) Quantifying transport between the tropical and mid-latitude lower stratosphere. *Science*, **272**, 1,763–1,768.

Volkovitsky, O. A., Gavrilov, A. A., Kaidalov, O. V. (1996) Is man really destroying the ozone layer? *Vestnik Russian Akad. Sci.*, **66**, N9, 783–786.

Vömel, H., Hofmann, D. J., Oltmans, S. J., Harris, J. M. (1995) Evidence for midwinter chemical ozone destruction over Antarctica. *Geophys. Res. Lett.*, **22**, N17, 2,381–2,384.

Vupputuri, R. K. R., Blanchet, J.-P. (1984) The possible effects of El Chichón eruption on atmospheric thermal structure and surface climate. *Geofís. Int.*, **23**, N3, 423–447.

Wallace, L., Livingston, W. (1992) The effects of the Pinatubo cloud on hydrogen chloride and hydrogen fluoride. *Geophys. Res. Lett.*, **19**, N12, 1209.

Wang, W.-Ch., Zhuang, Yi.-Ch., Bojkov, R. (1993) Climate implications of observed changes in ozone vertical distributions in the middle and high latitudes of the Northern Hemisphere. *Geophys. Res. Lett.*, **20**, N15, 1,567–1,570.

Wauben, W. M. F., Fortuin, J. P. F., van Velthoven, P. F. J., Kelder, H. M. (1998) Comparison of modeled ozone distributions with sonde and satellite observations. *J. Geophys. Res.*, **103**, ND3, 3,511–3,530.

Weinheimer, A. J., Campos, T. L., Walega, J. G., Grahek, F. E., Ridley, B. A., Baumgardner, D., Twohy, C. H., Gandrud, B., Jensen, E. J. (1998) Uptake of NOy on wave cloud ice particles. *Geophys. Res. Lett.*, **25**, N10, 1,725–1,728.

Weisenstein, D. K., Ko, M. K. W., Rodriguez, J. M, Sze, N.-D. (1993) Effects on stratospheric ozone from high speed civil transport: Sensitivity to stratospheric aerosol loading. *J. Geophys. Res.*, **98**, 23,133–23,140.

Weisenstein, D. K., Ko, M. K. W., Sze, N.-D., Rodriguez, J. M. (1996) Potential impact of SO_2 emissions from stratospheric aircraft on ozone. *Geophys. Res. Lett.*, **23**, N_2, 161–164.

Weisenstein, D. K., Ko, M. K. W., Dyominov, I. G., Pitari, G., Ricciardully, L., Visconti, G., Bekki, S. (1998) The effects of sulfur emissions from HSCT aircraft: A 2-D model intercomparison. *J. Geophys. Res.*, **103**, ND1, 1,527–1,547.

Wenneberg, P. O., Brault, J. W., Hanisco, T. F., Salawitch, R. J., Mount, G. H. (1997) The atmospheric column abundance of IO: Implications for stratospheric ozone. *J. Geophys. Res.*, **102**, ND7, 8,887–8,898.

Wilson, R. C. (1984) Measurements of total solar irradiance and its variability. *Space. Sci. Rev.*, **38**, 203–242.

Wilson, R. C., Hudson, H. S. (1991) The sun's luminosity over a complete solar cycle. *Nature*, **351**, 42–44.

World Meteorological Organization (1995) *1994 Assessment of Ozone Depletion*, Report No. 27, Ozone Secretariat, Geneva, Switzerland.

Wuebbles, D. J. (1996) Three-dimensional chemistry in the greenhouse. An Editorial comment. *Climatic Change*, **34**, N3–4, 397–404.

Wuebbles, D. J., Wei, C.-F., Patten, K. O. (1998) Effects on stratospheric ozone and temperature during the Maunder Minimum. *Geophys. Res. Lett.*, **25**, N4, 523–526.

Yamshanov, V. A. (1989) Nitrocompounds in the environment: A danger for the ozone layer. *Doklady Russian Akad, Sci.*, **309**, N3, 659–662 (in Russian).

Yordanov, D. L., Penenko, V. V., Aloyan, A. E. (1978) Parameterization of the stratified planetary boundary layer for numerical modelling of atmospheric processes. *Proc. Russian Akad. Sci.: Physics of the Atmos. and Ocean.*, **14**, N8, 815–823.

Yu, F., Turco, R. P. (1998) Contrail formation and impacts on aerosol properties in aircraft plumes: Effects of fuel sulfur content. *Geophys. Res. Lett.*, **25**, N3, 313–316.

Yung, Y. L., Allen, M., Crisp, D., Zurek, R. W., Sander, S. P. (1990) Spatial variation of ozone depletion rates in the springtime Antarctic polar vortex. *Science*, **248**, 721-724.

Zerefos, C. S., Tourpali, K., Isaksen, I. S. A., Rognerud, B., Bojkov, B. R., Schuurmans, C. J. E. (1998) The solar output and total ozone. In: *Atmospheric Ozone*, R. D. Bojkov, G. Visconti (eds.), Parco Scientifico e Tecnologico, Abruzzo, Italy, Vol. 1, pp. 279–282.

Zhang, R., Leu, M.-T., Molina, M. J. (1996) Formation of polar stratospheric clouds on preactivated background aerosols. *Geophys. Res. Lett.*, **23**, N13, 1,669–1,672.

Zhao, X., Turco, R. P., Kao, C.-Y. J., Elliot, S. (1996) Numerical simulation of the dynamical response of the Arctic vortex to aerosol-associated chemical perturbations in the lower stratosphere. *Geophys. Res. Lett.*, **23**, N12, 1,525–1,528.

Zhou, S., Rottman, G. J., Miller, A. J. (1997) Stratospheric ozone response to short- and intermediate-term variations of solar UV flux. *J. Geophys. Res.*, **102**, ND7, 9,003–9,012.

Ziemke, J. R., Chandra, S., Thompson, A. M., McNamara, D. P. (1996) Zonal asymmetries in southern hemisphere column ozone: Implications of biomass burning. *J. Geophys. Res.*, **101**, ND9, 14,421–14,427.

Ziereis, H., Schlager, H., Schulte, P., Köhler, I., Marquardt, R., Feigl, C. (1999) In situ measurements of the NO_x distribution and variability over the eastern North Atlantic. *J. Geophys. Res.*, **104**, ND13, 16,021–16,032.

Zillman, J. W. (1997) Atmospheric science and public policy. *Science*, **276**, N5,315, 1,084–1,086.

Zou, H., Gao, Y. (1997) Long-term variation in TOMS ozone over 60°–70°S. *Geophys. Res. Lett.*, **24**, N18, 2,295–2,298.

4

Surface ultraviolet radiation change because of total ozone variation

Recognition of the great importance of total ozone (TOZ) depletion has led to the realization of the significance of subsequent surface UV radiation (UVR) decreases which might be dangerous for man and the biosphere. UV radiation that reaches the Earth's surface can be arbitrarily divided into two subregions (*UNEP* ..., 1994): UVB (280–315 nm), which is strongly absorbed by ozone; and UVA (315–400 nm), which is only weakly absorbed by ozone. Less than 2% of extra-terrestrial solar energy falls within the UVB range, and only a small fraction of this reaches the surface. Figure 4.1 illustrates wavelength dependence of surface UV irradiance corresponding to values of TOZ (*UNEP* ..., 1994). The very rapid drop in irradiance with decreasing wavelengths indicates how difficult identification of UV trends is, which is further complicated by instrumental problems (primarily, precise calibration is absolutely necessary) and UV irradiance variability because of different factors such as aerosols, clouds and surface albedo. As mentioned in *UNEP* ... (1994), it will probably be necessary to use very short wavelengths in UVB that have a high sensitivity to ozone change to detect trends in UV because of ozone depletion.

Increasing attention to UVR problems has resulted in many recent publications including the *Proceedings of the NATO Advanced Research Workshop on Stratospheric Ozone Depletion/UVB Radiation in the Biosphere* (Biggs, Joyner, 1994), the *UNEP Scientific Assessment of Ozone Depletion* (*UNEP* ..., 1994), the NATO ASI monograph *Solar Ultraviolet Radiation* (Zerefos and Bais, 1997), the *European Conference on Atmospheric UV Radiation* (*Abstracts* ..., 1998), Krupa *et al.* (1998) and the WMO/UNEP *Scientific* ... (1999). Note, however, that there is no consensus in assessments of observed UVR trends and relevant implications (Michaels *et al.*, 1994). To avoid repetition we shall limit our discussion mainly to the most recent results.

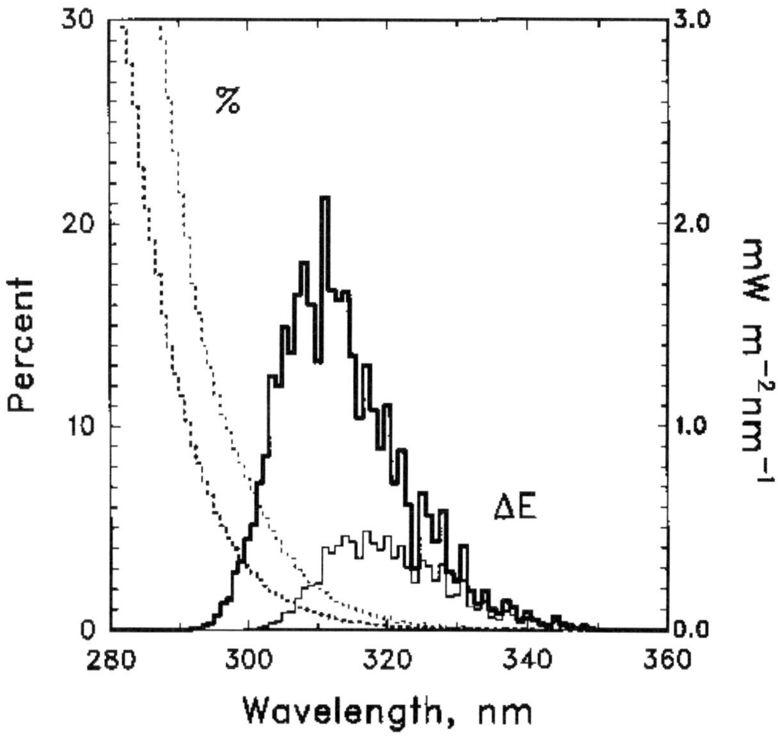

Figure 4.1. Increases in UV radiation in response to a 1% decrease in total ozone column near 300 DU (1 DU = 2.69×10^{16} mol cm^{-2}). Solid lines (right scale) give spectral irradiance changes, dotted lines (left scale) give % changes. Values are for overhead Sun (thick lines) and for a solar zenith angle of 70° (thin lines).

4.1 OBSERVED VARIABILITY OF SURFACE SOLAR UV RADIATION

Although global-scale UVR data are still in short supply (see Weatherhead *et al.* 1997a for a review) significant progress in the development of UVR observations has been made in recent years. Madronich (1998) discussed, in this context, activities of the WMO Scientific Steering Committee on UV monitoring (SSC), which was organized in 1995. The key efforts of the SSC have been directed towards solving problems of calibration, instrument comparison, archiving and analysis of UVR observational data as well as developing techniques to calculate surface UVR and retrieve UVR from satellite observational data.

A principal scientific aim has been to obtain representative observational data on UVR climatology, characterizing both mean values and variability with special emphasis on high-latitude observations. Already existing data for the three stations in the Antarctic indicate an enhancement trend in surface UVR over recent years that takes place in late spring–early summer. On the other hand, there was a UVR decrease towards the South Pole. Although it has been assumed that the UV

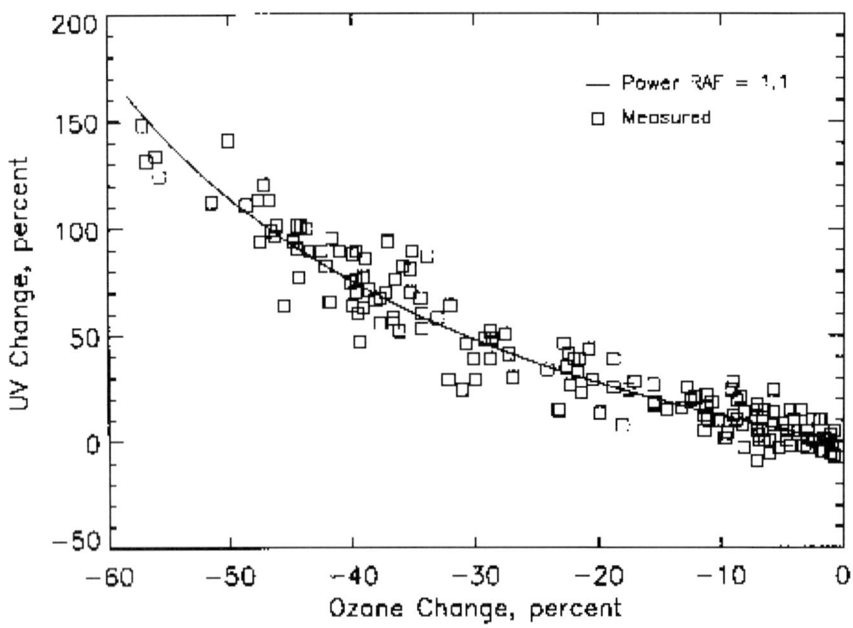

Figure 4.2. Dependence of erythemally weighted UV radiation on column ozone change. Measurements from South Pole, 1 February 1991 to 12 December 1992.

insolation level was lower before the time when regular formation of ozone holes started, there are still no observational data, however, to verify this.

As mentioned in *UNEP* ... (1994), the unique geometry at South Pole stations (90°S), where there are no diurnal cycles in solar-zenith angle, simplifies investigation of the relationship between UV, ozone and other parameters. Figure 4.2 illustrates strong anticorrelation between UV irradiance and TOZ. It is typical for Antarctica, however, that the highest instantaneous UV irradiances (erythema or UVB) take place not during the time of TOZ minimum but at a time closer to the summer solstice, combining the effects of higher solar-zenith angles with relatively low ozone. In contrast, visible radiation increases steadily as the solar-zenith angle decreases over the observation period (Figure 4.2) Another typical feature of South Pole stations is that perturbations of UV level by clouds are relatively small, probably because of the high surface albedo and extremely cold temperatures, which keep clouds from becoming optically thick. Huge year-to-year variations in UV have been measured at the South Pole. UV has a maximum in the spring, whereas longer wavelength radiation peaks near the summer solstice.

At Palmer Station (64.5°N, 64°W), the highest biologically weighted UV doses for the 6-year period before 1994 were observed in late October of 1993 when biologically damaging UV even exceeded the summer maximum measured at San Diego (32°N) (*UNEP* ..., 1994).

To characterize the interrelationship between TOZ change (ΔO_3) and UV variability (ΔE) the Radiation Amplification Factor (RAF) was introduced:

$$\text{RAF} = -(\Delta E/E)/(\Delta O_3/O_3)$$

Madronich (1993) showed, however, that a power low is more adequate for large TOZ depletions observed recently:

$$\text{R}AF = \ln(E^*/E)/\ln(O_3/O_3^*)$$

where E^* and E are two UV irradiances, and O_3^* and O_3 are corresponding TOZ values. Figure 4.2 illustrates the adequacy of the power low in the case of observations at the South Pole.

Quite a different situation was observed in the Northern Hemisphere. For instance, the UVR level in Garmisch-Partenkirchen (Germany, 47°N) was much lower than at the same latitude in the Southern Hemisphere. Such interhemispheric contrast could take place because of variability of several factors: TOZ, tropospheric pollution (especially by aerosols and tropospheric ozone), Earth–Sun distance. More complete observational data and simulation modelling are necessary to quantitatively explain the phenomenon mentioned.

According to observations by Cappelani and Kochler (1999) at Ispra from 1993 to 1997, there was TOZ decrease of 0.21%/yr (with a mean monthly value 3.19 ± 2 DU) and UVR (at the 305-nm wavelength) increase of about 2.0%/yr.

There have been a number of successful attempts at developing retrieval techniques for obtaining surface UVR from satellite observations, but further improvements are necessary as far as retrieval errors and spatio-temporal resolution are concerned. Presumably, we are still far from reaching 5% retrieval errors for monthly means or 20% errors for instant values. There are no reliable-enough assessments of retrieval errors as yet because of the absence of adequate comparisons with direct surface measurements. It is clear, however, that satellite observational data can serve as a reliable source of information on various input parameters, such as TOZ, aerosol and cloudiness characteristics, which are necessary to calculate surface UVR. Frederick *et al.* (1989) emphasized the need to take into account tropospheric ozone (TrOZ) trends which have, in general, the opposite sign to TOZ trends.

Calculated mean zonal mean annual values of erythema UVR indicate the presence of a positive trend (twice as high as MSDs (mean-square deviations)) between 1975 and 1992 for latitudes higher than 35–45° in both hemispheres (these trends were calculated using satellite data on TOZ and albedo). Forecast various scenarios of future ozone-depleting substance (ODS) growth, both with and without taking into account of the Montreal Protocol recommendations concerning reductions in ODS emissions, has allowed preliminary estimates of potential erythemal UVR growth in the future to be obtained. A principal conclusion is that it is very important to meet recommendations of the Montreal Protocol. Under these recommendations it should be possible to return to the pre-industrial UVR level by the middle of the 21st century. Future changes (maximum TOZ depletion has not been reached as yet) depend on such unpredictable (or difficult to predict) factors as volcanic eruptions, solar activity, climate change, rates of reductions in ODS emissions. Under such circumstances, accomplishing adequate UVR monitoring must take centre stage (Seckmeyer *et al.*, 1995)

A number of recent efforts have been made to solve monitoring problems. For instance, the US Department of Agriculture has been developing a UV Radiation Monitoring Program since 1994 (Bigelow *et al.*, 1998). The initial network of 12 stations employed broadband radiometers to measure UVB irradiance and included ancillary measurements of temperature, humidity and irradiance at several wavelengths in the visible produced by Multi-Filter Rotating Shadowband Radiometers (MFRSR). The network has now been expanded to more than 20 stations and broadband radiometers have been supplemented with a seven-wavelength Ultraviolet Multi-Filter Rotating Shadowband Radiometer (UV-MFRSR). The network has been designed to include 30 stations, each with a full complement of instrumentation. Broadband instruments are shown to be quite stable and collocated instruments are shown to agree to within 2.3% for zenith angles less than 80° under all sky conditions.

The US National Science Foundation supported the establishment in 1988 of the UV Monitoring Network for Polar Regions to collect data about the impact of ozone depletion gathered at Palmer Station, McMurdo and the South Pole (Antarctica) as well as at Ushuaia, Argentina. In 1990, a site was added at Barrow, Alaska, which is the first long-term spectral UV-irradiance-monitoring site in the US. In 1992, a sixth site was added in San Diego, California (Booth *et al.* in *Abstracts* ..., 1998). This network makes routine spectral measurements of UV irradiance in the 290–600-nm region at 1-nm resolution at hourly intervals during daylight (later increased to 4/hr).

UVR observational data are still rather scarce and can be illustrated by various fragmentary observational programmes. Kerr and McElroy (1993) were the first to show the effects of ozone depletion on integrated daily UVB at mid-latitudes, including cloudy conditions. Bernhard *et al.* (1997) were the first to carry out spectral UV irradiance measurements in the tropics of Australia (Townsville: 19.33°S, 146.76°E, 30 m a.s.l., during December 1995 and January 1996) using the mobile spectroradiometer developed at the Fraunhofer Institute for Atmospheric Environmental Research (Germany). Over this period, variation in daily erythemal radiation was found to be $\pm 24\%$ ($\pm 1\sigma$). An important finding was that TOZ variation contributed less than $\pm 3\%$ to this variability. The principal cause of observed UVR variability are clouds which, however, attenuate UV radiation less than total irradiation (300–3000 nm). The maximum erythemally weighted irradiance was 425 mW/m^2, and the average daily erythemal irradiation for the observation period was 6.06 kJ/m^2. These high radiation levels exceed the corresponding values for Garmisch-Partenkirchen, Germany (47.5°N, 11.0°E, 730 m a.s.l.) by between 55 and 70%. Such a difference between Australia and Germany is explained by the higher solar elevation and lower TOZ in the tropics. Van Weele *et al.* (1995) carried out combined measurements of UVA actinic flux, UVA irradiance and global radiation in connection with assessment of photodissociation rates in the troposphere.

Table 4.1. Summary of statistical results relating to variability of UV erythemal solar irradiance of 70° solar-zenith angle, the ratio UVB/UVA and TOZ in March 1995. DU is Dobson Units, σ is the standard deviation of the mean and σ_r is the coefficient of variation.

Sites	Erythernal irradiance (mW/m^2)			UVB/UVA (%)			Total ozone (DU)			Number of observations
	Mean	σ	σ_r	Mean	σ	σ_r	Mean	σ	σ_r	
Thessaloniki	11.45	4.37	0.38	0.123	0.014	0.11	360.1	35.5	0.10	28
Reading	12.99	3.50	0.27	0.136	0.021	0.15	346.3	46.6	0.13	30
Reykjavik	13.73	2.46	0.18	0.118	0.008	0.07	406.2	27.4	0.07	31
Esrange	12.08	3.92	0.32	0.124	0.020	0.16	345.2	54.1	0.16	23

Bais *et al.* (1996) discussed the results of UVR measurements over Europe within the Second European Stratospheric Arctic and Midlatitude Experiment (SESAME) campaign, which performed (in March 1995) a wide variety of measurements performed at several locations in the Arctic region and at selected sites at mid-latitudes of the Northern Hemisphere (Esrange, Sweden: 67.9°N; Reykjavik, Iceland: 64.1°N, Reading, England: 51.5°N; Lerwick, Scotland: 60.1°N; Thessaloniki, Greece: 40.6°N). Table 4.1 illustrates statistical results of observations. The ratio UVB/UVA was considered to exclude the interference of clouds and other wavelength-independent factors. An important result of statistical analysis is that the variability of UV erythemal dose, because of clouds, can be at least of similar level to that induced by the variability of TOZ. In the case of Thessaloniki the effect of clouds becomes even more important, being three to four times larger than that of ozone. As can be seen from Table 4.1, the mean levels of the erythemal solar irradiance measured at various sites were of the same order, since all irradiances were measured at the same solar-zenith angle (70°). The daily integral of erythemal solar irradiance at all sites varied between 0.2 and 2.2 kJ/m^2 in March 1995. The lowest values were observed at Reykjavik, dominated by increased TOZ and cloudiness, and the highest values were at Thessaloniki, because of its latitudinal difference from other sites.

Weatherhead *et al.* (1997b) undertook detailed statistical analysis of the UV irradiance datasets, obtained from several Robertson–Berger (RB) meters, to investigate their long-term trend behaviour between 1974 and 1991 for 14 sites in the United States. Results obtained indicate a significant negative trend of the order of −6% per decade overall, in agreement with similar data for the period 1974–1985 considered earlier (Scotto *et al.*, 1988). If, however, allowance was made for mean-level shift in the data for several of the stations around 1979, which may be due to calibration and other instrument-related problems, the resulting overall trend was found to be of the order of 2% per decade and not statistically significant. However, Weatherhead *et al.* (1997a) emphasized that, in spite of the lack of statistical significance of the results mentioned, they do rule out the possibility of a significant downward trend in UVR at the ground during the 1980s. More detailed analysis of the data from four of the stations between 1979 and 1991 has been done including radiative transfer model-based calculations of UVR, based on satellite TOMS data. Comparison between RB meter measurements and calculations uncovered substantial inconsistencies, with RB data showing significant downward movement relative to the calculations for three of the four sites. The final conclusion has to be made that existing RB meter measurements are unreliable for long-term trend detection.

In view of the importance of using numerical modelling results, Mayer *et al.* (1997) conducted systematic long-term comparison of spectral UVR measurements in Garmisch-Partenkirchen and modelling results obtained with the UVSP model package which allows calculation of diffuse and direct UV and visible irradiances and radiances (176–850 nm, 1.0 nm resolution) at any altitude (only cloudless conditions and low surface albedo were considered). Comparison of 2 years' observational data showed very good agreement between measurements and model. Systematic deviations in spectral irradiance ranged between −11 and +2% for wavelength interval 295–400 nm and solar-zenith angles up to 80°. These deviations are fully explained by uncertainty of both measurements and model. The comparison demonstrated that knowledge of TOZ and spectral aerosol depth, parametrized by the Ångstrom formula, is sufficient to achieve the mentioned level of agreement for a cloudless sky.

Gantner *et al.* (*Abstracts* ..., 1998) analysed results of UVB radiation change over three decades (1968–1996) at the Hohenpeissenberg Observatory (Germany). At this site, a trend showing stratospheric ozone decrease of 0.4% per year has been observed since 1968 but, on the other hand, a trend of TrOZ increase of 2% per year was observed until the mid-1980s, which resulted in net TOZ decrease of 3% per decade since 1968. Although it was natural to expect relevant UVB radiation increase, this has not happened in reality, because the decrease in global radiation of about 10% in 43 years was probably due to an increase in cirrus frequency. Spectral measurements of UV radiation in the wavelength interval 290–320 nm (with 0.5 nm spectral resolution) made since 1990 indicate the huge variability in UV radiation and therefore the difficulty of reliable trend assessments. Application of the technique of UVB extrapolation allowed, however, retrieval of trends between 1968 and 1996. The results demonstrate a strong increase in UV radiation since 1968 in spring, especially in the short wavelengths. For instance, UV radiation at 300 nm in March and April approximately doubled in 30 years. The trends later in the year and at longer wavelengths are much smaller.

The UVB monitoring network, consisting of three Solar Light UV-Biometers 501, was established in June 1993 along N–S cross-sections through Poland to represent the three types of climate found in the Baltic: coast, midland and Tatra Mountains (Litynska *et al.*, *Abstracts* ..., 1998). Statistical analysis of observational data June 1993 and December 1997 revealed distinct seasonal and geographical variability depending on the stations' altitude, TOZ, weather conditions and surface albedo. Large negative correlation was found between TOZ and maximum hourly UVB dose between April and August. The highest UVB values correspond to clear sky and minimum TOZ values.

On the basis of long-term surface UV irradiance observations at the Meteorological Observatory of Moscow State University and numerical simulations, Nezval *et al.* (*Abstracts* ..., 1998; Geogdzaev *et al.*, 1997) investigated effects of broken cloudiness. Satellite and surface observational data were used by Oikarinen *et al.* (*Abstracts* ..., 1998) to study how change in stratospheric ozone concentration or aerosols affects the surface UV radiance spectrum calculated using two radiative transfer models.

Sobolewski and Krzyscin (*Abstracts* ..., 1998) took measurements of the surface UV radiation in Hornsund Fjord (77°N, 15.5°E, 11 m a.s.l.) at the Polish Arctic station on Svålbard, Norway using the Robertson–Berger meter. For time periods March–October of 1996 and 1997, seasonal patterns and day-to-day variations in UV daily dose (erythemally weighted) and UV Index (UVI) were studied. Simultaneous observational data on global solar radiation, Dobson and TOMS TOZ data as well as radiative transfer calculations were also examined. Results indicate that the maximum value of UV daily dose and UVI during the late spring were 9 MED (1 MED = 210 J/m^2 day) and about 2, respectively. UV irradiance at midnight was about 3–4 mW/m^2 in late spring days. The seasonal pattern of the UV daily dose is asymmetric relative to the summer solstice. All sky UV measurements taken in early spring provided much higher UV daily dose and index than those in late summer because of the cloud/aerosol and snow impact on UV radiation. In early spring the measured UV doses under clear-sky conditions were about 30–40% higher than those in late summer (no snow cover). During summer days with heavy clouds and haze the UV daily dose dropped to a few percent of its modelled clear-sky values. Because of relatively small day-to-day TOZ variations over Hornsund Fjord, the impact of TOZ change on the UV level is much smaller than that related to short-term cloud/aerosol change.

Borkowski (*Abstracts* ..., 1998) discussed the results of long-term (since 1976) series of UV radiation observations at Belsk, Poland (52°N, 21°E). Results of the RB measurements indicate that the UVB trend for the autumn was not statistically significant, but for cold and warm seasons an increase in UV daily doses of about 6% per decade was observed. For the whole time period considered (1976–1997) the trend was $5.6 \pm 1.2\%$. Umkehr data on vertical ozone profiles, analysed by Krzyscin (*Abstracts* ..., 1998) indicate that temporal variations of the ratio of ozone content in Umkehr layers 1–4 (about 0–24 km) to that in Umkehr layers 5–10 (about 23–48 km) can be used as an index of ozone profile changes over Belsk. The linear trend of the index, equal to $-5.2 \pm 1.4\%$ (one standard deviation), demonstrates stronger ozone depletion in the lower stratosphere than in the upper stratosphere. Application of the multiple regression model allowed assessment of the influence of the ozone vertical profile on the daily erythemal dose. It was found that a 10% decrease (increase) in the index results in $\sim 1\%$ increase (decrease) in UV daily dose (TOZ kept constant). Both observations and radiative transfer calculations showed that ozone profile changes over Belsk between 1976 and 1996 resulted only in small long-term variation in surface UV radiation.

Krzyscin and Puchalski (1998) discussed observed UV radiation variability at Belsk because of changes in aerosol optical depth (AOD) at 550 nm between 1980 and 1996. Observational data indicate that the long-term decrease in AOD (−7.4% per decade) yields an increase in surface UV dose (erythemally weighted) of about 1% per decade. Observations between 1993 and 1997 show that the changes in UV radiation because of AOD variations are, in the mean, comparable with those owing to total ozone changes. A 10% increase (decrease) in AOD, relative to its reference value, manifests as ~1.5% decrease (increase) in erythemal daily dose at ground level. The mean cloud signal (characterized by the level of global solar radiation) in the UV data was found to be a few times larger than the mean ozone signal. Extreme AOD values in the Belsk record (0.1 and 0.7) correspond to 20–30% changes in erythemally weighted UV daily dose. Similar assessments were obtained on the basis of radiative transfer calculations.

Between 1994 and 1997, Van Weele and Tunc (*Abstracts* ..., 1998) performed simultaneous observations of surface total, UVB and UVA irradiances in de Bilt (52.10°N, 5.18°E). In all spectral ranges, global irradiances reach their maximum values during broken cloud conditions. Maximum values during total solar irradiance are found to be up to 30% higher than on a cloud-free day, in UVA the maximum values can even be up to 40% higher, while maximum enhancement over cloud-free values in UVB are less than 5%.

Weihs *et al.* (*Abstracts* ..., 1998) studied the influence of TOZ, aerosols and surface albedo on clear-sky UVA and UVB irradiance making high-mountain observations at the Sonnblick Observatory (3106 m a.s.l.) as well as using radiative transfer calculation results. Observed changes in UV irradiance because of TOZ variation indicate enhancement of UV irradiances of 560% at the 305-nm wavelength and 69% at 315 nm in the case of TOZ change from 380 to 230 DU. Combined use of measured and calculated values led to the conclusion that changes in aerosol optical depth could induce changes in UV irradiance of up to 5%. Cloud cover may enhance the UV irradiance on average by 6% in wintertime. UV values in winter are on average 4.5% higher than in summer (probably because of changes in surface albedo).

In accord with UVB irradiance observations at Thessaloniki (Greece) made using well-calibrated Brewer spectroradiometers between 1991 and 1997, which were discussed by Zerefos *et al.* (*Abstracts* ..., 1998), long-term changes in solar irradiance below 320 nm at ground level are mainly controlled by total ozone. This conclusion is supported by the similar long-term variability under clear-sky and all sky conditions during the period mentioned. Zerefos *et al.* emphasized that calculated trends in solar irradiance (in % per year) depend on the length of the observation period and thus any speculations on future changes in UVB solar irradiance should always be treated with caution. It was estimated, however, that a change of about 2% per year in solar irradiance at 305 nm at Thessaloniki is a rate that could possibly continue in view of the expected continuation of the ozone decline resulting in an increase of more than 20% per decade, which is comparable with the amplitude of the annual cycle at Thessaloniki. Weatherhead (*Abstracts* ..., 1998) also emphasized that detecting trends in UVB radiation has so far been difficult, despite the efforts that have gone into measuring UV radiation for the past several years. An important task in this context is to identify those areas that are better suited for detecting trends than others.

Within the project of the Photochemical Activity and Solar Ultraviolet Radiation (PAUR), Kylling *et al.* (1998) measured surface UV irradiances at two different sites in Greece (Athenes-Tatoi and a small Aegean island, Agios Efstratios) during June 1996 under non-cloudy conditions using the Brewer spectroradiometer. Comparison between observations and radiative transfer calculations (a discrete ordinate algorithm) indicates that UVB irradiance is reduced with respect to aerosol-free conditions by 5 to 35% depending on the aerosol optical depth and single-scattering albedo. During the period of PAUR observations, changes in aerosol loading gave larger variation in surface UV irradiance than the changes in the ozone column.

Sabziparvar *et al.* (1998) assessed UVR global change, since pre-industrial times, as a result of anthropogenic stratospheric and tropospheric ozone changes. Results are based on using the discrete–ordinate radiative transfer model with four streams to calculate changes in diurnally integrated surface UVB and erythemally weighted irradiances (EER). Input information on stratospheric ozone changes since 1964 were taken from TOMS and Dobson data. Tropospheric ozone changes since 1850 are estimated from the University of Oslo 3-D chemical transport model. Calculations show that in low latitudes a significant (up to 5% for UVB and 9% for EER) decrease in surface UV may have occurred, because of tropospheric ozone increases. Large UV increases are found at high latitudes, with up to 60% increases in October and 20% increases in April. Increased aerosol amounts (soot and sulphate aerosols were considered) may have decreased surface UVB and EER by up to 7% locally and about 2% on a global average. Aerosols may be a cause of the interhemispheric differences in UV trends, acting in the same way as tropospheric ozone that decrease UV more in the Northern Hemisphere than in the Southern Hemisphere (the interhemispheric differences are more pronounced in July and October, with the Northern Hemisphere showing very little trend and the Southern Hemisphere showing large increases in UVB). Table 4.2 illustrates annual average changes in AOD (in units of optical depth at 550 nm), UVB and EER (in %)

Sabzipavar *et al.* (1998) pointed out that owing to uncertainty in ozone trends at latitudes lower than 50°N most calculated UVB trends must be treated with some caution.

Since uncertainty in UV measurements is no less important, Seckmeyer *et al.* (1995) analysed results of intercomparisons of five UV spectroradiometers conducted in August 1994 at Garmisch-Partenkirchen (see also Zerefos *et al.*,

Table 4.2. Changes in surface UVB and EER because of aerosol increases.

Types of aerosol	AOD	UVB	EER
Soot	+0.003	−1.10	−1.00
Dust	+0.004	−0.60	−0.55
Sulphate	+0.019	−0.14	−0.15
Biomass burning	+0.017	−0.13	−0.14
Total	+0.043	−1.97	−1.84

1997). The general conclusion is that agreement between measurement systems is sufficiently good for investigation of geographical differences in UV, under all observing conditions. In accord with results mentioned above, Seckmeyer *et al.* (1995) pointed out that UV doses measured at sites in the Southern Hemisphere are systematically larger than those measured at corresponding northern latitudes. During the summer months daily doses at the South Pole exceed those at mid-latitudes in the Northern Hemisphere.

Erlick and Frederick (1998) studied effects of aerosols on the wavelength dependence of atmospheric transmission in the ultraviolet and visible in the cases of continental and urban aerosols under clear sky conditions (aerosol models suggested by d'Almeida *et al.* (1991) were used). Consideration of calculated normalized (with respect to aerosol-free atmosphere) transmission revealed the existence of two wavelength-dependent aerosol effects. The first is a dip in normalized transmission (NT) at wavelengths below about 320 nm because of coupling between aerosol multiple scattering and absorption by ozone and the rapidly increasing absorption coefficient of tropospheric water-soluble aerosols below 340 nm. The second effect is expressed as NT increase with wavelength, from 320 nm through the visible, which is caused by the decrease with wavelength of Mie scattering coefficients of tropospheric water-soluble, soot and stratospheric sulphate aerosols. For average summer humidity conditions the continental aerosol model predicts an increase in optical depth to 1.26 at 310 nm and to 0.65 at 550 nm, leading to a reduction in transmission of 15.2% at 310 nm and 9.7% at 550 nm, as compared with aerosol-free atmosphere (the significant contribution of aerosol absorption to the reduction of transmission was pointed out). The urban aerosol model predicts (again for summer humidity conditions) an increase in optical depth to 3.22 at 310 nm and to 1.65 at 550 nm, leading to a reduction in transmission of 40.0% at 310 nm (this is why a 5–18-fold decrease in surface UVB since the beginning of the industrial revolution has been observed in rural areas of industrialized countries) and to 25.3% at 550 nm, as compared with an aerosol-free atmosphere. Comparison of calculation results with groundbased observational data was carried out by Erlick and Frederick (1998), the results of which indicate that estimated optical depths are somewhat higher than observed in the case of continental aerosols, being more typical of industrial than rural areas. It must be emphasized that the results considered were restricted to continental mid-latitude regions typically associated with high atmospheric aerosol loadings.

McKenzie *et al.* (1998) adequately emphasized that clouds are important modulators of UV radiation, but quantification of their effects has been limited. This was clearly demonstrated by Lubin *et al.* (1998). An important aspect of UV radiation transfer is that most of the energy in the UVB region at the Earth's surface is in the form of scattered radiation, whereas in the visible the direct solar radiation dominates. This is why cloud effects are smaller in the UV region than in the visible.

Since 1989 New Zealand's National Institute of Water and Atmospheric Research (NIWA) has maintained a UV research programme at Lauder, Central Otago (45.04°S, 169.68°E, 370 m a.s.l.), mainly to measure and understand the causes of

UV variability (McKenzie *et al.*, 1991, 1996). According to groundbased observations, for annual averages, clouds typically reduce clear-sky irradiance in New Zealand by 25–30% but, under some conditions, the presence of clouds can result in a significant enhancement in UVB radiation.

With the purpose of studying cloud impacts on surface UV radiation, McKenzie *et al.* (1998) assessed the relationship of satellite-observed fractional cloud cover with near-simultaneous coincident groundbased observations of cloud cover. The correlation between fractional cloud cover (cloud amount) and measured UV was then examined. It was found that although there is a good correlation between measurements of cloud cover from satellite and from the ground, their correlation with surface irradiance in either the visible or UV parts of the spectrum is weaker. Using cloud amount as the only parameter is insufficient to parametrize cloud effect on incoming radiation. It is also important to take into account whether or not the Sun is obscured as well as to consider cloud optical depth values. McKenzie *et al.* (1998) demonstrated that, in view of the strong correlation between UV and visible radiation, the use of groundbased pyranometer data to infer cloud effects on UV radiation is more accurate than the use of presently available satellite data for process studies of the relationship between cloud cover and instantaneous UV measurements at a particular location.

Forster *et al.* (*Abstracts* ..., 1998) presented global changes in UV radiation since pre-industrial times, resulting from anthropogenic stratospheric and tropospheric ozone changes. Calculations (using a sophisticated radiative transfer scheme) of changes in diurnally integrated UVB and erythemally weighted (EER) irradiance reaching the Earth's surface were made. TOMS and Dobson TOZ data were used together with calculated TrOZ values. Calculations indicate that in low latitudes a significant (up to 5% for UVB and 9% for EER) decrease in surface UV may have occurred because of tropospheric ozone increases. Large UV increases are found at high latitudes, with up to 60% increases in October and 20% increases in April. Assessment of potential aerosol impact (soot and sulphate aerosols) on the UV shows that increased aerosol amounts may have decreased surface UVB and EER by up to 6.6% locally and about 2% on a global average.

As mentioned earlier, during recent years a number of attempts have been made to retrieve surface UV irradiances from satellite observations (Lubin and Jensen, 1997). For instance, Krotkov *et al.* (1998) demonstrated that, with the knowledge of extraterrestrial solar flux in the absence of clouds and UV absorbing aerosols, TOMS measurements of total column ozone and 380 nm (or 360 nm) radiances, combined with radiation transfer calculations, can provide estimates of surface spectral irradiances to accuracies comparable with typical groundbased instruments. Relationships between TOMS reflectivity at 380 nm (R380) and cloud/aerosol attenuation factor (CF) have been developed for various types of clouds and aerosols. The results show that for non-absorbing clouds and aerosols the R380 is highly correlated with CF at wavelengths between 290 and 490 nm (UV-absorbing aerosols may be a complication). A new technique using TOMS aerosol index data, developed by Krotkov *et al.* (1998), allows estimation of UV flux transmission by strongly absorbing aerosol. The results indicate that over certain parts of the Earth, aerosols can reduce UV flux at the surface by more than 50%. The most important need for reducing errors in TOMS-derived surface UV spectra is to improve the understanding of UV aerosol attenuation.

Qu *et al.* (*Abstracts* ..., 1998) applied the delta-Eddington approximation to simulate UVB irradiances at the Earth's surface using satellite ozone, cloud and aerosol measurements. Satellite data include TOZ from TOMS instrumentation; ISCCP cloud products; AVHRR and SAGE aerosol observation. For testing, two-stream model surface UVB measurements in North America and Russia were used. The deviation of calculated monthly UVB radiation from measured values was within $\pm 6\%$ for 25% of the time. Model results were sensitive, in descending order, to clouds, ozone, Rayleigh scattering and aerosols. Further improvements in simulating UVB radiation will require greatly improved remote-sensing data for clouds and aerosols.

Matthijsen *et al.* (*Abstracts* ..., 1998) developed a method to derive UV radiation from remote-sensing data and radiative transfer calculations, which allows for UV trend analysis and assessments of UV changes with a high spatial and temporal resolution. Between 1980 and 1991, relative UV increases were found to be largest over NW Europe (about 8%). Menkhaus *et al.* (*Abstracts* ..., 1998) applied three different models to derive UVB spectra and UVB index from GOME data. A similar study was undertaken by Peeters *et al.* (*Abstracts* ..., 1998).

Meerkoetter and Deguenther (*Abstracts* ..., 1998) developed a method for mapping erythemally weighted surface UV irradiances with a high spectral resolution, based on using GOME (ERS-2 satellite) and NOAA/AVHRR data (see also Meerkoetter *et al.*, 1997). Application of the method to selected areas in Europe indicates the extreme spatial variability of UV irradiances in the case of inhomogeneous cloud cover. For further improvement of the method, consideration of a 3-D radiative transfer is necessary to adequately take into account the impact of broken cloud fields. Verdebout (*Abstracts* ..., 1998) was the first to discuss results about mapping surface UV irradiance using the MVIRI sensor data from Meteosat satellites (the problem is that the MVIRI is not a UV instrument which makes it necessary to use other additional data). Comparison of the instantaneous UVB dose rate shows that the modelling performs well in the absence of clouds. On cloudy days, the modelling process is unable to reproduce the rapid dynamics seen in point ground observations. However, when comparing daily doses the ground measurements are themselves averaged and the agreement is then quite satisfactory.

Mayer and Madronich (*Abstracts* ..., 1998) proposed a method to derive the actinic flux (the amount of radiation incident on a sphere rather than on a flat surface as in the case of irradiance) from satellite measurements of TOZ and cloud reflectivity *R*. Methods for the calculation of surface irradiance from TOMS observations of TOZ and cloud reflectivity were earlier developed by Eck *et al.* (1995). It was shown by Mayer and Madronich that actinic flux transmittance is closely related to irradiance transmittance, when considering averages over area or time instead of point data. This allows earlier suggested methods of irradiance calculations of average actinic fluxes to be used. The uncertainty of such computations in the UV wavelength range is less than 20% (two standard deviations).

The most complete study of global surface UV radiation climatology was conducted by Lubin *et al.* (1998) who used satellite information on TOZ from Nimbus-7 TOMS data and solar reflectance from Earth–atmosphere system (NASA Earth Radiation Budget Experiment (ERBE)) data for the time period from the beginning of 1985 to early 1990 as well as ISCCP water-vapour climatology data as input information for relevant radiative transfer calculations (delta-Eddington approximation) to estimate climatological cloud opacity and, respectively, the dependence of UVR variability on geography and season (a technique for calculating surface UVR was discussed earlier by Frederick and Lubin (1988)). Tropospheric aerosol impact was not explicitly taken into account, but any significant aerosol opacity, over the large spatial scale (100 km) considered, was implicitly contained within cloud opacity retrievals. Surface UVR fluxes are spectrally resolved (for the entire shortwave spectrum with 179 narrow spectral bands) to enable weighted integration with any biological action spectrum. Absorption by ozone, CO_2 and water vapour was taken into account.

The results of numerical modelling, carried out by Lubin *et al.* (1998), indicate that solar elevation changes are more important than total column ozone in governing the variability of surface UVR over large geographical areas, although some regions with pronounced local minima in ozone (30 DU or more) will cause noticeable enhancements in integrated UVB flux relative to UVA. The greatest variability in surface UVR within a given climate zone is induced by cloud cover. In summertime, regions that show lower surface UVR fluxes relative to their surrounding regions include the eastern United States (versus the western United States), India, China (in the vicinity of the Yangtze River) and Japan (relative to the surrounding oceans). Cloud cover over tropical rainforest reduces surface UV irradiance relative to ocean areas at the same latitudes.

Analysis of zonally averaged data shows that there is high correlation between zonally averaged minimum daily UVB cloud transmissions and the clear-sky fraction. For instance, zonally averaged cloud transmission over land for January 1988 and 1989 and clear-sky fraction both have very similar variability with latitude, exhibiting a pronounced local maximum at 30°S and a pronounced local minimum at 10°S. In the Northern Hemisphere, the local maximum in clear-sky fraction occurs at 25°N, while the local maximum in UVB cloud transmission occurs just slightly south, between 10°N and 20°N. However, zonally averaged sky coverage shows different behaviour over land versus the ocean, with generally more pronounced local extremes over land. Lubin *et al.* (1998) emphasized that there is no single global relationship between cloud transmission and cloud coverage, although in most cases there is a statistical relationship between transmission and cloud amount for a given region and season. The cloud transmission considered depends on the cloud liquid-water path, cloud geometry and sky coverage, and varies between regions with different prevailing cloud types.

Plates 1a, b present global TOZ distributions for July and October 1989 (monthly averages) while Plates 2a, b illustrate global distributions of relevant monthly averaged ratios of F_{UVB}/F_{UVA} at local noon from TOMS and ERBE data. The ratio of UVB to UVA irradiance is largely independent of cloud opacity.

Lubin *et al.* (1998) pointed out that this ratio is an important quantity in photobiology.

Comparison of Plates 1a and 2a indicates that there is a coincidence between the lack of TOZ and F_{UVB}/F_{UVA} longitudinal variability in the equatorial belt 0–30°S. A local TOZ maximum off the coast of Canada and South Alaska corresponds to a local minimum of F_{UVB}/F_{UVA} of $\sim$0.025, as opposed to the average value of 0.028 in the surrounding areas. There is a local TOZ minimum over western China, which results in the highest values of F_{UVB}/F_{UVA} anywhere in the world. The maximum of this ratio is because of: (1) low TOZ values resulting in stronger increase of UVB than UVA; (2) high Sun elevation, which reduces the optical path for UVB through the ozone layer; and (3) the high-terrain elevation in western China, which reduces the contribution of Rayleigh scattering.

The range of TOZ variability in July 1989 is between 250 and 350 DU, or about 30–40% (it is mainly a latitude-dependent variability: the F_{UVB}/F_{UVA} global distribution is characterized by the high level of zonality). However, October 1989 data (Plates 1b and 2b) reveal a rather different picture showing springtime ozone depletion over the Antarctic Peninsula and southern parts of Argentina and Chile. Longitudinal F_{UVB}/F_{UVA} variability shows itself also in high northern latitudes.

Plates 3a and b illustrate the existence of strong spatial and temporal variability of monthly averaged cloud effective optical depth τ_c retrieved from ERBE data using the CCM2 algorithm. As can be seen, during the Northern Hemisphere summer (July 1989), regions with the lowest τ_c include the south-eastern United States (including all of California), Spain and Portugal, the eastern Mediterranean and the Middle East. Over most of the tropical oceans there is a local τ_c maximum just north of the equator. During the Southern Hemisphere spring (October 1989), τ_c is consistently large (of the order of 10) over the Southern Ocean and is small (less than 3) over most of Australia, Southern Africa and Northern Chile (Lubin *et al.*, 1998).

Basically, zonal distribution is typical for monthly-averaged local noon UVB (280–315 nm) irradiances (Plates 4a–c). Much of the regional variability can be understood in terms of cloud climatology (Plate 3). Global UVB distribution in the Southern Hemisphere summer (January) is characterized by maxima in the desert regions of central Australia, Namibia and Northern Chile (because of primarily sparse cloud cover). In tropical rainforest, such as the northern part of South America and southeastern Africa, surface UVB irradiance is 10–25% smaller than in other regions at the same latitudes. During the Northern Hemisphere spring and summer (April and July) UVB maxima are also observed over tropical ocean regions with low cloud opacity (Plate 3a). An example of cloud-induced UVB variability is India, where in April nearly as large UVB values (2.1–2.3 W/m^2) are found as over the deserts, whereas in July surface UVB irradiances over India drop to about 1.5 W/m^2. Comparison of Plates 4d, 2b and 1b helps analyses of how both cloud cover and solar position tend to offset the large-scale impact of springtime ozone depletion.

Plates 5a–d illustrate the highly inhomogenous global distribution of diurnal UVB amplitude variability the specific features of which define the time-integrated UVB dose (the most important characteristic from the viewpoint of assessing UVR

Colour plate section

A. JULY 1989

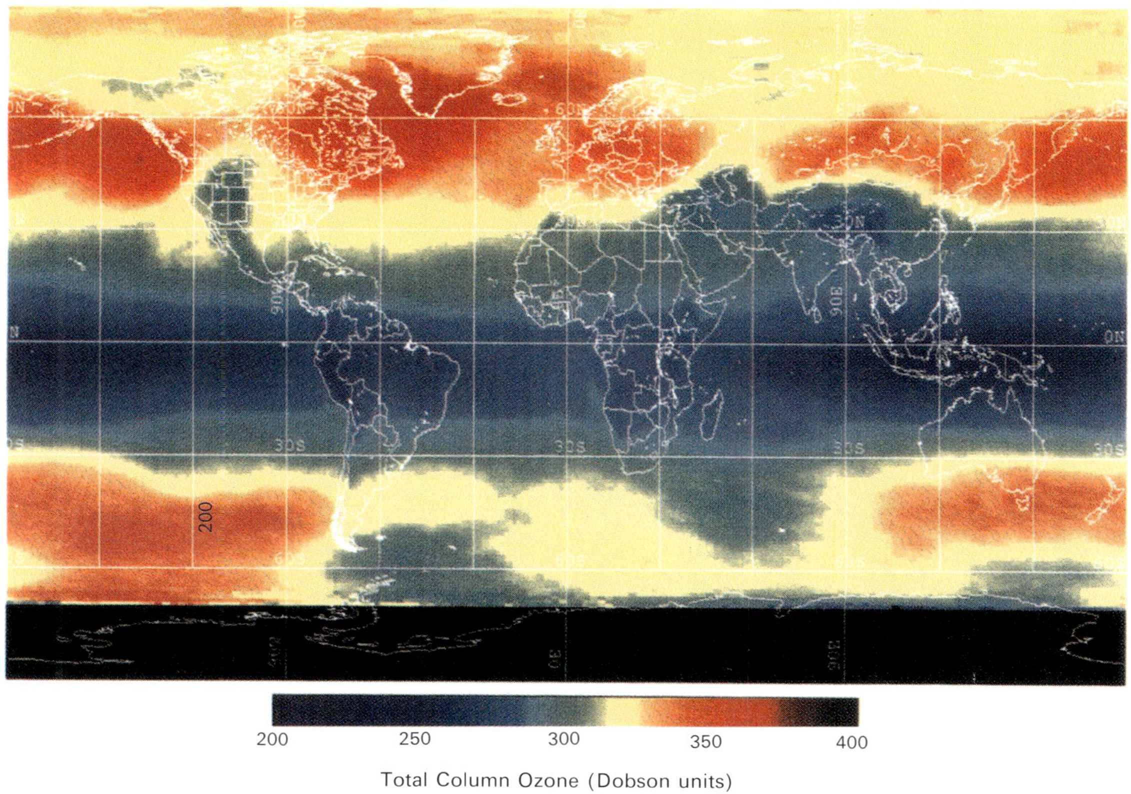

Plate 1a. Global distribution of total cclumn ozone abundance during 1989 as determined by the total ozone mapping spectrometer (TOMS): (a) monthly average for July, (b) monthly average for Oxtober.

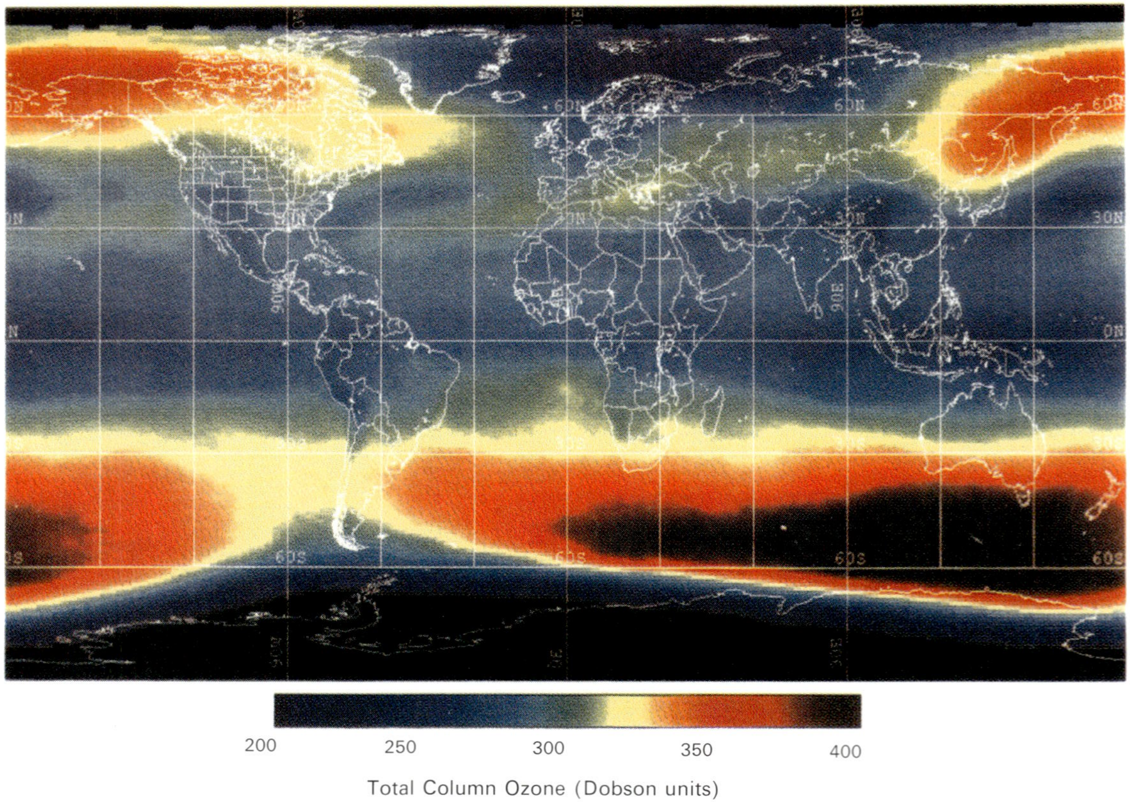

Plate 1b (*continued*)

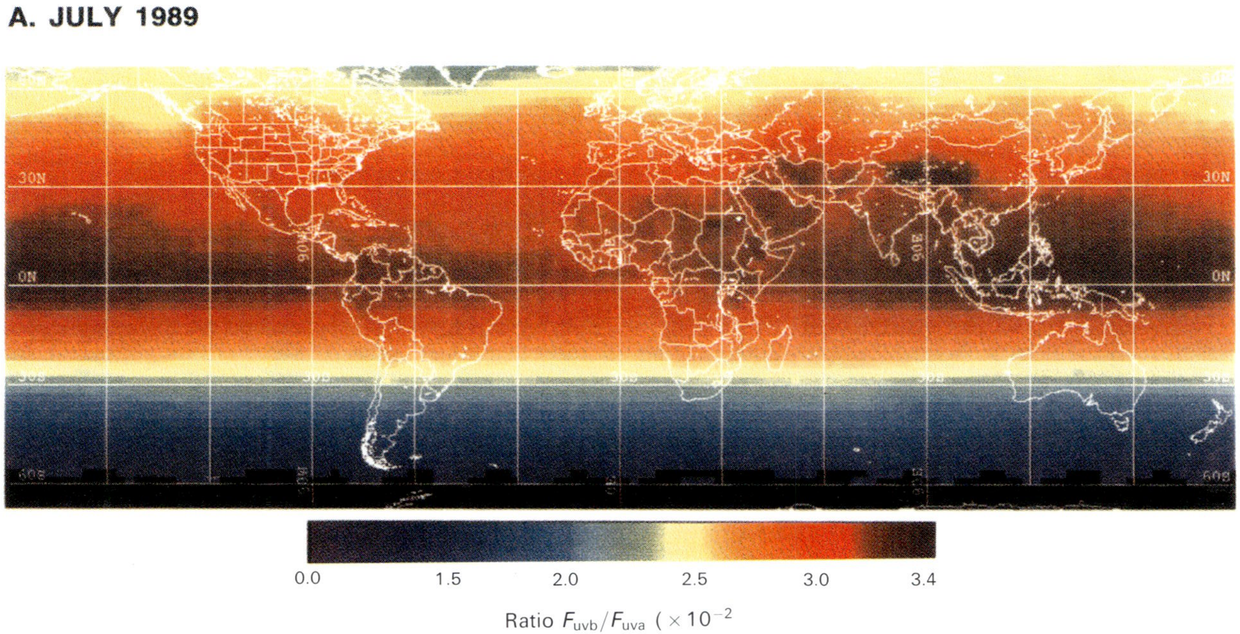

Plate 2a. Monthly averaged ratio F_{uvb}/F_{uva} at local noon estimated from the TOMS and ERBE data for the year 1989 for (a) January, (b) April, (c) July and (d) October.

B. OCTOBER 1989

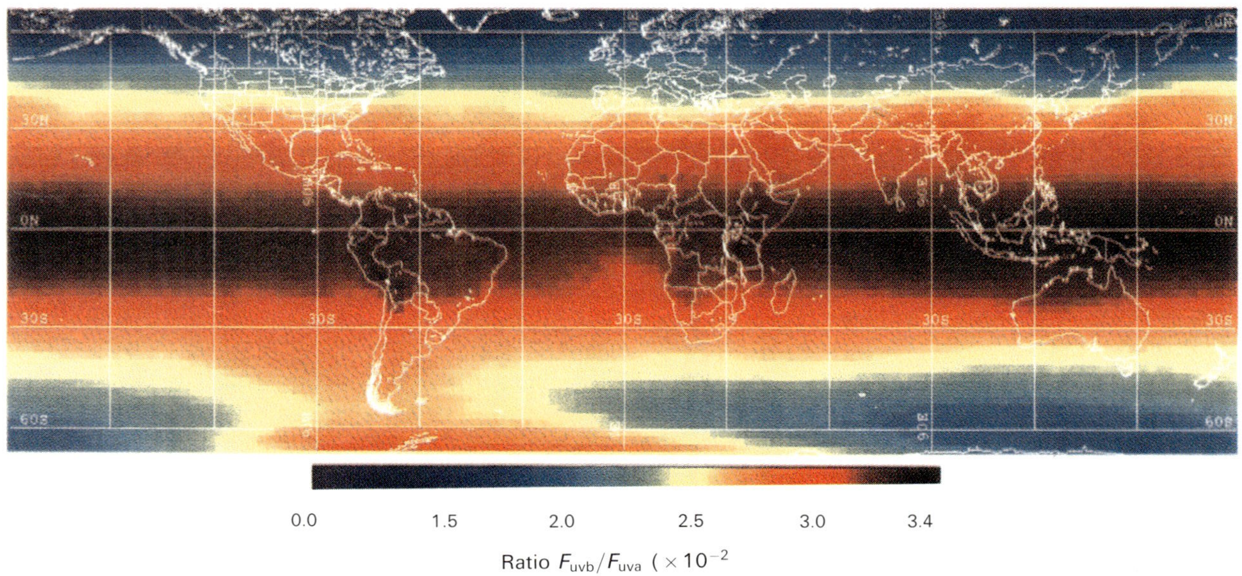

Plate 2b (*continued*)

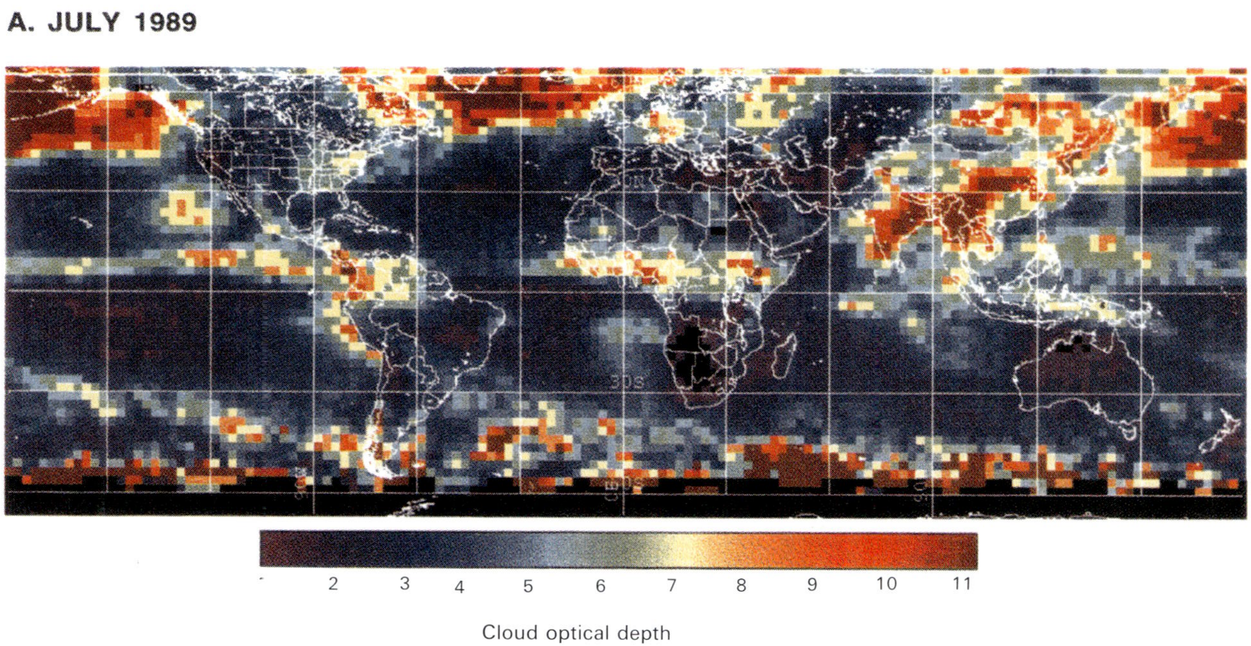

Plate 3a. Monthly averaged cloud effective optical depth τ_c retrieved from the ERBE data and the CCMR radiation algorithm for (a) July 1989 and (b) October 1989. The deepest shade of red denotes cloud optical depths 30 and greater.

B. OCTOBER 1989

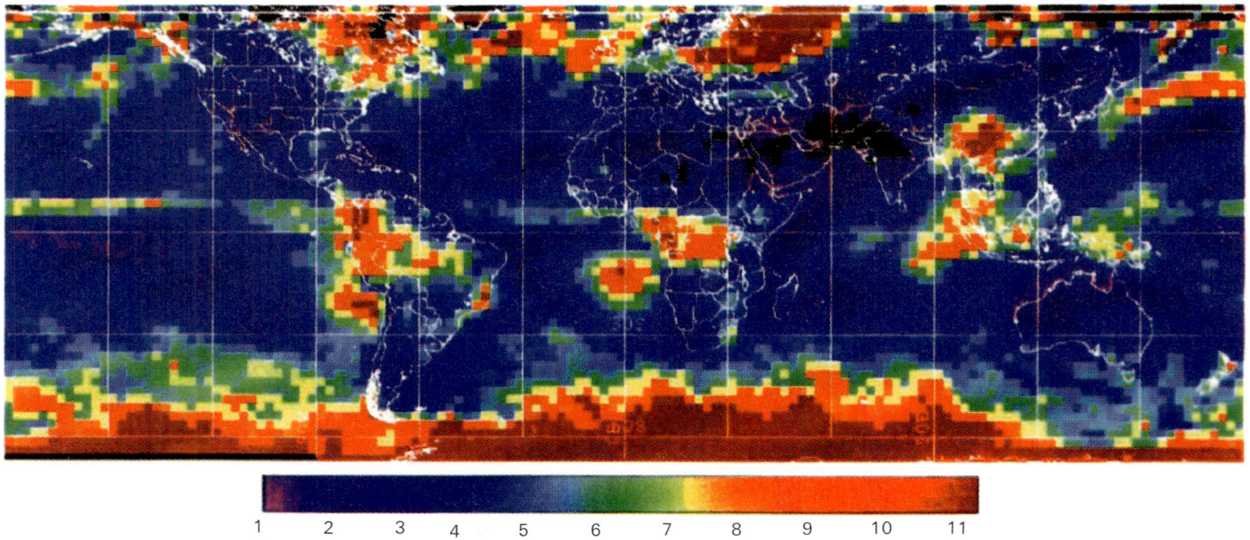

Plate 3b (*continued*)

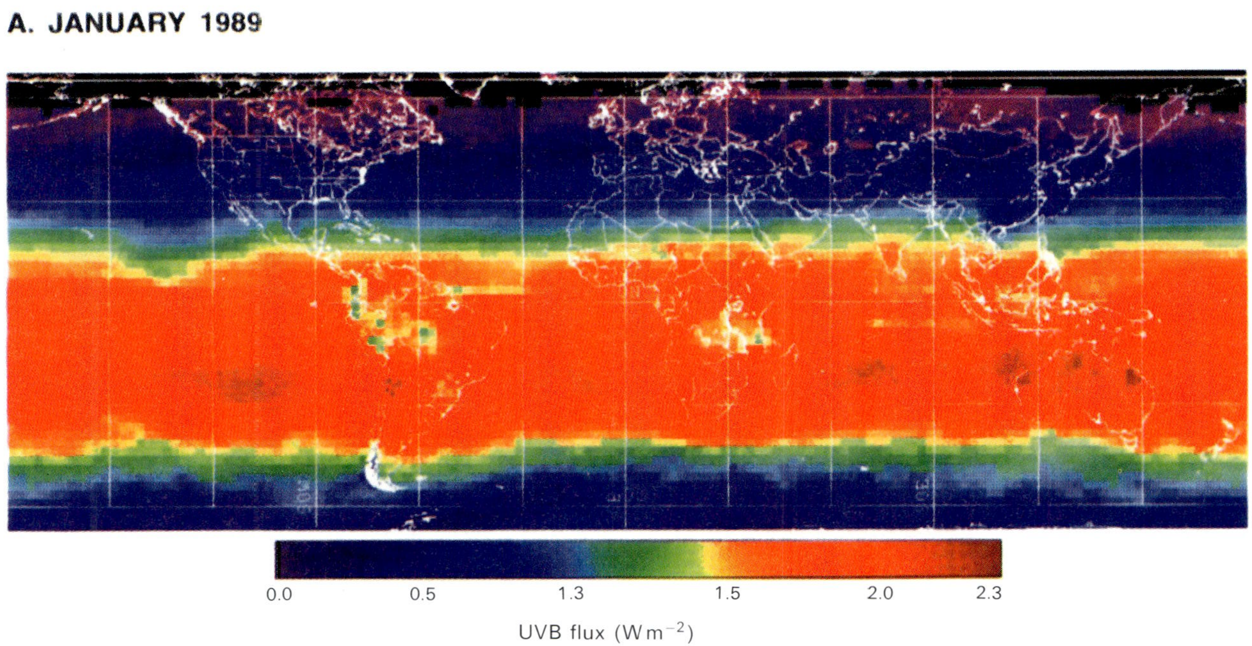

Plate 4a. Monthly veraged local noon s.irface UVB flux (wavelength integral 280–315 nm) estimated from the TOMS and ERBE data for the 1989 for (a) January, (b) April, (c) July and (d) October.

B. APRIL 1989

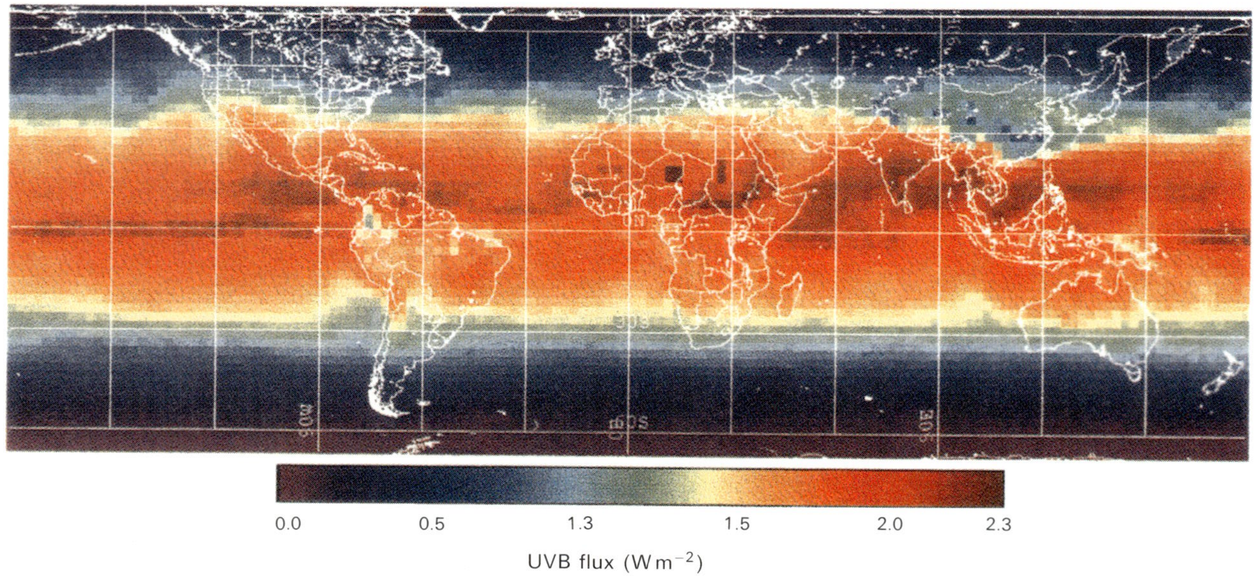

Plate 4b (*continued*)

C. JULY 1989

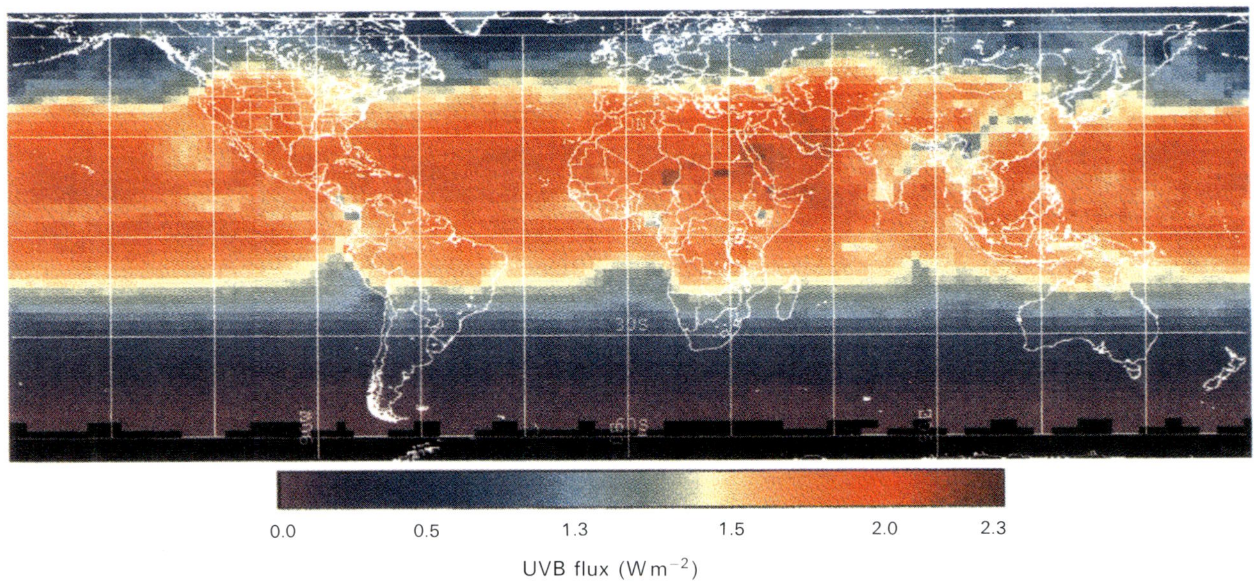

Plate 4c (*continued*)

D. OCTOBER 1989

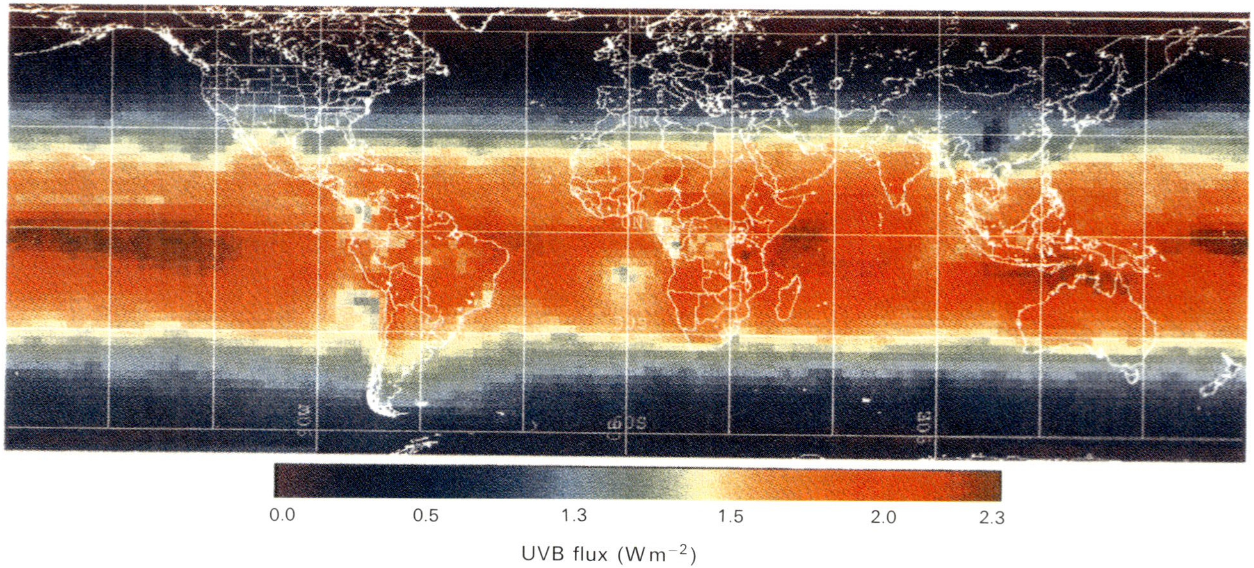

Plate 4d (*continued*)

A. JANUARY 1989

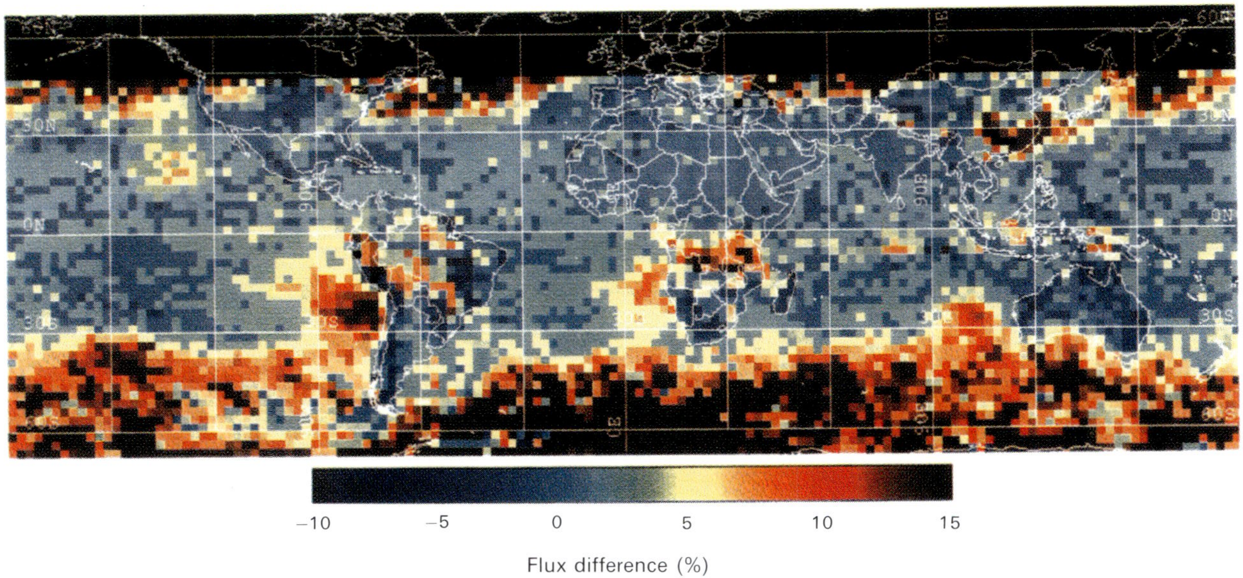

Plate 5a. Percentage difference between the instantaneous surface UVB fluxes at 15 : 00 and 09 : 00 (the afternoon flux less the morning flux) for the year 1989 for (a) January, (b) April, (c) July and (d) October.

B. APRIL 1989

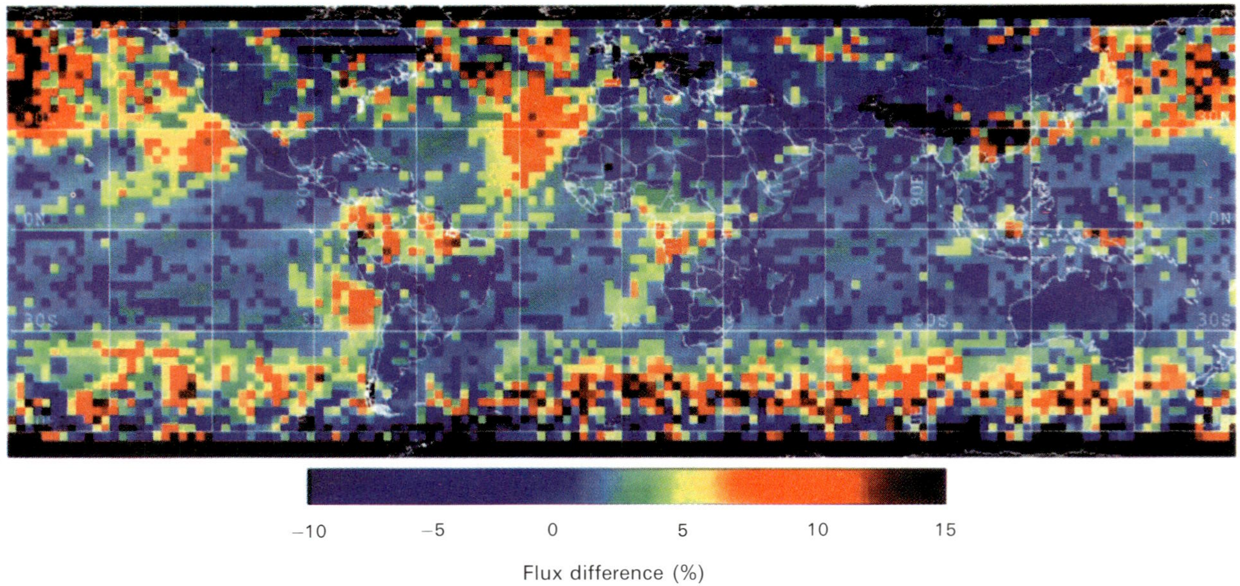

Plate 5b (*continued*)

C. JULY 1989

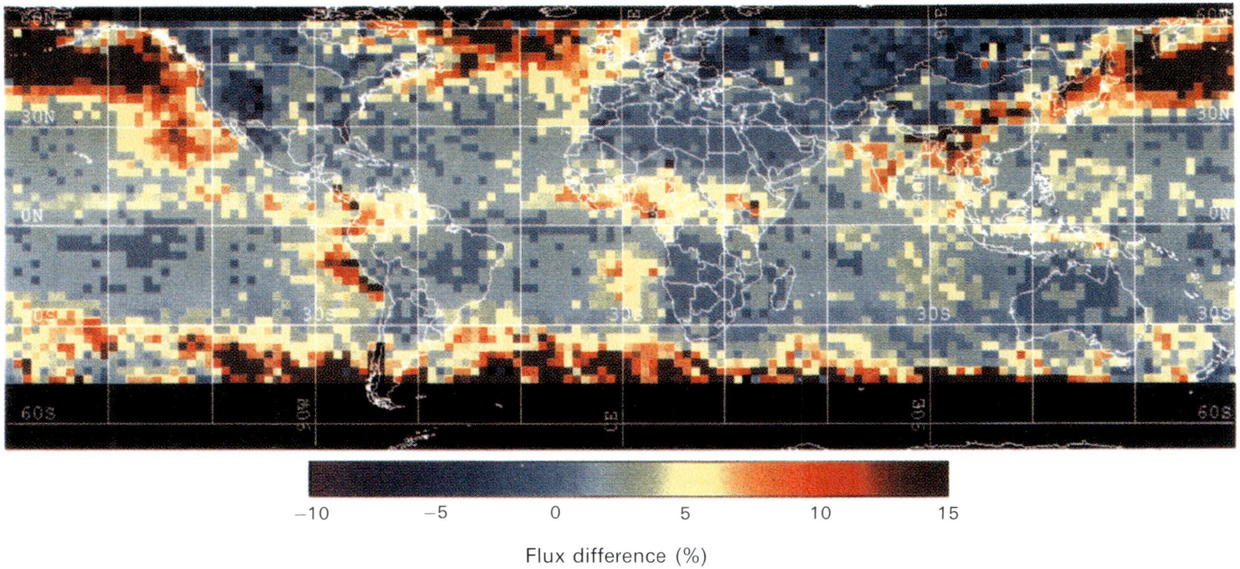

Plate 5c (*continued*)

D. OCTOBER 1989

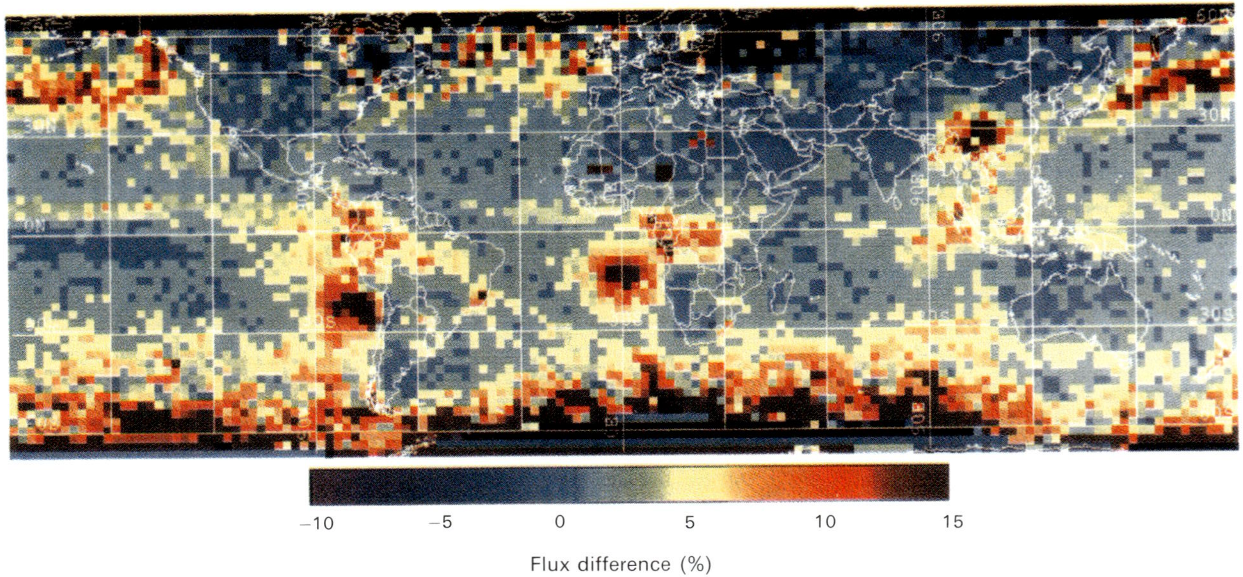

Plate 5d (*continued*)

significance for biological processes and ecological applications). Daily dose D is calculated through integration over the entire daylight period (Lubin *et al.*, 1998):

$$D = \int_0^{24\text{ hours}} \int_{280}^{400\text{ nm}} A(\lambda)F(\lambda)\, d\lambda\, dt \tag{4.1}$$

where $A(\lambda)$ is a biological action spectrum and $F(\lambda)$ is the instantaneous spectral irradiance. The two major factors of diurnal UVR variability are solar-zenith angle and diurnal variability in cloud cover. If diurnal UVB amplitude variability is defined as percentage difference between the instantaneous surface UVB irradiance at 15:00 and 09:00, then it becomes possible to filter out the solar-zenith-angle dependence (because the solar-zenith angle is the same for both of these times). Under such conditions the principal cause of diurnal variation is tropospheric opacity (mainly because of cloud cover change).

Plates 5a–d indicate that there are several temperate and tropical regions over the oceans where the afternoon UV irradiance is consistently 5–15% higher than the morning irradiance in all seasons (these regions include the south-eastern and north-eastern Pacific, the south-eastern Atlantic and parts of the tropical western Pacific). However, over most of the world's temperate and tropical oceans there is little difference between morning and afternoon irradiances. Over China and South-east Asia, and over equatorial Africa, the afternoon irradiance is consistently 5–15% higher than the morning irradiance in all seasons. During summer there are some land areas where the morning irradiance is more than 5% larger than the afternoon irradiance (mountainous parts of America, much of Russia, and most of coastal Australia) (Lubin *et al.*, 1998). It is easy to identify the principal cause of this variability, it is cloud-cover variability.

The situation is quite different, however, in the case of the time integral in surface UVR over an entire day, when the changing solar-zenith angle becomes the dominant factor, whereas cloud opacity variations play a much smaller role.

Despite the importance of the results obtained by Lubin *et al.* (1998), it is necessary to verify such results by comparison with direct surface UVR measurements. Unfortunately, such problems cannot as yet be solved in view of the fragmentary nature of existing UVR observations. For the time being, only a few limited-scale efforts to compare UV calculations with measurements have been carried out, as was done by Weihs and Webb (1997).

A very important step forward in the use of satellite information was taken by Ziemke *et al.* (1999), who examined the distribution of long-term trends in groundlevel erythemally weighted UVB exposures in northern latitudes between 1979 and 1991 using the Nimbus-7 TOMS instrument (a new erythemal UV dataset was produced recently by NASA). In the case of zonal mean erythemal UV data, results indicate ~3–7% per decade increases in mid to high latitudes. Detection of regional patterns in trends in erythemal UV favours summer months, when surface UV is strongest and noise factors such as clouds and aerosols are not as influential. Most of the regional increases in summer months (exceeding 6% per decade) in the Northern Hemisphere in the latitude range 30–40°N originate from the Pacific and

Atlantic oceanic regions. Increases (also exceeding 6% per decade) in latitudes 40–60°N appear to originate from the North American and Asian continents and also central Europe. Trends over East Asia at high latitudes indicate increases exceeding 10% per decade for May–August. Positive trends in subtropical latitudes originate mostly over oceanic regions, whereas positive trends at higher latitudes originate mostly over landmasses. Such increases may be partly due to decadal decreases in cloudiness.

Note that, although observational data on surface UVR indicate the existence of the complex picture characterized by UVR dependence on the factors mentioned above, it is undoubtedly true that, all things being equal, TOZ decreases result in surface UVR increases, in quantitative agreement with predictions by radiative transfer models (*UNEP* ..., 1994; *Scientific* ..., 1999). Specifically, large UVB increases have been observed in association with the ozone hole at high southern latitudes and also at northern mid and high latitudes. At the present time only a few studies have monitored UVB over timescales of decades, and these have yielded conflicting results on magnitude and even sign of the trends. Further improvement of instrumentation is necessary as well as continuation of regular UVR observations.

Among the factors affecting the amount of UV-irradiance is elevation above sea level. With increasing height, the overhead atmosphere becomes thinner and attenuation of UV radiation from atmospheric constituents decreases. Therefore, increase in solar UV radiation with height is expected.

Recent studies reveal vertical gradients of UV radiation with altitude of the order of 4–40%, mainly because of the different amount of atmospheric constituents which modify the transparency and the optical properties of the atmosphere (Piazena, 1996; *UNEP* ..., 1994; Varotsos and Kondratyev, 1998). Bener (1972) found an increase in UVB of about 2–16% per km at heights between 0 and 5 km. In the Swiss Alps an increase of 9–23% per km was measured (Blumthaler, 1993). Cabrera *et al.* (1994) showed an increase of 4–10% per km in the Chile region, while near Santiago, Chile this increase was 40% per km, approximately.

In the tropical Chilean Andes, Piazena (1996) observed an increase in global irradiance in UVB of about 8–10% per km as well as a change from 15 to 7% per km for UVA and for solar elevation between 20 and 90°. Changes in the direct component with height showed an increase from 60 to 20% per km for solar elevation between 20 and 90°, while diffuse irradiation showed a decrease of 2–4% per km for solar elevations of about 30–90°.

Kelley *et al.* (1995) conducted direct airborne measurements of both upwelling and downwelling irradiance from 0.2 to 7.6 km for clear-sky conditions and 58° solar-zenith angle. These observations indicate downwelling irradiance increase by 33%, while upwelling irradiance within the 0.2–7.6-km layer increased by a factor of 4. Excellent agreement was found with models of radiative transfer in cases having detailed solutions or the six-stream approximation, whereas a number of two-stream computer models agreed within 10 to 25%. Compared with clear-sky measurements, cloud enhancement of up to 58% was observed.

Varotsos and Kondratyev (1998) carried out aircraft observations of the vertical gradient of the biologically effective downwelling UV irradiance from sea level up to

10 km over Greece. These results are used to study the relation between the effective UV and overhead ozone content as obtained from vertical ozone profiles at the Athens station.

Radiometric data were obtained during the experimental campaign of the project entitled Radiation Field in the Troposphere (RAFT), in the framework of the EU project called Scientific Training and Access to Aircraft for Atmospheric Research Throughout Europe (STAAARTE). This campaign took place between 7 and 14 June 1997 and covered a large part of Greek territory. The research aircraft used in STAAARTE-RAFT was a Falcon 20-E5, D-CMET from DLR (German Aerospace Research Establishment).

To obtain the data, a UV-501 biometer (SOLAR) was used. The UV-biometer is a broadband instrument with a spectral response following the erythema action spectrum. The output of the detector is in MED/hr (Minimum Erythema Dose per hour). The effective power of UV radiation is calculated as follow: 1(MED/hr) = 0.05863 W/m^2.

The detector was mounted on top of the research aircraft and with an inclination of about 6° relative to the aircraft horizon, in order to compensate the aircraft angle of attack. Thus, the UV detector maintains horizontality during the flight. Recording frequency was set to a 1-min integrating interval. Measurements were taken throughout the flights at altitudes from sea level up to 10 km. Radiometric data were corrected relative to the solar-zenith angle, considering perpendicular incidence of the solar beam, in order to make the data comparable independently of the time of day in which the measurements were taken.

For evaluation of the altitude effect on measured effective UV radiation, only data during ascent and descent of the aircraft were considered in order to avoid latitudinal effects in the used data. In addition, while some ascents or descents of the aircraft were made in certain altitude steps, the UV dependence on the altitude range was examined. It should be noted that all the data were taken under clear-sky conditions. Measurements that were affected by the presence of clouds were excluded from the dataset. The geometric altitude is calculated in differential manner from the barometric height formula.

Ozonesondes used for vertical sounding were the Electrochemical Concentration Cells (ECC, Scientific Pump) after Komhyr (1969). For measurements of temperature and humidity, as well as for data transmission, RS-80 radiosondes (Vaisala system) were used. The interface cards for digitization of the signals and the receiving system were also from Vaisala. Data were taken during the ascent of the balloon (TOTEX 1200) at a rate of $5\,m\,s^{-1}$. Ozone sounding was performed on 13 June 1997 and the launch time was at 09:50 UTC.

In order to examine the relation between effective UV radiation and the ozone content above the measurement height level, ozone data were integrated from the considered height up to 10 hPa. Over 10 hPa, ozone residuals for the Athens Station were used.

On 7 June 1997, measurements were taken during an ascent of the aircraft between 900 and 6,300 m, at time intervals between 10 : 02 and 10 : 18 UTC. The vertical gradient of the effective UV was 7.9% per km. In the same region between

11:20 and 11:30 UTC, and at an altitude range between 6,300 and 12,200 m, the vertical gradient was 9.1% per km. Later on (11:52–12:03 UTC), at an altitude range of 6,300 to 1,200 m and at a latitude of 37.5°N, the vertical gradient was 8.8% per km.

The second experimental day was 9 June 1997. Measurements were taken during the ascent of the aircraft between 2,000 and 4,700 m. The time interval was from 08:49 to 08:55 UTC. The vertical gradient of the effective UV was 10.1% per km. The same day, during the descent of the aircraft from 4,700 to 370 m and between 09:13 and 09:28 UTC, the vertical gradient of the effective UV radiation was 11.4% per km.

On 10 June 1997, measurements were taken during the descent of the aircraft between 6,300 and 200 m, between 12:45 and 13:21 UTC. The vertical gradient of the effective UV was 9.5% per km.

In the case of 12 June 1997, measurements were taken during the descent from 6,300 to 430 m and between 10:12 and 10:27 UTC. The vertical gradient of the effective UV was 12.6% per km.

Finally, on 13 June 1997, measurements were taken during the ascent from 24 to 4,900 m and between 11:07 and 12:26 UTC. The vertical gradient of the effective UV was 9.5% per km.

All the results are summarized in Table 4.3.

Differences in vertical gradients of effective UV radiation during the flights may have arisen from variability in the amount of atmospheric constituents that modify the transparency and optical properties of the atmosphere.

In order to calculate the mean vertical gradient of effective UV radiation with height during the entire experimental period (7 days), the whole dataset was used. This is valid because measurements were taken in the same area and during similar meteorological conditions. Results are shown in Figure 4.3. The mean vertical gradient of effective UV radiation was 7.2% per km.

On 13 June 1997, at the same time when measurements of effective UV irradiance were taken over the greater city area of Athens, an ozonesonde was launched from the Athens ozonesounding station. Effective UV radiation and integrated ozone

Table 4.3. A summary of aircraft observations of UV downwelling vertical profiles.

Day	Altitude range (km)	Variation of the effective UV gradient (%/km)
7 June 1997	0.9–6.3	7.9
	6.3–12.2	9.1
	6.3–1.2	8.8
9 June 1997	2.0–4.7	10.1
	4.7–0.4	11.4
10 June 1997	6.3–0.2	13.8
12 June 1997	6.3–0.4	12.6
13 June 1997	0.1–4.9	9.5

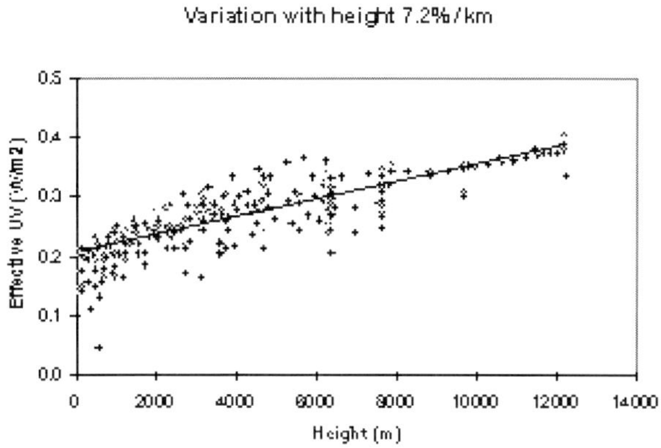

Figure 4.3. Mean variation of effective UV radiation.

content above each height level shows a strong anticorrelation, and effective UV-radiation increases of about 0.24% for each decrease of 1 DU of the overhead integrated ozone content. This analysis covered the height range from ground level up to 5,000 m. The R^2 value of the linear regression is 0.93.

4.2 CALCULATIONS OF SURFACE UV RADIATION

Calculation of radiative transfer in the atmosphere is a very broad scientific field which has continued to rapidly develop in recent years and has resulted in the creation of many techniques from very simple to very complex when 3-D atmospheric inhomogeneity and polarization are taken into account. Here we briefly discuss only those simplified techniques that are used in surface UV irradiance calculations (see also Isaksen, 1997; Madronich and Flocke, 1997; Zerefos, 1997).

As shown above, in the absence of other changes reduction in stratospheric ozone will result in increase of solar ultraviolet radiation (UVR), the enhancement of which in the 290–315-nm spectral region (UVB) is of special interest. Measurement of UVB trends is difficult because of instrumental complexity and stability. Therefore, some controversial results of long-term studies have been obtained. Nevertheless, at rural sites, the observed increase in UVB is comparable with that expected from total ozone change.

Frouin *et al.* (1989) were among the first to develop simplified techniques for UVR calculation, although many models and formulas had been earlier proposed for estimating total solar irradiance. They proposed a simple yet sufficiently accurate analytical formula to complete total and photosynthetically available solar irradiance (PAR) at the ocean surface under a clear sky, taking into account scattering by molecules and aerosols as well as absorption by water vapour, ozone and aerosols. Comparison of calculated total irradiance and PAR with observational data for both tropical and mid-latitude oceanic regions show 39 and 14 Wm^{-2} r.m.s. errors (6.5 and 4.7% of the average measured values) on an hourly timescale, respectively. Frouin *et al.* (1989) demonstrated the possibility of using an empirical approach to retrieve PAR from total irradiance values. In the case of a cloudy sky, it is generally sufficient to multiply clear-sky irradiance by cloud transmittance (for PAR the problem is reduced to determining cloud albedo, since clouds do not absorb in the PAR spectral interval).

Bodeker and McKenzie (1996) developed a semiempirical algorithm for inferring surface UV irradiance including cloud effects. Input information includes available meteorological variables and TOZ (Meteor-3 TOMS data). For substantiation of the algorithm, measurements made at Lauder (45.04°S, 169.68°E) between November 1993 and October 1994 were used to establish the relationship between cloud-induced reductions of erythemal UV and broadband irradiance, as a function of solar-zenith angle – termed cloud cover modifier functions. Application of these functions was illustrated by derivation of 10-min surface erythemal UV irradiances (EUV) for each day in November 1994. Error analysis of the November 1994 surface EUV indicates an accuracy of $0.12 \pm 0.69\,\mu W\,cm^{-2}$. Percentage error increases with solar zenith angle, with an accuracy of $2 \pm 10\%$ for solar-zenith angles less than 55°. However, such errors contribute little to the error in calculated erythemal daily doses in winter where the daily doses are small.

Ziemke *et al.* (1998) emphasized that even large daily changes in measured or predicted clear-sky UVR are usually smaller than daily UVR changes associated with transient clouds: surface UVR variability is dominated by local cloudiness conditions. They pointed out in this context that high-resolution TOMS-like side-scanning TOZ measurements are required for ozone and UVB monitoring and prediction, rather than SBUV-type nadir observations.

Since currently no standard results exist for radiation transfer codes that simulate surface UVR, Van Weele *et al.* (*Abstracts* ..., 1998) undertook an intercomparison of 12 numerical models to substantiate standard UVR spectra to be representative of various potential atmospheric conditions, which could serve as benchmarks. An important result of this intercomparison was that even for the most simple conditions (clear sky, high solar elevation) individual model calculations can still differ by up to 10–20% depending on the wavelength and on the considered case.

Koepke *et al.* (*Abstracts* ..., 1998) compared 18 radiative transfer models with different complexity that are in use for calculation of UV index. Calculations were made for more than 100 cloud-free atmospheres. A number of other approaches were discussed at the European Conference on Atmospheric UV Radiation (*Abstracts* ..., 1998).

Recently, Varotsos *et al.* (1995) proposed a simple parametric model for calculation of direct and diffuse UVR reaching the Earth's surface under different atmospheric conditions (including ozone content in the atmosphere) depending on time (including daily and seasonal variations) and geographic location. This model has been checked against field UV measurements on a daily, monthly and annual basis (Varotsos *et al.*, 1995).

Direct UVR propagating through the atmosphere and reaching the ground can be calculated by means of the following expression:

$$I_d(\lambda) = H_o(\lambda) D \tau_r(\lambda) \tau_a(\lambda) \tau_w(\lambda) \tau_o(\lambda) \tau_u(\lambda) \tag{4.2}$$

where $I_d(\lambda)$ is the direct normal irradiance at the Earth's surface at wavelength λ; $H_o(\lambda)$ is the extraterrestrial solar irradiance at the mean Earth–Sun distance, at a given wavelength λ; and D is the correction factor for the change in Earth–Sun distance throughout the year. It is given by the following simplified equation:

$$D = 1.000\,11 + 0.034\,221 \cos y + 0.001\,280 \sin y + 0.000\,719 \cos(2y) + 0.000\,077 \sin(2y) \tag{4.3}$$

where y is the day angle ($2\pi(d\text{-}1)/365$, d represents the Julian day that is the number of days elapsed from the beginning of the year). Extraterrestrial solar spectral irradiance was obtained by Neckel and Labs (1981). The parameters $\tau_r(\lambda)$, $\tau_a(\lambda)$, $\tau_w(\lambda)$, $\tau_o(\lambda)$ and $\tau_u(\lambda)$ denote transmission functions of the atmosphere at wavelength λ for molecular (Rayleigh) scattering, aerosol attenuation (scattering and absorption), water-vapour absorption, ozone absorption and uniformly mixed minor gases (CO_2, CO, CH_4, N_2O, O_2) absorption, respectively.

Various transmission functions used for calculating the components of formula (4.2) are:

- Rayleigh scattering transmission function (Bird and Riordan, 1986):

$$\tau_r(\lambda) = \exp\{-M'[\lambda^4(115.6406 - 1.335/\lambda^2)]\} \tag{4.4}$$

Aerosol attenuation transmission function (Bird and Riordan, 1986):

$$\tau_a(\lambda) = \exp(-\beta\lambda^{-a}M) \tag{4.5}$$

- Water-vapour transmission function (Leckner, 1978):

$$\tau_w(\lambda) = \exp[-0.2385 a_w(\lambda) WM/(1 + 20.07 a_w(\lambda) WM)^{0.45}] \tag{4.6}$$

- Ozone absorption transmission function (Leckner, 1978):

$$\tau_o(\lambda) = \exp[-a_o(\lambda) O_3 M_o] \tag{4.7}$$

- Transmission function for uniformly mixed gases (Leckner, 1978):

$$\tau_u(\lambda) = \exp[-1.41 a_u(\lambda) M'/(1 + 118.3 a_u(\lambda) M')^{0.45}] \tag{4.8}$$

In the above equations, λ is the wavelength in μm, M is the relative airmass and M' is the pressure-corrected relative airmass. The relative airmass is given by the following formula:

$$M = [\cos z + 0.15(93.885 - z)^{-1.253}]^{-1} \tag{4.9}$$

where z is the apparent solar-zenith angle in rad. It can be calculated by means of a computer algorithm. That algorithm can compute solar position with a precision to

Table 4.4. Temporal variation of Angstrom's turbidity coefficient throughout the year, based upon observations.

Month	β
January	1.22
February	1.197
March	1.0826
April	0.89
May	0.82
June	0.71
July	0.82
August	0.89
September	–
October	1.083
November	1.14
December	1.22

1′ of arc. The pressure-corrected relative optical airmass is given by the equation:

$$M' = MP/P_o \tag{4.10}$$

where P is the measured surface pressure in hPa and $P_o = 1013\,\text{hPa}$. Also, in equation (4.5) β is Angstrom's turbidity coefficient (Table 4.4) and a is the wavelength exponent. Angstrom's turbidity coefficient is given by the following expression:

$$\beta = [0.025 + 0.1\cos^2(\text{lat})]\exp(-0.7h) \tag{4.11}$$

where lat is the latitude of the area of interest and h is the altitude of the same area.

The latter expression gives the annual average of Angstrom's turbidity coefficient versus latitude. In order to account for temporal variation in this magnitude throughout the year, we considered a number of cases, relying on our own observations.

Note that in this model the dependence of Angstrom's turbidity coefficient on local conditions has not been taken into account.

The wavelength exponent α in equation (4.5) has the following values (Bird and Riordan, 1986):

$$\text{if } \lambda < 0.5\,\mu\text{m, then } \alpha = 1.0274 \quad \text{and} \quad \text{if } \lambda \geq 0.5\,\mu\text{m, then } \alpha = 1.206$$

Also, in equations (4.6), (4.7) and (4.8)

$a_w(\lambda)$ is the water-vapour absorption coefficient at wavelength λ and W is the precipitable water vapour (in cm) in a vertical path. Because $a_w(\lambda) = 0$ in the spectral region 0.28–0.75 μm, $\tau_w(\lambda) = 1$. $a_o(\lambda)$ is the ozone absorption coefficient at wave-

Table 4.5. Linear dependence of h_o on latitude (lat)

Month	h_o
January	$h_o = 35.919 + 0.0069$ (lat -30)
February	$h_o = 36.043 + 0.0858$ (lat -30)
March	$h_o = 33.799 + 0.1770$ (lat -30)
April	$h_o = 34.048 + 0.6680$ (lat -30)
May	$h_o = 31.610 + 0.1680$ (lat -30)
June	$h_o = 31.667 + 0.1881$ (lat -30)
July	$h_o = 34.238 + 0.1031$ (lat -30)
August	$h_o = 34.186 + 0.0986$ (lat -30)
September	$h_o = 34.069 + 0.1284$ (lat -30)
October	$h_o = 33.909 + 0.1789$ (lat -30)
November	$h_o = 33.734 + 0.2527$ (lat -30)
December	$h_o = 35.959 + 0.0672$ (lat -30)

length λ, O_3 is the total ozone amount in atm-cm and M_o is the relative ozone atmospheric mass. It is given by the following equation (Bird and Riordan, 1986):

$$M_o = (1 + h_o/6.370)/[\cos^2 z + 2h_o/6.370]^{0.5} \tag{4.12}$$

In the latter equation, h_o is the height of the maximum ozone concentration. In this model, h_o is calculated via the 12 expressions in Table 4.5, which resulted from the examination of h_o values for different latitudes and the whole year. A linear dependence of h_o on latitude was identified and is presented in Table 4.5.

The equations in Table 4.5 are valid for both hemispheres up to 70° latitude.

Also, $a_u(\lambda)$ is a combination of an absorption coefficient and gas amount. The values of $H_0(\lambda)$, $a_0(\lambda)$ and $a_u(\lambda)$ are given by Bird and Riordan (1986) for various wavelengths.

In order to approximately take into account the percentage cloud amount in the sky, we inserted the multiplying factor SK into expression (4.2). Three different cases have been specified (Diffey, 1982):

Clear sky	SK = 1
Partially cloudy sky	SK = 0.7
Overcast sky	SK = 0.2

In order to calculate direct solar irradiance at the Earth's surface on a horizontal plane, we must multiply the direct normal irradiance at the Earth's surface $I_d(\lambda)$ by $\cos z$ where z is the apparent solar-zenith angle.

In general, total diffuse irradiance on a horizontal surface at the ground is given by the equation:

$$I_s(\lambda) = I_r(\lambda) + I_a(\lambda) + I_g(\lambda) \tag{4.13}$$

where $I_s(\lambda)$ is the total diffuse irradiance on a horizontal surface at the ground, at wavelength λ; $I_r(\lambda)$ is the Rayleigh scattering component; $I_a(\lambda)$ is the aerosol

Table 4.6. Ground Albedo values.

Fresh snow	0.85
Ice	0.35
Sand	0.25
Greasy field	0.2
Dry ploughed field	0.125
Water	0.1
Forest	0.065

scattering component, and $I_g(\lambda)$ is the component that takes into account multiple reflected radiation between the air and the ground.

The model, modified by Bird and Riordan (1986), of Justus and Paris (1985) is employed for the calculation of diffuse irradiance. To consider percentage cloud amount in the sky, we use the following equation instead of equation (4.13)

$$I_s(\lambda) = [I_r(\lambda) + I_a(\lambda) + I_g(\lambda)]\,\mathrm{SK} \tag{4.14}$$

Special care is given to the use of appropriate ground albedo values. Albedo values are important for determination of multiple reflected radiation. For this purpose, the values in Table 4.6 are used. In the case of Athens we used the value 0.3 for mean ground albedo.

Total solar irradiance on a horizontal surface at the ground varies with the apparent solar-zenith angle z as follows:

$$I_t(\lambda) = I_d(\lambda)\cos z + I_s(\lambda) \tag{4.15}$$

where $I_t(\lambda)$ is total solar irradiance on a horizontal surface at the ground, at wavelength λ.

Application of the above-described model for the case of Athens was done by employing meteorological data collected from daily observations made at Athens University Ozone Station.

Daily column ozone observations made by TOMS (version 7) onboard the Nimbus-7 satellite (January 1979–April 1993) and groundbased TOZ measurements by Dobson No. 118 spectrophotometer at Athens, Greece (November 1991–December 1996) were considered as input values in the model. Very good agreement of daily TOZ values as deduced from the observations made by the two instruments, during common periods of their operation, justifies the assumption that this time series of 18 years of data is homogeneous.

Results from linear trend analysis, applied to the monthly mean values of TOZ, as deduced from daily measurements over Athens, for the mid-wintertime month (January) and the mid-springtime month (April) are shown in Figure 4.4a, b. It is evident from the figure that percentage total ozone depletion varies from about 15% in January to 10% in April. Recent studies, discussed in Chapter 2, showed that interannual variability has strong influence on TOZ depletion rates. For instance, Chandra *et al.* (1996) showed that the dynamically induced component of the trend

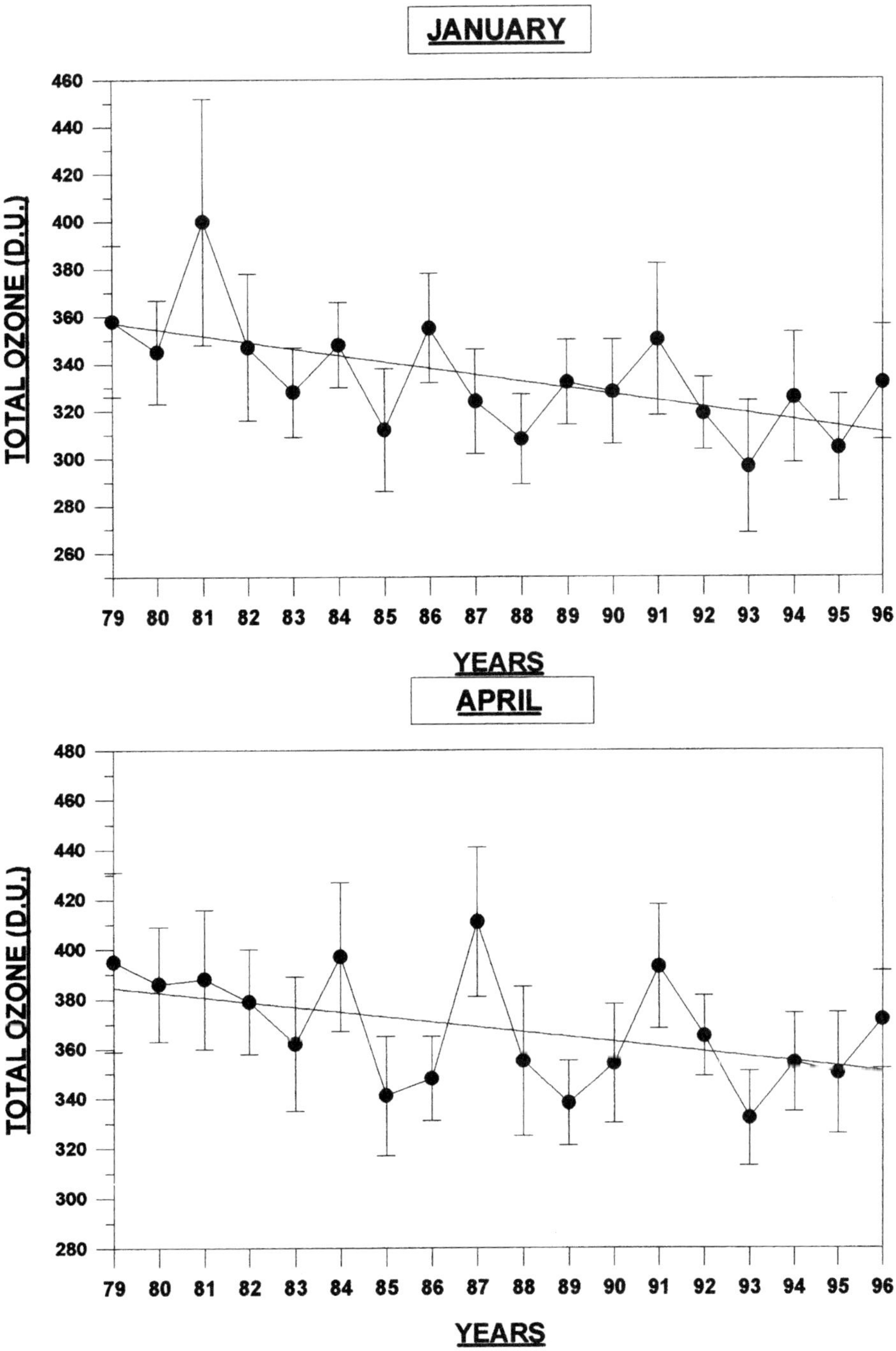

Figure 4.4. Monthly means and standard deviations in TOZ values deduced from TOMS and Dobson measurements at Athens (Greece) for January (a) and April (b).

in TOZ is about 5% and 3.3%, for January and April, respectively. Considering this influence, net ozone loss is significantly smaller than that shown in Figure 4.4. Inspection of the figure indicates that an extreme total ozone minimum occurred during the transition period from winter to summertime of 1993. This event was observed on a global scale.

It is therefore of great importance to use all available information to assess relevant increased levels of UVB at the Earth's surface at that time.

Use of daily total ozone values as input information for the above-mentioned parametric model allows us to detect fluctuations in monthly mean UVB irradiance for the summer months between 1993 and 1996 relative to the long-term mean (time period 1979–1996). Results are shown in Figure 4.5 where UVB fluctuations are plotted together with corresponding TOZ variations. Here the ratios of both TOZ and broadband UVB (290–315 nm) to the long-term means are shown. It is evident from Figure 4.5 that the reduction in TOZ of the order of $3.3 \pm 1.9\%$ resulted in an increase in calculated broadband UVB of $3.6 \pm 1.4\%$. Although the real measurements of broadband UVB made at the Athens station are influenced by local air pollution, very good agreement between observed and theoretically derived UVB values took place.

The 1993 total ozone content change were characterized by the lowest values observed above Athens since 1978. This sudden total ozone decline is consistent with the linear trend defined in earlier studies. The theoretically deduced UVB estimates

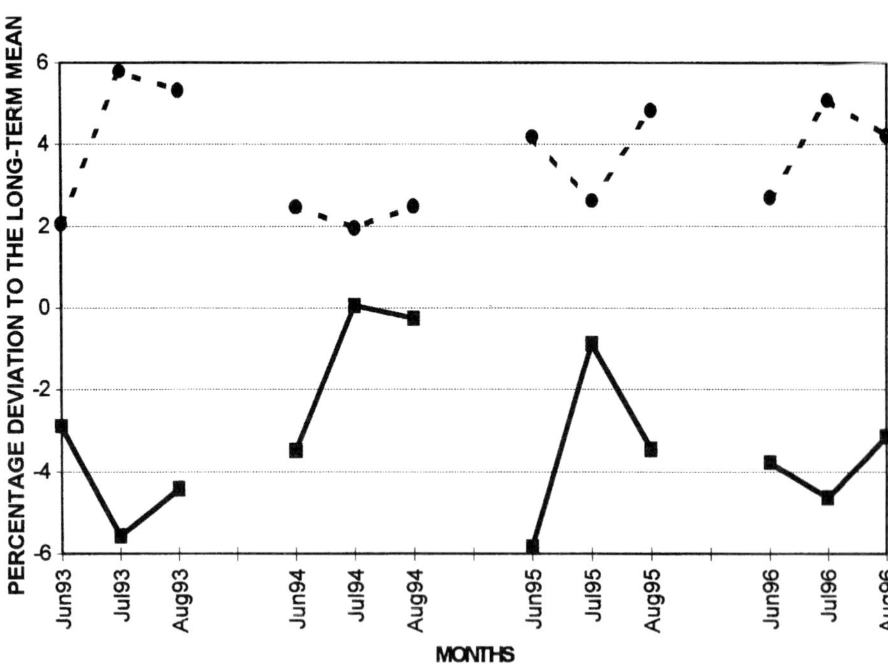

Figure 4.5. TOZ and UVB at Athens, expressed as % deviation in long-term mean, for summer.

Table 4.7. Atmospheric optical models for selected UV wavelength.

Model	λ, μm	0.28	0.30	0.32	0.34	0.36	0.40
	τ_{Rel}	1.545	1.222	0.927	0.717	0.564	0.364
	$\tau_{a\,scatt}$	0.387	0.371	0.355	0.339	0.339	0.276
	$\tau_{a\,abs}$	0.04	0.04	0.04	0.04	0.04	0.04
I	τ_{oz}	36.14	3.445	0.306	0.022	0.61(−3)	0.0
II	$\tau_{oz} - \tau_{oz\,max}$	13.56	1.110	0.117	0.007	0.16(−3)	0.0
III	$0.9\tau_{oz}$	32.43	3.100	0.275	0.019	0.55(−3)	0.0
I	τ_{I}	38.11	5.078	1.628	1.118	0.944	0.680
II	τ_{II}	24.55	2.744	1.439	1.103	0.943	
III	τ_{III}	34.50	4.733	1.597	1.116	0.944	
I	ω_{0I}	0.051	0.314	0.788	0.945	0.957	0.942
II	ω_{0II}	0.079	0.581	0.891	0.957		
III	ω_{0III}	0.056	0.337	0.802	0.947		
	S_λ, mW cm$^{-2}\,\mu^{-1}$	31.2	53	82	110	100	160

λ – wavelength; τ_{Rel} – scattering optical thickness of molecular atmosphere; $\tau_{a\,scatt}$ – scattering optical thickness of atmospheric aerosols; $\tau_{\alpha\,abs}$ – absorption optical thickness of atmospheric aerosols; $\tau_{oz} - T_{oz\,max}$ – absorption optical thickness of total ozone content without ozone maximum; $0.9\tau_{oz}$ – absorption optical thickness of total ozone content decreased for 10%; τ_I, τ_{II} and τ_{III} – optical thicknesses for three atmospheric models; ω_{0I}, ω_{0II} and ω_{0III} – single scattering albedo for three atmosphere models; S_λ incident solar irradiance on top of the atmosphere.

are in close agreement with observations. This fact supports the conclusion that severe ozone loss is accompanied by substantial enhancement of UVB doses.

Kondratyev *et al.* (*Abstracts* ..., 1998) carried out calculations of UV solar radiation incoming to the Earth's surface for three simple optical models of the atmosphere which differ in total ozone content. Three models were constructed for different atmospheric ozone content: (1) standard TOZ; (2) without ozone maximum (~0.3 TOZ); (3) less than standard by 10% (0.9 TOZ). Relevant values of optical thickness and single scattering albedo are presented in Table 4.7.

Surface irradiance calculations were carried using the delta-Eddington approximation for cosines of solar-zenith angles equal to 1, 0.5 and 0.1. Phase function corresponds to Rayleigh atmosphere. Ground albedo A is equal to 0 and 0.8. Results of numerical simulation are presented in absolute units in Table 4.8. It is clear that variations connected with changes of seasons and ground albedo are much stronger than variations induced by different ozone content. Impact of ground albedo grows with wavelength. For comparison with observations, spectral radiation measurements were taken from aircraft at different altitudes. The diffraction spectrometer, used in the aircraft experiments for measurements of solar fluxes, had been designed and constructed at the Atmospheric Physics Department of Leningrad State University. The spectral region of the monochromator is 0.250–1.000 μm. For irradiance measurements, frosted quartz or opal glass plate is used as the integrating element. The spectral resolution of the spectrometer is 0.002 μm. The recording time of one spectrogram is ~7 s, the time of inverse motion is 3 s. We use here the data at lowest altitude level 1,000 hPa for three dates with different values of ground albedo

Table 4.8. Downwelling solar irradiance to horizontal surface ($mW\,cm^{-2}\,\mu^{-1}$).

Model	λ, μm / ζ	0.28	0.30	0.32	0.33	0.34	0.36	0.40
	1.0	0	0.53	36.5	55.5	76.0	73.9	127.5
I, $A = 0$	0.5	0	0.016	10.7	22.2	28.0	28.2	50.8
	0.1	0	0.001	1.0	2.2	3.05	3.1	6.1
$A = 0.8$	0.5	0	0.016	12.9	27.5	37.3	36.8	62.0
	1.0	0	7.69	46.7	70.0	77.3		
II	0.5	0	1.25	15.7	25.0	28.9		
	0.1	0	0.011	1.6	2.8	3.2		
	1.0	0	0.80	38	66.6	76.0		
III	0.5	0	0.03	11.4	23.3	28.1		
	0.1	0	0.002	1.07	2.6	3.07		

Table 4.9. Measurement results for the selected wavelength corresponding to numerical simulation.

λ, μm / ζ	0.33	0.34	0.36	0.40
0.629, sand, November, desert	0.023	0.108	25.9	64.6
0.5	0.018	0.086	20.6	51.6
0.731, water, May, Ladoga	7.84	18.6	23.6	54.8
0.5	5.36	12.7	16.1	37.5
0.670, snow, April, Ladoga	19.9	30.5	38.3	82.1
0.5	14.9	22.8	28.6	61.3

and for different seasons as presented in Table 4.9. For comparison of measured irradiance with calculations, the data were reduced to solar incident angle cosine equal to 0.5.

Spectral measurement data and calculated values are shown in Figure 4.6. It can be seen from Table 4.9 and Figure 4.6 that surface UV irradiance is extremely variable because of dependence on wavelength and observation conditions. Variations connected with changes of seasons and surface albedo are much stronger than changes induced by different ozone contents. The impact of surface albedo grows with wavelength. Retrieval of ozone content from UV measurement data gives the approximate value of τ_{oz} of about 1.6 for $\lambda = 0.33$ μm. It corresponds to 440 DU of total ozone season maximum for latitude 60°N. The strong decrease of UV irradiance at $\lambda = 0.33$ μm on 16 October 1983 may be explained by strong influence of aerosol absorption and turbidity of the atmosphere.

As mentioned above, an important factor in UVR variability are clouds. It can be assumed that surface UV irradiance may increase slightly in the case of a small amount of cloud compared with the case of clear sky because of reflection and

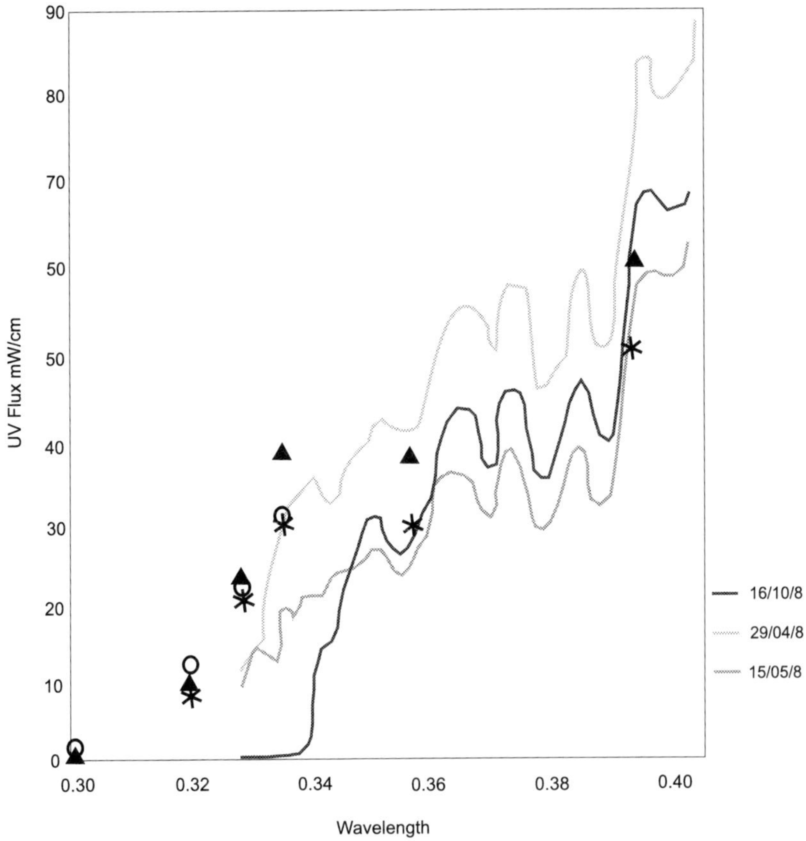

Figure 4.6. Results of spectral measurements (lines) and numerical simulations (points) of surface UV irradiance: asterisks – model *I*, $A = 0$; circles – model I, $A = 0.8$; triangles – model II, $A = 0$; rhombus – model III, $A = 0$.

scattering of solar radiation from cloud sides. Then, as cloud amount surface grows UV-irradiance is supposed to decrease. In this context, Kondratyev *et al.* (*Abstracts* ..., 1998) undertook an analysis of a 3-yr dataset for UV irradiance measured in St. Petersburg and Athens. The influence of solar zenith angle is taken into account. The dependence of surface UV irradiance on cloud amount and solar-zenith angle was considered. Observations indicate that cloudiness is essential and is the most changeable feature to have an impact on surface UV irradiance. This is what makes the study of the cloud influence in UV solar irradiance very important. It is known that clouds decrease UV radiation penetration to the surface. But it was recently observed that under certain conditions cumulus clouds increase UV flux by radiation reflection from their sides of 20% (Mims and Frederick, 1994). Aircraft observations by Hayasaka *et al.* (1995) revealed the same effect in broken cumulus.

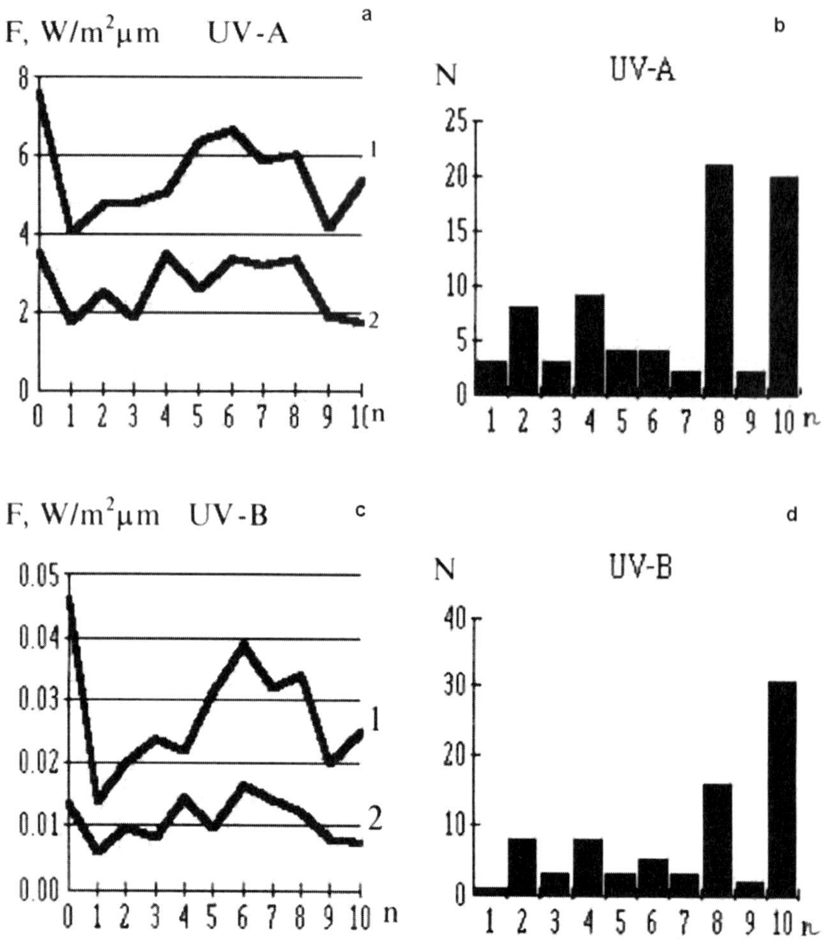

Figure 4.7. Dependence of surface UV irradiance on cloud amount. 1 – for mean values, 2 – for maximum values. *N* – number of cases when surface irradiance in cloudy conditions exceeds that clear-sky conditions. Surface albedo corresponds to wet soil; *n* – cloud amount.

For the purpose mentioned, routine UV-radiation observations over 3 years were processed to evaluate the role of broken clouds in reflection, transmission and scattering solar radiation. Measurement data were divided into groups of 10° range in solar elevation angle. In each group, dependence of UVB and UVA irradiance on cloud amount was considered. The results are shown in Figure 4.7–4.8. The number of cases when surface irradiance in cloudy conditions exceeds that in clear-sky conditions are presented in Figure 4.7b (UVA) and d (UVB). Two kinds of surfaces are presented: wet soil and snow for solar elevation ranges 0–10°.

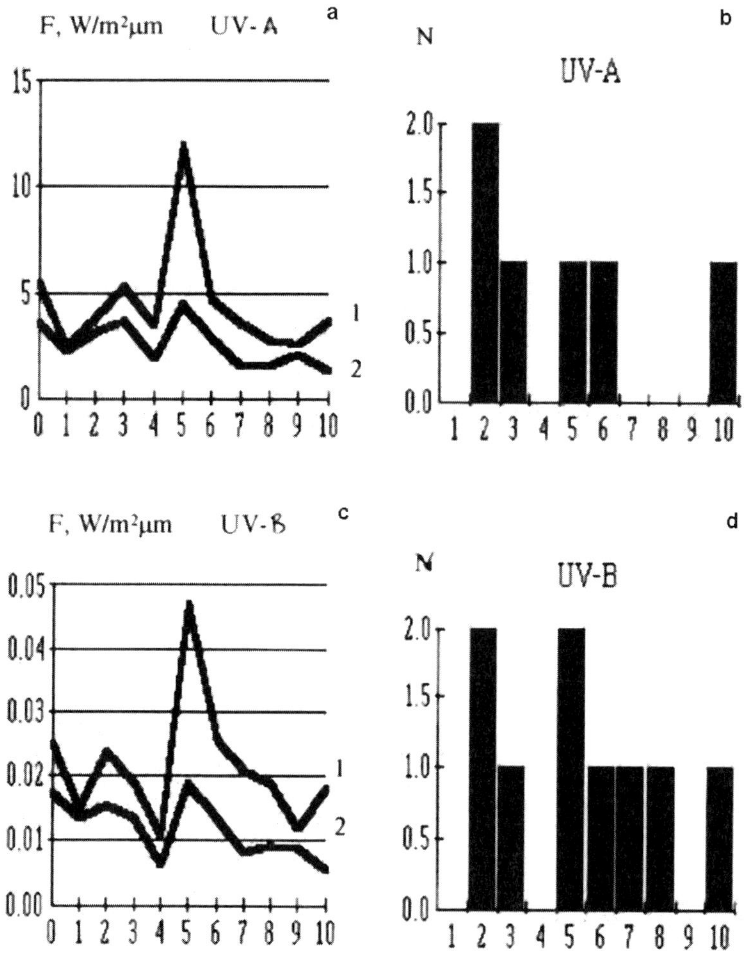

Figure 4.8. As Figure 4.7, but surface albedo corresponds to snow.

4.3 UNCERTAINTY IN MODELLED UV UP-AND-DOWN FLUXES BECAUSE OF LIMITED AVAILABILITY OF INPUT DATA[1]

4.3.1 Introduction

As discussed in Section 4.1, the main consequence of ozone depletion now apparent over much of the globe, is the increase in biologically damaging UV irradiance at

[1] This section was written in collaboration with E. V. Shalina.

groundlevel. Since biologically damaging irradiance is expected to increase in the near future, efforts have been made to create a global network of UV monitoring. Because UV irradiances have to be registered with high accuracy, spectral measurements of UV are expensive and difficult and therefore are restricted to a few locations. Radiative transfer calculations can provide UV levels for locations where measurements are not available, but the question remains open about UV irradiance accuracy that model calculations can deliver, having been discussed in a number of papers (Forster *et al.*, 1995; Schwander *et al.*, 1997; Zeng *et al.*, 1994). The aim of further consideration is, in the framework of model calculations, to assess the possibility of solving the problem of calculating surface UV irradiance levels using UV radiances measured at the top of the atmosphere by satellites.

The quality of the results of radiative transfer model calculations depends, on the one hand, on the complexity and precision of the used radiative transfer algorithm and, on the other hand, on the accuracy of the values for input atmospheric parameters. With respect to the latter, difference exists between internal constraints of a model and the variable input parameters used to describe the actual state of the atmosphere. For provision of vertical profiles of atmospheric constituents, especially great measurement efforts are required. Therefore, a common approach towards describing the state of the atmosphere for modelling is to prescribe of standard profiles for vertical distributions of atmospheric constituents, in combination with actual values for vertical integrals of the relevant parameters (total ozone amount, aerosol optical depth, etc.). We shall consider the question of whether provision of integral values is sufficient for modelling UV up-and-down radiances and what are the uncertainties in modelled UV radiance with respect to limited accuracy or non-availability of input vertical profiles for input parameters.

4.3.2 Method and model description

First, a sensitivity study was carried out to answer the question of how strongly uncertainty, because of limitations of measurement, observation or non-availability of atmospheric input parameters, affects modelled UV radiances. Second, analysis was done to compare sensitivities of up-and-down radiance. Calculations were restricted to cloud-free conditions. The delta-Eddington approximation was used for calculations of radiative transfer. The radiance was calculated as a function of solar-zenith angle for atmospheres consisting of 33 homogeneous layers above a reflecting surface, taking into account multiple molecular and aerosol scattering as well as absorption caused by aerosol particles and most important gases.

Atmospheric input parameters are divided into two classes: (1) bulk parameters (total amounts of ozone and other atmospheric gases, aerosol optical depth for one wavelength, surface albedo and others); (2) non-bulk parameters (vertical profiles, optical properties of aerosol particles and others). In the case of bulk parameters, only one value (the integral) is necessary to describe atmospheric conditions, while non-bulk parameters contain more detailed information. We have concentrated our efforts on parameters of ozone and aerosol. Uncertainty in the modelled radiance is calculated separately for every parameter. This means that the effect of one parameter is studied under the assumption that all other bulk and non-bulk values are error free. For bulk and non-bulk parameters which are related to the same atmospheric constituent (for example, ozone amount and ozone profile) the uncertainty in the profile is simulated by using different profiles but is normalized to the same total amount. UV radiance strongly depends on wavelength. Calculations indicate, however, that within a narrow wavelength interval, around 0.3 μm, radiance is equally sensitive to ozone and aerosol parameters. Therefore, most of the calculations were done for this wavelength. Four solar-zenith angles were prescribed with cosines equal to 1.0, 0.65, 0.5 and 0.3, the first being a theoretical reference, the second and third characterize mid-latitude conditions, and the last represents polar conditions. Solar incoming irradiance is taken to be unity, which means that all results of calculations are weighted to the solar irradiance.

4.3.3 The effect of uncertainty in model input parameters on UV radiances

Together with the solar-zenith angle, which can be determined for a given time and location, bulk parameters represent the minimum subset of input parameters for calculation of UV radiance. The accuracy of bulk values depends largely on how they are obtained.

Uncertainty for measurements of TOZ by Brewer or Dobson spectrometers is estimated to be about 3% (Schwander *et al.*, 1997). However, such measurements, as mentioned in Chapter 1, are restricted to a few locations, which means that for other places interpolation or other assessment procedures are to be used, and this is the main reason for taking larger values of uncertainty into account in our study. Therefore, we have calculated how fluxes change for TOZ depletion of up to 20%.

Figures 4.9–4.11 correspond to an atmosphere heavily loaded with aerosols, which means that aerosol and ozone have nearly the same optical depths for the 0.3-μm wavelength. The ordinate of the plots is scaled from 0 to 0.5 units of weighted flux to ensure easy comparison of the effects. These three figures show changes in up-and-down fluxes caused by TOZ depletion (bulk parameter) of up to 20%. These dependencies were calculated for three different compositions of aerosol (non-bulk parameters) to show mutual dependencies of ozone bulk and aerosol non-bulk parameters, total aerosol loading being unchanged. Figure 4.9 shows the results of calculations for non-absorbing aerosol in the atmosphere. Figure 4.11 relates to volcanic aerosol in the stratosphere with albedo for single scattering equal to 0.9.

Analysis of the results of calculations reveals the correlation between up-and-down irradiances when they change with ozone depletion. The correlation coefficient between the sets shown in Figure 4.9a and Figure 4.9b was obtained within the range −0.976–−0.985 for different solar angles. The correlation between sets of Figure 4.10 is even higher: -0.99. In the case shown in Figure 4.11 the upward flux dependence changes its character so appreciably that correlation becomes insignificant.

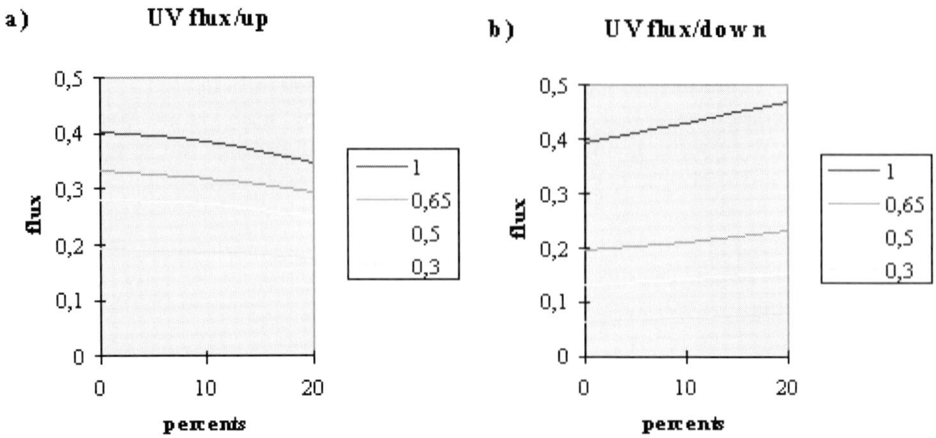

Figure 4.9. Dependence of up-and-down fluxes on % of ozone depletion. Four curves correspond to four solar-zenith angles, their cosines being shown to the right of the diagram. The albedo for single scattering of aerosols in the boundary layer is taken to be 0.999 and the same for stratospheric aerosols. Aerosol optical depth is 2.9.

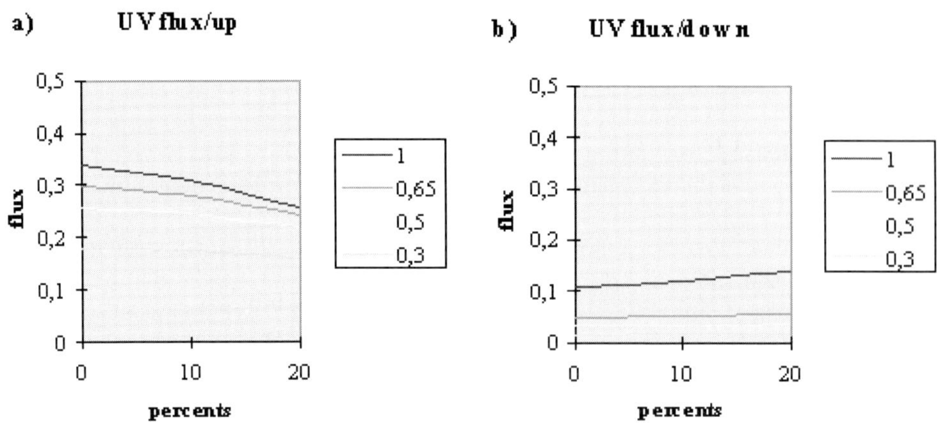

Figure 4.10. Dependence of up-and-down fluxes on % of ozone depletion. Four curves correspond to four solar-zenith angles, their cosines being shown to the right of the diagram. The albedo for single scattering of aerosols in the boundary layer is taken to be 0.5 and for stratospheric aerosols –0.999.

It can be seen that dependence of upward irradiance on ozone depletion is mostly sensitive to changes in characteristics of stratospheric aerosol: changes of 7–14% in upward flux with ozone depletion of 20% in Figure 4.9a turn out to be 1–3% in Figure 4.11a, where parameter ω of stratospheric aerosol decreases from 0.999 to 0.9. Downward flux dependence is also very sensitive to the mentioned changes. In this

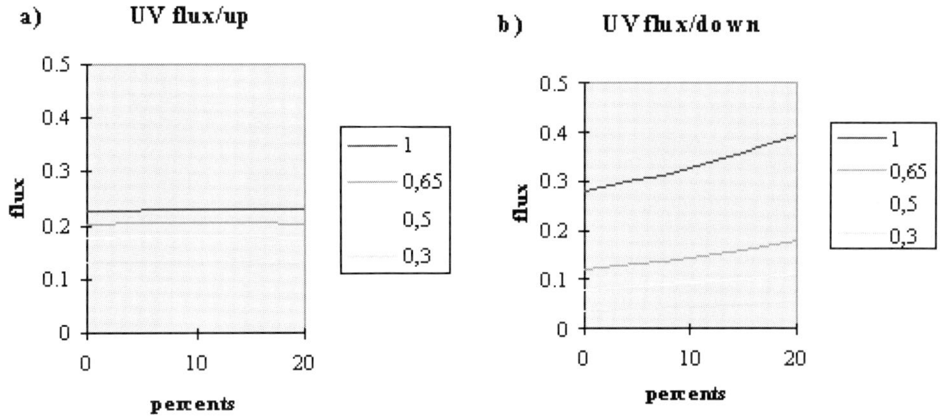

Figure 4.11. Dependence of up-and-down fluxes on % ozone depletion. Four curves correspond to four solar-zenith angles, their cosines being shown to the right of the diagram. The albedo for single scattering of aerosols in the boundary layer is taken to be 0.999 and for stratospheric aerosols –0.9.

case, changes of 17–19% in downward flux with 20% ozone depletion (Figure 4.9b) become 40–45% in Figure 4.11b. Regarding the influence of non-bulk aerosol parameters on values of up-and-down irradiances, it can be said that downward flux is mostly sensitive to boundary-layer aerosol properties: the value of the flux is four times as small when the value of ω decreases from 0.999 (Figure 4.9b) to 0.5 (Figure 4.10b). Upward flux is sensitive to changes in the stratosphere: it decreases by 44–31% when the value of ω of stratospheric aerosol changes, as indicated in Figure 4.9a through Figure 4.11a. Taking into account the overall goal of this study, a note should be made of the different (even opposite) sensitivity of up-and-down fluxes to non-bulk aerosol parameters, which complicates the task of estimation of UV irradiances from measured upward flux.

The next step in simulation modelling was investigation of the sensitivity of up-and-down fluxes to changes in aerosol bulk parameter – its optical depth. Figure 4.12 shows the results of radiative transfer calculations for aerosol optical depth which equals one-half of the optical depth taken for calculations shown in the Figure 4.9–4.11. Dependencies shown in Figure 4.12 are obtained for the same non-bulk parameters of aerosol as for Figure 4.9. Though the values of fluxes change (downward flux increases by 33% and upward flux decreases by 7–10% with decrease in optical depth from 2.92 to 1.45), trends of increase (decrease) with TOZ changes remain nearly the same as in Figure 4.9: increase in downward flux is 17–18% (it has been 17-19% in Figure 4.9) and decrease in upward flux is 10–16% (it was 8–10% in Figure 4.9) when ozone depletion reaches 20%. Statistical analysis shows very high correlation between the results of calculations shown in Figure 4.9 and Figure 4.12: it is about 0.99.

Figures 4.13–4.15 show the sensitivities of up-and-down irradiances to changes in aerosol optical depth for three sets of non-bulk aerosol parameters, the same as

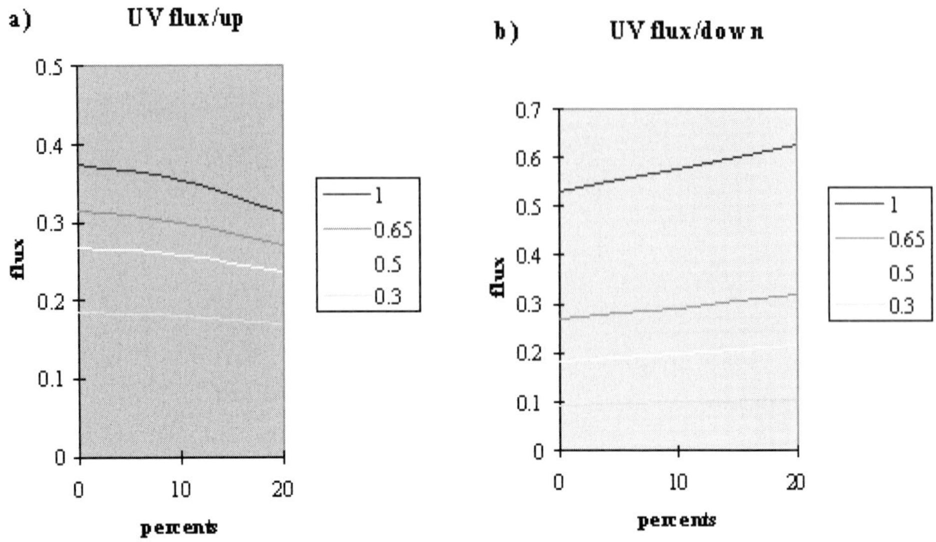

Figure 4.12. Dependence of up-and-down fluxes on % of ozone depletion. Four curves correspond to four solar-zenith angles, their cosines being shown to the right of the diagram. The albedo for single scattering of aerosols in the boundary layer is taken to be 0.999 and the same for stratospheric aerosols. Aerosol optical depth is 1.45.

prescribed for Figures 4.9–4.11. It is apparent that downward flux is more sensitive to aerosol changes than upward flux. Analysis of the results presented in Figures 4.13–4.15 shows that transformations caused by changes in vertical distribution of absorption properties of atmospheric aerosol are quite similar to those in Figures 4.9–4.11. Correlation between a and b sets in all these figures (i.e. between up-and-down irradiances) is close to 0.9.

Figures 4.16 and 4.17 demonstrate the dependencies of up-and-down irradiances on cosine of solar-zenith angle. The upper curves are for downward fluxes, the lower for upward ones. The difference in the sets presented in Figure 4.16 is caused by ozone depletion (three curves in every set correspond to 0.10 and 20% decrease in ozone concentration). in Figure 4.17 the difference between the two curves is generated by the change in asymmetry factor of atmospheric aerosol phase function. Dependencies are not very strong, but sets of up-and-down fluxes are highly correlated.

4.3.4 Conclusion

The most important result of sensitivity calculations is that not only the magnitude, but also even the sign of the effect obtained, by changing just a single input parameter, depends on the values of many of the others. This makes it difficult to

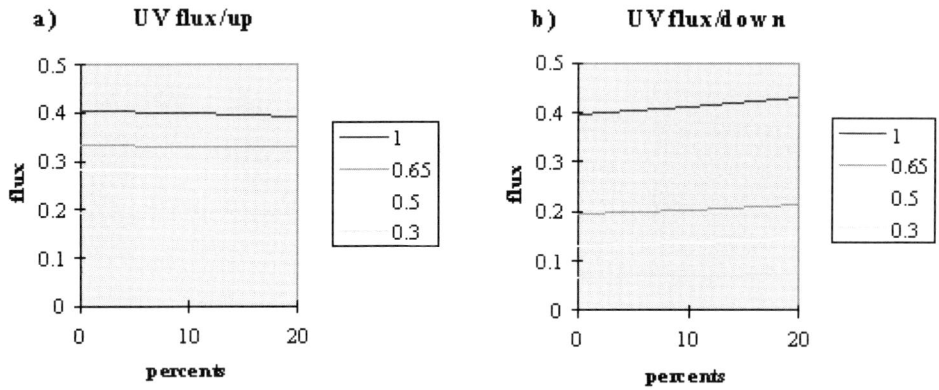

Figure 4.13. The dependence of up-and-down fluxes on % ozone depletion. Four curves correspond to four solar-zenith angles, their cosines being shown to the right of the diagram. The albedo for single scattering of aerosols in the boundary layer is taken to be 0.999 and the same for stratospheric aerosols. Ozone optical depth is 2.9.

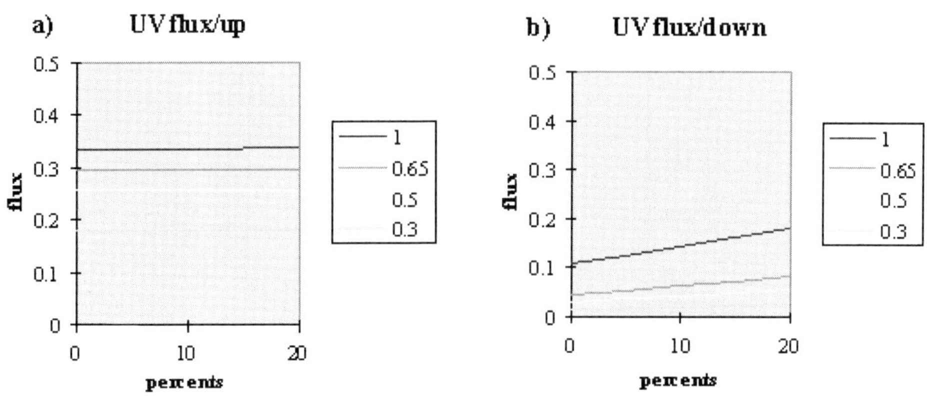

Figure 4.14. Dependence of up-and-down fluxes on % ozone depletion. Four curves correspond to four solar-zenith angles, their cosines being shown to the right of the diagram. The albedo for single scattering of aerosols in the boundary layer is taken to be 0.5 and for stratospheric aerosols it is 0.999. Ozone optical depth is 2.9.

determine the consequences of, for example, a slight vertical redistribution of ozone or change in aerosol properties (in the worst case of both) in different layers.

Sensitivity studies show appreciable dependence of up-and-down irradiances on solar-zenith angle, while statistical analysis demonstrate the strong correlation between up-and-down radiances as they change with solar-angle variations. Dependence on the aerosol asymmetry factor is rather weak, but correlation is high. Sensitivity of irradiances to variations in ozone and aerosol content strongly depend

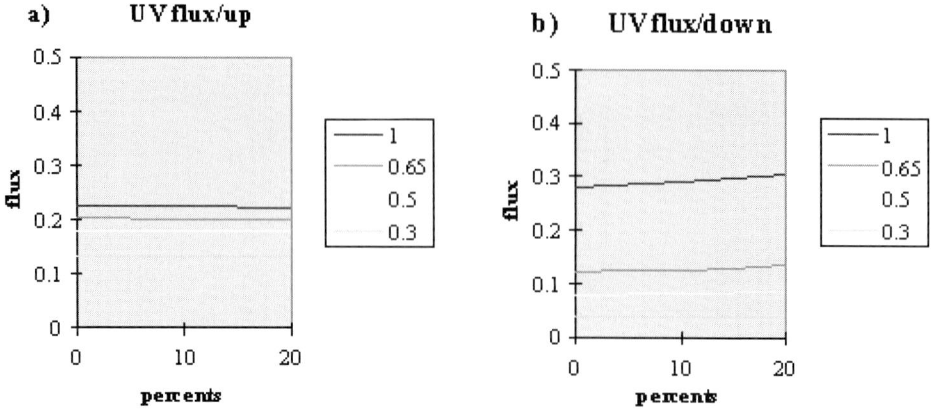

Figure 4.15. Dependence of up-and-down fluxes on % of ozone depletion. Four curves correspond to four solar-zenith angles, their cosines being shown to the right of the diagram. The albedo for single scattering of aerosols in the boundary layer is taken to be 0.999 and for stratospheric aerosols it is 0.9. Ozone optical depth is 2.9.

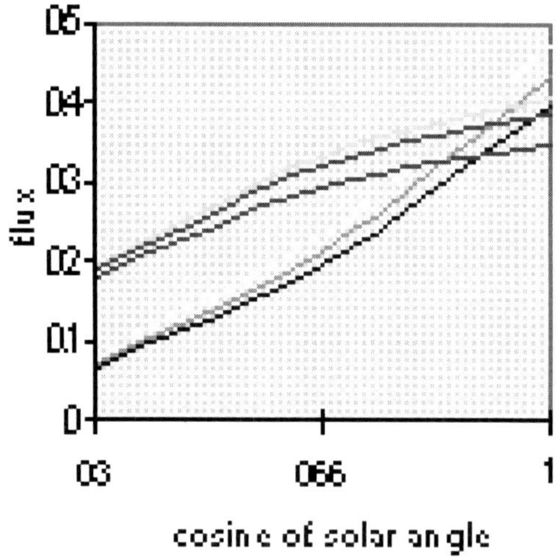

Figure 4.16. Dependence of up-and-down fluxes on the cosine of solar-zenith angle. The diagram is based on the same results as in Figure 4.9.

on other bulk and especially non-bulk parameters. Summarizing the results of statistical analysis, it can be stressed that changes in bulk parameters (ozone and aerosol integrals) cause variations in up-and-down irradiances which in many cases are highly correlated, but some combinations of non-bulk parameters totally modify the situation. Therefore carefully planned field experiments and numerical simulations are needed to improve accuracies of UVR calculations.

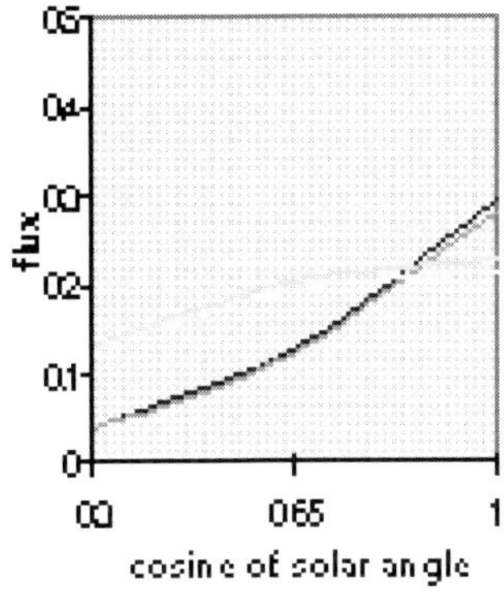

Figure 4.17. Dependence of up-and-down fluxes on the cosine of solar-zenith angle. Difference between the two curves in two groups (up-and-down) is caused by change in aerosol asymmetry factor from 0.7 to 0.75.

4.4 POSSIBILITIES TO PREDICT TOZ AND UVR CHANGES[2]

4.4.1 Introduction

Borisenkov (1963, 1980) undertook consideration of potential approaches to the solution of problems of long-term (one month to one season) and superlong-term (two seasons to 1–2 years) predictions of basic meteorological parameters. It was shown that the most realistic approaches are various kinds of physicomathematical methods of long-term and superlong-term meteorological forecasts. Then the possibility was substantiated of extending such approaches to the solution of the problem of long-term (superlong-term) forecasting of monthly (seasonal) mean anomalies of total ozone content (TOZ).

There were various efforts to develop TOZ prediction techniques through the use of statistical relationships (see, for instance, Spänkuch and Schulz, 1995) and regression models. Stanford and Ziemke (1996) used the latter approach. They examined two regression models to predict TOZ (Ω) from dynamical variables. One of the models is based on the following relationship between Ω (DU) and 50–100 hPa ($\sim$16–20-km layer) temperature T as the dependent variable:

$$\Omega_{\mathrm{t}} = \Omega_{\mathrm{to}} + \gamma(T_{\mathrm{t}} - T_{\mathrm{to}})$$

where t denotes forecast day, to is initialization day, γ is a local (latitude, longitude) day-indexed regression coefficient evaluated for each month.

The second model is based on backward spatial regressions (2-day means) using global hemispheric fields (30–60° latitude bands) of 100 and 500-hPa geopotential heights and 50-hPa temperatures as the database. Results discussed by Ziemke *et*

[2] This section was written in collaboration with E. P. Borisenkov.

al. (1998) indicate that both models provided improved 24-hour predictions of TOZ persistence for both hemispheres (Nimbus-7 TOMS and NOAA-II SBUV-2 data on TOZ were used). An important circumstance is that TOZ measurements from nadir-viewing SBUV instruments are seriously degraded by horizontal sampling which causes aliasing of the detected signal. This is not the case with new NASA earth probe and ADEOS high-density side-scanning TOMS spectrometer measurements.

A number of approaches to predict surface UVR were discussed at the European Conference on Atmospheric UV radiation in Helsinki (*Abstracts* ..., 1998). Schauberger and Schmalwieser pointed out, in this context, that since October 1995 daily forecasts of the UV index for the following day is published for Austria, Europe and worldwide. Assuming UV exposure over the whole day, the necessary Sun protection factor (SPF) is calculated to assess measures that are necessary to reduce the UV dose on a horizontal plane to 3 MED (Minimum Erythema Dose) and to 1 MED on a vertical plane. In 1997, routine forecasting of UV Index was implemented for the Czech Republic between May and September by the Solar and Ozone Observatory (SOO) of the Czech Hydrometeorological Institute (CHMI) in Hradec Kralove (Vanicek: *Abstracts* ..., 1998). Similar efforts are being made in Belgium, Japan and other countries, including development of satellite remote-sensing techniques to retrieve surface UVR spectra.

A version of the scheme of long-term and superlong-term forecast of monthly mean anomalies of TOZ for nine stations located in Western Europe was treated by Borisenkov and Kondratyev: *Abstracts* ... 1998. The scheme is based on representation of the fields of monthly mean TOZ anomalies as expansions into natural orthogonal function series. This makes information convenient for solving a number of problems, such as identifying quasi-periodic components in the expansion coefficients of basic time-dependent components; substantiation, on the basis of regular features revealed, of the quasi-cyclic prognostic model. Preliminary assessments were obtained for forecasts of monthly mean TOZ anomalies, performed using dependent and, where possible, independent information, for four central months (January, April, July and October).

Substantiation of a selected algorithm for constructing a prognostic scheme for all months of the year is now considered. Results of statistical analysis of the spatial–temporal structure of the fields of monthly mean TOZ anomalies for Western Europe for each month of the year will also be discussed. Predictions of monthly mean TOZ anomalies using both dependent and independent material are also considered. Results obtained indicate that the chosen approach is adequate for greater amounts of information. The TOZ observation series, embracing a 23-year period from 1973 to 1995, reveals no marked trends or sudden change in TOZ relevant to anthropogenic impact on ozone. This means that the results obtained are encouraging and make it possible to have confidence in the selected method of solving the problem of long-term and superlong-term TOZ forecasting, based as it is on detection of interannual quasiperiodic processes in the analysed series for the previous years. The main quasiperiodicities detected are apparently relevant to solar activity cycles and dynamic (circulation) processes.

4.4.2 General regular features of large-scale spatial–temporal structure of TOZ fields

Analysis of the literature, relevant to the spatial–temporal TOZ structure discussed in Chapter 2, makes it possible to single out two very pronounced zones, which differ one from the other, but each of which is homogeneous within its limits (Bekoriukov, 1990). These are (1) a zone of polar and temperate latitudes and (2) a zone of tropical ozone and between these two there is a zone with strong horizontal TOZ gradients. Inside each of these zones, TOZ latitudinal differences are comparable with longitudinal ones. Non-zonality of the ozone field is very pronounced there, especially in winter, and it reaches 40% (Bekoriukov, 1990).

Specific features of TOZ spatial distribution and intra-annual variability make it possible to conclude that variability in TOZ, especially during the cold period, is mostly due to solar-activity cycles and atmospheric dynamics.

It has been shown that, in latitude band 30–50°N where the region selected for examination is located, the zone of least variability in TOZ has been observed. These and other regularities make it possible to draw a general conclusion that development of physicostatistical models of long-term (superlong-term) TOZ prediction should be based, first of all, on taking account of observed spatial–temporal large-scale TOZ patterns in certain comparatively similar zones.

The analysis allows the following conclusions:

1. In order to eliminate the effect of very considerable annual TOZ variation, analysis and prediction not of instantaneous TOZ values but of monthly or seasonal mean TOZ anomalies has to be made.
2. Models of predicted TOZ anomalies should be developed separately for regions with similar large-scale spatial–temporal structures of TOZ fields and for certain months (seasons). One such region is Western Europe and the results presented in Table 4.10 refer to this region.

Description and substantiation of a cyclic model of long-term TOZ forecasting

Table 4.10 presents the numbers, geographic locations and coordinates of West European stations, whose data were used in further analysis. Initial data were borrowed from ozone data catalogues published by the World Ozone Data Center in Canada. Table 4.11 presents data about TOZ monthly mean values for each station for the period 1973 to 1994 (monthly mean values) and gives TOZ root-mean-square deviation values within each month for the same period.

The stations in Tables 4.11 and 4.12 have been placed in order of decreasing latitude. It can be seen that, for a given region, no significant dependence on latitude can be observed either in the mean values or in the MSDs. On the other hand, a pronounced intra-annual variation in TOZ is observed. TOZ maximum, averaged over the nine stations, is observed in April, when it reaches 358 DU; its maximum is

Table 4.10. Geographic names, numbers and coordinates ozonemetric stations chosen for the development of TOZ prediction models.

No	Station number	Station name	Coordinates	
			Latitude (°)	Longitude (°)
1	50	Potsdam	52.3N	13E
2	68	Belsk	51.5N	20.5E
3	53	Uccle	50.5N	4.2E
4	96	Hradec Kralove	51.1N	15.5E
5	99	Hohenpeissenberg	47.5N	11E
6	10	Budapest	47.3N	19.1E
7	35	Arosa	46.5N	9.4E
8	55	Vigna di Valle	42N	12.1E
9	82	Lisbon	38.5N	9.1W

noted in the middle and at the end of autumn, when it reaches 293 and 291 DU, respectively. The mean amplitude of the annual variation is 94 DU.

A similar pattern exists for root-mean-square TOZ deviations (Table 4.12). The maximum of TOZ monthly mean values variability is shifted, however, to the end of winter/beginning of spring, when it reaches 26 and 28 DU. The minimum of MSDs is shifted to the end of summer, when the root-mean-square deviation reaches 12 DU and to autumn, when its value varies within 11–13 D.U, i.e. at least twice that in autumn. The winter and spring seasons are characteristic of maximum TOZ variability, whereas the summer and autumn seasons reveal its minimum variability.

To eliminate annual variation in further analysis, monthly mean TOZ (ΔQ_{ijk}) have been calculated for every station:

$$\Delta Q_{ijk} = Q_{ijk} - \bar{Q}_{ij} \tag{4.16}$$

Here i is the number of the station ($i = 1, 2, \ldots, 9$); j is the number of the month of the year ($j = 1, 2, \ldots, 12$); k is the number of the year ($k = 1, 2, \ldots, 23$) ($k = 1$ corresponds to 1973). In some cases, where relevant data are lacking, they have been replaced by mean values.

The fields of monthly mean anomalies of TOZ (ΔQ) for every individual month have been presented as expansions into empirical orthogonal functions (EOFs).

$$\delta Q = \sum_{h=1}^{M} T_h(t) Q_h(x) \tag{4.17}$$

where $X_h(x)$ is a function (vector) for the spatial coordinate x directed from one

Table 4.11. TOZ monthly mean values for nine stations in Western Europe in Dobson Units (monthly mean values for 1974–1995).

Number		Station name	Months											
			12	1	2	3	4	5	6	7	8	9	10	11
1	$N = 50$	Potsdam	308	335	366	388	398	381	366	346	329	307	300	290
2	$N = 68$	Belsk	315	344	379	389	399	382	367	351	330	308	295	294
3	$N = 53$	Uccle	317	343	369	379	394	380	362	343	324	302	297	295
4	$N = 96$	Hradec Kralove	310	340	372	380	393	378	363	347	329	306	295	295
5	$N = 99$	Hohenpeissenberg	303	340	365	369	382	368	352	336	323	300	289	287
6	$N = 100$	Budapest	320	351	385	369	379	365	350	334	320	299	292	297
7	$N = 35$	Arosa	292	324	354	354	372	356	340	321	310	294	284	280
8	$N = 55$	Vigna di Valle	310	342	359	369	384	364	351	332	325	310	298	295
9	$N = 82$	Lisbon	297	315	333	340	368	366	349	331	320	309	302	288
	Regional mean value		308	337	361	365	385	385	356	338	323	304	293	291
			Winter XII–II			Spring III–V			Summer VI–VIII			Autumn IX–XI		
	Seasonal mean value		335			374			339			296		

Table 4.12. Data on root-mean-square deviations of TOZ for each month and for nine stations in Western Europe.

Number		Station Name	Months											
			12	1	2	3	4	5	6	7	8	9	10	11
1	$N = 50$	Potsdam	19	26	26	28	22	20	18	17	14	13	15	14
2	$N = 68$	Belsk	18	28	28	24	22	18	15	14	13	11	14	11
3	$N = 53$	Uccle	16	29	29	26	25	20	16	15	11	3	15	16
4	$N = 96$	Hradec Kralove	16	28	27	25	20	18	16	14	12	10	13	11
5	$N = 99$	Hohenpeissenberg	18	27	27	30	26	20	16	11	12	10	13	13
6	$N = 100$	Budapest	19	24	41	29	23	22	17	13	14	12	14	13
7	$N = 35$	Arosa	12	22	26	21	23	16	12	10	10	13	17	12
8	$N = 55$	Vigna diValle	19	21	31	27	23	22	16	11	10	12	12	13
9	$N = 82$	Lisbon	16	18	17	24	21	18	13	13	10	13	8	16
	Regional mean value		17	25	28	26	23	19	15	13	12	11	13	13
				Winter XII–II			Spring III–V			Summer VI–VIII			Autumn IX–XI	
	Seasonal mean value			23			23			13			12	

station to another, according to their number; $T_h(t)$ is the time function corresponding to $X_h(x)$; $h = 1, 2, \ldots, M$ is the dimension of the vector ($M = 9$).

The mean square of relative expansion error (4.17), related to full MSD, can be determined from the relationship:

$$\delta^2(\Delta Q) = 1 - R^2 = 1 - \frac{\sum_{h=1}^{H} \lambda_h}{\sum_{h=1}^{M} \lambda_h} = 1 - \frac{\delta_H^2}{\delta_M^2} \tag{4.18}$$

Here $\delta^2(\Delta Q)$ is the mean square of the expansion error; R is the multiple correlation; (ΔQ) is the total expansion MSD when H expansion terms are used ($H \leq M$); H is the number of terms in the expansion; λ_h are eigenvalues of a relevant correlation matrix which are usually placed in order of their decrease.

It can be seen that at $H = M$, $(\Delta Q) = 0$ and $R = 1$. Unlike the Fourier expansion series which is, in a certain sense, a formal procedure, an advantage of expansion (4.17) is the fact that it makes it possible to single out natural components, assess their contribution to the MSD field and their natural variation in time. Apart from this, a characteristic feature of this expansion is the circumstance that the very first expansion vectors $X_h(x)$, for which the eigenvalues λ_h are maximum in value, contain basic information about MSDs of the analysed fields; in our case, the fields of monthly mean TOZ anomalies. It appeared that the first four vectors contained basic information about the field of TOZ variability.

Table 4.13 presents eigenvalues λ_h of a relevant correlation matrix for the first four vectors $X_h(x)$ for each of the 12 months of the year.

Table 4.13. Eigenvalues λ_h corresponding to the first four vectors $X_h(x)$ for every month of the year.

Month		λ_h			
		$h = 1$	$h = 2$	$h = 3$	$h = 4$
1	January	6.943	0.991	0.523	0.314
2	February	6.740	0.828	0.587	0.375
3	March	6.996	1.159	0.424	0.261
4	April	6.687	1.185	0.368	0.338
5	May	7.307	0.765	0.463	0.192
6	June	6.715	1.021	0.437	0.366
7	July	5.772	1.098	0.835	0.648
8	August	6.476	1.126	0.494	0.474
9	September	4.824	1.515	0.976	0.689
10	October	6.007	1.414	1.013	0.234
11	November	5.444	1.039	0.989	0.660
12	December	5.388	1.301	0.864	0.708

Table 4.14. Contributions of vectors $X_h(x)$ to the MSD field for every month of the year: (a) contribution of each vector in %; (b) total contribution of vectors in order of their increase.

Month		$h = 1$	$h = 2$	$h = 3$	$h = 4$
January	a	77.144	11.011	5.811	3.489
	b	77.144	88.155	93.966	97.455
February	a	74.889	9.2	6.522	4.167
	b	74.889	84.089	90.611	94.778
March	a	77.733	12.878	4.711	2.9
	b	77.733	90.611	95.322	98.222
April	a	74.300	13.167	4.089	3.756
	b	74.300	87.476	91.556	95.312
May	a	81.189	8.5	5.144	2.133
	b	81.189	89.689	94.833	96.966
June	a	74.611	11.344	9.278	4.067
	b	74.611	85.955	85.611	94.878
July	a	64.133	12.2	5.489	7.2
	b	64.133	76.333	89.956	92.811
August	a	71.956	12.511	10.844	5.267
	b	71.956	84.467	81.277	95.223
September	a	53.600	16.833	10.844	7.656
	b	53.600	70.433	81.277	88.933
October	a	66.744	15.711	11.256	2.6
	b	66.744	82.455	93.711	96.311
November	a	60.489	11.544	10.989	7.333
	b	60.489	72.033	83.022	90.355
December	a	59.867	14.456	9.6	7.867
	b	59.867	74.323	83.923	91.790

It can be seen from Table 4.13 that the values of λ_1 are half an order or an order of magnitude greater than those of $\lambda_{2,3,4}$ for practically all months of the year. Therefore, as follows from (4.18), the very first vector already contains basic information about the TOZ field of anomalies. Comparison of maps for vectors $X_1(x)$ shows that every month has its own peculiar features characterizing the TOZ distribution.

Tables 4.13 and 4.14 indicate that the contribution of the first vector to total MSD reaches 74–77% in wintertime, 60–70% in summer and 53% in October. The contribution of the second vector amounts to 11–16%. On the whole, the first four vectors contain the following amount of information about the MSD field: 97–98% in January–February, 89–90% in September and November. Therefore, it would be fully justified to filter out the residual high-frequency component and base the strategy for developing a prognostic scheme for forecasting monthly mean TOZ anomalies on data about the first four coefficients $T_h(a)$ under the assumption of constancy of vectors $X_h(x)$ within the limits of the time interval considered.

Table 4.14 contains data on the contributions of each vector and the sum of vectors to the MSD field for every month of the year expressed in %. The prediction model can be expressed as:

$$\Delta Q_{np} = \sum_{h=2}^{H} T_{hn}(t) X_h(x) \tag{4.19}$$

where $T_{hn}(a)$ are of $T_h(a)$ predicted values. One of the conditions for prediction of $T_h(a)$ is the presence in these functions of quasiperiodic constituents, or of correlation of $T_h(a)$ with some predictors during previous time intervals. Taking account of the above-described specific features of the spatial–temporal structure of TOZ, the following approach was selected.

For every month of the year, each of the four vectors $T_h(a)$ was subject to statistical analysis to find out whether they had quasiperiodic components and to determine periods, confidence intervals, phase shifts, harmonic amplitudes and signal-to-noise ratio. The spectral analysis technique, usually employed for this purpose, is inapplicable in this case because of limited sampling duration (23 years).

The method of maximum entropy, described by Tzvetkov (1987), is used here. The method is based on application of a cyclic model of the following form:

$$A(t) = \sum_{k=1}^{K} H_k \cos(2\pi t/T_k + \varphi_k) + e(t) \tag{4.20}$$

Where $A(t)$ is a time function (in a given case, the function $T_h(a)$); T_k is the period of the kth harmonic; H_k is the amplitude of the kth harmonic; φ_k is the phase shift of the kth harmonic; $e(t)$ is the residue. Analysis provides for consecutive singling out of basic harmonics; it makes it possible to detect existing long-period harmonics with T_k period somewhat exceeding the sampling length and to separate the noise component $e(t)$.

Table 4.15 presents periods T_k, amplitudes H_k, signal-to-noise ratios S_k and phase shifts φ_k for each of the most significant harmonics. It can be seen that for the first eigenvalues λ_1 (corresponding to function $T_1(a)$), the first long-period component for practically all months of the year is a component with mean period $T_k = 28.7 \pm 1.6$ yr, the mean phase shift $16.1 \pm 16.6°$ and mean signal-to-noise ratio 0.85 ± 0.28. It is interesting to note that for a number of months (January, March, April, July, August) the signal-to-noise ratio is close to unity or much larger. Amplitudes of this harmonic vary from month to month within sufficiently broad limits. However, there is variability dependence of maximum amplitudes at the end of winter and in spring and minimum amplitudes in summer and autumn.

The next harmonic gives an average over all the months of the year of 10.1 ± 1.2 yr. These two harmonics probably reflect Brueckner and quasi-11-year cycles of solar activity. Then shorter periods follow: 6.3 ± 1, 7.9 ± 0.6 and

Table 4.15. Characteristics of the frequency spectrum of the first four functions $T_h(a)$ for every month of the year. T_k is the harmonic period, H_k is the amplitude, φ_k is the phase angle (in degrees), S_k is the signal-to-noise ratio.

Month				λ_1					λ_2		
December	T_k	27.8	13.7	8.3	4.4	3.5	30.3	10.6	7.6	6.2	4.2
	H_k	23.0	10.9	25.1	24.3	9.1	5.6	1.2	7.1	8.6	10.2
	S_k	0.8	0.4	0.8	0.8	0.3	0.4	0.1	0.5	0.5	0.6
	φ_k	45.8	−43.2	141.2	−150.9	−54.1	24.8	−159.4	62.3	−38.6	100.3
				λ_3					λ_4		
December	T_k	29.4	11.1	8.1	4.5	3.0	27.0	9.9	5.0	3.5	2.8
	H_k	2.7	10.9	11.1	6.8	5.3	11.7	7.6	8.7	3.7	7.3
	S_k	0.4	1.5	1.5	0.9	0.7	2.0	1.3	1.5	0.6	1.3
	φ_k	−157.5	−66.5	23.1	99.7	130.3	−56.5	88.7	−127.8	−7.6	164.1
				λ_3					λ_2		
January	T_k	24.4	9.3	5.7	4.5	3.3	26.3	9.3	5.2	4.0	3.6
	H_k	47.8	25.9	39.2	40.8	18.4	7.8	16.9	8.0	10.4	10.4
	S_k	1.1	0.6	0.9	1.0	0.4	0.4	0.8	0.4	0.5	0.5
	φ_k	27.2	−132.8	−117.1	136.2	−173.5	−111.8	68.4	63.0	−62.4	166.1
				λ_3					λ_4		
January	T_k	28.6	11.5	7.8	4.9	4.1	27.0	15.6	6.1	5.0	4.4
	H_k	11.0	5.3	4.9	3.2	5.0	4.4	2.9	4.3	10.6	12.9
	S_k	0.7	0.4	0.3	0.2	0.3	0.4	0.3	0.4	1.1	1.3
	φ_k	−84.2	−89.6	−11.3	−108.4	123.8	−15.8	20.0	177.8	126.6	−118.9
				λ_1					λ_2		
February	T_k	21.3	10.5	5.8	4.4	3.6	21.7	11.0	9.2	5.0	4.3
	H_k	45.3	15.5	20.4	48.1	28.9	12.2	5.6	16.4	7.4	14.4
	S_k	0.8	0.3	0.3	0.8	0.5	0.6	0.3	0.9	0.4	0.8
	φ_k	21.7	−142.0	−87.3	98.1	−160.2	−33.8	105.9	−30.5	33.7	−57.9
				λ_3					λ_4		
February	T_k	13.7	9.4	6.7	5.6	3.9	23.3	8.1	5.5	4.3	3.6
	H_k	3.3	8.9	5.8	12.0	12.8	21.4	3.8	5.6	5.1	6.7
	S_k	0.1	0.4	0.2	0.5	0.5	1.6	0.3	0.4	0.4	0.5
	φ_k	20.0	40.8	−85.8	46.1	−61.4	−55.0	21.7	−85.8	−153.8	151.5

3.8 ± 0.3 yr. The fact that these quasi-periodicities are stably present in all months of the year makes it possible to suppose that they are due to natural causes, perhaps to quasi-cyclic circulation processes. This makes it possible to consider these processes as quasi-stationary, which is very important for the development of a prognostic TOZ model.

Quasiperiodicities characteristic for $T_1(a)$ can also be found in the second vector, namely: 24.2 ± 2.8, 10.9 ± 0.9, 7.2 ± 1, 5.1 ± 0.5, 4 ± 0.3 yr.

Similar quasiperiodicities, but with more considerable month-to-month differences in amplitudes and phase shifts, are observed for vectors $T_3(a)$ and $T_4(a)$ (eigenvalues λ_3 and λ_4).

Analysis of data obtained makes it possible to draw at least three conclusions, which are of importance for the development of a scheme of long and superlong-term forecast of TOZ monthly mean anomalies:

1. During the entire 23-yr period of observations (1973–1994), we have not discovered in either of the vectors any trend components or sudden changes that could be interpreted as evidence of anthropogenic impact on TOZ.
2. The second conclusion, which partially confirms the previous one, is that in functions $T_h(a)$ a relatively stable presence of quasi-periodic constituents, close to the solar-dependent Brueckner cycle (of about 28 years) and the 11-year solar activity cycle, has been discovered. This leads to the assumption that TOZ quasi-rhythms, caused by solar activity, are of natural origin. Shorter periods (6–7, 5 or 4 years) are apparently because of dynamic (circulation) processes. Similar quasi-periodicities were observed earlier in the fields of other meteorological parameters and in a number of atmospheric processes.

The observed spatial–temporal structure of TOZ monthly mean anomalies over the chosen (rather homogeneous) region suggests a comparatively simple prognostic model for forecasting TOZ monthly anomalies, based on physicostatistical principles; that is, on the basis of accounting for the spatial–temporal structure of analysed fields. Here, at least for current and long-term forecasting, it is unnecessary to take potential anthropogenic impact on the ozone layer into account.

4.4.3 Results of testing a cyclic model for long and superlong-term forecast of TOZ monthly mean anomalies

As can be seen from (4.19), a cyclic model for forecasting TOZ monthly mean anomalies is based on the extrapolation prediction of the first four functions $T_h(a)$ for each month, under the assumption that the processes involved are quasi-stationary and the function $X_h(x)$ is constant. The model is assumed to be adjusted to new information. By analogy to (4.20), the following formula can be derived:

$$T_{hn}(t + \Delta t) = \sum H_k \cos(2\pi/T_k(t + \Delta t) + \varphi_k) \tag{4.21}$$

where Δt denotes that the forecast is made 1 or 2 years in advance; the values of H_k, T_k and φ_k are taken from (4.21).

As a rule, to assess the quality of prediction, a model is tested on independent and dependent observational data. To check the model on dependent samples, which is represented in our case by data for 1973–1994, all values of $T_h(a)$ for a given period were retrieved and simultaneously the forecast of $T_h(a)$ was made for 1995, 1996, 1997 and 1998. Since 1995 was initially omitted in the analysis, calculations were repeated taking this particular year into account. Then prognostic models could be compared with actual calculations of $T_h(a)$.

For quantitative assessment of the accuracy of retrieval of vectors $T_h(a)$ (forecasting of dependent data), we chose estimation criteria, which are usually employed in long-term forecasting:

$$\rho = (n^+ - n^-)/(n^+ + n^-) \tag{4.22}$$

$$Q^* = \frac{1}{9}\sum_{k=1}^{9}\left(\frac{\delta Q_k}{\sigma_k}\right)^2 \tag{4.23}$$

where n^+ is the number of points where the anomaly sign was correctly retrieved (predicted); n^- is the number of points where the anomaly sign was opposite; δQ_k is the absolute difference between the true value of the anomaly and its prognostic (or retrieved) value; σ_k is the root-mean-square deviation of Q.

If the criterion $\rho = 0$, this means that the number of successful and unsuccessful (of anomaly sign) predictions is the same. $\rho = 1$ means that all predictions were correct in sign. $\rho < 0$ means that predictions (retrieved values) were wrong in sign for the majority of points.

The criterion Q^* is a more rigid estimate of the justification of predictions (retrievals) of values. If $Q^* < 1$, it means that the anomaly value is predicted (retrieved) with an error which is less than the root-mean-square variability of the predicted value.

Table 4.16 presents estimates of ρ and Q^* averaged over seasons and all the nine locations, which shows whether the model was successful in retrieving monthly mean TOZ anomalies on a dependent sample.

As can be seen from Table 4.16, estimates of ρ and Q^* were averaged over seasons and are sufficiently high.

Table 4.16. Average estimates of how successfully retrieval of TOZ anomalies was using a cyclic model for the 1973–1994 period.

	Seasons				
	Spring	Summer	Autumn	Winter	Annual Average
ρ	0.452	0.431	0.494	0.492	0.467
Q^*	0.412	0.338	0.373	0.425	0.387

Table 4.17. Verification of monthly mean TOZ anomalies for an independent period in 1995; prediction on the basis of a cyclic model made 1 year in advance.

Month	5	6	7	11	12
ρ	−0.75	1	1	0.1	0.4
Q^*	5.1	3	1.4	1.7	1.6
Number of prediction check-up points	8	8	6	7	7

To assess the model on independent data, we used TOZ data for 1995 which had previously not been included in data processing. Regrettably, we did not possess data for all locations in 1995; therefore, only part of the reliable data for individual months in 1995 were used in our evaluations (for no less than six locations). These months were May, June, July, November and December.

Table 4.17 presents estimates of monthly mean TOZ anomalies for these 5 months of 1995.

It follows from Table 4.17 that for all months, except May (when forecast was unsuccessful for six out of eight points), prediction of sign of TOZ anomalies proved successful ($\rho < 1$). For June and July, justifiability in sign was 100%. With regard to the value of the anomaly, it is with an error greater than the value of σ_k (the criterion $Q^* > 1$).

All predictions were made 1 year in advance. All forecasts of the first four vectors $T_h(a)$ were made up to 1998 and can be checked by independent experts.

Thus the results of testing of the model, using dependent and independent data, have revealed that the approach was successful in solving the problem. Further efforts should be directed towards a more detailed check-up of the model, using independent data. This can be done as more initial data are accumulated. Other efforts should be oriented towards development of models for other regions which are characterized by different spatial–temporal structures of TOZ fields. Moreover, it would also be important (irrespective of TOZ prediction) to perform a similar statistical analysis of TOZ fields for other regions in order to detect quasi-cyclic processes and the presence of anthropogenic impact.

It can also be concluded that there are sound grounds to investigate the possibility of developing prognostic models that would take into account regular features of not only interannual TOZ variability but also its intra-annual variability. As noted above, in the observation series there were gaps for some months. These gaps were filled by mean values. Since the model allows reconstruction of the observation series, it is reasonable to forecast TOZ in two stages. At the first stage, missing observation points should be reconstructed. After that, all calculations should be repeated, replacing mean values (zero anomalies of TOZ) at the missing points by their retrieved values. In the case considered, taking account of the fact that gaps were not numerous, this was not done. However, the model makes it possible to do it in principle.

4.5 EFFECTS OF UVB RADIATION ON AQUATIC ECOSYSTEMS[3]

4.5.1 Introduction

Recent observations leave no doubt that in many regions of the globe at least a seasonal increase in solar ultraviolet (UV) radiation is taking place at the level of the Earth's surface, mainly because of anthropogenically induced depletion of total ozone (WMO, 1995). The nature of this phenomenon is associated with the combined effect of both natural and also (predominantly) anthropogenic factors. Research into solar physics points to the existence of a firm connection between levels of solar activity and UV radiation emitted by the Sun, which may be the result of variations in magnetic fluxes in the solar atmosphere and/or in the temperature and diameter of the star (Kondratyev, 1998). In particular, according to model extrapolated estimates, the period from 1645 to 1715 was noted for its exceptionally low solar activity (Maunder minimum); therefore, the fraction of UV radiation in solar-radiation flux (in comparison with the current level, which relates to a period of high solar activity) has fallen by 64% (at wavelength $\lambda = 121.6\,\text{nm}$), 8% (at $\lambda = 200\,\text{nm}$) and 3.5% in the spectral band 210–50 nm. Reduction in the level of UV radiation in extra-atmospheric solar flux inevitably leads to a decrease in concentration of ozone (C_{oz}) in the stratosphere and, correspondingly, to an increase in UV radiation at the Earth's surface in the absence of cloud-cover effects. Taking into account the fact that we are currently in a falling phase of the 11-year cycle of solar activity, we should expect (in cloud-free conditions) a relative rise in the level of UV radiation close to the Earth's surface.

An even more important role in relation to the increase in UV radiation is played by effects of anthropogenically caused (primarily because of emissions of chlorofluorocarbons into the troposphere) global trend for depletion in total ozone (TOZ), together with strong regional and seasonal decreases (so-called ozone holes). Calculations show, in particular, that for anthropogenically caused decrease in stratospheric ozone, the level of UVA radiation increases very insignificantly, while UVB radiation at the Earth's surface in cloud-free conditions may grow by six orders of magnitude. It should be borne in mind, however, that the effect of both natural and anthropogenic factors on the level of UV radiation at the Earth's surface is effectively modulated by cloud cover.

In assessing the effect of increase in UV radiation on natural ecosystems, the index of the amount by which its variations at the Earth's surface exceed corresponding variations because of variability in optical parameters of cloud cover over a specific point on the globe is of fundamental importance (Madronich, 1995). According to calculated figures, if anthropogenically caused destruction of the stratospheric ozone layer were to continue at its characteristic rate for the 1990s, this index would inevitably exceed 1.96, the particular threshold above which increased levels of UV radiation have a harmful effect on the biosphere. It is extremely significant that model estimates (Lubin and Jensen, 1995) predict that this threshold will be exceeded

[3] This section was written in collaboration with D. V. Pozdnyakov.

not only in high but also in mid-latitudes. These are reasons to think that the average-monthly threshold is being exceeded (not to mention the occurrence of short-term rises) in the summer months in many regions of Europe, Central Asia and North America, and also in the 0–30° latitudinal belt in the Southern Hemisphere.

Areas with the most favourable conditions at the current time include both tropical latitudes, where increase in surface UV radiation (against background levels) has not yet been significant, and several regions in higher latitudes which are characterized by statistically higher temporal variability in cloud-cover characteristics. It should be emphasized, however, that in light of the expected further destruction of the stratospheric ozone layer over the next decade and the increase in surface UV radiation, the tropical biosphere may suffer from more difficult conditions by virtue of the typically insignificant cloud cover in the so-called dry period of the year.

Together with the trend for increase in UV radiation at the surface in vast regions of both hemispheres, local falls in UV radiation are recorded, because of increase in the content of anthropogenically produced ozone in the lower troposphere, in large urban areas and industrial centres. For this and other reasons, global distribution of surface UV radiation is characterized by extremely significant heterogeneity. Against the background of expected continued decrease in TOZ and increase in UV radiation at the Earth's surface, the question of space–time dynamics of surface fields of UV radiation is becoming particularly important in the context of analysis of the mechanisms and scales of its effect on the biosphere. It is especially important to take full and adequate account of the whole diversity of direct and inverse relationships, inherent in both the biosphere as a whole and its components and, in particular, in water ecosystems (Kondratyev and Pozdnyakov, 1997a, b; Pozndyakov *et al.*, 1998; Kondratyev and Filatov, 1999).

UV radiation is usually subdivided into near ($\lambda = 400$–300 nm), middle ($\lambda = 300$–200 nm) and far ($\lambda = 200$–100 nm) bands. Solar UV radiation, passing through the atmosphere, actively interacts with the gaseous environment and, at the level of the Earth's surface, UV radiation with $\lambda \leq 280$ nm is totally eliminated from the solar spectrum as a result of its absorption by the atmosphere. It is extremely significant that the position of the minimum wavelength (λ_{min}) of light reaching the Earth's surface, and also the intensity of the UV radiation reaching the Earth's surface, are controlled to a decisive degree by the absorption of solar radiation by stratospheric ozone. Consequently, a decrease in C_{oz} in the stratosphere leads to a shift of λ_{min} to the shortwave interval of wavelengths and to a growth in UV irradiance in the interval $\lambda_{min} \sim 400$ nm.

Depending on the wavelength, UV radiation has different effects on various photochemical and biological processes. In this connection, it is traditional to single out wavelength bands 280–315 and 315–400 nm, which are referred to as UVB and UVA respectively. The effective UV irradiance, or exposure dose E is determined from the simple equation (Madronich *et al.*, 1995a, b):

$$E = \int F(\lambda) W(\lambda) \, d\lambda$$

where $W(\lambda)$ is the weighting function, also called the action spectrum and $F(\lambda)$ is the spectral distribution of incident light. The action spectrum is specific to each biological or photochemical process.

Exposure doses can be integrated over time. The need to introduce the weighting function results from the fact that UVB increases with wavelength, while its biological effect is most often intensified as λ decreases. Since the solar irradiance spectral distribution of E is largely determined by the ozone content in the stratosphere (i.e. C_{oz}), the so-called radiation amplification factor (RAF) is frequently introduced; this is calculated in %. For small variations of C_{oz}, this function is linear:

$$\Delta E/E = -\mathrm{RAF}(\Delta C_{oz}/C_{oz})$$

where $\Delta C_{oz}/C_{oz}$ is the relative change in ozone content and $\Delta E/E$ is the corresponding relative increase (decrease) in effective irradiance.

It should be emphasized that although the introduction of the RAF enables us to estimate the change in the exposure dose, it cannot be used to calculate the resulting biological effect since the latter is usually non-linearly dependent on the size of the exposure dose and on a number of other factors, such as the ability to recover, the phase of the lifecycle at which the effect occurred, etc. In particular, in the case of phytoplankton, subject to inhibition of the ability of cells to move by UVB fluxes, the RAF was 1.9 in January and 1.5 in July (Madronich *et al.*, 1995a). At the same time, the inhibition of the photosynthesis process of *Phaeodactyluni* spp. (alga), *Prorocentrum micans* and *Nodularia spunigena* (cyanobacteria), corresponding RAF values were 0.2 (0.2), 0.3 (0.4) and 0.2 (0.2) (the first value is for January, the second in brackets is for July). Thus, the introduction of the RAF enables us to compare the susceptibility of particular processes (effects) to variability in atmospheric ozone content. Action spectra $W(\lambda)$, characterized by a rapid fall as the wavelength increases in the UVB band, yield larger RAF values than spectral $W(\lambda)$ relating to the UVA band. In the latter case the RAF values amount to a few tenths of a percent only.

From the above example it can be seen that small RAF values are typical for processes of photosynthesis. This is true not only for phytoplankton but also for higher vegetation (Bronman *et al.*, 1984). For larger variations in TOZ, the preceding equation is modified to:

$$E_2/E_1 = [C_{oz}(1)/C_{oz}(2)]^{\mathrm{RAF}}$$

where E_1 and E_2 are the weighted values of the irradiance corresponding to the TOZ values $C_{oz}(1)$ and $C_{oz}(2)$.

In the case of analysis of the reaction of biological objects, weighting functions are essentially biological action spectra. The weighting function of the resulting effect is the convolution of a number of weighting functions at the molecular level, where the spectral absorption characteristics of the pigments have a critical effect. It is important to bear in mind that, in practice, weighting functions are determined by illuminating the biological material at each moment in time at a single wavelength and then summing the effects for all wavelengths. However, in natural conditions biological objects are simultaneously illuminated by all the solar light that reaches them. In this case, the intensity of the exposure in the UVA band and in the visible band is several orders greater than the level in the UVB band.

Radiation in the UVA band and in the visible band of the spectrum may cause interactive effects and, in particular, reduce the effectiveness of the influence of UVB on living organisms. Moreover, the action spectra are usually determined for limited time periods (one or more hours), while in natural conditions the exposure to sunlight covering the full spectral range may continue for several days or even months which, as one might expect, may lead to the onset of interactive effects, which are in no way a direct result of the nature of the corresponding action spectra. All the factors listed, together with other factors not mentioned here, which limit the adequacy of the weighting functions, must be borne in mind when analysing the reactions of specific biosystems to the effects of light. However, even in the absence of the necessary detailed information, it is preferable to use weighted fluxes rather than unweighted ones (Williamson *et al.*, 1994a, b). An illustration of this is provided by calculations given in Bukata *et al.* (1995), according to which DNA-weighted variations in the daily totals of incoming UVB are significantly greater than merely the percentage variations in the surface illumination in the same spectral interval. In particular, for the case of a decrease in ozone content in the stratosphere to 0.087 atm cm (as opposed to 0.350 atm cm, which is considered the 'background' value of C_{oz}), the overall flux of incoming solar radiation varies from $\sim$20 to 45% while for the DNA-weighted value for this flux (∇I) these variations amount to 90–$\sim$110%; moreover, which is particularly significant, the latter exhibit a significant degree of latitudinal and seasonal invariance. Bukata *et al.* (1995) proposed the following parametrization of ∇I by an exponential series:

$$\nabla I = 0.5 + 0.864(\Delta C_{oz}) - 0.000\,675(\Delta C_{oz})^2 + 0.000\,0530(\Delta C_{oz})^3$$

where ΔC_{oz} is the decrease in the content of stratospheric ozone, measured in Dobson Units (1,000 DU = 1 atm cm).

Aquatic ecosystems cover a wide spectrum of habitat from the oceans to large and small ponds, rivers, estuaries and wetlands whose state varies from the level of undisturbed ecosystems to practically complete anthropogenic transformation. Despite all diversity, aquatic systems have a feature in common: they are involved in the cycles of energy and substance, with the decisive role of varying incoming solar radiation (as a source of energy) and its spectral distribution. Primary production, whose level is bound up with the dynamics of nutritional chains and biochemical cycles of various compounds, determined by a wide range of phenomena from chemistry of aquatic domains to regional and global weather and climate, is the focus of the processes of photosynthesis taking place owing to solar energy. UVB-induced changes in the spectral regime signify the appearance of substantial environmental disturbances (first of all, connected with photoprocesses in aquatic ecosystems), which have existed over the long period the evolution of aquatic ecosystems.

Numerous facts have been found out which testify to the destructive impact of UVB radiation on aquatic organisms. In particular, the effect of UVB increase on primary production in the Antarctic. This not only affects phytoplankton resistivity but also breaks the polychromatic balance of the light field, which affects the interaction of the processes at molecular, biochemical, physiological and trophic levels and apparently causes changes in the evolution of natural communities. However, the consequences of the UV effects for regional and global aquatic primary production and related processes in the ecosystems remain unclear. In this connection, satellite and surface observational data to be obtained in the near future are crucially important: they will provide information about global UVB distribution over the ocean surface and enable estimation of the depth of penetration. In this context, the problem of obtaining data on photosynthetically active radiation (PAR) is also urgent.

One problem that is scarcely solved is analysis of the effect of UVB increase on communities and ecosystems (a community is defined as part of the ecosystem, an ensemble of organisms that exist within a certain ecological region), though it is clear that the response of aquatic ecosystems may vary in a wide range, from zero to catastrophic. This determines the need to:

- obtain more complete and reliable databases on UVB increase on local, regional and global scales;
- reveal the aquatic ecosystems potentially more sensitive to UVB forcing;
- investigate the relationships between the response of ecosystems to forcing and their subsequent recovery;
- make quantitative estimates of the effects manifested on various spatial–temporal scales;
- reveal the processes in ecosystems affected by increasing UVB radiation;
- investigate possible global positive feedback taking account of changes in cloud cover, temperature and processes of biological adaptation which can be determined by global changes in the level of the UVB radiation reaching the Earth's surface.

In view of the great importance of marine biosphere dynamics, we shall discuss effects of UVR on aquatic ecosystems in detail.

4.5.2 Algae primary production

Mainly because of phytoplankton photosynthetic activity, the global ocean stores about 50 times more carbon than the atmosphere, and each year oceans and atmosphere exchange around 15 times as much carbon dioxide as human activities produce. Phytoplankton produce more biomass than all other terrestrial ecosystems combined, and take up more than half the atmospheric carbon dioxide (Noever *et al.*, 1994).

Since the launch of the Coastal Zone Color Scanner (CZCS) in 1978, mapping primary production for the global ocean has been the focus of programmes such as JGOFS (1988), IGBP (1988), LOICZ (*Report* ..., 1996). Recently, this ever-persistent interest in assessing algae primary production on local, regional and global scales has received a new powerful momentum with putting into orbit a new sensor – SeaWiFS (the Sea-viewing Wide Field of View Sensor). One of the major challenges of this new ambitious global ocean colour mission is assessing global carbon fluxes between the atmosphere and global ocean (SeaWiFS, 1987). Carbon fluxes are believed to be among the principal players controlling the planet's climate status (Kondratyev, 1998). On the other hand, anthropogenically driven atmospheric ozone variations (Kondratyev *et al.*, 1995a, b; Kondratyev and Varotsos, 1996a, b) which are conducive to variations in solar ultraviolet radiation incident on the Earth's surface, are considered as being potentially capable of resulting in serious damage to aquatic ecosystems (Kondratyev and Pozdnyakov, 1996a).

Regardless of essential controversy in assessing photosynthetic rates of carbon production by phytoplanktonic algae in the euphotic zone of natural water bodies, it nevertheless is thought that total primary production in the global ocean is comparable in magnitude with that produced by the terrestrial phytosphere (Platt and Sathyendranath, 1988). Temporal and spatial requirements for global monitoring of aquatic photosynthetic activity can only be met by chlorophyll spatial distributions inferred from data downstreamed from ocean color sensors onboard satellites (Kondratyev *et al.*, 1996; Pozdnyakov *et al.*, 1998).

With the extended number of spectral channels in the visible spectrum and improved positioning of these channels in new sensors (Kondratyev and Tanaka, 1997; SeaWiFS, 1987), and with concomitant refinement of phytoplankton chlorophyll retrieval algorithms (Gordon and Wang, 1994; McClain and Yen Eueng-nan, 1994) it is reasonable to anticipate that accurate determinations of chlorophyll will be attainable in clear oceanic/marine waters (case-I waters according to the Morel (1988) classification). Serious progress (see e.g. a review by Pozdnyakov and Kondratyev, 1998; Pozdnyakov *et al.*, 1998), has been achieved in developing tools for remote sensing of chlorophyll content in case of turbid waters in marine coastal zones and inland water bodies which are usually referred to as non-case-I or else case-II waters. However, the perspective of chlorophyll content operational retrieval in case of turbid waters remains rather obscure and requires further efforts. Moreover, it is worth mentioning that the problem proves to be additionally complicated since there are no reliable atmospheric correction techniques applicable to case-II waters as yet, although first steps in this direction have been already made (see e.g. Land and Haigh, 1996; Heineman and Fisher, 1997).

The progress achieved in the areas of chlorophyll remote-sensing hydro-optical modeling (Bukata *et al.*, 1995; Morel, 1988; Pozdnyakov and Kondratyev, 1998), algae primary-production formulations (see e.g. Sakshaug and Slagstad (1991); Lee *et al.* (1996a, b); Berthon and Morel (1992)) and increasing concern prompted by reports about extending ozone holes stimulate analysis of feasibilities of remote sensing in assessing algae primary production *per se*, as well as its vulnerability to variation in the subsurface ultraviolet radiation budget occurring in natural waters because of the stratospheric ozone depletion.

Model formulations of algae primary production

Two basic approaches are widely used to numerically simulate algae primary production. The first approach relies on the use of production–light-intensity curves describing photosynthetic rate as a function of photosynthetically active irradiance propagating through the water surface down to the lower boundary of the photic zone over the time interval of interest, usually a day (see e.g. Vollenweider, 1966). The other approach depends on physiological (quantum yield of photosynthesis) and optical properties of phytoplankton (pigment-specific absorption coefficient α^*_{chl}). as well as the chlorophyll biomass in the photic zone (Kirk, 1986).

Production-light intensity functions

As noted above, production–light-intensity functions describe the photosynthetic response of phytoplankton to irradiance. The simplest formulation is a straight line:

$$P' = \alpha' E \tag{4.24}$$

where P' is the biomass-normalized photosynthetic rate, E is the irradiance, α' is the slope of the line.

As detailed in Talling (1957) and later in Jassby and Platt (1976), the relationship between the rate of photosynthesis P and solar photosynthetically active radiation (PAR) can be calculated from the extant chlorophyll concentration C_{chl} and two physiologically meaningful photosynthesis parameters (transfer coefficients) P^*_{max} and α^* (the asterisk indicates that both coefficients are normalized to the aquatic chlorophyll concentration, a surrogate for biomass). P^*_{max} is the maximum rate of photosynthesis and α^* is the slope of the production versus irradiance curve. The initial slope α^* is a direct consequence of the photochemical aspect of algal photosynthesis (a function of the type and quantity of indigenous photosynthetic pigments). Unlike P^*_{max}, α^* is not sensitive to water temperature. However, this parameter proves to be highly space and time variable. According to thorough investigations performed (see Lee *et al.*, 1996), α^* can vary by a factor of 4 for the same season for different regions, a factor of 4 for the same region for different seasons, or a factor of 5 for different regions and seasons.

The near-horizontal asymptotic portion of the curve at high irradiance levels represents the maximum rate of the enzymatic quantity of cellular enzymes and the specific temperature at which primary production is determined.

Because the quantum yield of photosynthesis decreases with increasing irradiance, the photosynthetic rate is a saturation function of irradiance. An empirical and simple saturation function was suggested by Webb *et al.* (1974):

$$P' = P^*_{max}[1 - \exp(\alpha' E / P^*_{max})] \tag{4.25}$$

This formulation, however, does not account for primary-production dependence on some factors, such as water temperature and nutrition depletion, as well as closely related effects of wind-induced mixing in the euphotic zone (Sakshaug and Slagstad, 1992). Moreover, it ignores the possible variability in P along the vertical in optically highly inhomogeneous waters.

Platt *et al.* (1991) express the depth dependence of primary production on available subsurface irradiance by the relation:

$$P(z) = \frac{C_{\text{chl}}(z)\alpha^* E(z)}{[1 + (\alpha^* E(z)/P^*_{\max})^2]^{1/2}} \tag{4.26}$$

where $P(z)$, $P^*_{\max}$ α^* and $E(z)$ are as previously defined. Although not obligatory, it is not unreasonable to assume, at least for case-I waters, that coefficients α^*, $P^*_{\max}$ are depth invariant. However, Denman and Gargett (1988) observed that the physical barrier imposed by a shallow thermocline resulted in disparate phytoplanktonic photoadaption profiles within each of the layers (i.e. distinct slopes in P versus PAR curves). They also observed that the diurnal cycle of vertical motion of algae within the upper layer generally proceeds slowly enough to enable them to adapt to high irradiances near the air–water interface, yet rapidly enough to redistribute the photosynthesis per unit chlorophyll roughly evenly throughout the epilimnion.

Equation (4.26) is based on the assumption that slope α^* is linear and that $E(z)$ is PAR. For the general case of a known spectral depth distribution of PAR $E(z, \lambda)$ and a known non-linear slope function $\alpha^*(z, \lambda)$, such wavelength dependency should also be taken into account. Further, the subsurface irradiance field $E(z, \lambda)$ is a function of in-water refracted solar-zenith angle θ' and can be partitioned into a direct sunlight component $E_{d_{\text{sun}}}(z, \lambda, \theta')$ and a diffuse skylight component $E_{d_{\text{sky}}}(z, \lambda)$. It is the daylight progressions of these functions $E_{d_{\text{sun}}}$ and $E_{d_{\text{sky}}}$ that must be considered.

To incorporate a depth-dependent and non-linear slope function, a spectral and angular dependence of the subsurface PAR light field, and a solar and sky partitioning of $E(z, \lambda)$ into the photosynthesis–light consideration, equation (4.26) may be rewritten as (Platt *et al.*, 1991):

$$P(z) = \frac{C_{\text{chl}}(z)\Lambda(z)}{[1 + (\Lambda(z)/P^*_{\max}{}^2]^{1/2}} \tag{4.27}$$

where

$$\Lambda(z) = \sec\theta' \int \alpha^*(z, \lambda, \theta') E_{d_{\text{sun}}}\, d\lambda + 1.2 \int \alpha^*(z, \lambda)\, E_{d_{\text{sky}}}\, d\lambda \tag{4.28}$$

Integration of equation (4.27) over the photic zone for daylight hours will then yield phytoplanktonic photosynthesis (approximate primary production) for the aquatic region being remotely monitored.

Vollenweider showed (1966) that the ratio $P/P_{\max}$, where $P_{\max}$ is the maximum rate of photosynthesis per unit volume ($P_{\max} = P^*_{\max} C_{\text{chl}}$), could be parametrized as follows:

$$\frac{P}{P_{\max}} = \frac{(E/E_k)}{\{[1 + (E/E_k)^2][1 + (\alpha E/E_k)^2]^n\}^{1/2}} \tag{4.29}$$

where a and n are parameters. An increase in α brings about a general lowering of the curve (Figure 4.18) at all irradiances, while an increase in n causes the curve slope to change more sharply about the point of maximum curvature. Except in the special case where either α or n equals zero, the curves exhibit three main features: a nearly

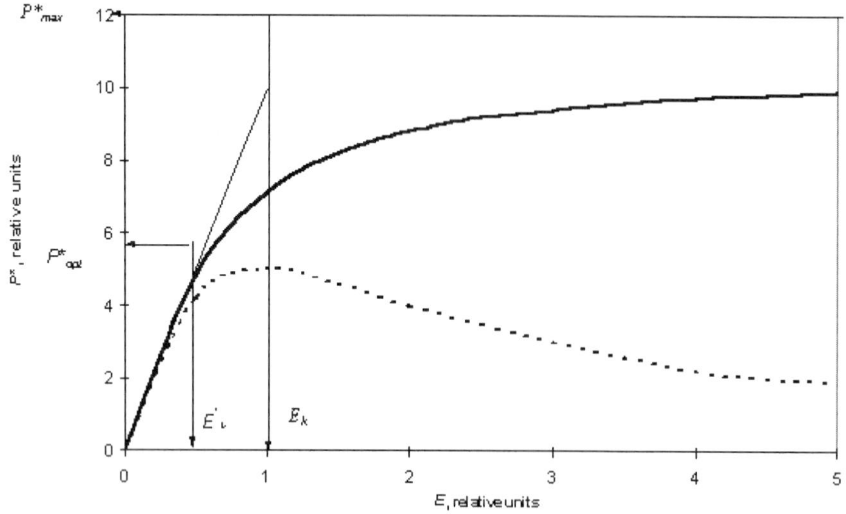

Figure 4.18. Production–light-intensity diagram illustrating the notations of E_k, E'_k, P^*_{max} and P^*_{opt}. E_k is the light intensity at which the extrapolation of the nearly linear portion of the production–light-intensity curve would intersect the asymptote. $P^* = P^*_{max} E'_k$ is the light intensity at which the extrapolated line would intersect the line $P^* = P^*_{opt}$.

linear relationship between photosynthetic rate and irradiance at low light intensities, a peak rate for an optimum irradiance and inhibition at higher irradiances. E is the downwelling irradiance integrated over the PAR spectral region (400–700 nm). The initial and asymptotic horizontal slopes intersect (Figure 4.18) is designated as E_k.

As can easily be seen from Figure 4.18, parameters P_{max} and E_k are not directly measurable if there is any kind of photosynthetic production inhibition by intense light. The measurable parameters are P^*_{opt} (normalized to the aquatic chlorophyll concentration) and E'_k which are related to the basic parameters via the following expressions:

$$P_{max} = \delta P^*_{opt} \tag{4.30}$$

and

$$E_k = \delta E'_k \tag{4.31}$$

where δ is a constant whose magnitude depends on α and n:

$$\delta = [(1 + E^2_{opt})(1 + \alpha^2 E^2_{opt})^n]^{1/2} / E_{opt} \tag{4.32}$$

$$E_{opt} = \left\{ \left[(1 - n)\alpha + \sqrt{\alpha^2 (n-1)^2 + 4n} \right] / 2n\alpha \right\}^{1/2} \tag{4.33}$$

The Fee (1969) integration of the Vollenweider primary model over time and depth yields:

$$P_\Sigma = P_{\text{opt}}\delta \int_{-\zeta}^{+\zeta} \int_{0.01E_d(z=-0,t)/E_k}^{E_d(z=-0,t)/E_k} \frac{dy\, dt}{K_d(z,t)\{(1+y^2)(1+\alpha^2 y^2)^n\}^{1/2}} \tag{4.34}$$

where $E_d(z = -0, t)$ is the photosynthetically active radiation (PAR) just below the water surface at time t; ζ is half the daylength; $K_d(z, t)$ is the downwelling irradiance attenuation coefficient at time t and depth z; $y = E(0, t)/E_k$; E_k is the light-saturation parameter ($E_k = P^*_{\max}\alpha^*$) when α or $n = 0$. The limits of integration in equation (4.34) are, for t, the times of local sunrise and local sunset and, for y, the values corresponding to the surface and the depth of the 1% irradiance level, below which depth no significant contribution to primary production is considered to occur. Note that the irradiance at a depth z for a subsurface refracted angle θ' and the incident radiation comprising both a direct and diffuse component may be written:

$$E_d(z, \theta' \equiv t) = E_d(z = -0, \theta' \equiv t) \exp(-K_d(z, \theta' \equiv t)z \tag{4.35}$$

where $\theta' = t$ underscores the identity between the local time and incoming radiation incidence angle, and points to the possibility that integration of equation (4.34) over time be replaced with integration over θ'.

Consequently, application of this approach for assessing primary productivity of a given algae community requires information on parameters determining the production–light-intensity curve (α, n, E_k and $P_{\max}$) appropriate for this specific assemblage of phytoplankton growing under concrete environmental conditions (water temperature, nutrients availability, etc.).

Explicitly physiological approach

As mentioned above, the second approach defines primary production P at a depth z as a function of chlorophyll specific absorption $\alpha^*_{\text{chl}}(\lambda)$, its biomass at depth z (concentration of chlorophyll C_{chl} in milligrams per cubic metre), quantum yield of photosynthesis $\phi(z)$ (in mol C per Eisteins (Ein) absorbed)[4] and quantum scalar irradiance $E_0(\lambda, z)$ (in Ein·m^{-2} nm^{-1}) at depth z:

$$P(z) = \phi(z) \int_\lambda \alpha^*_{\text{chl}}(\lambda) C_{\text{chl}}(z) E_0(\lambda, z)\, d\lambda \tag{4.36}$$

(the wavelength range of integration of equation (4.31) is 400–700 nm).

The value of ϕ is assumed to be independent of wavelength, but affected by scalar-irradiance level: at low, vanishing irradiance, it reaches a maximal value $\phi_{\max}$, and then decreases when irradiance decreases. Several formulations have been suggested. According to Kiefer and Mitchel (1983):

$$\phi(E_0) = \phi_{\max} \frac{E_\phi}{E_\phi + E_0} \tag{4.37}$$

[4] To convert irradiance expressed in W/m^2 into quanta/m^2 s, it should be multiplied by λ/hc, where c is the speed of light in m/s, h is Planck's constant equal to 6.63×10^{34} joules/quanta, λ is the wavelength in metres (1 Ein = 6.02×10^{-23} quanta).

where ϕ_{max}, a constant $= 0.06$ mol C (Ein absorbed)$^{-1}$, is a maximum quantum efficiency; E_ϕ, a constant $= 10$ Ein/day, is irradiance at which the quantum efficiency is equal to $\phi_{max}/2$; E_0 is the spectrally integrated (400–700 nm) scalar irradiance.

Platt *et al.* (1980) introduced a dimensionless function $f(x)$ describing the evolution of ϕ with irradiance $[\phi = \phi_{max} f(x)]$:

$$f(x) = x^{-1}(1 - e^{-x})e^{-\beta x} \tag{4.38}$$

where β is a parameter controlling light inhibition of photosynthesis ($\beta = 0.01$); x is a dimensionless irradiance = (photosynthetically usable radiation/photosynthetically usable radiation for which $\phi = \phi_{max}/2$). Function $f(x) = 1$ when $x = 0$, and tends toward 0 when x tends towards ∞. Importantly, this function could be tuned for temperature (Berthon and Morel, 1992).

Zaneveld and Kitchen (1993) suggested an alternative representation:

$$\phi(E_0) = \phi_1 - \phi_2 \ln[E_0(\text{PAR})] \tag{4.39}$$

where ϕ_1 and ϕ_2 are constants: $\phi_1 = 0.055$ and $\phi_2 = 0.0109$.

Equation (4.36) can be further rewritten in the form:

$$P(z) = \int \alpha(\lambda, z) C_{chl}(z) E_0(\lambda, z)\, d\lambda \tag{4.40}$$

where $\alpha(\lambda, z) = \phi(z)\, \alpha^*_{chl}(\lambda)$ is the rate of photosynthesis (in mol C mg chl (Ein m^{-2})$^{-1}$) which is commonly referred to as the *action spectrum* (Kirk, 1983).

The effect of photoinhibition on primary production could be formulated by introducing in (4.37) an exponential term:

$$\phi(E_0) = \phi_{max} \frac{E_\phi}{E_\phi + E_0} \exp(-\nu E_0) \tag{4.41}$$

where ν is a constant $= 0.01$ (Ein/m^2/day)$^{-1}$ – a value based on Platt *et al.* (1980).

Consequently, equation (4.40) can be finally presented in the following form:

$$P(z) = \frac{E_\phi \exp(-\nu E_0)}{E_\phi + E_0} \int \alpha(\lambda) C_{chl} E_0(\lambda, z)\, d\lambda \tag{4.42}$$

or else:

$$P(z) = \phi_{max} \frac{E_\phi \exp(-\nu E_0)}{E_\phi + E_0} \int_\lambda \alpha(\lambda) C_{chl} E_0(\lambda, z)\, d\lambda \tag{4.43}$$

4.5.3 Optical properties of and light transfer through water column as related to primary production

Downwelling irradiance attenuation and the depth of the photic zone

The general expression for attenuation of subsurface irradiance (E_d) downwelling along the pathlength (Z_θ) of the principal direction (θ') of subsurface propagation to a particular irradiance level is Beer's Law, which can be:

$$E_d(Z_{\theta'}, \theta', \lambda) = E_d(Z_{\theta'} = -0, \theta', \lambda)\exp(-K_d(\lambda)Z_{\theta'}(\theta'')) \quad (4.44)$$

where K_d is the irradiance attenuation coefficient.

Replacing Z_θ by the vertical depth z in equation (4.44) yields:

$$E_d(z, \theta', \lambda) = E(z = -0, \theta', \lambda)\exp(-K_d(\theta', \lambda)z) \quad (3.45)$$

where $K_d(\theta')$ is the vertical irradiance attenuation coefficient.

Since $Z_\theta = z$ only when the Sun is directly overhead (i.e. $\theta' = 0$), the experimentally determined value of the vertical irradiance attenuation coefficient varies with the time at which the subsurface irradiance profile is performed. Extending the use of an invariant $K_d(\theta')$ throughout the daylight hours (assuming the water column vertically homogeneous), therefore, defines an appropriate diurnal variation in subsurface irradiance levels E_d. Obviously, this is a factor capable of potently affecting a model evaluation of primary production when resorting to the first approach specified above (equation (4.34)).

Using Monte Carlo simulation of photon propagation through natural waters, Kirk (1984) presented empirical relationships relating the vertical irradiance attenuation coefficient to time of day and water type:

$$K_{d_{\text{sun}}}(\theta') = (\cos\theta')^{-1}[a^2 + (0.425\cos\theta' - 0.190)ab]^{1/2} \quad (4.46)$$

$$K_{d_{\text{sky}}} = 1.168[a^2 + 0.162ab]^{1/2} \quad (4.47)$$

where $K_{d_{\text{sun}}}(\theta')$ and $K_{d_{\text{sky}}}$ are the vertical irradiance attenuation coefficients for, respectively, direct and diffuse components of incident solar radiation; a and b are, respectively, absorption and scattering coefficients of the water (for simplicity, the wavelength stipulation, while understood, is not explicitly expressed).

For incident irradiation comprised of a diffuse sky fraction η, the resultant vertical irradiance attenuation coefficient $K_d(\theta')$ could be formulated as:

$$K_d(\theta') = \eta_w K_{d_{\text{sky}}} + (1 - \eta_w)K_{d_{\text{sun}}}(\theta') \quad (4.48)$$

where η_w is the fraction of the subsurface downwelling irradiance that is diffuse:

$$\eta_w = \eta(1 - \rho_{\text{sky}})/[\eta(1 - \rho_{\text{sky}}) + (1 \quad \eta)(1 - \rho_{\text{sun}}(\theta))] \quad (4.49)$$

where θ is the solar-zenith angle prior to refraction at the atmosphere – water interface; ρ_{sky} and $\rho_{\text{sun}}(\theta)$ are Fresnel reflectivities of sky and solar irradiance (direct propagating from the zenith angle θ), respectively.

Bukata *et al.* (1989) showed that for a watermass defined by a single scattering albedo $\omega[\omega = b/c;\ (c = a + b)]$, vertical irradiance attenuation coefficient K_d variations with θ exhibit two features: (i) as ω increases, the solar-zenith angle dependence of K_d decreases, and (ii) the relative value of K_d, for all values of ω, increases with increasing solar-zenith angles up to ca. 70° at which point the relative value of K_d (normalized to $K_d(0°)$) starts rapidly decreasing because of the increasing percentage of diffuse radiation in the total incident radiation observed for large solar-zenith angles. As a result, the depth of 1% subsurface irradiance level (defining the lower

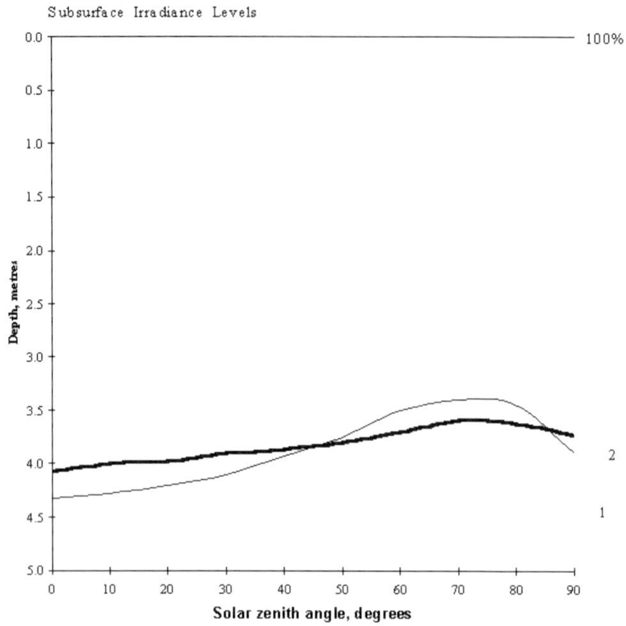

Figure 4.19. Depth of the 1% subsurface irradiance level as a function of solar-zenith angle for a watermass defined by $\omega = 0.60$, $K_d(0^\circ) = 1\,\mathrm{m}^{-1}$ (1) and $\omega = 0.90$, $K_d(0^\circ) = 1\,\mathrm{m}^{-1}$ (2).

boundary of the photic zone) reveals variations with θ similar to those of $K_d(\theta)$, and these variations degrade as ω increases (Figure 4.19) As can be seen from Figure 4.19, in relatively non-turbid waters ($\omega = 0.60$) for which photic depth is ca. 4 m, vertical migrations of the photic depth during daylight hours can attain an amplitude of about 1 m (i.e. 25%), although in the case of more turbid waters ($\omega = 0.90$) they prove to be less pronounced. These results strongly point to the necessity of applying equations (4.44–4.48) when evaluating primary production not only through modelling (e.g. equation (4.34) but also via *in situ* measurements (for a correct lowering of incubation bottles down to the photic depth level during daylight hours).

Relationship between vector and scalar irradiance throughout the photic zone

Global radiation, comprising a direct solar and diffuse sky irradiance downwelling on the air-water interface, possesses inherent geometrical properties related to the arrival directions of the impinging radiation.

The subsurface downwelling irradiance $E_d(z)$ determined from such global radiation values will retain corresponding directional biases, and is often referred to as *vector* irradiance.

A photosynthesizing phytoplanktonic cell suspended in water receives light energy not only from the downwelling irradiance but also from the upwelling irradiance to which it is subjected. *Scalar* irradiance is the total energy per unit area arriving at a point from all directions when all directions are equally weighted. Thus, $E_0(z)$ is the integration of the radiance distribution at a point over 4π space. When divided by the speed of light in water, E_0 becomes a measure of the energy density at a given point in the aquatic medium. Consequently, it is the *scalar* irradiance $E_0(z)$ rather than the *vector* irradiance $E_d(z)$ which is more representative of the cell's light regime.

According to Gershun (1936), the scalar irradiance is related to vector upwelling and downwelling irradiances $E_u(z)$ and $E_d(z)$ and the total absorption coefficient $\alpha(z)$ by the relationship:

$$E_0(z) = -\frac{1}{\alpha(z)}\frac{d}{dz}[E_d(z) - E_u(z)] \tag{4.50}$$

Prieur and Sathyendranath (1981) express the E_0/E_d ratio as:

$$\frac{E_0}{E_d}(R) = \frac{1}{\bar{\mu}_d} + \frac{R}{\bar{\mu}_u} \tag{4.51}$$

where $\bar{\mu}_d$ and $\bar{\mu}_u$ are the mean cosines for the downwelling and upwelling irradiance fields, respectively.

By curve fitting to Monte Carlo computer simulations for a homogeneous water column, Jerome *et al.* (1988, 1990) determined the ratio E_0/E_d as a function of volume reflectance $R = E_u/E_d$ and solar-zenith angle θ for various depths z within the water column.

For the depths of the 100% downwelling irradiance $z_{100} \equiv z = -0$:

$$\frac{E_0}{E_d}(R,\theta) = \left(\frac{1.068}{\mu'} - 0.068\right)(1 + 3.13\mathrm{R}(0^\circ)) \tag{4.52}$$

where $R(0^\circ)$ is the volume reflectance for the vertical incident radiation; μ_d is the reverse of the cosine of the in-water angle of refraction θ'. According to Bukata *et al.* (1995):

$$R(0^\circ) = R(\theta', \eta_w)/(1.165\eta_w + (1 - \eta_w)/\cos\theta') \tag{4.53}$$

and η_w is attainable from equation (3.49).

For the depths of the 10% downwelling irradiance (z_{10}):

$$\frac{E_0}{E_d}(R,\theta) = [1 + (28.0 - 50.5R(0^\circ))R(0^\circ)]^{0.5} + \left(\frac{\cos 48.6^\circ}{1 - \cos 48.6^\circ}\right)\left(\frac{1-\mu'}{\mu'}\right) \times \left[\frac{E_0}{E_d}(R, 89^\circ) - (1 + (28.0 - 50.5R(0^\circ))^{0.5}\right] \tag{4.54}$$

where

$$\frac{E_0}{E_d}(R, 89^\circ) = 1.512[1 + (7.39 - R(89^\circ) + 1376R^2(89^\circ))R(89^\circ)] \tag{4.55}$$

for $0 \le R \le 0.055$, and

$$\frac{E_0}{E_d}(R, 89^\circ) = 1.60 + 3.43R(89^\circ) \tag{4.56}$$

for $0.055 \le R \le 0.14$.

For the depth of the 1% downwelling irradiance (z_1):

$$\frac{E_0}{E_d}(R,\theta) = [1 + (39.1 - 176R(0^\circ))R(0^\circ)]^{0.5} + \left(\frac{\cos 48.6^\circ}{1 - \cos 48.6^\circ}\right)\left(\frac{1-\mu'}{\mu'}\right)$$
$$\times [1.512[1 + (-2.10 + 117R(89^\circ) - 582R^2(89^\circ))R(89^\circ)]^{0.8}$$
$$- (1 + (39.1 - 176R(0^\circ))R(0^\circ))^{0.5}] \tag{4.57}$$

where

$$\frac{E_0}{E_d}(R, 89^\circ) = 1.512[1 + (7.39 - R(89^\circ) + 1376R^2(89^\circ))R(89^\circ)] \tag{4.58}$$

for $0 \leq R \leq 0.055$, and

$$\frac{E_0}{E_d}(R,\theta) = 1.37 + 4.93R(\theta) \tag{4.59}$$

for $0.055 \leq R \leq 0.14$.

In the case of totally diffuse illumination of the water surface, for the depth of the 100% downwelling irradiance (z_{100}) the sought-for ratio takes the form:

$$\frac{E_0}{E_d}(R) = 1.177(1 + 3.13R) \tag{4.60}$$

For the depths of the 10% (z_{10}) and 1% (z_1) downwelling irradiance:

$$\frac{E_0}{E_d}(R) = 1.177[1 + (11.8 + 96.8R)R]^{0.5} \tag{4.61}$$

Expressing the E_0/E_d ratio via R offers an opportunity to evaluate it using water-column inherent hydro-optical properties. Numerous formulations for R have been suggested and a number of models of inherent optical properties of natural waters were developed for oceanic/marine and inland basins (see e.g. Bukata *et al.*, 1995). However, the above expressions are rather compound and calculations of the scalar irradiance vertical profile can be substantially simplified if, after having determined the E_0/E_d ratio at z_{100}, the coefficient of scalar irradiation attenuation is applied (Bannister, 1990):

$$K_0 = (\alpha/\mu')[1 + (0.247\mu')(b_b/\alpha)]^{1/2} \tag{4.62}$$

where α and b_b are absorption and back-scattering coefficients; and μ' is, as above, the cosine of the incident photon flux as measured just beneath the air/water interface. Hence:

$$E_0(z) = E(z_{100} \equiv z = -0)\exp(-K_0 z) \tag{4.63}$$

The coefficient of scalar irradiation attenuation K_0 can be assumed either vertically invariant or defined by specific vertical variations in α and b_b. The inherent optical properties of water (to which α and b_b belong) are additive in nature and hence are basically controlled by all major optically active components. However, in many cases the vertical inhomogeneity in α and b_b arises from a non-uniform distribution of chlorophyll in the vertical. The latter could be represented in a generalized form (Platt and Sathyendranath, 1988):

$$C_{\text{chl}}(z) = C_{0_{\text{chl}}} + \frac{h}{\sigma(2\pi)^{1/2}} \exp\left[-\frac{(z - z_m)^2}{2\sigma^2}\right] \tag{4.64}$$

where $C_{0_{\text{chl}}}$ is the baseline pigment concentration; z_m is the depth of the chlorophyll minimum; σ defines the breadth of the peak; and h determines the total biomass above the baseline. Consequently, in this case, the vertical profile of the attenuation coefficient K_0 will exhibit a dependence which is analogous to (4.61) but, certainly, differently scaled. However, when the optical inhomogeneity of the water column in the vertical is not exclusively controlled by chlorophyll but also by, say, suspended minerals or/and dissolved organics, then the depth dependence of K_0 will be accordingly modified.

4.5.4 Estimating primary production from remote-sensing data

Remote sensing of natural waters is based on capturing, by a remote sensor, the radiance signal coming up from the water surface. After subtraction of the Fresnel reflection component, the resultant radiance signal is a convolution of processes of solar light interactions with the water column. Consequently, the legitimate optical signal reaching the remote sensor carries information not just about a surface layer, it also contains contributions from layers deeper in the water column, although strongly damped by aquatic medium attenuation.

Gordon and McCluney (1975) determined that for a sensor viewing the ocean at any wavelength λ, 90% of the recorded signal originates from within a so-called penetration depth given by the inverse of the downwelling irradiance attenuation coefficient $K_d(\lambda)$. The penetration depth corresponds to one attenuation length; that is, the depth at which the downwelling irradiance falls to ca. 37% of its value just beneath the air–water interface. Thus, satellite observations over natural waters are restricted to the uppermost attenuation length. As was assumed above, the photic zone is considered to extend to the depth of the 1% irradiance level. The photic zone, therefore, extends to the 4.605-m penetration depth.

In relatively clear case-II waters (using the North American Great Lakes as an example), the values of K_{PAR} are generally in the range 0.1–0.5 m^{-1} resulting in penetration depths of 10 to 2 m being commonly observed in open waters. The corresponding photic zones in these waters would extend to depths in the range ~46–9 m. For relatively turbid case-II waters, the values of K_{PAR} generally are in the range 0.5–1.2 m^{-1} resulting in penetration depths between 2 and 0.8 m^{-1} and photic depths between 9 and 4 m (Bukata *et al.*, 1995).

Algorithms are available that convert the optical signal to give a chlorophyll concentration that is representative of the upper attenuation depth (Pozdnyakov and Kondratyev, 1998). At the same time, satellite techniques were developed to evaluate the downward solar radiation at the surface (Gautier *et al.*, 1980; Dedieu *et al.*, 1987).

However, for adequate evaluation of primary production in the water column, information is required on the *total* chlorophyll content ($\langle C_{\text{chl}} \rangle_{\text{tot}}$) within the euphotic zone (i.e. the zone at the depth where (Z_{eu}) the downwelling irradiance equals 1% of the irradiance just beneath the water surface).

Since remote-sensing means, in principle, are incapable of providing the required information on $\langle C_{\text{chl}} \rangle_{\text{tot}}$, the latter could be found either by establishing a regression relationship between the remotely determined chlorophyll concentration (C_{chls}) and $\langle C_{\text{chl}} \rangle_{\text{tot}}$, or by applying to C_{chls} statistical profiles of C_{chl} characteristic of the waterbody under surveillance (Morel and Berthon, 1989; Brooks and Torke, 1977).

Berthon and Morel (1992) used the model to estimate phytoplanktonic primary production which rests on the photosynthetically stored radiant energy $\bar{E}$ content in the euphotic zone:

$$P = (1/39)\tilde{E} = (1/39)\bar{E}_{\text{PAR}}(z = +0)\xi\langle C_{\text{chl}} \rangle_{\text{tot}}\psi^* \tag{4.65}$$

where $\bar{E}_{\text{ow}}(z = +0)$ is the photosynthetically available irradiance just above the water surface and integrated over daylength (W/m^2); ψ^* is the cross-section for photosynthesis per unit of area C_{chl} ($\text{m}^2/\text{g chl}$);[5] ξ is the ratio of photosynthetically active pigments to total pigments (including pheopigments):

$$\xi = C_{\text{chl}}/[C_{\text{chl}} + C_{\text{pheo}}] \tag{4.66}$$

the conversion factor ($39\,\text{kJ g}^{-1}$ of C fixed) is taken from Platt (1969), $\langle C_{\text{chl}} \rangle_{\text{tot}}$ is as above:

$$\langle C_{\text{chl}} \rangle_{\text{tot}} = \int_0^{z_{eu}} C_{\text{chl}}(z)\, dz \tag{4.67}$$

To adjust this model to remote-sensing needs, the vertically integrated biomass $\langle C_{\text{chl}} \rangle_{\text{tot}}$ is replaced by an empirical non-linear expression which involves the chlorophyll concentration retrieved from remote-sensing data (C_{chlrs}), a factor F and an exponent γ derived from a regression analysis. In vertically homogeneous or well-mixed waters, γ is nearly 0.5, but the existence of non-uniform profiles (particularly with deep pigment maxima) results in somewhat lower values of γ.

A more simplified formulation for P_{eu} via remotely sensed parameters was suggested by Berthelot and Deschamps (1994). This formulation does not encompass parameters related to photosynthesis *per se*, but is based solely upon

[5] There is experimental evidence (Berthon and Morel, 1992) that the mean value of ψ^* for the euphotic zone is about $0.088\,\text{m}^2\,(\text{g chl-a})^{-1}$ as opposed to 0.07 (relevant units) previously reported by Platt (1986).

chlorophyll content in the upper layer and the daily PAR:

$$\lg P_{eu} = -4.286 - 1.390 \lg(R_1/R_3) + 0.621 \lg(0.43 E_d(z = +0.400 - 700\,\text{nm})) \tag{4.68}$$

where R_1 and R_2 are volume reflectances at λ equal to 440 and 550 nm, respectively. The ratio R_1/R_3 with appropriate (locally tuned) coefficients stands, in equation (4.68), for chlorophyll concentration in the first attenuation water column which is further regressed against euphotic zone chlorophyll concentration.

The second term in equation (4.68) estimates the amount of PAR coming down onto the water surface. Barthelot and Deschamps (1994) claim that using this algorithm the r.m.s. accuracy of the primary-production estimate is 0.17 on a logarithmic scale; that is a factor of 1.5.

Balch *et al.* (1989) analysed the reliability of relating euphotic zone primary production P_{eu} to the concentration of chlorophyll in the top optical depth (C_{chlrs}) leaving PAR beyond explicit formulation. The algorithms suggested for evaluating integral primary productivity in the euphotic zone do not explicitly incorporate the daily PAR at sea level:

$$P_{eu} = (210 C_{chlrs} + 383)0.01 \tag{4.69}$$

$$P_{eu} = 10^{(1.25+0.75 \lg C_{chlrs})}\{4.6(6.9 - 6.5 \lg C_{chlrs})\}0.001 \tag{4.70}$$

$$\lg P_{eu} = 0.48 \lg C_{chlrs} - 0.29 \tag{4.71}$$

$$\ln P_{eu} = 3.06 + 0.5 \ln C_{chlrs} - 0.24\text{PTA} + 0.25\text{DL} \tag{4.72}$$

where PTA is the local area temperature anomaly; and DL is daylength.

The validity of the above algorithms was tested against shipborne measurements. Results of comparison of adequacy of the above algorithms displayed that the first algorithm (4.68) could be applied for cases when $P_{eu} \geq 0.38\,\text{g}\,^\circ\text{C}\,\text{m}^{-2}\,\text{day}^{-1}$, the second algorithm (4.69) has an upper limit of applicability ($P_{eu} \leq 0.69\,\text{g}\,^\circ\text{C}\,\text{m}^{-2}\,\text{day}^{-1}$ and $C_{chlrs} \leq 2.9\,\text{mg}\,\text{m}^{-3}$), the third and fourth algorithms exhibited no upper or lower limits to predicted production, but they underestimated the values of P_{eu} by ca. 25%. However, the best results were obtained when the algorithm takes the dependence of the maximum photosynthesis P'_{max} on water temperature T (in °C) into account:

$$P'_{max} = 10^{(0.006T+0.308)} \tag{4.73}$$

Euphotic zone chlorophyll concentration was calculated via:

$$C_{chleu} = \frac{C_{chlrs}(6.9 - 6.5 \lg C_{chlrs})}{0.157} \tag{4.74}$$

Temperature at depth was approximated using an empirical relation between mixed-layer depth (ML, in meters) and the surface temperature T_{surf}:

$$\text{ML} = 133 - 7(T_{surf}) \tag{4.75}$$

Ultimately, the temperature profile over depth was used to construct the profile of P'_{max} which was then convolved with the chlorophyll profile to calculate P_{max} through the water column. Photosynthesis was assumed maximal at light intensities $\geq 10.5\%$ of the surface irradiance. Below this light intensity, P_{max} was weighted by a factor $F(E, z)$ which decreased photosynthesis according to light intensity at depth (using Beer's law). In a summarized form, the algorithm could be formulated as:

$$P_{eu} = \sum_{z=0}^{z=z_{eu}} P'_{max} C_{chl}(z) F(E, z) \tag{4.76}$$

Lee *et al.* (1996b) report that they developed an algorithm for estimating primary production at depth, from remote sensing, that uses an analytically derived phytoplankton (chlorophyll) absorption cross-section at 440 nm ($\alpha^*_{chl}(440)$) and the bulkwater absorption coefficient $\alpha(\lambda)$ based on remote measurements of above-water upwelling and downwelling radiances. The general scheme of the algorithm encompasses equations (4.18–4.19). The photoinhibition parameter ν in equations (4.42–4.43) was set at 0.01 Ein/m^2/day^{-1}, a value based on Platt *et al.* (1980).

However, instead of using equation (4.61) for the coefficient of scalar irradiation attenuation K_0, the algorithm in question uses the coefficient of downwelling vector irradiance attenuation $K_d \approx \alpha/\mu_d$, where α is the bulk water absorption coefficient and is a sum of three components: $\alpha_w + \alpha_{dom} + \alpha_{ph(chl)}$, α_w is the pure water absorption, α_{dom} is the dissolved organics and detritus absorption, $\alpha_{ph(chl)}$ is the absorption by phytoplankton (chlorophyllous pigments):

$$\alpha_{dom}(\lambda) = \alpha_{dom}(440) \exp[-S(\lambda - 440)] \tag{4.77}$$

S is a variable exponent and μ_d is the subsurface mean cosine for the downwelling light field.

To retrieve the bulk-absorption coefficient, the relationship between the ratio

$$E_u(z = +0, \lambda)/E_d(z = +0, \lambda)$$

which is generally referred to as remote-sensing reflectance and inherent water properties was used (Lee *et al.*, 1996b):

$$R_{rs}(\lambda) \approx \frac{0.17}{\alpha_w(\lambda) + \alpha_{dom}(\lambda) + \alpha_{ph(chl)}} \left[\frac{b_{b_w}(\lambda)}{3.4} + X \left(\frac{400}{\lambda} \right)^Y \right] \tag{4.78}$$

where X, Y are coefficients; b_{b_w} is the water-backscattering coefficient.

Averaging of K_d over the euphotic depth is attained via multiplying by an empirically found coefficient 1.08. Remotely assessed concentration of chlorophyll was decreased by a factor of 0.8 to account for the contribution of pheopigments. μ_d was given a value of 0.83 for high-latitude cloudy days. E_0 and α'_{chl} (averaged over the PAR spectral region) were given values of 10 Ein/day and 0.016 m^2/mg chl-a, respectively. Integration of $E_0(\lambda)$ over depth was performed down to the lower boundary of the euphotic zone (i.e. where $E_0(z, \lambda) = 0.01 E_0(z = 10, \lambda)$).

Kondratyev and Pozdnyakov (1994; see also Bukata *et al.*, 1995) suggested a protocol for estimating primary production in non-case-I waters (i.e. marine coastal and inland) from remote measurements based on the E_0-R formalism discussed above and a hydro-optical model developed for Lake Ladoga (Kondratyev *et al.*, 1990). A generalized flow diagram outlining parameters and sequential operations involved in remotely estimating primary production in turbid waters is presented in Figure 4.20.

The recorded upwelling radiance spectrum $L_u(z = +0, \lambda, \theta)$ is converted into the subsurface volume reflectance spectrum $R(z = +0, \lambda, \theta)$ at just below the air–water interface:

$$R(z = -0, \theta) = \frac{Q[L_u(+0, \theta) - 0.021\alpha E_{\text{sky}} - f_2 E_{\text{sun}}]}{0.544(0.944E_{\text{sky}} + f_4 E_{\text{sun}})} \tag{4.79}$$

where Q is the ratio of the upwelling irradiance below the water surface to the upwelling nadir radiance below the water surface; α is the spectrally dependent ratio of the downwelling zenith sky radiance to the downwelling sky irradiance; f_2 is reflection of the solar irradiance into the field of view of the sensor because of surface waves; f_4 is reflection of the solar irradiance because of surface waves; E_{sun} is spectrally dependent downwelling irradiance from the Sun; E_{sky} is spectrally dependent downwelling irradiance from the sky.

The subsurface volume reflectance just beneath the air–water surface interface $R(z = -0, \lambda, \theta)$ will, of course, be remotely estimated under variable conditions of solar-zenith angle θ, and relative fractions of diffuse and direct downwelling global radiation. Thus, it is necessary to normalize the subsurface volume reflectance to a standard solar angle and a standard downwelling radiation. Consequently, the estimated subsurface volume reflectance just beneath the air–water interface can be partitioned into components responsive to direct and diffuse components of global radiation:

$$R(z = -0, \theta, \eta_w) = (1 - \eta_w)R_{\text{sun}}(z = -0, \theta) + \eta_w R_{\text{sky}}(z = -0) \tag{4.80}$$

where η_w is described by equation (4.46) and is the fraction of total subsurface downwelling irradiance that is diffuse; the fraction that is direct would then be given by $(1 - \eta_w)$; $R_{\text{sun}}(z = -0, \theta)$ is the volume reflectance just beneath the air–water interface resulting from the direct fraction of global radiation; it is dependent on θ (or rather on θ' which is equal to the in-water refraction angle associated with a solar-zenith angle of θ); $R_{\text{sky}}(z = -0)$ is the volume reflectance just beneath the air–water interface, resulting from the diffuse fraction of global radiation; it is independent of θ.

According to Bukata *et al.* (1995), subsurface volume reflectance $R_{\text{sun}}(z = -0, 0°)$, observable when the sun is vertically overhead (i.e. $\theta = 0°$), may be related to remotely estimated subsurface volume reflectance $R(z = -0, \theta, \eta_w)$ through equation (4.52). From such zero-angle normalized subsurface volume reflectance values and appropriate cross-section spectra for major optically active components in turbid waters (Kondratyev *et al.*, 1990), the near-surface chlorophyllous pigment concentrations (along with co-existing concentrations of suspended minerals (sm) and dissolved organics (doc)) may be estimated making use of one of the available retrieval algorithms (see e.g. Pozdnyakov and

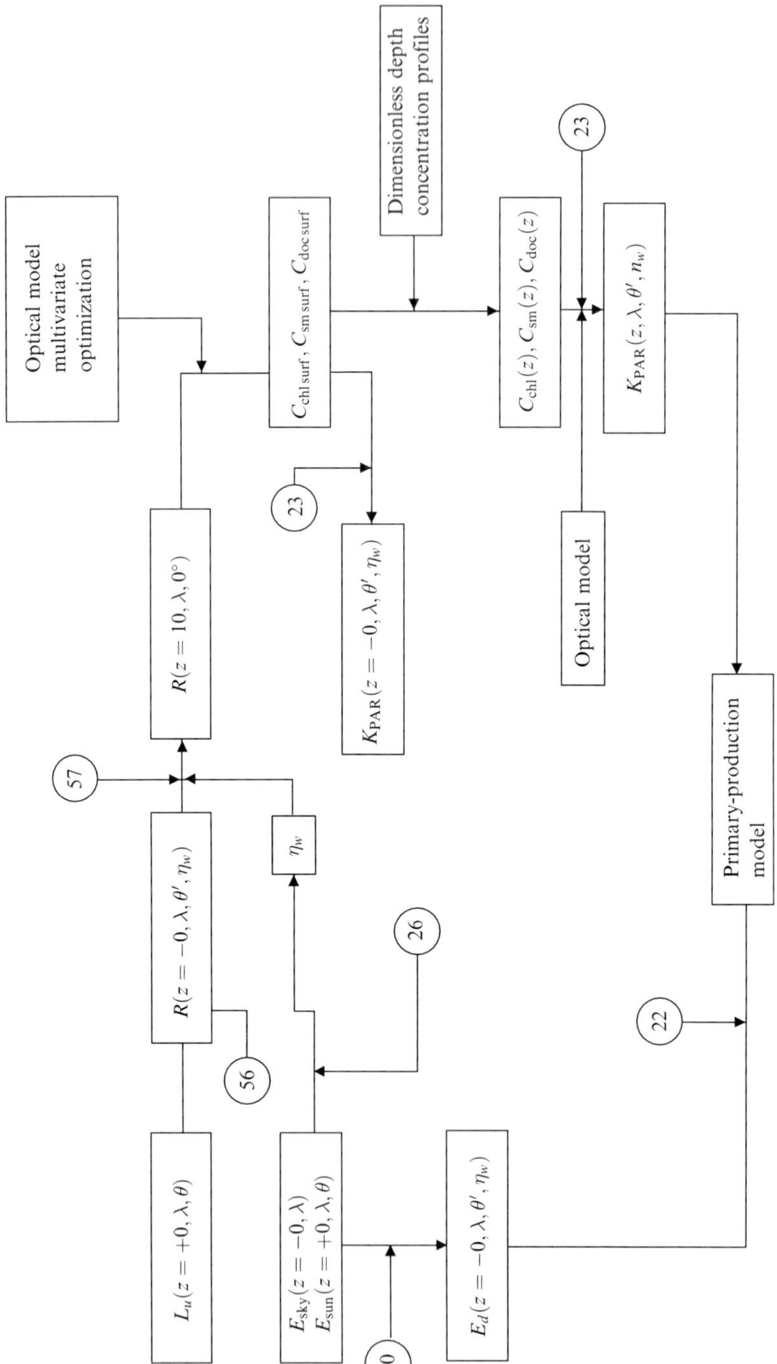

Figure 4.20. Generalized flow diagram outlining parameters and sequential operations involved in remotely estimating primary production in non-case-I waters. The mathematical formalism is described in the text.

Kondratyev, 1998) Multiplying the retrieved concentration of chlorophyll by the coefficient ξ (equation (4.65)) will result in evaluation of abundance of chl-a in the top layer of the water column.

When the water surface is calm, downwelling subsurface irradiance $E_d(z = -0, \theta, R)$ can be obtained either directly from on-site measurements or indirectly estimated from above-surface values of downwelling irradiance $E_d(z = +0, \theta, R)$ through the relationship:

$$E_d(z = -0, \theta', R) = \left[1 + \frac{R(\theta')\rho_u(\theta')}{1 - R(\theta')\rho_u(\theta')}\right](1 - \rho_d(\theta))E_d(z = +0, \theta, R) \quad (4.81)$$

where $\rho_d(\theta)$ is the coefficient of reflection of above-surface downwelling irradiance for solar-zenith angle θ; $\rho_u(\theta')$ is the coefficient of reflection of subsurface upwelling irradiance for solar-zenith angle θ'.

It was shown (Bukata *et al.*, 1995) that:

$$\rho_u(\theta') = 0.271 + 0.249\cos(\theta') \quad (4.82)$$

for direct incident solar radiation, and $\rho_u = 0.561$ for cardioidal diffuse incident radiation distribution. Values of $\rho_d(\theta)$ may be calculated from Fresnel's equation, while ρ_d for diffuse cardioidal distribution can be taken as 0.066 (Yerlov, 1976).

Baker and Smith (1990) suggest a simplified expression for $E_d(z = -0)$:

$$E_d(z = -0) = \frac{1}{1 - \rho_u R(\theta')} t(\theta)E_d(z = +0, \theta) \quad (4.83)$$

where $t(\theta)$ stands for global transmittance through the air–water interface:

$$t(\theta) = 1 - [\rho_{\text{sun}}(1 - \eta) + \rho_{\text{sky}}\eta] \quad (4.84)$$

The value of ρ_{sky} is estimated at 0.066 for uniform sky-radiance distribution, and 0.052 for completely overcast skies (assuming cardioidal distribution); ρ_w is given the value of 0.47. Fresnel's formula for unpolarized light gives the reflectance of a direct solar beam.

Evaluation of sunlight propagation right down to the euphotic-zone lower limit (z_{eu}), requires information on vertical profiles of the major optically active components, such as chl, sm and doc.

Surficial concentrations of chl, sm and doc can be retrieved in non-case-I waters from the upwelling irradiance by employing sophisticated techniques, such as the multivariate optimization, simplex method, neural networks and some others described elsewhere (Atkinson, and Tatnall, 1997; Pozdnyakov and Kondratyev, 1998). However, the problem is difficult because, for adequate assessment of column primary production, first a vertical profile of chl concentration is required and second, for determining z_{eu}, information on column-content variation in other optically active components is mandatory.

In an attempt to overcome variability in depth profiles of chlorophyll concentrations in inland and coastal waters, it might be worth considering the approach, suggested for case-I waters, by Morel and Berthon (1989). This approach considered two extreme aquatic configurations that could accommodate

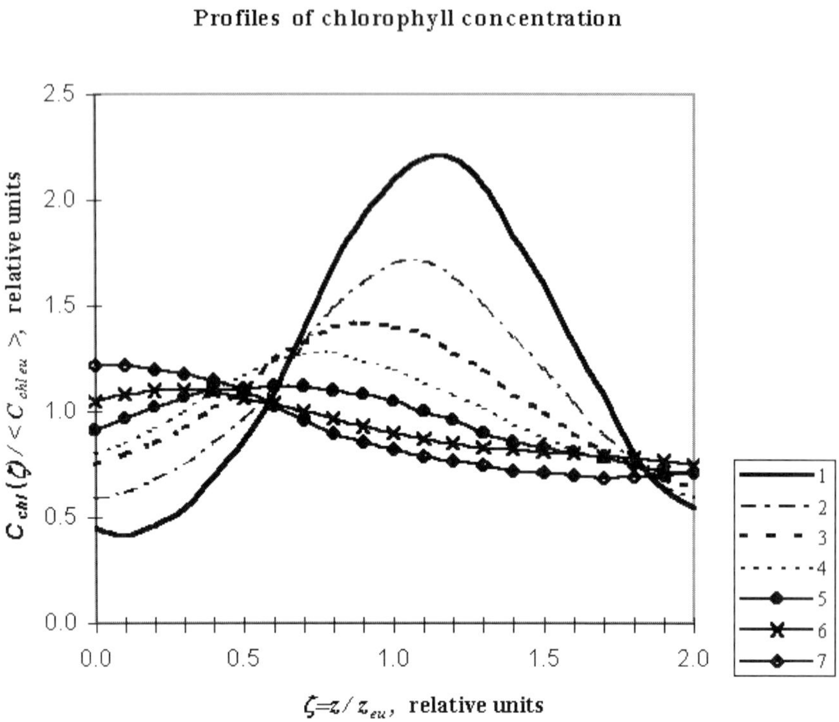

Figure 4.21. Seven generic dimensionless chlorophyll depth profiles for case-I waters.

chlorophyll concentration profiles of oceanic waters. The first extreme was taken to be stratified oligotrophic oceanic waters wherein nutrient depletion is observed in the near-surface layer. In this case, maximum chlorophyll concentration would be downward driven. The second extreme was taken to be well-mixed waters in which sufficient irradiance levels have established eutrophic conditions. In this case, vertical distribution of chlorophyll $C_{\text{chl}}(z)$ throughout the photic zone would be quite uniform. Understandably, a wide range of intermediate chlorophyll profiles within these extremes are encountered in natural waters. It is perhaps not unreasonable, however, to reduce this wide range of profiles to a limited number of generic chlorophyll depth distributions pertinent to specific types of water columns (and on through specific water bodies), depending on trophic status of the water, aquatic vegetation growth cycle and intensity and duration of wind-induced vertical mixing.

Statistically analysing over 4,000 oceanic-station datasets, Morel and Berthon (1989) illustrated the presence of seven generic classifications of oceanic waters. These profiles are schematically illustrated in Figure 4.21. The seven profiles are plotted in terms of a dimensionless chlorophyll concentration $C_{\text{chl}}(\zeta)/\langle C_{\text{chl eu}}\rangle$ and a dimensionless depth $\zeta = z/z_{\text{eu}}$; that is depth reckoned as a fraction of photic depth and chlorophyll concentration reckoned as a fraction of total photic-zone chlorophyll:

$$\langle C_{\text{chl}} \rangle = z_{\text{eu}}^{-1} \int_0^{z_{\text{eu}}} C_{\text{chl}}(z)\, dz \tag{4.85}$$

Numerical analysis of $C_{\text{chl}}(\zeta)/\langle C_{\text{chl eu}} \rangle$ profiles in Figure 4.4, as a function of remotely determined (i.e. surficial) concentration of chl (denoted c in equation (4.86)), suggests the following parametrization (Morel and Berthon, 1989):

$$C_{\text{chl}}(\zeta)/\langle C_{\text{chl eu}} \rangle = C_b + C_{\max} \exp\{-[(\zeta - \zeta_{\max})/\Delta\zeta]^2\} \tag{4.86}$$

where $C_b = 0.768 + 0.087 \lg c - 0.179 (\lg c)^2 - 0.025 (\lg c)^3$; $C_{\max} = 0.299 - 0.289 \lg c + 0.579 (\lg c)^2$; $\zeta_{\max} = 0.600 - 0.640 \lg c + 0.02 (\lg c)^2 + 0.115 (\lg c)^3$; $\Delta\zeta = 0.710 + 0.159 \lg c + 0.02 (\lg c)^3$. C_b is a background over which a Gaussian curve is superimposed with a maximum value given by $C_{\max}$ occurring at $\zeta_{\max}$, and having a broadness controlled by $\Delta\zeta$.

Brooks and Torke (1977) seemingly found the same kind of regularities in chl vertical profiles in Lake Michigan. They report that for a multi-year period of observations in Lake Michigan chlorophyll concentration increased uniformly at all depths during spring reaching 3–4 mg/m^3 by late May. Thermal stratification was followed by development of a subthermocline chlorophyll peak between 10 and 30 m that reached 8.5 mg/m^3 by late July. The major subthermocline peak collapsed in mid-August but was followed by two lesser peaks at depths of 10 and 30 m. Autumn mixing dispersed these peaks in the mixed layer, increasing chlorophyll content of the epilimnion at a time when integral chlorophyll levels were declining. At the end of autumn, chlorophyll concentrations were uniformly distributed at approximately 1 mg/m^3.

Unfortunately, in non-case-I waters, the levels of high energy available for photosynthesis and the depth of the photic zone are not controlled only by chlorophyll content ($\langle C_{\text{chl eu}} \rangle$), but also by the presence of other responsive aquatic components that do not necessarily co-vary with $C_{\text{chl}}(z)$. However, analysis of field data in large temperate lakes in the Northern Hemisphere points to the fact that vertical homogeneity of suspended minerals and dissolved organics may very often be a not-unreasonable assumption (Petrova, 1990).

Based on these data, Kondratyev and Pozdnyakov (1994) suggested using three types of chl vertical profiles for large temperate lakes: uniform in spring, submerged peak in mid-summer and a near-surface peak in early to mid-autumn (profiles 6, 1 and 7 in Figure 4.21). Distributions of sm and doc were believed to be vertically homogeneous.

So, vertical profiles of bulk-water absorption and backscattering coefficients can be determined down to z_{eu}. In its turn, the vertical profile of K_d or K_0 can be calculated (equations (4.46–4.48), (4.52)), and ultimately scalar or vector irradiance vertical profiles (equations (4.35), (4.52), (4.64)). Finally, irradiance profiles applied to any chosen primary production model (i.e. equations (4.27–4.28), (4.34), (4.30–4.43)) can be calculated.

This protocol was further used to process field data and evaluate primary production in Lake Ladoga. Although no remote sensing *per se* was performed during this campaign, optical measurements performed above and below the air–water interface are pertinent to support the suggested procedure.

Daily spectral measurements of the incident global PAR were made in late 1989 aboard a small research vessel in the northern coastal zone of Lake Ladoga. Concurrent with these actinometric observations were *in situ* measurements of the vertical distribution of phytoplanktonic chlorophyll concentration throughout the photic zone, and subsurface upwelling and downwelling irradiances. Field data were collected under a cloudless sky and an almost perfectly calm air–water interface at local noon ($\theta = 35°$). The diurnal value of the incident global PAR was calculated from continuously recorded data to be 27 Ein/m^2. The subsurface volume reflectance spectrum displayed a peak value of 0.040 at $\lambda = 580$ nm.

Use of optical cross-section spectra for Lake Ladoga and multivariate optimization technique (Kondratyev *et al.*, 1990) resulted in retrieval from the volume reflectance spectrum of 7.1 mg/m^3 of chl-a, 0.8 g/m^3 of sm and 8 g C/m^3 of doc as estimates of co-existing near-surface aquatic concentrations. The value of $K_{PAR}(z = -0)$, deduced on the basis of the bio-optical model and retrieved concentration values of chl, sm and doc, was 1.8 m^{-1}. This $K_{PAR}(z = -0)$ corresponds to a penetration depth of ca. 0.6 m.

The following input values of the Vollenweider model parameters of equation (4.34) were selected as appropriate to Lake Ladoga: $a = 1.0$, $n = 0.5$, $\delta = 0.26$, $\zeta = 12$ hours, $P_{opt} = 80$ mg C/m^3, $E_k = 800$ μEin/m^2 s. It was assumed that the depth of sm and doc remained invariant from their estimated near-surface values throughout the photic zone. The chlorophyll depth profile, as mentioned above, was taken from *in situ* measurements.

Integration of the Vollenweider equation (4.34) yielded a daily phytoplanktonic photosynthesis value of 346 mg C/m^{2}day. This estimate was in good agreement with direct measurements of primary production in Lake Ladoga performed during the same vegetation season (relevant values in the range 320–380 mg C/m^2 day were reported by Petrova (1990)). The primary-production value, determined from volume reflectance estimates, was not corrected for scalar irradiance since the direct measurements reported in Petrova were values for downwelling irradiance.

As a conclusion to this section, some remarks are made about persisting problems in remote sensing of primary production in natural turbid waters. Leaving aside the issue of quantitative evaluation of chlorophyll and other optically active substances in surficial concentrations (which constitute a separate topic), emphasis should be placed here on parameters of two primary models that describe the photosynthetic activity of algae.

When employing models that depend either on algal physiological parameters (chlorophyll quantum yield of photosynthesis ϕ and PAR specific absorption coefficient α^*_{chl}) or production–light-intensity curves (initial slope α' and P^*_{max} normalized to chlorophyll content), special attention should be drawn to the space and time variability of locally determined values of these parameters to other water areas or/and water bodies.

Pozdnyakov and Kondratyev (1998) analysed the available data on α^*_{chl} variability as affected by phytoplankton volumetric concentration and taxonomic composition. These variations can be very substantial and some typical values of α^*_{chl}, which would be readily applicable to watermasses/waterbodies that differ greatly in geographical location and vegetation cycle, are hardly possible. Moreover, even vertical invariability in α^*_{chl} should be considered in many cases (particularly in case-I-waters) as a rather rough approximation.

There is direct evidence (Berthon and Morel, 1992) that the physiological status of algae, as well as their taxonomic composition, render extensive departures of ϕ from some typical value. The same could be stated about production – light-intensity curve parameters (Platt and Sathyendranath, 1988).

This indicates that, first, more research is needed to establish probabilistic characteristics of the above parameters for a variety of water basins and vegetation seasons and, second, it is considerably more reliable, at least to date, to perform remote sensing of aquatic primary production on a local scale, and much caution should be used in applying primary-production assessment results to water areas/waterbodies other than those for which photosynthesis parameters have actually been obtained.

This last conclusion may be illustrated by results of the analysis of specific features of Antarctic phytoplanktonic photosynthesis inhibited by UVR during incubation (Neale *et al.*, 1998). It was discovered in this case that inhibition is worse in regions beneath the Antarctic ozone hole. Neale *et al.* (1998) emphasized that direct extrapolation to the conditions of the open waters of the Antarctic is impossible, because vertical mixing of phytoplankton alters UV exposure and has significant effects on integrated inhibition through the water column. In this context, a model of UV-influenced photosynthesis in the presence of vertical mixing was developed. Calculations made by Neale *et al.* (1998) indicate that photosynthesis is strongly inhibited by near-surface UV radiation. This inhibition can be either enhanced or decreased by vertical mixing, depending on the depth of the mixed layer.

Voytex (1990) discussed biological implications of TOZ depletion for the Antarctic in a much broader context and emphasized that the most serious consequences of UVR enhancement, because of TOZ depletion, are connected with changes in energy transfer in food chains. As far as phytoplanktonic dynamics is concerned, Voytex (1990) mentioned that marine phytoplankton in the Antarctic was acreage in a stress situation.

4.5.5 UVB radiation variation impact on primary production in turbid waters

Kondratyev and Pozdnyakov (1996a, b) analysed possible physiological responses of algae to enhancement of incident UV radiation and related hydro-optical aspects of this problem.

The first reliable evidence for the UVB radiation effect on Antarctic phytoplankton was published in 1994. Based on analysis of UVB spatial and temporal variability caused by the Antarctic ozone hole, it was shown (Prézelin *et al.*, 1994) that UVB enhancement reduced phytoplankton productivity by 6–23% in the marginal zone of the ice sheet.

The damaging effect of UVB radiation on the process of primary production takes place at the DNA level, the normal processes of intercellular metabolism, photosynthesis and nitrogen fixation being broken. Pigments get bleached, the ability of the cells of phytoplankton to orientate and their motor ability become inhibited. UV radiation is believed to damage the reactive centres of photosystem II in the cells of phytoplankton, which leads to decrease in the efficiency of the electron-transport mechanism and to breaking the normal functioning of that part of the photosynthesizing apparatus of the cell which is responsible for decomposition of water molecules.

The photoinhibition mechanism consists apparently in photo-oxidation of reactive centres and mediating molecules (such as plastoquinone), which provide the normal functioning of electron-transport paths. The presence of both UVA and UVB in incident radiation flux intensifies photoinhibition and slows down processes of self-repair of the normal functioning of the cell. However, it is still unclear whether the effect of UVB radiation is reduced only by the enhancing effect of UVA, or whether some other damaging mechanism is equally important.

However, it was found that, under the influence of UV radiation, the total content of lipids in an algal cell decreases, which disturbs the normal functioning of membranes. Provoked by UVB impact, inhibition of photosynthesis can also hyperbolically depend on the level of total irradiance: a relatively short-period exposure to large doses of UVB is more damaging than longer exposure to less intensive radiation fluxes. There are reports suggesting that the damaging effect combines with initiation of algal-cell-protective functions. Thus, the resulting action spectra are of a composite character, which is complicated by interactive effects of enhanced protective and repairing abilities of an algal cell under the influence of UVA and visible light. In this connection, it should be emphasized that the threshold of photoinhibition for many algae is different (e.g. there are direct indications that flagellates are more sensitive to UV radiation than diatoms), and there are several taxons not affected by high-level solar irradiance, whereas small-sized and ultra-small planktonic organisms[6] prove to be particularly vulnerable to UVB radiation impact.

Apart from the purely physiological impact, UVB radiation is believed to reduce primary-production rates through a host of indirect mechanisms. One of the options is related to influencing phytoplankton vertical positioning. The space of productive activity of phytoplankton is confined to the euphotic zone; that is, the upper layer of the water column whose bottom is determined by the level at which photosynthic energy production in a cell is balanced by its losses on respiration. Vertical profiles of plankton concentration in the euphotic zone are determined by both solar irradiance conditions and physical processes, such as wind-driven mixing, temperature-density variation and many others.

Many phyta of phytoplankton can migrate vertically on their own. Planktonic organisms, deprived of original mobility functions (such as cyanobacteria, diatoms

[6] It is presently believed that nano and picoplankton account for at least 40% of total plankton biomass (see Kondratyev and Pozdnyakov, 1996a, b).

and bacteria), use this facility to regulate buoyancy and, with the help of gas vacuoles (spaces filled with oily liquids), also move up and down within the euphotic zone (and outside it).

Phytoplankton use light at different wavelengths in the visible and UVA spectral regions to orientate. Incidentally, their sensor systems cannot functionally respond to variation in UVB radiation and hence do not provide cells with proper information about the necessity to move towards deeper (radiation protected) levels of the water column. This additional effect is bound to further enhance the injurious effect of UVB radiation on primary production.

However, there is evidence that, being exposed to enhanced radiation doses, many floristic types of phytoplankton can either produce protective (screening) pigments (microsporines) or intensify the use of carotenoids to take up excess PAR from the photosynthesizing pigments. Moreover, as was pointed out above, the net result of UVB incident radiation increase at the air–water interface is complicated by interactive effects of enhanced protective and repairing abilities of an algal cell under the influence of UVA and visible light.

Poor understanding of physiological consequences for natural aquatic environments notwithstanding, the problem could be considered purely in terms of changes in short-wave radiation propagation through the water column. Obviously, such an approach is only capable of analysing the increment in photosynthetic available light within the euphotic depth as well as in the euphotic zone *per se* because of increased intensity in solar short-wave irradiance.

Since penetration depth has a pronounced dependency on λ, eutrophic depth is also wavelength dependent. It means, first, that spectrally mean euphotic depth, being controlled by permeating photosynthetically active irradiance, is theoretically amenable to some extension down to lower depths because of enhanced levels of incident radiation at wavelengths around 400 nm. Second, for the same reason, an increase in PAR should be expected throughout the total photic zone.

The actual importance of the effects mentioned above is controlled by the optical properties of the particular water mass. As can be seen, attenuation rates of light propagating though the water column are determined by coefficients of irradiance attenuation (equations (4.44)–(4.48)), which are functions of bulkwater absorption and scattering (backscattering) coefficients. Since we have concentrated on turbid non-case-I waters, it should be remembered that such waters are rich in optically active components. Chlorophyll and dissolved organics strongly absorb in the shortwavelength portion of PAR. It means that the expected increase in z_{eu} and PAR levels in the photic zone will be considerably subdued.

Based on the hydro-optical model for Lake Ladoga (Kondratyev *et al.*, 1990) and the formalism presented above, assessments were made regarding increase in z_{eu} because of 10% increase in radiation at 400–410 nm in a water column characterized by concentrations typical of Lake Ladoga: $C_{chl} = 5\,\mu g/l$, $C_{doc} = 8\,mg\,C/l$, $C_{sm} = 0.8\,mg/l$. Calculations yielded the value of $\Delta z_{eu} = 1.2\,mm$. Obviously, no real consequences for primary-production volume could be expected from such an increment in z_{eu}. Increase in primary production at depths ranging from $z = -0$ to z_{eu} because of a 10% rise in above-water radiation at 400–410 nm produced changes in primary production varying from less than 1% at $z = -0.5$ m to zero at $z \geq 0.75$ m. Again, the influence produced by increased irradiance levels is insignificant. However, in less turbid and less rich chl or doc waters (also pertaining to case-II waters), the consequences of shortwave incident PAR increase could be more substantial although not considerably and the increase would be restricted to uppermost layers of the water column, still leaving gross production in the photic zone practically intact.

4.5.6 Estimates of the effect of UVB on water ecosystems

Marine-water ecosystem

It is known that over 30% of the animal protein in food consumed by the inhabitants of our planet comes from the sea (Hader *et al.*, 1995); therefore, research into the effect of changes in UVB on hydrobiota is very important. In addition, the World Ocean plays a key role in the formation of the global climate: phytoplankton are the main consumers of atmospheric carbon dioxide (CO_2), and this determines their effect on trends in CO_2 content in the atmosphere. Any reduction in oceanic assimilation of CO_2 from the atmosphere because of reduction in its consumption by phytoplankton (as a result of decreasing phytoplankton numbers and biomass) immediately results in enhancement of the atmospheric greenhouse effect.

The exceptional role of phytoplankton in the dynamics of the biosphere stems from their fundamental position in the hierarchy of food interactions in water ecosystems: phytoplankton are primary producers (according to available estimates (*UNEP* ..., 1994), they are responsible for around 50% of annual production of organic carbon on the planet); they form the base of the food chain for all other levels of trophic interaction. According to current estimates, a 5% reduction in the amount of phytoplankton in the Pacific Ocean could reduce the annual fish catch by 7 million tonnes (Karantz, 1994). In the light of this and certain other data, obtained during experiments on short-term incubation periods, extremely alarming conclusions have been drawn about a 30% reduction in fish stocks in the Pacific Ocean. Nevertheless, these estimates do not take into account the possibility that fish populations may move their habitat zones to deeper layers of the ocean, may activate intracellular processes to restore DNA or may produce photoprotective substances such as mycosporin aminoacids. When these adaptive mechanisms are taken into account such pessimistic estimates about reduction in fish stocks apparently have less justification (Vincent and Roy, 1993). The spatial distribution of phytoplankton in the Pacific Ocean is extremely uneven. The largest concentrations occurs in high latitudes with infrequent maxima in low latitudes, confined to upwelling zones on continental shelves (Hader, 1993)

In addition to phytoplankton, nano and picoplankton, bacterio and zooplankton and also fish, crabs, shrimps and amphibians (at early stages in their development) are subject to the harmful effect of UVB. Higher-water vegetation is no exception to the list. It should be emphasized that in recent years a significant revision has been made to estimates of the relative significance of nano and picoplankton to primary production in the Pacific Ocean. There is now reason to suppose that at least 40% of total plankton biomass comprises nano and picoplankton. According to Karantz (1994), it is precisely ultra-small organisms which are most vulnerable to the effect of UVB. The effect of UV radiation on phytoplankton is of selective nature, even at the level of its floral composition; in particular, there are direct indications that flagellates are more sensitive to UV radiation that diatoms (Helbling *et al.*, 1994). Possible depletion of a specific trophic level as a result of intensification of UVB penetrating the sea inevitably leads to change in size and constitution of populations at higher trophic levels.

The harmful effect of UVB on the primary-production process occurs at the level of DNA. Normal processes of intracellular metabolism, photosynthesis and nitrogen fixation are disrupted. Bleaching of pigments also occurs and the ability of phytoplanktonic cells to move and orient themselves is inhibited (Dohler *et al.*, 1991). It is conjectured that UV radiation damages the reaction centres of photosystem II in phytoplanktonic cells, leading to reduction in the effectiveness of the electron-transport mechanism and to disruption of the normal functioning of that part of the cell's photosynthesizing apparatus which is responsible for breaking down water molecules (Hader, 1994). In addition, it has been shown that under exposure to UV a reduction in the total content of lipids in an algal cell occurs, leading to disruption of functioning of the membranes. Many studies show, however, that inhibition of photosynthesis may be caused not only by exposure to UVB but also by hyperbolic dependence on the level of full illumination (Leser and Cullen, 1994); a relatively short exposure to larger doses of UVB is more harmful than a more prolonged exposure to less intense radiation fluxes. These data confirm that the resulting effect of exposure is a combination of a harmful effect and the activation of the plant cell's protective functions. Thus, the resulting action spectra have a composite nature, which is complicated by the interactive effects of the increase in the plant cell's ability to protect itself and recover from the effect of UVA and visible light. In this connection, it is important to note that the threshold for the onset of photo-inhibition varies greatly for many algae and there exist a number of taxa which are not harmed in any way by the effect of high levels of solar illumination (Lewis *et al.*, 1984). The mechanism for photo-inhibition apparently involves photo-oxidation of the reaction centres and intermediate molecules, such as plastoquinones, which ensure the normal functioning of electron-transfer paths. Here, the presence of both UVA and UVB in the incident flux increases photo-inhibition and retards the process of self-recovery of the normal work of the cell. However, it still remains unclear whether the effect of UVB merely reduces intensification of the inhibitory effect of UVA or whether some other mechanism with a harmful effect is involved in this.

Spatially, the productive activity of phytoplankton is bounded by the euphotic zone; that is, by the upper layer of the water column, whose lower boundary is determined by the position of the level where photosynthetic production of energy in a cell is balanced by its losses in respiration. Usually, the lower boundary of the euphotic zone corresponds to the level at which incoming photosynthetically active radiation (PAR) amounts to 1% of its value on the water surface. Vertical profiles of plankton concentration in the euphotic zone are determined both by the conditions of solar illumination and by physical processes such as wind enhancement, variations in temperature and density, etc.

Many forms of phytoplankton can migrate independently in the vertical direction. Plankton organisms devoid of independent motor functions (such as cyanobacteria, diatomic algae and bacteria) take advantage of their ability to regulate buoyancy and, with the aid of gas vacuoles or volumes filled with oily liquids, also move up and down within the boundaries of the euphotic zone and outside it.

Phytoplankton use light at various wavelengths in the visible and UVA bands of the spectrum to orient themselves in space. However, their sensor systems are not capable of reacting functionally to variations in the intensity of radiation in the UVB band and, consequently, do not provide phytoplanktonic cells with corresponding information about the need for them to move to deeper (protected from radiation) levels of the water column (Donkar *et al.*, 1993). In this connection, note that higher-water vegetation, unlike plankton in being anchored, is unable to change its position in terms of its depth in the water and thus has no possibility of countering the effect of UVB by getting the water layer to act as a filter. Note that in this case it is more vulnerable than plankton. This property applies to macrophytes in tidal regions which are fully or partially devoid of a protective water filter for many hours. The most vulnerable of all these are benthic macrophytes (Larkum and Wood, 1993).

Returning to plankton organisms, evidence is now appearing in the literature that many forms of phytoplankton, subject to high doses of exposure, are able either to manufacture protective (screening) pigments (micosporins) or to increase the effectiveness of the use of carotenoids to divert excess PAR from photosynthesizing pigments (Karantz *et al.*, 1991). In 1994, the first reliable evidence of the effect of UVB on Antarctic phytoplankton was published. Based on an analysis of the space–time variability of UVB because of the Antarctic ozone hole, it was shown (Weiler and Penhale, 1994) that an increase in UVB led to reduction in phytoplankton productivity of 6–23% in the marginal zone of the ice sheet.

The susceptibility of marine invertebrates to UVB varies greatly from species to species. Some crustaceans, having risen to the surface, are liable to a 50% mortality rate at current levels of UVB. However, larval shrimps were able to withstand the typical 16% increase in the spring level of UVB in 1982–1988 (Dey, 1989). Field observations show that UVB kills most species of crustaceans and immature crabs in the north-eastern area of the Pacific Ocean. At latitudes above 40°N similar phenomena can also be expected in connection with anthropogenically caused seasonal variation of C_{oz} (Krzyscin, 1995). Benthic marine organisms such as sea sponges, confined to coral ecosystems and consequently to waters with high translucence, exhibit a low threshold of vulnerability to UVB. However, those which are able to produce melanins as protective pigments are more resistant to UVB exposure. Coral systems vary significantly in their susceptibility to UVB. These variations are particularly clearly manifested as a function of the depth at which the corals live. UVB apparently affects corals via a photostimulated oxidizing effect on these invertebrate mechanisms (Shick *et al.*, 1991; Gleason and Wellington, 1993).

High levels of UVB significantly retard growth and reduce survival rate of the fry of a number of seafish. According to available estimates, a 16% increase in UVB over the level for 1982–1983 at a depth of 0.5 m leads to 50, 80 and even 100% mortality rate for anchovy fry at ages 2, 4 and 8 days, respectively (Hunter *et al.*, 1982). Similar results were also obtained for rainbow trout and two forms of salmon (Little and Fabacher, 1994).

Reduction in the amphibian population noted in recent years may have been caused by growing levels of UVB. Eggs of certain species of frog laid in open (and consequently clear) waters are very strongly affected by the rise in UVB and react to it with increased embryo-mortality rate. It is interesting to note, however, that certain amphibians have already been able to adapt to the changed conditions of solar illumination; while the fry are developing the eggs become coated with a coloured slime (Hader *et al.*, 1995).

Freshwater ecosystems

The increase in UVB (amounting to 10–20% over 10 years) observed in mid-latitudes of both hemispheres (Kerr and McElroy, 1993), where there is a large number of freshwater lakes, has raised the urgent matter of assessment of the damage to these ecosystems by processes destroying the ozone layer. According to Feister and Grewe, (1995), in the summer months of 1982–1983 (when there was a depletion in TOZ of the order of 10% of the average multiyear level) the increase in erythemal exposure around Potsdam (Germany) amounted to 20–30% in comparison with preceding years (at the same time of year), while the level of UV radiation harmful to higher vegetation exceeded the corresponding average multiyear level by 20–40%. For all the scantiness of the results of research into marine ecosystems discussed earlier, the reaction of freshwater ecosystems to growing UV radiation at the surface has been studied in an even more fragmentary manner (Williamson, 1995).

Based on the general consideration that the depth of penetration of UVB (e.g. 10% of the level of illumination at the water surface) in most freshwater lakes (with the rare exception of particularly clear lakes) usually amounts to several metres and most frequently to significantly less, it can be conjectured that the biota of freshwater lakes are less vulnerable to UVB. However, such conjecture should be treated with extreme caution. It is known that the global average depth of freshwater lakes is less than 10 m (Wetzel, 1990), while the warm mixed layer, in which conditions for life and development of living organisms are most favourable, is less than a few metres thick. This is absolutely true for rivers of different depths too. Thus, it is precisely the first millimetres or, in the extreme case, the first few metres of the upper layer of the water, that changes in spatial variation of the UVB field and the majority of biota are combined. This means that potentially UVB may have a simultaneous and direct effect not only on primary producers but also on the entire chain of trophic interactions in freshwater biota (Bothwell *et al.*, 1994). It is evident, and we were reminded of this earlier, that hydrobiota most vulnerable to variability in UVB are those which are unable to detect variations in UVB or actively respond to them; for example, by developing protective pigments or moving to deeper layers of the lake. In the literature, in particular, it is noted that the freshwater cladoceran *Daphnia* and the larvae of the flat sea-urchin have receptors sensitive to UVB and can react adequately to an increase in UVB. However, invertebrates such as the euphausiid *Thysanoessa* and the freshwater midge *Cricotopus* do not have this ability (Bothwell *et al.*, 1994). Note that in mountain areas, zooplankton in clear translucent lakes are, as a rule brightly coloured with protective (screening) pigments. At the same time, photoreceptors in the majority of aquatic invertebrates have a spectral sensitivity limited to the UVA band and, like land biota, are not able to react adequately in a behavioral or organismal sense to the increase in UVB (Siebeck, 1994).

Unlike in seas and oceans, in freshwater lakes the increase in UVB is also a function of the height of the lake above sea level, which is an exceptionally important factor as far as the problem in hand is concerned. In this context, it is extremely significant that the height of the lake above sea level determines not only the level of incoming UVB at its surface, but also the levels of UVA and radiation in the visible band, whose effect, as noted earlier, may be either of a neutralizing nature as far as UVB is concerned or may lead, in certain special cases, to increase in the latter (Siebeck and Bohm, 1994).

Increase in UVB may also have serious consequences for freshwater ecosystems from the point of view of reduction of the most suitable habitat and associated population problems. In particular, in oligotrophic lakes at the time of the summer solstice small zooplankton experience serious difficulties because of two competing processes in the lake. Migration of predatory invertebrates and large zooplankton to the bottom of the lake during daylight hours (in a bid to hide from the ichthyofauna of the lake (Lampert, 1989)) frees living space for small zooplankton in the upper well-heated layers. Reduction in the turbidity of these upperwater layers because of this migration means that they become more transparent to solar radiation and to UVB in particular. As a result, it turns out that small zooplankton are concentrated in a zone of increased radiation danger which leads to a rise in their deathrate and a reduction in the overall population size (Ringelberg *et al.*, 1995). It is clearly appropriate to stress here that the effect of UV radiation on certain forms of zooplankton can lead to the latter being forced to sink to deeper, less translucent and less suitable layers of the lake. This is possible in conditions of both high and 'normal' levels of incoming UV radiation. The main factor here is the presence of a stable stratification in the lake which, to a certain degree, depends on weather and climatic factors. Thus, for individual oligotrophic stratified lakes, especially in their pelagic part, the stress experienced by the ecosystem may be primarily the result of climatic change and variations because of depletion of the stratospheric ozone layer and intensification of incoming UV radiation (although the close interdependence of the two processes should not be forgotten) (Williamson *et al.*, 1994a, b).

It is safe, however, to assume that this effect is not characteristic of eutrophic lakes (Zepp *et al.*, 1992), where the vertical gradient of UVB absorption is primarily determined not by the turbidity of the water but its absorbance in a given wavelength interval. In addition to the direct radiation effect, enhancement in UVB may alter the relationship between the level of food resources available to a species and the number of predators *per capita* significantly; this phenomenon is known in limnology as an indirect effect of cascaded euphotic interaction (Carpenter and Kitchell, 1993). In particular, if for some reason UVB had a stronger effect on the immediately superior trophic levels than on a given population, the resulting reaction of the latter might be significantly more serious (negative) than the direct radiation-related oppression. For example, susceptibility of phytoplankton to harm by UVB, is noted above, results in a significantly more negative effect on zooplankton through the latter than the corresponding direct effect. Note that there exist hypotheses based on the results of studying land vegetation (Smith, 1991), according to which pigments such as carotenoids and mycosporin, when exposed to UVB, may change the nutritional qualities of phytoplankton in its trophic interactions with zooplankton. Zooplankton digesting phytoplankton with a changed relationship between the pigments and possibly also in their properties, may also be subject to certain changes in their pigmentation. However, the question of the extent to which such changes can affect the vulnerability of plankton to predators of a higher trophic level remains open (Byron, 1992). Potentially, predators such as amphibians and fish, which regulate the spatial distribution and composition of species of zooplankton in the lake, may themselves experience the oppressive effect of UVB, especially at the stage of development of embryos in eggs (fish roe) and also (later) at the fry stage. Note that convincing evidence of the oppressive effect of UVB on the dynamics of the amphibian population has already been obtained (Herndl *et al.*, 1993).

When zooplankton populations are most vulnerable to the effect of UVB (and the development of phytoplankton is limited by the rate of their consumption by the latter) a growth in the biomass of phytoplankton should be expected (Bothwell *et al.*, 1994). This indicates the need for longer, more representative experiments capable of covering timespans and ranges of trophic levels which would permit adequate research into the effect of UVB, not only at the level of individual organisms but also at the level of the ecosystem. At the same time, bacterial communities should also be taken into account. It is known that UVB inhibits the heterophase fixation of carbon by bacterioplankton (Lindell and Rai, 1994). In addition, it has been shown that UV radiation, by breaking down the inhibitory fraction of the humic substances of the low-molecular DOM fraction and transforming them into forms more easily assimilated by heterotrophic bacterioplankton, thereby stimulates the productivity of the latter. In turn, the photostimulated breakdown of humic substances leads to a fall in the concentration of dissolved oxygen, because of formation of secondary toxic products as hydroxy radicals, singlet oxygen and hydrogen peroxide (Visscher *et al.*, 1994). The effect of these secondary products at various levels of the trophic pyramid has been studied in an even more fragmentary manner.

The problem of fluorocarbon compounds is very closely related to the problems discussed. Some HFC and HCFC compounds (primarily HFC 134, HCFC 123 and

HFC 124) photo-oxidize when exposed to UVB, forming trifluoroacetic acid (TFA), CF_3COOH. TFA is not subject to photolysis and does not undergo further physicochemical transformations. Evidently, TFA is not involved in the metabolic processes of higher vegetation. However, it is known that micro-organisms such as sulphur and soil bacteria break it down (Chumley, 1994). Consequently, TFA may be a previously disregarded form of long-lived pollutant. According to available estimates, in the air over the southern parts of Germany concentrations of $C_{\mathrm{TFA}} \cong 0.01$–$0.05\,\mathrm{ng\,m^{-2}}$. By 2010, concentrations of C_{TFA} could increase 40-fold. For inland lakes with relatively small water volumes and highly eutrophic state, a 5–10-fold increase in bio-accumulation may occur because of water-blossoming phenomena (with the mass breakdown of phytoplankton) (Groeneveld *et al.*, 1992). TFA is mildly toxic for the majority of freshwater forms of phytoplankton. At the same time, there exist forms of freshwater phytoplankton, such as the green algae *Selenastrum capricornutum*, for which EC_{50} is less than $\sim 4.8\,\mathrm{mg\,l^{-1}}$ (Baker *et al.*, 1991). Currently, we know of no data about bio-accumulation of TFA in phytoplankton, or on other levels of inland-lake trophic interactions. However, this possibility should clearly be taken into account, especially in the case of shallow inland lakes.

Another mechanism behind the pollution of freshwater lakes is over-acidity resulting from dry and wet precipitation of pollutants from the atmosphere. The waters of over-acid lakes usually have lower concentrations of humic substances and therefore greater translucence (Schindler, 1990). This means that shortwave solar radiation on such lakes may penetrate to greater depths. As a result, reconstruction of the composition of a population by species at different trophic levels resulting from the effect of over-acidity may potentially also be 'modulated' by the effect of UVB.

In summary, it should be emphasized that the existence of a large number of uncertainties about the nature of direct, inverse and indirect interactions in water ecosystems makes forecasting the results of the effects of change in UVB extremely difficult, both at a particular individual trophic level (e.g. the reduction/increase in primary productivity of phytoplankton) and at the level of the system as a whole. At the same time, as previously mentioned, the importance of the effect is itself very significant for the lifecycle of each type of hydrobiota (e.g. the beginning of the seasonal reconstruction of the structure of communities of different trophic levels or the established state of a community (ecosystem)).

Finally, we consider the nature of the indirect effect of UVB on hydrobiota, including climatic changes because of variations in TOZ and UVB. For example, there exists information on the real (as opposed to results from various model numerical experiments) influence of climatic change on the hydrologic regime of lakes of the Canadian Shield, relating to increase in the depth of the thermocline and transmission of the waters. There is no doubt that the effect of variations in UVB is potentially very diverse and profound. Existence of a large number of direct and indirect interactions broadens the physical boundaries of the influence of ecosystems of freshwater lakes considerably and makes the latter an inherent integral part of global processes.

4.6 EFFECTS OF UVB RADIATION ON TERRESTRIAL ECOSYSTEMS AND BIOGEOCHEMICAL CYCLES[7]

After detailed analysis of effects of UVR on aquatic ecosystems we shall now briefly dwell on similar problems relevant to terrestrial ecosystems and biogeochemical cycles.

4.6.1 Terrestrial ecosystems

Few studies have so far, been carried out on the effect of UVB radiation (wavelengths 280–320 nm) on various crops (mainly grown in special greenhouses and in just a few cases in outdoor field conditions). Mainly individual plants were studied and not populations, communities or ecosystems. However, there is no doubt that long-term investigations of the response of integrated ecosystems to UVB increase are needed (*Effects* ..., 1992; Kondratyev and Varotsos, 1996a, b). Interaction and feedback between individual species of ecosystems cannot be assessed reliably with the data available; but for individual species they can. Therefore, study of individual species should be considered only as an initial stage of subsequent more complete investigations. One of the unexpected and surprising results of the considered consequences consists in detecting a very strong variability in sensitivity of individual numbers various species (different samples of plants within a given species) to changes in UVB radiation. Characteristics such as vegetation growth and biomass distribution were used as quantitative criteria of sensitivity. As a rule, the sensitivity of plants to UVB changes is weaker in field conditions than in greenhouses.

Decrease in biomass because of ceasing processes of photosynthesis, when UVB radiation intensifies, is one of the concrete manifestations of the plant's sensitivity. In some cases, the biomass of plants shrinks as a result of the decrease in leaf-cover extent, but in other cases some morphological changes appear, manifesting themselves as an increasing number of branches, changing thickness of leaves and growing foliage, or shortened trunks. Response to UVB increase may weaken owing to various interacting chromophores (e.g. nucleic acids, phytohormones and pigments). In some cases, functioning of specific UVB receptors has been discovered, which absorb light mainly at wavelengths ~290 nm.

A few studies of tree species have shown, for example, that increasing UVB radiation suppresses the development of pines (*Pinus taeda*), with an accumulation of negative consequences for the subsequent three years.

As there is a distinct likelihood in the future of not only UVB increase but also increase in CO_2 concentration and climatic warming, a combined account of these factors becomes urgent. From the data available, increase in CO_2 concentration mainly affects C3 species, whereas C4 species are much less sensitive from the point of view of photosynthetic reaction. Experiments with this combined account revealed in the case of C3 species, such as rice and wheat, that when CO_2 concentration increases, the biomass does not grow unless the plants are additionally

[7] This section was written in collaboration with D. V. Pozdnyakov.

irradiated with the UVB lamps, whereas soyabeans do not react to CO_2 when the UVB increases. With solar irradiation, temperature increase in the case of maize (C4) or CO_2 increase in the case of sunflower (C3) could compensate for negative forcing because of increasing UVB radiation. If, however, UVB increase is combined with that of CO_2 over saline marshes, these two factors do not practically interact. Thus, the response of plants to the combined forcing of CO_2 and UVB depends substantially on individual plant species.

The effect of UVB increase on plant chemistry shows itself mainly through the increasing concentration of UVB-absorbing pigments (first of all, flavonoids), whose function is to protect the plants from the harmful effect of UVB. So, for example, an accumulation of such pigments in the epidermis of wheat protects the leaves from photosynthetic suppression. An increase in UVB may also lead to changes in the concentrations of minor components such as alkaloids and coumarines, which can affect the quality of agricultural produce, the rate of organic disintegration and the intensity of emissions of trace gases to the atmosphere.

The effect of UVB on suppressed plants (plants lacking minerals or moisture) from the ecosystem is a separate aspect of the problem. It has been established, for example, that plants lacking moisture react weaker to UVB increase than those that are well watered. Plants suffering from deficient phosphorus are also less sensitive to a UVB increase (the effect of other deficient nutrients has not been studied). Understanding interactions between the effects of soil contamination and UVB increase is very important. The following unsolved problems connected with the effect of increased UVB radiation should be given priority:

- Assessment of the impact on primary production of various agricultural and non-agricultural ecosystems.
- Assessment of the impact on morphological changes in the architecture of vegetation cover and biomass distribution as well as the respective consequences.
- Assessment of the direct and indirect impacts on users of the ecosystem's produce and of the possible effect of feedback to the ecosystem.
- Analysis of the genetic basis of resistance to the effects of UVB change.

4.6.2 Biogeochemical cycles

Cyclic variations in solar radiation, manifested as daily and annual courses, as well as solar-activity-induced long-range variability (an 11-year cyclicity is its most clear manifestation) substantially affect biogeochemical cycles determined by a complicated interaction of biological, chemical and physical processes of the substance and energy transformation. A well-studied example of cyclicity of the biogeochemical cycle is the annual course of atmospheric CO_2 concentration determined by global biomass cyclicity under the influence of annual change in incoming solar radiation and, respectively, vegetation processes. In this connection, an important contribution is made by anthropogenic CO_2 emissions which, however, despite their considerable volume (~6 Gt C/yr), are by an order of magnitude less than CO_2 exchange between the atmosphere and terrestrial biosphere (~60 Gt C/yr), and the close-in-value CO_2 exchange between the atmosphere and the ocean. The latter means that even small variations in biosphere fluxes can cause substantial changes in atmospheric CO_2 concentrations.

Analysis of the mean global CO_2 trend reveals the interesting fact that, despite slowing down of anthropogenic CO_2 emissions, the ratio of the observed rate of CO_2 increase to that of fossil-carbon-fuel burning has substantially increased. This may be interpreted as a result of increasing UV radiation at the Earth's surface over the last 15 years, causing suppression of photosynthetical activity and hence decrease in primary production and weakening of assimilation of atmospheric CO_2 (the effect of other factors, e.g. changes in land use in the tropics, is not excluded).

UVB increase cannot only suppress photosynthetic activity but also intensify the disintegration of non-living organic substances. Over long time periods, variation in the productivity of ecosystems can affect the dynamics of the nutrient cycles. In particular, the rate of the cycle of compounds resisting disintegration processes (lignin, humus) can increase. Observational data show, for example, that increase in UVB radiation favours disintegration of aquatic humus substances.

The close interaction of biogeochemical cycles determines inevitable accompanying changes in cycles of nitrogen, oxygen, sulphur, halogens and metals which play an important role in cellular metabolism. Three major manifestations of the effect of UVB increase on the cycles of these elements are possible:

- Direct effect on biological processes can happen because of intensified destruction of DNA and other biological molecules (first of all, chlorophyll) and, as a result, changes take place in photosynthetic activity and respiration of terrestrial vegetation as well as in processes either directly or indirectly connected with marine phytoplankton, such as microbe respiration, nitrification, sulphur cycle and metallic redox reactions with the participation of metals. These effects may be connected with such phenomena as global climate warming and CO_2 increase, and PAR variation.
- UVB increase strongly affects geochemical processes in the sea and on land. The processes taking place both in the sea and on land include: production of 'greenhouse' and chemically active gases (CO_2, CO, NO, COS (carbonnylsulphide), NMHC (non-methane hydrocarbons)); mineralization of nitrogen and phosphorus; transformation of organic matter to biologically simple organic compounds with smaller molecular weight; deeper penetration of UV radiation to seawater thickness; changes in redox state of the upper oceanic layer, and others.
- UVB-induced chemical processes in the atmosphere manifest themselves through changes in: oxidizing property of the troposphere; formation (more intensive) of greenhouse gases, such as methane and ozone; concentrations of aerosol and cloud condensation nuclei; concentrations of ozone-depleting gases (including various halogenized compounds). Such phenomena should strongly interact with chemically active gases emitted by the burnt biomass (organic substances) and other anthropogenic processes.

All these changes may be followed by appearance of feedback which either intensifies or reduces formation of greenhouse gases and aerosol in the atmosphere, which will affect global climate dynamics.

To obtain reliable estimates of regional and global consequences of the effect of UVB increase on biogeochemical cycles and respective feedback, it is necessary to develop numerical simulation models and to obtain adequate data of observations of the dynamics of atmospheric chemistry. Deeper insight into the correlation of climate change and biogeochemical cycles of atmospheric gases (TG), such as CO_2, CO, N_2O, DMS and OCS, highlights the increasing urgency of developments in the sphere of numerical climate modeling, taking account of remotely-sensed data on cycles of respective TG obtained from satellites. As an example, Erickson and Eaton (1993) discussed the possibility of using CZCS data to estimate global distribution of the phytoplankton-induced input of carbonylsulphide (OCS) from surface waters to the atmosphere.

Interactivity of chemical processes in the atmosphere, affected by increased UV radiation, highlights the need to accumulate a global database characterizing the dynamics of atmospheric chemistry caused by anthropogenic impact. At present, it has been stated that anthropogenic emissions of nitrogen oxides (NO_x) and sulphur dioxide because of industrial emission, burning of fossil fuel and biomass substantially affect chemical processes in the global troposphere.

In this connection, a project was initiated (Penner *et al.*, 1993) to draw up an inventory of global emissions (GEIA) within the key IGBP project on global atmospheric chemistry (IGAC), bearing in mind data archiving over the $1 \times 1°$ latitude–longitude grid. Such an archive, together with models of photochemical processes in the atmosphere, can be used to forecast the dynamics of most important atmospheric trace gases (TG).

Comparison of the results thus obtained with observational data will make it possible to assess both the adequacy of present models of photochemical processes in the atmosphere and also numerical modelling of the budget, transport and other processes. All this can provide the basis for reliable estimates of possible consequences of anthropogenic TG emissions to the atmosphere.

As mentioned above, two aspects of the discussed problems are of particular interest:

(1) The oxidizing property of the atmosphere and its possible changes under the influence of emitted methane, nitrogen oxides, carbon monoxide and non-methane hydrocarbons (NMHC). The oxidizing property of the atmosphere is known to be determined mainly by the concentration of hydroxyl (Table 4.19), but ozone and hydrogen peroxide also play an important role. If hydroxyl concentration decreases on global scales, the concentration of TG removed by hydroxyl may increase even in the absence of change in intensity of their sources. As for hydroxyl, it forms from O_3 photolysis (under the influence of solar radiation at wavelengths $< 310\,nm$) giving $O(^1D)$. The subsequent reaction of $O(^1D)$ with a water-vapour molecule gives two hydroxyl radicals. Since OH concentration in the background atmosphere is mainly determined by concen-

Table 4.19. Atmospheric trace gases either removed or oxidized by chemical reactions in the presence of hydroxyl.

Component	Value of the component considered
CH_4	Greenhouse gas; affects the content of tropospheric and stratospheric ozone and OH.
CO	Urban atmospheric pollutant; affects the tropospheric O_3 cycle and OH.
CH_3CCl_3, CH_3Br, CH_3Cl	Greenhouse gases; their disintegration in the stratosphere gives free chlorine which destroys ozone.
HCFCs	Hydrochlorofluorocarbons are greenhouse gases; CFC substitutes giving free chlorine and destroying ozone in the atmosphere.
NMHCs	Urban atmospheric pollutants; determine increase of tropospheric ozone in the regions with increased concentrations of NO_x, responsible for sink reactions of tropospheric OH.
DMS, H_2S, SO_2	Lead to the formation of sulphate aerosols which contribute to the formation of acid rain and affect the climate (because of increased backscattering of solar radiation); formation of sulphate condensation nuclei affects optical properties of clouds.

trations of CO, CH_4, NO_x and H_2O, to assess the effect of these TG on O_3 and OH, data are needed on the power not only of their natural but also of their anthropogenic emissions.

(2) The effect on climate of aerosol formed from gas-to-particle reactions with the participation of anthropogenic emissions of sulphate compounds. Various types of thus-formed anthropogenic aerosols are possible depending on their chemical composition (dissolved in water non-organic components, elemental or soot carbon, mineral dust). An important fact is that processes involved in formation of sulphate aerosols depend on the oxidizing property of the atmosphere, and this determines in particular the importance of data about anthropogenic emissions of NO_x and SO_2, as well as CH_4 and CO.

The global database that characterizes the spatial–temporal variability of NO_x and SO_2 emissions (over the $1 \times 1°$ grid), the estimates of global propagation of these TG, as well as the formation of sulphate aerosols and the effect of nitrogen oxides on tropospheric ozone were discussed earlier (Penner *et al.*, 1993).

Calculations of the global concentration of sulphate aerosols were made with due regard to five natural and anthropogenic sources of sulphate compounds: formation (mainly) of dimethylsulphide (DMS) by oceanic phytoplankton; volcanic emissions of sulphur dioxide (SO_2); terrestrial emissions of DMS and sulphurous hydrogen (H_2S) by soils, forests and crops; SO_2 emissions due to fossil fuel and biomass burning as well as industrial emissions of SO_2. The data in Table 4.20 are characteristic of global emissions of sulphur compounds.

The forecast emissions in the model of global transport and chemical transformation have led, on the whole, to a realistic simulation of the observed concentration of

Table 4.20. The power of global sources of chemically active sulphur compounds and their possible changes (Mt S/yr).

Source	Total emissions	Probable range
Fossil fuel and industry	66	60–100
Biomass burning	2.2	1–4
Soils, forests, crops	0.9	1–10
Volcanoes: degassing	3.4	
Volcanoes: eruptions	6.4	
Oceans	23.7	12–40

Table 4.21. The power of global sources of chemically active nitrogen compounds and their possible ranges in changes (MtN/yr).

Sources	Total emissions	Possible range
Anthropogenic		
Fossil fuel burning	22.4	15–5
Biomass burning	10	2–20
Aircraft wastes	0.23	0.1–0.7
Forest fires	0.2	0.1–0.8
Natural		
Emissions due to microbiological processes in soils	5	1–20
Lightning discharges	10	2–100
Formation in the stratosphere	1.0	0.5–1.5
Total	48.8	20–168

sulphates in the atmosphere and their wet and dry deposition onto the underlying surface. In the numerical modelling of ozone concentration in the troposphere determined by oxidation of CO, CH_4 and NMHC in the presence of NO_x (and insolation), the sources of chemically active nitrogen compounds were prescribed as shown in Table 4.21.

Tropospheric ozone number density turned out to be overestimated, though the near-surface concentration is close to that observed. This highlights the necessity for further specified calculations (e.g. NMHCs have not been specified and the formation of PAN has not been considered). It is important to increase the reliability of the initial database on emissions, both natural and anthropogenic. Means of spaceborne remote sounding, such as those used on the UARS satellite and planned for EOS, open up new prospects for obtaining a representative global database characterizing the dynamics of atmospheric chemistry.

The main unsolved problem is undoubtedly poorly studied effects of UVB radiation on ecosystems. So, it is very important to investigate the effect of UVB radiation on:

1. Terrestrial ecosystems, including:
   1.1 Primary productivity of individual agricultural and non-agricultural ecosystems.
   1.2 Morphology (architecture) and its spatial structure on distribution of the vegetation biomass.
   1.3 The structure of populations and its stability.
   1.4 The effect on chemical composition and related studies at the level of ecosystems.
   1.5 The effect of long-term evolution of vegetation, taking account of genetic dynamics.
2. Aquatic ecosystems, including:
   2.1 Variability in UVB radiation and of the relationship UVB, UVA, PAR at different depths and the respective effects on ecosystems.
   2.2 Spatial–temporal sensitivity of ecosystems (primary productivity, balances of substance and energy adaptivity) and UVB changes.
   2.3 Effect of physical, chemical and biological factors on manifestation of the impact of UVB increase on ecosystems.
   2.4 Revealing the role of potential feedbacks (determined, e.g., by changes in clouds and temperature) which can show itself in the process of the effect of increased UVB radiation on the ecosystems.
3. Biogeochemical cycles and global changes (Table 4.22), including:
   3.1 Cycles of carbon and nitrogen in terrestrial ecosystems.
   3.2 Cycles of carbon and nitrogen in aquatic ecosystems.
   3.3 The cycle of sulphur in aquatic ecosystems.
   3.4 Cycles of biologically important metals and phosphorus in the aquatic domain.
   3.5 Formation and degradation of organogenes.
   3.6 Disturbances introduced to atmospheric biogeochemical cycles, which is of interest from the viewpoint of the effect on climate (CO_2, CH_4, N_2O, SO_2, DMS, O_3), the quality of tropospheric air (NO_x, SO_2, CH_4, CO, O_3, NMHC) and stratospheric chemistry (COS, CH_4, halogenometals).
   3.7 Development and application of remote sensing techniques to assess the effect of increased UVB radiation on ecosystems on global scales, taking account of other related processes.
   3.8 Development and application of techniques for simulation numerical modelling of environmental dynamics (mainly global biogeochemical cycles) in conditions of anthropogenic effects with complex account of interactive factors such as decrease in total content of stratospheric and increase in tropospheric ozone, increase in UVB radiation, increase in concentration of CO_2 and other greenhouse gases, intensification of the greenhouse effect of the atmosphere, climate warming and others.

Table 4.22. Direct and indirect forcings of UVB radiation on biogeochemical cycles.

Carbon	Terrestrial ecosystems	(1) Photosynthesis, shifting of plant species. (2) Emission of non-methane hydrocarbons. (3) Composition of the substratum, disintegration of organic substance.
Carbon	Aquatic ecosystems	(1) Degradation and cycle of organic substance. (2) Spectral properties of aquatic domain. (3) Synthesis of organic carbon and CO_2 assimilation by phytoplankton and macrophytes.
Carbon	Atmosphere	(1) Exchange of carbon-containing gases. (2) The rate of chemical transformation, especially in the formation of radicals.
Nitrogen	Terrestrial ecosystems	(1) Fixation of nitrogen. (2) Nitrification, emissions of N_2O and NO. (3) Composition and disintegration of the substratum. (4) Emissions of inorganic N as a result of degradation of organic matter and subsequent effect on soil fertility, CH_4/CO sinks.
Nitrogen	Aquatic ecosystems	(1) Nitrification, emissions of N_2O and NO. (2) Formation of inorganic N as a result of degradation of organic matter and subsequent effect on accessibility of nutrients. (3) Formation of radicals from NO_3^- and NO_2^-.
Nitrogen	Atmosphere	(1) Exchange of nitrogen-containing gases (N_2O, NO, NH_3) between the atmosphere and other natural domains. (2) The rate of chemical transformation, especially in the formation of radicals.
Sulphur	Terrestrial ecosystems	(1) Assimilation and emission of COS by vegetation.
Sulphur	Aquatic ecosystems	(1) Assimilation and emission of the precursors DMS, COS and S by aquatic organisms. (2) Formation and consumption of DMS. (3) Photo-oxidation of DMS.
Sulphur	Atmosphere	(1) Exchange of sulphur-containing gases (COS, DMS, SO_2) between the atmosphere and other natural domains. (2) The rate of chemical transformation, especially in the formation of radicals.
Metal-phosphorus compounds	Terrestrial ecosystems	(1) Emissions of inorganic P as a result of degradation of organic matter and subsequent effect on soil fertility.

Metal-phosphorus compounds	Aquatic ecosystems	(1) Existence of small admixtures of metals (e.g. Fe, Cu, Mn). (2) Formation of OH through reaction in the presence of Fe.
Oxygen	Terrestrial ecosystems	(1) Photosynthesis. (2) Disintegration of the substratum.
Oxygen	Aquatic ecosystems	(1) Photosynthesis and oxidation of DOM. (2) Formation and disintegration of H_2O_2. (3) Formation of HO_2 and OH.
Oxygen	Atmosphere	(1) Transformation $O_2 \rightarrow O_3$.
Halogens	Terrestrial ecosystems	(1) Formation of organohalogens.
Halogens	Aquatic ecosystems	(1) Formation of nutrients and consumption of organohalogens.
Halogens	Atmosphere	(1) Disintegration of methylhalides, chloroflourocarbons and their substitutes with the subsequent destruction of ozone.

4.7 IMPLICATIONS OF UVR CHANGE FOR HUMAN AND ANIMAL HEALTH

Observational data over the past two decades indicate that stratospheric ozone has already decreased and, according to theoretical assessments, further decrease is expected in the coming years (*Scientific* ..., 1999).

Increased doses of solar ultraviolet radiation (SUVR) because of ozone-layer depletion may result in serious consequences for living organisms. A 10% reduction in total ozone column (TOZ) could lead to as much as a 15–20% increase in UV exposure, depending on the biological process being considered (WHO, 1994).

UV absorption by a biomolecule leads to the formation of generally reactive species or free radicals, which can produce effects that can last for hours, days or even years. For instance, DNA which carries the genetic information for the synthesis of proteins is the most critical target for damage by UVB radiation.

This section focusses on the serious consequences of excessive levels of SUVR at the Earth's surface, associated with depletion of stratospheric ozone, on human and animal health.

4.7.1 Human health

It is well known that enhanced UVB radiation has adverse effects on human health, like skin damage (including sunburn and skin cancer), eye abnormalities (including snowblindness and cataracts) and changes in the immune system (which may alter the course of cancer or infectious disease). UVR penetrates the skin by less than 1 mm and is absorbed by ocular tissues (mainly the cornea and the lens) before it reaches the retina. However, while the risk of skin cancer can be most reliably estimated, present knowledge of the quantitative role of UV radiation in damage to the eye or immunological effects is limited (Diffey, 1982, 1992). For instance, skin cancer incidence in the United Kingdom is increasing rapidly but there is no scientific proof that stratospheric ozone depletion has contibuted to this. In the following, a brief description of the existing information regarding incidence of beneficial and harmful effects on human health, caused by UVB radiation, is presented.

4.7.1.1 Skin sensitivity to UV exposure

Human skin develops specific mechanisms for photoprotection and biological responses to SUVR exposure that involve absorption, reflection or scattering. UV absorption takes place in wavelength bands, where molecules have characteristic absorption spectra and the absorbed energy is dissipated as heat or re-emission of light or is spent on photochemical reactions. Because of scattering, the direction of UV propagation is altered especially at boundaries of refractive index change. For example, in white skin, the change in refractive index causes 5% of normally incident light to be reflected. The remaining 95% is absorbed or reflected at the dermis, the second layer of the skin (below the epidermis), which gives the epidermis its elasticity and supportive strength. The backscattering component is reflected again at different interfaces resulting in multiplication of the strength of incident light. Another important adaptive response of the skin to UV exposure is epidermal thickening or hyperplasia, which begins around 72 hours after exposure. This occurs because of the increased rate of cell division in the lower epidermis which results in thickening of both epidermis and the outermost layer of the epidermis (stratum corneum). Six phototypes of skin sensitivity to UV have been defined: types I to IV are characteristic of caucasoid populations; type V represents mongoloid Middle Eastern populations; and type VI represents African and American negroid populations (WHO, 1994).

UV exposure of the skin has been shown to have both beneficial and harmful effects. The beneficial effects include production of vitamin D. The harmful effects include sunburn, phototoxicity, photoallergy, photoageing and skin cancer. Brief information in this respect is given below.

4.7.1.1.1 Beneficial effects of UV exposure

It has been suggested that one (and the only one well-established) of the beneficial effects of UV exposure is synthesis of vitamin D_3. For vitamin D_3 production the amount of UV radiation reaching the skin is critical. Continual UVB exposure causes the provitamins D_3 to be isomerized to pre-vitamin D_3 which forms by photoisomerization of various biologically inert photoproducts. Vitamin D_3 is required for intestinal absorption of calcium (Davies, 1985). Widespread use of sunscreens could lead therefore to vitamin-D deficiency and inadequate fixation of calcium, resulting in weakening of the bones. Only short outdoor exposures are required for synthesis of vitamin D_3 in the skin. For instance, from spring until autumn 15 min exposure between 9 am and 4 pm is adequate to synthesize the needed vitamin D_3. Increased melanin pigmentation in the skin can limit the production of vitamin D_3 with increasing age (WHO, 1994).

Another beneficial effect of UV exposure is maintenance of the skin's ability to sustain further UV exposures.

It has also been suggested that beneficial effects of UV exposure may occur such as improvement in cardiopathy and better micro-organism defence.

4.7.1.1.2 Acute effects of UVR exposure

Skin exposure to UVR creates two distinct tannings, one immediate and one delayed.

The first begins immediately on exposure to UVR and is called immediate pigment darkening (IPD). The main cause of IPD is darkening of the pigment melanin caused by UVA and begins to fade a few hours after exposure.

The second type of tanning is more persistent than IPD and takes about 3 days to develop. It is more effectively produced by UVB and results in increase in thickness and scattering properties of the epidermis and increases the number, size and pigmentation of melanin granules. Note that the tan obtained cosmetically from UVA (which does not produce thickening of the epidermis) is not as effective in protecting against further exposure to UV as the pigmentation induced by exposure to UVB. Multiple use of certain agents may produce a photosensitizing effect on exposure to UV. Reactions to photosensitizing agents involve both photoallergy and phototoxicity which are accelerated by UV exposure (WHO, 1994).

Skin exposure to UVR usually leads to reddening (erythema) that lasts about 8 hours after exposure and gradually fades after a few days. Reddening occurs 3–5 hours after sun exposure and reaches a maximum between 8 and 24 hours, fading over 3 days. In the case of too much exposure, the skin becomes painful and oedematous, and blistering may result. The degree of reddening depends on skin type, the time of exposure to UV and the wavelength of the UV radiation (UVA, UVB and UVC are all able to induce reddening). The most active wavelengths for reddening are in the 250–290-nm range with a decrease in effectiveness as the wavelengths increase. Although UVA radiation is much less erythemogenic than UVB radiation – broadly speaking by a factor of 1,000 – the much higher UVA irradiance present in sunlight means that in summertime UVA radiation contributes about 15–20% of the sunburn reaction. Exposure time to UVB solar radiation in order to elicit erythema increases after about 60 years of age (WHO, 1994).

The best documented effect of increasing UVB irradiance is the increase in incidence of skin cancer, which is briefly described in the next section.

4.7.1.1.3 Skin cancer

Recent studies (Madronich, 1993) showed that skin-cancer incidence in Australia (close to the Antarctic ozone hole) is the highest reported in the world, and in The Netherlands skin cancer is among the most frequently occurring cancers.

In principle, exposure to sunlight even without ozone depletion is the main cause of non-melanoma skin cancer. Already-available epidemiological evidence of carcinogenicity of UV in humans is outlined in the following (Kricker *et al.*, 1993):

- cancer occurs more frequently in people who are sensitive to solar UV;
- cancer occurs mainly at sun-exposed body sites;
- cancer is increased in residents of areas of high ambient solar UV;
- cancer is increased in people with high personal outdoor exposure.

Although direct epidemiological evidence linking sun exposure and skin cancer is weak, it could be inferred indirectly that two main categories of skin cancer induced from solar UV can be distinguished: cutaneous malignant melanoma and non-melanocytic cancer. The first type of skin cancer originates from pigment cells in the skin while the second is from normal epithelial cells in the epidermis. Epidemiological evidence shows that their incidence is less in dark-skinned populations than in those with light skins, residing in the same geographic area. Also, risk of skin cancer increases with decreasing pigmentation of the skin and reduced ability to produce a protective tan. However, the main uncertainty is about malignant melanoma, which is the type with the highest mortality (WHO, 1994).

Nowadays, there is little doubt in the scientific community that increased UVB irradiance (other things being equal) will lead to increased incidence of non-melanoma skin cancerns. These skin cancers have high incidence but comparatively low mortality (about 1%).

Non-melanoma skin cancer can be subdivided into two major types: basal cell carcinoma (BCC) and squamous cell carcinoma (SCC). There is now growing evidence that 1% decrease in ozone will lead to 1.2–1.4% increase in carcinogenic-effective UV radiation. Considering that for every 1% increase in carcinogenic-effective UVB radiation there will be 2.5% increase in SCC incidence and corresponding 1.5% increase in BCC, we finally arrive at the conclusion that 1% decrease in ozone leads to 3.5% increase in SCC and 2.1% increase in BCC. Note that BCC is the most common. According to recent literature for BCC and malignant melanoma, neither the wavelengths of SUVR, nor the exposure pattern that results in risk, have been established with certainty; whereas for SCC, both UVB and UVA are implicated. Furthermore, whilst SCC is strongly related to cumulative lifetime exposure to SUVR, this relationship is not so convincing for BCC (Madronich, 1993).

As far as skin-cancer incidence is concerned, SCC is more clearly related to SUV exposure. In particular, these tumours occur in face, neck, back of hands (i.e. skin regularly exposed to the Sun). According to Vitasa *et al.* (1990), the SCC risk rises proportional to total dose of sunlight received over a lifetime. It should be stressed that BCC hardly ever occurs on the regularly Sun-exposed backs of the hands, but it occurs more frequently than SSC on the trunk of the body (Madronich, 1993).

Among the possible effects of exposure to UV radiation, only the induction of non-melanoma skin cancers is understood sufficiently well. Madronich (1993) predicted significant increases in the steady-state incidence of two types of non-

Table 4.23. Expected increases (%)[a] in annual UV doses and skin-cancer incidence because of stratospheric ozone depletion from 1979 through 1992.

Latitude	Total ozone[a]	Erythema induction dose	DNA damage dose	Skin cancer dose	Basal cell carcinoma incidence[b]	Squamous cell carcinoma incidence
85N	-8.8 ± 3.2	7.1 ± 1.7	14.8 ± 3.6	10.6 ± 2.5	15.1 ± 5.6	28.5 ± 11.2
75N	-9.0 ± 2.9	7.6 ± 1.7	14.9 ± 3.3	10.8 ± 2.4	15.4 ± 5.8	29.1 ± 11.5
65N	-7.4 ± 1.7	7.6 ± 1.7	14.1 ± 3.2	10.3 ± 2.3	14.7 ± 5.6	27.7 ± 11.0
55N	-7.4 ± 1.3	7.2 ± 1.7	12.9 ± 3.2	9.5 ± 2.3	13.5 ± 5.3	25.4 ± 10.3
4N	-6.6 ± 1.2	6.5 ± 1.7	10.9 ± 3.0	8.1 ± 2.2	11.6 ± 4.7	21.6 ± 9.0
35N	-4.8 ± 1.4	5.0 ± 1.8	8.2 ± 3.0	6.1 ± 2.2	8.6 ± 4.0	16.0 ± 7.6
25N	-2.7 ± 1.5	3.0 ± 1.9	4.8 ± 3.1	3.5 ± 2.2	5.0 ± 3.5	9.0 ± 6.4
15N	-1.5 ± 1.1	1.7 ± 1.4	2.6 ± 2.3	1.9 ± 1.6	2.7 ± 2.4	4.8 ± 4.4
5N	-0.6 ± 1.6	0.7 ± 1.9	1.2 ± 3.0	0.8 ± 2.2	1.2 ± 3.1	2.1 ± 5.5
5S	-1.1 ± 1.4	1.2 ± 1.7	2.0 ± 3.0	1.4 ± 2.0	2.0 ± 2.8	3.6 ± 5.2
15S	-1.9 ± 1.3	2.2 ± 1.5	3.5 ± 2.4	2.5 ± 1.7	3.6 ± 2.6	6.5 ± 4.8
25S	-2.6 ± 1.6	3.1 ± 1.7	5.0 ± 2.4	3.6 ± 1.9	5.1 ± 3.1	9.2 ± 5.8
35S	-4.0 ± 1.6	4.8 ± 1.6	7.9 ± 2.6	5.7 ± 1.9	8.1 ± 3.6	14.9 ± 6.8
45S	-5.6 ± 1.4	7.2 ± 1.6	12.5 ± 2.7	8.9 ± 1.9	12.7 ± 4.8	23.9 ± 9.2
55S	-9.0 ± 1.5	10.9 ± 2.0	19.7 ± 3.6	14.2 ± 2.6	20.4 ± 7.4	39.3 ± 15.1
65S	-15.0 ± 2.0	16.3 ± 3.0	30.5 ± 5.8	21.9 ± 4.1	31.9 ± 12.2	64.0 ± 26.2
75S	-19.5 ± 2.6	24.1 ± 5.4	49.8 ± 12.0	34.0 ± 7.7	50.6 ± 21.4	107.7 ± 52.0
85S	-21.1 ± 3.0	31.0 ± 6.8	72.0 ± 17.6	46.5 ± 10.5	70.6 ± 31.2	159.6 ± 83.6

[a] Evaluated over the 14-year data record, expressed as % relative to the 1979 intercept. Uncertainty is one standard deviation.
[b] Based on skin-cancer dose increase and relevant biological amplifications factors.

melanoma skin cancer among fair-skinned populations under the assumptions that (1) UVB levels increased between 1979 and 1992 in accordance with observed ozone changes, (2) no further UV increases or decreases will occur and (3) human exposure habits will remain unaltered (see Table 4.23).

The aforementioned carcinogenic-effective exposure to sunlight is often expressed in 'minimal erythema dose' or MED. One MED is the minimum dose of SUVR which produces a reddening of white skin 8–24 hr after exposure. It is usually taken to be equivalent to an erythemally weighted radiant exposure of $200\,\mathrm{J\,m^{-2}}$ (McKinley and Diffey, 1987). As UVR decreases with increasing distance from the equator, the annual number of MED on an unshaded, horizontal surface at mid-latitudes can be roughly estimated as (Diffey, 1990):

$$\text{Annual MED} = 2 \times 10[4] \exp(-\text{latitude}/20)$$

In the absence of ozone depletion, adult indoor workers receive an annual facial exposure dose of 93 MED, children receive 157 MED up to the age of 18, and thereafter 93 MED (Diffey, 1992).

It has generally been observed that an inverse relationship with latitude and incidence of both non-melanocytic skin cancer and cutaneous melanoma exists (WHO, 1994). However, in Europe the incidence and mortality rates in Scandinavia are considerably higher than those in Mediterranean countries. Migration studies showed that sun exposure in childhood is a crucial period for melanoma risk. Similar studies showed that, in the case of immigrants coming from a temperate climate to a subtropical one, the risk of MSC later in life rises strongly with people arriving before adolescence.

Together, the indirect and direct evidence is sufficient to conclude that Sun exposure causes both melanoma and non-melanocytic skin cancer.

Like skin cancer, cancer of the lip is much more common in white than black populations. However, cancer of the lip and the ocular melanoma are rare and there is not the clear relationship with geographical latitude that is seen with skin cancers.

Recent studies on risk analysis of human skin cancer have shown that age and environmental UV exposure are the two most important factors in determining the relative risk. These findings have led to the following simple power low relationship for non-melanoma skin cancer:

$$\text{risk} \propto (\text{annual UV dose})^{\beta}\,(\text{age})^{\alpha}$$

where α and β are numerical constants.

When taking account of changes in lifestyle with age the following more complex model risk was developed (Diffey, 1992):

$$\text{risk} \propto (\text{cumulative UV dose at age } T)^{\beta-1} \sum_{t=0}^{T} (\text{annual dose at age } [T-t])\,t^{\alpha-\beta}$$

Another important factor in assessing the risk of skin cancer from SUVR is the so-called 'action spectrum' that denotes wavelength dependency. An action spectrum is the relative effectiveness of radiation at different wavelengths to produce a given biological effect. Comparison of experimentally determined action spectra with absorption spectra of appropriate molecules can sometimes give insight into the molecules primarily responsible for the effect (Diffey, 1992). Based on the information given by both the action spectrum $A(\lambda)$ and total ozone, the annual carcinogenic effective SUV dose D can be determined by the formula:

$$D = \sum_{\lambda=100}^{400} \left[\sum_{m=1}^{12} S(\lambda) \right] A(\lambda)$$

where $S(\lambda)$ is the total monthly spectral dose for a month m. Note that the action spectrum expresses the relative effective weight factor at wavelength λ. Considering that β denotes the exposure fraction received by the most exposed parts of the skin, then βD represents the actual effective dose for the skin (Slaper, 1987). Of course, this exposure fraction depends on individual lifestyles with regard to SUV exposure. For instance, for The Netherlands and Australia this fraction is equal to 0.02 and 0.05, respectively. If the effects of clouds were also taken into account, the UV doses would be reduced to 65% of clear-sky values (Diffey, 1992).

Because incidence of non-melanoma skin cancers is correlated with cumulative lifetime UV exposure, calculation of biologically accumulated dosage (BAD) of UVB radiation over days, months and years is very useful. The slope of annual BAD as function of latitude was found to correspond to epidemiological data for non-melanoma skin cancers between 30 and 50°N. It was also found that a small ozone loss in the tropics can provide as much additional BAD as a much larger ozone loss at higher latitudes. For ozone depletions of $> 5\%$ BAD increases exponentially with decreasing ozone levels.

In conclusion, skin cancer is the most investigated ramification of increased UVB irradiance and has led to such well-developed concepts as the radiation amplification factor which is briefly discussed in the next section.

4.7.1.1.4 Radiation Amplification Factor

In general, the Radiation Amplification Factor (RAF) is a number which denotes differences in the response of a biological system under different scenarios of UV exposure. For instance, this number varies between 2 and 3 for non-melanoma skin cancer, between 1 and 3 for cataracts and between 1 and 2 for certain immune responses (Varotsos, 1994). However, it is very important to consider the limitations of the data that generated the RAF, because RAFs values are not linear over a wide range of ozone-depletion rates. Therefore, it is preferable to consider ranges of RAF numbers instead of single numbers.

Total ozone depletion over the entire time period is characterized by strong variations from month to month reaching 13% for December but with a slightly positive trend for September. Recently, Varotsos *et al.*, (1995) calculated spectral SUVR by employing the above-mentioned parametric model and taking into account total ozone-loss rates over St. Petersburg.

Let us now attempt to estimate the relationship between the SUVR (calculated by Varotsos *et al.*, 1995) and ozone change in order to detect damage to the human eye.

Effective UV irradiance E or dose rate is given by

$$E = \int F(\lambda) W(\lambda)\, d(\lambda)$$

where $W(\lambda)$ is the weighting function or action spectrum for a specific biological effect and $F(\lambda)$ is the spectral irradiance for a given time and location. Although solar UVB decreases steeply towards longer wavelengths, biological effectiveness often increases towards shorter wavelengths. Therefore, weighted UV irradiances computed with different action spectra have different responses to atmospheric ozone change. A commonly used measure of this dependence is the radiation amplification factor (RAF) defined by

$$\Delta E/E = -\mathrm{RAF}(\Delta O_3/O_3)$$

where $\Delta O_3/O_3$ is the % change in ozone and $\Delta E/E$ is the corresponding % increase in weighted irradiance.

This simple formula is valid only for small O_3 change. For large O_3 change, a more accurate relation is

$$E_1/E_2 = [(O_3)_2/(O_3)_1]^{RAF}$$

where E_1 and E_2 are weighted irradiances corresponding to ozone $(O_3)_1$ and $(O_3)_2$, respectively.

In this study, radiation amplification factor is calculated by considering monthly deviation from the 15-year mean monthly value for total ozone content and corresponding monthly deviation from the 15-year mean monthly value for UVB radiation reaching the ground. These calculations were carried out for four different months (January, April, July and October) to gain insight into RAF seasonal variability.

RAF values for January varied from 0.67 in 1990 to 2.98 in 1981 with a mean value of about 2.0. For April, the minimum RAF value was –0.7 in 1987 and the maximum 1.1 in 1985 with an average value of 0.7. Corresponding RAF values for the July varied from –0.4 in 1988 to 2.0 in 1991 with an average value of about 0.8. Finally, for October the minimum RAF value was 0.6 in 1979 and the maximum value 3.2 in 1988. The average value was about 1.1 approximately.

From the above analysis it is evident that relative maximum RAF values appear during the winter (for January the mean RAF value was 2.0), which can be attributed to the relative larger deviation in UVB than the ozone deviation from the corresponding mean values. During the spring and summer, RAF values are low (mean RAF values for April and July are 0.7 and 0.8, respectively). Springtime RAF values are higher than those for fall and summer and relatively lower than that of the winter period.

4.7.1.1.5 Implication for ocular damage

It is a truism that increase in UVB radiation can have various direct and indirect effects on human health. UVB exposure can affect the incidence of skin cancer and ocular damage and can potentially affect incidence and severity of certain infectious diseases. The potential immunological impacts of UV radiation are not limited to fair-skinned populations; effects on the immune system are observed in deeply pigmented individuals as well.

Cataracts are responsible for 17 million cases of blindness each year (50% of the world's cases). A 1% stratospheric ozone loss is predicted to result in 0.6% increase in cataract incidence or 100,000 new cases per year. Known ocular effects because of increased UVB radiation include photokeratitis (snowblindness), early presbyopia, lens-capsule deformation and ocular melanoma.

From seasonal analysis of RAF, discussed in the section 4.1, an assessment can be made for eye damage resulting from SUVR exposure.

For instance, RAF at 40°N in January (300 DU) and July (310 DU) is 1.2 and 1.1, respectively, for damage to the cornea. Another example is that, at 40°N, RAF is 0.8 and 0.7 in January and July, respectively, for cataract damage (Varotsos and Feretis, 1997).

4.7.1.2 *Eye sensitivity to UVR exposure*

Cataracts are opacities of the eye lens that lead to impaired vision which can be surgically treated, but is a leading cause of blindness in the world.

Although ophthalmologists suggested long ago that there exists an association between UVR and development of cataracts and other ocular defects (Widmark, 1889; Duke-Elder, 1926), it was only during the last three decades that relevant studies provided a scientific link through epidemiological studies. For instance, there is sufficient evidence to link photokeratitis to acute ocular exposure to UV. Ambient levels of SUVR are such that acute phototoxic corneal disorder is largely confined to snowfields and deserts. It has been estimated that 2-h exposure outdoors around noon on snow-covered terrain is sufficient to induce photokeratitis, and that an exposure of 6–8 h may be needed in sandy terrain. It has also been shown that outdoor workers with high solar exposure run a threefold risk of developing pterygium and a sixfold risk of having climatic droplet keratopathy. Furthermore, experimental studies have confirmed that development of certain types of cataract are associated with ocular UVB exposure. Doubling of cumulative exposure also increased the risk of cortical cataract by a factor of 1.6. Exposure to sunlight, especially in early life, may be important in the development of intraocular malignant melanoma. Finally, the macula in the aphakic eye may be at risk from SUVR exposure. The natural crystalline lens absorbs all incident UVB and almost all UVA radiation. Concern has been expressed that some types of implanted plastic intraocular lenses may not provide adequate UV protection for the retina (Diffey, 1990). A 46-year-old Caucasian man with no other known risk factors developed bilateral posterior subcapsular cataracts soon after intense exposure to UVR.

Despite these studies, many uncertainties still remain. So, there is limited evidence:

- To link sunlight exposure of the eye to the development of pterygioum. It is still unclear whether the observed association is specific for UV.
- To associate climatic droplet keratopathy with UV exposure.
- To link ocular UV exposure and acute solar retinitis, age-related macular degeneration, acceleration of pigmentary retinopathies and exfoliation syndrome.

4.7.1.3 *Immune responses to UVR exposure*

The immune system plays an essential role in defending the body against invading organisms and provides resistance against the development of certain types of cancer.

It has been known that UVR exposures at environmental levels suppress immune responses. This immune suppression results in enhanced susceptibility to certain skin infections as well as some systematic infections.

According to laboratory studies, the most common immune responses to UVB radiation are suppression of contact hypersensitivity (CHS), or delayed-type hypersensitivity response (DTH), and the suppression of the rejection of transplanted organs. It is believed that a natural component of skin, a deanimated form of histidine, which may be involved in skin-cancer development is responsible for suppression of CHS. As far as which infectious diseases might be exacerbated by UVB is concerned, two criteria regarding the infectious agent are suggested (1) it must come into contact with the skin and (2) the skin exposed to UVB must be susceptible to immune suppression. The combination of these two processes result in the formation of suppressor T cells which accumulate in such a way that rejection of the infectious agent is down-modulated or switched off. Infectious agents such as *Leishmania* parasite and *Herpes* virus are affected by UVB radiation. Also diabetes and multiple sclerosis (not infectious diseases) have shown a latitudional gradient, which denotes that UVB may play a role. Note that both diabetes and multiple sclerosis are considered to have a strong autoimmunological component (De Fabo, 1983). Another infectious disease believed to be related to UVB exposure is vaccination. This means that there is the possibility that efficacy of a particular vaccination with live or attenuated organisms, following exposure to UVB, may be affected (WHO, 1994).

The following studies are needed to assess the consequent risk to human health:

- To determine whether systemic effects occur in humans as a result of UVR exposure.
- To examine whether currently available sunscreens provide protection and whether dietary supplements are protective.
- To explore the spectral effectiveness of UVR in producing immune-function defects.
- To determine the role of UVB exposure in allergic reactions.

4.7.2 Animal health

4.7.2.1 Skin sensitivity to UVR exposure

From animal studies about UVR exposure of the skin the following can be concluded. Among domestic and food animals an often-occurring phenomenon is carcinogenesis and in particular squamous cell carcinomas, mainly caused by UV radiation. Effectiveness for induction of these carcinomas is found to peak in the UVB range. In addition, UVA is also carcinogenic, but at a much lower level, similar to that found for erythema and tanning. Melanomas are much less common among animals, while basal-cell carcinomas are uncommon (WHO, 1994).

As mentioned in section 4.7.1.1.3, information on wavelength dependency is very important for assessment of skin-cancer risk. However, such information cannot be obtained directly for humans and has to be determined by other methods, such as measurement of the action spectrum in hairless mice. Conversion of this result to humans is done by correcting for optical differences between human and mouse skin.

It is well known from animal studies that UVB immune suppression plays a major role in suppressing destruction of skin tumours (De Fabo, 1983). Using animal models to study the mechanism of metastasis, and the effect of diet and nutrition on it, exemplify new direction to take to assess the overall impact of UVB on animal and human health.

4.7.2.2 *Eye sensitivity to UVR exposure*

Studies of effects such as cataract, retinopathy and UV damage mechanisms in the cornea, lens and retina are only possible in animal models. Analogous to acute sunburn of the skin is photokeratitis which is an acute reversible radiation-induced injury of the corneal epithilium. Current knowledge from animal studies is that exposure to UVB can lead to acute photokeratitis. More specifically, the mean threshold of UVB for photokeratitis was estimated at $35\,J\,m^{-2}$ at 270 nm (artificial source of UVB). Oldenburg *et al.* (1990) demonstrated that using UVB filtering contact lenses on rabbits protects against photokeratitis. According to studies of the last decade, shortwavelength visible non-coherent and coherent light, as well as UVA, can cause photochemical, mechanical and thermal damage to the retina and the pigment epithelium (Organisciak and Winkler, 1993). The impact on important farm animals can lead to an impact on food production, especially in countries that depend heavily on raising such species as cattle and sheep (De Fabo, 1983).

4.7.2.3 *Immune responses*

It is generally accepted in the literature that exposure of the skin to UVB radiation can result in suppression of certain types of immune responses in laboratory animals (Jeevan and Kripke, 1990). Among these are (a) rejection of UV-induced cancers and melanomas, (b) contact-allergy reactions to chemicals, (c) delayed-type-hypersensitivity responses to microbial and other antigens and (d) phagocytosis and elimination of certain bacteria from lymphoid tissues. Recent studies, based on mycobacterial infection of mice, showed that exposure to UVB radiation decreased the delayed hypersensitivity response to mycobacterial antigens, increased the severity of infection and reduced the survival rate of mice from lethal infections (WHO, 1994).

4.8 REFERENCES

Abstracts of the European Conference on Atmospheric UV Radiation, 29 June – 2 July 1998, Helsinki, Finnish Meteorological Instituted, Helsinki, 127 pp.

Anav, A., Moriconi, M. L., Giannoccolo, S., Di Menno, M., (1996) Field measurements of the global UV-B radiation: A comparison between a broad-band radiometer and a Brewer spectrophotometer. *Nuovo Cimento*, **19C**, Ser. 1, N4, 505–516.

Atkinson, P. M., Tantall, A. R. L. (1997) Neural networks in remote sensing. *Int. J. Remote Sensing*, **18**, N4, 699–709.

Bais, A. F. (1997) Variability of UV-B at four stations in Europe. *Geophys. Res. Lett.*, **24**, 1,363–1,366.

Bais, A. F., Zerefos, C. S., McElroy, C. T. (1996) Solar UVB measurements with the double- and single-monochromator Brewer Ozone Spectrophotometers, *Geophys. Res. Lett.*, **23**, 833–836.

Bais, A., Blumthaler, M., Webb, A. R., Groeber, J., Kirsch, P. J., Gardiner, B. G., Zerefos, C. S., Svenoe, T., Martin, T. J. (1997) Spectral UV measurements over Europe within the SESAME activities. *J. Geophys. Res.*, **102**, ND, 8,731–8,736.

Baker, K. S., Smith, R. C. (1990) Irradiance transmittance through the air–water interface. SPIE Proc., **1,302**, 556–565.

Baker, L. A., Herlihy, A. T., Eilers, J. M. (1991) Acidic lakes and streams in the United States: The role of acidic deposition. *Science*, **252**, 1,151–1,154.

Balachandran, N. K., Rind, D. (1995) Modeling the effects of UV variability and the QBO on the troposphere-stratosphere system. Part II. The Middle Atmosphere. *J. Climate*, **8**, N8, 2,058–2,079.

Balch, W. M., Abbott, M. R., Eppley, R. W. (1989) Remote sensing of primary production. I. A comparison of empirical and semi-analytical algorithms. *Deep-Sea Res.*, **36**, N2, 281–295.

Bannister, T. T. (1990) Empirical equations relating solar irradiance to a, b/a, and solar zenith angle. *Limnol. Oceanogr.*, **35**, 173–177.

Beaglehole, D., Carter, G. G. (1992a) Antarctic skies. 1. Diurnal variations of the sky irradiances and UV effects of the ozone hole. *J. Geophys. Res.*, **97**, ND2, 2,589–2,596.

Beaglehole, D., Carter, G. G. (1992a) Antarctic skies. 2. Characterization of the intensity and polarization of skylight in a high albedo environment. *J. Geophys. Res.*, **97**, ND2, 2,597–2,600.

Beaglehole, D., Ramanathan, B., Rumberg, J. (1998) The UV to IR transmittance of Antarctic snow. *J. Geophys. Res.*, **103**, ND8, 8,849–8,858.

Bégué, A., Myneni, R. (1996) Operational relationships between NOAA-Advanced Very High Resolution Radiometer vegetation indices and daily fraction of absorbed photosynthetically active radiation, established for Sahelian vegetation canopies. *J. Geophys. Res.*, **101**, ND16, 21,275–21,290.

Bekoriukov, V. I. (1990) *An Atlas of Climatic Characteristics of Total Content and Partial Pressure of Ozone*. Gidrometeoizdat, Moscow, 180 pp (in Russian).

Bener, P. (1972) *Final Technical Report No. DASA 37-68-C-1017*, European Research Office, US Army, 98 pp.

Bernhard, G., Mayer, B., Seckmeyer, G., Moise, A. (1997) Measurements of spectral solar UV irradiance in tropical Australia. *J. Geophys. Res.*, **102**, ND7, 8,719–8,730.

Berthelot, B., Deschamps, P.-Y. (1994) Evaluation of bio-optical algorithms to remotely sensed marine primary production from space. *J. Geophys. Res.*, **99**, NC4, 7,979–7,989.

Berthon, J.-F., Morel, A. (1992) Validation of a spectral light-photosynthesis model and use of the model in conjuction with remotely sensed pigment observations. *Limnol. Oceanogr.*, **37**, 781–796.

Bigelow, D. S., Slusser, J. R., Beaubien, A. F., Gibson, J. H. (1998) The USDA Ultraviolet Radiation Monitoring Program. *Bull. Amer. Meteorol. Soc.*, **79**, N4, 601–615.

Biggs, R. H., Joyner, M. E. B. (eds.) (1994) *Stratospheric Ozone Depletion/UV-B Radiation in the Biosphere*, NATO ASI Series Vol. 18, Springer-Verlag, Berlin, 358 pp.

Bird, R., Riordan, C. (1986) Simple solar spectral model for direct irradiance on horizontal and tilted planes at the earth's surface for cloudless atmosphere. *J. Clim. Appl. Meteorol.*, **25**, 87–97.

Bjarnason, G. G., Rögnvaldsson, Ö. E. (1997) Coherency between solar UV radiation and equatorial total ozone. *J. Geophys. Res.*, **102**, ND11, 13,009–13,018.

Blaustein, A. R. (1994) UV repair and resistance to solar UV-B in amphibians eggs: A link to population declines? *Proc. Natl. Acad. Sci.*, **91**, 1,791–1,795.

Blumthaler, M. (1993) UV-radiation and ozone depletion. In: *Environmental Effects of Ultraviolet Radiation*, M. Tevini (ed.), Lewis Publ., Boca Raton, FL, pp. 79–94.

Bodeker, G. E., McKenzie, R. L. (1996) An algorithm for inferring surface UV irradiance including cloud effects. *J. Appl. Meteorol.*, **35**, 1,860–1,877.

Bodhaine, B. A. (1996) Central Antarctica: Atmospheric chemical composition and atmospheric transport. In: *Proceedings NATO Advanced Research Workshop 'Process Chemical Exchange between Atmosphere and Polar Snow', 19–23 March 1995, Ciocco, Italy*. Springer, Berlin, pp. 145–172.

Bodhaine, B. A., Barrie, L. A. (1996) Chemical reactions in the polar troposphere relevant to C, S and N components. In: *Proceedings NATO Advanced Research Workshop 'Process Chemical Exchange between Atmosphere and Polar Snow', 19–23 March 1995, Ciocco, Italy*. Springer, Berlin, pp. 201–224.

Bodhaine, B. A., Dutton, E. G., Hofmann, D. J., McKenzie, R. L., Johnston, P. V. (1997) UV measurements at Mauna Loa: July 1995 to July 1996. *J. Geophys. Res.*, **102**, ND15, 19,265–19,274.

Bojkov, R. D., Fioletov, V. E. (1995) Estimating the global ozone characteristics during the last 30 years. *J. Geophys. Res.*, **100**, 16,537–16,551.

Booth, C. R., Lucas, T. B., Tusson, IV J. R., Morrow, J. H., Mestechkina, T. (1994) Antarctic UV spectroradiometer monitoring program: Contrasts in UV radiance at the South Pole and Barrow, Alaska. *CLDM Summer Report No. 22*, Boulder CO, pp. 98–110.

Bordewijk, J. A., Slaper, H., Reiner, H. A. J. M., Schlamann, E. (1995) Total solar radiation and the influence of clouds and aerosols on the biologically effective UV. *Geophys. Res. Lett.*, **22**, 2,151–2,154.

Borisenkov, E. P. (1963) *Physico-Statistical Methods of Analysis and Prediction of Meteorological Fields*. Marine Transport Publ., Leningrad, 243 pp. (in Russian).

Borisenkov, E. P. (1980) Principles of constructing physico-statistical schemes of long-range meteorological forecasts and their optimisation. In: *Proceedings*

WMO Symposium on Probabilistic and Statistical Methods in Weather Forecasting. WMO, Geneva, pp. 37–47.

Bothwell, M. L., Sherdot, D. M., Pollock, C. M. (1994) Ecosystem response to solar ultraviolet-B radiation: Influence of trophic-level interaction. *Science*, **265**, 97–100.

Bronman, J. F. L., Bjoern, L. O., Akerlund, H.-E. (1984) Action spectrum for inhibition by UV radiation of photosystem II activity in spinach thylakoids. *Photobiochem. Photobiophys*, **8**, 305–313.

Brooks, A. S., Torke, B. G. (1977) Vertical and seasonal distributions of chlorophyll α in Lake Michigan. *J. Fish. Res. Board Can.*, **34**, 2,280–2,287.

Brühl, C., Crutzen, P. J. (1989) On the disproportional role of tropospheric ozone as a filter against UV-B radiation. *Geophys. Res. Lett.*, **16**, 703–706.

Bukata, R. P., Jerome, J. H., Bruton, J. E. (1989) Determination of irradiation and primary production using a time-dependent attenuation coefficient. *J. Great Lakes Res.*, **15**, 327–338.

Bukata, R. P., Jerome, J. H., Kondratyev, K. Ya., Pozdnyakov, D. V. (1995) *Optical Properties and Remote Sensing of Inland and Coastal Waters*. CRC Press, Boca Raton, FL, 360 pp.

Bull, K. R., Sutton, M. A. (1998) Critical loads and the relevance of ammonia to an effects-based nitrogen protocol. *Atmos. Environ.*, **32**, N2, 565–572.

Byron, E. R. (1992) The adaptive significance of calanoid copepod pigmentation: A comparative and experimental analysis. *Ecology*, **63**, 1,817–1,886.

Cabrera, S., Bozzo, S., Fuenzalida, H. (1995) Variations in UV radiation in Chile. *Photochem. and Photobiol.*, **28**, 137–142.

Cappelani F., Kochler, C. (1999) Ozone and UV-B variations at Ispra from 1993 to 1997. *Atmos. Environ.*, **33**, 3,787–3,790.

Carpenter, S. R., Kitchell, J. E. (1993) The trophic cascade in lakes. *BioScience*, **45**, 634–639.

Cebula, R. P., DeLand, M. T., Hilsenrath, E. (1998) NOAA 11 Solar Backscatter Ultraviolet, model 2 (SBUV/2) instrument solar spectral irradiance measurements in 1989–1994. 1. Observations and long-term calibration. *J. Geophys. Res.*, **103**, ND13, 16,235–16,245.

Chandra, S, Varotsos, C., Flynn, L. E. (1996) The midlatitude total ozone trend in the northern hemisphere. *Geophys. Res. Lett.*, **23**, 555–558.

Chumley, F. G. (1994) *Workshop on the Environmental Fate of Trifluoroacetic Acid (TFA), Miami Beach, FL*, SPA-AFEAS Inc., Washington, DC.

Correll, D. L., Clark, C. O., Goldberg, B., Goodrich, V. R., Hayes, D. R., Klein, W. H., Schecher, W. D. (1992) Spectral ultraviolet-B radiation fluxes at the earth's surface: Long term variations at 39°N, 77°W. *J. Geophys. Res.*, **97**, 7,579–7,591.

Cullen, J. C., Neale, P. J., Lesser, M. P. (1992) Biological weighting function for the inhibition of phytoplankton photosynthesis by ultraviolet radiation. *Science*, **258**, 646–650.

d'Almeida, G. A., Koepke, P., Shettle, E. P. (1991) *Atmospheric Aerosols: Global Climatology and Radiative Characteristics*. A. Deepak Publ., Hampton, VA.

Damakaer, D. M., Dey, D. B. (1983) UV damage and photo-reactivation potentials of larval shrimp, *Pandalus platyceros* and adult euphausiidis, *Thysanoessa raschif. Oecologia*, **60**, 169–175.

Davies, D. M. (1985) Calcium metabolism in healthy men deprived of sunlight. *Ann. N. Y. Acad. Sci.*, **453**, 21–27.

Dedieu, G., Deschamps, P. Y., Kekk, Y. H. (1987) Satellite estimation of solar irradiance at the surface of the Earth and of surface albedo using a physical model applied to Meteosat data. *J. Clim. Appl. Meteorol.*, **26**, 79–87.

De Gruijl, F. R. (1995) Impacts of a projected depletion of the ozone layer. *Consequences*, **1**, N2, 13.

DeLand, M. T., Cebula, R. P. (1998) NOAA 11 Solar Backscatter Ultraviolet, model 2 (SBUV/2) instrument solar spectral irradiance measurements in 1989–1994. 2. Results, validation, and comparisons. *J. Geophys. Res.*, **103**, ND13, 16,251–16,273.

Denman, K. L., Gargett, A. E. (1988) Multiple thermoclines are barriers to vertical exchange in the subarctic Pacific during SUPER, May 1984. *J. Mar. Res.*, **46**, 77–103.

Diffey, B. (1982) The consistency of studies of ultraviolet erythema in normal human skin. *Phys. Med. Biol.*, **27**, 715–720.

Diffey, B. L. (1990) Human exposure to ultraviolet radiation. *Semin Dermatol.*, **9**, 2–10.

Diffey, B. L. (1992) Stratospheric ozone depletion and the risk of non-melanoma skin cancer in a British population. *Phys. Med. Biol.*, **37**, N12, 2,267–2,274.

Dlugokencky, E. J., Dutton, E. G., Novelli, P. C., Tans, P. P., Masarie, K. A., Lantz, K. O., Madronich, S. (1996) Changes in CH_4 and CO growth rates after the eruption of Mt. Pinatubo and their link with changes in tropical tropospheric UV flux. *Geophys. Res. Lett.*, **23**, N20, 2,761–2,764.

Dohler, G., Hagmeier, E., Grigoleit, E. (1991) Impact of solar UV radiation on uptake of ^{15}N-ammonia and ^{15}N-nitrate by marine diatoms and natural phytoplankton. *Biol. Phys. Pfl.*, **187**, 293–303.

Donkar, V. A., Damian, H. A. K., Hader, D. P. (1993) Effects of tropical solar radiation on the velocity and photophobic behaviour of filamentous gliding cyanobacteria. *Acta Protozoa*, **32**, 67–72.

Duke-Elder, W. S. (1926) The pathological action of light upon the eye. I. Action on the outer eye: Photophthalmia. *Lancet*, **1**, 1,137–1,141.

Eck, T., Bhartia, P., Kerr, J. (1995) Satellite estimation of spectral UVB irradiance using TOMS derived total ozone and UV reflectivity. *Geophys. Res. Lett.*, **22**, 611–614.

Effects of Increased Ultraviolet Radiation on Global Ecosystems (1992) SCOPE/UNEP, Paris, 47 pp.

Erickson, D. J., Eaton, R. E. (1993) Global biochemical cycling estimates with CZCS satellite data and general circulation models. *Geophys. Res. Lett.*, **20**, N8, 683–686.

Erlick, C., Frederick, J. E. (1998) Effects of aerosols on the wavelength dependence of atmospheric transmission in the ultraviolet and visible. 2. Continental and urban aerosols in clear skies. *J. Geophys. Res.*, **103**, ND13, 23,275–23,285.

De Fabo, E. C., Noonan, F. P. (1983) Mechanism of immune suppression by ultraviolet radiation in vivo. I. Evidence for the existence of a unique photoreceptor in skin and its role in photoimmunology. *J. Exp. Med.*, **158**, 84–98.

Fee, E. J. (1969) A numerical model for the estimation of photosynthetic production, integrated over time and depth in natural waters. *Limnol. Oceanogr.*, **14**, 906–911.

Feister, U., Grewe, R. (1995) Higher UV radiation inferred from low ozone levels at northern midlatitudes in 1992 and 1993. *Global and Planetary Change*, **11**, Nos. 1/2, 25–34.

Fioletov, V. E., Griffioen, E., Kerr, J. B., Wardle, D. I., Uchino, O. (1998) Influence of volcanic sulphur dioxide on spectral UV irradiance as measured by Brewer spectrophotometers. *Geophys. Res. Lett.*, **25**, N10, 1,665–1,668.

Fioletov, V. E., Kerr, J. B., Wardle, D. I. (1997) The relationship between total ozone and spectral UV irradiance from Brewer observations and its use for derivation of total ozone from UV measurements. *Geophys. Res. Lett.*, **24**, N23, 2,997–2,300.

Forster, P. M. de F. (1995) Modeling ultraviolet radiation at the Earth's surface. Part I: The sensitivity of ultraviolet irradiance to atmospheric changes. *J. Appl. Meteorol.*, **34**, N11, 2,412–2,425.

Forster, P. M. de F., Shine, K. P. (1995) A comparison of two radiation schemes for calculating ultraviolet radiation. *Quart. J. Roy. Meteorol. Soc.*, **121**, Part A, N525, 1,113–1,132.

Forster, P. M. de F., Shine, K. P., Webb, A. R. (1995) Modeling ultraviolet radiation at the Earth's surface. Part II. Model and instrument comparison. *J. Appl. Meteorol.*, **34**, N11, 2,426–2,439.

Frederick, J. E., Lubin, D. (1988) The budget of biologically active ultraviolet radiation in the earth-atmosphere system. *J. Geophys. Res.*, **93**, 3,825–3,832.

Frederick, J. E., Snell, H. E., Haywood, E. K. (1989) Solar ultraviolet radiation at the earth's surface. *Photochem. and Photobiol.*, **50**, N4, 443–450.

Frederick, J. E., Erlick, C. (1995) Trends and interannual variations in erythemal sunlight, 1978-1993. *Photochem. and Photobiol.*, **62**, 476–484.

Frouin, R., Linger, D. W., Gautier, C., Baker, K. S., Smith, R. C. (1989) A simple analytical formula to compute clear sky total and photosynthetically available solar irradiance at the ocean surface. *J. Geophys. Res.*, **94**, NC7, 9,731–9,742.

Gautier, C., Diak, G., Masse, S. (1980) A simple physical model to estimate incident solar radiation at the surface from GOES satellite data. *J. Appl. Meteorol.*, **19**, 1,005–1,012.

Geogdzaev, I. V., Kondranin, T. V., Rublev, A. N., Chubarova, N. E. (1997) UV radiation transfer through broken cloud fields: Modeling and comparisons with measurements. *Physika atmosphery i okeana*, **33**, N5, 680–686 (in Russian).

Gershun, A. (1936) On the photometry of turbid media. *Trudy Gos. Okeanogr. Inst.*, **11**, 99–125 (in Russian).

Gleason, D. F., Wellington, G. M. (1993) Ultraviolet radiation and coral bleaching. *Nature*, **365**, 836–838.

Gordon, H. R., McCluney, W. R. (1975) Estimation of the depth of sunlight penetration in the sea for remote sensing. *Appl. Opt.*, **14**, 413–416.

Gordon, H. R., Wang, M. (1994) Retrieval of water-leaving radiance and aerosol optical thickness over the ocean with SeaWiFS: A preliminary algorithm. *Appl. Opt.*, **33**, N3, 443–452.

Grant, R. H., Heisler, G. M. (1997) Obscured overcast sky radiance distribution for ultraviolet and photosynthetically active radiation. *J. Appl. Meteorol.*, **36**, N10, 1,336–1,345.

Groeneveld, A. H. C., de Kok, H. A. M., van den Berg, G. (1992) The toxicity of sodium trifluoroacetate to the alga *Selenastrum capricornutum. AFEAS Report No. 56635/52/92*, pp. 156–159.

Gruzdev, A. N. (1995) Possible changes in the dose of biologically active ultraviolet radiation received by the biosphere in the summertime Arctic due to total ozone interannual variability. *Sci. Total Environ.*, **160/161**, 669–675.

Hader, D. P. (1993) Risks of enhanced solar ultraviolet radiation for aquatic ecosystems. *Progr. Phycholog. Res.*, **9**, 1–45.

Hader, D. P. (1994) UV-B effects on aquatic systems. *Stratospheric Ozone Depletion/UV-B Radiation in the Biosphere*, R. H. Biggs, M. E. B. Joyner (eds.), (Springer: Heidelberg), pp. 155–162.

Hader, D. P., Worrest, R. C., Kumar, H. D., Smith, R. C. (1995) Effects of increased solar ultraviolet radiation on aquatic ecosystems. *UNEP Environmental Effects of Ozone Depletion. 1994 Assessment*. WMO, Geneva, Switzerland, pp. 1–19.

Hayasaka, T., Kikuchi, N., Tanaka, M. (1995) Absorption of solar radiation by stratocumulus clouds: Aircraft measurements and theoretical calculations. *J. Appl. Meteorol.*, **34**, 1,047–1,055.

Heineman, Th., Fisher, J. (1997) Simultaneous retrieval of oceanic and atmospheric properties using satellite remote sensing measurements. *SPIE Proc.*, **2,963**, 691–638.

Helbling, E. W., Villafane, V., Holm-Hansen, O. (1994) Effects of utraviolet radiation on antarctic marine phytoplankton photosynthesis with particular attention to the influence of mixing. *Ultraviolet Radiation in Antarctica: Measurements and Biological Effects* (Antarctic Research Series vol 62), C. S. Weiler, P. A. Penhale, American Geophysics Union, Washington, DC, pp. 207–227.

Henriques, D. V. (1997) The Portuguese ozone and UV-B monitoring stations. In: *Atmospheric Ozone Dynamics: Observations in the Mediterranean Region*, C. Varotsos (ed.), NATO ASI Series Vol. 153, Springer-Verlag, Berlin, pp. 147–156.

Herman, J. R., Bhartia, P. K., Torres, O., Hsu, S., Seftor, C., Celarier, F. (1997) Global distributions of UV-absorbing aerosols from Nimbus 7/TOMS data. *J. Geophys. Res.*, **102**, ND14, 16,911–16,922.

Herndl, G. J., Mueller-Niklas, G., Frick, J. (1993) Major role of ultraviolet-B in controlling bacterioplankton growth in the surface layer of the ocean. *Nature*, **361**, 717–719.

Hunter, J. R., Kaupp, S. E., Taylor, J. H. (1982) Assessment of effects of UV radiation on marine fish larvae. *The Role of Solar Ultraviolet Radiation in Marine Ecosystems*, J. Calcins (ed.), Plenum, New York, pp. 459–497.

IGBP (1988) The International Geosphere-Biosphere Programme: A study of global change. *IGBP Rep. No. 4*, IGBP Publ., Paris, 200 pp.

Impacts of projected depletions of the ozone layer (1995) *Consequences*, N2, 12–21.

Isaksen, I. S. A. (1997) Stratospheric ozone depletion and UV-B changes. In: *Solar Ultraviolet Radiation*, NATO ASI Series Vol. 52, C. S. Zerefos, A. F. Bais (eds.), Springer-Verlag, Berlin, pp. 13–21.

Jassby, A. D., Platt, T. (1969) Mathematical formulation on the relationship between photosynthesis and light for phytoplankton. *Limnol. Oceanogr.*, **21**, 540–547.

Jeevan, A., Kripke, M. L. (1990) Alteration of the immune response to *Mycobacterium bovis* BCG in mice exposed chronically to low doses of UV radiation. *Cell Immunol.*, **130**, 32–41.

Jerome, J. H., Bukata, R. P., Bruton, J. E. (1988) Utilizing the components of vector irradiance to estimate the scalar irradiance in natural waters. *Appl. Opt.*, **27**, N19, 4,012–4,018.

Jerome, J. H., Bukata, R. P., Bruton, J. E. (1990) Determination of available subsurface light for photochemical and photobiological activity. *J. Great Lakes Res.*, **16**, N3, 436–443.

JGOFS (1988) *Report of the First Session of the SCOR Committee for JGOFS. Sci. Comm. Ocean. Res.*, IGBP Publ., Stockholm, 50 pp.

Justus, C. G., Paris, M. V. (1985) A model for solar spectral irradiance at the bottom and top of a cloudless atmosphere. *J. Clim. Appl. Meteorol.*, **24**, 193–502.

Karantz, D., Cleaver, J. E., Mitchell, D. L. (1991) Cell survival characteristics and molecular responses of Antarctic phytoplankton to ultraviolet-B radiation. *J. Phycol.*, **27**, 326–341.

Karantz, D. (1994) Impact of UV-B radiation on pelagic ecosystems: report of working group on bacteria and phytoplankton. *Arch. Hydrobiol*, **35**, 56–65.

Katsambas, A., Antoniou, Ch., Stratigos, J., Arvanitis, I., Zolota, F., Varotsos, C., Cartalis, C., Asimakopoulos, D. N. (1991) A simple algorithm for simulating the solar ultraviolet radiation at the earth's surface: An application in determining minimum erythema dose. *Earth, Moon and Planets*, **53**, 191–204.

Kelley, P., Dickerson, R. R., Luke, W. T., Kok, G. L. (1995) Rate of NO_2 photolysis from the surface to 7.6 km altitude in clear-sky and clouds. *Geophys. Res. Lett.*, **22**, N19, 2,621–2,624.

Kerr, J. B., McElroy, C. T. (1993) Evidence for large upward trends of ultraviolet-B radiation linked to ozone depletion. *Science*, **262**, 1,032–1,034.

Kiefer, D. A., Mitchel, B. G. (1983) A simple, steady state description of phytoplankton growth based on absorption cross section and quantum efficiency. *Limnol. Oceanogr.*, **28**, N4, 770–776.

Kirk, J. T. O. (1983) *Light and Photosynthesis in Aquatic Ecosystems*. Cambridge University Press, Melbourne, Australia, 401 pp.

Kirk, J. T. O. (1984) Dependence on relationship between inherent and apparent optical properties of water on solar altitude. *Limnol. Oceanogr.*, **29**, N2, 350–356.

Kirk, J. T. O. (1986) *Light and Photosynthesis in Aquatic Ecosystems*. Cambridge University Press, London, 239 pp.

Köhler, U., Vandersee, W., Winkler, P., Gontner, L. (1998) *Spectral and Integral Observations of UV-B-Radiation and Ozone Measurements*, Berichte des Deutchen Wetterdienstes München, N204, 56 pp.

Komhyr, W. D. (1969) Electrochemical concentration cells for gas analysis. *Ann. Geophys.*, **25**, 203–210

Kondratyev, K. Ya., Pozdnyakov, D. V., Isakov, V. Yu. (1990) *Radiation and Hydrooptical Experiments on Lakes*, Nauka, Leningrad, 115 pp. (in Russian).

Kondratyev, K. Ya., Pozdnyakov, D. V. (1994) Some aspects of optical remote sensing applications in contemporary limnological studies. *Water Resources*, **21**, N5, 563–574.

Kondratyev, K. Ya., Varotsos, C., Katsambas, A., Stratigos, G., Antoniou, C. (1994) On the risk for human skin from the solar ultraviolet radiation. *Doklady Russian Acad. Sci.*, **338**, 262–263 (in Russian).

Kondratyev, K. Ya., Varotsos, C. A., Fedchenko, P. P. (1995) Global dynamics of total ozone content, its impact on the solar ultraviolet radiation variations and ecosystems. *Studying the Earth from Space*, N4, 105–117.

Kondratyev, K. Ya., Pokrovsky, O. M., Dalyuk, I. U., Varotsos, C. A. (1995) Atmospheric ozone trends and other factors of surface ultraviolet radiation variability. *Environmental Conservation*, N6, 17–22.

Kondratyev, K. Ya., Buznikov, A. A., Pokrovsky, O. M. (1996) *Global Change and Remote Sensing*, Wiley/Praxis, Chichester, UK, 389 pp.

Kondratyev, K. Ya., Pozdnyakov, D. V. (1996a) The influence of UV solar radiation on aquatic ecosystems. I. Hydrobiological aspects of the problem. *Studying the Earth from Space*, N5, 94–106 (in Russian).

Kondratyev, K. Ya., Pozdnyakov, D. V. (1996b) The influence of UV solar radiation on aquatic ecosystems. I. Hydrooptical aspects of the problem. *Studying the Earth from Space*, N6, 105 (in Russian).

Kondratyev, K. Ya., Varotsos, C. A. (1996a) Global total ozone dynamics – Impact on surface solar ultraviolet radiation variability and ecosystems. Part I. Global ozone dynamics and environmental safety. *Environ. Sci. Pollut. Res.*, **3**, N3, 153–157.

Kondratyev, K. Ya., Varotsos, C. A. (1996b) Global total ozone dynamics – Impact on surface solar ultraviolet radiation variability and ecosystems. Part II.

Dynamics of atmospheric chemical composition: The role of remote sensing. *Environ. Sci. Pollut. Res.*, **3**, N4, 205–209.

Kondratyev, K. Ya., Pozdnyakov, D. V. (1997a) The effect of solar ultraviolet radiation on water ecosystems due to the fall of the total ozone. 1. Hydrobiological aspects. *Earth Obs. Rem. Sens.*, **14**, 805–822.

Kondratyev, K. Ya., Pozdnyakov, D. V. (1997b) The effect of solar ultraviolet radiation on water ecosystems due to the fall of the total ozone. 1. Hydro-optical aspects. *Earth Obs. Rem. Sens.*, **14**, 997–1,009.

Kondratyev, K. Ya., Tanaka, T. (1997) Advanced Earth Observing Satellite-II (ADEOS-II): New perspectives of global environmental monitoring. *Studying the Earth from Space*, N1, 105–122.

Kondratyev, K. Ya. (1998) *Multidimensional Global Change*. Wiley/Praxis, Chichester, UK, 761 pp.

Kondratyev, K. Ya., Filatov, N. N. (eds.) (1999) *Limnology and Remote Sensing*, Springer/Praxis, Chichester, UK, 375 pp.

Kricker, A., Armstrong, B. K., Jones, M. E., Burton, R. C. (1993) Health, solar UV radiation and environmental change. *IARC Technical Report No. 13*, Lyon, International Agency for Research on Cancer.

Kripke, M. L. (1985) Health effects of stratospheric ozone depletion. In: *Atmospheric Ozone Research and Its Policy Implications*, T. Schneider *et al.* (eds.), Elsevier, Amsterdam, pp. 795–801.

Krotkov, N. A., Bhartia, P. K., Herman, J. R., Fioletov, V., Kerr, J. (1998) Satellite estimation of spectral surface UV irradiance in the presence of tropospheric aerosols. 1. Cloud-free case. *J. Geophys. Res.*, **103**, ND8, 8,779–8,794.

Krotkov, N. A., Krueger, A. J., Bhartia, P. K. (1997) Ultraviolet optical model of volcanic clouds for remote sensing of ash and sulfur dioxide. *J. Geophys. Res.*, **102**, ND18, 21,891–21,904.

Krupa, S. V., Kickert, R. N. (1989) The greenhouse effect: Impacts of UV-B, CO_2 and O_3 on vegetation. *Environ. Poll.*, **61**, 263–293.

Krupa, S. V., Kickert, R. N. (1993) The greenhouse effect: The impact of carbon dioxide (CO_2), Ultraviolet-B (UV-B) radiation and ozone on vegetation (crops). *Vegetatio*, **104/105**, 223–238.

Krupa, S. V., Kickert, R. N., Jäger, H. J. (1998) *Elevated Ultraviolet (UV)-B Radiation and Agriculture*, Springer, Berlin, Vol. XII, 296 pp.

Krzyscin, J. (1995) Variations of the total ozone trend over the northern hemisphere mid-latitudes in winter–spring seasons 1963–1964 through 1992–1993. *J. Geophys. Res.*, **100**, N(D10), 20,927–20, 936.

Krzyscin, J. W., Puchalski, S. (1998) Aerosol impact on the surface UV radiation from the ground-based measurements taken at Belsk, Poland, 1980–1996. *J. Geophys. Res.*, **103**, ND13, 16,175–16,181.

Kylling, A., Bais, A. F., Blumthaler, M., Schreder, J., Zerefos, C. S., Kosmidis, E. (1998) Effect of aerosols on solar UV irradiances during the Photochemical Activity and Solar Ultraviolet Radiation Campaign. *J. Geophys. Res.*, **103**, ND20, 26,057–26,060.

Lampert, W. (1989) The adaptive significance of vertical migration of zooplankton. *Funct. Ecol.*, **3**, 21–27.

Land, P. E., Haigh, J. D. (1996) Atmospheric correction over case 2 waters with an iterative fitting algorithm. *Appl. Opt.*, **35**, N27, 5,443–5,451.

Larkum, A. W. D., Wood, W. C. (1993) The effect of UV-B radiation on photosynthesis and respiration of phytoplankton, benthic microalgae and seagrasses. *Photosynth. Res.*, **36**, 17–23.

Leckner, B. (1978) The spectral distribution of solar radiation at the earth's surface: Elements of a model. *Solar Energy*, **20**, 143–150.

Lee., Z. P., Carder, K. L., Marra, J., Steward, G. G., Perry, M. J. (1996a) Estimating primary production at depth from remote sensing. *Appl. Opt.*, **35**, N3, 463–474.

Lee., Z. P., Carder, K. L., Peakock, T. G., Davis, C. O., Mueller, J. (1996b) Method to derive ocean absorption coefficients from remote sensing reflectances. *Appl. Opt.*, **35**, N3, 453–462.

Leser, M. P., Cullen, J. J. (1994) Carbon uptake in a marine diatom during acute exposure to ultraviolet-B radiation: Relative importance of damage and repair. *J. Phycol.*, **30**, 183–192.

Lewis, M. L., Cullen, J. J., Platt, T. (1984) Relationships between vertical mixing and photoadaptation of phytoplankton: Similarity criteria. *Mar. Ecol. Progr. Ser.*, **15**, 141–149.

Lindell, M. J., Rai, H. (1994) Photochemical oxygen consumption in humic waters. *Ergeb. Limnol.*, **43**, 145–155.

Little, E. E., Fabacher, D. L. (1994) Combative sensitivity of rainbow trout and two threatened salmonids to UV-B radiation. *Arch. Hydrobiol.*, **35**, 34–43.

Lubin, D., Mitchell, B. G., Frederick, J. E., Alberts, A. D., Booth, C. R., Lucas, T. B., Neuschuler, D. A. (1992) A contribution toward understanding the biospherical significance of Antarctic ozone depletion. *J. Geophys. Res.*, **97**, 7,817–7,828.

Lubin, D., Gautier, C., Ricchiazzi, P., Whritner, R. M. (1994) A method for mapping Antarctic surface ultraviolet radiation using multispectral satellite imaging. In: *Ultraviolet Radiation in Antarctica: Measurements and Biological Effects*, Antarc. Res. Sci. Vol. 62, C. S. Weiler, P. A. Penhale (eds.), AGU, Washington, DC, pp. 53–81.

Lubin, D., Holm-Hansen, O. (1995) Atmospheric ozone and the biological impact of solar ultraviolet radiation. In: *Encyclopedia of Environmental Biology*, Academic Press, San Diego, CA, Vol. 1, pp. 147–167.

Lubin, D., Jensen, E. H. (1995) Effects of clouds and stratospheric ozone depletion on ultraviolet radiation trends. *Nature*, **377**, 710–713.

Lubin, D., Jensen, E. H. (1997) Satellite mapping of the solar ultraviolet radiation at the earth's surface. In: *Solar Ultraviolet Radiation: Modeling, Measurements and Effects*, NATO Adv. Study Inst. Ser. Vol. 52, C. S. Zerefos, A. F. Bais (eds.), Springer-Verlag, New York, pp. 95–118.

Lubin, D., Jensen, E. H., Gies, H. P. (1998) Global surface ultraviolet radiation climatology from TOMS and ERBE data. *J. Geophys. Res.*, **103**, ND20, 26,061–26,091.

Madronich, S. (1992) Implications of recent total ozone measurements for biologically active ultraviolet radiation reaching the earth's surface. *Geophys. Res. Lett.*, **19**, 37–40.

Madronich, S. (1993) UV radiation in the natural and perturbed atmosphere In: *Environmental Effects of Ultraviolet Radiation*, M. Tevini (ed.), Lewis Publ., Boca Raton, FL, pp. 1769.

Madronich, S. (1995) The radiation equation. *Nature*, **377**, 680.

Madronich, S., McKenzie, R. L., Caldwell, M. M., Bjoern, L. O. (1995a) Changes in ultraviolet radiation reaching the Earth's surface. *WMO Global Ozone Research and Monitoring Project Report No. 37*, WMO, Geneva, Switzerland, pp. 1–30.

Madronich, S., McKenzie, R., Caldwell, M. M., Bjorn, O. (1995b) Changes in ultraviolet radiation reaching the Earth's surface. *Ambio*, **24**, 143–152.

Madronich, S., Flocke, S. (1997) Theoretical estimation of biologically effective UV radiation at the Earth's surface. In: *Solar Ultraviolet Radiation*, NATO ASI Series Vol. 52, C. S. Zerefos, A. F. Bais (eds.), Springer-Verlag, Berlin, pp. 23–48.

Madronich, S. (1998) Scientific Steering Committee on UV monitoring: Activities and scientific issues. *SPARC Newsletter*, N10, 19–21.

Matthijsen, J., Suhre, K., Rosset, R., Eisele, F. L., Mauldin, R. L., III, Tanner, D. I. (1998) Photodissociation and UV radiative transfer in a cloudy atmosphere: Modeling and measurements. *J. Geophys. Res.*, **103**, ND13, 16,665–16,676.

Mayer, B., Seckmeyer, G., Kylling, A. (1997) Systematic long-term comparison of spectral UV measurements and UVSPEC modeling results. *J. Geophys. Res.*, **102**, ND7, 8,755–8,768.

McClain, C. R., Yen Eueng-nan (1994) CZCS bio-optical algorithm comparison. Case studies for the SeaWiFS calibration and validation. *P.I./NASA Tech. Memo. 104566*, NASA Publ., Greenbelt, MD, 52 pp.

McElroy, M. (1983) *Global Change: A Biogeochemical Perspective*, JPL Publ. 83-51, Jet Prop. Lab., Pasadena, CA, vi + 33 pp.

McKenzie, E. L., Matheus, W. A., Johnson, P. V. (1991) The relationship between erythemal UV and ozone, derived from spectral irradiance measurements. *Geophys. Res. Lett.*, **18**, 2,265–2,272.

McKenzie, R. L., Bodeker, G. E., Keep, D. J., Kotkamp, M., Evans, J. M. (1996) UV radiation in New Zealand: Measured North to South differences, and relationship to other latitudes. *Weather and Climate*, **16**, 17–26.

McKenzie, R. L., Paulin, K. J., Bodeker, G. E., Liloy, J. B., Sturman, A. P. (1998) Cloud cover measured by satellite and from the ground: Relationship to UV radiation at the surface. *Int. J. Remote Sensing*, **19**, N15, 2,969–2,985.

McKinley, A. F., Diffey, B. L. (1987) A reference action spectrum for ultraviolet induced erythema in human skin. *CIE J.*, **6**, 17–22.

Meerkoetter, R., Wissinger, B., Seckmeyer, G. (1997) Surface UV from ERS-2/GOME and NOAA/AVHRR data: A case study. *Geophys. Res. Lett.*, **24**, 1,939–1,942.

Michaels, P. J., Singer, S. F., Knappenberger, P. C. (1994) Analyzing ultraviolet-B radiation: Is there a trend? *Science*, **264**, 1,341–1,342.

Mims, F. M., III, Ladd, J. W., Blaha, R. A. (1995) Increased solar ultraviolet-B associated with record low ozone over Texas. *Geophys. Res. Lett.*, **22**, N3, 227–230.

Mims, F. M., Frederick, J. E. (1994) Cumulus clouds and UV-B. *Nature*, **371**, 291–293.

Min, Q., Harrison, L. C. (1998) Syntyhetic spectra for terrestrial ultraviolet from discrete measurements. *J. Geophys. Res.*, **103**, ND4, 17,033–17,039.

Morel, A. (1988) Optical modelling of the upper ocean in relation to its biogenous matter content (case-I waters). *J. Geophys. Res.*, **93**, 10,749–10,768.

Morel, A., Berthon, J.-F. (1989) Surface pigments, algal biomass profiles, and potential production of the euphotic layer: Relationships reinvestigated in view of remote sensing applications. *Limnol. Oceanogr.*, **34**, N8., 1,545–1,562.

Neale, P. J., Davis, R. F., Cullen, J. J. (1998) Interactive effects of ozone depletion and vertical mixing on photosynthesis of Antarctic phytoplankton. *Nature*, **392**, N6,676, 585–589.

Nekel, H., Labs, D. (1981) Improved data of solar irradiance from 0.33 to 1.25 μm. *Solar Physics*, **74**, 231–249.

Nezval, E. I. (1996) Statistical characteristics of UV radiation in Moscow due to the data for 1968–1992. *Meteorol. and Hydrol.*, N8, 61–71 (in Russian).

Noever, D. A., Matsos, H. C., Cronise, R. J. (1994) Ocean–atmosphere CO_2 exchange: An accessible lab simulation for considering biological effects. *Climate Change*, **27**, 299–320.

Nuñez, M., Chen, D. (1998) A comparison of cloudless sky erythemal ultraviolet radiation at two sites in Southwest Sweden. *Int. J. Climatolog.*, **18**, N8, 915–930.

Oldenburg, J. B., Gritz, D. C., McNonnell, P. J. (1990) Tropical ultraviolet light-absorbing chromophore protects against experimental photokeratitis. *Arch. Opthalmol.*, **108**, 1,142-1,144.

Organisciak, D. T., Winkler, B. S. (1993) Retinal light damage: Practical and theoretical considerations. In: *Progress in Retinal Research*, N. Osborne, G. Chader (eds.), Pergamon Press, Oxford.

Otten, C., Ferlemann, F., Platt, U., Wagner, T., Pfeilsticker, K. (1998) Ground-based DOAS UV/visible measurements at Kiruna (Sweden) during SESAME winters 1993/94 and 1994/95. *J. Atmos. Chem.*, **30**, N1, 141–162.

Penner, J. E., Atherton, C. S., Graedl, T. E. (1993) *Global Emissions and Models of Photochemically Active Compounds*, Lawrence Livermore National Laboratory Preprint UCLA-JC-112273, Los Angeles, CA, 28 pp.

Petrova, N. A. (1990) *Successions of Phytoplankton with the Anthropogenic Eutrophication of Large Lakes*. Nauka, Leningrad, 200 pp. (in Russian).

Piazena, H. (1996) The effect of altitude upon the solar UV-B and UV-A irradiance in the tropical Chilean Andes. *Solar Energy*, **57**, 139–140.

Platt, T. (1969) The concept of energy efficiency in primary production. *Limnol. Oceanogr.*, **14**, 653–659.

Platt, T., Gallegoes, C. L., Harrison (1980) Photoinhibition of photosynthesis in natural assemlages of marine plankton. *J. Marine Res.*, **38**, 687–701.

Platt, T. (1986) Primary production of the ocean water column as a function of surface light intensity: Algorithms for remote sensing. *Deep-Sea Res.*, **33**, 149–163.

Platt, T., Sathyendranath, S. (1988) Oceanic primary production: Estimation by remote sensing at local and regional scales. *Science*, **241**, 1,613–1,620.

Platt, T., Caverhill, C., Sathyendranath, S. (1991) Basin-scale estimates of oceanic primary production by remote sensing: The North Atlantic. *J. Geophys. Res.*, **96**, 15,147–15,159.

Pozdnyakov, D. V., Kondratyev, K. Ya. (1998) Remote sensing of natural waters in the visible band of the spectrum. 1. Formation of the water–surface radiance. *Earth. Obs. Rem . Sens.*, **15**, 1–26.

Pozdnyakov, D. V., Kondratyev, K. Ya., Bukata, R. P., Jerome, J. H. (1998) Numerical modelling of natural water colour: Implications for remote sensing and limnological studies. *Int. J. Remote Sensing*, **19**, N10, 1,913–1,932.

Prieur, L., Sathyendranath, S. (1981) An optical classification of coastal and oceanic waters based on the specific spectral absorption curves of phytoplankton pigments, dissolved organic matter and other particulate materials. *Limnol. Oceanogr.*, **26**, 671–685.

Prézelin, B. B., Boucher, N. P., Schofield, O. (1994) Evaluation of field studies of UVB radiation effects on Antarctic marine primary productivity. In: *Stratospheric Ozone Depletion/UV-B Radiation in the Biosphere*, H. Biggs, M. E. B. Joyner (eds.), NATO ASI Series Vol. 18, Springer-Verlag, Berlin, pp. 181–194.

Report on the International Workshop on Continental Shelf Fluxes of Carbon, Nitrogen and Phosphorus (1996) J. Hall, S. V. Smith, P. R. Boudreau (eds.), LOICZ/R&S/96-9, LOICZ Publ., Texel, 50 pp.

Rind, D., Balachandran, N. K. (1995) Modeling the effects of UV variability and the QBO on the troposphere–stratosphere system. Part II. The Troposphere. *J. Climate*, **8**, N8, 2,080–2,095.

Ringelberg, J., Keyser, A. L., Flick, J. G. (1995) The mortality effect of ultraviolet radiation in a translucent and in a red morph of *Acanthodiaptomus denticornis* (Crustacea Copepoda) and its possible ecological revealence. *Hydrobiologia*, **112**, 217–222.

Sabziparvar, A. A., Forster, P. M. de F., Shine, K. P. (1998) Changes in ultraviolet radiation due to stratospheric and tropospheric ozone changes since preindustrial times. *J. Geophys. Res.*, **103**, ND20, 26,107–26,113.

Sakshaug, E., Slagstad, D. (1991) Light and productivity of phytoplankton in polar marine ecosystems: A physiological view. *Polar Res.*, **10**, N1, 69–85.

Sakshaug, E., Slagstad, D. (1992) Sea ice and wind: Effects on primary productivity in the Barents Sea. *Atmosphere-Ocean*, **30**, N4, 579–591.

Sathyendranath, S., Platt, T. (1997) Analytical model of ocean colour. *Appl. Opt.*, **36**, N12, 2,620–2,629.

Schindler, D. W. (1990) Effects of climatic warming on lakes in the central boreal forests. *Science*, **250**, 967–970.

Schwander, H., Koepke, P., Rugabber, A. (1997) Uncertainties in modelled UV irradiances due to limited accuracy and availability of input data. *J. Geophys. Res.*, **102**, ND8, 9,419–9,429.

Scientific Assessment of Ozone Depletion: 1998. WMO Global Ozone Research and Monitoring Project Report No. 44 (1999) WMO, Geneva, Switzerland.

Scotto, J., Cotton, G., Urbach, F., Berger, D., Fears, T. (1988) Biologically effective ultraviolet radiation: Surface measurements in the United States, 1974 to 1985. *Science*, **239**, 762–764.

SeaWiFS (1987) *Report of the joint EOSTA/NASA SeaWiFS Working Group*, NASA/EOSC Publ., Washington, 91 pp.

Seckmeyer, G., Mayer, B., Bernhard, G., McKenzie, R. L., Johnston, P. V., Kotkamp, M., Bluth, C. R., Lucas, T., Mestechkina, T., Roy, C. R., Gies, H. P., Tomlinson, D. (1995) Geographical differences in the UV measured by intercompared spectroradiometers. *Geophys. Res. Lett.*, **22**, N14, 1,889–1,892.

Semenov, S. M., Kunina, I. M. (1993) An experience of modelling ozone impact on vegetation in orthogenesis. In: *Problems of Ecological Monitoring and Ecosystem Model.*, Gidrometeoizdat, Moscow, pp. 283–289 (in Russian).

Shick, J. M., Lesser, M. P., Stochaj, W. R. (1991) Ultraviolet radiation and photooxidative stress in zooxanthelate anthozoa: The sea anemone *Phyllodiscus semini* and the octrocoral *Clavularia* sp., *Symbio.*, **10**, 145–173.

Siebeck, O. (1994) Impact of UV-B radiation on zooplankton and fish in pelagic freshwater ecosystems. *Ergeb. Limnol.*, **43**, 101–114.

Siebeck, O., Bohm, U. (1994) Challenges for an appraisal of UV-B effects upon planktonic crustacean under natural radiation conditions with non-migrating and migrating cladoceran. *Ergeb. Limnol.*, **43**, 197–206.

Sinha, A., Toumi, R. (1997) Tropospheric ozone, lightning, and climate change. *J. Geophys. Res.*, **102**, ND9, 10,667–10,672.

Slaper, H. (1987) Skin cancer and UV exposure: Investigation on the estimation of risks. PhD thesis, University of Utrecht, The Netherlands.

Smith, R. C., Prezelin, B. B., Baker, K. S., Bidigare, R. R., Boucher, N. P., Coley, T., Karentz, D., MacIntyre, S., Matlick, H. A., Menzies, D., Ondrusek, M., Wan, Z., Waters, K. J. (1992) Ozone depletion: Ultraviolet radiation and phytoplankton biology in Antarctic waters. *Science*, **255**, N5,047, 952–959.

Smith, V. H. (1991) Competition among consumers. *Limnol. Oceanogr.*, **36**, 820–823.

Spänkuch, D., Schulz, E. (1995) Diagnosing and forecasting total column ozone by statistical relations. *J. Geophys. Res.*, **100**, 18,873–18,885.

Stanford, J. L., Ziemke, J. R. (1996) A practical method for predicting midlatitude total column ozone from operational temperature fields. *J. Geophys. Res.*, **101**, 28,769–28,773.

Talling, J. F. (1957) The phytoplankton population as a compound photosynthetic system. *New Phytol.*, **56**, 287–295.

Tzvetkov, A. V. (1987) Variations of intensity and action centres of the atmosphere in the Atlantic and Pacific Oceans. *Proc. Main Geophys. Obs.*, iss. 513, 48–54 (in Russian).

UNEP Scientific Assessment of Ozone Depletion; WMO Global Ozone Research and Monitoring Project Report No. 37 (1994) WMO, Geneva, Switzerland.

Urbach, F. (1989) The biological effects of increased ultraviolet radiation: An update. *Photochem. and Photobiology*, **50**, N4, 439–441.

Van Weele, M., de Arellano, J. V.-G., Kuik, F. (1995) Combined measurements of UV-A actinic flux, UV-A irradiance and global radiation in relation to dissociation rates. *Tellus*, **47B**, 353–364.

Varotsos, C. (1994) Solar ultraviolet radiation and total ozone, as derived from satellite and ground-based instrumentation. *Geophys.Res. Lett.*, **21**, 1,787–1,790.

Varotsos, C., Kondratyev, K. Ya. (1994) Solar ultraviolet radiation changes at the earth's surface due to ozone content variations in the troposphere and stratosphere. *Studying the Earth from Space*, N1, 3–10 (in Russian).

Varotsos, C., Kondratyev, K. Ya. (1995) The tropospheric pollution and ultraviolet solar radiation. *Optics Atmos. Ocean*, **8**, N4, 614–618.

Varotsos, C., Kondratyev, K. Ya., Katsikis, S. (1995) On the relationship between total ozone and solar ultraviolet radiation at St. Petersburg, Russia. *Geophys. Res. Lett.*, **22**, 3,481–3,484.

Varotsos, C. (1995) On the association between the column ozone and the spectral solar ultraviolet radiation. *Toxicol. Environ. Chem.*, **50**, 119–130.

Varotsos, C., Feretis, E., Kondratyev, K. Ya. (1999) Impact of total ozone variability on surface solar ultraviolet radiation change. *Toxicol. Environ. Chem*, **71**, N13–19, 1,999.

Varotsos, C., Feretis, E. (1997) Health effects on human eye resulting from the increased ambient solar ultraviolet radiation. *Toxicol. Environ. Chem.*, **61**, 43–68.

Varotsos, C., Kondratyev, K. Ya. (1998) Vertical profiles of ultraviolet solar radiation in the troposphere. *Doklady Russian Acad. Sci.*, **36**, N2, 254–256 (in Russian).

Vincent, W. F., Roy, S. (1993) Solar ultraviolet-B radiation and aquatic primary production: Damage, protection and recovery. *Environ. Rev.*, **1**, 1–12.

Visscher, P. T., Culbertson, C. W., Dremland, R. S. (1994) Degratation of trifluoroacetate in oxic and anoxic sediments. *Nature*, **369**, 729–731.

Vitasa, B. C., Taylor, H. R., Strickland, P. T., Rosenthal, F. S., West, S., Abbey, H., Ng, S. K., Muñoz, B., Emmett, E. A. (1990) Association of nonmelanoma skin cancer and actinic keratosis with cumulative solar ultraviolet exposure in Maryland watermen. *Cancer*, **65**, 2,811–2,817.

Vollenweider, R. A. (1966) Calculation models for estimating day rates phytoplankton photosynthetic primary production. *Primary Productivity in Aquatic Environments*, C. Goldman (ed.), University of California Press, Berkeley, CA, pp. 425–457.

Voytex, M. A. (1990) Addressing the biological effects of decreased ozone on the Antarctic environment. *Ambio*, **19**, N2, 52–61.

Weatherhead, E. C., Reinsel, G. C., Tiao, G. C., Frederick, J. E. (1997a) Trends in UV irradiance in the 1980's. Monthly Update. *Atmospheric Chemistry Program*, **8**, N1, 1–4.

Weatherhead, E. C., Tiao, G. C., Reinsel, G. C., Frederick, J. E., DeLuisi, J. J., Choi, D., Tam, W. (1997b) Analysis of long-term behaviour of ultraviolet radiation measured by Robertson-Berger meters at 14 sites in the United States. *J. Geophys. Res.*, **102**, ND7, 8,738–8,754.

Webb, A. R. (1992) Spectral measurements of solar Ultraviolet-B radiation in Southern England. *J. Appl. Meteorol.*, **31**, 212–216.

Webb, W. L., Newton, M., Starr, D. (1974) Carbon dioxide exchange of *Alnus rubra*: A mathematical model. *Oecologia*, **17**, 281–291.

Weihs, P., Webb, A. R. (1997) Accuracy of spectral UV model calculations. 2. Comparison of UV calculations with measurements. *J. Geophys. Res.*, **102**, 1,551–1,560.

Weiler, C. S., Penhale, P. A. (eds.) (1994) *Ultraviolet Radiation in Antarctica: Measurements and Biological Effects*, American Geophysics Union, Washington, DC, 315 pp.

Wetzel, R. G. (1990) Land–water interfaces: Metabolic and limnological regulators. *Theor. Angew. Limnol. Verh.*, **24**, 6–24.

WHO (1994) *Ultraviolet Radiation. An Authoritative Scientific Review of Environmental and Health Effects of UV, with Reference to Global Ozone Layer Depletion*, Environmental Health Criteria No. 160, WHO, Geneva, Switzerland.

Widmark, J. (1889) Über den einfluss des Lichtes auf die vorderen medien des auges. *Skand. Arch. Physiol.*, **1**, 264–330.

Williamson, C. (1995) What role does UV-B radiation play in freshwater ecosystems? *Limnol. Oceanogr.*, **40**, 386–392.

Williamson, C. E., Mislivits, M., Orr, P. (1994) The vulnerability of Daphnia to UV-B radiation in the northern United States. *Ergeb. Limnol.*, **43**, 207–216.

Williamson, C. E., Zagarese, H. E., Seva, J. (1994) The impact of short-term exposure to UV-B radiation on zooplankton communities in north temperate lakes. *J. Plankton Res.*, **16**, 205–218.

(1995) *WMO Global Ozone Research and Monitoring Project Report No. 37*, WMO, Geneva, Switzerland, 292 pp.

Yerlov, N. G. (1976) *Marine Optics*, Elsevier Oceanography Series 14, Elsevier, Amsterdam, 321 pp.

Yung, Y. L., Jiang, Y., Liao, H., Gerstell, M. F. (1997) Enhanced UV penetration due to ozone cross-section changes induced by CO_2 doubling. *Geophys. Res. Lett.*, **24**, N24, 3,229–3,232.

Zaneveld, J. R., Kitchen, J. C. (1993) Vertical structure of productivity and its vertical integration as derived from remotely sensed observations. *Limnol. Oceanogr.*, **38**, N7, 1,384–1393.

Zeng, J., McKenzie, R., Stamnes, K., Wineland, M., Rosen, J. (1994) Measured UV spectra compared with discrete ordinate method simulation. *J. Geophys. Res.*, **99**, ND11, 23,019–23,030.

Zepp, R. G., Faust, B. C., Hoigne, J. (1992) Hydroxyl radical formation in aqueous reactions (pH 3–8) of iron (II) with hydrogen peroxide: The photofention reactions. *Environ. Sci. Technol.*, **26**, 313–319.

Zerefos, C. S., Bais, A. F. (eds.) (1997) *Solar Ultraviolet Radiation. Modelling, Measurements and Effects*, NATO ASI Series Vol. 52, Springer-Verlag, Berlin, 324 pp.

Zerefos, C. S., Balis, D. S., Bais, A. F., Gillotay, D., Simon, P. C., Mayer, B., Seckmeyer, G. (1997) Variability of UV-B at four stations in Europe. *Geophys. Res. Lett.*, **24**, N11, 1,363–1,366.

Zerefos, C. (1997) Factors influencing the transmission of solar ultraviolet irradiance through the earth's atmosphere. In: *Solar Ultraviolet Radiation*, C. S. Zerefos, A. F. Bais (eds.), NATO ASI Series Vol. 52, Springer-Verlag, Berlin, pp. 133–141.

Ziemke, J. R., Herman, J. R., Stanford, J. L., Bhartia, P. K. (1998) Total ozone/UVB monitoring and forecasting: Impacts of clouds and the horizontal resolution of satellite retrievals. *J. Geophys. Res.*, **103**, ND4, 3,865–3,871.

Ziemke, J., Chandra, S., Herman, J., Varotsos, C. (2000) Erythemal weighted UV trends over northern latitudes derived from Nimbus-7 TOMS measurements. *J. Geophys. Res.* (in press).

5

Tropospheric ozone

5.1 INTRODUCTION

Stratospheric and tropospheric ozone trends observed during recent years are of great concern from the viewpoint of their impact on climate and human health as well as on ecosystems. The significance of this problem was recognized fairly long ago (Fishman and Crutzen, 1978; Fishman *et al.*, 1979)

Studies of tropospheric ozone variability are motivated by its significance as a greenhouse gas that influences climate formation as well as its serious impact on human health, under conditions of increasing surface ozone concentration because of smog photochemical reactions in the presence of growing atmospheric pollution (nitrogen oxides, hydrocarbon compounds, etc.).

As far as stratospheric ozone dynamics is concerned the main problem (as we have seen in earlier chapters) is anthropogenically induced depletion of total ozone content which has been most dramatically expressed by formation of spring total ozone content minimum in Antarctica (the so-called 'ozone hole') as well as miniholes in high latitudes of the Northern Hemisphere. Even in low and mid-latitudes the decreasing trend of total ozone content is of serious concern, because of the impact on human health (skin cancer, etc.). That is why the Montreal Protocol and further similar documents have been approved and the use of the ozone-destroying chlorofluorocarbons CFC-11 and CFC-12, as well as other ozone-destroying gases, prohibited (Chapter 3). There are also substantial climatic impacts of ozone variability in the stratosphere, which deserve serious attention (Chapter 6).

The complex nature of environmental ozone impact requires application of a systematic approach for studying ozone change and the relevant impact. The purpose of this chapter is to discuss the situation in general and to treat as a case study the experience of complex ozone studies (tropospheric ozone dynamics) in the Athens region.

As is well known (Hales, 1996; Høv, 1997; Isaksen, 1998; McKee, 1994; Prinn, 1994a, b), tropospheric ozone originates from two sources: intermittent engulfment of stratospheric ozone in the troposphere and *in situ* formation from chemical processes among tropospheric trace gases. An increasing number of ozone observations on the ground and in the free troposphere has resulted in substantial improvement of the tropospheric ozone database. Many recent studies in the field of simulation modelling have led to much better understanding of tropospheric ozone dynamics. Since relevant results have been discussed in a number of recent fundamental publications (Bojkov, 1999; Høv, 1997; Isaksen, 1998; Lefohn, 1992; McKee, 1994; Prinn, 1994a; *UNEP* ..., 1994; Wang and Isaksen, 1995; *Scientific* ..., 1999) only a few recent studies will be mentioned here.

5.2 AN EXPERIENCE OF STUDYING TROPOSPHERIC OZONE DYNAMICS IN ATHENS, GREECE

A fundamental effort of studying tropospheric ozone dynamics in the area of Greater Athens was undertaken during the summer of 1994 (20 August–20 September) within the programme of the Mediterranean Campaign of Photochemical Tracers–Transport and Chemical Evolution (MEDCAPHOT–TRACE) experiment (Duriex *et al.*, 1998; Suppan *et al.*, 1998; Ziomas, 1998; Ziomas *et al.*, 1998a, b). Sixteen European scientific groups participated in this experiment, which was coordinated by the Laboratory of Atmospheric Physics, Aristotle University of Thessaloniki, Greece. The principal aim of the experiment was development of further understanding of the emissions, distribution and transformation of ozone related trace gases in a typical Mediterranean coastal densely populated urban area. The project included both experimental and modelling work and addressed the following general research tasks (Angeletti, 1998):

(1) investigate the evolution of ozone related trace gases (over land and sea), modulated by the various synoptic and local meteorological conditions (sea/land breeze circulation, local winds, inversion layers, etc.) in densely populated urban Mediterranean regions;
(2) quantify the background levels of ozone in the Mediterranean urban (semi-urban) atmosphere;
(3) contribute to the improvement/evaluation of available mesoscale photochemical dispersion models and to design national strategies for abatement of severe air-pollution episodes in coastal urban/industrial Mediterranean areas (see also Asimakopoulos *et al.*, 1992; Christakos and Vyas, 1998; Guesten *et al.*, 1997; Kalabokas and Bartzis, 1998).

We start by discussing some earlier surface ozone observational data.

5.2.1 Surface ozone observations

There have been few long-term surface ozone observations. The unique historical ozone series, which is considered below, consists of measurements of surface ozone mixing ratios taken at the National Observatory in Athens between 1901 and 1940, using De James colorimetric papers. Paper strips were exposed to atmospheric air for varying periods and were shielded from sun and rain. Depending on the ozone mixing ratio, there was a corresponding colour change, which was then correlated to a chromatic scale from 0 to 21.

Measurements were influenced by relative humidity, an effect which was found to be significant in several studies (Varotsos and Cartalis, 1991). This was taken into account using a conversion chart which provides correlation between the De James scale and the surface ozone mixing ratio and involves relative humidity (Varotsos and Cartalis, 1991). Note that construction of the conversion chart was based on test runs for Schönbein paper at varying relative humidity levels (Kley *et al.*, 1988, Varotsos *et al.*, 1991).

Recently, Pavelin *et al.* (1999) discussed ozone measurements made using Schönbein's method during the late nineteenth and early twentieth centuries. These data were presented from sites in Europe, Asia, Africa, Australia and South America. The values obtained lie in the range 5–15 ppbv for all sites.

The 1987–1990 ozone record consists of daily averages of measurements taken at 30-min intervals at several monitoring stations of the National Service for Air Pollution Monitoring, located at the perimeter of the National Observatory of Athens.

Daily mean ozone mixing ratios were analysed to yield:

- the annual cycle of surface ozone mixing ratio from 1901 to 1940 and from 1987 to 1990.
- the amplitude, phase and % contribution to variance of the first three harmonics for both periods.

Note that harmonic analysis between 1901 and 1940 was applied only for corrected ozone mixing ratios.

The annual cycle of corrected mean surface ozone between 1901 and 1940 shows that maximum and minimum corrected surface ozone values were observed from April to June and from October to December, respectively. Uncorrected surface ozone for these time periods demonstrated that there exist large differences, which result from ignoring the effect of relative humidity and exposure time on measurements. Note here that, for several subperiods (1901–1903, 1907–1908, 1934–1935 and 1938–1940), the 4-month cycle was stronger in magnitude and had a larger % contribution to total variance than the 6-month cycle. Between 1987 and 1990 the 6-month cycle was comparable with the 4-month cycle.

In the following, annual cycles of daily surface ozone mixing ratios at Montsouris (1877–1907), Hohenpeissenberg (1971–1988) and Arosa (1954–1958) are compared with the annual cycle of daily surface ozone mixing ratios in Athens for the periods 1901–1940 and 1987–1990. One interesting feature is the relatively small differences in ozone levels between the 1901–1940 and 1987–1990 Athens records. This is because daily (24 h) averaged ozone values are used for the interpretation. If average morning ozone values are used instead, differences increase and the 1987–1990 ozone values are almost double the 1901–1940 values. The large differences are directly attributed to the increase in concentrations of precursors of ozone, namely

NO_x and hydrocarbons (Varotsos and Kondratyev, 1994). Another interesting feature is the relatively small differences in ozone levels between the 1901–1940 and 1987–1997 Athens night-time records (Varotsos *et al.*, 1999).

5.2.1.1 Surface ozone variation because of changing meteorological conditions

Many efforts have been made to interpret daily and seasonal variations in the amount of surface ozone, because of the influence of various meteorological phenomena. In India, for example, surface ozone concentration variations have been correlated with rainfall in a non-polluted area and the conclusion was made that daytime rainfall is accompanied by decrease in surface ozone concentration, while nocturnal rainfall is accompanied by increase in surface ozone concentration.

Galbally (1968) conducted efforts to explain the correlation between surface ozone concentration and variations in surface air temperature during the day and attributed this correlation to thermal turbulence and mixing in the lower troposphere. Later Tiwari and Sreedharan (1973), Shreffler and Evans (1982) and Evans *et al.* (1983) proved that a direct relationship exists between daily variations in air temperature and surface ozone concentration. We believe, however, that since temperature variations are usually very well correlated with solar-radiation changes the latter produce variations in photochemical activity, which lead to ozone variability.

Guesten *et al.* (1988) observed variations in surface ozone concentration during vertical motions of airmasses. Short-term fluctuations, which often appear in surface ozone concentration daily variations have also been studied. Their investigation in relation to various meteorological phenomena is particularly difficult, since the simultaneous interaction with more than one meteorological parameter often leads to inconsistent conclusions. For instance, the association between surface ozone concentration and the occurrence of rain has led to formulation of contradictory conclusions among researchers (Sanhuenza *et al.*, 1985; Muralidharan *et al.*, 1989).

We considered hourly surface ozone concentration measurements obtained by the existing network of groundbased stations in the Greater Athens area during 1987 (Varotsos and Kondratyev, 1995). These measurements were performed at four points within the Greater Athens area. Hourly values of rainfall, air–surface temperature and wind speed were obtained by the National Observatory of Athens, which is located on the top of a small hill, 107 m a.s.l., in the centre of the urban area.

From the set of rainy days in 1987 we selected cases with total rainfall greater than 5 mm. For these cases we studied variations in hourly surface ozone concentration values in combination with hourly rainfall amounts, surface air temperature and wind-speed variations. We chose finally three rainfall events. Two of them occurred on 13 October 1987 and one on 11 March 1987. Two of these events were associated with the passing of a cold front and they were diurnal events, while the third was a nocturnal case without a cold-front passage.

During all rainfall events, decrease in surface ozone concentration was observed. In the third case (11 March 1987) rain occurred during the night and the early morning hours. Simultaneously, a significant decrease in air temperature was observed. As a result of this overnight cooling, temperature inversion formed which lasted throughout the day. The very low air temperatures were due to the existence of a low pressure system with an associated cold front in the Aegean Sea. The observed frontal rainfall occurred just before the passing of the cold front.

What we observed during this rainfall event was a significant decrease in surface ozone concentration at all stations. The station in the centre of the city (P) showed unusually high ozone concentrations at the beginning of the event, while surface ozone concentration minima appeared after the passage of the cold front and not during the rain event.

In the first and second case (13 October 1987), rain occurred during the morning hours 06:00–08:00 LT and heavy rain at 16:00 LT. Decrease in air temperature during the morning rain cannot be considered as important: substantial decrease in surface ozone concentration was observed only at one station. This decrease, however, cannot easily be attributed to the rain, because it coincided with maximum values of the normally observed daily course of surface ozone concentration in Athens (Guesten *et al.*, 1988). Moreover, during the maximum intensity of the rain at 08:00 LT a small increase in surface ozone concentration was observed. On the contrary, during the heavy evening rain (17.3 mm per hour) which was accompanied by very significant decrease in air temperature (4.3°C per hour) and which was due to the passing of a cold front over Athens, we observed a substantial decrease in the surface ozone concentration.

Increased ozone values during midday hours which are observed before the passing of a cold front are due mainly to the diurnal photochemical activity cycle, which is also enhanced by the southern airflow coming from the Saronicos Gulf, which is considered to be a zone of ozone formation as a result of sea-breeze circulation. Downward ozone transport from the upper troposphere before passage of a cold front could also contribute to this increase.

According to Muralidharan *et al.* (1989), surface ozone concentration during rain events is influenced by two physical mechanisms: vertical mixing and advection of ozone from higher altitudes accompanied by its partial destruction at the ground. The first mechanism reveals itself during daytime rain, as a result of air temperature decreases near the ground causing attenuation of vertical mixing and therefore, a decrease in ozone concentration. The influence of the mixing mechanism usually exceeds that of rain.

It seems as though variations in ozone concentration during a rain event vary significantly, depending on the type of rain. A usual rain event in the Athens region is associated with the passage of a cold front. It is known that before the passage of a cold front downward motions of airmasses prevail. These airmasses are usually richer in ozone at higher altitudes because in general over urban areas we observe increase in ozone concentration with altitude. This may be partially due to ozone destruction at the ground especially during the night but the main reason are NO emissions because of human activities which rapidly destroy ozone.

In both frontal rain events, we observed a strong ozone decrease. It should be mentioned, however, that unusually high values were observed at the Patission

Station (in the city centre), which were about three times higher than the average value for that period of the year and for that hour. In this connection, it is especially interesting to note the heavy evening rain of short duration on 13 October 1987, when very little ozone increase was observed as could be expected from the decrease in photochemical activity. This is also in agreement with the literature, which reports very low solubility of ozone in water.

The rain event which occurred during the morning hours of 13 October 1987 was not associated with a cold front. It cannot give us significant information because ozone concentrations for all stations were already very low. Nevertheless, a slight ozone increase for all stations when the maximum rain rate was observed can be seen. In this case, the change in horizontal wind direction may be responsible for the ozone increase because, as mentioned earlier, southern winds in this area usually transport air rich in ozone.

The phenomenon of strong surface ozone concentration which decreases during the passage of a cold front has been very frequently observed in Central Europe. The suggested explanation is that gas-phase reactions of ozone with nitrogen oxides, in combination with the fast heterogeneous ozone removal under cloudy conditions, characteristic of airmass travelling behind a cold front, may lead to a sharp decrease in surface ozone concentration. This decrease is enhanced during the night. Under such conditions the ozone concentration may be reduced to one-quarter of its initial value. These observations were made in a rural area of Central Europe (Hohenpeissenberg), which is often influenced by polluted air coming from Munich. It is remarkable though to observe a similar strong surface ozone concentration decrease in the middle of a dense urban area in the Athens region, which can be even more pronounced (reduction to one-seventh of surface ozone concentration initial value). If we examine absolute surface ozone concentration values in Athens they can be even lower than in Central Europe, because fresh nitrogen oxide emissions destroy the remaining ozone in the cold front zone of the airmass.

Note, however, that cloudy conditions decelerate photochemical ozone formation, which makes it impossible to observe any increasing ozone signal within 2–3 hours after passage of the cold front.

As far as the overnight rains impact is concerned, surface ozone concentration in a polluted area does not increase as it does in a non-polluted area, although such an increase was mentioned in recent literature. This may, possibly, be due to the fact that the mechanism of vertical ozone advection by rain during the night is of smaller order of magnitude than other mechanisms or, alternatively, to the fact that the vertical gradient of ozone concentration in the area is not important. The second alternative appears to be supported by measurements obtained by Guesten *et al.* (1988).

5.2.1.2 Surface ozone and its precursors

Photochemical pollution is a very complex process involving meteorological topographic emissions and chemical parameters. It is characterized by atmospheric formation of photochemical oxidants (mainly ozone), which are produced by irradiation of mixtures of hydrocarbons and nitrogen oxides (precursors). The most important chemical mechanisms involved in these atmospheric processes have already been identified and studied. But many unknown parameters still exist because of the large number of participating chemical reactions with a great variety of atmospheric hydrocarbons (Logan *et al.*, 1981, Logan, 1985; Finlayson-Pitts and Pitts, 1986).

For investigation of this complex atmospheric phenomenon, various mathematical models called Air Quality Simulation Models have been developed. These models fall into two main categories: prognostic models, which are based on physicochemical principles determining photochemical pollution and diagnostic models, which are statistical descriptions of observed air-pollution data.

Significant progress has recently been made in development of prognostic Air Quality Simulation Models, which represent the most fundamental approach relating concentrations of photochemical oxidants to those of their precursors (Kuntasal and Chang, 1987). Nevertheless, many uncertainties remain regarding such models because of the difficulties encountered in representing real atmospheric processes and because of the extensive field data requirement for their verification.

Substantiation of statistical models from data already available, which is the second modelling approach (diagnostic models), contains one important weakness. Since photochemical oxidant formation involves complex spatial and temporal characteristics of emissions, meteorological variables and photochemical reactions, it is difficult to choose realistic databases for statistical substantiation of the O_3-HC-NO_x relation (Lonneman *et al.*, 1986). Meteorological variables which were usually used for development of statistical models were inversion-base height, wind speed and direction and temperature at the 850-hPa level. It is difficult to deal, however, with inversion-base height and wind speed and direction because their fluctuations appear complex and uncertain. But temperature at 850 hPa, which is considered to be a good indicator of airmass stability, has been shown to correlate well with ozone levels in California. For this reason it is used as a meteorological index to characterize ozone-forming potential (Chock *et al.*, 1982; Davidson *et al.*, 1985).

It is of great importance to investigate whether similar statistical models could be applied in Athens, one site in the Mediterranean region where very frequent photochemical pollution episodes occur. Therefore, five-station composite monthly averages of daily concentrations in the Greater Athens area have been used for regression analysis, in order to examine the broad features of basin-wide O_3-HC-NO_x relations based on monthly variation in ozone and concentrations of its precursors.

Daily maximum hour measurements of surface O_3 and NO_x concentration between 1986 and 1990, at five stations of the operating network in the Greater Athens Area (GAA), have been considered. These measurements were performed by using conventional methods of UV absorption for O_3 and chemiluminescence for NO_x, while background detection levels of the instruments were 0.01 ppm and 0.04 ppm for O_3 and NO_x, respectively. Use has also been made of

Table 5.1. Regression results for O_3-NO_x-NMHC relations. Correlation coefficients r are significant at the 5% level.

	a (ppm)	b (ppm/°C)	c	d	r
(1) $[O_3] = a + b\,T_{850}$	−0.25	0.62			0.82
(2) $[O_3] = a + b\,T_{850} + c\,[NO_x] + d\,[NMHC]$	−5.87	0.68	0.14	0.012	0.90

radiosonde-derived temperature data at the 850-hPa pressure level, these data being held in archive by the Hellenic National Meteorological Service from 1986 to 1990.

Kuntasal and Chang (1987) found that linear correlation exists between surface ozone concentration and temperature at the pressure level of 850 hPa. Correlation coefficients are shown in Table 5.1. Kuntasal and Chang also proposed a linear relation between surface concentrations of O_3, NO_x, non-methane hydrocarbons (NMHC) and T_{850}. Coefficients of this relation are also shown in Table 5.1. Annual variation in surface ozone concentration derived by averaging mean monthly values at five stations at various locations within the GAA and the expected variation in surface ozone concentration applying equation (5.1), proposed by Kuntasal and Chang, indicates that between April and November observed surface ozone concentration is lower than could be predicted on the basis of the Kuntasal and Chang model, whereas observed values exceed those predicted for the winter months. The discrepancy may be attributed to different initial levels of chemical precursors NO_x and HC as well as to different NO_x/HC ratios because of the different profile of human activities in the two urban areas. Seasonal variation in solar radiation may be another factor because solar ultraviolet radiation exerts a direct influence on surface ozone concentration. We should also keep in mind that the Kuntasal and Chang model is derived from measurements in an urban region (Los Angeles), which experiences much higher levels of air pollution than Athens (ozone concentrations are three times as high). Topography and meteorological conditions in both areas may also be other factors causing differences.

Annual variation in $[O_3]$ and [NMHC], which equally represents annual variation in surface ozone concentration as a function of $[NO_x]$ and of T_{850}, was also calculated taking into account mean monthly values of T_{850} and corresponding mean monthly values of $[NO_x]$. Comparison of these data with those given by equation (5.1) indicates that the latter is systematically lower. This relation shows that the presence of NMHC exerts a positive influence on ozone concentration in an urban region, which is in agreement with theoretical expectations (Finlayson-Pitts and Pitts, 1986).

In order to define the relation between T_{850} and surface ozone concentration for the GAA, linear correlations have been studied between:

(i) Mean monthly values of T_{850} and corresponding mean monthly values of daily maximum hour surface ozone concentration. (These measurements were averaged over the five stations for the period 1986–1990.)

Table 4.2. Values of constants a, b in the equation: $[O_3] = a + b\,T_{850}$ and correlation coefficients between O_3 and T_{850} for various stations.

Station	Year	a (ppm)	b (ppm/°C)	r
Athens Basin	1987	0.11	0.20	0.92
Athens Basin	1988	1.62	0.18	0.94
Nea Smirni	1987	0.99	0.26	0.81
Nea Smirni	1988	1.07	0.25	0.81
Liossion	1987	2.56	0.11	0.50

(ii) Mean monthly values of T_{850} and corresponding mean monthly values of (daily maximum hour) surface ozone concentration at the Nea Smirni Station and at the Liossion Station.

The Nea Smirni and Liossion Stations were selected for further analysis because in these stations the highest and lowest values of surface ozone concentration have been observed.

If we examine Table 5.2, we observe that the strongest correlation appears to be for the set of five stations and for 1988, whilst the weakest is that for the Liossion Station.

By applying values for T_{850}, a and b in equation (5.1) are obtained from the five stations for 1987; the expected value of surface ozone concentration can now be calculated.

Finally, mean monthly values of NO_x concentrations for the five stations were used in order to permit linear correlation of T_{850}, surface ozone concentration and NO_x concentration:

$$[O_3] = a + b\,T_{850} + c\,[NO_x] \tag{5.1}$$

where $a = -1.72$ ppm, $b = 0.2$ ppm/°C and $c = 0.6$. The result is that constant b is comparable with that in the equation $[O_3] = a + b\,T_{850}$ while constant a is lower than the average corresponding value in this equation.

In conclusion, we can say that significant linear correlation exists between surface ozone concentration in the Athens basin and temperature at the 850-hPa pressure level. A linear relation, whose coefficients are estimated by processing local data, permits improved forecasting of surface ozone concentrations in the GAA in comparison with coefficients chosen for the Kuntasal and Chang (1987) model. This is probably due to differences in precursor concentrations and meteorological conditions between the two areas. Detailed prognostic models should be developed for the Athens Basin in order to be able to assess the reasons for this discrepancy.

5.2.2 Tropospheric ozone concentration at Athens, Greece (see also Section 2.2)

5.2.2.1 Vertical ozone distribution in the troposphere

Long-term vertical soundings of ozone in Central Europe have shown an increase in tropospheric ozone over the last 25 years (Brühl and Crutzen, 1989). This effect could be attributed to either tropospheric photochemical production and transport, or to variations in the troposphere–stratosphere exchange pattern (Reid and Vaughan, 1991; Varotsos *et al.*, 1992).

In order to investigate the origin of the free tropospheric ozone, systematic long-term measurements should be carried out in the widest possible latitudinal and longitudinal ranges. These measurements should be taken at the surface and in the free troposphere.

The ozone sensors that have been used for vertical ozone soundings were of the ECC type. The Electrochemical Concentration Cell (ECC) ozonesonde is an ozone measuring device designed to be flown with and on the same balloon as a standard National Weather Service meteorological radiosonde. The sensing unit for the ozonesonde is an iodine/iodide redox concentration cell composed of two platinum electrodes, immersed in neutral buffered iodine solutions of different concentrations in the anode and cathode chambers. The sensor operates as a galvanic cell that causes a current to flow in an external circuit while it drives to equilibrium. In operation, a continuous stream of ambient air is bubbled through the cathode cell of the sensor by a constant volume pump. Ozone oxidizes iodide in the cathode to iodine unbalancing the equilibrium of the cell.

As the cell returns to equilibrium, the resulting current in the external circuit gives a measure of the amount of ozone in the air sample. The ozonesonde is connected to the meteorological radiosonde (RS-80, Väisälä). Data were taken during the ascent of the balloon (Totex, 1,200 gr). Ascent velocity was about $5\,m\,s^{-1}$. Soundings were performed at about 10:00 UTC. Observational results showed that the average vertical profile of tropospheric ozone up to about 7 km was around 50 ppt. We also observed no vertical gradient of ozone-mixing ratios in the troposphere, which is in accordance with other observations in the Northern Hemisphere.

If we compare mean values with average winter values, which were measured at Jülich, Germany (50.6°N, 6.2°E) 2 years before (Volz-Thomas *et al.*, 1991) we see that between 1 and 4 km our values are about 20% higher than those obtained at Jülich, while between 4 and 8 km the difference is less than 10% and is considered insignificant, since it is below the relative standard error (10–15%). The above remarks should be handled with caution because our dataset does not cover the winter season completely and because there are also year-to-year variation in seasonal average of ozone-mixing ratios.

We also tried to examine whether, from our limited data, we could observe any sign of pollution transport in the free troposphere coming from Western Europe to our site. In order to minimize the influence of local pollution, we examined transport at the 700-hPa level. Advection from the north-western sector was dominant, as was expected for our region covering 60% of the cases. The ozone mean value is then

(51 ± 4) ppb. Advection from the south-south-western sector was also relatively frequent covering 25% of the cases. The average ozone value is (47 ± 2) ppb.

Difference becomes much more significant if we examine the same behaviour within the same month in order to eliminate month-to-month variation. March serves as a good example because for three out of the eight soundings we performed, the wind at the 700-hPa level came from the south-south-eastern sector, while on the other days the wind was blowing from the north-western sector. The average value in the first case is (46 ± 2) ppb. The decrease observed in ozone-mixing ratio at 3 km in the monthly curve of March could be attributed to this influence.

5.2.2.2 *Cross-tropopause ozone exchange during a cut-off low event and the polar vortex*

Many studies show that there is an increasing trend in background values of ozone in the troposphere (Bojkov, 1986). This creates serious concern because of the damage caused by high ozone concentrations to human beings, animals and plants and also because of possible climatic effects. Stratospheric ozone leaks into the troposphere contributing to a natural background. In addition, man-made and natural emissions of nitrogen oxides and hydrocarbons lead to production of ozone in the sunlit troposphere.

In order to assess the anthropogenic component of free tropospheric ozone, the nature of stratosphere–troposphere exchange must be well understood. It was believed, until recently, that the main transfer of stratospheric ozone occurs in tropopause folding events associated with rapid cyclogenesis.

Major tropopause folding events occur at the flanks of large-scale northern flow. A threefold tropopause structure model has been proposed, according to which our sounding station in Athens is situated at such a latitudinal band (36–38°N) that stratospheric air intrusions should occur by means of tropopause foldings between the polar and the mid-latitude troposphere. This fact makes the Athens region subject to the influence of the polar front and sometimes to the impact of the subtropical jetstream. Influence of the meteorological situation has been revealed by observations as the laminated structure of the vertical profile of ozone over Athens.

Observations made near a cut-off low in the upper troposphere suggest that a cut-off low may also contribute to removal of ozone from the stratosphere. But only a limited number of studies have so far been conducted and there still remain many uncertainties about this type of stratosphere-troposphere exchange.

In the following, we examine the stratosphere–troposphere ozone exchange over Athens, Greece, during the winter 1991–1992 EASOE campaign, in relation to the presence of a cut-off low in the upper troposphere or a possible tropopause folding, when the station was close to the flank of the upper-level trough. We should mention that this was the first time such an intensive sounding campaign was performed in Athens. We examined four ozone soundings, which correspond to the above meteorological situations. It is also important to emphasize that the nearest ozone-sounding station to Athens (Hohenpeissenberg, Germany: 48°N, 11°E) is too far away to obtain information about the situation over our region, which seems quite interesting from the atmospheric dynamics point of view, as mentioned above.

5.2.2.2.1 Cross-tropopause exchange during a cut-off low event

A cut-off low system developed over Athens on 18 and 20 March. On the first day (18 March) the lowest height for the pressure level of 200 hPa during our campaign was observed. This resulted in a very low tropopause (approaching the 400-hPa level).

We also observed very high ozone values with a maximum of 190 nb and a strong secondary ozone maximum (about 120 nb) in the lower stratosphere (about 5 km deep). The synoptic map shows that our site is influenced by northern circulation, which brings ozone-rich air to our region. From another point of view, we may say that a cut-off low event is combined with high ozone partial pressure at low levels. Because stratospheric air is then at higher pressure than normal, ozone partial pressure will be similarly increased. So the remarkable fact is the minimum in partial pressure near 15 km, which our ozone data clearly shows (G. Vaughan, personal communication). The ozone profile on that day does not show any characteristics of detectable cross-tropopause exchange.

On 20 March (the day of the next ozone sounding), we observed that the cut-off low system was moving eastwards and that our sounding site was influenced by polar air. The intensity of the low had weakened and consequently the height of the tropopause had increased.

On 18 March, the already observed laminated ozone structure in the region of 100–300 hPa was much more pronounced. This is in agreement with the laminar structure in ozone profile over Athens, which is strongly associated with establishment of north-western circulation over the region. If we examine the tropospheric temperature profile of both days, we find that tropospheric temperature on 20 March is higher by 3–5°C. This agrees fairly well with findings of similar studies of cut-off low systems. This might be associated with cooling of the lower stratosphere and lifting of the tropopause, which was in fact observed on that day. The most important feature observed on 20 March was the significant tropopause peak in ozone (about 10 nb) at 600 mb associated with sharp decrease in relative humidity, suggesting that this air might originate in the stratosphere. It was also seen on the same day that ozone partial pressure increases sharply before arriving at the tropopause, which demonstrates another example of stratospheric ozone intrusion into the tropopause.

5.2.2.2.2 Troposphere–stratosphere exchange at the flank of the polar vortex

On 27 March and 8 April our sounding site was situated on the eastern edge of the upper-level trough. The strong laminae observed on those days was also an indication that we were close to the vortex boundaries. As mentioned above, in this type of situation tropopause folding occurs which creates the potential for troposphere–stratosphere exchange. In fact, we observed tropospheric ozone peaks of 5–10 nb in the region of 400–600-hPa level. These peaks are less pronounced than the broad peak observed on 20 March under a cut-off low regime. Also, these peaks are not associated with sharp decrease in relative humidity. In addition, on 8 April we observed a sharp tropospheric ozone increase before the beginning of cross-tropopause exchange. A slight positive temperature gradient right above the tropopause, as observed on the same day, was another indication of ongoing troposphere–stratosphere exchange (Dobson, 1973).

Further analysis of the synoptic situation and the history of relevant air parcels is essential. For instance, plots of potential vorticity and cross-sections of potential temperature should strongly indicate the exchange that occurred.

5.3 GLOBAL TROPOSPHERIC OZONE-OBSERVED DYNAMICS

Although *in situ* tropospheric ozone measurements are rather fragmentary as are observations of tropospheric ozone precursors, there has been substantial progress in obtaining relevant information because of the development of satellite remote sensing (Fishman *et al.*, 1986, 1990, 1996, 1997, 1988; Hudson *et al.*, 1993; Hudson and Thomson, 1998; Kondratyev, 1998; *UNEP* ..., 1994; *Scientific* ..., 1999). Apart from that, very important progress has been achieved in development and application of *in situ* measurements and surface remote sounding (Chapter 2). A persuasive illustration of such progress has been development of the Network for Detection of Stratospheric Change (NDSC). In this context, McDermid *et al.* (1998a, b) discussed the results of intercomparison of ozoneprofiling instruments: two differential absorption lidars (DIAL), a microwave radiometer, ozonesondes and the SAGE-II satellite instrument. These results indicate sufficiently good agreement between various kinds of measurements (within 15% for single profiles and within 10% for the campaign average, at altitudes between 20 and 40 km; outside this region the differences were greater). It has now become possible to analyse global-scale TrOZ dynamics as well as specific features of ozone concentration vertical profile. We shall first discuss, however, some fragments of existing information obtained by application of both *in situ* and remote-sensing techniques.

Bojkov and Reinsel (1984) analysed ozonesonde data (11 stations) and Umkehr profiles for trend detection in the troposphere. The weighted average trend estimate for the Umkehr layer is $+(0.65 \pm 0.69)\%$ per year between 1970 and 1981. Because of ozonesonde data over the Northern Hemisphere the trend in the free troposphere (850–400 hPa) was $+(1.12 \pm 0.54)\%$ per year over 15 years. Bojkov (1988) pointed out that ozone in the upper part of the troposphere (500–300 hPa) usually shows a lower rate of increase than ozone in the lower part (850–500 hPa).

Naja and Lal (1996) conducted surface ozone concentration observations at Ahmedabad (23°N) during 1954–1955 and 1991–1993 periods. Results indicate that higher ozone concentrations, including larger amplitudes of diurnal and seasonal variations, were observed in 1991–1993 than measurements made during 1954–1955. Observed seasonal maximum during 1954–1955 was in spring months, whereas during 1991–1993 it was in winter months. Average ozone concentration increased from 14.7 ppbv during 1954–1955 to 25.3 ppbv during 1991–1993, resulting in a linear increase of 1.45%/yr. Brace and Peterson (1998) studied spatial patterns of tropospheric ozone in the Mount Rainier region of the Cascade Mountains.

Vukovich (1998) analysed surface ozone concentration (SOC) changes over the US from the Mid-West towards the eastern part of the country. SOC increase was observed towards the east with daily maximum SOC values (DMOC) changing (on average) from <40 ppb to >60 ppb. Temporal DMOC variability between 1981 and 1990 was presented in the form:

$$O_3(t) = \bar{O}_3(t) + A(t') + I(t'') + S(t)$$

where $\bar{O}_3(t)$ is long-term DMOC value; $A(t')$ is long-term average interannual DMOC variability; $I(t'')$ is interannual variability; $S(t)$ is interday (synoptic) DMOC change. The results of observations indicate the presence of substantial spatial variability in all $O_3(t)$ components. There is correlation of interannual DMOC variability with emissions of pollutants, while interannual variations correlate with climate changes.

Theodorsen *et al.* (1996) considered surface ozone changes at high latitudes because of atmospheric pollution on the basis of observational data for Tromsø (70°N) and Longyerbyen Svålbard (78°N) from November 1992 and throughout 1993, using ozonometers with high temporal resolution. It has been shown that at high latitudes in Norway, car exhaust and air pollution essentially destroy surface ozone.

An important source of information on tropospheric ozone are lidar soundings. Applying an airborne differential absorption lidar (DIAL) system, Browell *et al.* (1988, 1985) measured ozone and aerosol distributions during July–August 1985 over the tropical rainforest of Brazil as part of the NASA Global Tropospheric Experiment to study the Amazon boundary layer (*Global* ..., 1984). As pointed out by Crutzen *et al.* (1985), three specific processes are particularly important in the region considered: (1) photochemical production of O_3 associated with biomass burning; (2) O_3 destruction in the boundary layer over the rainforest by reactions with isoprene (C_5H_8) and other hydrocarbons, and (3) dry deposition on the forestry canopy.

Results obtained by Browell *et al.* (1988), using aircraft and ozonesonde data, indicate the presence of large-scale variation in vertical and horizontal distributions of O_3 and aerosols in the Amazon region, with O_3 mixing ratio exceeding 50 ppbv in some regions. Both negative and positive correlations were observed$_3$3 and aerosols were negatively correlated, this represented clean mid-tropospheric air, and when they were positively correlated, the airmass had undergone photochemical O_3 production as a result of biomass during.

Reichardt *et al.* (1996) discussed Raman lidar measurements of ozone, water vapour and cirrus cloud optical properties over northern Germany (Geesthacht: 53.4°N, 10.4°E) between October 1995 and January 1996. Pronounced minima in ozone vertical profiles, with values close to zero, were found several times in the presence of ice-cloud layers, in contrast to smooth ozone profiles with mixing ratio values of 50 to 100 ppbv in a cloud-free and dry upper troposphere. The cause of such minima may be heterogeneous chemical reactions on ice particles and changed conditions for homogeneous chemical processes in the presence of cirrus clouds. Further studies of cause and effect interrelationships are necessary, however.

Durieux *et al.* (1998) discussed results of 3-D lidar measurements (DIAL system at wavelengths 272.2 and 291.65 nm) of tropospheric ozone conducted during the MEDCAPHOT–TRACE field experiment in the Greater Athens area (Ziomas *et al.*, 1998a, b). Lidar soundings were conducted, up to 1,500 m above groundlevel, to monitor vertical ozone profile evolution over four successful days on the west side of the Ymittos Mountain. During the last day, development of photochemical smog was observed. Fiorani *et al.* (Ziomas *et al.*, 1998a, b) used these data (in combination with wind observations) to calculate ozone fluxes. Ziomas *et al.* (1998a, b) applied MEDCAPHOT–TRACE observational data to verify the Urban Airshed model and showed that from the viewpoint of the ozone abatement strategy, the first priority was VOC emission controls rather than controlling NO_x. Reiter *et al.* (1987) developed an initiative highlighting the significant contribution of stratospheric ozone to tropospheric ozone budget. Implications of the correlation between tropospheric ozone and potential vorticity (PV) in the western Pacific for Asian pollution were discussed by Newell *et al.* (1997) using aircraft lidar sounding data. Langford *et al.* (1998) used a 5-year time series of lidar profiles from a site in Colorado to investigate sources and scales of variability in ozone in the free troposphere.

Hudson *et al.* (1995) developed a new algorithm to retrieve tropospheric column ozone from satellite TOMS data, with the purpose of using this algorithm to assess the contribution of biomass burning to tropospheric ozone budget. They demonstrated that, over the region bounded by 20°W and 60°E longitude and 20° and 0° latitude during the 1989 biomass-burning season, tropospheric ozone content approached 20% of total ozone.

Biomass burning (as we saw earlier and will discuss later as well) is a very significant contributor to tropospheric ozone change. Pillay *et al.* (*Atmospheric* ..., 1998) illustrated this conclusion by satellite observations off the east coast of South Africa (extent of impacts of unplanned fires in the sugar cane and forestry industries on the eastern seaboard of South Africa was assessed). A clear relationship between the occurrence of fire and enhancement in TOMS total ozone was detected.

The very important progress in studying tropospheric ozone was marked by the recent symposium on Atmospheric Ozone (*Atmospheric* ..., 1998). Bisling *et al.* (*Atmospheric* ..., 1998) applied lidar sounding data in combination with the $5F_6$ tracer-dispersion experiment 'Schauinsland '95' to investigate vertical ozone profile (VOP) and long-range transport of ozone in the planetary boundary layer (up to 1,200 m above groundlevel). Maximum O_3 values of about 270 μg/m^3 were found at noon in the mixed layer and correlate with the arrival of the trace gas close to the lidar site. Bonasoni *et al.* (*Atmospheric* ..., 1998) studied the impact of Saharan dust transport on surface ozone at Mt. Cimone, Italy. Monitoring of ozone has been conducted since 1991 using an ultraviolet Dasibi 1108 photometric analyser. Results of observations indicate that Saharan dust transport was accompanied by clear reduction in surface ozone concentration. Reduction in ozone concentration was probably due to gas absorption by dust particles and/or to the lower altitude origin area of the airmass.

Surface ozone measurements, using two Dasibi 1008 type instruments, were made by Brunke and Scheel (*Atmospheric* ..., 1998), at the coastal site Cape Point (34°S, 18°E) between 1983 and 1995. The results showed annual average mixing ratio of about 21 ppb in clean air without indications of a long-term trend. Seasonal variations were registered with average peak-to-peak amplitude of about 14 ppb between a maximum in July/August and a minimum in January. In polluted airmasses (with elevated CO concentration), photochemical O_3 production in summer and O_3 loss in winter was observed.

Elansky and Smirnova (*Atmospheric* ..., 1998) discussed surface observations of O_3, NO, NO_2 in the centre of Moscow between 1990 and 1993, using gas analysers Dasibi 1008-AH and AC-30M. The results indicate that, at all seasons, mean O_3 concentration is lower and NO_x concentration is significantly higher in urban air than in the incoming rural flow. Seasonal variations show that O_3 and NO_x are in a state close to photochemical equilibrium. As a rule, during night-time the air around Moscow is almost cleaned of pollution, and O_3 and NO_x concentrations at sunrise are close to those observed in neighbouring rural areas.

Long-term series of surface ozone observations (1980–1991) at the background station Preila, Lithuania (55°20′N, 21°13′E, 5 m a.s.l.) was discussed by Girgzdiene (*Atmospheric* ..., 1998). An electrochemical analyser was used for these observations. Analysis of observational data revealed the trend of increase in ozone concentration by 2.5% per year from 1980 to 1995. During this time period, not only increase in maximum hourly ozone concentration was observed but also more frequent (than before) ozone values, above the threshold, of 60, 80 and 100 $\mu g/m^3$. Considerable change has been noticed in the character of seasonal variations. Earlier results (Girdziene, 1991) indicated that highest ozone values were connected with efficient stratosphere–troposphere exchange in spring. Summer peak is because of anthropogenic pollution and photochemical processes.

Chemical airmass transformation during long-range transport of ozone and other reactive species became the focus of the PEM–WEST(B) campaign between February and March 1994; O_3 and CO observations from March 1994 to February 1996 were also studied. Kajii *et al.* (1998), as well as Pochnart *et al.* (*Atmospheric* ..., 1998) conducted simultaneous measurements of ozone, CO, SO_2, HCl and formic (HCOOH) and nitric acids (HNO_3) as well as aerosols at Oki, a remote island site in the Sea of Japan. The results indicate that the cleanest atmosphere was observed in the northerly airmass coming from the Bering Sea, whereas the highest concentrations of all species were observed in the airmass of westerly flow passing through a lower boundary layer over the Yellow Sea and South Korea. O_3 and CO reached highest concentrations in August. Strong diurnal variations were found in July and August. Koike *et al.* (*Atmospheric* ..., 1998) made similar aircraft observations of NO_y, O_3 and CO over the Western Pacific in February and March 1994. Enhancements in mixing ratios of NO_y and CO were observed in the lower and mid-troposphere. Positive correlation was found between these two species suggesting that high NO_y values were due to anthropogenic emissions over the continental surface.

Lam *et al.* (*Atmospheric* ..., 1998) carried out O_3 and CO surface *in situ* observations at the coastal site in Hong Kong in 1994 and 1995. It was found that continental winter airmasses carried much higher O_3 ($\sim$40 ppbv) and very high CO ($\sim$400 ppbv). Generally, O_3 and CO are positively correlated in summer and negatively correlated in winter.

A number of attempts have been directed towards quantitative assessments of various mechanisms of tropospheric ozone changes. Combrink *et al.* (*Atmospheric* ..., 1998) studied stratospheric–tropospheric exchange (STE) of ozone using ozone-profile characteristics over a 4-day period, when a particularly deep westerly low-pressure system or cut off low (COL) passed over southern Africa. (The STE problem was discussed in detail in *Stratospheric* ... (1998).) For this purpose, ozonesonde data were collected at Irene (Pretoria: 25°53′S, 28°13′E, 1,523 m a.s.l.) over the 4-day period 27 to 30 May 1994, during the South African Atmospheric Research Initiative (SAFARI). Meteorological parameters and potential vorticity were used as STE indicators. Combrink *et al.* pointed out that lowering of the radiosonde and PV tropopause at Irene and intrusion of high PV and high stratospheric air to tropospheric altitudes, coupled with downward flow near the tropopause, provides dynamic evidence of coupling of tropospheric and stratospheric air in the upper troposphere. PV data do not provide conclusive evidence that tropopause folding took place in the case study considered. In the absence of chemical tracer measurements, it is recognized that advection may play a role. Turbulent mixing between upper and lower tropospheric air is facilitated by up and downdraughts within the COL system. Vertical mixing and strong downward motion, in the vicinity of the upper tropospheric disturbance, promoted downward transport of ozone-rich stratospheric air producing a fairly well-mixed upper tropospheric layer on 25 May. As the COL system weakened and moved away from the country, the tropopause was restored at the higher altitude.

Diab *et al.* (*Atmospheric* ..., 1998) outlined the build-up of ozone, other trace gases and aerosols because of anticyclonic recirculation over the African Continent. For this purpose, they compared mean seasonal (summer and spring) ozone profiles for Irene and Reunion, using ozonesonde data. Such comparison underscored the role of activities, such as biomass burning on the African Continent and Madagascar, on observed ozone vertical profiles above Reunion and highlighted the role of long-distance transport from Africa to the Indian Ocean islands.

Eisele and Trickl (*Atmospheric* ..., 1998) analysed results of a long- term series of DIAL soundings in Garmisch-Partenkirchen with an emphasis on stratosphere–troposphere exchange. Various episodes of stratospheric air transport into the troposphere were discussed (one of the episodes lasted for 4 days). Hintsa *et al.* (1998) used NASA ER-2 aircraft data to assess troposphere-to-stratosphere transport in the lowermost stratosphere by measurements of H_2O, CO_2, N_2O and O_3.

Cooper and Moody (*Atmospheric* ..., 1998) analysed effects of atmospheric dynamics on O_3 mixing ratio using hourly O_3 measurements from the Big

Meadows mountain-top site in Virginia, USA (an ultraviolet photometric analyser was used with detection limit of 2 ppb) between 16 April and 4 May 1993. Three consecutive cold-front passages, each associated with a deep trough, resulted in similar O_3 fluctuations. Six back trajectories per day were calculated using a 3-D kinematical model. Airmasses delivering peak O_3 values to the site passed through regions potentially influenced by tropopause folding. O_3 peaks were probably caused by an enhanced pool of O_3 residing above the boundary layer in the anticyclones following the cold fronts. Vertical mixing brought enhanced O_3 to the surface during daytime, while O_3 destruction reduced boundary-layer mixing ratios at night. Cooper and Moody emphasized, in conclusion, the significance of a stratospheric source for enhancing background O_3 values above the boundary layer. The O_3 events considered are not unusual and potentially occur under similar atmospheric conditions.

Ebel *et al.* (*Atmospheric ...*, 1998) discussed vertical airmass fluxes in the troposphere and lowest stratosphere and their role for regional ozone budgets at mid-latitudes, using a 3-D regional chemical transport model, with emphasis on possible impact of downward flux of free troposphere ozone into lower layers on oxidation capacity of these layers. Ebel *et al.* showed that ozone generation through gas-phase chemistry significantly depends on such downward flux.

To accomplish an empirical analysis of extra-tropical cross-tropopause transport between the stratosphere and the troposphere via quasi-isentropic processes in the 'middle-world' (a notion introduced by Hoskins (1991) for that part of the atmosphere in which isentropic surfaces intersect the tropopause and where stratospheric and tropospheric air can mix horizontally), Wang *et al.* (1998a, b) studied seasonal variations in ozone latitudinal distribution in the isentropic layer between 300 and 380 K (approximately corresponding to the middle-world), based on measurements from the Stratospheric Aerosol and Gas Experiment (SAGE) II.

Wang *et al.* (1998b) pointed out that in the tropics stratosphere–troposphere exchange (STE) involves airmass transport across isentropic surfaces mostly via strong moist convections, while at high latitudes STE is primarily induced by radiative cooling, particularly during winter, leading to large-scale airmass descending circulation. Ascending tropical air motion, together with high-latitude descending circulation, constitute the so-called Brewer–Dobson (diabatic) circulation.

The results obtained by Wang *et al.* (1998a, b) are consistent with: (1) build-up of ozone-rich air in the extra-tropical middle-world through large-scale descending mass circulation during winter; (2) spread of ozone-rich air in the isentropic layer from mid-latitudes to subtropics via quasi-isentropic transport during spring; (3) significant photochemical ozone removal and absence of ozone-rich supply of air to the latter in summer, and (4) airmass exchange between the subtropics and the extra-tropics in summer. Thus, SAGE-II observational data are consistent with the conceptual role of the eddy quasi-adiabatic transport in the stratosphere–troposphere exchange reported earlier in the literature (see also Van Velthoven and Kolder, 1996).

Vaughan and Timmis (1998) pointed out that, during the last week of January 1952, ozonesonde lidar observations over Europe revealed a layer of extremely low ozone (<100 ppbv) and aerosol concentrations in the stratospheric layer 360–380 K (potential temperature). The layer corresponded well to low PV in ECMWF analysis fields. Back-trajectory calculations at 370 K confirmed the subtropical origin for the layer and suggested that it was formed in a stream of low-PV air drawn from the troposphere over North Africa, as the subtropical jetstream turned northwards on 25 January.

Timmis and Vaughan (*Atmospheric* ..., 1998) emphasized that their study demonstrated that troposphere–stratosphere exchange can occur in winter as well as in summer, and that a single event can be responsible for transporting a large mass ($\sim 5 \times 10^{15}$ kg) of air. An important point about the example considered is that it occurred when a blocking anticyclone was established in the upper troposphere and lower stratosphere – an event which was particularly pronounced in January 1952, but which occurs on a smaller scale in most winters. Further work is required to determine whether the transfer considered is associated with all blocking anticyclones or only the more prominent ones.

Vaughan *et al.* (*Atmospheric* ..., 1998) described ozone observations by lidars and ozonesondes which were conducted at OHP (44°N, 5.7°E) and Aberystwyth (52°N, 4°W) during the week 16–21 March 1994. They pointed out that during the nights of 15–20 March the lidar measured tropospheric air rich in ozone with tropospheric values of static stability, implying low PV. Ozonesonde observations confirmed these findings. ECMWF assimilations showed a layer of dry stratospheric air passing over the observing sites in good agreement with observations. This layer had a mass about 3.6×10^{15} kg – corresponding to 6×10^{33} ozone molecules. So, the principal conclusion is that the air observed by the ozonesonde was aged stratospheric air, which descended into the troposphere 4 days previously at the base of a tropopause fold over N. America. The air experienced diabatic cooling, which lowered the potential temperature and decreased the PV, but retained its stratospheric chemical signatures. Maximum ozone concentrations of ~80 ppbv were intermediate between lower stratospheric and background tropospheric values, showing that some mixing occurred in the extruded stratospheric air. Vaughan *et al.* emphasized that the long lifetime in the troposphere of a layer of unmixed stratospheric air has implications for tropospheric chemistry, and in particular for estimates of mix-down times for chemical species.

Kalabokas *et al.* (*Atmospheric* ..., 1998) discussed rural ozone data (Dasibi 1008), collected at the air pollution station of the NCSR at Demokritos located about 10 km east of the centre of Athens at 280-m altitude, to evaluate inter-annual variability during the 4 years of operation of the station as well as seasonal diurnal ozone variations. The results indicate clear seasonal variation in rural ozone mixing ratios with maximum average seasonal values in summer (60 to 65 ppb) and minimum values in winter (~35 ppb). Average diurnal variation is relatively small in comparison with average summer values through the day, which vary between 55 and 70 ppb.

Preliminary analysis of tropospheric ozone in the lower, mid and upper troposphere at five ozonesonde sites in Canada between 1980 and 1995 was presented by Tarasick *et al.* (*Atmospheric* ..., 1998), who discovered decreases in concentrations at all levels, below 50 hPa, including the surface, at all stations except Alert, which shows an increase in the surface layer. Therefore, the average of all five Canadian sites shows statistically significant (at 2σ level) negative trends at all levels, including the surface. Tropospheric and lower stratospheric long-terms trends are similar, and stratospheric variations on shorter timescales appear to be correlated with those in the troposphere. Tarasick *et al.* concluded that stratospheric ozone levels appear to be the dominant influence in tropospheric ozone amounts in remote northern regions, possibly through transport of photochemical forcing.

Tsutsumi *et al.* (*Atmospheric* ..., 1998) considered latitudinal distribution of tropospheric ozone in the Western Pacific Ocean (within the meridional band ~120–140°E from Australia to Japan) in different seasons using measurement data from three aircraft campaigns (Fokker F-27 research aircraft): PACE-1 (23–31 January 1994), PACE-2 (20–28 October 1994), PACE-3 (7–17 July 1994). Observational data indicate that the most distinctive feature of tropospheric ozone changes in the region were high ozone concentration at about 20°S and low ozone concentration around the equator. The principal causes of the high ozone concentration were biomass burning (PACE-2) and transport of stratospheric ozone into the troposphere (PACE-3). The low ozone airmass in the subtropics extended to 30°N during PACE-3, which has been ascribed to airmass arrival from the north.

Photochemistry in biomass-burning plumes and implications for tropospheric ozone were studied by Mauzerall *et al.* (1998) using aircraft observational data for the tropical South Atlantic, during the dry season in September 1992, and results of simulation modelling. Observational data for various trace components were used to calculate the ratio $\Delta Y/\Delta X$ (Δ is concentration excess above the background level). Minor component Y includes CO, O_3, N_2O, HNO_3, PAN, CH_2O, acetone, H_2O_2, CH_3OOH, HCOOH, CH_3COOH and aerosols. X is CO or CO_2 mixing ratio. The results indicate that O_3 formation in an old plume continues for at least 1 week at $\Delta O_3/\Delta CO = 0.7$. O_3 formation in a plume was limited by NO_x level and took place mainly within plume zones. Assessment of minimum regional level of O_3 formation has given the value of 17×10 molecules O_3/cm^2 s, which is high enough to explain observed 30% O_3 increase over the tropical South Atlantic during the dry season, as a result of biomass burning. O_3 formation during the dry season is approximately three times more intensive that during the wet season.

Poppe *et al.* (1998) studied the influence of atmospheric dilution in the case of ozone formation on biomass-burning plumes. It was shown that formation of photo-oxidants in biomass-burning plumes is an important source of photo-oxidants in the tropical atmosphere, as well as of organic compounds and nitrogen oxides. The contribution of photo-oxidants to tropospheric ozone budget is about 15%. The implications of biomass burning have recently attracted great attention, especially in connection with the IGAC's Biomass Burning Experiment (BIBEX) activity (Andreae, 1998; Andreae and Fishman, 1998; Delmas and Guenther, 1998; Helas *et al.*, 1998; Swap *et al.*, 1998). Christopher *et al.* (1998a, b) used satellite information to study the 1985 biomass-burning season in South America. Formaldehyde may serve as an indirect indicator of VOC (Lee *et al.*, 1998a, b).

Ziemke and Chandra (*Atmospheric* ..., 1998) investigated tropospheric ozone dynamics in the context of the well-known phenomenon of TOZ variability in the southern tropics, which is characterized by presence of year-round stationary wave-1 pattern with maximum peak-to-peak amplitudes ~20–30 DU during the southern spring (September–October) (Chapters 2, 3). The crest of this wave occurs in the South Atlantic region with larger amplitudes occurring in the months of intense biomass burning in South America and Africa (July–October). Ziemke and Chandra underlined that, since the wave exists in all months of the year (weakest in May–June with peak-to-peak amplitude 10–12 DU), this indicates the presence of a persistent dynamical forcing mechanism.

Ziemke and Chandra (1998b) studied the persistence of the wave-1 pattern, combining ozonesonde data, from several sites in the tropics over 14 years (1975–1992), from both Nimbus-7 TOMS total ozone data and National Centers for Environmental Prediction (NCEP) geo-potential height analysis. Analysis of the limited number of ozonesonde profiles in the tropics indicates that the primary contribution to both TrOZ and the TOZ ozone wave-1 lies in the low to mid-troposphere, maximizing around 4–5-km altitudes. Near this maxima, 500 hPa (~5-km altitude) geo-potential heights show a persistent (year-round) wave-1 pattern in the South Atlantic similar to TOMS total ozone, indicating both equatorward transport of subtropical ozone into the region and potential for planetary-scale subsidence effects. It was thus substantiated by Ziemke and Chandra, for the first time, that lower tropospheric ozone and dynamical transport have major roles in establishing persistence of the wave-1 anomaly in total ozone.

Gros *et al.* (1998) discussed the O_3 time series for a 3-year period (1994–1996) using the results of continuous surface O_3 measurements carried out at Amsterdam Island (37°S, 77°E) in the southern Indian Ocean (Thermo-Electron analysers were used). The purpose was to observe and understand seasonality and long-term trends in surface ozone concentration in the remote marine atmosphere. The results indicate the presence of seasonal variation with maxima around 30 ppbv during winter (July–September) and minima around 13 ppbv during summer (December–February). These O_3 values are close to those observed at Cape Grim (Tasmania: 41°S, 145°E) between 1991 and 1995.

Application of the global 3-D Climatological Model of the Global Universal Tracer Transport in the Atmosphere (MOGUNTIA) for simulation calculations showed that the model is adequate enough to reproduce observed O_3 mixing ratios and their seasonal cycle. The seasonal variation of stratospheric origin O_3 calculated by MOGUNTIA shows a maximum of 18 ppbv in September and a maximum of 13 ppbv in April. Oxidation of continentally emitted O_3 precursors during their transport to Amsterdam Island contributes to photochemical production of O_3 by up to 12 ppbv during the austral summer. Gros *et al.* (1998) found that, of this amount, 5–7 ppbv are directly linked non-methane hydrocarbon oxidation chemistry under relatively high NO_x conditions. In particular, biomass-burning emissions contribute up to 5.5 ppbv to observed O_3 levels in September. Both photochemical production by biomass burning and stratospheric influx of O_3 lead to a maximum in O_3 mixing ratios during late winter to early Spring in agreement with observations. Minimum O_3 mixing ratios observed during the austral summer are related to photochemical O_3 depletion. These conclusions agree with simulation modeling carried out by Roelofs *et al.* (1997), who showed that intrusion of stratospheric air might be a significant source of surface ozone. Contribution of O_3 from the stratosphere to the surface O_3 is maximum in winter and spring, when the photochemical lifetime of O_3 is relatively long. According to Roelofs *et al.*, contributions of stratospheric ozone to the tropospheric content of O_3 in the Southern Hemisphere are 35 and 50% during the summer and winter, respectively. The principal conclusions made by Gros *et al.* (1998) are that the annual cycle of surface O_3, observed at Amsterdam Islands from 1994 to 1996, is consistent (1) with predominant photochemical destruction of O_3 mainly of stratospheric origin during summer, (2) with long-range transported O_3 (because of biomass burning in Southern Africa and Madagascar) and enhanced O_3 levels of stratospheric ozone during winter.

Bodeker *et al.* (1998) discussed the results of the first analysis of trend and variability in vertical ozone and temperature profiles, measured by ozonesonde at Lauder, New Zealand (45.045°S, 165.684°E) between August 1986 and July 1996. The results indicate wintertime upper tropospheric ozone decrease of more than $-30 \pm 24\%$ per decade (2σ), post-vortex breaking trends in a narrow altitude region above the tropopause of $-20 \pm 20\%$ per decade (2σ) and positive trends of up to $30 \pm 14\%$ per decade (2σ) in the lower stratosphere during late winter, spring and early summer. A significant QBO component was found in ozone variability throughout the lower stratosphere during winter, spring and early summer, but little or no dependence on the solar cycle. On the contrary, however, temperature changes are characterized by little dependence on QBO and are more influenced by the solar cycle. Although the Mt. Pinatubo eruption had little influence on Lauder ozone, it resulted in significant cooling of the troposphere. The ENSO cycle in both ozone and temperature was weak except at uppermost analysis levels.

The important field experiment SOAPEX was conducted at Cape Grim in north-western Tasmania (41°N) under conditions of remote marine atmospheric boundary layer (Monks *et al.*, 1998). The results obtained allowed analysis of photochemical processes of ozone formation and destruction in 'clean' atmosphere, where three principal factors contribute to tropospheric ozone variability:

$$\frac{d[O_3]}{dt} = C + E_v([O_3]_{ft} - [O_3])/H + v_d[O_3]/H$$

where C is photochemical formation and destruction of ozone; E_v is involvement velocity; $[O_3]_{ft}$ is ozone concentration in the free troposphere; v_d is dry deposition velocity; H is height of the boundary layer. Because of estimates of daytime photochemical ozone destruction its value is about 2.5 ppb/day.

Correlations between O_3, N_2O and NO_y (the sum of total reactive nitrogen species) concentrations were considered by Michelsen *et al.* (1998), using ATMOS data, which covered a wide range of altitudes and latitudes, including the tropics. The results obtained are in good agreement with aircraft and balloonborne observations. Correlations between the mentioned components demonstrate rapid production of NO_y in tropical and O_3 in the tropical mid-stratosphere and episodic export of air from this region to higher latitudes.

Michelsen *et al.* (1998) studied isolation of air within developing polar vortices during the autumn. Arctic vortex data from April 1993 appear to indicate denitrification of 20–30%, which is evident as a 3.0–4.5 ppb deficit in NO_y concentration. Mixture of air from above 40 km with air from lower altitudes can fully account for this deficit, in addition to approximately one-half of apparent Arctic ozone loss, as inferred on vortex and extravortex $[O_3]:[N_2O]$ correlations. Similar comparison for the Antarctic indicates 60–80% deficit in $[NO_y]$ and 80–100% deficit in $[O_3]$; at least half of this apparent denitrification and ozone loss can be attributed to mixing of air from higher altitudes with air from lower altitudes.

In June 1990, Ridley *et al.* (1998a) conducted aircraft observations, within the Rural Oxidants in the Southern Environmental (ROSE 1) experiment, with the aim of investigating photochemical and dynamical processes involving anthropogenic and biogenic emissions that determine air quality, such as ozone abundance, in the South-eastern United States (over Mississippi and Alabama). Urban, biomass burn, electric power plant and paper-mill plumes are examples of sources encountered during these flights. In the case of the urban plume from Mobile, Alabama, the upper limit of ozone production efficiency was 6–7 ppbv of ozone per ppbv of NO_x oxidized. Emissions from several biomass fires had lower efficiencies of 2–4 ppbv/ppbv, but the average rate of production of ozone was as high as 58 ppbv/h for one fire, where the plume was isolated from mixing at the surface. Ridley *et al.* (1998a) pointed out that close encounters with paper mill and power-plant emissions showed chemical consumption of O_3, while far-field encounters with power-plant plumes showed that maximum mixing ratios of O_3 in similarly aged (6–7 hours) plumes, from a large power plant to the north of Mobile, at 85 ppbv were 10 ppbv larger than those in the Mobile urban plume. Ridley *et al.* (1998a) emphasized the need for considerably more detailed aircraft studies of chemical and vertical evolution of power plant, biomass burn, pulp mill and urban plumes to assess their relative contributions to regional O_3 production.

Analysis of long-term trends in groundlevel ozone over the contiguous United States (1980–1995), carried out by Fiore *et al.* (1998), led to the conclusion that no region of the United States experienced significant increase in O_3 concentrations during the 1980–1995 period. Decreasing trends were predominately clustered in the three largest metropolitan areas: New York, Los Angeles and Chicago.

Klemm *et al.* (1995) observed several ozone plumes (with enrichment of O_3 of 10–45 ppb above levels of surrounding airmasses) in the southern part of Eastern Germany during a photosmog period in summer 1993. Synchronously with O_3, other species were enriched (SO_2, NO_y, CO and photochemically formed CH_2O). O_3 production in the plumes probably resulted from emissions from surface-burning sulphur rich in coal under non-optimum conditions, such as old power plants and domestic water-heating systems. Co-emissions of VOCs with SO_2 led to formation of O_3 and CH_2O. Most of the plumes seemed to be less than 1 day old.

Observations within the North Atlantic Regional Experiment (NARE) have shown that chemical processes at the surface and in the free troposphere often develop independently. Therefore, surface observational data may not be representative for the free troposphere. In view of these circumstances, Berkowitz *et al.* (1998) studied formation mechanisms and chemical characteristics of elevated photochemical smog over north-east USA in 1995, using both surface meteorological observational data and aircraft soundings of atmospheric chemical composition in the eastern part of Connecticut and central part of Massachusetts. The principal purpose was to assess the impact of turbulent mixing in the atmosphere boundary layer (ABL) on chemical process kinetics under conditions of clear sky and light wind, during ABL transfer from nocturnal stable stratification to afternoon convection.

Trevino and Mendez (1999) studied the transient ozone-generation process in polluted urban areas using a simplified set of chemical reactions for NO_x and hydrocarbon photolysis. The subject of urban impact on surface and tropospheric ozone has recently attracted great attention. Starn *et al.* (1998) showed, under conditions of an urban-impacted forest site, the importance of night-time isoprene chemistry (isoprene is emitted to the atmosphere by vegetation).

An unprecedented cooperative study, the Southern Oxidants Study Nashville/Middle Tennessee Ozone Study (SOS), was carried out in the USA (Cowling *et al.*, 1998; Meagher *et al.*, 1998; Huebler *et al.*, 1998; Starn *et al.*, 1998; Lee *et al.*, 1998c; Novaime *et al.*, 1998; Roberts *et al.*, 1998c; Frost *et al.*, 1998a; Gillani *et al.*, 1998a, b; Sillman *et al.*, 1998; Kasibhatla *et al.*, 1998). The SOS was initiated in 10 states in 1998 to address many national and regional scientific and policy concerns. The programme has two general foci (Cowling *et al.*, 1998):

(1) using the southern United States as a natural laboratory for policy-relevant scientific investigations to improve scientific and public understanding of chemical and meteorological processes that cause ozone and other photochemical oxidants to accumulate in the atmosphere near the ground;
(2) evaluating alternative strategies that would allow federal, state, municipal, industrial and commercial organizations to better manage ozone and other photochemical oxidants in the lower atmosphere and decrease the injurious effects of these airborne chemicals in urban and rural areas.

Cowling *et al.* (1998) emphasized that, prior to the late 1980s, biogenic hydrocarbons were believed to play little or no role in ozone formation in either the rural or urban atmosphere. It has, however, been shown that biogenic hydrocarbons played a major role in rural ozone episodes in the eastern United States, as well as represented a significant VOC source in Atlanta. These findings were a major driving force behind the inception of SOS and, as a result, biogenic VOC were identified as a prominent research theme for the SOS programme. The results of the studies demonstrated the great value of the 3-D information provided by aircraft-based measurement platforms (Huebler *et al.*, 1998).

Some of the more significant findings of the 1994–1995 studies include the following (Meagher *et al.*, 1998):

(1) Ozone production in Nashville (Tennessee) was found to be close to the transition between NO_x-sensitive and VOC-sensitive chemistry.
(2) Ozone production efficiency (OPE) in power-plant plumes, molecules of ozone formed per molecule of NO_x emitted (from 1 to 7 molecules), was found to be inversely proportional to NO_x emissions exhibiting the lowest OPE.
(3) During stagnant conditions, night-time winds dominated pollution transport and represented the major mechanism for transporting urban pollutants to rural areas (Bante *et al.*, 1998).
(4) Biogenic emissions of soil NO_x in the south-east was small (4.1%).
(5) Isoprene chemistry dominated ozone production in the forested rural areas of the Nashville/middle Tennessee region (Roberts *et al.*, 1998c).
(6) Production and loss of peroxide compounds were examined in urban and power-plant plumes (Jobson *et al.*, 1998a, b).

The results obtained by Berkowitz *et al.* (1998) showed that strong daily variation in surface ozone concentration was connected with the presence of elevated layers of higher ozone concentration as well as concentrations of some other trace gases. Such layers were characterized by both strong vertical (within ABL) and horizontal (between urban and rural regions) NO_x gradients. This demonstrates the decisive role of local anthropogenic emissions in North America in ozone formation under urban conditions.

As discussed earlier in the book, tropospheric ozone is known to have increased in Central Europe since the end of the last century. Observational data indicated the increasing trend of 0.5–3.0% per year up to the early 1980s, with some temporal and spatial variations (Bojkov and Reinsel, 1984; Bojkov, 1986, 1988; Logan, 1985). However, this increasing trend has continued over the last decade, which may be ascribed to NO_x emission leveling off in Western European countries during this period (Lee *et al.* 1998c). In contrast to Europe, very little data have been available for long-term ozone measurement in East Asia.

Lee *et al.* (1998c) analysed 8-year-long ozonesonde data obtained between 1989 and 1997 at Naha (Okinawa Island, Japan), as well as results of backward trajectory calculations, in order to elucidate recent tropospheric ozone trends in North-east Asia. Earlier data of 22 year sets of ozone soundings at the stations of Kagoshima, Tsukuba, and Sapporo showed an average statistical increase of about 2% in the planetary boundary layer during the 1970s and 1980s. New ozone data from Naha have been considered with focus on the regionally polluted continental outflow (airmasses that reached Naha from northern, north-western and western directions during autumn/winter/early spring seasons (from October to March). The results indicate an ozone increase of about $2.5 \pm 0.6\%$ per year for the 0–2-km layer, which is in accord with earlier results and demonstrates the increase in tropospheric ozone in Northern Asia in the 1990s. Lee *et al.* (1998c) suggested that this tropospheric ozone trend could be related to increasing emissions of NO_x from North-east Asia (China, Japan, South Korea, Taiwan) over this period at a rate of 3.9% per year.

Carmichael *et al.* (1998) pointed out that most surface ozone measurements in the Northern Hemisphere show a spring maximum, but emphasized that the roles of photochemical production from anthropogenic emissions of precursors, and intrusion of stratospheric air with high concentrations of ozone, in producing those springtime maxima remain to be quantified. In this context, Carmichael *et al.* studied ozone transport and chemistry in spring in East Asia using the STEM-II (Sulphur Transport Eulerian Model) regional-scale transport/chemistry model. Relevant 3-D calculations were made between 1 and 15 May 1987, a period characterized by strong downward transport of ozone in East Asia, associated with travelling low-pressure systems. Elevated ozone levels were observed at high-altitude surface sites in Japan during this period.

In order to assess relative contributions of transport and chemical sources of tropospheric ozone, model simulations were performed by Carmichael *et al.* (1998) both with and without photochemical processes. Comparison of model results with observational data for Japan has shown that calculations accurately capture most of the important observed features. Simulation modelling indicates that near-surface ozone levels were strongly influenced both by continental outflow of precursors occurring behind the cold fronts as they moved out over the Pacific Ocean and by strong downward transport of ozone-rich air from the upper troposphere, which occurred in association with the weather systems.

To simulate the effects of meteorology on ozone in Houston (USA), Davis *et al.* (1998b) applied cluster analysis and generalized additive models. Seven statistically significant different meteorological regimes were identified which differ mainly by maximum values of daytime ozone concentration. The highest ozone concentration was determined by wind conditions.

Cárdenas *et al.* (1998) studied correlations between CO, NO_y, O_3 and NMHC and their relationships with meteorological conditions during winter 1993 on the North Norfolk Coast, UK. Special emphazis was placed on carbon monoxide as one of the important lower-troposphere-influencing processes of ozone formation and destruction. Correlation between CO and O_3 happened to be strong and negative (with background concentrations in the boundary layer equal to 122 ppb and 33 ppb, respectively). Simultaneous observations of O_3, NO, NO_2 concentrations, NO_2 photodissociation rate coefficient $j(NO_2)$ and air temperature were made during the winter 1993 field experiment WAOWE-93 at the observational station in North Norfolk, with the principal purpose of studying the photochemically stable state (PSS) of a moderately polluted atmosphere (the atmosphere was strongly polluted by NO_x, however). Allowing for observational errors, the ratio $\Phi = j(NO_2)[NO_2]/k_3[NO][O_3]$ (k_3 is a relevant reaction rate), which is a PSS indicator, was practically equal to unity (this means the absence of peroxide radicals). During WAOWE-94 (October 1994), $\Phi \sim 2$ because of the impact of peroxide radicals whose concentration could reach 60 ppt.

Bremaud *et al.* (1998) described the diurnal cycle of ozone in the marine atmospheric boundary layer, measured at Reunion Island (21°S, 55°E) in the western part of the Indian, between 22 August to 4 September 1995 (an electrochemical concentration cell was used as sensor). There was a sharp decrease in ozone concentration between sunrise and the late afternoon, with a variation of 3–4 ppbv. Observational data were compared with box chemistry model calculations. The principal conclusion concerning the nature of the diurnal cycle was that the most important mechanism was the exchange between the ozone-rich free troposphere and the ozone-poor marine ABL. This exchange is supposed to be more important during the night than during the day, allowing ozone night recovery. Bremaud *et al.* (1998) underlined that this is the key point of the observed diurnal cycle, since daytime ozone photochemistry is well described by the model.

Sun *et al.* (1998) investigated transport of carbon dioxide, water vapour and ozone by turbulence and local circulations using data collected by Canadian Twin Otter aircraft and a deck boat which traversed Candle Lake (Saskatchewan, Canada) during the Boreal Ecosystem–Atmosphere Study (BOREAS), under conditions of nocturnal land breezes and daytime lake breezes. It was shown that nocturnal vertical transport over the lake consists of two parts: (1) mesoscale rising motion associated with land-breeze convergence and (2) significant turbulence and vertical mixing driven by buoyancy in the lower part of the internal boundary layer and shear generation in the top part of the internal boundary layer.

A 5-year time series of lidar profiles from a site at Peak Observatory near Rollinsville, Colorado (39.9°N, 105.5°W, 2.68 km a.s.l.) were used by Langford *et al.* (1998) to investigate sources and scales of variability for ozone in the free troposphere. Analysis of 475 daily profiles derived from more than 25,000 individual profiles showed that about 66% of total variance is due to fluctuations on timescales from 1 day to 1 month, 30% on timescales from 1 month to 1 year and 4% on timescales greater than 1 year. It was also found that low-frequency variability, correlating with ENSO-related changes, can also lead to apparent ozone trends that range from −0.2 to +0.5 ppbv/yr for different 10-year intervals between 1970 and 1990. These results demonstrate the important impact of low-frequency fluctuations on trend analyses.

As mentioned earlier, an important source of tropospheric ozone (TrOZ) information are satellite data. At the beginning of the 1980s, the NASA Global Tropospheric Experiment began studying the chemical composition of the troposphere, as well as its interaction with the stratosphere, land and ocean, which determines global biogeochemical cycles (McNeal *et al.*, 1983). Special emphasis was placed, in this context, on the necessity to investigate global ozone dynamics and develop relevant satellite remote-sensing techniques.

Fishman *et al.* (1986) discussed three independent examples which suggest that photochemical ozone production in the troposphere can be observed in the tropics from analysis of TOZ data. According to this example, the seasonal TOZ cycle is dominated by the seasonal TrOZ cycle, even though tropospheric ozone accounts for only 5–15% of total ozone. The second example, a case study presented by Fishman *et al.* (1986), showed that enhanced TOZ observed over the Amazon Basin can be associated with the presence of biomass burning. As we saw earlier, the problem of biomass burning as a cause of global biogeochemical cycle change is acquiring more and more attention. Crutzen *et al.* (1979) were the first to put forward the hypothesis about the importance of biomass burning in photochemistry of the troposphere, which has been confirmed by further studies. *In situ* measurements have demonstrated that biomass burning results in the production of photochemically generated ozone, analogous to the formation of smog in industrialized regions.

The third example, discussed by Fishman *et al.* (1986), was discovery of strong correlation between carbon monoxide and concurrent TOZ distribution between 5°S and 10°N, which was revealed by analysis of Space Shuttle observational data. Since all sources of carbon monoxide are located in the troposphere, this means that TOZ gradients at low latitudes must also reflect processes occurring in the troposphere.

Later, Fishman *et al.* (1990, 1996) (see also Fishman, 1998) analysed tropospheric ozone distribution obtained from satellite data using various kinds of retrieval techniques. For example, analysis was made (Fishman *et al.*, 1990) of more than 2,000 ozone profiles from SAGE-I (1979–1981) and SAGE-II (1984–1987) between 50°S and 50°N in combination with 9 years (1979–1987) of Nimbus-7 TOMS total ozone data, which allowed investigation of the spatial distribution and seasonal cycle of the integrated amounts of ozone in the troposphere (TrOZ). Comparison with ozonesonde data has shown that TrOZ retrieval accuracy is probably better than 10% in the tropics and 15% at mid-latitudes. The use of satellite data has allowed production of the first global (50°N–50°S) TrOZ climatology.

The results obtained indicate that, in the tropics, highest concentrations are found in the eastern Atlantic Ocean downwind (west) of Africa and maximize at the time when biomass burning is most prevalent, between July and October. A different seasonal cycle is observed in the tropics over Indonesia, where a relative maximum is present in March–April, but in this case the maximum is also connected with the season of biomass burning. At mid-latitudes, highest ozone concentrations are found downwind of Asia and maximize in the summer. Relatively high amounts of tropospheric ozone are observed downwind of North America and Europe.

More recently, Fishman *et al.* (1996) combined Nimbus-7 TOMS TOZ data with vertical ozone profiles from the NOAA-11 SBUV/2 from TrOZ retrievals. Ziemke and Chandra (1998a) pointed out, however, that such an approach is not correct: significant errors arise in stratospheric column ozone in the tropics derived from SBUV because of the combination of vertical weighting functions and the inversion algorithm.

Using ozonesonde data, Hofmann *et al.* (1992) investigated tropospheric ozone variations in the Arctic in January 1990. They discovered episodes of near-complete ozone destruction beneath temperature inversions in the first few kilometres above the surface and emphasized that tropospheric ozone depletion, in association with the 'Arctic haze' phenomenon, may be a more prominent feature of the Arctic than previously appreciated. This conclusion has been confirmed by further studies (*UNEP...*, 1994).

Mauzerall *et al.* (1996) quantified the tropospheric ozone budget over remote high northern latitudes in summer using chemical and meteorological measurements between 0 and 6 km, made during the summer of 1990 by the Arctic Boundary Layer Expedition (ABLE 3b). It was found that, in the 0–6-km column, net photochemical ozone production is near zero because regional photochemical production and loss were approximately equal. Manzerall *et al.* (1996) discovered, however, that dispersed *in situ* photochemical production, driven by background NO levels (5–10 pptv), is the largest source of the ozone budget (62%). The influx of stratospheric ozone is of secondary importance (27%), long-range transport of pollution ozone makes a small contribution (9%) and photochemical production of ozone within biomass wildfire plumes is relatively negligible (2%) in the budget. It was found that *in situ* photochemistry, within the 0–6-km column, accounts for nearly 90% of the ozone mixing ratio within the boundary layer, while above 5 km it accounts for only about 40%.

Kim and Newchurch (1998) pointed out that TrOZ maps, derived from SAGE, SBUV and TOMS satellite data, show elevated tropospheric ozone from South America across the Atlantic to South Africa during the biomass-burning season (see also Fishman *et al.*, 1996; Fishman and Brackett, 1997). To further explore the problem, Kim and Newchurch used monthly averaged Nimbus-7 version-7 TOMS total ozone data on a 1° latitude by 1.25° longitude grid from January 1979 to April 1993. One technique of TrOZ retrieval was based on the use of TOZ values over two nearby regions with topographic contrast. Lower-tropospheric ozone amounts were derived over western South America and over the western Pacific Ocean near New Guinea, from the column difference between two nearby regions with topographic contrast, mountain and sea-level (New Guinea and The South America Andes provide places near the equator that best meet necessary requirements for reliable retrievals). Retrieved TrOZ seasonal variation agrees with relevant ozonesonde data.

The results obtained by Kim and Newchurch (1998) suggest that, in the presence of persistent easterly winds throughout the year over equatorial South America, favourable conditions exist for transporting biomass-burning products from burning regions to western South America. A difficult seasonal TrOZ variation was observed over New Guinea with maximum in July–September and minimum in January–February. Less lower-tropospheric ozone was observed over western New Guinea ($6.6 + 3.4/-2.0$ DU). Although there was no significant ozone trend, west of New Guinea (downwind of the biomass-burning regions), a statistically significant increase was found of 0.06 ± 0.04 DU/yr (at 95% confidence interval), corresponding to 1.0 ± 0.6% per year. This is comparatively close to the assessment obtained by Jiang and Yung (1996) for the latitude range 10–23°S in the eastern Pacific Ocean immediately west of South America: 0.21 ± 0.06 DU/yr or 1.48 ± 0.40% per year. Thus, the study considered found strong evidence linking biomass burning to enhanced production of tropospheric ozone in the western Pacific Ocean near New Guinea.

Ziemke *et al.* (1998) developed two new approaches for determining TrOZ from satellite data. The first was based on the combined rise in UARS HALOE and MSL ozone measurements. Tropospheric column ozone is obtained by subtracting stratospheric ozone amounts from TOZ values. In all cases considered total ozone included retrievals from Nimbus-7 (November 1978 to May 1993) and Earth Probe (July 1996 to December 1997) total ozone mapping spectrometers (TOMS). Data from HALOE are used in the first method to extend the vertical span of MLS (highest pressure level 46 hPa) using simple regression. Such assimilation allows high-resolution daily maps of stratospheric and tropospheric ozone to be drawn, which is not possible from solar occultation measurements alone, such as HALOE or SAGE.

The second new approach, developed by Ziemke *et al.* (1998), yields TrOZ values directly from TOMS high-density footprint measurements in regions of high convective clouds. This technique is defined as the convective cloud differential (CCD) technique since it is based on subtracting column ozone above high cloud tops from TOZ. It has been shown that the CCD method provides long-time series (essentially late 1978 to December 1997) of tropospheric ozone in regions dominated by persistent high tropopause-level clouds, such as maritime tropical Pacific and within or near mid-latitude continental landmasses.

The results covering tropical latitudes (the eastern Pacific tropics), between January 1979 and December 1997, reveal that dominant interannual variability in stratospheric ozone is driven by the quasi-biennual oscillation (QBO), whereas for tropospheric ozone it is driven by El Niño events. Comparison of the results with ozonesonde data from several southern tropical stations has demonstrated good qualitative agreement. Ziemke *et al.* (1998) pointed out that further analysis will include ozone data for extratropics.

Chandra *et al.* (1998) investigated effects of the 1997–1998 El Niño on tropospheric ozone and water vapour using TOMS data from Earth Probe and MLS data from UARS. This El Niño was considered as one of the strongest events of the century (even in comparison with the 1982–1983 El Niño). The major impact of the SST changes has been the shift in the convection pattern from the western to the eastern Pacific affecting the response of rain-producing cumulonimbus. There has been, therefore, a significant increase in rainfall over the eastern Pacific and a decrease over the western Pacific and Indonesia. The dryness in the Indonesia region, because of El Niño, has contributed to large-scale burning by uncontrolled wildfires in the tropical rainforest of Sumatra and Borneo. The meteorological situation has resulted in TrOZ decrease by 4–8 DU in the eastern Pacific and increase by about 10–20 DU in the western Pacific largely as a result of the eastward shift in tropical convective activity. Chandra *et al.* (1998) emphasized that TrOZ changes, during the 1997–1998 El Niño, appear to be caused by combination of large-scale circulation processes associated with the shift in the tropical convection pattern surface/boundary layer processes, associated with forest fires in the Indonesian region.

Muro *et al.* (1998) carried out direct measurements of tropospheric ozone distributions from space using Global Ozone Monitoring Experiment (GOME) data from the ERS-2 satellite in April 1995. Data on backscattered radiation for bands 1A (237–307 nm) and 2B (312–405 nm) were chosen. The retrieval technique was non-linear optimal estimation. Comparison with ozonesonde data shows good agreement.

A new TrOZ retrieval technique was developed by Spänkuch *et al.* (1998), who used groundbased Fourier-transform infrared emission measurements, applying the concept of differential optical emission spectroscopy analogous to the concept of differential optical absorption spectroscopy. Since the strength of the spectral signature in the chosen spectral interval ($\sim$1,043 cm^{-1}) depends not only on the amount of column ozone between the instrument site and a cloud base (the detection limit) is shown to depend on cloud-base height. An approach was suggested by Spänkuch *et al.* (1998) to assess multiple scattering of emitted radiation within clouds.

5.4 TROPOSPHERIC OZONE PRECURSORS

Schultz *et al.* (1998) emphasized that local tropospheric ozone balance is largely influenced by the UV radiation field and the concentration of H_2O and nitrogen oxides ($NO_x = NO + NO_2$).

As pointed out in *UNEP...* (1994), the production of ozone in a polluted troposphere is accomplished through a complex series of reactions referred to as the photochemical smog mechanism (see Section 5.5). This well-known mechanism (Figure 5.1) involves photo-oxidation of volatile organic compounds (VOC) and carbon monoxide (CO) in the presence of NO_x ($= NO + NO_2$). Typical of this mechanism are reactions (5.1) through (5.7):

$$RH + OH \longrightarrow R + H_2O \tag{5.1}$$

$$R + O_2 + M \longrightarrow RO_2 + M \tag{5.2}$$

$$RO_2 + NO \longrightarrow RO + NO_2 \tag{5.3}$$

$$RO + O_2 \longrightarrow HO_2 + R'CHO \tag{5.4}$$

$$HO_2 + NO \longrightarrow OH + NO_2 \tag{5.5}$$

$$NO_2 + h\nu \longrightarrow NO + O \tag{5.6}$$

$$O + O_2 + M \longrightarrow O_3 + M \tag{5.7}$$

Net reaction:

$$RH + 4O_2 + 2h\nu \longrightarrow R'CHO + H_2O + 2O_3$$

Here an initial reaction between hydrocarbon (RH) and hydroxyl radical (OH) results in the production of two O_3 molecules and an aldehyde $R'CHO$ or a ketone. Additional ozone molecules can then be produced from degradation of $R'CHO$.

Ozone can also be generated from CO oxidation via (5.8) and (5.9) followed by (5.5), (5.6) and (5.7):

$$CO + OH \longrightarrow CO_2 + M \tag{5.8}$$

$$H_2 + O_2 + M \longrightarrow HO_2 + M \tag{5.9}$$

In contrast to hydrocarbons and CO, which provide the fuel for production of tropospheric ozone and are consumed in the process, NO_x is conserved in the process of ozone production and this acts as a catalyst in ozone formation.

The central step is conversion of NO to NO_2 by peroxy radicals (HO_2 and RO_2), since rapid photolysis of NO_2 yields the oxygen atom required to produce ozone

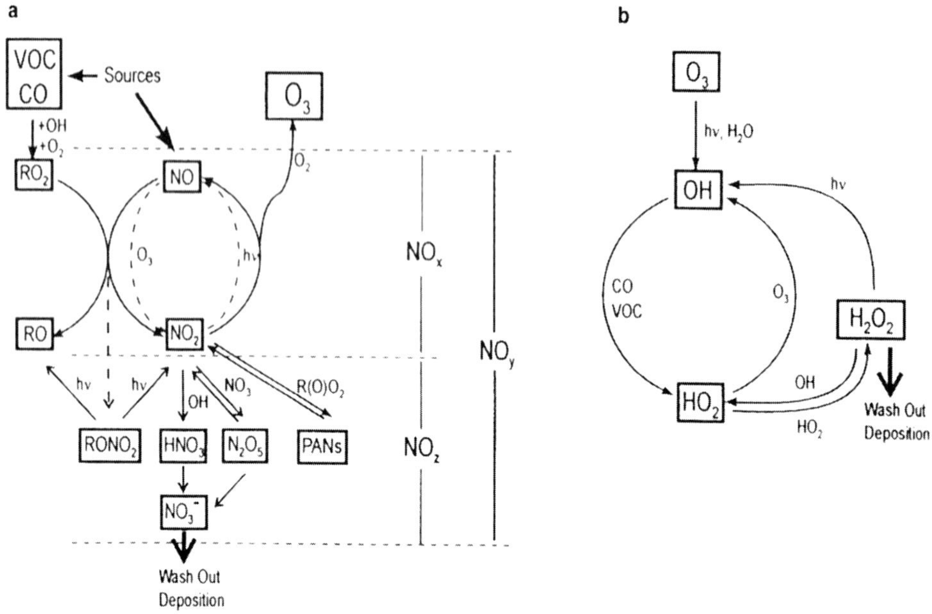

Figure 5.1. The photochemical smog mechanism.

(5.7). The conversion of NO to NO_2 occurs to a large extent through reaction with O_3 itself:

$$NO + O_3 \longrightarrow NO_2 + O_2 \tag{5.10}$$

Because of the rapid interconversion between NO and NO_2 during daylight, the quantity $NO_x = NO + NO_2$ is arrived at and, moreover, so is the quantity $O_x = O + NO_2$.

Photochemical loss of tropospheric ozone is accomplished through photolysis followed by reaction of the $O(^1D)$ atom with water vapour:

$$O_3 + h\nu \longrightarrow O(^1D) + O_2 \tag{5.11}$$

$$O(^1D) + R_2O \longrightarrow 2OH \tag{5.12}$$

$$HO_2 + O_3 \longrightarrow OH + 2O_2 \tag{5.13}$$

$$OH + O_3 \longrightarrow HO_2 + O_2 \tag{5.14}$$

The hydroxyl radical formed in (5.12) oxidizes CO, methane and other organic compounds (RH) and thereby produces peroxy radicals HO_2 and RO_2:

$$OH + CO + O_2 \longrightarrow CO_2 + HO_2 \tag{5.15}$$

$$OH + CH_4 + O_2 \longrightarrow CH_3O_2 + H_2O \tag{5.16}$$

$$OH + RH + O_2 \longrightarrow RO_2 + H_2O \tag{5.17}$$

In the absence of NO_x, self-reactions and cross reactions of HO_2 and RO_2 and the reaction of HO_2 with ozone are important, leading to further ozone loss:

$$HO_2 + HO_2 \longrightarrow H_2O_2 + O_2 \quad (5.18)$$

$$HO_2 + RO_2 \longrightarrow ROOH + O_2 \quad (5.19)$$

$$HO_2 + O_3 \longrightarrow OH + 2O_2 \quad (5.20)$$

The primary photochemical source of ozone in the troposphere is photolysis of NO_2 in the presence of molecular oxygen (reactions (5.6)–(5.7)). Most of NO reacts with O_3 to re-form as NO_2:

$$NO + O_3 \longrightarrow NO_2 + O_2 \quad (5.21)$$

Schulz *et al.* (1998) underlined that the ozone-production rate is approximately proportional to NO concentration, whereas the loss rate is almost independent of NO. Thus, local net chemical balance of ozone depends critically on the concentration of NO_x, H_2O and the UV radiation flux.

Apart from photochemistry, the budget of ozone in a given region is governed by transport of ozone into and out of the region.

The simplified scheme of tropospheric ozone formation, considered above, gives an idea of what priorities need to be taken into account when considering tropospheric ozone precursors. The problem of precursors is very complex because of a great number of relevant substances (one example of the complexity are VOCs). The consideration below is limited by just a few examples (see also Cox and Jones, 1998; Frost *et al.*, 1998, 1999).

Luecken *et al.* (1999) established that the atmospheric behaviour of relative oxidized nitrogen (NO_y) must be understood to predict the production of ozone and other oxidants as well as fine particulates. The components of NO_y that participate in the complex cycle of atmospheric chemistry include nitrogen oxide (NO), nitrogen dioxide (NO_2), nitrous acid (OHNO), peroxynitric acid (HNO_4), nitrate radical (NO_3), nitrate aerosol (NO_3^-), dinitrogen pentoxide (N_2O_5), nitric acid (HNO_3), peroxyacetyl nitrate (PAN) as well as other PAN analogues and organic nitrates (ORNI).

Luecken *et al.* (1999) pointed out that individual NO_y species can be classified as either short or long-lived (lifetimes <1 h) NO_y species that rapidly interconvert (Dentener and Crutzen, 1993):

$$\left.\begin{aligned} NO_x &= NO + NO_2 \\ NO_y &= NO_x + NO_3 + 2 \times N_2O_5 + HNO_4 + PAN + MPAN + nitrates \end{aligned}\right\} \quad (5.22)$$

Free-radical reactions convert NO_x to relatively long-lived (lifetime >4 h) forms of NO_y:

$$NO_z = HNO_3 + NO_3^- + ORNI + PAN \quad (5.23)$$

Thus, $NO_y = NO_x + NO_z$. PAN is an intermediate compound with a time constant that varies with temperature from hours to days.

Table 5.3. Initial concentrations, emissions and speciation of VOCs.

Species	Low rural	High rural	Urban
Initial concentrations			
VOC (ppb C)	48	60	400
NO (ppb NO_2)	1.2	3.0	50.0
CO (ppb)	125	200	500
PAN (ppb)	0.05	0.4	2.0
Emissions (kg/km^2 h)			
VOC	0.64	1.2	2.7
NO_x	0.08	0.3	1.7
CO	0.001	1.7	1.0
Fraction of total VOCs (by ppbC)			
Ethane	0.24	0.09	0.05
Propane	0.12	0.11	0.05
Butane	0.1	0.13	0.16
Pentane	0.06	0.07	0.18
Hexane	0	0.02	0.07
Ethene	0.03	0.02	0.05
Propene	0.02	0.02	0.03
Butene	0	0.02	0.03
Acetylene	0.05	0.02	0.02
Benzene	0.05	0.04	0.02
Toluene	0.03	0.04	0.1
Xylene	0.05	0.04	0.1
Formaldehyde	0.05	0.14	0.02
Acetaldehyde	0.01	0.03	0.01
Isoprene	0.14	0.11	0.01
Non-reactive	0.1	0.1	0.1

Table 5.3, compiled by Luecken *et al.* (1999), shows initial concentrations, emissions and speciation of VOCs, whereas Table 5.4 contains simulation modelling results obtained by application of three models; RADM (Regional Acid Deposition Model 2), CB4 (the regulatory version of the carbon bond 4 mechanism), HARW (2-D global model) for low rural, high rural and urban conditions, respectively. A location was chosen at 38.75°N, 88.0°W (St. Louis, Missouri) on 20 June from 06 : 00 (sunrise) to 20 : 00 (sunset). The most important reaction under rural conditions is oxidation of isoprene, which influences the production of PAN and organic nitrate (see also Gillani *et al.*, 1998a, b; Sillman *et al.*, 1998).

An important initiative, developed by NASA, is the Pacific Exploratory Mission in the Western Pacific Ocean (PEM–West), which is a major component of the East Asia/North Pacific Regional Study (APARE), an IGAC project. Its broad objectives, discussed by Hoell *et al.* (1997), are to study chemical processes and long-range transport of atmospheric trace species over the north-west Pacific Ocean and to estimate the extent of human impact on these species over this region. The first phase of PEM–West (PEM–West A) was conducted in September–October 1991, a period

Table 5.4. Predicted final or maximum concentrations (ppb) of O_3, NO_y and related species.

Concentration	Low rural			High rural			Urban		
	RADM	CB4	HARW	RADM	CB4	HARW	RADM	CB4	HARW
O_3 (max)	40.1	41.7	41.1	74.6	80.8	81.0	120	124	114
O_3 (final)	36.4	37.7	37.1	67.1	72.7	72.5	105	109	99.0
NO_y (final)	1.99	2.37	2.07	6.16	6.93	6.00	42.3	42.9	45.4
O_3/NO_x (final)	31.6	41.4	34.7	21.2	26.7	22.7	4.04	4.44	2.90
PAN (max)	0.76	0.50	0.43	2.44	1.80	1.58	6.80	4.52	3.34
ORNI (max)	0.09	1.00	0.58	0.26	2.25	0.98	2.38	6.89	1.08
HNO_3 (max)	0.41	0.33	0.41	1.65	1.51	1.77	12.6	12.6	11.6
NO (final)	0.003	0.003	0.003	0.006	0.005	0.005	0.01	0.009	0.01
NO_x (final)	1.16	0.91	1.07	3.17	2.73	3.20	26.0	24.5	34.1

characterized by minimum outflow from the Asian continent. The second phase of this mission (PEM–West B) was conducted during February–March 1994, a period characterized by enhanced outflow from the Asian continent.

In a later study, Hoell *et al.* (1999) examined the impact of human and natural processes on the chemistry of the troposphere over the Pacific basin using PEM–Tropics results, whose principal objectives included improving understanding of the oxidizing power of the tropical atmosphere (HO_x and NO_x photochemistry), as well as investigating oceanic sulphur compounds and their conversion to aerosols. The PEM–Tropics field campaign provided a unique set of atmospheric measurements in a heretofore data-sparse region (OH, H_2, SO_4, NO, NO_2 and actinic fluxes were measured from two NASA DC-8 and P-3B aircraft).

Frost *et al.* (1998) performed comparisons between a steady-state photochemical model and the first airborne measurements of hydroxyl radical concentrations [OH], in the lower free troposphere and marine boundary layer, taken during the First Aerosol Characterization Experiment (ACE1). The model quantifies the sources and sinks of OH for low levels of nitric oxide (NO). Predicted concentrations of peroxy radicals HO_2 and CH_3O_2 and total radical production and loss rates have been presented. It has been shown that model overpredictions of observed [OH] within clouds can be corrected by including a loss in gas phase OH to cloud droplets. Low [NO] conditions only preclude HO_2 uptake from explaining observed depletions of OH within clouds.

Hendricks *et al.* (1999) developed a box model to simulate heterogeneous reactions of nitrogen chlorine and bromine compounds on and in sulphate aerosols under conditions encountered in the lowermost stratosphere and upper troposphere. The simulations revealed that heterogeneous transformations in the upper troposphere lead to detoxification. The upper troposphere appears to show a large potential for heterogeneous activation of ClO_x and BrO_x. Heterogeneous chemistry induces ozone loss in the entire tropopause region. In the upper troposphere, heterogeneous reactions of chlorine compounds on and in sulphate aerosols have a major impact on ozone chemistry, when enhanced aerosol loading is assumed.

Spatial and temporal variations in particulate precursor gases and photochemical reaction products, during SJVAQS/AUSPEX ozone episodes, were discussed by Chow *et al.* (1998). Both precursors (NH_3, SO_2, NO, NO_2) and final products of photochemical reactions (HNO_3, PAN, O_3) were observed in San Joaquin Valley (California) in 1990. Five ozone episodes were examined.

Straume *et al.* (1998) discussed observations of perfluorocarbon background concentrations in Europe. Strong space–time variability in concentrations was observed, which, however, does not exclude use of perfluorocarbons as passive tracers.

Isaac *et al.* (1998) discussed the results of the Eulerian Model Evaluation Field Study (EMEFS), which was conducted in the summer of 1988 and the spring of 1990 in Central Ontario, Canada.

A major part of field research was connected with the North Atlantic Regional Experiment (NARE) (Penkett at al., 1998a, b). This experiment is responsible for the most extensive dataset on chemical composition of the North Atlantic atmosphere. A very important finding of these field studies is observation of multilayered structures of pollutants in the atmosphere above the ocean both close to Antarctic and European source regions, coupled with suggestions of long-range transport across the Atlantic. NARE observational data come from groundbased measurements (mainly in the Canadian Maritime Provinces, Izaña in the Canary Islands, in the Azores, on Bermuda and Spitzbergen, and at Mace Head, Ireland) as well as from four aircraft measurements.

The principal aim of NARE has been to understand the processes responsible for either *in situ* production of transport of ozone to all regions of the atmosphere over the Atlantic, including the more remote areas. The most typical and well-known feature of tropospheric ozone dynamics is the strong spatio-temporal variability in ozone concentration. Parrish *et al.* (1998) discovered, for example, that a maximum in ozone was clearly observed at the remote mid-Atlantic site in spring. The behaviour of ozone is clearly correlated to the behaviour of carbon monoxide, an excellent long-lived tracer of background anthropogenic pollution at that time of year. These observations suggest that the ozone spring maximum has two major contributions, photochemical production in the troposphere and subsidence of ozone formed in the stratosphere.

The overall conclusion reached by Penkett *et al.* (1998a) is that in spring peroxy-radical chemistry produces ozone in the presence of NO_x, whereas in the summer the same chemistry preferentially diminishes ozone and produces peroxides. Summer ozone loss over the ocean is perturbed by the presence of substantial layers of polluted air 1–2-km thick at latitudes up to 8 km. The sources of pollution may be located in Europe, North Africa or North America (Knapp *et al.*, 1998a, b; Penkett *et al.*, 1998b). Formation of polluted layers by various processes was investigated by Bethan *et al.* (1998). Modelling studies by Liang *et al.* (1998) led to the conclusion that eventual ozone production in the global troposphere from US emissions is about twice as large as direct export of ozone from the US boundary layer. This is primarily due to the transport from the continents of NO_x and PAN along with ozone. Munger *et al.* (1998) demonstrated, in this context, that isoprene chemistry in the United States continental boundary layer is a substantial factor in formation and export of PAN and other organic nitrates. The results of numerical modelling conducted by Schultz *et al.* (1998) indicate that the presence of NO_x in the air at the end of the trajectory was probably associated with decomposition of PAN during descent of the airmass, which served to limit photochemical destruction of transported ozone.

NARE publications contain extensive observational data on ozone precursors. According to Peterson *et al.* (1998), in the clean marine boundary layer on the Azores the concentrations of NO_x and NO_y were below 100 pptv (NO_y ~73 pptv, NO ~16 pptv maximum). At these levels, ozone destruction will exceed ozone production (Schultz *et al.* 1998). Above the boundary layer, the NO_y levels were substantially higher (Peterson *et al.*, 1998) reaching an average 300–400 pptv, with NO_x ~50 pptv. This is sufficient for photochemical production to substantially offset ozone destruction. Similar NO_y and NO_x levels were observed in the free troposphere at Izaña (400–500 pptv NO_y and ~50 pptv NO_x), just below concentrations required for the ozone budget to be in balance according to photochemical models (Schultz *et al.*, 1998).

Aircraft and surface measurements of *in situ* ozone production rates indicate positive values in the range from 1–4 ppbv/day observed at the surface in the Canadian Maritime Provinces (Duderstadt *et al.*, 1998). In the case of aircraft studies, limitation in rate of O_3 production was dominated by NO_x levels, but peroxy-radical concentrations were critical 10% of the time (Kleinman *et al.*, 1997).

Apart from the ozone precursors mentioned above, NARE observations included such components as non-methane hydrocarbons (Jobson *et al.*, 1998b), organic nitrates (Roberts *et al.*, 1998a), total non-methane organic carbon (Roberts *et al.*, 1998b) and oxygenates such as formaldehyde (Fischer *et al.*, 1998a). It has been shown that organic nitrates correlate well with carbon monoxide and as such are excellent tracers of the impact of anthropogenically driven ozone production from precursors such as NO_x and VOCs. Observations of formaldehyde suggest that non-methane hydrocarbon chemistry must play a significant role in the chemistry of the free troposphere over the remote Atlantic. These findings show that sources of HO_x radicals, other than ozone photolysis, should be taken into account in the overall chemistry of the North Atlantic atmosphere. The principal conclusion from NARE observations is that the atmosphere over the North Atlantic Ocean is heavily impacted by emissions of pollutants from continents on either shore (Penkett *et al.*, 1998b). Overall impact on free-radical chemistry and ozone budget of this pollution is far from being settled, however. Significantly more data need to be collected throughout the year in future field experiments. Novaime *et al.* (1998) conducted the measurements of peroxyacetyl nitrate (PAN), peroxypropionyl nitrate (PPN), and peroxymethacryloyl nitrate (μPAN) during the Southern Oxidants Study.

Particular attention should be paid to hydroxyl radical OH which is mainly responsible for the oxidizing capacity of the atmosphere. This radical is produced as a result of O_3 photolysis in the presence of water vapour:

$$O_3 + h\nu\ (\lambda < 320\ \text{nm}) \longrightarrow O(^1D) + O_2 \tag{5.24}$$

$$O(^1D) + H_2O \longrightarrow 2OH \tag{5.25}$$

OH controls removal and, therefore, concentrations of many man-made gaseous pollutants as well as natural gases (Ehhalt, 1998; Ehhalt *et al.*, 1998). Reduction of OH with molecules such as CO or hydrocarbons returns another radical HO_2, which eventually is recycled to OH. So, oxidation by OH is catalytic; one OH can destroy several molecules. The rate of HO_2 recycling, and therefore OH concentration, is greatly enhanced by the presence of NO. At the same time, the other nitrogen radical NO_2, which is always present along with NO, reacts with OH to form nitric acid (HNO_3), removing OH from the atmosphere. The opposing actions of both processes lead to a highly non-linear dependence of OH concentration on NO_x, which not only influences the rate of which $HO_x(=OH + HO_2)$ is recycled between OH and HO_2, but also controls how much O_3 is produced in each cycle. The resulting net production rate of O_3 depends on NO_x concentration in a highly non-linear manner. This non-linear dependence makes model predictions about changes, even the sign of the change, in local OH concentration and net O_3 production very difficult. Moreover, it is necessary to include an additional input of NO_x dependent on pre-existing levels of NO_x (Penner *et al.*, 1998).

Ehhalt (1998) emphasized that, given its role as a driver of tropospheric chemistry, measurement of OH provides a crucial test of our understanding of atmospheric chemistry. However, only recently have accurate and fast measurements of OH and HO_2 in the troposphere been possible and only by a few groups worldwide. Among them were Wennberg *et al.* (1998), who measured concentration of OH and HO_2 in the mid and upper troposphere simultaneously with NO, O_3, CO, H_2O, CH_4, non-methane hydrocarbons and with the ultraviolet and visible radiation field. All observations were obtained between October 1995 and August 1996 with instruments aboard NASA ER-2 aircraft near Mountain View, California (37°N, 122°W) and Banders Point Naval Air Station, Hawaii (21°N, 158°W).

Wennberg *et al.* (1998) emphasized that these data allow direct examination of the processes that produce O_3 in the mid and upper troposphere. Comparison of measured concentrations of OH and HO_2 with calculations, taking account of their production from H_2O, O_3 and CH_4, have shown that these sources are insufficient to explain observed radical concentrations in the upper troposphere. Photolysis of carbonyl and peroxide components, transported to this region from the lower troposphere, may provide the source of HO_x required to sustain measured abundances of these radical species. Assessments of thc production rate for ozone have led to the conclusion that this production rate is faster than previously thought. This implies that anthropogenic activities that add NO to the upper troposphere, such as biomass burning and aviation, lead to production of more O_3 than expected.

Krol *et al.* (1998) pointed out that on a global scale the most important sink for

OH is its reaction with either CO or CH_4, which are relatively well-mixed gases that limit the lifetime of OH to 1–10 s. In the planetary boundary layer, however, non-methane hydrocarbons may be the dominant OH sink, particularly in polluted and forested areas. The lifetime of many trace gases, and most notably those of CH4 and CO, are determined by reaction with OH.

In this regard, Krol *et al.*, (1998) used methylchloroform (MCF) measurements taken at Atmospheric Lifetime Experiment/Global Atmospheric Gases Experiment (ALE/GAGE) measurement stations to deduce tropospheric OH concentration and its linear trend between 1978 and 1993. Global 3-D OH concentration fields have been calculated by a transport model that included background photochemistry. Optimized OH fields (adjusted to fit the measurements optimally) implied MCF lifetime in the troposphere of 4.7 years in 1978 and 4.5 years in 1993. For CH_4 these lifetimes (only because of OH destruction) are 9.2 and 8.6 years in 1978 and 1993, respectively. In an earlier study, Krol and von Weele (1997) examined the consequences of variations in global photodissociation rates for tropospheric chemistry. Calculations of daily-integrated photodissociation rates (DIPRs), in combination with a global chemistry transport model, have shown that the globally averaged lifetime of methane in an atmosphere with clouds is 8.8 years (removal of all clouds reduces the lifetime by only 2.4%).

Optimized OH concentration is sensitive to the strength of other MCF sinks in the model and is constrained to $1.00(+0.05\ \ -0.15) \times 10^{16}$ mol/cm^3 in 1978 and to $1.07(+0.09\ \ -0.17) \times 10$ mol/cm^3 in 1993. Deduced OH trend is sensitive to the trend in MCF emissions and is confined to the interval between −0.1 and +1.1% per year with a most likely value of 0.46% per year. Krol *et al.* (1998) discussed possible causes of global increase in OH. Calculations indicate that the positive trend was due to stratospheric ozone depletion, declining CO concentrations, increased water-vapour abundance and enhanced NO_x emissions. Despite the high degree of uncertainty about atmospheric composition changes, it can be concluded that observed changes are consistent with significant increases in OH over the past decades.

The hydroxyl radical is a tropospheric component that regulates methyl chloride (CH_3Cl) concentration. Khalil and Rasmussen (1999) pointed out that even though CH_3Cl is present at only about 600 pptv, methyl chloride is one of the largest reservoirs of gaseous chlorine in the Earth's atmosphere (5 Tg). It is thought to be mostly of natural origin and, as such, is probably the largest natural source of chlorine in the stratosphere. Analysis of observational results has led to the discovery of small mostly decreasing trends. On average, there was 4% less CH_3Cl during the time period 1994–1996 than in 1985–1988. Observations indicate the presence of continental CH_3Cl sources, mostly confined to the tropics.

Since reactions of ozone with alkanes in the gas phase are believed to be important processes in tropospheric chemistry, Marston *et al.* (1998) measured relative rate constants for the reactions of eight pairs of organic compounds with reactive species, formed by reaction of ozone with 2-methylbut-2-ene. The results obtained are consistent with the assumption that the reactive intermediate is the hydroxyl radical OH.

Table 5.5. Means, medians and number of samples for the Mauna Loa Observatory Photochemical Experiment intensives.

	Autumn 1991	Winter 1992	Spring 1992	Spring* 1988	Summer 1992
NO_x (pptv)					
Means ± SD	26.9 ± 9.9	32.6 ± 12.0	40.5 ± 16.6	32.4 ± 12.5	29.8 ± 12.7
Median	26.1	31.1	37.0	31.0	29.2
Number of samples	8.034	11.767	10.136	12.436	5.077
PAN (pptv)					
Means ± SD	11.5 ± 10.5	29.1 ± 24.7	36.4 ± 30.6	17.0 ± 14.3	8.9 ± 11.7
Median	8.2	25.0	28.3	13.1	4.1
Number of samples	567	777	670	514	411
O_3 (ppbv)					
Means ± SD	36.0 ± 8.7	42.8 ± 6.0	63.0 ± 9.8	43.4 ± 12.4	38.8 ± 16.1
Median	35.2	43.5	64.0	43.6	35.6
Number of samples	16.087	19.646	16.532	15.833	9.950

*The MLOPEX 1 spring intensive (1 May to 4 June 1988) was later in the year than the spring intensive of 1992 (15 April to 15 May). SD = standard deviations.

In order to estimate O_3 production over the seasons, Ridley *et al.* (1998b) measured peroxyacetyl nitrate (PAN) and NO_x, as well as a variety of other constituents, over approximately 1-month-long periods of intensive observations in the autumn of 1991, and the winter, spring and summer of 1992, during the second Mauna Loa Observatory Photochemistry Experiment (MLOPEX 2). Ridley *et al.* pointed out that it has been known for some time that O_3 shows a maximum in spring at the Mauna Loa Observatory, but there were no sufficiently reliable assessments of relative contributions to this maximum of flow stratospheric exchange and photochemical production in the troposphere. The principal purpose of the MLOPEX 2 has been to contribute to the solution of the problem. The results indicate that PAN and NO_x in the free troposphere had maximum abundances in spring in concert with the well-known maximum in O_3 (Table 5.5). The ratio of spring-to-summer averages was a factor of 4.1 for PAN, a factor of 1.6 for O_3 and only a factor of 1.4 for NO_x. The characteristic feature of minor gas changes was that during most intensities, variations in periods of a few days to a week were often larger than the average seasonal amplitude.

Ridley *et al.* (1998b) emphasized that in free tropospheric airmasses local to Hawaii, average PAN/NO_x ratios were maximum in winter through spring but in the range of 0.25–0.86 in all intensives. PAN decomposition is unlikely to be the major net source of NO_x local airmasses in summer and fall. The low HNO_3/NO_x ratios, determined during MLOPEX 1, were confirmed during MLOPEX 2. Intensive average ratios of 1.6–3.8 over the year are lower than some model predictions. Ridley *et al.* (1998a) underlined that both the low ratio and magnitude of NO_x imply a shortcoming in our understanding of transformations and sources of NO_x constituents in the central Pacific.

The 3-to-4 km altitude region near Hawaii was a net importer of O_3, on average, over the year. Average net rate of O_3 production in the free troposphere was near zero in winter, −0.4 to −0.8 ppbv/day in spring, −1.4 ppbv/day in summer and −0.6 ppbv/day in autumn. These figures demonstrate that the spring O_3 maximum is not due to local photochemistry. Ridley *et al.* (1998a) believe that variation in ozone precursors over the year and on shorter timescales of a few days to a week is controlled predominantly by changes in large-range transport.

Human activities have resulted in substantial increase of impacts of nitrogen oxides on the atmosphere. To obtain the necessary information, Munger *et al.* (1998) have conducted measurements of nitrogen deposition and concentrations of NO, NO_2, NO_y (total oxidized N), and O_3 at Harvard Forest in central Massachusetts (USA) since 1990. The results of observations between 1990 and 1996 indicate that total (wet plus dry) reactive N deposition averages 47 mmol/m^2 day (126 mmol/m^2 day, 3.6 kg N/ha . yr), with 34% contributed by dry deposition. Additional input from the atmosphere amounts to about 12% in comparison with the N made available annually by mineralization in the forest soil. The corresponding deposition rate at a distant site, Schefferville, Quebec (Canada), was 20 mmol/m^2 day during summer 1990. In the boundary layer at the atmosphere, both heterogeneous and homogeneous reactions efficiently convert NO_x to HNO_3, which is subsequently removed rapidly by either dry deposition or precipitation.

According to data obtained by Munger *et al.* (1998), characteristic (e-folding) time for NO_x oxidation ranges from 0.30 days in summer, when OH radical is abundant, to ~1.5 days in the winter, when heterogeneous reactions are dominant and O_3 concentration are lowest. The characteristic time for removal of NO_x oxidation products (defined as NO_y minus; $NO_y = NO_x + NO_3 + N_2O_5 + HNO_3 + PAN +$ other organic nitrates + aerosol nitrate) from the boundary layer by wet and dry deposition is ~1 day, except in winter when it decreases to 0.6 day. Substantial contribution to N deposition is made by biogenic hydrocarbons, through formation of organic nitrates, but precursors of reservoir species such as PAN may also be exported from the region. According to a simple model assuming pseudo-first-order rates for oxidation of NO_x followed by deposition, about 45% of NO_x in the north-eastern Q.S. boundary layer is removed in 1 day during summer and 27% is removed in winter. It takes 3.5 and 5 days for 95% removal in summer and winter, respectively.

Hoffman and Sulkowski (1998) analysed surface data from the air-monitoring network in Kedzierzyn-Kozle (Poland) from June 1992 to July 1995 to investigate the influence of local NO_x emission on surface ozone concentration during the night and day. Results show that the night-time depletion effect was more apparent than the daytime effect.

De Mazière *et al.* (1998) studied post-Mount Pinatubo NO_2 reduction and recovery, based on 10 years of Fourier transform infrared and UV-visible spectroscopic measurement (FTIR) at Jungfraujoch in the Swiss Alps (46.6°N, 8.0°E, 380 m a.s.l.). Analysis of the FTIR data has not revealed any long-term trends in NO_2 observations over 11 years. A sudden drop was registered, however, at the beginning of January 1992, which can be ascribed to the impact of the Pinatubo eruption in June 1991.

During recent years, quite a number of studies have been devoted to natural emissions of various minor gas components (volatile organic compounds – VOC) by vegetation. In this context, Fantechi *et al.* (1998) investigated atmospheric oxidation of methylbutenol (MBO) by HO radicals, ozone and NO_3 radicals. Laboratory studies (in glass tubes) were performed of mechanisms and MBO reactions with hydroxyl radicals, O_3 and NO_3. The results obtained demonstrate that the atmosphere may be considered as a gigantic chemical reactor under strongly varying conditions. MBO reactions with OH, O_3 and NO_3 contribute most as MBO sinks.

There still remains the very important and still-unresolved problem of solar activity impact on tropospheric ozone. Using TrOz values obtained from Nimbus-7 TOMS measurement from 1975 to 1992, Chandra *et al.* (1999) provided the first observational evidence of changes in tropospheric ozone in the marine atmosphere of the tropics, which are out of phase with stratospheric ozone changes over an 11-year solar cycle (a retrieval technique was described by Ziemke *et al.* (1998)). Estimated changes in tropospheric and stratospheric column ozone over the solar cycle are respectively -2.98 ± 1.31 and $+8.63 \pm 1.99$ DU or -12.6 ± 5.6 and $3.65 \pm 0.86\%$ from solar minimum to solar maximum (these values are statistically significant at the 2σ level). Chandra *et al.* (1999) pointed out that, by comparison, linear trends in tropical TrOZ are not statistically significant.

The results considered are qualitatively consistent with a modulation effect on troposphere ozone photochemistry by UV-induced changes in stratospheric ozone. However, in the low NO_x regime of the marine atmosphere, observed changes are significantly larger than estimated by a photochemical model. Chandra *et al.* (1999) emphasized that explanation for the solar signal may include subtle solar-induced changes in transport in the troposphere involving low boundary-layer ozone and ozone precursors. If the NO_x mixing ratio is low relative to that of O_3, a likely reaction is:

$$HO_2 + O_3 \longrightarrow OH + 2O_2 \tag{5.26}$$

which means a net ozone loss. However, at larger NO_x mixing ratios the reaction:

$$HO_2 + NO \longrightarrow NO_2 + OH \tag{5.27}$$

will dominate leading to net O_3 production. Box-model calculations made by Chandra *et al.* (1999), taking into account 64 reactions among 25 variable species (basic O_x, HO_x, NO_x chemistry along with CH_4, C_2H_4 and C_2H_6 oxidation sequences were considered), confirmed this assumption.

Chandra *et al.* (1999) underlined the potential importance of TrOZ variability over a decade from the viewpoint of climate change. Assuming a scaling factor of 0.05–0.08 $W/m^2 \cdot DU$, solar-cycle related change in TrOZ implies an anomaly of about 0.15–0.25 W/m^2.

Richter *et al.* (1998) pointed out that long-term TrOZ observations in polar regions have repeatedly shown periods of low tropospheric ozone values in spring in the Arctic and Antarctic, which are believed to be mostly of natural origin (see also Kieser *et al.*, 1993; Solberg *et al.*, 1996), since surface observations revealed enhanced Br levels before and during such events. Richter *et al.* (1998) analysed measurements from GOME for tropospheric BrO in the Northern Hemisphere spring and summer 1997. Tropospheric excess TrOZ values have been determined by subtracting measurements from a longitude range assumed to represent background conditions. From February until the end of May, enhanced tropospheric BrO columns were observed over the Hudson Bay area and parts of the Canadian Arctic (satellite mapping of enhanced BrO concentrations was carried out by Wagner and Platt (1998a, b)). This large and persistent event has not been reported before and can only be explained by the large local source of bromine. Apart from that, from March to May other smaller and shorter tropospheric BrO events were detectable along the coastlines of the Arctic Sea and over the polar ice. Assuming that the layer of BrO was 2 km high then tropospheric BrO mixing ratios of up to 20 ppt are implied.

Richter *et al.* (1998) noted that several theories have been proposed to explain the production of enhanced tropospheric bromine at high latitudes (40–90°) in spring, including assumptions concerning both biologic and organic mechanisms. GOME measurements of large amounts of BrO in spring but not in summer or autumn, and the correlation of BrO with the movement of sea ice across the seasons are consistent with a biological source releasing bromine at sea-ice cracks.

However, there is also some evidence of an inorganic source mechanism. McElroy *et al.* (1999) pointed out that, according to earlier studies, a potential origin of BrO could be the autocatalytic release of bromine from sea salt accumulated on the Arctic snow pack, followed by photolytic and heterogeneous reactions, which produce and recycle the oxide.

Since there are no data about the vertical BrO concentration profile, McElroy *et al.* (1999) conducted observations from high-altitude aircraft (NASA ER-2) that suggest the presence of significant amounts of BrO not only in the planetary boundary layer (PBL) but also in the free troposphere above it. Assuming that all the BrO resides within the PBL at 1-km height, BrO mixing ratios of 50–100 pptv are obtained. This is substantially larger than the maximum BrO mixing ratio previously observed within the PBL. Assuming that the BrO resides above the PBL (in 1–5-km layer), mixing ratios of 20–30 pptv are obtained. If a uniform mixed 0–5-km layer is assumed then slightly lower BrO mixing ratios, are obtained. Note that BrO mixing ratios of 50 pptv in the 0–1-km PBL are required to destroy most of PBL ozone within a day, if bromine chemistry alone is considered.

McElroy *et al.* (1999) believe that BrO is transported from the PBL into the free troposphere through convection over large Arctic ice leads (openings in the pack ice). Convective transport also lifts ice crystals and water droplets well above the PBL, thus providing surfaces for heterogeneous reactions that can recycle BrO from less-reactive forms and thereby maintain its ability to affect the chemistry of the free atmosphere. Such heterogeneous reactions play a very important role, since otherwise Br atoms, produced by photolysis of BrO, will rapidly react with such species as HO_2, HCHO and C_2H_2 to produce relatively inactive HBr. So, in the absence of heterogeneous recycling, BrO will rapidly disappear.

An important role is played in the chemistry of both the stratosphere and troposphere by nitrate radical NO_3, which is formed by reactions of NO_2 with O_3 through the strongly temperature-dependent reaction (Aliwell and Jones, 1998):

$$NO_2 + O_3 \longrightarrow NO_3 + O_2 \tag{5.28}$$

NO_3 reacts further with NO_2 to produce N_2O_5, some of which thermally decomposes to produce NO_3 and NO_2 again:

$$NO_3 + NO_2 + M \longrightarrow N_2O_5 + M \tag{5.29}$$

$$N_2O_5 + M \longrightarrow NO_3 + NO_2 + M \tag{5.30}$$

Since these reactions are also very strongly temperature dependent it is expected that NO_3 concentration must be higher during warmer days. In the polluted boundary layer there is a number of reactions which destroy NO_3 at night in the absence of photolysis. One of the most significant is reaction (5.31) and this is the main sink for NO at night:

$$NO_3 + NO \longrightarrow NO_2 + NO_2 \tag{5.31}$$

NO also reacts with O_3:

$$O_3 + NO \longrightarrow O_2 + NO_2 \tag{5.32}$$

NO_3 also reacts with DMS, which is emitted by algae, though this reaction is only potentially significant at the marine boundary layer. Reactions with isoprene and various terpenes, which are emitted by plants, are rapid and may also be a significant sink in rural areas. A number of reactions with alkanes, alkenes and aldehydes may also become significant in the polluted urban-boundary layer at night and lead to formation of peroxy radicals and therefore OH radicals.

Having in view the processes mentioned, Aliwell and Jones (1998) measured the column abundance of NO_3 in the troposphere just prior to sunrise at Cambridge, England (52.2°N, 0.1°E) during the summer and autumn (July to October) of 1995. The zenith sky viewing spectrometer was used to apply a 'twilight' retrieval technique, which allows separation of stratospheric and tropospheric components of total NO_3 vertical column at sunrise. The results of observations indicate that, if it is assumed that all tropospheric NO_3 is present in a 1-km thick atmospheric boundary layer, then an average NO_3 mixing ratio is 89 ppt. The characteristic feature of NO_3 variability is one, however, of considerable atmospheric day-to-day change.

Aliwell and Jones (1998) concluded, from NO_3 measurements together with point measurements of O_3, NO_2, NO and air temperature, that NO_3 lifetime is significantly longer than reported earlier. This may be explained using surface observational data in previous assessments, which cannot be considered representative for the whole boundary layer, since vertical gradient in NO_3 concentration would be expected in the boundary layer.

Röckmann *et al.* (1998) underlined the significant role of carbon monoxide in tropospheric chemical processes, which has both natural and anthropogenic sources, while its mixing ratio varies from 40 nmol/mol in remote parts of the atmosphere up to several μmol/mol under conditions of polluted urban atmosphere. The principal CO sink is determined by its oxygenation reaction to CO_2 with hydroxyl participation (about one-half of hydroxyls in the atmosphere are spent for CO removal). CO impact strongly on a number of minor gas components; in particular, on such a greenhouse gas as CH_4. Carbon monoxide is characterized by strong spatio-temporal variability, because of its relatively short lifetime (from about 3 weeks in the tropics up to several months at high latitudes) and the complex combination of its sources and sinks. This is why assessment of CO global budget is still highly uncertain.

The important role of HNO_3 in the troposphere is determined by its functioning as a sink for NO_x and substantial components of acid deposition. Analysis of NO_x measurements in an unpolluted atmosphere has shown that the observed $[NO_x]/[HNO_3]$ ratio is much higher than the calculated ratio (using a photochemical model), which indicates substantial uncertainty in assessments of tropospheric ozone formation under regional and global scales. In order to obtain more reliable data on HNO_3 concentration in the gas phase, Fehsenfeld *et al.* (1998) developed the necessary instrumentation and undertook an intercomparison of two ionization mass-spectrometers (CIMS) and filter probes (FP) during winter and early summer 1996 at a site close to Denver (USA). Although the results obtained show good enough agreement, the FP technique requires further studies of its reliability.

Flocke *et al.* (1998) conducted continuous measurement of alkyl nitrates during the Tropospheric Ozone Research (TOR) subproject of the European Experiment on Transport and Transformation of Environmentally Relevant Trace Constituents in the Troposphere over Europe (EUROTRAC) at the Schaunsland Station in the Black Forest (Germany) between June 1990 and May 1991, using an automated gas chromatograph for NO_y detection.

Flocke *et al.* (1998) pointed out that alkyl nitrates are formed as a minor product in the reaction of alkylperoxy radicals with NO:

$$NO + RO_2 \longrightarrow NO_2 + RO \tag{5.33a}$$

$$NO + RO \longrightarrow RNO_2 \tag{5.33b}$$

The major channel of this reaction leads to formation of NO_2 and constitutes the rate-limiting step in tropospheric ozone production.

Another potential source of alkyl nitrates is the reaction of alkoxy radicals with NO_2:

$$NO_2 + RO \longrightarrow RONO_2 \tag{5.34}$$

Multifunctional nitrates may also be formed during oxidation of hydrocarbons in the presence of NO_x. Because of their formation from (5.33), which also represents the only known significant source for ozone in the troposphere, alkyl nitrates can be utilized as tracers of photochemical ozone production from anthropogenic precursors.

The results obtained by Flocke *et al.* (1998) indicate that the mixing ratio of the sum of alkyl nitrates (e_1–C_8) ranged between 30 and 630 ppt and averaged 120 ppt over the whole measurement period. Average contribution of alkyl nitrates to total odd nitrogen was about 3%, with little seasonal variation. The most abundant individual nitrates were 2-propyl, 2-butyl and methylnitrate. Mixing ratios of nitrates exhibited summer maxima and pronounced daily variations, with maxima in the early evening. Abundance increased from C_3 to C_5. In clean, photochemically aged airmasses, alkylnitrate mixing ratios showed summer minimum and no pronounced diurnal variations. Decrease in abundance from C_3 toward higher carbon chainlengths was observed, as is expected for air that is characterized by photochemical destruction. Flocke *et al.* (1998) emphasized that the observed behaviour and mixing ratios of methyl and ethylnitrate cannot be explained solely by formation from the reaction of peroxy radicals and NO.

The interesting idea of monitoring atmospheric chemical composition changes along the Trans-Siberian Railroad from Moscow to Vladivostok was realized by Bergamaschi *et al.* (1998a). Between 5 and 12 August 1996 measurements of isotropic composition of CO (^{13}C, ^{14}C, ^{18}O) and CH_4 (^{13}C, D) were made. The results obtained indicate that, apart from short-time CO and CH_4 fluctuations because of local emission sources, certain large-scale (hundreds of kilometres) and stable anomalies of enhanced concentrations have been observed. This may be explained by the impact of long-range transport from powerful sources of CO and CH_4. Large-scale CH_4 enhancement has been observed, for example, over West Siberian lowland. Methane isotropic signatures ($\delta^{13}C = -62.5 \pm 4.7‰$ V-PDB; $\delta D = -311 \pm 14‰$ V-SMOW) clearly indicate its biogenic origin. The second important feature was very strong CO enhancement (up to 1,500 nmol/mol to the east of Chita (Eastern Siberia) along the Amur River at a distance of about 2,000 km). Measured ^{14}C concentrations and calculated back-trajectories indicate that the cause of this anomaly was biomass burning.

Semenov (1996) substantiated a general ‘genetic’ approach to assessment of contributions of various components into total budgets of relevant components. This approach has been illustrated by calculations for concentrations of six trace gases (NO, NO_2, O_3, HNO_3, NO_3^- and PAN), which take part in the reactions determining, in particular, tropospheric ozone change because NO and NO_2 emissions.

5.5 NUMERICAL MODELLING OF TROPOSPHERIC OZONE VARIABILITY

Various causes of tropospheric changes have been considered above in connection with analysis of observational data. It is clear, however, that the principal instrument for understanding the numerous and often interacting causes of ozone changes is numerical modelling (Berntsen *et al.*, 1996; Berntsen and Isaksen, 1997, 1998; Bottenheim and Barrie, 1996; Hauglustaine *et al.*, 1998; Kley, 1997; Kritz *et al.*, 1998; Wang *et al.*, 1998a, b).

To assess impacts of increased emissions in Asia on tropospheric ozone and climate, Berntsen *et al.* (1996) used a global 3-D chemical transport model (CTM) developed at the NASA Goddard Institute for Space Studies (GISS). An extended

Table 5.6. Global and Asian surface-emission rates applied in the reference run of the model.

Species	Emissions (E_R)		
	Global (Tg/yr)	Asian (Tg/yr)	F_a
NO_x	34.6 (N)	7.5 (N)	1.0
CO	1.575	294	0.50
C_2H_4	36.7	6.3	0.61
C_2H_6	37.0	7.4	0.61
C_3H_6	27.0	4.8	0.73
C_4H_{10}	29.9	5.7	0.52
C_6H_{14}	24.3	5.0	0.70
m-Xylene	22.7	5.8	0.68
HCHO	1.11	0.2	0.61
Isoprene	205	27	0.0
CH_4	394	169	0.0

set of chemical compounds (including such principal ozone precursors as hydrocarbons, nitrogen oxides, carbon monoxide) has been taken into account in the model, of which 25 are transported. The Asian region was selected because anthropogenic emissions are rapidly increasing in this region and because there is a huge potential for future increases.

Three model experiments have been conducted to study the impact of tropospheric ozone from surface emission changes: (1) a reference run (R) with global emissions of NO_x and hydrocarbons representative for 1980 (Table 5.6); (2) two additional experiments with increased emissions (E_{A2}, E_{A4}), where anthropogenic sources were increased by factors of 2 and 4, respectively. New emissions applied in the Asian region are then defined by equations:

$$E_{A2} = 2 \times F_a \times E_R + (1 - F_a) \times E_R$$
$$E_{A4} = 4 \times F_a \times E_R + (1 - F_a) \times E_R$$

where F_a (Table 5.6) defines the fraction of the emissions assumed to be of anthropogenic origin for each component. The Asian region is defined as the region between the equator and 80°N, and between 57.5°E and 172.5°W. The factor F_a gives the fraction of the emissions, which are perturbed in the runs with increased emissions relative to the reference run (A2 and A4).

The basic-model case (R), approximately covering emissions in 1980, was run for 15 months. Perturbations were then run for 4 months each starting 1 May with initial conditions equal to the concentrations in the reference case at that time. Final results are averages for the month of August (after a 3-month spin-up period).

Berntsen *et al.* (1996) pointed out that reference results lead to surface NO_x concentrations in the Asian region with maxima over Japan and north-east China of about 1 ppbv as a monthly averaged concentration in August. This level is about

Table 5.7. Calculated surface concentrations of NO_x and total reactive carbon for the three model runs, and corresponding changes in surface ozone concentrations.

Region	Case	NO_x (pptv)	Total *C* (ppbv)	ΔO_3 (ppbv %)	PE_{O3} (ppbv/Tg (N) yr^{-1})	$\Delta O_3/\Delta NO_x$ (ppbv/ppbv)
Japan	R	970	16.3			
	A2	1,710	24.2	13.2/28	23	18
	A4	3,260	40.4	22.5/47	8.4	6.0
China	R	1,160	13.8			
	A2	2,110	17.3	7.9/29	17	8.4
	A4	4,060	25.1	14.0/52	6.5	3.1
India	R	290	16.6			
	A2	490	20.7	12.8/43	119	64
	A4	790	29.3	3.2/114	100	71

The changes (A2 versus R and A4 versus A2) in ozone production efficiency $PE_{O3} = \Delta E_{NO_x}$ (ppbv/Tg (N) yr^{-1}) and in the ratio between increases in ozone and NO_x concentrations ΔO_3 (ppbv)/ΔNO_x (ppbv) are given in the last two columns.

45% and 30% of the corresponding levels predicted over the eastern United States and Central Europe, respectively. Over south-east Asia and India, NO_x concentrations in the model are significantly lower (300 pptv or less).

The results obtained by Berntsen *et al.* (1996) indicate that in 3-D model calculations, the adopted perturbations lead to significant increases in surface ozone in the Asian region (up to 70–90% in the case of A2). Table 5.7 illustrates some results.

Note that enhanced short-term ozone values during episodes are becoming more pronounced in cases with high emissions. Significant ozone increases have also been obtained in the upper troposphere because of efficient convective transport. Doubling NO_x emissions results in ozone increases up to 30% in the upper troposphere. Estimates of regionally positive radiative forcing (RF) because of increases in tropospheric ozone give values up to 0.5 W/m^2, which is 30–50% of negative RF owing to the direct effect of sulphate aerosols in the region. Averaged RF values for the Northern Hemisphere, the Southern Hemisphere and their total are calculated to be 0.13, 0.016 and 0.073 W/m^2, respectively. Longwave RF dominates total RF, which is highly inhomogeneous. These conclusions demonstrate the important role of tropospheric ozone as a greenhouse gas (Wang and Isaksen, 1995).

The calculations presented in Table 5.7 demonstrate the importance of background concentration impact on the increase in emissions.

The value of the ozone production efficiency, defined by the term $PE_{O3} = \Delta O_3/\Delta E_{NO_3}$, is maximum over north-east India because of significantly lower NO_x concentrations in this region. Table 5.8 shows the contribution of various processes to change in tropospheric ozone in the Northern Hemisphere.

As can be seen, in the middle free troposphere (approximately 1.5–7.5-km altitude) there is an increase in the flux of ozone from the boundary layer of

Table 5.8. Contribution to ozone change in troposphere by various processes in the model for the whole Northern Hemisphere.

	R (Tg/month)		A2 (Tg/month)		Diff. (A2−R) (Tg/month)	
Process	May	August	May	August	May	August
Strat. flux	96.7	33.4	95.8	31.9	−0.9	−1.5
Deposition	80.5	71.0	85.2	77.8	4.7	6.8
Net chemical product (PBL)	50.7	68.5	60.5	79.8	9.8	11.3
Net chemical product (MFT)	−66.4	−74.2	−66.5	−81.8	−0.1	−7.6
Net chemical product (UFT)	18.3	30.1	21.9	35.2	3.6	5.1
Upword flux at top of PBL	−24.6	5.5	−20.4	10.0	4.1	4.4

PBL: planetary boundary layer ($\sigma = 1.0$–0.87, surface to approximately 1.5 km); MFT: middle free troposphere ($\sigma = 0.87$–0.39, approximately 1.5–7.5 km); UFT: upper free troposphere ($\sigma = 0.39$–0.14, approximately 7.5–13.7 km).

about 4.4 Gt/month for August when anthropogenic emissions are doubled over Asia. There is also increase in downward transport from the upper free troposphere.

Berntsen *et al.* (1996) concluded that, with sustained economic growth in the Asian region and increased use of coal as the primary source of energy, local pollution effects from sulphur emissions will probably lead to emission controls on emissions. Emissions of NO_x, CO and hydrocarbons leading to ozone production are more difficult to control and will probably continue to increase. The net effect of anthropogenic emissions of sulphur dioxide and ozone precursors on the radiation budget over Asia is therefore likely to become positive in the future.

Flatøy *et al.* (1996) applied the 3-D Mesoscale Chemistry Transport (MCT) model, driven by meteorological data from the Numerical Weather Prediction (NWP) model, to calculate ozone and other chemical species over Europe over a 10-day period (1–10 July 1991) which was characterized by warm weather and frequent occurrence of cumulus convection. Since photolysis rates are highly dependent on cloud amount and cloud optical depth, the meteorological model uses cloud water as one of the prognostic variables and has an advanced treatment of stratiform and convective cloud and precipitation processes, which allows sufficiently reliable simulation of cloud properties. The results obtained by Flatøy *et al.* (1996) show that: (1) physical processes, especially convection, may dominate vertical distribution of ozone in the free troposphere; (2) sinking air which compensates for convective updrafts is important for the tropospheric ozone budget; (3) transport of ozone precursors from the boundary layer into the free troposphere by convection enhances the rate of O_3 photochemical production significantly.

A case study has been considered of photochemical development and transport of ozone and other chemical components in the troposphere over Europe between 00:00

UT on 1 July 1991 and 06:00 UT on 11 July 1991. Model values were compared to ground measurements and vertical soundings. Comparison with ground-level measurements shows that the model performance is reasonable, but some phase and magnitude errors can be found (there are various reasons, why this happened). Photochemical production rates as high as 1.5 ppb/h were calculated when the boundary layer contained about 3 ppb NO_x, 130 ppb CO and 20 ppb NMHC (non-methane hydrocarbons), which agrees with some earlier results.

Flatøy *et al.* (1996) emphasized the need to further improve the model as well as more thorough comparisons with observational data. Future modelling will include a study of how North American tracer emissions of NO_x, NH_3, SO_2 and VOCs (volatile organic compounds), and the secondary products formed, can be identified and in what chemical form. In addition, the chemical impact of subsonic aircraft emissions of NO_x over the North Atlantic on upper tropospheric NO_x and ozone budgets will be calculated. New assessments will be obtained of how the flux of stratospheric ozone into the troposphere in tropopause-folding events can influence oxidation of NO_x and VOCs.

Mauzerall *et al.* (1996) quantified the tropospheric ozone budget over remote high northern latitudes in summer using chemical and meteorological measurements between 0 and 6 km, made during the summer of 1990 by the Arctic Boundary Layer Expedition (ABLE 3b). All components in the ozone budget were considered: both sinks (*in situ* chemical loss and deposition) and sources (*in situ* photochemical production, advection of pollution ozone into the region, production in biomass wildfire plumes and downwards transport from the upper troposphere/stratosphere).

Observations and 1-D photochemical modelling indicate that regional photochemical production and loss in the 0–6-km column are approximately equal. Therefore, net photochemical production is near zero. Influx of stratospheric ozone is of secondary importance (27%), long-range transport of pollution ozone makes a small contribution (9%) and photochemical production of ozone within biomass wildfire plumes is a relatively negligible term (2%) in the budget. Mauzerall *et al.* (1996) found that *in situ* photochemistry within the 0–6-km column accounts for nearly 90% of the ozone mixing ratio within the boundary layer, while above 5 km it accounts for only 40%.

With the purpose of assessing the level of agreement between observed ratios of NO_2 to NO and photochemical modelling, Crawford *et al.* (1996) compared measurements made during the NASA Global Tropospheric Experiment (GTE) Pacific Exploratory Mission-West A (PEM-West A) with photostationary-state model calculations. The results indicate that the value of the ratio $(NO_2)_{expt}/(NO_2)_{calc} = R_e/R_c$ showed a general trend of increasing magnitude with increasing altitude and decreasing latitude. In order to understand causes of discrepancies, various assumptions were made about different components in atmospheric chemical composition. The general conclusion is that incompleteness in the model's chemistry was probably not the major cause of the discrepancy. Measurement errors were probably the major cause, which highlights the need to further develop measurement techniques for NO_2. Chemical adequacy of current photochemical models also requires careful analysis.

Analysis of numerical modelling results indicates that about 300 organic compounds contribute to photochemical processes of surface ozone formation. Derwent *et al.* (1998a) applied a Master Chemical Mechanism (MCM) approach to assess photochemical ozone creation potentials for organic compounds in North-west Europe under conditions of polluted atmospheric boundary layer ($>$2,400 compounds and $>$7,100 chemical reactions were considered to assess the decay of 120 organic compounds). The results obtained are significant for regional planning of measures for pollution prevention.

The most recent development of simulation modelling was marked by outstanding progress in substantation and application of global 3-D photochemical tracer transport models (CTM) to analyse causes and potential future changes in tropospheric ozone (Berntsen *et al.* 1998; Hauglustaine *et al.*, 1998; Wang *et al.*, 1998a, b).

To study photochemistry of the troposphere, Berntsen and Isaksen (1997) developed the model which is based on the NASA Goddard Institute for Space Studies (GISS) CTM with incorporation of an extensive photochemical scheme. Berntsen and Isaksen emphasized that, in view of the strong non-linearities in photochemistry of the troposphere, it has become increasingly evident that 3-D resolution of the atmosphere is crucial in models aimed at studying distribution of ozone and its precursors. In the case considered, model resolution is 8° latitude × 10° longitude with nine vertical levels below 10 hPa. The model includes 3-D transport by transient weather systems and convection coupled with an extensive chemical scheme. The basis for this model was developed by Prather *et al.* (1987). One year of meteorological data from a free-running GCM (NASA/GISS), including advective winds and convection frequency with 8-hour time resolution, is used as input information. The chemical scheme includes 49 components, 85 chemical reactions, and 16 photolytic reactions.

Berntsen and Isaksen (1997) pointed out that their model simulates well the lower tropospheric distribution of key species like carbon monoxide, NMHCs and ozone. The model is also able to simulate the important potential background NO_x distribution. Some discrepancies have been found in the upper troposphere (because of coarse model resolution) between calculated and observed ozone concentration, especially at high latitudes. Numerical modelling, carried out by Berntsen and Isaksen, indicates that net ozone production takes place in the boundary layer and in the upper troposphere, but in the mid-troposphere there is generally a loss of ozone except in the European and North American regions. Globally, net loss in the middle free troposphere is of the same magnitude as net production in the boundary layer. Table 5.9 shows basic features in ozone-budget variability.

It has been found that, for the troposphere as a whole, ozone content is mainly controlled by tropospheric photochemical processes and, to a lesser extent, by injections from the stratosphere.

There are various uncertainties in calculated concentrations but it is difficult to give their general quantification, because it will be different from species to species and also in space and time.

Table 5.9. Global and regional tropospheric ozone budgets calculated in the model.

Region	Stratospheric flux		Deposition (Tg/yr)	Net chemistry (Tg/yr)			
	(10^{10} mol'cm^2) (s)	(Tg/yr)		Boundary layer	Mid-troposphere	Upper troposphere	Total
Global	6.6	846	−1178	797	−840	338	295
Arctic	5.9	39	−70	12.5	−4.2	3.5	11.8
Pacific	−14.3	−7.6	−23	0.6	−24.4	−0.4	−24.2
North America	9.6	44	−76	61.4	6.8	8.9	77.1
Atlantic	−13.7	−44	−17	−0.2	−14.0	5.2	−9.0
Europe	37.6	149	−75	55.3	28.5	13.1	96.9
Asia	17.4	115	−103	72.3	−12.7	4.9	64.5
Northern Hemisphere Tropics	13.1	445	−413	342.4	−495	141.1	−11.1
Southern Hemisphere Tropics	2.7	172	−395	252.4	−325	161.7	88.9

Berntsen and Isaksen (1997) identified the following areas (in particular) toward which model development will be directed in the near future:

1. Aqueous phase chemistry.
2. Improved data on distribution of clouds and improved parametrizations of the impact of clouds on photolysis rates.
3. Improved description of the chemistry of alkyl nitrates.
4. Improved vertical resolution in the boundary layer and around the tropopause.
5. Insufficient knowledge of emissions both with respect to total source strength and geographical distribution needs to be resolved.

A new fundamental effort to develop and apply a global 3-D chemical-transport model (MOZART, version 1) was undertaken by Brasseur *et al.* (1998a) and Hauglestaine *et al.* (1998). This model has been based on the use of the NCAR Climate Community Model and studies the distribution and budget of tropospheric ozone and its precursors. The model has horizontal resolution of 2.8° in latitude and longitude, includes 25 levels in the vertical between the surface and upper boundary, located at approximately 35-km altitude (3 hPa). In its present configuration, the model calculates global distribution of 56 chemical constituents (with a timestep of 20 min) and accounts for surface emission and deposition, large-scale advective transport, subgrid convective and boundary-layer exchanges, chemical and photochemical transformations, as well as wet scavenging. The chemical scheme includes 140 chemical and photochemical reactions and considers the photochemical oxidation schemes of methane (CH_4), ethane (C_2H_6), propane (C_3H_8), ethylene (C_2H_4), propylene (C_3H_6), isoprene (C_5H_8), terpenes (as α-pinene, $C_{10}H_{16}$) and a lumped compound *n*-butane (C_4H_{10}) used as surrogate for other C_4 hydrocarbons (Hauglustaine *et al.*, 1998). Transport is simulated 'off-line' from CCM with dynamical variables provided every 3 hours from pre-established 'history targets'.

Table 5.10. Fields provided by NCAR CCM2 and used to drive MOZART.

Field	Description	Units
PS	Surface pressure	Pa
D_s	Water equivalent snow depth	m
U	Zonal wind component	m/s
V	Meridional wind component	m/s
T	Temperature	K
q	Specific humidity	kg/kg
K_c	Diffusivity of heat in the boundary layer	m^2/s
γ_c	Boundary layer countergradient factor	$1/m^3m$
f_c	Cloud fraction	Fraction
W_{H_2O}	Water-vapour tendency from rainout	kg/kg/s
L	Liquid-water content	kg/kg
η_c	Convective mass flux	$kg/m^2/s$
β_c	Convection overshoot parameter	Fraction

Table 5.10 lists relevant meteorological parameters. Analysis of spatial and temporal variability in the chemical fields predicted by the model suggests that regional events such as summertime ozone episodes in polluted areas can be simulated by MOZART.

Table 5.11 contains a list of chemical species considered in MOZART. As pointed out above, one very important aspect of modelling is adequate prescription of *a priori* information on surface emissions of biogenic and anthropogenic components (fossil-fuel combustion and biomass-burning contributions were specifically taken into account). Table 5.12 shows global emissions for each chemical compound expressed by category (relevant global maps for January and July were drawn).

The chemical scheme used in MOZART (Tables 5.13 and 5.14) is very similar to the scheme implemented previously in the IMAGES model (Müller and Brasseur, 1995). The scheme includes 107 gas phase and 5 heterogeneous reactions and 28 photochemical reactions.

Hauglustaine *et al.* (1998) discussed in detail the application of the MOZART model to simulate global distributions of tropospheric ozone and its precursors. The general conclusion is that the model is able to simulate the distribution of species in the troposphere with fair agreement. MOZART reproduces accurately important transport-chemistry processes, including ventilation of the planetary boundary layer, convective transport and the Walker circulation in the tropics, the impact of monsoon circulation on chemical tracers, and the vertical and seasonal gradients in ozone and nitrogen species in the vicinity of the tropopause. The model simulates distribution of methane, NMHCs and CO successfully enough. However, model evaluation in the tropics stresses the need for better representation of biomass-burning emissions in order to evaluate the budget of carbon monoxide, nitrogen species and ozone with more accuracy in this region. MOZART reproduces NO observations in most parts of the troposphere. Nitric acid, however, is overestimated

Table 5.11. Chemical species considered in MOZART.

Number	Species formula
1	
$O_x = 0(^3P) + O(^1D) + O_3$	
2	$NO_x = N + NO + NO_2$
3	OH
4	HO_2
5	N_2O
6	NO_3
7	HNO_3
8	HO_2NO_2
9	N_2O_5
10	PAN
11	MPAN
12	ONIT
13	CH_4
14	CO
15	H_2
16	C_2H_6
17	C_3H_8
18	C_2H_4
19	C_3H_6
20	C_4H_{10}
21	$C_{10}H_{16}$
22	ISO
23	MVK
24	MACR
25	CH_3COCH_2O
26	CH_3O_2
27	$C_3H_7O_2$
28	$C_2H_5O_2$
29	$C_3H_6OHO_2$
30	$CH_3COCH_2O_2$
31	ISO1
32	MCO_3
33	$MOHO_2$
34	H_2O_2
35	CH_3OOH
36	C_2H_5OOH
37	C_3H_7OOH
38	C_3H_6OHOOH
39	CH_3COCH_2OOH
40	CH_3COOOH
41	CH_2O
42	CH_3CHO
43	CH_2OHCHO
44	CH_3COCHO

45	CH_3CO_3
46	CH_3COCH_3
47	Rn 222
48	Pb 210
49	O_3 stratosphere
50	O_3 inert

PAN, peroxyacetyl nitrate; MPAN, peroxymethacrylic nitrate; C_4H_{10}, surrogate for $\geq C_4$ hydrocarbons; $C_{10}H_{16}$, α-pinene; ISO, isoprene; ISO1, surrogate for peroxy radicals formed from isoprene oxidation; ONIT, surrogate for organic nitrates; MVK, methylvinyl ketone;

Table 5.12. Global surface emissions of trace gases considered in MOZART.

Species	Industrial	Agricultural	Biomass burning	Biogenic	Oceans	Total
CH_4 (Tg/yr)	130.9	179.0	39.4	110.0	9.8	469.1
CO (Tg/yr)	381.6	0.0	661.8	162.1	13.0	1218.5
N_2O (Tg N/yr)	1.1	4.3	0.6	5.6	3.8	15.4
NO_x (Tg N/yr)	21.4	0.0	7.5	6.6	0	35.5
CH_3COCH_3 (Tg/yr)	1.5	0.0	14.0	18.0	0.0	33.5
CH_2O (Tg/yr)	2.3	0.0	0.0	0.0	0.0	2.3
C_2H_6 (Tg C/yr)	6.0	0.0	4.4	0.8	0.8	12.0
C_3H_8 (Tg C/yr)	3.8	0.0	2.3	1.6	1.4	9.1
C_2H_4 (Tg C/yr)	4.3	0.0	13.6	4.3	8.2	30.4
C_3H_6 (Tg C/yr)	1.7	0.0	6.0	0.8	9.9	18.4
C_4H_{10} (Tg C/yr)	63.7	0.0	11.3	73.2	58.2	206.4
Isoprene (Tg C/yr)	0.0	0.0	0.0	220.0	0.0	220.0
Terpenes (Tg C/yr)	0.0	0.0	0.0	129.5	0.0	129.5

above the Pacific up to a factor of 10 and over continental regions by a factor of 2–3. The seasonal cycle of ozone in the troposphere is generally well reproduced by the model in comparison with ozone soundings. MOZART tends, however, to underestimate O_3 at higher latitudes and specifically above 300 hPa. Global mean OH concentration is probably underestimated by 10% in the model. Distribution of radicals (OH, RO_2, HO_2/RO_2) are generally consistent with available measurements. This is an important positive feature of the model because recent studies stressed the significant role played by aldehydes, peroxides and acetone as potential sources of radicals in the upper stratosphere (Prather and Jacob, 1997). Hauglustaine *et al.* (1998) underlined that the global budget of nitrogen species in the troposphere is still an important area for future model improvement.

Now, let us consider some of the results in more detail. Table 5.15 shows components of CO annual budget. Hauglustaine *et al.* (1998) emphasized that the importance of CO is connected with the fact that the reaction of carbon monoxide with hydroxyl radicals provides the major sink of OH in the free troposphere and leads to ozone photochemical production in the presence of NO_x. The most

Table 5.13. Chemical reactions considered in MOZART.

	Reaction	Rate
(R1)	$O(^1D) + N_2 \longrightarrow O + N_2$	$k_1 = 1.80\text{E-}11 \exp(110/T)$
(R2)	$O(^1D) + O_2 \longrightarrow O + O_2$	$k_2 = 3.20\text{E-}11 \exp(70/T)$
(R3)	$O(^1D) + H_2O \longrightarrow 2\,OH$	$k_3 = 2.20\text{E-}10$
(R4)	$O + O_2 + M \longrightarrow O_3 + M$	$k_4 = 6.00\text{E-}34[M](300/T)^{2.3}$
(R5)	$O + O_3 \longrightarrow 2O_2$	$k_5 = 8.00\text{E-}12 \exp(-2,060/T)$
(R6)	$O + OH \longrightarrow HO_2 + O_2$	$k_6 = 2.20\text{E-}11 \exp(120/T)$
(R7)	$O + HO_2 \longrightarrow OH + O_2$	$k_7 = 3.00\text{E-}11 \exp(200/T)$
(R8)	$OH + O_3 \longrightarrow HO_2 + O_2$	$k_8 = 1.60\text{E-}12 \exp(-940/T)$
(R9)	$HO_2 + O_3 \longrightarrow OH + 2O_2$	$k_9 = 1.10\text{E-}14 \exp(-500/T)$
(R10)	$HO_2 + HO_2 \longrightarrow H_2O_2$	$k_{10} = (k_a + k_b)k_c$
		$k_a = 2.30\text{E-}13 \exp(600/T)$
		$k_b = 1.70\text{E-}33[M] \exp(1,000/T)$
		$k_c = 1 + 1.40\text{E-}21\ [H_2O] \exp(2,200/T)$
(R11)	$H_2O_2 + OH \longrightarrow H_2O + HO_2$	$k_{11} = 2.90\text{E-}12 \exp(-160/T)$
(R12)	$OH + HO_2 \longrightarrow H_2O + O_2$	$k_{12} = 4.80\text{E-}11 \exp(250/T)$
(R13)	$OH + OH \longrightarrow H_2O + O$	$k_{13} = 4.20\text{E-}12 \exp(-240/T)$
(R14)	$H_2 + OH \longrightarrow H_2O + HO_2$	$k_{14} = 5.50\text{E-}12 \exp(-2,000/T)$
(R15)	$H_2 + O(^1D) \longrightarrow HO_2 + OH$	$k_{15} = 1.00\text{E-}10$
(R16)	$N_2O + O(^1D) \longrightarrow 2NO$	$k_{16} = 6.70\text{E-}11$
(R17)	$N_2O + O(^1D) \longrightarrow N_2 + O_2$	$k_{17} = 490\text{E-}11$
(R18)	$N + O_2 \longrightarrow NO + O$	$k_{18} = 1.50\text{E-}11 \exp(-3,600/T)$
(R19)	$N + NO \longrightarrow N_2 + O$	$k_{19} = 2.10\text{E-}11 \exp(100/T)$
(R20)	$NO + HO_2 \longrightarrow NO_2 + OH$	$k_{20} = 3.50\text{E-}12 \exp(250/T)$
(R21)	$NO + O_3 \longrightarrow NO_2 + O_2$	$k_{21} = 2.00\text{E-}12 \exp(-1,400/T)$
(R22)	$NO_2 + O \longrightarrow NO + O_2$	$k_{22} = 6.50\text{E-}12 \exp(120/T)$
(R23)	$NO_2 + O_3 \longrightarrow NO_3 + O_2$	$k_{23} = 120\text{E-}13 \exp(-2,450/T)$
(R24)	$NO_3 + HO_2 \longrightarrow 0.4HNO_3 + 0.6OH + 0.6NO_2$	$k_{24} = 2.30\text{E-}12 \exp(170/T)$
(R25)	$NO_2 + NO_3 + M \longrightarrow N_2O_5 + M$	$k_0 = 2.20\text{E-}30\ (300/T)^{3.9}$
		$k_\infty = 1.50\text{E-}12\ (300/T)^{0.7}$
		$F_c = 0.6$
(R26)	$N_2O_5 + M \longrightarrow NO_2 + NO_3 + M$	$k_{26} = k_{25}\,3.7\text{E}26 \exp(-11,000/T)$
(R27)	$N_2O_5 + H_2O \longrightarrow 2\,HNO_3$	$k_{27} = 2.00\text{E-}21$
(R28)	$NO_2 + OH + M \longrightarrow HNO_3 + M$	$k_0 = 2.50\text{E-}30\ (300/T)^{4.4}$
		$k_\infty = 1.60\text{E-}11\ (300/T)^{1.7}$
		$F_c = 0.6$
(R29)	$HNO_3 + OH \longrightarrow NO_3 + H_2O$	$k_{29} = k_a + 7.20\text{E-}15 \exp(785/T)$
		$k_a = k_b/(1 + k_b/k_c)$
		$k_b = 1.90\text{E-}33 \exp(725/T)[M]$
		$k_c = 4.10\text{E-}16 \exp(1,440/T)$
(R30)	$NO_3 + NO \longrightarrow 2NO_2$	$k_{30} = 1.50\text{E-}11 \exp(170/T)$
(R31)	$NO_2 + HO_2 + M \longrightarrow HO_2NO_2 + M$	$k_0 = 4.80\text{E-}31\ (300T)^{3.2}$
		$k_\infty = 4.70\text{E-}12\ (300T)^{1.4}$
		$F_c = 0.6$
(R32)	$HO_2NO_2 + M \longrightarrow HO_2 + NO_2 + M$	$k_{32} = k_{31}\,4.7\text{E}26 \exp(-10,900/T)$
(R33)	$HO_2NO_2 + OH \longrightarrow H_2O + NO_2 + O_2$	$k_{33} = 1.30\text{E-}12 \exp(380/T)$

(R34)	$CH_4 + OH \longrightarrow CH_3O_2 + H_2O$	$k_{34} = 2.45E\text{-}12 \exp(-1,775/T)$
(R35)	$CH_4 + O(^1D) \longrightarrow CH_3O_2 + OH$	$k_{35} = 1.40E\text{-}10$
(R36)	$CH_4 + O(^1D) \longrightarrow H_2 + CH_2O$	$k_{36} = 1.40E\text{-}11$
(R37)	$CH_3O_2 + NO \longrightarrow CH_2O + NO_2 + HO_2$	$k_{37} = 3.00E\text{-}12 \exp(280/T)$
(R38)	$CH_3O_2 + CH_3O_2 \longrightarrow 1.4CH_2O + 0.8HO_2$	$k_{38} = 2.50E\text{-}13 \exp(190/T)$
(R39)	$CH_3O_2 + HO_2 \longrightarrow CH_3OOH + O_2$	$k_{39} = 3.80E\text{-}13 \exp(800/T)$
(R40)	$CH_3OOH + OH \longrightarrow 0.58CH_3O_2 + 0.42OH + 0.42CH_2O$	$k_{40} = 3.80E\text{-}12 \exp(200/T)$
(R41)	$CH_2O + OH \longrightarrow CO + H_2O + HO_2$	$k_{41} = 1.10E\text{-}11$
(R42)	$CH_2O + NO_3 \longrightarrow CO + HO_2 + HNO_3$	$k_{42} = 6.00E\text{-}13 \exp(-2,058/T)$
(R43)	$CO + OH \longrightarrow CO_2 + HO_2$	$k_{43} = 1.50E\text{-}13\ (1 + 0.6P)$
(R44)	$C_2H_6 + OH \longrightarrow C_2H_5O_2 + H_2O$	$k_{44} = 8.70E\text{-}12 \exp(-1,070/T)$
(R45)	$C_2H_5O_2 + NO \longrightarrow CH_3CHO + HO_2 + NO_2$	$k_{45} = 2.60E\text{-}12 \exp(365/T)$
(R46)	$C_2H_5O_2 + HO_2 \longrightarrow C_2H_5OOH + O_2$	$k_{46} = 7.50E\text{-}13 \exp(700/T)$
(R47)	$C_2H_5O_2 + CH_3O_2 \longrightarrow 0.7CH_2O + 0.8CH_3CHO + HO_2$	$k_{47} = 3.75E\text{-}13\ \exp(-40/T)$
(R48)	$C_2H_5O_2 + C_2H_5O_2 \longrightarrow 1.6CH_3CHO + 1.2HO_2$	$k_{48} = 6.50E\text{-}14$
(R49)	$C_2H_5OOH + OH \longrightarrow 0.5C_2H_5O_2 + 0.5CH_3CHO + 0.5OH$	$k_{49} = 3.80E\text{-}12 \exp(200/T)$
(R50)	$C_3H_8 + OH \longrightarrow C_3H_7O_2 + H_2O$	$k_{50} = 1.00E\text{-}11 \exp(-660/T)$
(R51)	$C_3H_7O_2 + NO \longrightarrow 0.82CH_3COCH_3 + NO_2 + HO_2$	$k_{51} = 4.20E\text{-}12 \exp(180/T)$
(R52)	$C_3H_7O_2 + HO_2 \longrightarrow C_3H_7OOH + O_2$	$k_{52} = 7.50E\text{-}13 \exp(700/T)$
(R53)	$C_3H_7O_2 + CH_3O_2 \longrightarrow 0.5CH_2O + HO_2 + 0.82CH_3COCH_3$	$k_{53} = 3.75E\text{-}13\ \exp(-40/T)$
(R54)	$C_3H_7OOH + OH \longrightarrow H_2O + C_3H_7O_2$	$k_{54} = 3.80E\text{-}12 \exp(200/T)$
(R55)	$CH_3COCH_3 + OH \longrightarrow CH_3COCH_2O_2 + H_2O$	$k_{55} = 2.20E\text{-}12 \exp(-685/T)$
(R56)	$CH_3COCH_2O_2 + NO \longrightarrow CH_3COOH_2O + NO_2$	$k_{56} = 4.20E\text{-}12 \exp(180/T)$
(R57)	$CH_3COCH_2O_2 + HO_2 \longrightarrow CH_3COCH_2OOH + O_2$	$k_{57} = 7.50E\text{-}13 \exp(700/T)$
(R58)	$CH_3COCH_2OOH + OH \longrightarrow CH_3COOH_2O_2 + H_2O$	$k_{58} = 3.80E\text{-}12 \exp(200/T)$
(R59)	$CH_3COCH_2O + M \longrightarrow CH_3CO_3 + CH_2O + M$	$k_{59} = 8.70E + 13 \exp(-7,300/T)$
(R60)	$CH_3COCH_2O + O_2 \longrightarrow CH_3COCHO + HO_2$	$k_{60} = 6.50E\text{-}14 \exp(-1,400/T)$
(R61)	$C_4H_{10} + OH \longrightarrow 0.8ISO1$	$k_{61} = 1.55E\text{-}11 \exp(-540/T)$
(R62)	$C_3H_6 + OH + M \longrightarrow C_3H_6OHO_2 + M$	$k_0 = 8.00E\text{-}27\ (300/T)^{3.5}$; $k_\infty = 3.00E\text{-}11$; $F_c = 0.5$
(R63)	$C_3H_6 + O_3 \longrightarrow 0.532CH_2O + 0.585HO_2 + 0.4275OH + 0.08CH_4 + 0.37CO + 0.5CH_3CHO + 0.2875CH_3O_2$	$k_{63} = 6.50E\text{-}15 \exp(-1,900/T)$
(R64)	$C_3H_6 + NO_3 \longrightarrow ONIT$	$k_{64} = 4.00E\text{-}15$
(R65)	$C_3H_6OHO_2 + NO \longrightarrow CH_3CHO + CH_2O + HO_2 + NO_2$	$k_{65} = 4.20E\text{-}12 \exp(180/T)$
(R66)	$C_3H_6OHO_2 + HO_2 \longrightarrow C_3H_6OHOOH + O_2$	$k_{66} = 6.50E\text{-}13 \exp(650/T)$
(R67)	$C_3H_6OHOOH + OH \longrightarrow 0.5C_3H_6OHO_2 + 0.5OH + H_2O$	$k_{67} = 3.80E\text{-}12 \exp(200/T)$
(R68)	$CH_3OHO + OH \longrightarrow CH_3CO_3 + H_2O$	$k_{68} = 5.60E\text{-}12 \exp(270/T)$
(R69)	$CH_3OHO + NO_3 \longrightarrow CH_3CO_3 + HNO_3$	$k_{69} = 1.40E\text{-}12 \exp(-1,900/T)$
(R70)	$CH_3CO_3 + NO \longrightarrow CH_3O_2 + CO_2 + NO_2$	$k_{70} = 5.30E\text{-}12 \exp(360/T)$
(R71)	$CH_3CO_3 + NO_2 + M \longrightarrow PAN + M$	$k_0 = 9.70E\text{-}29\ (300/T)^{5.6}$; $k_\infty = 9.30E\text{-}12\ (300/T)1.5 = 0.6$; $F_c = 0.6$
(R72)	$PAN + M \longrightarrow CH_3CO_3 + NO_2 + M$	$k_{72} = k_{71}\ 1.11E28 \exp(-14,000/T)$

Table 5.13 (*cont.*)

	Reaction	Rate
(R73)	$CH_3CO_3 + HO_2 \longrightarrow 0.67CH_3COOOH$	$k_{73} = 4.30E\text{-}13 \exp(1{,}040/T)$
	$+ 0.33CH_3COOH + 0.33O_3$	$k_{73} = 4.30E\text{-}13 \exp(1040/T)$
(R74)	$CH_3CO_3 + CH_3O_2 \longrightarrow CH_3O_2 + CH_2O + HO_2 + CO_2$	$k_{74} = 1.80E\text{-}09 \exp(-1{,}800/T)$
(R75)	$CH_3CO_3 + CH_3O_2 \longrightarrow CH_2O + CH_3COOH + O_2$	$k_{75} = 4.10E\text{-}15 \exp(2{,}100/T)$
(R76)	$CH_3CO_3 + CH_3CO_3 \longrightarrow 2CH_3O_2 + 2CO_2$	$k_{76} = 2.90E\text{-}12 \exp(500/T)$
(R77)	$CH_3COOOH + OH \longrightarrow CH_3CO_3 + H_2O$	$k_{77} = 1.00E\text{-}11$
(R78)	$C_2H_4 + OH + M \longrightarrow 0.6667C_3H_6OHO_2 + M$	$k_0 = 1.00E\text{-}28\ (300/T)^{0.8}$
		$k_\infty = 8.80E\text{-}12$
		$F_c = 0.6$
(R79)	$C_2H_4 + O_3 \longrightarrow CH_2O + 0.52HO_2 + 0.42CO$	$k_{79} = 1.20E\text{-}14 \exp(-2{,}630/T)$
(R80)	$ISO + OH \longrightarrow ISO1$	$k_{80} = 2.50E\text{-}11 \exp(450/T)$
(R81)	$ISO + O_3 \longrightarrow 0.66MACR + 0.26MVK + 0.45O$	$K_{81} = 1.20E\text{-}14 \exp(-2{,}013/T)$
	$+ 0.7825CH_2O + 0.06HO_2 + 0.68OH$	
	$+ 0.07C_3H_6 + 0.19CO$	
(R82)	$ISO + NO_3 \longrightarrow ONIT$	$k_{82} = 3.03E\text{-}12 \exp(-466/T)$
(R83)	$ISO1 + NO \longrightarrow 0.42MVK + 0.58MACR + CH_2O$	
	$+ HO_2 + NO_2$	$k_{83} = 3.70E\text{-}12 \exp(180/T)$
(R84)	$ISO1 + NO \longrightarrow 0.59CH_3OO_3 + CH_2O$	$k_{84} = 5.00E\text{-}13 \exp(180/T)$
	$+ 0.59CH_2OHCHO + 0.59NO_2$	
	$+ 0.41ONIT$	
(R85)	$ISO1 + HO_2 \longrightarrow 0.126MVK + 0.174MACR + 0.3CH_2O$	$k_{85} = 7.50E\text{-}13 \exp(750/T)$
	$+ 0.3HO_2$	
(R86)	$ISO1 + CH_3O_2 \longrightarrow MACR + 2CH_2O + 2HO_2$	$k_{86} = 1.30E\text{-}14$
(R87)	$ISO1 + CH_3CO_3 \longrightarrow MACR + HO_2 + CH_2O$	$k_{87} = 4.90E\text{-}14$
(R88)	$MVK + OH \longrightarrow MOHO_2$	$k_{88} = 4.13E\text{-}12 \exp(452/T)$
(R89)	$MVK + O_3 \longrightarrow 0.05CO + 0.06HO_2 + 0.04CH_3OHO$	$k_{89} = 4.00{=}\text{-}15 \exp(-2{,}000/T)$
	$+ 0.82CH_3COCHO + 0.8CH_2O + 0.2O$	
	$+ 0.8OH$	
(R90)	$MACR + OH \longrightarrow 0.5MOHO_2 + 0.5MCO_3$	$k_{90} = 1.86E\text{-}11 \exp(175/T)$
(R91)	$MACR + O_3 \longrightarrow 0.275HO_2 + 0.215OH$	$k_{91} = 4.40E\text{-}15 \exp(-2{,}500/T)$
	$+ 0.8CH_3COCHO + 0.7CH_2O + 0.2O$	
	$+ 0.2CO$	
(R92)	$MOHO_2 + NO \longrightarrow CH_2OHCHO + CH_3CO_3 + NO_2$	$k_{92} = 3.50E\text{-}12 \exp(180/T)$
(R93)	$MOHO_2 + NO \longrightarrow ONIT$	$k_{93} = 4.50E\text{-}13 \exp(180/T)$
(R94)	$MOHO_2 + HO_2 \longrightarrow 0.3CH_3CO_3 + 0.3CH_2OHCHO$	$k_{94} = 7.50E\text{-}13 \exp(700/T)$
(R95)	$CH_2OHCHO + OH \longrightarrow 0.5CH_3CO_3 + 0.5HO_2$	$k_{95} = 6.00E\text{-}12 \exp(250/T)$
(R96)	$CH_2OHCHO + NO_3 \longrightarrow CH_3CO_3 + HNO_3$	$k_{96} = 1.40E\text{-}12 \exp(-1{,}900/T)$
(R97)	$MCO_3 + NO \longrightarrow CH_3COCH_2O + NO_2$	$k_{97} = 2.40E\text{-}11$
(R98)	$MCO_3 + NO_2 \longrightarrow MPAN$	$k_0 = 9.70E\text{-}29\ (300/T)^{5.6}$
		$k_\infty = 9.30E\text{-}12\ (300/T)^{1.5}$
		$F_c = 0.6$
(R99)	$MPAN + M \longrightarrow MCO_3 + NO_2 + M$	$k_{99} = k_{86}\ 1.11E28 \exp(-14{,}000/T)$
(R100)	$MCO_3 + HO_2 \longrightarrow MOHO_2$	$k_{100} = 4.50E\text{-}13 \exp(1{,}040/T)$
(R101)	$MCO_3 + CH_3O_2 \longrightarrow CH_2O + HO_2 + CH_3COCH_2O$	$k_{101} = 2.20E\text{-}12 \exp(490/T)$
(R102)	$MCO_3 + CH_3CO_3 \longrightarrow CH_3COCH_2O + CH_3O_2$	$k_{102} = 5.00E\text{-}12 \exp(550/T)$

(R103)	$MCO_3 + MCO_3 \longrightarrow 2CH_3COCH_2O$	$k_{103} = 2.50E\text{-}12 \exp(550/T)$
(R104)	$ONIT + OH \longrightarrow NO_2 + MOHO_2$	$k_{104} = 6.80E\text{-}13$
(R105)	$C_{10}H_{16} + OH \longrightarrow 1.64ISO1 + CH_3COCH_3$	$k_{105} = 1.20E\text{-}11 \exp(444/T)$
(R106)	$C_{10}H_{16} + O_3 \longrightarrow 1.122MACR + 0.442MVK + 0.765O + 1.156OH + 0.119C_3H_6 + 1.326CH_2O + 0.323CO + 0.102HO_2$	$k_{106} = 9.90E\text{-}16 \exp(-730/T)$
(R107)	$C_{10}H_{16} + NO_3 \longrightarrow 1.7ISO1 + NO_2$	$k_{107} = 5.60E\text{-}11 \exp(-650/T)$
(R108)	$N_2O_5 + SO_4^{=} \longrightarrow HNO_3 + HNO_3$	$k_{108} = f([SO_4^{=}], T, [H_2O])$
(R109)	$NO_3 + SO_4^{=} \longrightarrow HNO_3$	$k_{109} = f([SO_4^{=}], T, [H_2O])$
(R110)	$MVK + SO_4^{=} \longrightarrow$ products	$k_{110} = f([SO_4^{=}], T, [H_2O])$
(R111)	$MACR + SO_4^{=} \longrightarrow$ products	$k_{111} = f([SO_4^{=}], T, [H_2O])$
(R112)	$CH_3COCHO + SO_4^{=} \longrightarrow$ products	$k_{112} = f([SO_4^{=}], T, [H_2O])$

Read 1.80E-11 as 1.80×10^{-11}. T is temperature (K), $[M]$ atmospheric density (cm^{-3}), $[H_2O]$ water-vapour density (cm^{-3}), P pressure (hPa), $[SO_4^{=}]$ sulphate density. The three body reaction rates are calculated from:

$$k = \frac{k_0[M]}{1 + k_0[M]/k_\infty} F_c \{1 + [\lg(k_0[M]/k_\infty)]^2\}^{-1}$$

Table 5.14. Photolytic reactions considered in MOZART.

(R1)	$O_2 + h\nu \longrightarrow O + O$
(R2)	$O_3 + h\nu \longrightarrow O(^1D) + O_2$
(R3)	$O_3 + h\nu \longrightarrow O + O_2$
(R4)	$N_2O + h\nu \longrightarrow O(^1D) + N_2$
(R5)	$NO + h\nu \longrightarrow N + O$
(R6)	$NO_2 + h\nu \longrightarrow NO + O$
(R7)	$N_2O_5 + h\nu \longrightarrow NO_2 + NO_3$
(R8)	$HNO_3 + h\nu \longrightarrow NO_2 + OH$
(R9)	$NO_3 + h\nu \longrightarrow 0.89NO_2 + 0.11NO + 0.89O_3$
(R10)	$HO_2NO_2 + h\nu \longrightarrow NO_2 + HO_2$
(R11)	$CH_3OOH + h\nu \longrightarrow CH_2O + HO_2 + OH$
(R12)	$CH_2O + h\nu \longrightarrow CO + 2HO_2$
(R13)	$CH_2O + h\nu \longrightarrow CO + H_2$
(R14)	$H_2O + h\nu \longrightarrow OH + HO_2$
(R15)	$H_2O_2 + h\nu \longrightarrow OH + OH$
(R16)	$CH_3CHO + h\nu \longrightarrow CH_3O_2 + CO + HO_2$
(R17)	$C_3H_6OHOOH + h\nu \longrightarrow CH_3CHO + CH_2O + HO_2 + OH$
(R18)	$CH_3COOOH + h\nu \longrightarrow CH_3O_2 + OH$
(R19)	$PAN + h\nu \longrightarrow CH_3CO_3 + NO_2$
(R20)	$MPAN + h\nu \longrightarrow MCO_3 + NO_2$
(R21)	$CH_2OHCHO + h\nu \longrightarrow CH_2O + CO + 2HO_2$
(R22)	$MACR + h\nu \longrightarrow MCO_3 + CO + 0.6CH_3COCHO + 0.4CH_3CO_3 + 0.4CH_2O + 2.6HO_2$
(R23)	$MVK + h\nu \longrightarrow CH_3CO_3 + HO_2 + 0.25CH_2O + 0.25CO$
(R24)	$C_2H_5OOH + h\nu \longrightarrow CH_3CHO + HO_2 + OH$
(R25)	$C_3H_7OOH + h\nu \longrightarrow 0.82CH_3COCH_3 + OH + HO_2$
(R26)	$CH_3COCH_2OOH + h\nu \longrightarrow CH_3COCH_2O + OH$
(R27)	$CH_3COCH_3 + h\nu \longrightarrow CH_3CO_3 + CH_3O_2$
(R28)	$CH_3COCHO + h\nu \longrightarrow CH_3CO_3 + CO + HO_2$

Table 5.15. Annual budget of carbon monoxide in the troposphere (below 250 hPa) calculated by MOZART (Tg CO/yr).

	Northern Hemisphere	Southern Hemisphere	Global
Surface emission	847	372	1,219
Photochemical production	539	342	881
Total source	1,386	714	2,100
Photochemical destruction	−1,092	−638	−1,730
Dry deposition	−129	−61	−190
Total sink	−1,221	−99	−1,920
Burden (Tg CO)	196	125	321
Lifetime (month)	1.9	2.1	2.0

important source is direct emission from incomplete combination of fossil fuel and biomass burning. In addition, carbon monoxide is produced by oxidation of methane and NMHCs. Calculations of CO global distribution indicate highest concentrations, resulting from anthropogenic activities, over the industrial regions of Europe, Asia and North America. CO mixing ratios over these areas are typically 300–400 ppb during winter and 200–300 ppb during summer. The effect of biomass burning over Africa and South America is also visible, with CO surface concentrations of typically 200–300 ppb (biomass burning is most intense during the dry season). Comparison with NASA MAPS (measurement of air pollution from satellites) instrumentational data (MAPS flew aboard the Space Shuttle in November 1981, October 1984, and in April and October 1994) shows good enough agreement (the data cover latitudes from 57°N to 57°S), although the model tends to overestimate CO concentration over Africa (probably because either biomass burning CO emissions or CO production from isoprene are overestimated in the model or simulated vertical transport is too vigorous).

As can be seen in Table 5.13, surface primary emissions constitute 58% of total CO source on the global scale, while photochemical production associated with degradation of methane and NMHCs accounts for the remainder. Approximately 70% of CO surface emission and 60% of its photochemical production take place in the Northern Hemisphere. The apparent imbalance of the CO budget (180 Tg/yr, or 9% of the total source) is associated with transport to the stratosphere. Annual average tropospheric burden of CO is 321 Tg (60% located in the Northern Hemisphere).

Analysis of model data on nitrogen species ($NO_x = NO + NO_2$; $NO_y = NO_x + NO_3 + 2N_2O_5 + HNO_3 + HNO_4 + PAN + MPAN + \text{nitrate}$) made by Hauglustaine *et al.* (1998) indicates (in the case of the monthly mixing ratio of NO_x at the surface and at the 500-hPa level for January and July conditions) the large impact of atmospheric sources in the boundary layer in the Northern Hemisphere. The contribution of biomass burning in the tropics is visible. Maximum NO_x concentrations (its lifetime is of the order of a few days or less in the boundary layer) were found over industrial and urbanized regions of the eastern United States,

Table 5.16. Annual budget of ozone in the troposphere (below 250 hPa) calculated by MOZART (Tg O_3/yr).

	Northern Hemisphere	Southern Hemisphere	Global
Photochemical production	2,023	995	301
Stratospheric influx	170	221	391
Total source	2,193	1,216	3,409
Photochemical destruction	−1,540	−971	−2,511
Dry deposition	−625	−273	−898
Total sink	−2,165	−1,244	−3,409
Burden (Tg)	114	79	193
Lifetime (days)	19	23	21

California, western and central Europe, China and Japan. Typical mixing ratios in these regions are in the range 1–5 ppb (according to Horowitz *et al.* (1998), this is in the higher range of observations).

Hauglustaine *et al.* (1998) pointed out that the simulated total ozone distribution is characterized by low tropical values (typically about 250 DU) and high polar values, particularly in the winter hemisphere (of the order of 380–450 DU in the Northern Hemisphere and 280–320 DU in the Southern Hemisphere during January), in agreement with observations. Table 5.16 illustrates calculations of annual ozone-budget components.

Global ozone photochemical production is 3,018 Tg/yr (NH contribution dominates). Stratospheric ozone intrusion (391 Tg/yr) contributes 11% to the global source of ozone in the troposphere (8% and 11% for the NH and SH, respectively). Photochemical destruction is 2,511 Tg/yr. Dry deposition at the surface (898 Tg/yr) contributes 26% to the global sink of tropospheric ozone (about 30% and 20% in the NH and SH, respectively). These data clearly demonstrate that the budget of tropospheric ozone is dominated by photochemistry. Typically, net photochemical production (production – destruction) is the small difference between two large terms. Calculations show that net tropospheric ozone production exhibits a strong seasonal cycle, varying from about 200 Tg/yr in January to 600–700 Tg/yr in June–September. Strong seasonal variability also takes place in the case of ozone influx from the stratosphere across the 250-hPa upper boundary, with annual maximum of about 1,200 Tg/yr in March-April. Dry deposition is characterized by a strong seasonal cycle in the Northern Hemisphere associated with larger vegetation uptake during summer.

Approximately 60% of the global tropospheric burden (163 Tg) is found in the Northern Hemisphere. The calculated lifetime of ozone is about 1 month in global and annual average. When surface dry deposition is considered, the global lifetime decreases by about 20 days. This lifetime exhibits strong seasonal and geographical variability. In the Northern Hemisphere, lifetimes range from 15 days in summer to about 30 days in winter on average.

Table 5.17. Annual and global budget of tropospheric ozone calculated by various CTMs (Tg O_3/yr).

	Photochemistry			Influx		
	Production	Destruction	Net	Stratosphere	Deposition	Net
Müller and Brasseur (1995)	4,550	−4,000	550	550	−1,100	−550
Lelieveld and van Dorland (1995)	3,609	−3,183	426	528	−953	425
Roelofs and Lelieveld (1995)	3,206	−3,037	169	575	−740	−165
Levy *et al.* (1997)			128	696	−825	−129
Bernsten and Isaksen (1997)			295	846	−1,178	−332
Roelofs *et al.* (1997)	3,415	−3,340	75	459	−534	−75
Wang *et al.* (1998)	4,300	−3,810	490	400	−890	−490
Hauglustaine *et al.* (1998)	3,018	−2,511	507	391	−898	−507

The level of uncertainties in TrOZ simulation modelling are shown in Table 5.17, which demonstrate significant differences between models, depending on model assumptions as well as on definition of the domain under consideration. It is, of course, particularly interesting that significant disagreement exists between the results of the most recent 3-D global tropospheric ozone modelling (Hauglustaine *et al.*, 1998; Wang *et al.*, 1998a, b).

Background tropospheric columns calculated by Hauglustaine *et al.* (1998) are typically 15–25 DU. Enhanced TrOZ values are calculated as a result of anthropogenic activities, including fossil-fuel combustion in the Northern Hemisphere and biomass-burning emissions in the tropics. The following basic features of TrOZ global distribution are in the agreement with observational data considered by Fishman and Brackett (1997): NH high ozone values; evidence of the plume structure emanating from North America, Asia, Europe and Africa; relatively lower TrOZ over the Rockies, Himalayas and Andes because of surface elevation; low ozone amount over the tropical Pacific Ocean where convective transport is dominant, and high TrOZ values over the tropical South Pacific associated with biomass burning in Africa and South America. For example, between September and October calculated TrOZ values reach a maximum of 40 DU in Northern Hemisphere plumes, 35 DU over the tropical Atlantic, 30–35 DU in the southern part of Australia (African plume), and a minimum of less than 20 DU in the central Pacific. At mid-latitudes, observed TrOZ values reach 30–35 DU, whereas MOZART produces values of only 15–20 DU (the disagreement may be partly due to differences in definition of the tropopause height).

Hauglustaine *et al.*, (1998) emphasized that, although many of results provided by MOZART are encouraging, the model will continuously require improved formulations and parametrizations of physical and photochemical processes. Future research needs to address the problem of obtaining more adequate observational data, which are very important for both improvement and validation of global models (Kritz *et al.*, 1998; Stockwell *et al.*, 1998). More complete and precise laboratory data on various reaction rates are also necessary.

The important phenomenon of surface ozone depletion at polar sunrise fuelled by sea-salt halogens may be used for model-verification purposes. Barrie *et al.* (1998a, b, c) pointed out that at polar sunrise the Arctic troposphere (0 to ~8 km) becomes a unique chemical reactor, influenced by human activity and the Arctic Ocean. At this time, a relatively high abundance of photochemically reactive compounds and aerosols is found in the troposphere, which results in O_3 depletion in the surface-based inversion layer over the Arctic Ocean (a similar phenomenon probably develops in the Antarctic as well). From mid-March to the end of May, when sunlight is present and the Arctic Ocean is still frozen, groundbased O_3 mixing ratios decrease from mid-winter mean of 30 to 40 ppb to < 1 ppb ('surface ozone hole' is formed). It can be assumed that the involvement in polar-sunrise chemistry of BrO/Br, ClO/Cl and possibly IO/I, acting individually or in combination, as well as the important role of heterogeneous reactions on ice or acid aerosols, is responsible for observed O_3 destruction. In contrast to stratospheric O_3 depletion, Br atoms rather than Cl are responsible for most of the O_3 destruction. Strong evidence is now emerging to show that atmospheric reactions involving oxidants lead to autocatalytic liberation of reactive bromine from sea salt. In this context, the ARC Troposphere O_3 Chemistry (ARC-TOC) project was organized by European scientists.

Although complex 3-D global transport/chemistry models are undoubtedly decisive tools in simulation modelling, it would be unwise to ignore the need for further development of simpler models. Fortunately, such efforts are successfully continuing.

Klonecky and Levy (1997) used a photochemical box model with CO-CH_4-NO_y-H_2O chemistry to calculate the diurnally averaged net photochemical rate of change in ozone (called the chemical ozone tendency) in the troposphere for different values of parameters: NO_x and ozone concentration, temperature, humidity, CO concentration, and surface albedo. Relevant sensitivity studies indicate that the magnitude of the tendency decreases rapidly with height mostly as a result of lower absolute humidity and temperature. In the upper troposphere (at 190 hPa), maximum tendencies are below 2 ppbv/day Lower temperature and specific humidity cause a shift in NO_x values. In the upper troposphere, the net tendency is at least as sensitive to variations in H_2O concentration as to NO_x. The latter conclusion, reached by Klonecky and Levy (1997), suggests a possible synergism between direct NO_x pollution by aircraft and indirect modification of H_2O by climate change.

Klonecky and Levy (1997) used box-model-calculated rates for the chemical-tendency terms of ozone in a simulation conducted with the 3-D global chemistry/transport model (GCTM). The results indicate that chemical-tendency fields show strong correlation with NO_x fields. In contrast with the lower and mid-troposphere, where tendencies are negative in remote regions over the oceans, in the upper troposphere, where NO_x is generally greater than 50 pptv and the balance is low, the tendencies are generally small but positive. GCTM simulations of preindustrial ozone show that in the upper troposphere present-day ozone tendencies are greater than simulated preindustrial tendencies. Since in the mid-troposphere NO_x sources are strongest over land, there is generally production of ozone over land and destruction over the sea. In the upper troposphere this land/sea contrast is generally absent.

Tie and Hess (1997) used GCTM to study ozone mass exchange between the stratosphere and troposphere under conditions of background and volcanic sulphur aerosols in the northern and southern extratropical lowermost stratosphere and in the tropical upper troposphere. Calculations showed that the stratospheric ozone mass flux into the troposphere is highest during the spring months in both hemispheres but higher in the NH than in the SH. During background aerosol conditions, the net modelled stratospheric flux of ozone to the troposphere is 792 Tg/yr. The results obtained indicate a significant impact on ozone-mass flux across the tropopause of heterogeneous chemical reactions occurring on the surface of sulphate aerosols, leading to a possible 15% reduction in the annual global ozone mass transported from the stratosphere to the troposphere after a large volcanic eruption. Most of the reduction in ozone-mass flux occurs in the Northern Hemisphere. In the Southern Hemisphere winter and spring, ozone concentration is already significantly perturbed by heterogeneous reactions occurring on the surface of polar stratospheric cloud particles over Antarctica.

A number of recent studies have been devoted to improvement in various aspects of tropospheric chemistry modelling.

Kindler *et al.* (1998) pointed out that one of the substantial uncertainties in simulation and prediction of minor species change is the poor mechanistic understanding of the role of advective and subgrid scale transport in redistribution of key species. This requires further improvement in GCTMs with the aim of accurately simulating current distributions of all major atmospheric species. In this context, Kindler *et al.* (1998) undertook an evaluation using ^{14}C and N_2O simulations of 3-D transport driven by UK Meteorological Office and Goddard Space Flight Center assimilated winds. Their principal conclusion was that, because of significant discrepancies between model simulations, using both wind-field sets, and the observations, it is necessary to be careful when using these assimilated winds for long-term model studies.

Brunner *et al.*, (1998) used aircraft observational data over the Northern Hemisphere to examine the role of large-scale nitrogen-oxide plumes in the tropopause for ozone. It was found that upper-tropospheric nitrogen-oxide distribution is strongly influenced by large-scale plumes, extending about 100 to 1,300 km along flight tracks. Because the plumes were frequently observed downwind of thunderstorms and frontal systems, it can be assumed that they were formed because of upward transport of polluted air from the continental boundary layer. Some contribution may also belong to nitrogen-oxide production by lightning strikes.

The origin of NO_x in the upper troposphere over the central United States has been studied by Jaeglé *et al.* (1998a), using aircraft observations obtained during the SUCCESS campaign in April–May of 1996. They discovered correlations between NO_y and CO at the 8–12-km altitude, which indicates that NO_x originates primarily from convective transport of polluted boundary-layer air. According to Jaeglé *et al.*

(1998a), lightning and aircraft emissions appear to be only minor sources of NO_x. The principal conclusion is that the NO_x/NO_y ratio is maintained above the chemical steady state by frequent convective injections of fresh NO_x from the polluted boundary layer and by the long lifetime of NO_x in the upper troposphere (5–10 days). In contrast to previous studies, no evidence for fast heterogeneous recycling from HNO_3 to NO_x in the upper troposphere was found.

Using observational data obtained during the same SUCCESS aircraft campaign in April–May 1996, Jaeglé *et al.* (1998b) assessed the sources of NO_x (out peroxy radicals) and the associated production of ozone at 8–12 km over the United States, by examining modelled observational data of OH, HO_2, NO and other species. The results indicate that HO_x concentrations measured in SUCCESS were up to a factor of 3 higher than can be calculated from oxidation of water vapour and photolysis of acetone. The highest discrepancy was found in the outflow of a convective storm. It has been shown that convective injection of peroxides (CH_3COOH and H_2O_2) and formaldehyde (CH_2O) from the boundary layer to the upper troposphere could resolve this discrepancy. A general conclusion is that, in the case considered, local convection was a major source of HO_x and NO_x to the upper troposphere. Average net ozone production of 2 ppbv/day between 8 and 12 km over the continental United States in the spring was found. This ozone production was NO_x limited. High levels of HO_x present in the upper troposphere stimulate ozone production and increase the sensitivity of ozone to NO_x emissions from aircraft and other sources.

In view of the important role of non-methane hydrocarbons (NMHC) in global-scale tropospheric chemistry, Houweling *et al.* (1998) suggested a new chemical scheme (for NMHC impact parametrization). Results of global Tracer Model 3 (TM3) runs, with and without NMHC chemistry (compared with observations), illustrate the effects of NMHC on key compounds of tropospheric photochemistry, such as ozone, the hydroxyl radical, carbon monoxide and NO_x. It has been shown, in particular, that the simulation of ozone over polluted region improves when NMHC chemistry is taken into account. Globally, the contribution of NMHC to net photochemical ozone production has been estimated at about 40%, leading to a 17% increase of tropospheric ozone column. OH is depleted over the continents owing to reactions with NMHC, which is most evident in regions with strong biogenic emissions. Although NMHC significantly influence global OH distribution, their effect on its total tropospheric content appears to be marginal.

Results of sensitivity calculations, with and without organic peroxy nitrates, show that this N reservoir may significantly change global NO_x distribution, leading to NO_x increase over the oceans of the order of 50%. Although the results obtained by Houweling *et al.* (1998) reproduce NMHC impact on the main features of tropospheric chemistry, some large discrepancies with observations still remain.

Using a 3-D continental-scale model of tropospheric O_3-NO_x-NMHC chemistry for conditions in North America in summer, Horowitz *et al.* (1998) demonstrated that the global influence on tropospheric chemistry of nitrogen oxides ($NO_x = NO + NO_2$) emitted by fossil-fuel combustion may be strongly modulated by non-methane hydrocarbon chemistry taking place in the continental boundary layer. A specific feature of the model is consideration of the isoprene-oxidation mechanism. The chemical mechanism used in the model includes detailed photo-oxidation schemes for four NMHCs: propane, butane, propene and isoprene (butane is considered as a surrogate for all alkanes $\geq C_4$, while propene is used to represent all alkenes $\geq C_3$). Model calculations have been compared with observations for ozone, reactive nitrogen species and photochemical tracers at a number of sites in North America. The principal objective of simulation modelling was improvement in understanding of the export of NO_x and its reservoirs out of the continental boundary layer over the United States in summer (the US accounts for about 30% of the global NO_x source from fossil-fuel combustion).

Budget analysis for the United States carried out by Horowitz *et al.* (1998) indicates that 9% of NO_x emitted from fossil-fuel combustion is exported out of the continental boundary layer as NO_x, 3.5% is exported as peroxyacetyl nitrate (PAN) and 3.7% is exported as other organic nitrates. Isoprene is the principal NMHC responsible for formation and export of organic nitrate, which eventually decompose to provide a source of NO_x in the remote troposphere. It has been shown that export of NO_x and organic nitrates from the US boundary layer is a major source of NO_x on the scale of the Northern Hemisphere troposphere and on the scale of the upper troposphere at northern mid-latitudes. Therefore, adequate representation of isoprene chemistry in the continental boundary layer is important for simulation of NO_x in global tropospheric chemistry models.

Liang *et al.* (1998) applied the same 3-D continental-scale photochemical model to investigate seasonal budgets of O_3 and NO_y species (including NO_x and its oxidant products) in the boundary layer of the United States to assess the export of these species from the US boundary layer to the global atmosphere. Model results have been verified by comparison with year-round observations for O_3, CO and NO_y species at non-urban sites. The model solves the 3-D continuity equations for 21 chemical tracers (Table 5.18) over a domain including North America and large portions of the neighbouring oceans.

Table 5.19 illustrates calculations of the budget of fuel combustion NO_y in the US boundary layer made by Liang *et al.* (1998). Seasonal transition from NO_x to hydrocarbon-limited conditions for O_3 production over the eastern United States has been found to take place in the autumn, with reverse transition taking place in the spring. The mean NO_x/NO_y molar ratio in the US boundary layer ranges (according to the model) from 0.2 in summer to 0.6 in winter, in accord with observations, and largely reflecting seasonal variation in the chemical lifetime of NO_x.

According to Liang *et al.* (1998), formation of hydroxy organic nitrates during oxidation of isoprene, followed by decomposition of these nitrates to HNO_3, is estimated to account for 30% of the chemical sink for NO_x in the US boundary layer in summer. Model results indicate that PANs are most abundant in spring (25% of total NO_y), reflecting a combination of active photochemistry and low temperature. About 20% of the NO_x emitted from fossil-fuel combustion in the US is exported out of the US boundary layer as NO_x or PANs (15% in summer, 25% in winter).

All quantities are 3-month averages for the US boundary layer extending horizontally over the area mentioned and vertically to the top of model layer 3

Table 5.18. Chemical tracers in the model.

Tracer	Component(s)
O_x (odd oxygen)	$O_3 + NO_2 + 2NO_3 + 3N_2O_5 + HNO_4$
H_2O_2	Hydrogen peroxide
NO_x	$NO + NO_2 + NO_3 + HNO_2 + HNO_4 + 2N_2O_5$
RNO_3	HNO_3 + aerosol + NO_3^-
PAN	Peroxyacetyl nitrate
PMN	Peroxymethacryloyl nitrate
PPN (lumped)	Peroxyacyl nitrates from RCHO
ISN2 (lumped)	Isoprene nitrates
R4N2 (lumped)	$C_{4,5}$ alkylnitrates
C_3H_8	Propane
ALK4 (lumped)	$>C_3$ alkanes
ALKE (lumped)	$>C_2$ alkenes, except isoprene
C_5H_8	Isoprene
CR_2O	Formaldehyde
CR_3CHO	Acetaldehyde
RCRO (lumped)	$>C_2$ aldehydes except MACR
CH_3COCH_3	Acetone
MEK (lumped)	$>C_3$ ketones produced from ALK4
MVK	Methylvinyl ketone
MACR	Methylacrolein

Lumping of organic compounds is done on a per carbon basis. Uniform concentrations of 1.7 ppmv CH_4 and 1.2 ppbv C_2H_6 are assumed over the model domain.

(about 740 hPa). The budgets are for NO_y emitted by fossil-fuel combustion within the region, as determined by subtracting background NO_y concentrations and fluxes from zero-NO_x simulation. PANs includes PAN, PMN and PPN; RNO_3 includes isoprene nitrates and alkylnitrates; HNO_3 (I) denotes HNO_3 of inorganic origin, namely, formed from oxidation of NO_2 by OH and hydrolysis of N_2O_5 in aerosols; HNO_3 (O) denotes HNO_3 produced by decomposition of hydroxy organic nitrates (mainly isoprene nitrates). Aerosol nitrate is lumped with HNO_3 in the model. Lifetimes are defined as concentrations divided by the sum of chemical and deposition loss rates. JJA, June, July and August; SON, September, October and November; DJF, December, January and February; MAN, March, April and May.

Liang *et al.* (1998) estimated annual mean export of NO_x and PANs from the US boundary layer to be equal to 1.4 Tg N/yr, representing an important source of NO_x on the scale of the Northern Hemisphere troposphere. Eventual O_3 production in the global troposphere because of exported NO_x and PANs has been estimated to be twice as large, on an annual basis, as the direct export of O_3 pollution from the US boundary layer. Combustion of fossil fuel in the United States into the total source of NO_x in the NH troposphere (on an annual basis) is about 10%.

A very important and rapidly developing branch of environmental numerical modelling is simulation of urban environmental dynamics including tropospheric

Table 5.19. Budget of fossil-fuel combustion NO_x in the US boundary layer.

Season	Summer (JJA)	Fall (SON)	Winter (DJF)	Spring (MAM)
Concentration (ppbv)				
NO_x	0.21	0.51	0.80	0.36
PAN	0.13	0.11	0.08	0.20
PMN	0.012	0.006	0.003	0.007
PPN (lumped	0.023	0.020	0.003	0.024
Isoprene nitrates	0.028	0.020	0.001	0.005
Alkylnitrates	0.015	0.015	0.015	0.031
HNO_3	0.55	0.60	0.35	0.29
Total NO	1.0	1.3	1.3	0.9
Lifetime (days)				
NO_x	0.27	0.64	1.2	0.47
PAN	0.086	0.23	0.95	0.47
HNO_3	0.93	1.1	0.67	0.50
NO_x emission (Gmol day^{-1})	1.34	1.34	1.34	1.34
Dry deposition (Gmol day^{-1})				
NO_x	0.077	0.094	0.064	0.092
PANs	0.030	0.015	0.0039	0.020
RNO_3	0.034	0.023	0.0022	0.010
HNO_3 (I)	0.35	0.39	0.39	0.41
HNO_3 (O)	0.12	0.051	0.0099	0.037
Total dry deposition	0.62	0.57	0.47	0.57
Wet deposition (Gmol day^{-1})				
HNO_3 (I)	0.23	0.30	0.39	0.37
HNO_3 (O)	0.14	0.048	0.016	0.052
Total wet deposition	0.37	0.35	0.41	0.42
Export (Gmol day^{-1})				
NO_x	0.10	0.14	0.27	0.15
PANs	0.122	0.108	0.054	0.12
RNO_3	0.027	0.022	0.0090	0.019
HNO_3 (I)	0.073	0.13	0.11	0.061
HNO_3 (O)	0.026	0.016	0.0021	0.0036
Total export	0.35	0.42	0.46	0.35

ozone dynamics in a polluted atmosphere. Although we do not intend to consider this special subject here, we mention some recent publications.

Ziomas *et al.* (1998a, b) discussed in a thematic issue of *Atmospheric Environment* the results of the Mediterranean Campaign of Photochemical Tracers–Transport and Chemical Evolution (MEDCAPHOT–TRACE) experiment conducted in the Athens area in 1994–1995. One of the important results of this project was development of the Urban Airshed Model to simulate environmental dynamics in the Greater Athens area, including assessment of ozone-episode formation. This model studied the effect on ozone of 36 possible scenarios of VOC and NO_x reduction across the entire area. Comparison of model results with observations has proven the adequacy of the model. It has been shown that ozone abatement strategy should focus mainly on VOC emission control rather than controlling NO_x. VOC reduction should be of the order of 30–40% to effectively reduce ozone levels at the northern part of Greater Athens, down to the European Union limit of 90 ppb.

Similar studies have recently been carried out by a number of other research groups (Berkowitz *et al.*, 1998; Calbó *et al.*, 1998; Duncan and Chameides, 1998; Jacobson, 1998), although it must be pointed out that some earlier results are still very important, such as those from analysis of sensitivity of urban/regional chemistry to climate change (Gery *et al.*, 1989; Wuebbles and Penner, 1988).

To adequately plan measures to reduce surface ozone concentration, it is necessary to assess the sensitivity of ozone concentration to various perturbations. Summarizing relevant results, Lu and Chang (1998) pointed out, first of all, that the dependence of ozone formation on the change in concentration of precursors (VOC and NO_x) is highly non-linear. Since comparable ozone levels can be associated with diverse combinations of VOC and NO_x concentrations, the ozone level by itself is not sufficient to foretell its sensitivity to precursor change. So, Lu and Chang (1998) extended the concept of indicator species to combinations of observable species that are consistently associated with different site characteristics. They illustrated the results by application of the SARMAR Air Quality Model (SAQM) to simulate the situation in San Joaquin Valley, California.

The results obtained by Lu and Chang (1998) show that the use of three indicator ratios ($O_3/(NO_y\text{-}NO_x)$, $HCHO/NO_y$ and H_2O_2/HNO_3) along with SAQM-derived criteria are most effective for identifying VOC or NO_x-sensitive regimes. An important result is that SAQM-derived threshold criteria for assessing ozone sensitivities differ from threshold criteria proposed by previous studies using different models and under different conditions (Table 5.20). The conclusion is therefore that threshold criteria are likely to be dependent on location and environmental conditions, including emission patterns and rates. Thus, local observed indicator species and ratios can be used to determine ozone sensitivities only if appropriate threshold criteria have been derived for the local conditions.

Table 5.20. Threshold values suggested by various simulations.

	SAQM base case	SAQM perturbed case	Sillman (1995)	Milford *et al.* (1994)
NO_y (ppb)	4.0–5.0	3.0–4.0	7.0–31.0	10.0–25.0
$HCHO/NO_y$	0.5–0.7	0.6–0.9	0.2–0.39	NA
H_2O_2/HNO_3	0.8–1.0	0.9–1.2	0.25–0.67	NA
$O_3/(NO_y\text{-}NO_x)$	25.0–30.0	25.0–30.0	6.1–14.0	NA
$(NO_y\text{-}NO_x)/NO_y$	0.5–0.6	0.5–0.6	NA	NA

NA – not available.

As mentioned earlier, an important aspect of modelling tropospheric ozone dynamics is parametrization of the effects of clouds on tropospheric chemical dynamics. In this regard, Kreidenweis *et al.* (1997) carried out a modelling study of the effects of cumulus convection on transformation and redistribution of chemical species in the troposphere. Clouds play significant roles in tropospheric chemistry via several mechanisms by which they influence gas and aerosol composition of the atmosphere, chemical conversion rates, transport and radiative transfer. Many observations and modelling studies indicate, in particular, that convective clouds can redistribute species from the boundary layer to the mid and upper troposphere. Since redistributed gases and aerosols generally have longer lifetimes in the upper troposphere, their regional and global cycles can be modified by convection.

Kreidenweis *et al.*, (1997) emphasized that slightly insoluble species, such as ozone, are efficiently redistributed by cloud motions. However, most chemical species interact to some degree with cloud water, and the effectiveness of convective systems in transporting these species should therefore depend on cloud chemical and dynamic environment. Many chemical reactions proceed rapidly in the liquid phase, but aqueous-phase oxidation of sulphur dioxide is of particular importance. Dissolved oxidants important in SO_2 conversion include ozone, oxygen and hydrogen peroxide (in-cloud formation of hydroxymethane sulphonate HMS has also received attention recently).

In order to simulate these cloud effects, it is necessary to couple a cloud dynamics model with models describing microphysical and chemical processes. For this purpose, Kreidenweis *et al.* (1997) employed a sulphate chemistry model (including prognostic equations for CCN mass and number concentrations) coupled with a 2-D mixed-phase two-moment microphysics Eulerian cloud model. The aim was to simulate effects of convective transport on active chemical species redistribution in a small mid-latitude continental cumulonimbus. The results obtained by Kreidenweis *et al.* (1997) demonstrate significant redistribution of SO_2 from the boundary layer to the mid and upper troposphere with a larger percentage redistributed in the case of chemical fields that are representative of polluted continental conditions.

To conclude this section, it is proper to emphasize once more the coupled and non-linear nature of those processes that are responsible for tropospheric chemical changes. For example, Prather (1998) pointed out, in this context, that photochemical coupling of N_2O, NO_y and O_3 in the stratosphere leads to a natural mode, for coupled perturbations to all these species, that decays 10 to 15% faster than the N_2O steady-state lifetime implies. Similar situation exists in the troposphere as was shown in a number of the above cases.

Serious uncertainties still exist concerning surface-ozone trends (Roemer, 1999; Zanis *et al.*, 1999). Roemer (1999) analysed data of measured ozone between 1988 and 1995 for about 50 sites mainly in NW Europe. Ozone and oxidant data do not provide a uniform picture of ozone changes. Positive as well as negative trends were found (most of them are not significant at the 95% confidence level). Even cities located nearby (100–200 km) indicate the presence of relatively large differences in trends.

5.6 REFERENCES

A Global Strategy for European Atmospheric Interdisciplinary Research AIRES (1997) Air Pollution Research Report No. 61, EC Directorate-General Science, Research and Development, EUR 17645 EN, Luxembourg, 45 pp.

Aikin, A. C., Pesnell. W. D. (1998) Uptake coefficient of charged aerosols: Implications for atmospheric chemistry. *Geophys. Res. Lett.*, **25**, N9, 1,309–1,312.

Albold, A. (1997) Transmittance of a cloud is wavelength-dependent in the UV range: Physical interpretation. *Geophys. Res. Lett.*, **24**, N4, 397–400.

Aliwell, S. R., Jones, R. L. (1998) Measurements of tropospheric NO_3 at mid-latitudes. *J. Geophys. Res.*, **103**, N5, 5,719–5,727.

Anderson, B. E., Cofer, W. R., Bagwell, D. R., Barrick, J. W., Hudgins, C. M., Brunke, K. E. (1998) Airborne observations of aircraft aerosol emissions. I. Total nonvolatile particle emission indices. *Geophys. Res. Lett.*, **25**, N10, 1,685–1,692.

Anderson, B. E., Cofer, W. R., Barrick, J. W., Bagwell, D. R., Hudgins, C. M.. (1998) Airborne observations of aircraft aerosol emissions. II, Factors controlling volatile particle production. *Geophys. Res. Lett.*, **25**, N10, 1,693–1,696.

Andreae, M. O. (1998) The IGAC Biomass Burning Experiment (BIBEX): Rationale and evolution. *IGACtivities Newsletter*, N15, 2–3.

Andreae, M. O., Fishman, J. (1998) The Southern Tropical Atlantic Region Experiment (STARE): Transport and Atmospheric Chemistry near the Equator-Atlantic (TRACE-A) and Southern African Fire/Atmospheric Research Initiative (SAFARI). *IGACtivities Newsletter*, N15, 3–6.

Angeletti, G. (1998) Introduction. The Mediterranean Campaign of Photochemical Tracers–Transport and Chemical Evolution (MEDCAPHOT–TRACE). *Atmos. Environ.*, **32**, N12, 2,043.

Apel, E. C., Calvert, J. G., Reimer, D., Pos, W., Zika, R., Kleindienst, T. E., Lonneman, W. A., Fung, K., Fujita, E., Shepson, P. B., Starn, T. K., Roberts, P. (1998) Measurements comparison of oxygenated volatile organic compounds at a rural site during the 1995 SOS Nashville Intensive. *J. Geophys. Res.*, **103**, ND17, 22,295–22,316.

Arakati, T., Faust, B. C. (1998) Sources, sinks, and mechanisms of hydroxyl radical (OH) photoproduction and consumption in authentic acidic continental cloud water from Whiteface Mountain, New York: The role of the Fe(r) (r=II, III) photochemical cycle. *J. Geophys. Res.*, **103**, ND3, 3,487–3,504.

Aranda, A. *et al.* (1997) The BrO + CH3O2 reaction: Kinetics and role in the atmospheric ozone budget. *Geophys. Res. Lett.*, **24**, N22, 2,745–2,748.

Arctic Tropospheric Ozone Chemistry ARCTOC (1997) U. Platt, E. Lehrer (eds.), Air Pollution Research Report No. 64, EC Directorate-General Science, Research and Development, EUR 17783 EN, Luxembourg, 218 pp.

Armerding, W., Comes, F. J., Crawack, H. S., Forberich, O., Gold, G., Ruger, C., Spiekermann, M., Walters, J., Cuevas, E., Redondas, A., Schitt, R., Matuska, P. (1997) Testing the daytime oxidizing capacity of the troposphere: 1994 OH field campaign at the Izaña observatory, Tenerife. *J. Geophys. Res.*, **102**, ND9, 10,603–10,612.

Asimakopoulos, D., Deligiorgi, D., Drakopoulos, C., Helmis, C., Kokkori, K., Labas, D., Sikiotis, D., Varotsos, C. (1992) An experimental study of night-time air-pollutant transport over complex terrain in Athens. *Atmos. Envioron.*, **26B**, N1, 59–71.

Atmospheric Ozone, Proceedings XVIII Quadrennial Ozone Symposium, 12–21 September 1996, L'Aquila, Italy, R. D. Bojkov, G. Visconti (eds.) (1998) Parco Scientifico e Tecnologico, Abruzzo, Italy, Vol. 1, pp. 319–450.

Backer, H., Muer, D., Sadelaer, G. (1998) Comparison of ozone profiles obtained with Brewer-Mast and Z-ECC sensor during simultaneous ascents. *J. Geophys. Res.*, **103**, ND16, 19,641–19,648.

Bante, R. M., Senff, C. J., White, A. B., Trainer, M., McNider, R. T., Valente, R. J., Mayor, S. D., Alvarez, R. J., Hardesty, R. M., Parrish, D., Fehsenfeld, F. C. (1998) Daytime buildup and night-time transport of urban ozone in the boundary layer during a stagnation episode. *J. Geophys. Res.*, **103**, NC10, 22,519–22,544.

Barrie, L. A. (1996) Occurrence and trends of pollution in the Arctic troposphere. In: *Chemical Exchange Between the Atmosphere and Polar Snow*, E. W. Wolff, R. C. Bales (eds.) NATO ASI Series Vol. 143, Springer-Verlag, Berlin, pp. 93–129.

Barrie, L. A., Delmas, R. J. (1998) The Polar Atmospheric and Snow Chemistry (PASC) Activity. *IGACtivities Newsletter*, N14, 2–7.

Barrie, L. A., Bottenheim, J. W., Schnell, J. W., Crutzen, P. J., Rasmussen, R. A. (1998a) Ozone destruction and photochemical reactions at polar sunrise in the lower Arctic atmosphere. *Nature*, **334**, 138–141.

Barrie, L. A., Platt, U., Shepson, P. (1998b) Surface ozone depletion at polar sunrise fueled by sea-salt halogens. *IGAC Newsletter*, N14, 4–7.

Bastrup-Birk, A., Brandt, J., Zlatev, Z. (1997) Studying cumulative ozone exposures in Europe during a 7-year period. *J. Geophys. Res.*, **102**, ND20, 23,917–23,935.

Belan, B. D., Kovalevski, V. K., Plotnikov, A. P., Sklyadneva, T. K. (1998) Temporal change of ozone and nitrogen oxides in the surface layer of the atmosphere near Tomsk. *Optics of Atmos. Ocean*, **11**, N12, 1,325–1,327.

Bergamaschi, P., Brenninkmeijer, C. A. M., Hahn, M., Röckmann, T., Scharffe, D. H., Crutzen, P. J., Elansky, N. F., Belskov, I. B., Trivett, N. B. A., Worthy, D. E. J. (1998a) Isotope analysis based source identification for atmospheric CH_4 and CO sampled across Russia using the Trans-Siberian railroad. *J. Geophys. Res.*, **103**, ND7, 8,227–8,235.

Bergamaschi, P., Lubina, C., Königstedt, R., Fisher, H., Veltkamp, A. C., Zwaagstra, O. (1998b) Stable isotopic signatures (d13C, dD) of methane from European landfill sites. *J. Geophys. Res.*, **103**, ND7, 8,251–8,265.

Berkowitz, C. M., Fast, J. M., Springston, S. R., Larsen, R. J., Spicer, C. W., Doskey, P. V., Hubbe, J. M., Plastridge, R. (1998) Formation mechanisms and chemical characteristics of elevated photochemical layers over the northeast United States. *J. Geophys. Res.*, **103**, ND9, 10,631–10,647.

Berland, B. S., Foster, K. L., Tolbert, M. A., George, S. M. (1996) UV absorption spectra of H2O/HNO_3 films. *Geophys. Res. Lett.*, **23**, N20, 2,757–2,760.

Berntsen, T. K., Isaksen, I. S. A., Wang, W.-C., Liang, X.-Z. (1996) Impacts of increased atmospheric emissions in Asia on tropospheric ozone and climate. A global 3-D model study. *Tellus*, **48B**, 13–32.

Berntsen, T. K., Isaksen, I. S. A. (1997) A global three-dimensional chemical transport model for the troposphere. 1. Model description and CO and ozone results. *J. Geophys. Res.*, **102**, ND17, 21,239–21,280.

Berntsen, T. K., Isaksen, I. S. A., Fugelstvedt, J. S., Myhre, G., Larsen, T. A., Stordal. F., Freckleton, R. S., Shine, K. P. (1998) Effect of anthropogenic emissions on tropospheric ozone and its radiative forcing. *J. Geophys. Res.*, **102**, ND23, 28,101–28,126.

Bethan, S., Vaughan, G., Gerbig, C., Volz-Thomas, A., Richer, H., Tiddeman, D. A. (1998) Chemical air mass differences near fronts. *J. Geophys. Res.*, **103**, ND11, 13,413–13,434.

Blum, O., Bythnerowicz, A., Manning, W., Popovicheva, L. (1998) Ambient tropospheric ozone in the Ukrainian Carpathian mountain and Kiev region: Detection with passive samples and biodetector plants. *Environ. Pollut.*, **98**, N3, 299–304.

Bodeker, G. E., McKenzie, R. L. (1996) An algorithm for inferring surface UV irradiance including cloud effects. *J. Appl. Meteorol.*, **35**, N10, 1,860–1,877.

Bodeker, G. E., Boyd, I. S., Matthews, W. A. (1998) Trends and variability in vertical ozone and temperature profiles measured by ozonesondes at Lauder, New Zealand: 1986–1996. *J. Geophys. Res.*, **103**, ND22, 28,661–28,681.

Bojkov, R. D., Reinsel, G. C. (1984) Trends in tropospheric ozone concentration. *Proceedings of the Ozone Symposium in Greece*, D. Reidel, Dordrecht, pp. 775–781.

Bojkov, R. D. (1986) Surface ozone during the second half of the nineteenth century. *J. Clim. and Appl. Meteorol.*, **25**, 343–352.

Bojkov, R. (1988) Ozone changes at the surface and in the free troposphere. In: *Tropospheric Ozone*, I. S. A. Isaksen (ed.), D. Reidel, Dordrecht, pp. 83–96.

Bojkov, R. D. (1999) International Assessment of Ozone Depletion: 1998. *WMO Bull.*, **48**, 35–44.

Borrell, P., Builtjes, P., Grennfelt, P., Høv, Ø. (eds.) (1997) *Photo-oxidants, Acidification and Tools: Policy Applications of EUROTRAC Results*, Springer, Berlin, Vol. XXII, 216 pp.

Borrell, P., Builtjes, P., Grennfelt, P., Høv, Ø., van Alst, R., Fowler, D., Mégre, G., Moussiopoulos, N., Warneck, P., Volz-Thomas, A., Wayne, R. (1997) EUROTRAC: Applications to photo-oxidants and the future project. In: *Atmospheric Ozone Dynamics, Observations in the Mediterranean Region*, C. Varotsos (ed.), NATO ASI Series Vol. 153, Springer-Verlag, Berlin, pp. 1–7.

Borrell, P., Varotsos, C., Kondratyev, K. Ya. (1998) Key problems of chemistry and photochemistry of the troposphere: Completion of the first phase of EUROTRAC and future perspectives. *Ecological Chemistry*, **7**, N1, 1–12 (in Russian).

Bottenheim, J. W., Sirois, A. (1996) Long-term daily mixing ratios of O_3, PAN, HNO_3, and particle nitrate at a rural location in eastern Canada: Relationships and implied ozone production efficiency. *J. Geophys. Res.*, **101**, ND2, 4,189–4,204.

Bottenheim, J. W., Barrie, L. A. (1996) Chemical reactions in the polar stratosphere relevant to C, S and N compounds. In: *Chemical Exchange between the Atmosphere* and Polar Snow, *Proceedings NATO Advanced Research Workshop*, Springer-Verlag, Berlin, 201–224.

Boucher, O. (1999) Air traffic may increase cirrus cloudiness. *Nature*, **397**, N6,714, 30–31.

Brace, S., Peterson, D. L. (1998) Spatial patterns of tropospheric ozone in the Mount Rainier region of the cascade mountains, USA. *Atmos. Environ.*, **32**, N21, 3,629–3,637.

Bradshaw, J., Sandholm, S., Talbot, R. (1998) An update on reactive odd-nitrogen measurements made during recent NASA Global Tropospheric Experiment programs. *J. Geophys. Res.*, **103**, ND15, 19,129–19,148.

Brasseur, G. P., Müller, J.-F., Granier, C. (1996) Atmospheric impact of NO_x emissions by subsonic aircraft: A three-dimensional model study. *J. Geophys. Res.*, **101**, ND1, 1,423–1,428.

Brasseur, G. P., Hauglustaine, D. A., Walters, S., Rasch, P. J., Müller, J.-F., Granier, C., Tie, X. X. (1998a) MOZART, a global chemical transport model for ozone and related chemical tracers. 1. Model description. *J. Geophys. Res.*, **103**, ND21, 28,265–28,290.

Brasseur, G. P., Cox, R. A., Hauglustaine, D., Isaksen, I., Lelieveld, J., Lister, D. M., Sausen, D., Schumann, K., Wahner, A., Wiesen, P. (1998b) European scientific assessment of the atmospheric effects of aircraft emissions. *Atmos. Environ.*, **13**, N32, 2,329–2,418.

Bremaud, P. J., Taupin, F., Thompson, A. M., Chaumerliac, N. (1998) Ozone nighttime recovery in the marine boundary layer: Measurements and simulation of the ozone diurnal cycle at Reunion Island. *J. Geophys. Res.*, **103**, ND3, 3,463–3,474.

Broennimann, S., Neu, U. (1998) A possible photochemical link between stratospheric and near-surface ozone in Swiss mountain sites in late winter. *J. Atmos. Chem.*, **31**, N3, 299–319.

Browell, E. V., Gregory, G. L., Harriss, R. C., Kirchhoff, V. W. J. H. (1988) Tropospheric ozone and aerosol distribution across the Amazon Basin. *J. Geophys. Res.*, **93**, ND2, 1,431–1,451.

Browell, E. V. (1989) Differential absorption lidar sensing of ozone. *Proc. IEEE*, **89**, N3, 419–432.

Brühl, C., Crutzen, P. J. (1989) On the disproportionate role of tropospheric ozone as a filter against solar UV-B radiation. *Geophys. Res. Lett.*, 16, 703–706.

Brune, W. H., Faloona, I. C., Tan, D., Weinheimer, A. S., Campos, T., Ridley, B. A., Vay, S. A., Collins, J. E., Sachse, G. W., Jaeglé, L., Jacob, D. J. (1998) Airborne in-situ OH and HO_2 observations in the cloud-free troposphere and lower stratosphere during SUCCESS. *Geophys. Res. Lett.*, **25**, N10, 1,701–1,704.

Brunner, D., Staehelin, J., Jeker, D. (1998) Large-scale nitrogen oxide plumes in the stratopause region and implications for ozone. *Science*, **282**, N5,392, 1,305–1,307.

Burkholder, J. B., Orlando J. J. (1998) Rate coefficient upper limits for the $BrONO_2$ and $ClONO_2 + O_3$ reactions. *Geophys. Res. Lett.*, **25**, N19, 3,567–3,570.

Burnett, C. R., Minschwaner, K. (1998) Continuing development in the regime of decreased atmospheric column OH at Fritz Peak, Colorado. *Geophys. Res. Lett.*, **25**, N9, 1,313–1,316.

Calbó, J., Pan, W., Webster, M., Prinn, R. G., McRae, G. J. (1998) Parameterization of urban subgrid scale processes in global atmospheric chemistry models. *J. Geophys. Res.*, **103**, ND3, 3,437–3,452.

Caligirou, A., Larsen, B. B., Kotzias, D. (1999) Gas-phase terpene oxidation products: A review. *Atmos. Environ.*, **33**, 1,423–1,439.

Cammas, J.-P., Jacoby-Koaly, S., Suhre, K., Rosset, R., Marenco, A. (1998) Atlantic subtropical potential vorticity barrier as seen by Measurements of Ozone by Airbus Service Aircraft (MOZAIC) flights. *J. Geophys. Res.*, **103**, ND19, 25,681–25,693.

Cárdenas, L. M., Austin, J. F., Burgess, R. A., Clemitshaw, K. C., Dorling, S., Penkett, S. A., Harrison, R. M. (1998) Correlations between CO, NO_y, O_3 and non-methane hydrocarbons and their relationships with meteorology during winter 1993 on the North Norfolk Coast, UK. *Atmos. Environ.*, **32**, N19, 3,339–3,351.

Carmichael, G. R., Uno, I., Phandis, M. I., Zhang, Y., Sunwoo, Y. (1998) Tropospheric ozone production and transport in the springtime in east Asia. *J. Geophys. Res.*, **103**, ND9, 10,649–10,671.

Carpenter, L. J., Clemitshaw, K. C., Burgess, R. A., Penkett, S. A., Cape, J. N., McFadyen, G. G. (1998) Investigation and evaluation of the NO_x/O_3 photochemical steady state. *Atmos. Environ.*, **32**, N19, 3,353–3,365.

Carsey, T. P., Churchill, D. D., Farmer, M. L., Fisher, C. J., Pszenny, A. A., Ross, V. B., Salzman, E. S., Springer-Young, M., Bonsang, B. (1997) Nitrogen oxides and ozone production in the North Atlantic marine boundary layer. *J. Geophys. Res.*, **102**, ND9, 10.653–10,666.

Cerveny, R. S., Balling, R. C., Jr (1998) Weekly cycles of air pollutants, precipitation and tropical cyclones in the coastal NW Atlantic region. *Nature*, **394**, N6,693, 561–563.

Chadwick, D. R., Sneath, R. W., Phillips, V. R., Pain, B. F. (1999) A UK inventory of nitrous oxide emission from farmed livestock. *Atmos. Environ.*, **33**, N20, 3,345–3,354.

Chameides, W. L., Saylor, R. D., Cowling, E. B. (1997) Ozone pollution in the rural United States and the now NAAQS. *Science*, **276**, N5,315, 916–917.

Chan, L. Y., Chan, C. Y., Qin, Y. (1998) Surface ozone pattern in Hong Kong. *J. Appl. Meteorol.*, **37**, N10, 1,153–1,165.

Chandra, S., Ziemke, J. R., Min, W., Read, W. G. (1998) Effects of 1997–1998 El Niño on tropospheric ozone and water vapor. *Geophys. Res. Lett*, **25**, N20, 3,867–3,870.

Chandra, S., Ziemke, J. R, Stewart, R. W. (1999) An 11-year solar cycle in tropospheric ozone from TOMS measurements. *Geophys. Res. Lett*, **26** (in print).

Chen, J. L., Islam, S., Biswas, P. (1998) Nonlinear dynamics of hourly ozone concentrations: nonparametric short term prediction. *Atmos. Environment*, **32**, N11, 1,839–1,848.

Chew, A. A., Atkinson, R. (1996) OH radical formation yields from the gas-phase reactions of O_3 with alkenes and monoterpans. *J. Geophys. Res.*, **101**, ND22, 28,645–28,659.

Chin, M., Rood, R. B., Allen, D. J., Andreae, M. O., Thompson, A. M., Lin, S.-J., Atlas, R. M., Ardizzone, J. V. (1998) Processes controlling dimethylsulfide over the ocean: Case studies using a 3-D model driven by assimilated meteorological fields. *J. Geophys. Res.*, **103**, ND7, 8,341–8,353.

Chow, J. C., Watson, J. G., Lowenthal, D. H., Egami, R. T., Solomon, P. A., Thuillier, R. H., Magliano, K., Ranzie, A. (1998) Spatial and temporal variations of particulate precursor gases and photochemical reaction products during SJVAQS/AUSPEX ozone episodes. *Atmos. Environ.*, **32**, N16, 2,835–2,844.

Chock, D. P., Kumar, S., Hermann, R. W. (1982) An analysis of trends in oxidant air quality in the South Coast Air Basin of California. *Atmos. Environ.*, **16**, 2,615–2,624.

Christakos, G., Vyas, V. M. (1998) A composite space/time approach to studying ozone distribution over eastern United States. *Atmos. Environ.*, **32**, N16, 2,845–2,857.

Christoffersen, T. S., Hjorth, J., Horie, O., Jensen, N. R., Kotzias, D., Molander, L. L., Neeb, P., Ruppert, L., Winterhalter, R., Virkula, A., Wirtz, K., Larsen, B. R. (1998) *cis*-Pinic acid, a possible precursor for organic aerosol formation from ozonolysis of α-pinere. *Atmos. Environ.* **32**, N10, 1,657–1,661.

Christopher, S. A., Chen, J., Welch, R. M., Kliehe, D. V., Connor, V. S. (1998a) Satellite investigations of fire, smoke, and carbon monoxide during April 1994 MAPS mission: Case studies over tropical Asia. *J. Geophys. Res.*, **103**, ND15, 19,327–19,336.

Christopher, S. A., Wang, M., Berendes, T. A., Welch, R. M., Yang, S.-K. (1998b) The 1985 Biomass Burning Season in South America: Satellite remote sensing of fires, smoke, and regional radiative energy budgets. *J. Appl. Meteorol.*, **37**, N7, 661–678.

Chu, Sh.-H., Cox, W. M. (1995) Effects of emissions reductions on ozone predictions by the Regional Oxidant Model during the July 1988 episode. *J. Appl. Meteorol.*, **34**, N3, 679–693.

Cofer, W. R., III, Winstead, E. L., Stocks, B. J., Goldammer, J. G., Cahoon, D. R. (1998) Crown fire emissions of CO_2, CO, H_2, CH_4 and TNMHC from a dense jack pine boreal forest fire. *Geophys. Res. Lett.*, **25**, N21, 3,919–3,922.

Collins, W. J., Stevenson, D. S., Johnson, C. E., Derwent, R. G. (1997) Tropospheric ozone in a global-scale three-dimensional Lagrangian model and its response to NO_x emission controls. *J. Atmos. Chem.*, **26**, N3, 223–274.

Cooper, O. R., Moody, J. L., Davenport, J. C., Oltmans, S. J., Johnson, B. J., Chen, X., Shepson, P. B., Merrill, J. T. (1998) Influence of springtime weather systems on vertical ozone distributions over three North American sites. *J. Geophys. Res.*, **103**, ND17, 22,001–22,013.

Cordero, E. C., Nathan, T. R., Echols, R. S. (1998) An analytical study of ozone feedbacks on Kelvin and Rossby-gravity waves: Effects on the QBO. *J. Atmos. Sci.*, **55**, N6, 1,051–1,062.

Cotte, H., Devaux, C., Carlier, P. (1997) Transformation of irradiance measurements into spectral actinic flux for photolysis rates determination. *J. Atmos. Chem.*, **26**, N1, 1–28.

Cowling, E. B., Chameides, W. L., Kiang, C. S., Fehsenfeld, F. C., Meagher, J. F. (1998) Introduction to special section: Southern Oxidants Study Nashville/Middle Tennessee Ozone Study. *J. Geophys. Res.*, **103**, ND17, 22,204–22,212.

Cox, R. A., Jones, R. L. (1998) Tropospheric composition, lifetimes and photochemistry. In: *Tropospheric Chemistry and Space Observations*, N. Chaumberliac, G. Mégre, N Papineau, G. Angeletti (eds.), EC Directorate-General Science, Research and Development, EUR 17790, Luxembourg, pp. 1–13.

Coy, L. *et al.* (1997) Meteorology of the polar vortex: Spring 1997. *Geophys. Res. Lett.*, **24**, N22, 2,693–2,696.

Crawford, J., Davis, D., Chen, G., Bradshaw, J., Sandholm, S., Gregory, G., Sachse, G., Anderson, B., Collins, J., Blake, D., Singh, H., Heikes, B., Talbot, R., Rodriguez, J. (1996) Photostationary state analysis of the NO_2-NO system based on airborne observations from the western and central North Pacific. *J. Geophys. Res.*, **101**, ND1, 2,053–2,072.

Crutzen, P. J., Heidt, L. E., Krasnec, L. P., Pollock, W. H. Seiler, W. (1979) Biomass burning as a source of atmospheric gases CO, H_2, N_2O, NO, CH_3Cl, and COS. *Nature*, **282**, 253–256.

Crutzen, P. J., Delany, A. C., Greenberg, J., Haagenson, P., Heidt, L., Lueb, R., Pollock, W., Seiler, W., Wartburg, A., Zimmerman, P. (1985) Tropospheric chemical composition measurements in Brazil during the dry season. *J. Atmos. Chem.*, **2**, 233–256.

Crutzen, P. J., Zimmermann, P. H. (1991) The changing photochemistry in the troposphere. *Tellus*, **43(AB)**, 136–151.

de la Cruz, J., Martin-González, J. M., González, B., Sancho, P. (1997) Surface ozone measurements at Taliarte, Gran Canaria (Canary Islands, Spain). *Il Nuovo Cimento*, **20C**, Ser. 2, N2, 181–194.

Davidson, A., Hoggan, M., Wong, P. (1985) *Air Quality Trends in the South Coast Air Basin 1975–1984*, South Coast Air Quality Management, El Monte, CA, pp. 41–46.

Davis, D., Chen, G., Kasibhatla, P., Jefferson, A., Tanner, D., Lenschow, D., Neff, W., Berresheim, H. (1998a) DMS oxidation in the Antarctic boundary layer:

Comparison of model simulations and field observations of DMS, DMSO, $DMSO_2$, H_2SO_4(g), MSA(g), and MSA(p). *J. Geophys. Res.*, **103**, ND1, 1,657–1,678.

Davis, J. M., Eder, B. K., Nychka, D., Yang, Q. (1998b) Modeling the effects of meteorology on ozone in Houston using cluster analysis and generalized additive models. *Atmos. Environ.*, **32**, N14/15, 2,505–2,520.

Decker, C. E., Ripperton, L. A., Worth, J. J. B., Vukovich, F. M., Bach, W. D., Tomerdahl, J., Smith, F., Wagoner, D. E. (1976) *Formation and Transport of Oxidants along Gulf Coast and in Northern US*, Report. EPA-450/3-76-033, US Environment Protection Agency, Research Triangle Park, NC.

Delmas R., Guenther A. (1998) The Experiment for Regional Sources and Sinks of Oxidants (EXPRESSO). *IGACtivities Newsletter*, N15, pp. 9–11.

Dentener, F. J., Crutzen, P. J. (1993) Reaction of N_2O_5 on tropospheric aerosols: Impact on the global distributions of NO_x, O_3, and CH. *J. Geophys. Res.*, **98**, 7,149–7,163.

Derwent, R. G., Jenkin, M. E., Saunders, S. M., Pilling, M. J. (1998a) Photochemical ozone creation potentials for organic compounds in northwest Europe calculated with a master chemical mechanism. *Atmos. Environ.*, **32**, N14/15, 2,429–2,441.

Derwent, R. G., Simmonds, P. G., O'Doherty, S., Ciais, P., Ryall, D. B. (1998b) European source strengths and Northern Hemisphere baseline concentrations of radiatively active trace gases at Mace Head, Ireland. *Atmos. Environ.*, **32**, N21, 3,703–3,715.

Dimitroulopoulou, C., Ap Simon, H. M. (1999) The influence of the photolysis rates on modelled ozone concentrations. *Atmos. Environ.*, **33**, N1, 147–154.

Dobson, G. M. B. (1973) The laminated structure of the ozone in the atmosphere. *Quart. J. Roy. Meteorol. Soc.*, **99**, 599–607.

Donovan, D. P., Bird, J. C., Carswell, A. I. (1997) Ozone, column ClO, and PSC measurements made at the NDSC Eureka observatory (80°N, 86°W) during the spring of 1997. *Geophys. Res. Lett.*, **24**, N22, 2,709–2,712.

Dore, J. E., Popp, B. N., Karl, D. M., Sansone, F. J. (1998) A large source of atmospheric nitrous oxide from subtropical North Pacific surface waters. *Nature*, **396**, N6,706, 63–66.

Duderstadt, K. A., Carroll, M. A., Sillman, S., Wang, T., Albercook, G. M., Feng, L., Parrish, D. D., Holloway, J. S., Fehsenfeld, F. C., Blake, D. R., Blake, N. J., Forbes, G. (1998) Photochemical production and loss rates of ozone at Sable Island, Nova Scotia during the North Atlantic Regional Experiment (NARE) 1993 summer intensive. *J. Geophys. Res.*, **103**, ND11, 13,531–13,535.

Duncan, B. N., Chameides, W. L. (1998) Effects of urban emission control strategies on the export of ozone and ozone precursors from the urban atmosphere to the troposphere. *J. Geophys. Res.*, **103**, ND21, 28,159–28,179.

Durieux, E., Fiorani, L., Calpini, B., Flamm, M., Jaquet, L., van den Bergh, H. (1998) Tropospheric ozone measurements over the Great Athens area during the Medcaphot-Trace campaign with a new shot-per-shot dial instrument: Experimental system and results. *Atmos. Environ.*, **32**, N12, 2,141–2,150.

Ebel, A., Friedrich, R., Rodhe, H. (eds.) (1997) *Tropospheric Modelling and Emission Estimation*, Springer-Verlag, Berlin, Vol. XXIV, 450 pp.

Ebel, A., Memmesheimer, M., Jakobs, H. J. (1997) Regional modelling of tropospheric ozone distribution and budgets. In: *Atmospheric Ozone Dynamics Observations in the Mediterranean Region*, C. Varotsos (eds.), NATO ASI Series Vol. 153, Springer-Verlag, Berlin, pp. 37–57.

Ehhalt, D. H. (1998) Radical ideas. *Science*, **279**, 1,002–1,003.

Ehhalt, D. H., Rohrer, F., Wahner, A., Prather, M. J., Blake, D. R. (1998) On the use of hydrocarbons for the determination of tropospheric OH concentration. *J. Geophys. Res.*, **103**, ND15, 18,981–18,998.

Elliot, S., Zhao, X., Turko, R., Kao, Ch.-Y. J., Shen, M. (1995) Photochemical numerics for global-scale modeling: Fidelity and GCM testing. *J. Appl. Meteorol.*, **34**, N3, 694–718.

Evans, G., Finkelstein, P., Martin, B., Possiel, N., Graves, M. (1983), Ozone measurements from a network of remote sites. *J. Air Pollut. Control Assoc.*, **33**, 291–296.

Fantechi, G., Jensen, N. R., Hjorth, J., Peeters, J. (1998) Mechanistic studies of the atmospheric oxidation of methyl butenol by OH radicals, ozone and NO_3 radicals. *Atmos. Environ.*, **32**, N20, 3,547–3,556.

Fast, J. D., Zhing, S. (1998) Meteorological factors associated with inhomogeneous ozone concentrations within the Mexico City basin. *J. Geophys. Res.*, **103**, ND15, 18,927–18,946.

Fehsenfeld, F. C., Daum, P., Leatich, W. R., Trainer, M., Parrish, D. D., Hübler, G. (1996) Transport and processing of O_3 and O_3 precursors over North Atlantic: An overview of the 1993 North Atlantic Regional Experiment (NARE) summer intensive. *J. Geophys. Res.*, **101**, 28,877–28,891.

Fehsenfeld, F. C, Huly, L. G., Sueper, D. T., Norton, R. B., Williams, E. J. (1998) Ground-based intercomparison of nitric acid measurement techniques. *J. Geophys. Res.*, **103**, ND3, 3,343–3,353.

Finlayson-Pitts, B. J., Pitts, J. N., Jr. (1986) *Atmospheric Chemistry*, Wiley-Interscience, New York, pp. 25–32.

Finlayson-Pitts, B. J., Pitts, J. N., Jr. (1997) Tropospheric air pollution: Ozone, airborne toxics, polycyclic aromatic hydrocarbons, and particles. *Science*, **276**, N5,315, 1,045–1,052.

Fiorani, L., Calpini, B., Jaquet, L., Van der Bergh, H., Durieux, E. (1997) Correction scheme for experimental biases in differential absorption lidar tropospheric ozone measurements based on the analysis of shot per shot data samples. *Appl. Opt.*, **36**, N27, 6,857–6,863.

Fiore, A. M., Jacob, D. J., Logan, J. A., Yin, J. H. (1998) Long-term trends in ground level ozone over the contiguous United States, 1980–1995. *J. Geophys. Res.*, **103**, ND1, 1,471–1,480.

Fischer, H., Nikitas, C., Parchatka, U., Zenker, T., Harris, G. W., Matuska, P., Schmitt, R., Mihelcic, D., Muesgen, P., Paetz, H.-W., Schultz, M., Volz-

Thomas, A. (1998) Trace gas measurements during Oxidizing Capacity of the Tropospheric Atmosphere campaign 1993 at Izaña. *J. Geophys. Res.*, **103**, ND11, 13,505–13,518.

Fischer, H., Wagenbach, D., Kipfstuhl, J. (1998a) Sulfate and nitrate firn concentration on the Greenland ice sheet. 1. Large-scale geographical deposition changes. *J. Geophys. Res.*, **103**, ND17, 21,927–21,934.

Fischer, H., Wagenbach, D., Kipfstuhl, J. (1998b) Sulfate and nitrate firn concentration on the Greenland ice sheet. 2. Temporal anthropogenic deposition changes. *J. Geophys. Res.*, **103**, ND17, 21,935–21,942.

Fish, D. J., Shallcross, D. E., Jones, R. L. (1999) The vertical distribution of NO_3 in the atmospheric boundary layer. *Atmos. Environ.*, **33**, N5, 687–691.

Fishman, J., Crutzen, P. J. (1978) The origin of ozone in the troposphere. *Nature*, **274**, 855–858.

Fishman, J., Solomon, S., Crutzen, P. J. (1979) Observational and theoretical evidence in support of a significant in situ photochemical source of tropospheric ozone. *Tellus*, **31**, 432–446.

Fishman, J., Minnis, P., Reichle, H. G., Jr. (1986) Use of satellite data to study tropospheric ozone in the Tropics. *J. Geophys. Res.*, **91**, ND13, 14,451–14,465.

Fishman, J., Watson, C. E., Larsen, J. C., Logan, J. A. (1990) Distribution of tropospheric ozone determined from satellite data. *J. Geophys. Res.*, **95**, ND4, 3,599–3,617.

Fishman, J., Brackett, V. G., Browell, E. V., Grant, W. B. (1996) Tropospheric ozone derived from TOMS/SBUV measurements during TRACE A. *J. Geophys. Res.*, **101**, ND19, 24,069–24,082.

Fishman, J., Brackett, V. G. (1997) The climatological distribution of tropospheric ozone derived from satellite measurements using version 7 Total Ozone Mapping Spectrometer and Stratospheric Aerosol and Gas Experiment data sets. *J. Geophys. Res.*, **102**, ND15, 19,275–19,278.

Fishman, J. (1998) Global chemistry and distribution of tropospheric ozone. In: *Tropospheric Chemistry and Space Observations*, N. Chaumberliac, G. Mégre, N Papineau, G. Angeletti (eds.), EC Directorate-General Science, Research and Development, EUR 17790, Luxembourg, pp. 217–227.

Flatøy, F., Høv, Ø. (1996) Three-dimensional model studies of the effect of NO_x emissions from aircraft on ozone in the upper troposphere over Europe and North Atlantic. *J. Geophys. Res.*, **101**, ND1, 1,401–1,422.

Flatøy, F., Høv, Ø., Smit, H. (1996) Three-dimensional model studies of exchange processes of ozone in the troposphere over Europe. *J. Geophys. Res.*, **101**, ND6, 11,465–11,481.

Flocke, F., Volz-Thomas, A., Buers, H.-J., Patz, W., Garthe, H.-J., Kley, D. (1998) Long-term measurements of alkyl nitrates in southern Germany. 1. General behavior and seasonal and diurnal variation. *J. Geophys. Res.*, **103**, ND5, 5,729–5,746.

Flossman, A. I., Wobrock, W. (1996) Venting of gases by convective clouds. *J. Geophys. Res.*, **101**, 18,639–18,649.

Folkins, I., Wenneberg, P. O., Hanisco, T. F., Anderson, J. G., Salawitch, R. J. (1997) OH, HO_2, and NO in two biomass burning plumes: Sources of HO_x and implications for ozone production. *Geophys. Res. Lett.*, **24**, N24, 3,185–3,188.

Folkins, I., Chatfield, R., Singh, H., Chen, Y., Keikos, B. (1998) Ozone production efficiencies of acetone and peroxides in the upper troposphere. *Geophys. Res. Lett.*, **25**, N9, 1,305–1,308.

Frost, G. J., Trainer, M., Allwine, G. *et al.* (1998) Photochemical ozone production in the rural southeastern United States during the 1990 Rural Oxidants in the Southern Environment (ROSE) program. *J. Geophys. Res.*, **103**, NC10, 22,491–22,508.

Frost, G. J., Trainer, M., Mauldir, R. L., III, Eisele, F. L., Prevot, A. S. H., Flocke, S. S., Madronich, S., Kok, G., Schillawski, R. D., Baumgardner, D., Bradshaw, J. (1999) Photochemical modelling of OH levels during the First Aerosol Characterization Experiment (ACE 1.). *J. Geophys. Res.*, **104**, N13, 16,041–16,052.

Fruekilde, P., Hjorth, J., Jensen, N. R., Kotzias, D., Larsen, B. (1998) Ozonolysis at vegetation surfaces: A source of acetone, 4-ozopentanal, 6-methyl-5-hepten-2-one, and geranyl acetone in the troposphere. *Atmos. Environ*, **32**, N11, 1,893–1,902.

Fujiwara, M., Kita, K., Ogawa, T. (1998) Stratosphere–troposphere exchange of ozone associated with the equatorial Kelvin wave as observed with ozonesondes and rawinsondes. *J. Geophys. Res.*, **103**, ND15, 19,173–19,182.

Galbally, L E. (1968) Some measurements of ozone variation and destruction in the atmospheric-surface layer. *Nature*, **218**, 456–457.

Ganzeveld, L., Lelieveld, J., Roelofs, G.-J. (1998) A dry deposition parameterization for sulfur oxides in a chemistry and general circulation model. *J. Geophys. Res.*, **103**, ND7, 5,679–5,694.

Gao, R. S., Kurcher, B., Keim, E.R., Fathey, D.W. (1998) Constraining the heterogeneous loss of O_3 on soot particles with observations in jet engine exhaust plumes. *Geophys. Res. Lett.*, **25**, N17, 3,327–3,330.

Gerlach, T. M., McGee, K. A., Sutton, A. J., Elias, T. (1998) Rates of volcanic CO_2 degassing from airborne determinations of SO_2 emissions rates and plume CO_2/SO_2: Test study at Pu'u 'O'cone, Kilauea Volcano, Hawaii. *Geophys. Res. Lett.*, **25**, N14, 2,675–2,678.

Gery, M. W., Edmond, R. D., Whitten, G. Z. (1989) Potential effect of stratospheric ozone depletion and global temperature rise on urban photochemistry. In: *Atmosphere Ozone Research and Policy Implications Proceedings 3rd US–Dutch International Symposius, 9–13 May 1988, Nijmegen*, Amsterdam, pp. 365–375.

Ghosh, S., Varotsos, C. (1999) On the uptake of O_3 into aerosol and water droplets over Athens, Greece. *Toxicol. & Environ. Chem.*, **68**, 117–131.

Gillani, N. V., Luria, M., Valente, R. J., Tanner, R. L., Imhoff, R. E., Meagher, J. F. (1998a) Loss rate of NO_2 from a power plant plume based on aircraft measurements. *J. Geophys. Res.*, **103**, NC10, 22,585–22,592.

Gillani, N. V., Luria, M., Valente, R. J., Tanner, R. L., Imhoff, R. E., Meagher, J. F. (1998b) Relative production of ozone and nitrates in urban and rural power plant plumes. 1. Composite results based on data from 10 field measurement days. *J. Geophys. Res.*, **103**, NC10, 22,593–22,616.

Gilliband, A. B., Hartley, D. E. (1998) Interhemispheric transport and the role of convective parameterizations. *J. Geophys. Res.*, **103**, ND17, 22,039–22,045.

Girdziene, R. (1991) Surface ozone measurements in Lithuania. *Atmos. Environ.*, **25A**, N9, 1,791–1,794.

Glatthor, N., Blom, C. E., von Clarmann, T., Fisher, H., Gulde, T., Piesch, C., Erle, F., Pfeilsticker, K., Chipperfield, M. P., Lee, A. M., Pyle, J. A. (1998) Airborne remote sensing of NO_2 in the Arctic winter of 1994–1995 and comparison with a three-dimensional chemical transport model. *J. Geophys. Res.*, **103**, NC7, 13,315–13,326.

Glavas, S. (1999) Surface ozone and NO_x concentrations at a high altitude Mediterranean site, Greece. *Atmos. Environ.*, **33**, N23, 3,813–3,820.

Gleitsmann, G., Zellner, R. (1998) A modeling study of the formation of cloud condensation nuclei in the jet regime of aircraft plumes. *J. Geophys. Res.*, **103**, ND16, 19,543–19,556.

Global Tropospheric Chemistry. A Plan for Action (1984) National Academy Press, Washington, DC, 194 pp.

Gnauk, T., Rolle, W. (1998) A three-year study of non-methane hydrocarbons in surface air over Saxony (Germany). *J. Atmos. Chem.*, **30**, N3, 371–395.

Grant, K. E., Chuang, C. C., Grossman, A. S., Penner, J. E. (1999) Modelling the spectral optical properties of ammonium sulfate and biomass burning aerosols: Parameterization of relative humidity effects and model results. *Atmos. Environm.*, **33**, N17, 2,603–2,620.

Greenhut, G. K., Jochum, A. M., Neiniger, B. (1995) Boundary-layer turbulent transport and production/destruction of ozone during summertime smog episodes over the Swiss Plateau. *Boundary-Layer Meteorol.*, **73**, N4, 357–372.

Gregory, G. L., Anderson, B. E., Browell, E. V. (1996) Influence of lower tropospheric ozone as observed over the Pacific Ocean during the 1991 PEM–West A expedition. *J. Geophys. Res.*, **101**, ND1, 1,919–1,930.

Gros, V., Poisson, N., Martin, D., Kanakidou, M., Bonsang, B. (1998) Observations and modeling of the seasonal variation of surface ozone at Amsterdam Island: 1994–1996. *J. Geophys. Res.*, **103**, ND21, 28,103–28,109.

Grosjean, E., Grosjean, D. (1998) The gas phase reaction of alkenes with ozone: Formation yields of carbonyls from biradicals in ozone-alkene-cyclohexane experiments. *Atmos. Environ.*, **32**, N20, 3,393–3,402.

Grosjean, E., Grosjean, D. (1999) The reaction of unsaturated aliphatic oxygenates with ozone. *J. Atmos. Chem.*, **32**, N2, 205–232.

Güllük, T., Slemr, F., Stauffer, R. (1998) Simultaneous measurements of CO_2, CH_4, and N_2O in air extracted by sublimation from Antarctica ice cores: Confirmation of the data obtained using other extraction techniques. *J. Geophys. Res.*, **103**, ND13, 15,971–15,978.

Gurneg, K. R. (1998) Evidence for increasing ultraviolet irradiance at Point Barrow, Alaska. *Geophys. Res. Lett.*, **25**, N7, 903–906.

Guesten, H., Heinrich, G., Cvitas, T., Klasing I., Ruscic, B., Lalas, D., Petrakis, M. (1988) Photochemical formation and transport of ozone in Athens, Greece. *Atmos. Environ.*, **22**, 1,855–1,861.

Guesten, H., Heinrich, G., Muennich, E., Weppner, J., Cvitas, T., Klasing, I., Varotsos, C. A., Asimakopoulos, D. N. (1997) Thessaloniki '91 field measurement campaign-II. Ozone formation in the Greater Thessaloniki area. *Atmos. Environm.*, **32**, N8, 1,115–1,126.

Hales, J. (1996) Scientific background for AMS Policy Statement on Atmospheric Ozone. *Bull. Amer. Meteorol. Soc.*, **77**, N8, 1,249–1,253.

Harris, J. M., Oltmans, S. J. (1997) Variations in tropospheric ozone related to transport at American Samoa. *J. Geophys. Res.*, **102**, ND7, 8,781–8,792.

Hauglustaine, D. A., Brasseur, G. P., Walters, S., Rasch, P. J., Müller, J.-F., Emmons, L. K., Carroll, M. A. (1998) MOZART, a global chemical transport model for ozone and related chemical tracers. 2. Model results and evaluation. *J. Geophys. Res.*, **103**, ND21, 28,291–28,335.

Helas, G. *et al.* (1998) Domestic versus wild fires in Africa – A comparison between emissions from everyday domestic fire practices and wild fires. *IGACtivities Newsletter*, N15, 11–12.

Helmig, D., Greenberg, J., Guenter, A., Zimmerman, P., Geron, C. (1998) Volatile organic compounds and isoprene oxidation products at a temperate deciduous forest site. *J. Geophys. Res.*, **103**, NC10, 22,397–22,414.

Hendricks, J., Lippert, E., Petry, H., Ebel, A. (1999) Heterogeneous reactions on and in sulfate aerosols: Implications for the chemistry of the midlatitude tropopause region. *J. Geophys. Res.*, **104**, ND5, 5,531–5,550.

Herring, J. A., Jaffe, D. A., Beine, H. J., Madronich, S., Blake, D. R. (1997) High latitude springtime photochemistry. Part II. Sensitivity studies of ozone production. *J. Atmos. Chem.*, **27**, 155–178.

Highwood, E. J., Hoskins, B. J. (1998) The tropical tropopause. *Quart. J. Roy. Meteorol. Soc.*, **124**, Part A, 1,579–1,604.

Hintsa, E. J., Boering, K. A., Weinstock, E. M., Anderson, J. G., Gary, B. L., Pfister, L., Daube, B. C., Wofsy, S. C., Loewenstein, M., Podolske, J. R., Margitan, J. J., Bui, T. P. (1998) Troposphere-to-stratosphere transport in the lowermost stratosphere from measurements of H_2O, CO_2, N_2O, and O_3. *Geophys. Res. Lett.*, **25**, N14, 2,655–2,658.

Hoell, J. M., Davis, D. D., Liu, S. C., Newell, R. E., Akimoto, H., McNeal, R. J., Bendura, R. J. (1997) The Pacific Exploratory Mission–West Phase B: February–March 1994. *J. Geophys. Res.*, **102**, ND23, 28,223–28,239.

Hoell, J. M., Davis, D. D., Jacob, D. J., Rodgers, M. O., Newell, R. E., Fuelberg, H. E., McNeal, R. J., Raper, J. L., Bendura, R. J. (1999) Pacific Exploratory Mission in the tropical Pacific: PEM–Tropics A, August–September 1996. *J. Geophys. Res.*, **104**, ND5, 5,567–5,583.

Hoffman, S., Sulkowski, W. (1998) Influence of local NO_x emission on surface ozone concentration (South Poland, 1992–95). *Ecological Chemistry*, **7**, N1, 69–72.

Hofmann, D. J., Ferguson, E. E., Johnston, P. V., Mathews, W. A. (1992) Tropospheric ozone variations in the Arctic January 1990. *Planet. Space Sci.*, **40**, N2/3, 203–210.

Hopper, J. F., Barrie, L. A., Silis, A., Hart, W., Gallant, A. J., Dryfhout, H. (1998) Ozone and meteorology during the 1994 Polar Sunrise Experiment. *J. Geophys. Res.*, **103**, ND1, 1,481–1,492.

Horowitz, L. W., Liang, J., Gardner, G. M., Jacob, D. S. (1998) Export of reactive nitrogen from North America during summertime: Sensitivity to hydrocarbon chemistry. *J. Geophys. Res.*, **103**, ND11, 13,451–13,476.

Hoskins, B. J. (1991) Towards a PV-theta view of the general circulation. *Tellus*, **43AB**, 27–35.

Houweling, S., Dentener, F., Lelieveld, J. (1998) The impact of nonmethane hydrocarbon compounds on tropospheric photochemistry. *J. Geophys. Res.*, **103**, ND9, 10,673–10,696.

Høv, Ø. (ed.) (1998) *Tropospheric Ozone Research. Tropospheric Ozone in the Regional Sub-regional Context*, Springer-Verlag, Berlin, XXIV, 499 pp.

Høv, Ø. (ed.) (1997) *Tropospheric Ozone Research. Tropospheric Ozone in the Regional Sub-regional Context*, Springer-Verlag, Berlin, Heidelberg, 499 pp.

Høv, Ø., Kley, D., Volz-Thomas, A., Beck, J., Grenfelt, P., Penkett, S. A. (1997) An overview of tropospheric ozone research. In: *Tropospheric Ozone Research*, Ø. Høv (ed.), Springer-Verlag, Berlin, pp. 1–34.

Hubbard, M. C. and Cobourn, W. G., (1998) Development of a regression model to forecast ground level ozone in Louisville, Kentucky. *Atmospheric Environment*, **32**, 2,637–2,647.

Hudson, R. D., Kim, J.-H., Thompson, A. M. (1995) On the derivation of tropospheric column ozone from radiancies measured by the Total Ozone Mapping Spectrometer. *J. Geophys. Res.*, **100**, ND6, 11,137–11,145.

Hudson, R. D., Thompson, A. M. (1998) Tropical tropospheric ozone from total ozone mapping spectrometer by a modified residual method. *J. Geophys. Res.*, **103**, ND19, 22,129–22,145.

Huebler, G., Alvarez, R., Daum, P., Dennis, R., Gillani, R., Kleinman, L., Luke, W., Meagher, J., Rider, D., Trainer, M., Valente, R. (1998) An overview of the airborne activities during the Southern Oxidants Study (SOS) 1995 Nashville/Middle Tennessee Ozone Study. *J. Geophys. Res.*, **103**, ND17, 22,245–22,259.

Isaac, G. A., Banic, C. M., Leait, Ch., Anlauf, K. G., Couture, M. D., Liu, P., Macdonald, A., MacQuarrie, K., Pucket, K. J., Wiebe, H. A. (1998) Vertical profiles and horizontal transport of atmospheric aerosols and trace gases over central Ontario. *J. Geophys. Res.*, **103**, ND17, 22,015–22,038.

Isaksen, I. S. A. (ed.) (1998) *Tropospheric Ozone. Regional and Global Scale Interactions*, NATO ASI Series Vol. 227, Kluwer, Dordrecht, 425 pp.

Isidorov, V. A. (1992) *Organic Chemistry of the Atmosphere*, Chemistry, St. Petersburg, 288 pp. (in Russian).

Ivlev, L. S. (1991) Time variability of ozone density in the atmospheric lower layers. In: *Atmospheric Ozone*, Leningrad Hydrometeorological Institute, St. Petersburg, pp. 86–102 (in Russian).

Jacobson, M. Z. (1998) Studying the effects of aerosols on vertical photolysis rate coefficient and temperature profiles over an urban airshed. *J. Geophys. Res.*, **103**, ND9, 10,593–10,604.

Jacovides, C. P., Asimakopoulos, D. N., Steven, M. D. (1999) On the optical properties of the polluted Athens atmosphere during May 1995. *Atmos. Environm.*, **33**, N20, 3,427–3,431.

Jaeglé, L., Jacob, D. J., Wang, Y., Weinheimer, A. J., Ridley, B. A., Campos, T. L., Sachse, G. W., Hagen, D. E. (1998a) Sources and chemistry of NO_x in the upper troposphere over the United States. *Geophys. Res. Lett.*, **25**, N10, 1,705–1,708.

Jaeglé, J., Jacob, D. J., Brune, W. H., Tan, D., Faloona, I. C., Weinheimer, A. J., Ridley, B. A., Campos, T. L., Sachse, G. W. (1998b) Sources of HO_x and production of ozone in the upper troposphere over the United States. *Geophys. Res. Lett.*, **25**, N10, 1,709–1,712.

Jaffe, D., Herring, J., Madronich, S. (1998) Impact of large solar zenith angles, total column ozone and new O^1D quantum yields on atmospheric chemistry at high latitudes. *IGACtivities Newsletter*, N14, 8–10.

Jeffers, P. M., Wolfe, N. L., Nzengung, V. (1998) Green plants: A terrestrial sink for atmospheric CH3Br. *Geophys. Res. Lett.*, **25**, N1, 43–46.

Jiang, Y., Yung, Y. L. (1996) Concentrations of tropospheric ozone from 1979 to 1992 over tropical Pacific South America from TOMS data. *Science*, **272**, N5,262, 714–716.

Jobson, B. T., Parrish, D. D., Goldan, P., Kuster, W., Fehsenfeld, F. C., Blake, D. R., Blake, N. J., Niki, H. (1998a) Spatial and temporal variability of nonmethane hydrocarbon mixing ratios and their relation to photochemical lifetime. *J. Geophys. Res.*, **103**, ND11, 13,557–13,567.

Jobson, B. T., Frost G. J., McKeen, S. A., Ruerson, T. B., Buhr, M. P., Parrish, D. D., Trainer, M., Fehsenfeld, F. C. (1998b) Hydrogen peroxide dry deposition lifetime determined from observed loss rates in a power plant plume. *J. Geophys. Res.*, **103**, ND17, 22,617–22,628.

Jonguières, I., Marenca, A., Maalej, A., Rohrer, F. (1998) Study of ozone formation and transatlantic transport from biomass burning emissions over West Africa during the airborne Tropospheric Ozone Campaigns TROPOZ I and TROPOZ II. *J. Geophys. Res.*, **103**, ND15, 19,059–19,074.

Kajii, Y., Someno, K., Tanimoto, H., Hirokawa, J., Akimoto, H., Katsuno, T., Kawara, J. (1998) Evidence for the seasonal variation of photochemical activity of tropospheric ozone: Continuous observation of ozone and CO at Happo, Japan. *Geophys. Res. Lett.*, **25**, N18, 3,505–3,508.

Kalabokas, P. D., Bartzis, J. G. (1998) Photochemical air poluution characteristics at the station of the NCRS-Demokritos, during the MEDCAPHOT–TRACE campaign in Athens, Greece (20 Aug.–20 Sep., 1994). *Atmos. Environ.*, **32**, N12, 2,123–2,139.

Kalachev, L. V., Field, R. J. (1998) Absence of multiple steady states and a transition from steady state to monotonic growth behavior of CO and O_3 in a simple, nonlinear, oscillatory model of tropospheric photochemistry. *Geophys. Res. Lett.*, **25**, N24, 4,505–4,508.

Kalberer, M., Amman, M., Gaggeler, H.W., Baltensperger, U. (1999) Absorption of NO2 on carbon aerosol particles in the low ppb range. *Atmos. Environm.*, **33**, N17, 2,815–2,822.

Kallos, G., Kotroni, V., Lagouvardos, K. (1997) Photochemical air pollution: A regional scale problem. In: *Atmospheric Ozone Dynamics Observations in the Mediterranean Region*, C. Varotsos (ed.) NATO ASI Series Vol. 153, Springer-Verlag, Berlin, pp. 157–166.

Karol, I. L., Ozolin, Y. E., Rozanov, E. V. (1997) Box and Gaussian plume models of the exhaust composition evolution of subsonic transport aircraft in and out of the flight corridor. *Ann. Geophys.*, **15**, 88–96.

Kasibhatla, P., Chameides, W. L., Saylor, R. D., Olerud, D. (1998) Relationships between regional ozone pollution and emissions of nitrogen oxides in the eastern United States. *J. Geophys. Res.*, **103**, NC10, 22,663–22,670.

Khalil, M. A. K., Rasmussen, R. A. (1999) Atmospheric methyl chloride. *Atmos. Environ.*, **33**, 1,305–1,321.

Kieser, B. N., Bottenheim, J. W., Sideris, T., Niki, H. (1993) Spring 1989 observations of lower tropospheric chemistry in the Canadian high arctic. *Atmos. Environ.*, **27A**, 2,979–2,988.

Kim, J. H., Newchurch, M. J. (1998) Biomass-burning influence on tropospheric ozone over New Guinea and South Africa. *J. Geophys. Res.*, **103**, ND1, 1,455–1,462.

Kindler, T. P., Cunnold, D. M., Alyea, F. N., Chameides, W. L., Lou, G. P., Schwan, K. (1998) An evaluation using ^{14}C and N_2O simulations of three-dimensional transport driven by United Kingdom Meterological Office and Goddard Space Flight Center assimilated winds. *J. Geophys. Res.*, **103**, ND9, 10,827–10,847.

Kirchner, F., Stockwell, W. R. (1996) Effect of peroxy radical reactions on the predicted concentrations of ozone, nitrogenous compounds, and radicals. *J. Geophys. Res.*, **101**, ND15, 21,007–21,022.

Kirstine, W., Golbally, Ye. Y., Hooper, M. (1998) Emissions of volatile organic compounds (primarily oxygenated species) from pasture. *J. Geophys. Res.*, **103**, ND9, 10,605–10,619.

Kleinman, L. I. (1994) Low and high Nox tropospheric photochemistry. *J. Geophys. Res.*, **99**, ND8, 16,831–16,838.

Kleinman, L., Danm, P. H., Lee, J. H., Lee, Y.-N., Nunnermacker, L. J., Springston, S., R., Newman, L., Weinstein-Lloyd, J., Sillman, S. (1997) Dependence of ozone production on NO and hydrocarbons in the troposphere. *Geophys. Res. Lett.*, **24**, N18, 2,295–2,302.

Klemm, O., Werhahn, J., Shaller, E., Schlager, H., Krautstrunk, M. (1995) Characterization of ozone plumes in eastern Germany. *Geophys. Res. Lett.*, **22**, N15, 2,021–2,024.

Kley, D., Volz, A., Mulheims, F. (1988) *Ozone Measurements in Historic Perspective. Tropospheric Ozone*, I. S. A. Isaksen (ed.), D. Reidel, Dordrecht, pp. 63–72.

Kley, D. (1997) Tropospheric chemistry and transport. *Science*, **276**, N5,315, 1,043–1,045.

Kley, D., Smit, H. G. I., Vömel, H., Grassl, H., Ramanathan, V., Crutzen, P. J., Williams, S., Megwerk, J., Oltmans, S. J. (1997) Tropospheric water-vapour and ozone cross-sections in a zonal plane over the central equatorial Pacific Ocean. *Quart. J. Roy. Meteorol. Soc.*, **123**, PartA, N543, 2,009–2,040.

Klonecki, A., Levy, H., II (1997) Tropospheric chemical ozone tendencies in CO-CH_4-NO_y-H_2O system: Their sensitivity to variations in environmental parameters and their application to a global chemistry transport model study. *J. Geophys. Res.*, **102**, ND17, 21,221–21,237.

Knapp, K. G., Jensen, M. L., Balsley, B. B., Bognar, J. A., Oltmans, S. J., Smith, T. W., Birks, J. W. (1998a) Vertical profiling using a complementary kite and tethered balloon platform at Ferryland Downs, Newfoundland, Canada: Observation of a dry, ozone-rich plume in the free troposphere. *J. Geophys. Res.*, **103**, ND11, 13,389–13,397.

Knapp, K. G., Balsley, B. B., Jansen, M. L., Hanson, H. P., Birks, J. W. (1998b) Observation of the transport of polluted air masses from the northeastern United States to Cape Sable Island, Nova Scotia, Canada, during the 1993 NARE summer intensive. *J. Geophys. Res.*, **103**, ND11, 13,399–13,411.

Knoth, O., Wolke, R. (1998) An explicit–implicit numerical approach for atmospheric chemistry–transport modelling. *Atmos. Environ.*, **32**, N10, 1,785–1,797.

Kondratyev, K. Ya., Varotsos, C. (1997) A review of greenhouse effect and ozone dynamics over Greece. In: *Atmospheric Ozone Dynamics. Observations in the Mediterranean Region*, C. Varotsos (ed.), Springer-Verlag, Berlin.

Kondratyev, K. Ya. (1998) *Multidimensional Global Change*, Wiley/Praxis, Chichester, UK, 761 pp.

Kourtidis, K., Borchers, R., Fabian, P. (1998) Vertical distribution of methyl bromide in the stratosphere. *Geophys. Res. Lett.*, **25**, N4, 505–508.

Kreidenweis, S. M., Zhang, Y., Taylor, G. R. (1997) The effects of clouds on aerosol and chemical species production and distribution. 2. Chemistry model description and sensitivity analysis. *J. Geophys. Res.*, **102**, ND20, 23,867–23,882.

Kritz, M. A., Rosner, S. W., Stockwell, D. Z. (1998) Validation of an off-line three dimensional chemical transport model using observed radon profiles. 1. Observations. *J. Geophys. Res.*, **103**, ND7, 8,425–8,432.

Krol M. C., van Weele, M. (1997) Implications of variations in photodissociation rates for global tropospheric chemistry. *Atmos. Environm.*, **31**, N9, 1,257–1,273.

Krol, M., van Leeuwen, P. J., Lelieveld, J. (1998) Global OH trend inferred from methylchloroform measurements. *J. Geophys. Res.*, **103**, ND9, 10,697–10,711.

Kuhn, M., Builtjes, P. J. H., Poppe, D., Simpson, D., Stockwell, W. R., Anderson-Sköld, Y., Baart, A., Das, M., Fielder, F., Hov, Ø., Kirchner, E., Makar, P. A., Milford, J. B., Roemer, M. G. M., Ruhake, R., Strand, A., Vogel, B., Vogel, H. (1998) Intercomparison of the gas-phase chemistry in several chemistry and transport models. *Atmos. Environ.*, **32**, N4, 693–709.

Kuntasal, G., Chang, T. Y. (1987) Trends and relationships of O_3, NO_x and HC in the South Coast Air Basin of California. *J. Air Pollut. Control Assoc.*, **37**, 1,158–1,163.

Kunz, H., Speth, P. (1997) Variability of near-ground ozone concentrations during cold front passages – a possible effect of tropopause folding events. *J. Atmos. Chem.*, **28**, N1–3, 77–95.

Laaksonen, A., Hienola, J., Kulmala, M., Arnold, F. (1997) Supercooled cirrus cloud formation modified by nitric acid pollution of the upper troposphere. *Geophys. Res. Lett.*, **24**, N23, 3,009–3,012.

Laakso, K., Hattunen, S. (1998) Effects of the ultraviolet B-radiation (UV-B) on conifers: A review. *Environmental Pollution*, **99**, N3, 319–328.

Lal, S., Naja, M., Jayaraman, A. (1998) Ozone in the marine boundary layer over the tropical Indian Ocean. *J. Geophys. Res.*, **103**, ND15, 18,907–18,918.

Langford, A. O., O'Leary, T. J., Masters, C. D., Aikin, K. C., Profitt, M. H. (1998) Modulation of middle and upper tropospheric ozone at Northern midlatitudes by the El Niño/Southern Oscillation. *Geophys. Res. Lett.*, **25**, N14, 2,667–2,670.

Laurila, T., Lättilä, H. (1993) Surface ozone exposures measured in Finland. *Atmos. Environ.*, **28**, N1, 103–114.

Laurila, T., Joffre, S. M., Lättilä, H., Koskinen, T., Boij, H., Hakula, H. (1993) Photochemically active trace species over the Baltic Sea. In: EUROTRAC/TOR Annual Report 1993, D. Kley (ed.), Zerhock-Schmidt, Garmisch-Partenkirchen, pp. 101–107.

Laurila, T., Lättilä, H. (1994) Surface ozone exposures measured in Northern Europe. *Atmos. Environ.*, **28A**, 103–114.

Laurila, T. (1995) Ozone in Finland. *Finnish Air Pollution Prevention News*, N3, 5–11.

Lee, M., Heikes, B. G. (1998) Enhancements of hydroperoxides and formaldehyde in biomass burning impacted air and their effect on atmospheric oxidants cycles. *J. Geophys. Res.*, **103**, ND11, 13,201–13,212.

Lee, S. H., Akimoto, H., Nakane, H., Kurnosenko, S., Kinjo, Y. (1998a) Lower tropospheric ozone trend observed in 1989–1997 at Okinawa, Japan. *Geophys. Res. Lett.*, **25**, N10, 1,637–1,640.

Lee, M., Heiker, B. G., Jacob, D. J. (1998b) Enhancements of hydroxides and formaldehyde in biomass burning impacted air and their effect on atmospheric oxidant cycles. *J. Geophys. Res.*, **103**, ND11, 13,201–13,212.

Lee-Taylor, J. M., Doney, S. C., Brasseur, G. P., Müller, J.-F. (1998) A global three-dimensional atmosphere–ocean model of methyl bromide distributions. *J. Geophys. Res.*, **103**, ND13, 16,039–16,057.

Lee, Y. N., Zhou, X., Kleinman L. L. *et al.* (1998c) Atmospheric chemistry and distribution of formaldehyde and several multioxygenated carbonyl compounds during the 1995 Nashville/Middle Tennessee Ozone Study. *J. Geophys. Res.*, **103**, ND17, 22,449–22,462.

Lefohn, A. S. (ed.) (1992) *Surface Level Ozone Exposures and Their Effects on Vegetation*, Lewis Publ., Chelsea, MI, 366 pp.

Legrand, M., DeAngelis, M., Cachier, H., Gaudichet, A. (1995) Boreal biomass burning over the last 80 years recorded in a Summit-Greenland ice core. In: *Ice Core Studies of Global Biogeochemical Cycles*, NATO ASI Series Vol. 30, R. Delmas (ed.), Springer-Verlag, New York, 1995, pp. 347–360.

Leinonen, L. (ed.) (1993) *Air Quality Measurements*, Finnish Meteorological Institute, Helsinki, 1993, 250 pp.

Levy, H., II, Chameides, W. L. (1997) The global impact of human activity on tropospheric ozone. *Geophys. Res. Lett.*, **24**, N7, 791–794.

Liang, J., Horowitz, L. W., Jacob, D. J., Wang, Y., Fiore, A. M., Logan, J. A., Gardner, G. M., Munger, J. W. (1998) Seasonal budgets of reactive nitrogen species and ozone over the United States, and export fluxes to the global atmosphere. *J. Geophys. Res.*, **103**, ND11, 13,435–13,450.

Lin, C.-J., Pehkonen, S. Q. (1998) Oxidation of elemental mercury by aqueous chlorine (HOCl/OCl-): Implications for tropospheric mercury chemistry. *J. Geophys. Res.*, **103**, ND21, 28,093–28,102.

Linvill, D., Hooker, W. J., Olson, B. (1980) Ozone in Michigan's environment 1876–1900. *Monthly Review*, 1,883–1,891.

Liu, X., Jeffries, H. E., Sexton, K. G. (1999) Hydroxyl radical and ozone initiated photochemical reactions of 1,3-butadiene. *J. Atmos. Environ.*, **33**, N18, 3,005–3,022.

Logan, L. A., Prather, M. J., Wofsy, S. C., McElroy, M. B. (1981) Tropospheric chemistry: a global perspective. *J. Geophys. Res.*, **86**, 7,210–7,230.

Logan, L. A. (1985) Tropospheric ozone: Seasonal behaviour, trends, and anthropogenic influence. *J. Geophys. Res.*, **90**, 10.463–10,482.

Lonneman, W. A., Sella, R. L., Meeks, S. A. (1986) Non-methane organic composition in the Lincoln tunnel. *Environmental Science Technology*, **20**, 790–799.

Low, P. S., Kelly, P. M., Davies, T. D. (1992) Variations in surface ozone trends over Europe. *Geophys. Res. Lett.*, **19**, N11, 1,117–1,120.

Lu, Ch. H., Chang, J. S. (1998) On the indicator-based approach to assess ozone sensitivities and emission features. *J. Geophys. Res.*, **103**, ND3, 3,453–3,462.

Luria, M., Valente, R. J., Tanner, R. L., Gillani, N. V., Imhoff, R. E., Mueller, S. F., Olszyna, K. J., Meagher, J. F. (1999) The evolution of photochemical smog in a power plant plume. *J. Atmos. Environ.*, **33**, N18, 3,023–3,036.

Luecken, D. J., Tonnesen, G. S., Sickles, J. E., II (1999) Differences in NO_y speciation predicted by three photochemical mechanisms. *Atmos. Environ.*, **33**, N7, 1,073–1,084.

Lyons, W. A., Tremback, C. J., Pielke, R. A. (1995) Applications of the regional atmospheric modeling system (RAMS) to provide input to photochemical grid models for the Lake Michigan ozone studies (LMOS). *J. Appl. Meteorol.*, **34**, N8, 1,762–1,786.

Madronich, S., Granier, C. (1994) Tropospheric chemistry changes due to increased UV-B radiation. *Stratospheric Ozone Depletion/UV-B Radiation in the*

Biosphere, NATO ASI Series Vol. 18, R. M. Biggs, M. E. B. Joyner (eds.), Springer-Verlag, Berlin, pp. 3–10.

Madronich, S. (1998) WMO ad hoc Scientific Steering Committee on UV monitoring: Activities and Scientific issues. *SPARC Newsletter*, N10, 19–21.

Mahlman, J. D. (1997) Dynamics of transport processes in the upper troposphere. *Science*, **276**, N5,315, 1,079–1,083.

Mak, J. E., Southon, J. R. (1998) Assessment of tropical OH seasonality using atmospheric ^{14}CO measurements from Barbados. *Geophys. Res. Lett.*, **25**, N15, 2,801–2,804.

Marston, G., McGill, C. D., Rickard, A. R. (1998) Hydroxyl-radical formation in the gas-phase ozonolysis of 2-methylbut-2-ene. *Geophys. Res. Lett.*, **25**, N12, 2,177–2,180.

Mauldin, R. L., III, Madronich, S., Flocke, S. J., Eisell, F. L., Frost, G. S., Prevot, A. S. H. (1997) New insights on OH: Measurements around and in clouds. *Geophys. Res. Lett.*, **24**, N23, 3,033–3,036.

Mauzerall, D. L., Jacob, D. J., Fan, S.-M., Bradshaw, J. D., Gregory, G. L., Sachse, G. W., Blake, D. L. (1996) Origin of tropospheric ozone at remote high northern latitudes in summer. *J. Geophys. Res.*, **101**, ND2, 4,175–4,188.

Mauzerall, D. L., Logan, J. A., Jacob, D. J., Anderson, B. E., Blake, D. R., Bradshaw, I. D., Heikes, B., Sachse, G. W., Singh, H., Talbot, B. (1998) Photochemistry in biomass burning plumes and implications for tropospheric ozone over the tropical South Atlantic. *J. Geophys. Res.*, **103**, ND7, 8,401–8,423.

De Mazière, M., van Roosendael, M., Hermans, C., Simon, P. C., Demoulin, P., Roland, G., Zander, R. (1998) Quantitative evaluation of the post-Mount Pinatubo NO_2 reduction and recovery, based on 10 years of Fourier transform infrared and UV-visible spectroscopic measurements at Jungfraujoch. *J. Geophys. Res.*, **103**, ND9, 10,849–10,858.

McClenny, W. A., Daughtrey, E. H., Jr., Adams, J. R., Oliver, K. D., Kronmiller, K. G. (1998) Volatile organic compound concentration patterns at the New Handersonville monitoring site in the 1995 Southern Oxidants Study in the Nashville, Tennessee, area. *J. Geophys. Res.*, **103**, NC10, 22,509–22,518.

McDermid, I. S., Bergwerff, J. B., Bodeker G., Boyd, I. S., Brinksma, E. J., Connor, B. J., Farmer, R., Gross, M. R., Kimvilakani, P., Matthews, W. A., McGee, T. J., Ormel, F. T., Parrish, A., Singh, U., Swart, D. P. J., Tsou, J. J., Wang, P. H., Zawodny, J. (1998a) OPAL: Network for the Detection of Stratospheric Change Ozone Profiler Assessment at Lauder, New Zealand. 1. Blind intercomparison. *J. Geophys. Res.*, **103**, ND22, 28,683–28,692.

McDermid, I. S., Bergwerff, J. B., Bodeker, G., Boyd, I. S., Brinksma, E. J., Connor, B. J., Farmer, R., Gross, M. R., Kimvilakani, P., Matthews, W. A., MGee, T. J., Ormel, F. T., Parrish, A., Singh, U., Swart, D. P. J., Tsou, J. J. (1998b) OPAL: Network for the Detection of Stratospheric Change Ozone Profiler Assessment at Lauder, New Zealand. 2. Intercomparison of Revised Results. *J. Geophys. Res.*, **103**, ND22, 28,693–28,699.

McElroy, C. T., McLinden, C. A., McConnell, J. C. (1999) Evidence for bromine monoxide in the free troposphere during the Arctic polar sunrise. *Nature*, **397**, N6717, 338–341.

McKee, D. J. (1994) *Tropospheric Ozone*. CRC Press, Boca Raton, FL, 333 pp.

McNeal, R. J., Mugler, J. P., Jr., Harriss, R. C., Hoell, J. H., Jr. (1983) NASA Global Tropospheric Experiment. *EOS*, **64**, N38, 561–562.

Meagher, J. F., Cowling, E. B., Fehsenfeld, F. C., Parkhurst, W. J. (1998) Ozone formation and transport in southeastern United States: Overview of the SOS Nashville/Middle Tennessee Ozone Study. *J. Geophys. Res.*, **103**, ND17, 22,213–22,224.

Meywerk, J., Oltmans, S. J. (1998) Tropospheric water vapour and ozone cross sections in a zonal plane over the central equatorial Pacific. *Quart. J. Roy. Meteorol. Soc.*, **124**, Part B, 2,009.

Michelsen, H. A., Manney, G. L., Gunson, M. R., Zander, R. (1998) Correlations of stratospheric abundances of NOy, O_3, N_2O, and CH_4 derived from ATMOS measurements. *J. Geophys. Res.*, **103**, ND21, 28,347–28,359.

Miller, T. M., Grassian, V. H. (1998) Heterogenic chemistry of NO_2 on mineral oxide particles: Spectroscopic evidence for oxide-coordinated and water-solvated surface nitrate. *Geophys. Res. Lett.* **25**, N20, 3,835–3,838.

Molemaker, M. J., de Arellano, J. V.-G. (1998) Control of chemical reactions by convective turbulence in the boundary layer. *J. Atmos. Sci.*, **55**, N4, 568–574.

Monks, P. S., Carpenter, L. J., Penkett, S. A., Ayers, G. P., Gillett, R. W., Galbally, I. E., Meyer, C. P. (1998) Fundamental ozone photochemistry in the remote marine boundary layer: The SOAPEX experiment, measurement and theory. *Atmos. Environ.*, **32**, N21, 3,647–3,664.

Moulik, M. D., Milford, J. B. (1999) Factors influencing ozone chemistry in subsonic aircraft plumes. *Atmos. Environ.*, **33**, N6, 869–880.

Moussiopoulos, N. (1997) Air pollution models and their role in environmental policy. In: *Atmospheric Ozone Dynamics Observations in the Mediterranean Region*, NATO ASI Series Vol. 153, c. Varotsos (ed.), Springer-Verlag, Berlin, pp. 241–250.

Müller, J.-F., Brasseur, G. (1995) IMAGES: A three-dimensional chemical transport model of the global troposphere. *J. Geophys. Res.*, **100**, ND8, 16,445–16,490.

Munger, J. W., Fan, S.-M., Bakwin, P. S., Goulden, M. L., Goldstein, A. H., Colman, A. S., Wofsy, S. C. (1998) Regional budgets for nitrogen oxides from continental sources: Variations of rates for oxidation and deposition with season and distance from source regions. *J. Geophys. Res.*, **103**, ND7, 8,355–8,368.

Munro, R., Siddans, R., Reburn, W. J., Kerridge, B. J. (1998) Direct measurement of tropospheric ozone distributions from space. *Nature*, **392**, N6,672, 169–171.

Muralidharan, V., Mohan Kumar, G., Sampath, S. (1989) Surface ozone variation associated with rainfall. *Pageoph.*, **130**, 47–55.

Murayama, Sh., Yamazaki, K., Nakazawa, T., Aoki, Sh. (1998) Interpretation of high mixing ratios of O_3 observed in the upper troposphere over

Syowa, Antarctica, using a trajectory analysis. *Geophys. Res. Lett.*, **25**, N8, 1,177–1,180.

Naja, M., Lal, S. (1996) Changes in surface ozone amount and its diurnal and seasonal patterns, from 1954–1955 to 1991–1993, measured at Ahmedabad (23°N), India. *Geophys. Res. Lett.*, **23**, N1, 81–84.

Neftel, A., Blatter, A., Gut, A., Högger, D., Meixner, F., Ammann, C., Nathaus, F. J. (1998) NH3 soil and soil surface gas measurements in a triticale wheat field. *Atmos. Environ.*, **32**, N2, 499–505.

Newell, R. E., Browell, E. V., Davis, D. D., Lin, S. C. (1997) Western Pacific tropospheric ozone and potential vorticity: Implications for Asian pollution. *Geophys. Res. Lett.*, **24**, N22, 2,733–2,736.

Novaime, G., Bertman, S. B., Seaver, C., Elyea D., Huang, H., Shepson, P. B., Starn, T. K., Riemer, D. D., Zika, R. G., Olszyna, K. (1998) Sequential oxidation products from tropospheric isoprene chemistry: MACR and MPAN at a NO_x-rich forest environment in the southeastern United States. *J. Geophys. Res.*, **103**, NC10, 22,463–22,472.

Olcese, L. E., Taselli, B. M. (1998) Unexpected high levels of ozone measured in Cordoba, Argentina. *J. Atmos. Chem.*, **31**, N3, 269–279.

Oltmans, S. J., Levy, H., II (1994) Surface ozone measurements from a global network. *Atmos. Environ.*, **28**, 9–24.

Oltmans, S. J., Lefohn, A. S., Scheel, H. E., Harris, J. M., Levy, H., II, Galbally, I. E., Brunke, E.-G., Meyer, C. P., Lathrop, J. A., Johnson, B. J., Shadwick, D. S., Cuevas, E., Schmidlin, F. J., Tarasick, D. W. (1998) Trends of ozone in the troposphere. *Geophys. Res. Lett.*, **25**, N2, 139–142.

Oppenheimer, C., Francis, P., Stix, J. (1998) Depletion rates of sulfur dioxide in tropospheric volcanic plumes. *Geophys. Res. Lett.*, **25**, N14, 2,671–2,674.

Oum, K. W., Lakin, M. J., Finlayson-Pitts, B. J. (1998) Bromine activation in the troposphere by the dark reaction of O_3 with seawater ice. *Geophys. Res. Lett.*, **25**, N21, 3,923–3,926.

Parrish, D. D., Trainer, M., Holloway, J. S., Yee, J. E., Warshawsky, M. S., Fehsenfeld, F. C., Forbes, G. L., Moody, J. L. (1998) Relationships between ozone and carbon monoxide at surface in the North Atlantic region. *J. Geophys. Res.*, **103**, ND11, 13,357–13,376.

Pavelin, E. G., Johnson, C. E., Rughooputh, S., Toumi, R. (1999) Evaluation of pre-industrial surface ozone measurements made using Schönbein's method. *Atmos. Environ.*, **33**, 919–929.

Peleg, M., Luria, M., Sharf, G., Vanger, A., Kallos, G., Kotroni, V., Lagouvardos, K., Varinou, M. (1997) Observational evidence of an ozone episode of the Greater Athens Area. *Atmospheric Environment*, **31**, N23, 3,969–3,983.

Penkett, S. A., Reeves, C. E., Bandy, B. J., Kent, J. M., Richer, H. R. (1998a) Comparison of calculated and measured peroxide data collected in marine air to investigate prominent features of the annual cycle of ozone in the troposphere. *J. Geophys. Res.*, **103**, ND11, 13,377–13,388.

Penkett, S. A., Volz-Thomas, A., Parrish, D. D., Honrath, R. E., Fehsenfeld, F. C. (1998b) North Atlantic Regional Experiment (NARE II). Preface. *J. Geophys. Res.*, **103**, ND11, 13,353–13,355.

Penner, J. E., Bergmann, D. J., Walton, J. J., Kinnison, D., Prather, M. J., Rotman, D., Price, C., Pickering, K. E., Baughcum, S. L. (1998) An evaluation of upper tropospheric NO_x with two models. *J. Geophys. Res.*, **103**, ND17, 22,097–22,113.

Peters, K. L., Berkowitz, C. M., Carmichel, G. R., Easter, R. C., Fairweather, G., Ghan, S. J., Hales, J. M., Leung, L. R., Pennell, W. R., Potra, F. A., Saylor, R. D., Tsang, T. T. (1995) The current state and future direction of Eulerian models in simulating the tropospheric chemistry and transport of tracer species: A review. *Atmos. Environ.*, **29**, N2, 189–222.

Peterson, M. C., Honrath, R. E., Parrish, D. D., Oltmans, S. J. (1998) Measurements of nitrogen oxides and a simple model of NOy fate in the remote North Atlantic marine atmosphere. *J. Geophys. Res.*, **103**, ND11, 13,489–13,503.

Petzold A., Strom J., Schroeder, F. P., Karcher, B. (1999) Carbonaceous aerosol in the jet engine exhaust: Emission characteristics and implications for heterogeneous chemical reactions. *Atmos. Environm.*, **33**, N17, 2,689–2,698.

Pirjola, L., Laaksonen, A., Aalto, P., Kulmala, M. (1998) Sulfate aerosol formation in Arctic boundary layer. *J. Geophys. Res.*, **103**, ND7, 8,309–8,321.

Plant, R. A. J. (1998) GIS-based extrapolation of land use-related nitrous oxide flux in the Atlantic zone of Costa Rica. *Water, Air, and Soil Pollut.*, **105**, N1–2, 131–141.

Platt, U. *et al.* (1997) Influence of DMS on the O_x-NO_y partitioning and the NO_x distribution in the marine background atmosphere. *Geophys. Res. Lett.*, **24**, N15, 1,935–1,938.

Podgorny, I., Lubin, D. (1998) Biologically active insolation over Antarctic waters: Effects of highly reflecting coastline. *J. Geophys. Res.*, **103**, NC2, 2,919–2,928.

Poppe, D., Koppmann, R., Rudolph, J. (1998) Ozone formation in biomass burning plumes: Influence of atmospheric dilution. *Geophys. Res. Lett.*, **25**, N20, 3,823–3,826.

Prather, M., McElroy, M., Wofsy, S., Russel, G., Rind, D. (1987) Chemistry of the global troposphere: Fluorocarbons as tracers of air motion. *J. Geophys. Res.*, **92**, 6,579–6,613.

Prather, M. J., Jacob, D. J. (1997) A persistent imbalance in HO_x and NO_x photochemistry of the upper troposphere driven by deep tropical convection. *Geophys. Res. Lett.*, **24**, N24, 3,189–3,192.

Prather, M. J. (1998) Time scales in atmospheric chemistry: Coupled perturbations to N_2O, NO_y, and O_3. *Science*, **279**, 1,339–1,341.

Prinn, R. G. (ed.) (1994a) *Global Atmospheric–Biospheric Chemistry*, Plenum Press, New York, 261 pp.

Prinn, R. G. (ed.) (1994b) Global atmospheric–biospheric chemistry. In: *Global Atmospheric–Biospheric Chemistry*, Plenum Press, New York, pp. 1–18.

Rao, S. T., Zurbenko, I. G., Neagu, R., Porter, P. S., Ku, J. Y., Henry, R. F. (1997) Space and time scales in ambient ozone data. *Bull. Amer. Meteorol. Soc.*, **78**, N10, 2,153–2,166.

Rappenglueck, B., Fabian, P., Kalabokas, P., Viras, G., Ziomas, I. C. (1998) Quasi-continuous measurements of non-methane hydrocarbons (NMHC) in the Greater Athens area during MEDCAPHOT–TRACE. *Atmos. Environ.*, **32**, N12, 2,103–2,121.

Ravishankara, A. R. (1997) Heterogeneous and multiphase chemistry in the troposphere. *Science*, **276**, N5,315, 1,058–1,065.

Reichardt, J., Ansmann, A., Serwazi, M., Weitkamp, C., Michaelis, W. (1996) Unexpectedly low ozone concentration in midlatitude tropospheric ice clouds: A case study. *Geophys. Res. Lett.*, **23**, N15, 1,929–1,932.

Reid, S. T., Vaughan, G. (1991) Lamination in ozone profiles in the lower stratosphere, *Quart. J. Roy. Meteorol. Soc.*, **117**, 825–844.

Reiter, R., Sladkovic, R., Kanter, H.-J. (1987) Concentration of trace gases in the lower troposphere, simultaneously recorded at neighboring mountain stations. Part II: Ozone. *Meteorol. Atmos. Phys.*, **37**, 27–47.

Richter, A., Wittrock, F., Eisinger, M., Burrows, J. P. (1998) GOME observations of tropospheric BrO in Northern Hemisphere spring and summer 1997. *Geophys. Res. Lett.*, **25**, N14, 2,683–2,686.

Ridley, B. A., Walega, J. G., Lamarque, J.-F., Grahek, F. E., Trainer, M., Hübler, G., Lin, X., Fehsenfeld, F. C. (1998a) Measurements of reactive nitrogen and ozone to 5-km altitude in June 1990 over the southeastern United States. *J. Geophys. Res.*, **103**, ND7, 8,365–8,388.

Ridley, B., Walega, J., Hübler, G., Montzka, D., Atlas, E., Hauglustaine, D., Grahek, F., Lind, J., Campos, T., Norton, R., Greenberg, J., Schauffer, S., Oltmans, S., Whittlestone, S. (1998b) Measurements of NO_x and PAN and estimates of O_3 production over the seasons during Mauna Loa Observatory Photochemistry Experiment 2. *J. Geophys. Res.*, **103**, ND7, 8,323–8,340.

Roberts, J. M., Bertman, J. B., Parrish, D. D., Fehsenfeld, F. C., Jobson, B. T., Niki, H. (1998a) Measurements of alkyl nitrates at Chebogue Point, Nova Scotia during the 1993 North Atlantic Regional Experiment (NARE) intensive. *J. Geophys. Res.*, **103**, ND11, 13,569–13,580.

Roberts, J. M., Bertman, S. B., Jobson, T., Niki, H., Tanner, R. (1998b) Measurement of total nonmethane organic carbon (C_y): Development and application at Chebogue Point, Nova Scotia, during the 1993 North Atlantic Regional Experiment campaign. *J. Geophys. Res.*, **103**, ND11, 13,581–13,592.

Roberts, J. M., Williams, J., Baumann, K. *et al.* (1998c) Measurements of PAN, PPN, and MPAN made during the 1994 and 1995 Nashville Intensives of the Southern Oxidant Study: Implications for regional ozone production from biogenic hydrocarbons. *J. Geophys. Res.*, **103**, NC10, 22,473–22,490.

Röckmann, T., Brenninkmeijer, C. A. M., Neeb, P., Crutzen, P. J. (1998) Ozonolysis of nonmethane hydrocarbons as a source of the observed mass independent oxygen isotope enrichment in tropospheric CO. *J. Geophys. Res.*, **103**, ND1, 1,463–1,470.

Roelofs, G.-J., Lelieveld, J., Smit, H. G. J., Kley, D. (1997) Ozone production and transports in the tropical Atlantic region during the biomass burning season. *J. Geophys. Res.*, **102**, ND9, 10,637–10,652.

Roemer, M. (1999) There are no trends like ozone trends. In: *Proceedings EUROTRAC Symposium '98*, P. M. Borrell, P. Borrell (eds.), WIT Press, Southampton, UK, 208–213.

Roscoe, H. K., Clemitshaw, K. C. (1997) Measurements techniques in gas-phase tropospheric chemistry: A selective view of the past, present, and future. *Science*, **276**, N5315, 1,065–1,072.

Russel, A., Milford, J., Bergin, M. S., McBride, S., McNair, L., Yang, Y., Stockwell, W. R., Croes, B. (1995) Urban ozone control and atmospheric reactivity of organic gases. *Science*, **269**, N5,223, 491–495.

Ryerson, T. B., Buhr, M. P., Frost, G. J. *et al.* (1998) Emission lifetimes and ozone formation in power plant plumes. *J. Geophys. Res.*, **103**, NC10, 22,569–22,584.

Salardino, D. H., Carroll, J. J. (1998) Correlation between ozone exposure and visible foliar injury in ponderosa and Jeffrey pines. *Atmos. Environ.*, **32**, N17, 3,001–3,010.

Sandroni, S., Anfossi, D., Viarengo, S. (1992) Surface ozone levels at the end of the nineteenth century in South America. *J. Geophys. Res.*, **97**, ND2, 2,535–2,540.

Sanford, S., He, D., Pippin, M. R., Daum, P. H., Imre, D. G., Kleiman, L. I., Lee, J. H., Weinstein-Lloyd, J. (1998) Model correlations for ozone, reactive nitrogen, and peroxides for Nashville in comparison with measurements: Implications for O_3, NO_x-hydrocarbon chemistry. *J. Geophys. Res.*, **103**, NC10, 22,629–22,644.

Sanhuenza, E., Octavio, K. H., Arrocha, A., (1985) Surface ozone measurements in the Venezuelan Tropical Savannah. *J. Atmos Chem.*, **2**, 377–385.

De Santis, F. (1999) New Directions: will a new European vegetation ozone standard be fair to all European Countries? *Atmos. Environ.*, **33**, N23, 3,873–3,874.

Savarino, J., Legrant, M. (1998) High northern latitude forest fires and vegetation emissions over the last millennium inferred from the chemistry of a central Greenland ice core. *J. Geophys. Res.*, **103**, ND7, 8,267–8,279.

Scheel, K. E., Brunke, E.-G., Sladkovic, R., Seiler, W. (1998) In situ CO concentrations at the sites Zugspitze (47°N, 11°E) and Cape Point (34°S, 18°E) in April and October 1994. *J. Geophys. Res.*, **103**, ND15, 12,295–12,304.

Schoeberl, M. R., Jackman, C. H., Rosenfield, J. E. (1998) A Lagrangian estimate of aircraft effluent life time. *J. Geophys. Res.*, **103**, ND9, 10,817–10,825.

Schreiner, J., Voigt, C., Kohlmann, A., Arnold, F., Mauersberger, K., Larsen, N. (1999) Chemical analysis of polar stratospheric cloud particles. *Science*, **283**, N5,404, 968–970.

Schroeder, W. H., Barrie, L. A. (1998) Is mercury input to polar ecosystems enhanced by spring ozone depletion chemistry? *IGACtivities Newsletter*, N14, 7–8.

Schultz, M., Schmitt, R., Thomas, K., Volz-Thomas, A. (1998) Photochemical box modeling of long-range transport from North America to Tenerife during the North Atlantic Regional Experiment (NARE) 1993. *J. Geophys. Res.*, **103**, ND11, 13,472–13,488.

Scientific Assessment of Ozone Depletion: 1998. WMO Global Ozone Research and Monitoring Project Report No. 44 (1999) WMO, Geneva, Switzerland.

Seaman, N. L., Stauffer, D. R., Lario-Gibbs, A. M. (1995) A multiscale four-dimensional data assimilation system applied in the San Joaquin Valey during SARMAP. Part I. Modeling design and basic performance characteristics. *J. Appl. Meteorol.*, **34**, N8, 1,739–1,761.

Semenov, S. M. (1996) Genetic approach to assessments of contributions of various sources into total budget of environmental components. *Doklady Russian Acad. Sci.*, **351**, N2, 284–286 (in Russian).

Shreffler, J. H., Evans, R. B. (1982) The surface ozone record from the regional air pollution study. *Atmos. Environ.*, **16**, 1,311–1,321.

Sillman, S., He, D., Pippin, M. R., Daum, P. H., Imre, D. G., Kleiman, L. I., Lee, J. H., Weinstein-Lloyd, J. (1998) Model correlations for ozone, reactive nitrogen, and peroxides for Nashville in comparison with measurements: Implications for O_3-NO_x-hydrocarbon chemistry. *J. Geophys. Res.*, **103**, ND17, 22,629–22,644.

Sillman, S. (1999) The relation between ozone, NO_x and hydrocarbons in urban and polluted rural environments. *Atmos. Environ.*, **33**, Iss. 12, 1,821–1,852.

Simmonds, P. G., Cunnold, D. M., Weiss, R. F., Prinn, R. G., Fraser, P. J., McCulloch, A., Alyea, F. N., O'Doherty, S (1998a) Global trends and emission estimates of CCl4 from in situ background observations from July 1978 to June 1996. *J. Geophys. Res.*, **103**, ND13, 16,017–16,027.

Simmonds, P. G., O'Doherty, S, Huang, J., Prinn, R. G., Derwent, R. G., Ryall, D., Nickless, G., Cunnold, D. (1998b) Calculated trends and the atmospheric abundance of 1,1,1,2-tetrafluoroethane, 1,1-dichloro-1-fluoroethane, and 1-chloro-1,1-difluoroethane using automated in situ gas chromatography–mass spectrometry measurements recorded at Mace Head, Ireland, from October 1994 to March 1997. *J. Geophys. Res.*, **103**, ND13, 16,029–16,037.

Smit, H., Kley, D., McKeen, S., Voltz, A, Gilge, S. (1989) *Ozone in the Atmosphere*, R. Bojkov, P. Fabian (eds.), A. Deepak Publishing, Hampton, VA, pp. 419–422.

Smith, D. M., Chughtai, A. R. (1997) Photochemical effects in the heterogeneous reaction of soot with ozone at low concentrations. *J. Atmos. Chem.*, **26**, N1, 77–91.

Solberg, S., Schmidbauer, N., Semb, A., Stordal, F., Høv, Ø. (1996) Boudary-layer ozone depletion as seen in the Norwegian Arctic in spring. *J. Atmos. Chem.*, **23**, 301–332.

Solomon, P. A. (1995) Regional photochemical measurements and modeling studies: A summary of the A&WMA International Specialty Conference. *J. Air Waste Manage. Assoc.*, **45**, N4, 253–286.

Song, Q., Leighton, H. G. (1998) A parameterization of sulphate production in convective clouds for use in large scale models. *Atmos. Res.*, **47–48**, 555–572.

Spänkuch, D., Döhler, Güldner, J., Schulz, E. (1998) Estimation of the amount of tropospheric ozone in a cloudy sky by ground-based Fourier-transform infrared emission spectroscopy. *Appl. Optics.*, **37**, N15, 3,133–3,142.

Starn, T. K., Shepson, P. B., Bertman, S. B., White, J. S., Splawn, B. G., Riemer, D. D., Zika, R. G., Olszyna, K. (1998) Observations of isoprene chemistry and its role in ozone production at a semirural site during the 1995 Southern Oxidants Study. *J. Geophys. Res.*, **103**, NC10, 22,425–22,436.

Starn, T. K., Shepson, P. B., Bertman, S. B., Riemer, D. D., Zika, R. G., Olszyna, K. (1998) Night-time isoprene chemistry at an urban-impacted forest site. *J. Geophys. Res.*, **103**, NC10, 22,437–22,448.

St. John, J. C., Chameides, W. L., Saylor, R. (1998) Role of anthropogenic NO_x and VOC as ozone precursors: A case study from the SOS Nashville/Middle Tennessee Ozone Study. *J. Geophys. Res.*, **103**, ND17, 22,415–22,424.

Stockwell, D. Z., Kritz, M. A, Chipperfield, M. P., Pyle, J. A. (1998) Validation of an off-line three-dimensional chemical transport model using radon profiles. 2. Model results. *J. Geophys. Res.*, **103**, ND7, 8,433–8,445.

Stratospheric Processes and Their Role in Climate (SPARC) (1998) WCRP-99 (WMO/TD-N814), Vol. II, WMO, Geneva, Switzerland, pp. 351–428.

Straume, A. G., Dietz, R. N., Koffi, E. N., Nodop, K. (1998) Perfluorocarbon background concentrations in Europe. *Atmos. Environ.*, **32**, N24, 4,109–4,122.

Sumner, A. L., Shepson, P. B. (1999) Snowpack production of formaldehyde and its effect on the Arctic troposphere. *Nature*, **398**, N6,724, 230–233.

Sun, J., Desjardins, R., Mahrt, L., MacPherson, I. (1998) Transport of carbon dioxide, water vapor, and ozone by turbulence and local circulations. *J. Geophys. Res.*, **103**, ND20, 25,873–25,885.

Sun, J., Lenschow, D. H., Mahrt, L., Crawford, T. L., Davis, K., Oncley, S. P., MacPherson, J. I., Wang, Q., Dobosy, R. J., Desjardins, R. L. (1997). Lake-induced atmospheric circulations during BOREAS. *J. Geophys. Res.*, BOREAS Special Issue, **102**, ND24, 29,155–29,166.

Suppan, P., Fabian, P., Vyras, L., Gyning, S. E. (1998) The behaviour of ozone and peroxyacetylnitrate concentrations for different wind regimes during the MEDCAPHOT–TRACE campaign in the Greater Area of Athens, Greece. *Atmos. Environ.*, **32**, N12, 2,089–2,102.

Swap, B., Annagarn, H., Scholes, M., Justice, C. (1998) SAFARI-2000: A Southern African Regional Science Initiative. *IGACtivities Newsletter*, N15, 16–18.

Taalas, P., Koskela, T., Kyrö, E. (1994) The unusually low total ozone columns in Northern Europe during 1992 and 1993: Background and effects on solar UVB radiation. *Ann. Geophys.*, **12**, Part III, C603.

Taalas, P. (1994) UV-research and its results in Northern Europe and in Finland. In: *Proceedings WMO Meeting of Experts on UV-B Monitoring, Data Quality and Standardization of UV Indices, 25–28 July 1994, Les Diablerets, Switzerland.*

Talukdar, R. K., Longfellow, C. A., Gilles, M. R., Ravishankara, A. R. (1998) Quantum yields of $O(^1D)$ in the photolysis of ozone between 289 and 329 nm as a function of temperature, *Geophys. Res. Lett.*, **25**, N2, 143–146.

Tarasick, D. W., Wardle, D. I., Kerr, J. B., Bellefleur, J. J., Davies, J. (1995) Tropospheric ozone trends over Canada: 1980–1993. *Geophys. Res. Lett.*, **22**, N4, 409–412.

Taylor, G. R., Kreindenweis, S., Zhang, Y. (1997) The effects on aerosol and chemical species and distribution. 1. Cloud model formulation, mixing, and detainment. *J. Geophys. Res.*, **103**, ND20, 23,851–23,865.

The Oxidizing Capacity of the Troposphere (1997) B. Larsen, B. Versino, G. Angeletti (eds.), Air Pollution Research Report No. 17,482, EC Directorate-General Science, Research and Development, Luxembourg, 716 pp.

Theodorsen, A., Henriksen, K., Bersås, S., Sirota, V., Vasiliev, A. (1996) Surface ozone at high latitudes. *Geophys. Res. Lett.*, **23**, N1, 77–80.

Thompson, A. M., Pickering, K. E., McNamara, D. P., Schoeberl, M. R., Hudson, R. D., Kim, J. H., Browell, E. V., Kirchoff, V. W. J. H., Nganga, D. (1996) Where did tropospheric ozone over southern Africa and the tropical Atlantic come from in October 1992? Insights from TOMS, GTE TRACE A. and SAFARI 1992. *J. Geophys. Res.*, **101**, ND19, 24,251–24,278.

Tiao, G. C., Phadke, M. S., Box, G. E. P., (1976) Some empirical models for the Los Angeles photochemical smog data. *J. Air PollUT. Control Assoc.*, **26**, 485.

Tie, X., Hess, P. (1997) Ozone mass exchange between the stratosphere and troposphere for background and volcanic sulfate aerosol conditions. *J. Geophys. Res.*, **102**, ND21, 25,487–25,500.

Tilmes, S., Zimmerman, J. (1998) Investigation on the spatial scales of the variability in measured near-ground ozone mixing ratios. *J. Geophys. Res.*, **103**, ND17, 22,663–22,669.

Tiwari, V. S., Sreedharan, C. R. (1973) Ozone concentration studies and ozone flux mesurements near the ground at Poona. *Pageoph.*, **106–108**, 1,124–1,138.

Transboundary Photooxidation Air Pollution in Europe: Calculations of Tropospheric Ozone and Comparison with Observations (1998) The Norwegian Meteorological Institute Research Report No. 67/EMEP/MSC-W Report 2/98, Oslo, Norway, 74 pp., Appendices.

Tremmel, H. G., Schlager, H., Konopka, P., Schulte, P., Arnold, F., Klemm, M., Droste-Franke, B. (1998) Observations and model calculations of jet aircraft exhaust products at cruise altitude and inferred initial OH emissions. *J. Geophys. Res.*, **103**, ND9, 10,803–10,816.

Treviño, C., Méndez, F. (1999) Simplified model for the prediction of ozone generation in polluted urban areas with continuous precursor species emissions. *Atmos. Environ.*, **33**, 1,103–1,110.

Tropospheric Chemistry and Space Observations (1998) N. Chaumberliac, G. Mégre, N. Papineau, G. Angeletti (eds.), Air Pollution Research Report No. 65, EC Directorate-General Science, Research and Development, EUR 17790, Luxembourg, 277 pp.

Tsutsumi, Y., Igarashi, Y., Zaizen, Y., Makino, Y. (1998) Case studies of tropospheric ozone events observed at the summit of Mount Fuji. *J. Geophys. Res.*, **103**, ND14, 16,935–16,951.

Tulet, P., Maalej, A., Crassier, V., Rosset, R. (1999) An episode of photooxidant plume pollution over the Paris region. *Atmos. Environ.*, **33**, Iss. 11, 1,651–1,662.

Tuovinen, J.-P., Laurila, T. (1993) Ozone concentrations and exposures in Finland. In: *Proceedings EMEP Workshop on the Control of Photochemical Oxidants in*

Europe, 20–22 April, 1993, Porvoo, Finland, Finnish Meteorological Institute, pp. 15–24.

Tuovinen, J.-P., Laurila, T., Lättilä, H., Ryaboshapko, A., Brunkhanov, Korolev, S. (1993) Impact of the sulphur dioxide sources in the Kola peninsula on air quality in northernmost Europe. *Atmos. Environ.*, **27A**, N9, 1,379–1,395.

Tyson, D. D., Garstang, M., Thompson, A. M., D'Abreton, P., Diab, R. D., Browell, E. V. (1997) Atmospheric transport and photochemistry of ozone over central Southern Africa during the Southern Africa Fire–Atmosphere Research Initiative. *J. Geophys. Res.*, **102**, ND9, 10,625–10,636.

UNEP Scientific Assessment of Ozone Depletion, WMO Global Ozone Research and Monitoring Project Report No. 37 (1994) WMO, Geneva, Switzerland.

Uspensky, A. B., Trotsenko, A. N., Rublev, A. N., Romanov, D. Yu. (1998) Determination of column amounts of the atmospheric minor gas constituent using IASI IR-sounder. I. Information content analysis. *Studying the Earth from Space*, N2, 3–16 (in Russian).

Valente, E. J., Imhoff, R. E., Tanner, R. L., Meagher, J. F., Daum, P. H., Hardesty, R. M., Bante, R. M., Alvarez, R. J., McNider, R. T., Gillani, N. V. (1998) Ozone production during an urban air stagnation episode over Nashville, Tennessee. *J. Geophys. Res.*, **103**, NC10, 22,555–22,568.

Van Velthoven, P. F. J., Kelder, H. (1996) Estimates of stratosphere–troposphere exchange: Sensitivity to model formulation and horizontal resolution. *J. Geophys. Res.*, **101**, ND1, 1,429–1,434.

Varotsos, C., Cartalis, C., Vlassi, A. (1991) Comparing the intraseasonal variations of an historic surface ozone time series and recent surface ozone data for Athens, *Greece. 5th AMS Conference on Climate Variation, 14–18 October 1991, Denver, CO*, pp. 169–172, American Meteorological Society, Boston.

Varotsos, C., Cartalis, C. (1991) Re-evaluation of surface ozone over Athens, Greece for the period 1901–1940, *Atmos. Res.*, **26**, 303–310.

Varotsos, C., Kalabokas, P., Chronopoulos, G. (1992) On the laminated structure of ozone in the subtropical atmosphere. *Quadrennial Ozone Symposium, Charlottesville, VA*, R. D. Hudson (ed.), NASA Cont. Pub. 3,266, VA.

Varotsos, C., and Kondratyev, K. Ya. (1994) Athens environmental dynamics: From a rural to an urban region. *Optics Atmos. Ocean*, **7**, 3–17 (in Russian).

Varotsos, C. A., Kondratyev, K. Ya. (1995) Ozone dynamics over Greece as derived from satellite and in situ measurements. *Int. J. Remote Sensing*, **16**, N10, 1,777–1,798.

Varotsos, C. A., Kondratyev, K. Ya., Efstathiou, M. (2000) Seasonal variation of the surface ozone mixing ratio over Athens, Greece. *TOR-2 Workshop, 19–21 September, 1999, Athens*, Swedish Environmental Research Institute, Sweden.

Vasilyev, A. V., Rozanov, V. V., Timofeyev, Yu. M. (1998) Information content of outgoing reflected and scattered solar radiation in UV and visible spectral regions. *Studying the Earth from Space*, N2, 51–59 (in Russian).

Vaughan, G., Timmis, C. (1998) Transport of near-tropopause air into the lower midlatitude stratosphere. *Quart. J. Roy. Meteorol. Soc.*, **124**, Part A, 1,559–1,578.

Vogt, R., Crutzen, P. J., Sander, R. A. (1996) A mechanism for halogen release from sea-salt aerosol in the remote marine boundary layer. *Nature*, **383**, 327–330.

Vogt, R., Sander R., von Glasow R., Crutzen, P. J. (1999) Iodine chemistry and its role in halogen activation and ozone loss in the marine boundary layer: A model study. *J. Atmos. Chem.*, **32**, N3, 375–395.

Voltz, A., Geiss, H., McKeen, S., Kley, D. (1988) Correlation of ozone and solar-radiation at Mountsuris and Hohenpeissenberg: Indications for photochemical influence. In: *Ozone in the Atmosphere*, R. Bojkov, P. Fabian (eds.), A. Deepak Publishing, Hampton, VA.

Voltz-Thomas, A., Smit, H, Kley, D., Aschmutat, U., Buers, H., Floke, F., Garthe, H., Geiss, H., Gilge, S., Heil, T., Houben, N., Klemp, D., Loup, H., Mihelcic, D., Muesgen, P., Paetz, W., Pilwat, G., Sraeter, W., Su, Y. (1991) In: *Tropospheric Ozone Research (TOR) 1990 Annual Report, EUROTRAC Annual Report*, Part 9, P. Kley (ed.), International Scientific Secretariat, Munich.

Vukovich, F. M., Brackett, V., Fishman, J., Sickles, J. E., II (1997) A 5-year evaluation of the representativeness of the tropospheric ozone radical at nonclimatological periods. *J. Geophys. Res.*, **102**, ND13, 15,927–15,932.

Vukovich, F. M. (1998) Aspects of subregional ozone variations in the SOS region. *Atmos. Environ.*, **32**, N22, 3,881–3,889.

Wagner, T., Platt, U. (1998a) Satellite mapping of enhanced BrO concentrations in the troposphere. *Nature*, **395**, 466–470.

Wagner, T., Platt, U. (1998b) Satellite mapping of enhanced BrO concentrations in the troposphere. *Nature*, **396**, N6,701, 486–490.

Wakamatsu, S., Uno, I., Ohara, T., Schere, K. L. (1999) A study of the relationship between photochemical ozone and its precursor emissions of nitrogen oxides and hydrocarbons in Tokyo and surrounding areas. *Atmos. Environ.*, **33**, N19, 3,097–3,108.

Wang, P.-H., Cunnold, D. M., Zawodny, J. M., Pierce, R. B., Olson, J. R., Kent, G. S., Skeens, K. M. (1998) Seasonal ozone variations in the isentropic layer between 330 and 380 K as observed by SAGE II: Implications of extratropical cross-tropopause transport. *J. Geophys. Res.*, **103**, ND22, 28,647–28,659.

Wang, W.-C., Isaksen, I. S. A. (eds.) (1995) *Atmospheric Ozone as a Climate Gas: General Circulation Model Simulations*, Springer-Verlag, Berlin, 459 pp.

Wang, Y., Jacob, D. J., Logan, J. A. (1998a) Global simulation of tropospheric O_3-NO_x-hydrocarbon chemistry. 2. Model evaluation and global ozone budget. *J. Geophys. Res.*, **103**, ND9, 10,727–10,755.

Wang, Y., Jacob, D. J., Logan, J. A. (1998b) Global simulation of tropospheric O_3-NO_x-hydrocarbon chemistry. 3. Origin of tropospheric ozone and effects of nonmethane hydrocarbons. *J. Geophys. Res.*, **103**, ND9, 10,757–10,767.

Warneck, P. , Helas G., Slanina, J., Steinbrecher, R. (1998) Biogenic volatile organic compounds in the atmosphere. *J. Atmos. Chem.*, **30**, N3, 407–409.

Watanabe, A., Yoshida, M., Kimura, M. (1998) Contribution of rice straw carbon to CH4 emission from rice paddies using 13C-enriched rice straw. *J. Geophys. Res.*, **103**, ND7, 8,237–8,242.

Wauben, W. M. F. (1997) The effect of bandwidth on the retrieval of ozone from measurements of attenuated solar radiation with a Brewer spectrometer. *Beitr. Phys. Atmos.*, **70**, N3, 221–232.

Wauben, W. M. F., van Velthoven, P. F. J., Kelder, H. (1997a) A 3D chemistry transport model study of changes in atmospheric ozone due to aircraft NO_x emissions. *Atmos. Environ.*, **31**, N12, 1,819–1,836.

Wauben, W. M. F., Bintanja, R., van Velthoven, P. F. J., Kelder, H. (1997b) On the magnitude of transport out of the Antarctic polar vortex. *J. Geophys. Res.*, **102**, ND1, 1,229–1,238.

Wauben, W. M. F., van Velthoven, P. F. J., Kelder, H., Bintanja, R. (1997c) Reply. *J. Geophys. Res.*, **102**, ND23, 28,219–28,221.

Wauben, W. M. F., Fortuin, J. P. F., van Velthoven, P. F. J., Kelder, H. M. (1998) Comparison of modelled ozone distribution with sonde and satellite measurement. In: *Atmospheric Ozone*, R. D. Bojkov, G. Visconti (eds.), Parco Scientifico e Tecnologico d'Abruzzo, L'Aquila, Italy, Vol. 2, pp. 791–794.

van Weele, M., Vilà-Guerau de Arellano, J., Kuik, K. (1995) Combined measurements of UV-A actinic flux, UV-A irradiance and global radiation in relation to photodissociation rates. *TelLus*, **47B**, N3, 353–364.

Weinheimer, A. J., Montzka, D. D., Campos, T. L. *et al.* (1998a) Comparison between DC-8 and ER-2 species measurements in the tropical middle troposphere: NO, NO_y, O_3, CO_2, CH_4, and N_2O. *J. Geophys. Res.*, **103**, ND17, 22,073–22,085.

Weinheimer, A. J., Montzka, D. D., Campos, T. L. *et al.* (1998b) Comparison between DC-8 and ER-2 species measurements in the tropical middle troposphere: NO, NO_y, O_3, CO_2, CH_4, and N_2O. *J. Geophys. Res.*, **103**, ND17, 22,087–22,096.

Wennberg, P. O., Hanisco, T. F., Jaeglé L., Jacob, D. J., Hintsa, E. J., Lanzendorf, E. J., Anderson, J. G., Gao, R., Keim, E. R., Donnelly, S. G., Negro, L. A., Fahey, D. W., McKeen, S. A., Salawitch, R. J., Webster, C. R., May, R. D., Herman, R. L., Proffitt, M. H., Margitan, J. J., Atlas, E. L., Schauffler, S. M., Flocke, F., McElroy, C. T., Bui, T. P. (1998) Hydrogen radicals, nitrogen radicals, and the production of O_3 in the upper troposphere. *Science*, **279**, 49–53.

Wingenter, O. W., Wang, C. J.-L., Blake, D. R., Rowland, F. S. (1998) Seasonal variation of tropospheric methyl bromide concentrations: Constraints on anthropogenic input. *Geophys. Res. Lett.*, **25**, N15, 2,797–2,800.

Wolff, E. W., Martin, S. (1998) Sea ice as a source of sea salt aerosol in polar regions. *IGACtivities Newsletter*, N14, 10–12.

Wuebbles, D. J., Penner, J. E. (1988) *Sensitivity of Urban/Regional Chemistry to Climate Change Report of the Workshop*, Lawrence Livermore Nat. Laboratory, Preprint UCRL-99436, Livermore, CA, 17 pp., illustrations.

Yates, S. R., Wang, D., Gan, J., Ernst, F. F., Jury, W. A. (1998) Minimizing methyl bromide emissions from soil fumigation. *Geophys. Res. Lett.*, **25**, N10, 1,633–1,636.

Yu, F., Turco, R. P. (1998) Contrail formation and impacts on aerosol properties in aircraft plumes: Effects on fuel sulfur content. *Geophys. Res. Lett.*, **25**, N3, 313–316.

Zanis, P., Schuepbach, E., Scheel, H. E., Baudenbacher, M., Buchmann, B. (1999) Inhomogeneities and trends in the surface ozone record (1988–1996) at Jungfraujoch in the Swiss Alps. *Atmos. Environ.*, **33**, N23, 3,777–3,786.

Zhang, D., Iwasaka, Y. (1999) Nitrate and sulfate in individual Asian dust-storm particles in Beijing, China, in spring of 1995 and 1996. *Atmos. Environ.*, **33**, N19, 3,213–3,223.

Zhang, J., Rao, S. T., Duggupaty, S. M. (1998) Meteorological processes and ozone exceedances in the Northeastern United States during the 12–16 July 1995 episode. *J. Appl. Meteorol.*, **37**, N8, 776–789.

Zhang, Y., Bischof, C. H., Easter, R. C., Wu, P.-T. (1998) Sensitivity analysis of a mixed-phase chemical mechanism using automatic differentiation. *J. Geophys. Res.*, **103**, ND15, 18,955–18,979.

Ziemke, J. R., Chandra, S. (1998a) Comment on Tropospheric ozone derived from TOMS/SBUV measurements during TRACE A by J. Fishman *et al. J. Geophys. Res.*, **103**, ND12, 13,903–13,906.

Ziemke, J. R., Chandra, S. (1998b) On tropospheric ozone and the tropical wave 1 in total ozone. *Proceedings 18th Quadrennial Ozone Symposium*, Parco Scientifico e Tecnologico d'Abruzzo, L'Aquila, Italy, Vol. 1, pp. 447–450.

Ziemke, J. R., Chandra, S., Bhartia, D. K. (1998) Two new methods for deriving tropospheric column ozone from TOMS measurements: Assimilated UARS MLS/HALOE and convective cloud differential technique. *J. Geophys. Res.*, **103**, ND17, 22,115–22,127.

Ziomas, I. C. (1998) The Mediterranean campaign of photochemical tracer-transport and chemical evolution (MEDCAPHOT–TRACE): An outline. *Atmos. Environ.*, **32**, N12, 2,045–2,053.

Ziomas, I. C., Gryning, S.-E., Bornstein, R. O. (eds.) (1998a) The Mediterranean Campaign of Photochemical Tracer–Transport and Chemical Evolution (MEDCAPHOT–TRACE). *Atmos. Environ.*, **32**, N12, 2,043–2,326.

Ziomas, I. C., Tzoumaka, P., Balis, D., Melas, D., Zerefos, C. S., Klemm, O. (1998b) Ozone episodes in Athens, Greece. A modelling approach using data from the MEDCAPHOT–TRACE. *Atmos. Environ.*, **32**, N12, 2,313–2,321.

Zondlo, M. A., Barone, S. B. (1997) Uptake of HNO_3 on ice under upper tropospheric conditions. *Geophys. Res. Lett.*, **24**, N11, 1,391–1,394.

6

Ozone and climate

6.1 INTRODUCTION

Many interactive processes are responsible for climate formation and change (Brasseur, 1997; Houghton *et al.*, 1996; Kondratyev, 1989, 1998; Zillman, 1997). It is therefore difficult to single out impacts of individual causes of climate change. Nevertheless, many attempts have been made to assess the climatic impacts of various processes and phenomena, including processes in the stratosphere especially regarding the role of stratospheric ozone variability (Kondratyev, 1998, 1999). Long ago, a hypothesis was proposed concerning the important role of the stratosphere in climate change ('stratospheric climate regulation'). Although it was soon shown that such a hypothesis should be considered as a gross exaggeration, there is no doubt that the 'stratosphere and climate' problem deserves serious attention, especially in the context of anthropogenically induced ozone depletion (Bojkov, 1997; Hansen *et al.*, 1997; Kondratyev, 1981, 1985, 1998; *Ozone* ..., 1989; *Stratospheric* ..., 1998; Uchino, 1998; Varotsos, 1997; Wang and Isaksen, 1995). This is why the WCRP Core Project 'Stratospheric Processes and Their Role in Climate' (SPARC) has been approved and is now being successfully carried out (Chanin, 1993, 1998; Chanin and Geller, 1998; Geller, 1998; Shine *et al.*, 1995; *Stratospheric* ..., 1997). The following key scientific issues have been identified (Phillips, 1997):

- Influence of the stratosphere on climate.
- Physics and chemistry associated with stratospheric ozone decrease.
- Stratospheric variability and monitoring.
- UV irradiation changes.

Later, (*Stratospheric* ..., 1998) the principal SPARC research topics were classified into the three following categories: 'stratospheric indicators of climate change', 'stratospheric processes and their relation to climate', 'modelling stratospheric effects on climate'.

As far as stratospheric processes and their relation to climate is concerned, the following important topics have been emphasized:

- Gravity wave processes and parametrization of the effects of unresolved internal gravity waves in global numerical atmospheric models.
- Stratosphere–troposphere exchange.
- Chemistry climate of the upper troposphere–lower stratosphere.
- Changes in ozone and its influence on climate.
- Quasi-biennial oscillation and its possible role in coupling the stratosphere and troposphere.

To study upper tropospheric and lower stratospheric ozone, aerosols and climate the following priorities should be taken into account:

- Observations of stratospheric aerosols and PSCs.
- Heterogeneous chemistry of the upper troposphere.
- Gas-phase processes.
- Chemical–radiative coupling in the upper troposphere and lower stratosphere.

Nowadays, it has become clear that understanding basic climate features is impossible without consideration of the complexity of all interactive processes including chemistry and dynamics of the atmosphere (Wang *et al.* 1996), especially in the study of atmospheric greenhouse-effect changes (Wuebbles, 1996). Spatio-temporal variability in optically active minor gas and aerosol components in the atmosphere controls radiative-flux divergence (including greenhouse effect) and, on the other hand, climate dynamics is a principal factor of minor component changes (having to take into account biospheric processes makes the situation prohibitively complex).

An important step forward has been made, in this context, by Wang *et al.* (1998a, b) who studied climate chemistry interactions and their uncertainty using a coupled global-scale modelling system, consisting of a global climate model and an atmospheric chemistry model. This system has been designed to predict as function of time, latitude and altitude the zonally averaged concentrations of major chemically and radiatively important trace species in the atmosphere. Biogeochemical models and climate submodels in this system are fully interactive. The nine-level 2-D statistical–dynamical climate submodel has horizontal resolution of 7,826°. The biogeochemical submodel has 25 chemical species including CO_2, CH_4, N_2O, O_3, CO, H_2O, NO_x, HO_x, SO_2, sulphate aerosols and chlorofluorocarbons. The model does not include non-methane hydrocarbon chemistry, which leads to underestimation of ozone production, especially in polluted areas. Continuity equations for $CFCl_3$, N_2O, O_3, CO, NO, CO_2, NO_2, N_2O_5, HNO_3, CH_4, CH_2O, SO_2 and H_2SO_4 include convergence because of transport (advection, eddy diffusion and convection), whereas the remaining very reactive atoms, free radicals, or molecules are assumed to be unaffected by transport (photochemical steady states) because of their very short lifetimes.

There are 41 gas phase and 12 heterogeneous reactions in the chemistry model. Gas-phase chemistry has three major parts: tropospheric O_3-HO_x-NO_x-CO-CH_4 reactions; tropospheric SO_2-sulphate reactions and stratospheric chlorofluorocarbon and N_2O removal reactions (Table 6.1). Dissolution of soluble gases in precipitating water is included in the wet deposition term in the model (Table 6.2). Emissions of CO_2, CH_4, N_2O, $CFCl_3$, CF_2Cl_2, CO, NO and SO_2 are taken into account as functions of latitude. With the exception of two CFCs, all emissions are separated into natural and anthropogenic emissions.

Predicted concentrations of chemical species in the chemistry submodel are used interactively to calculate radiative forcing (RF) in the climate submodel. The results obtained indicate that model predictions of surface trends of several key species are close to observations over the past 10–20 years. The model provides reasonable simulations of seasonal variability and good agreement with observations at high latitudes. Predicted vertical distributions of climate-relevant species, as well as seasonal variations, are also in good agreement with observations. Prediction of future atmospheric chemical composition by 2100 shows that, if current trends persist, it will be quite different from the one presently observed. Wang *et al.* (1998a) emphasized that the differences involve not only higher concentrations of major long-lived trace gases such as CO_2 (745 ppm), N_2O (410 ppb) and CH_4 (4.4 ppm) but also about 20% lower concentrations of the major tropospheric oxidizer (OH free radicals), and almost double the current concentrations of short-lived air pollutants CO and NO_x. Quantification of two-way feedback between climate dynamics and atmospheric chemistry would be a very importan result. Wang *et al.* (1998a) underlined, however, that there are still many uncertainties in the model and therefore further studies are necessary.

Ozone in the stratosphere and troposphere affects the climate in various ways. In the stratosphere, ozone is a strong absorber of ultraviolet solar radiation as well as of upwelling thermal emission, while in the troposphere, ozone (although it absorbs ultraviolet radiation) plays the role of a greenhouse gas (Wang and Isaksen, 1995). In recent years, substantial progress has been made in assessing the contributions of the dynamics of ozone content in the stratosphere and troposphere to climate change (Mahlman *et al.*, 1994; Shine, 1986; van Dorland and Fortuin, 1994; Varotsos, 1997).

Calculations have shown (*UNEP* ..., 1994; *Scientific* ..., 1999) that the total ozone content decrease observed in the 1980s and 1990s ought to have determined the decrease of radiative forcing for the surface–atmosphere system at mid and high latitudes. Being globally averaged over 10 years, such a decrease turns out to be similar in magnitude, but opposite in sign, to radiative forcing enhancement because of CFC concentration growth over the same time period. Since, in this case, the regional inhomogeneities of the radiative forcing distribution are very substantial (especially in high latitudes), specification of these estimates based on using global 3-D models, is very urgent.

Assessing the possible effect of anthropogenic sulphate aerosol in the Northern Hemisphere (in the Southern Hemisphere the effect is negligibly small) in clear-sky conditions has led to a cooling estimate of about 1 Wm^{-2} on average for the Northern Hemisphere, which should be compared with greenhouse warming of about 2.5 Wm^{-2} for the last century. No doubt, estimates about the climate-forcing contribution of aerosols are very preliminary and require further verification, because more so the important role of aerosol as a factor affecting radiative properties of clouds, may be substantial (Kondratyev, 1999).

Table 6.1. Gaseous phase chemical reactions included in the model.

Number	Reaction
R1	$O_3 + h\nu \longrightarrow O(^1D) + O_2$
R2	$O(^1D) + H_2O \longrightarrow 2OH$
R3	$O(^1D) + N_2 \longrightarrow O + N_2$
R4	$O(^1D) + O_2 \longrightarrow O + O_2$
R5	$CO + OH \longrightarrow H + CO_2$
R6	$H + O_2 + M \longrightarrow HO_2 + M$
R7	$HO_2 + NO \longrightarrow OH + NO_2$
R8	$NO_2 + h\nu \longrightarrow NO + O$
R9	$O + O_2 + M \longrightarrow O_3 + M$
R10	$HO_2 + O_3OH + 2O_2$
R11	$OH + O_3 \longrightarrow HO_2 + O_2$
R12	$NO + O_3 \longrightarrow NO_2 + O_2$
R13	$NO_2 + OH + M \longrightarrow HNO_3 + M$
R14	$NO_2 + O_3 \longrightarrow NO_3 + O_2$
R15	$NO_3 + NO_2 + M \longrightarrow N_2O_5 + O_2$
R16	$HO_2 + HO_2 \longrightarrow H_2O_2 + O_2$
R17	$H_2O_2 + h\nu \longrightarrow 2OH$
R18	$H_2O_2 + OH \longrightarrow HO_2 + H_2O$
R19	$HO + HO_2 \longrightarrow H_2O + O_2$
R20	$HO + HO \longrightarrow H_2O + O$
R21	$HO + HO + M \longrightarrow H_2O_2 + M$
R22	$CH_4 + OH \longrightarrow CH_3 + H_2O$
R23	$CH_3 + O_2 + M \longrightarrow CH_3O_2 + M$
R24	$CH_3O_2 + NO \longrightarrow CH_3O + NO_2$
R25	$CH_3O + O_2 \longrightarrow CH_2O + HO_2$
R26	$CH_3O_2 + HO_2 \longrightarrow CH_3O_2H + O_2$
R27	$CH_3O_2H + h\nu \longrightarrow CH_3O + OH$
R28	$CH_3O_2H + OH \longrightarrow CH_3O_2 + H_2O$
R29	$CH_2O + h\nu \longrightarrow CHO + H$
R30	$CH_2O + OH \longrightarrow CHO + H_2O$
R31	$CHO + O_2 \longrightarrow CO + HO_2$
R32	$SO_2 + OH + M \longrightarrow HOSO_2 + M$
R33	$HOSO_2 + O_2 \longrightarrow HO_2 + SO_3$
R34	$SO_3 + H_2O \longrightarrow H_2SO_4$
R35	$CFCl_3 + O(^1D) \longrightarrow$ products
R36	$CFCl_3 + h\nu \longrightarrow$ products
R37	$CF_2Cl_2 + O(^1D) \longrightarrow$ products
R38	$CF_2Cl_2 + h\nu \longrightarrow$ products
R39	$N_2O + h\nu \longrightarrow N_2 + O(^1D)$
R40	$N_2O + O(^1D) \longrightarrow 2NO$
R41	$N_2O + O(^1D) \longrightarrow N_2 + O_2$

Table 6.2. Aqueous phase chemical reactions included in the model.

Number	Reaction
R42	$H_2SO_4(g) \longleftrightarrow H_2SO_4(aq)$
R43	$H_2SO_4(aq) \longleftrightarrow HSO_4^- + H^+$
R44	$HNO_3(g) \longleftrightarrow HNO_3(aq)$
R45	$HNO_3(aq) \longleftrightarrow NO_3^- + H^+$
R46	$CH_2O(g) \longleftrightarrow CH_2O(aq)$
R47	$SO_2(g) \longleftrightarrow SO_2(aq)$
R48	$SO_2(aq) + H_2O \longleftrightarrow HSO_3 + H^+$
R49	$HSO_3^- \longleftrightarrow SO_2^- + OH^-$
R50	$H_2O_2(g) \longrightarrow H_2O_2(aq)$
R51	$HO(g) \longrightarrow HO(aq)$
R52	$HO_2(g)HO_2(aq)$
R53	$HO_2(aq) \longrightarrow H^+ + O_2^-$

6.2 RADIATIVE FORCING CALCULATIONS

Most recent climatic assessments of ozone impacts have concentrated on radiative forcing (RF) calculations (e.g. Hansen *et al.*, 1997; Kondratyev, 1998; Portmann *et al.*, 1997; Solomon and Daniel, 1996; Stevenson *et al.*, 1998; *Stratospheric* ..., 1998; Zwick, 1998).

New assessments about radiative forcing caused by decreasing ozone content in the lower stratosphere at mid and high latitudes suggested the possibility of considerable negative radiative forcing (cooling) for the surface–troposphere system, contrary to the positive greenhouse radiative forcing because of the increasing amount of CFCs. However, initial calculations of anthropogenic climate changes did not take into account the recent growth in ozone content in the troposphere, which should cause greenhouse warming.

According to Hansen *et al.* (1997) there has been a net negative RF from all ozone changes, since 1979, which is larger than the changes from all non-CO_2 long-lived gases over the same period. This means that ozone layer recovery in the coming decades will be accompanied by enhancement in GHG warming. For the time being, it is assumed that the observed cooling of the stratosphere is mostly a result of ozone depletion (*Stratospheric* ..., 1998).

No doubt, when considering the effect of observed ozone changes on radiative forcing and climate, it is necessary to take into account both decrease in ozone content in the stratosphere and its increase in the troposphere. Moreover, specific features about the vertical ozone profile should be taken into account. In this connection, Wang *et al.* (1993) analysed results of observations of the vertical ozone profile at seven stations at mid and high latitudes (from Tateno, 36°N, to Resolute, 75°N) to study ozone variations over the last few decades and to assess climate change in the context of increased concentrations of other GHGs: CO_2, CH_4, CFC-11, CFC-12 and N_2O.

Table 6.3. Radiative forcing (W/m^2) because of changes in ozone content between 1972 and 1990.

	Stratosphere		Troposphere		Total	
Month	SW	LW	SW	LW	SW	LW
Hohenpeissenberg						
Fixed air temperature (January)	0.207	−0.019	0.062	0.160	0.270	0.140
Fixed dynamic heating (January)	0.207	−0.103	0.062	0.074	0.270	0.057
Fixed air temperature (July)	0.084	−0.004	0.098	0.392	0.182	0.388
Fixed dynamic heating (July)	0.084	−0.026	0.098	0.369	0.182	0.366
Payern						
Fixed air temperature (January)	0.141	−0.027	0.056	0.186	0.198	0.158
Fixed dynamic heating (January)	0.141	−0.193	0.056	0.018	0.198	−0.008
Fixed air temperature (July)	0.346	−0.045	0.027	0.07	0.373	0.024
Fixed dynamic heating (July)	0.346	−0.306	0.027	−0.194	0.373	−0.237

Results obtained by Wang *et al.* (1993) show that observed changes in the vertical ozone profile (especially in the lower stratosphere and upper troposphere) can markedly affect the climate. While causes of ozone decrease in the lower stratosphere are known to be connected with increasing concentrations of CFCs, there are still many uncertainties with regard to the increasing trend of tropospheric ozone at the Northern Hemisphere mid and high latitudes (up to 2–2.5% per year in the upper spring and fall troposphere), though it is probably determined by the effect of such ozone precursors as CO, NO_x and hydrocarbons. Much effort is needed to solve this problem, especially bearing in mind the importance of forecasts for tropospheric ozone trends.

Table 6.3 gives the results of calculations of radiative forcing because of shortwave (SW) and longwave (LW) radiation for the surface–atmosphere system, determined by changes in ozone content over Hohenpeissenberg and Payern for 1971–1980 to 1981–1990. Calculations were made using two techniques: at fixed air temperature and also adjusting the temperature profile by prescribed fixed dynamic heating.

Comparison of data for the troposphere and stratosphere makes analysis of the effect of the vertical ozone profile on formation of radiative forcing possible. As can be seen, the increase in ozone content in the troposphere determines positive radiative forcing for both SW and LW radiation; the troposphere contributes most to formation of total radiative forcing in July.

Table 6.4 contains estimates of relative contributions of ozone and other GHGs to radiative forcing. This table demonstrates the marked contribution of tropospheric ozone dynamics to formation of radiative forcing.

The strong sensitivity of ozone-induced radiative forcing to the lower stratospheric temperature points to the need for studies about the climatic consequences of ozone-content variations based on using atmospheric general-circulation models with adequate account of radiation–dynamics interaction.

Table 6.4. The relative role of contributions of ozone and other greenhouse gases to formation of radiative forcing (W/m^2).

	Changes in ozone content				Combined effect of ozone and other greenhouse gases	
	Total		Troposphere			
Season	FT	FD	FT	FD	FT	FD
Hohenpeissenberg						
January	0.410	0.327	0.222	0.136	0.794	0.710
July	0.570	0.548	0.490	0.467	1.050	1.027
Payern						
January	0.356	0.190	0.242	0.074	0.797	0.630
July	0.397	0.136	0.097	−0.167	0.879	0.617
Resolute						
Winter	−0.038	−0.085	0.012	−0.036	0.230	0.183
Summer	0.194	−0.009	0.052	−0.153	0.620	0.415
Goose Bay						
Winter	0.025	−0.066	0.013	0.012	0.379	0.288
Summer	0.043	−0.218	−0.153	−0.415	0.497	0.235
Churchill						
Winter	0.013	−0.086	−0.022	−0.122	0.209	0.110
Summer	−0.097	−0.425	−0.507	−0.839	0.155	−0.174
Edmonton						
Winter	0.080	−0.069	−0.008	−0.158	0.275	0.127
Summer	0.110	−0.002	0.078	−0.031	0.361	0.252
Tateno						
Winter	0.239	0.096	0.084	−0.060	0.491	0.348
Summer	0.304	0.287	0.295	0.279	0.585	0.569

FT is fixed air temperature and FD is fixed dynamic heating.

Schwartzkopf and Ramaswamy (1993) pointed out that observations from satelliteborne and groundbased instruments indicate that significant changes in atmospheric ozone concentrations took place during the 1980s. Decreases in total ozone in mid and high latitudes have been measured by TOMS instrument, while Stratospheric Aerosol and Gas Experiment (SAGE) data indicate ozone depletion in the lower stratosphere at all latitudes, including the tropics. On the other hand, observations at several locations in mid-latitudes of the Northern Hemisphere have shown significant decadal increases in tropospheric ozone. Earlier calculations led to the conclusion that decadal changes in lower stratospheric ozone produce substantial negative radiative forcing of the surface troposphere system at mid and high latitudes. The magnitude of ozone forcing is sensitive to both total ozone loss and altitude profile of ozone depletion.

Schwartzkopf and Ramaswamy (1993) undertook new calculations about the sensitivity of radiative forcing to ozone changes using SAGE-I and II data on ozone vertical profiles up to 17 km (below this level the presence of clouds complicates remote sounding) as well as data on increase in tropospheric ozone.

The following parameters were fixed: vertical temperature and humidity profiles, cloudiness, surface temperature and albedo. Three scenarios for annual-averaged profiles within the 17–50 km layer for decadal ozone losses in tropical (4.5°N) and mid (40.5°N) latitudes (similar in each latitude belt) were chosen:

i. No ozone depletion below 17 km (S1);
ii. Linear decrease with altitude of the percentage of ozone depletion between 17 km and the tropopause, where the depletion becomes zero (S2);
iii. Depletion percentage between the tropopause and 17 km is equal to the SAGE depletion percentage at 17 km (S3).

An additional standard profile (S4) was also used which employs percentage ozone depletion adjusted so that TOZ is equal to that obtained using profile S1. Tables 6.5 and 6.6 illustrate some of the results obtained by Schwartzkopf and Ramaswamy (1993).

These results show that the amount of stratospheric ozone radiative forcing during the 1980s depended principally on two factors: total ozone-content depletion and altitude profile of the ozone change. In the tropics, where stratospheric ozone change profiles are available almost right up to the troposphere, the principal question is about the amount of lower stratospheric ozone loss. The important

Table 6.5. Decadal stratospheric column ozone change in DU, stratospheric ozone forcing in W/m^2 and stratospheric radiative forcing gradient (SRFG) in W/m^2 $(DU)^{-1}$ for standard tropical (January, 4.5°N) profiles. The SRFG derived here is valid over stratospheric column ozone decrease of 0–10 DU.

Profile	S1	S2	S3	S4
Column change	−7.34	−8.07	−8.79	−7.34
Forcing	−0.052	−0.071	−0.090	−0.052
SRFG	0.0071	0.0087	0.0102	0.0071

Table 6.6. Same as Table 6.5, except for mid-latitude (January, 40.5°N). The SRFG derived here is valid over stratospheric total ozone decrease of 0–30 DU.

Profile	S1	S2	S3	S4
Column change	−10.38	−15.55	−20.72	−10.38
Forcing	−0.035	−0.084	−0.134	−0.078
SRFG	0.0034	0.0054	0.0065	0.0075

conclusion reached is that the magnitude of decadal Northern Hemisphere stratospheric ozone radiative forcing would significantly increase if ozone depletion had occurred in the tropics, as implied by SAGE observations.

In mid-latitudes, stratospheric ozone forcing depends strongly on the altitude profile of decadal ozone change in the mid-latitude lower stratosphere, especially in the layers nearest the tropopause (accurate measurements of ozone change in the vicinity of the tropopause are necessary to arrive at a precise estimate for stratospheric ozone radiative forcing). If the observed increase in tropospheric ozone in various regions of the Northern Hemisphere are hemispherically representative then it can be concluded that substantial positive tropospheric ozone radiative forcing occurred during the 1980s of magnitude possibly comparable with that produced by decadal stratospheric ozone decreases. Although the results considered require further confirmation (uncertainty in both magnitude and sign of decadal total ozone forcing is rather high), it may be concluded that changes in total ozone content and vertical ozone profiles (both in the stratosphere and troposphere) are attributable to significant climate-forming processes. This is why various radiative transfer calculations have recently been made to assess ozone-induced radiative forcing variability (e.g. Berntsen *et al.*, 1997; Kondratyev, 1998; Lelieveld and van Dorland, 1995; van Dorland *et al.*, 1997; Roelofs *et al.*, 1997; Wang and Isaksen, 1995; Wang and Sze, 1980; Wang *et al.*, 1993, 1995, 1998a, b).

An important aspect of ozone climatic impact is its interactive nature with regard to the contribution of other greenhouse gases (Kondratyev, 1998). To analyse the stratospheric contribution to such an interactivity, Schneider *et al.* (1993) assessed the response of stratospheric ozone to increased levels of CO, N_2O and CFC (such an analysis is undoubtedly needed for tropospheric ozone). The computations show that CO_2 doubling leads to 3.1% increase in global ozone content which results in an associated increase in solar heating, reducing maximum temperature drop (because of CO_2 doubling) by about 3°C. CFC increase from continuous emissions at the 1985 rate causes 4.5% loss of ozone. For combined perturbation (because of CO_2 doubling and CFC increase), net loss of total ozone of 1.3% has been calculated. All ozone perturbations, calculated by Schneider *et al.* (1993), exhibit asymmetry between Northern and Southern Hemispheres, relative ozone losses being larger at high latitudes of the Southern Hemisphere as against larger eddy mixing and larger horizontal advection in the Northern Hemisphere.

It became clear at an early stage of studying ozone impact on climate that climate changes, greenhouse gases (GHGs) and ozone are highly coupled. This was demonstrated by Nicoli and Visconti (1982), who applied a 1-D radiative–convective–photochemical model of the atmosphere (0–70 km) taking account of 60 chemical and 25 photochemical reactions to simulate the impact of N_2O, CH_4, CO_2 and Cl_y emissions to the atmosphere on TOZ and surface air temperature (SAT). Calculations were made for cloudless atmosphere the solar-zenith angle 60°. Table 6.7 gives various versions of prescribed trace gases dynamics (1 or 2 mean nominal or doubled concentration values, FB, NFB with (without) taking account of feedback). Table 6.8 illustrates the results of numerical modelling for SAT change (K) and TOZ change (%).

Table 6.7. Atmospheric composition changes.

Version	CO_2 (ppm)	Cl_y (ppb)	CH_4	N_2O
A	320	2.3	1	1
B	320	10(FB)	1	1
C	320	10(NFB)	1	1
D	640	2.3	1	1
E	640	10	1	1
F	320	2.3	2	1
G	320	2.3	1	2
H	640	10	1	2
I	640	10	2	2

Table 6.8. SAT and TOZ change.

Version	SAT changes (K)	Total ozone changes (%)
B		−9.3
C	NFB	−10.9
D	1.88	2.2
E	1.67	−9
F	0.27	1.7
G	0.35	−2.7
H	2.1	−9.7
I	2.3	−4.7

CO_2-doubling-induced tropospheric warming and stratospheric cooling of the Cl_y mixing ratio by 10 ppb results in TOZ depletion equal to 9% (NFB) or 10.9% (FB). Nitrous-oxide-doubling leads to ozone depletion and weak warming, whereas the result of CH_4 doubling is TOZ increase and climate warming.

Methane emissions weaken Cl_y impact on ozone because of Cl_y-induced HCl formation. Enhancement in the greenhouse effect because of combined N_2O and CH_4 emissions lead to SAT increase of about 2.3 K (instead of 1.67 K in the case of CO_2 doubling). In the case of the simple additive effect of the four components considered, TOZ depletion is 11.9%, while interactive consideration of their impacts results in only 4.7% ozone depletion (Table 6.8).

Increasing atmospheric CH_4 concentration results not only in greenhouse effect enhancement but also tends to increase two radiatively–chemically important gases: tropospheric O_3 by enhanced tropospheric chemical activities and stratospheric H_2O because of stratospheric CH_4 oxidation. These increases lead to RF increase and thus enhance the CH_4 future warming effect.

To examine the indirect effects of the increase in atmospheric CH_4^- induced by climate–chemistry interaction, Wang *et al.* (1998a) conducted model-sensitivity

Table 6.9. Minor gas concentrations used in the radiative forcing calculations.

Year	CO_2 (ppmv)	N_2O (ppmv)	$CFCl_3$ (ppbv)	CF_2Cl_2 (ppbv)	Cases	CH_4 (ppmv)	Troposphere mean O_3 (%)	Stratosphere H_2O (ppmv)
1990	354	0.31	0.28	0.48		1.72	–	–
2050	539	0.37	0.55	1.07	I	3.24	9.6	0–3
					II	2.69	−1.4	0–2

Values of tropospheric O_3 and stratospheric H_2O are annual and global mean changes; cases I and II correspond to 'without' and 'with' global warming, respectively.

Table 6.10. Global and annual radiative forcing (W/m^2).

Case	CO_2	N_2O	$CFCl_3$ CF_2Cl_2	CH_4	Tropospheric O_3	Stratospheric H_2O	Total
I	2.59	0.19	0.28	0.6	0.19	0.030	3.92
II	2.59	0.19	0.28	0.44	−0.04	0.019	3.47

experiments. A concrete aim of modelling has been to analyse climate implications of changes in tropospheric O_3 and CH_4, when increases in moisture and temperature caused by increasing GHGs (CO_2, N_2O, CH_4 and CFCs) are considered.

One of the principal conclusions made by Wang *et al.* (1998a) was that tropospheric O_3 can be substantially increased in response to CH_4 increase, which is, however, in sharp contrast to O_3 decrease, when increases in moisture and temperature are included. In addition, CH_4 increase can be substantially reduced when enhanced OH associated with global warming is taken into account. Wang *et al.* (1998a) predicted chemical composition changes, shown in Table 6.9, to calculate relevant radiative forcing changes (Table 6.10).

As can be seen in Table 6.10, the calculated annual and global mean total RF is $3.92\,W/m^2$ with CO_2 increase contributing $2.59\,W/m^2$, while CH_4 accounts for $0.65\,W/m^2$. The contributions of increases in tropospheric O_3 and stratospheric H_2O associated with CH_4 increase are 0.19 and $0.03\,W/m^2$, respectively. Numerical modelling has shown that the effect of global warming is to decrease zonal mean forcing in all latitudes with maximum value of about $0.5\,W/m^2$ at low latitudes. Annual and global mean total RF (Table 6.10) decreases to $3.47\,W/m^2$ (11.5% reduction) with CH_4, O_3 and stratospheric H_2O contributing 0.44, −0.04 and $0.02\,W/m^2$, respectively. There are much larger latitudinal effects, however. For example, RF because of O_3 changes in low latitudes decreases from positive $0.2\,W/m^2$ to negative $0.1\,W/m^2$.

Wang *et al.* (1998a) emphasized that the results considered indicate the importance of including climatic changes in tropospheric chemistry when studying CH_4-O_3 coupling. Undoubtedly, further studies using 3-D models are necessary.

Rosenfield and Schoeberl (1986) made calculations about radiative flux divergence (RFD) in the stratosphere and RFD-induced atmospheric circulation using observed temperature and ozone fields in the Southern Hemisphere for September–October 1979, 1980 and 1984. The principal purpose of numerical modelling was to analyse limitations imposed by the stratospheric radiative regime at the end of winter–beginning of spring on formation of atmospheric circulation because of RFD sensitivity to temperature and ozone variation. The results obtained indicate that:

(1) between September and October there was enhancement in total radiative warming of the lower stratosphere near the south pole (RFD > 0);
(2) an interannual depletion trend in total radiative cooling in the lower stratosphere of mid and subpolar latitudes took place because of decrease in mean zonal temperature;
(3) temperature changes because of observed ozone variations were highly improbable;
(4) RFD-induced atmospheric circulation in the lower stratosphere in spring revealed trends in upward motion enhancement near the pole in 1979–1980 but weakening of downward motions in mid and subpolar latitudes in 1979–1984 (this indicated annual enhancement in the upward branch of the circulation cell near the pole and downward branch in mid latitudes in September and October (this conclusion agrees with observed TOZ changes);
(5) assessments of the circulation of mass showed that support of upward motions near the pole was impossible without an assumption concerning the presence of an additional source of stratospheric warming that led to temperature increase of about 0.5 K/day (this source was more powerful in 1984 than in 1980).

Forster and Shine (1997) carried out new calculations about radiative forcing with improved radiative transfer schemes and temperature trends from stratospheric ozone changes (Nimbus-7 SBUV and SAGE-I, II data). The results are shown in Table 6.11. These results clearly indicate the importance of ozone-induced radiative forcing.

Van Dorland *et al.* (1997) carried out numerical modelling of the RF global distribution because of tropospheric ozone and sulphate aerosols. The 3-D concentration field for both components and their precursors was calculated on a $10 \times 10°$ horizontal grid (10 isobaric layers in the vertical) using a transport/chemistry model of the troposphere (MOGUNTIA) for pre-industrial (1850), contemporary (1990) and future atmosphere (2050).

As discussed earlier (Chapters 1 and 2), significant TrOZ growth has been observed since pre-industrial times. According to van Dorland *et al.* (1997), increases up to 19 DU are found in July in Europe. As far as the seasonal cycle of zonal mean ozone is concerned, the highest change of about 14 DU has been found at northern mid-latitudes (40–60°N) in summer with the principal contribution from Europe and North America. A second zonal mean maximum of about 8 DU occurs in September and October, mainly because of biomass burning in the tropics of the Southern Hemisphere (Africa and South America). This has been discussed in more detail by Levy *et al.* (1997).

Table 6.11. Adjusted annually and globally averaged net radiative forcing (W/m^2) from various ozone changes and greenhouse-gas changes over different time periods.

Ozone data	1979–1991	1979–1996	1964–1996
SBUV	-0.13 ± 0.02	-0.22 ± 0.03	-0.26 ± 0.05
SAGE-A (no depletion below 15.5 km)	-0.08 ± 0.02	-0.14 ± 0.03	-0.16 ± 0.05
SAGE-B (constant % depletion between 15.5 km and the tropopause)	-0.11 ± 0.02	-0.19 ± 0.03	-0.22 ± 0.05
SAGE-B normalized to SBUV total ozone	-0.14 ± 0.02	-0.23 ± 0.03	-0.28 ± 0.05
Dobson	-0.11 ± 0.03	-0.18 ± 0.04	-0.21 ± 0.06
Well-mixed greenhouse gases	$+0.45 \pm 0.05$	$+0.70 \pm 0.05$	$+1.7 \pm 0.10$

The results of RF calculations made by van Dorland *et al.* (1997) indicate (the extended Morcrette radiative transfer model was utilized) that global and annual average radiative forcing over the industrial period (1985–1990) is equal to 0.38 ± 0.15 and $-0.36\,W/m^2$ for tropospheric ozone (a TrOZ increase of about 8 DU) and sulphate aerosol changes, respectively; that is, an approximate balance between these two RF components takes place. Van Dorland *et al.* (1997) emphasized, however, that both constituents have characteristic seasonal cycles and their RF dependencies on several radiative parameters are quite different, being characterized by different space and time variability. Local RF values achieve $1.0\,W/m^2$ over northern Africa and Saudi Arabia. The pattern of sulphate RF shows higher variability than ozone.

It is quite natural for ozone-induced RF changes to be maximum downwind of industrialized regions and biomass-burning areas, resulting in some asymmetry between both hemispheres. Peak values of about $1\,W/m^2$ have been found over north Africa and Saudi Arabia. Sulphate aerosol increases are confined to industrial areas in the Northern Hemisphere and to biomass-burning regions in Africa and South America to an even greater degree than that of TrOZ, because of the shorter lifetime in the atmosphere. This leads to large variability in space and time with peak values up to $-5.5\,W/m^2$ over Europe and the United States in July, and $-4\,W/m^2$ over south-east Asia in January.

The significant difference between RF, because of ozone, and sulphate changes exists because ozone is optically active in both the longwave and shortwave spectral regions (on average, longwave RF is about 80% of the total), whereas sulphate aerosols mainly scatter shortwave radiation. Therefore, ozone RF is maximum over areas with high surface temperatures and its changes are strongest in the vicinity of the tropopause, in particular in summer and over regions with large convective activity.

Van Dorland *et al.* (1997) pointed out that clouds mask the greenhouse effect of tropospheric ozone, but enhance shortwave forcing. The net effect is a decrease of about 13% in global and annual mean. Since normalized shortwave RF is linearly dependent on the product of surface albedo and $\mu_0 = \sec\Theta_0$ (Θ_0 is solar-zenith angle), the ratio of longwave to shortwave RF is smallest over cold bright surfaces and in the case of large insolation (like Antarctica in January). The dependence of shortwave RF, because of sulphate aerosol, on surface albedo and angle of incidence is quite different from that of ozone.

Van Dorland *et al.* (1997) developed analytical fits for both ozone and sulphate RF which take into account the most important relevant radiative parameters: surface temperature for longwave RF, surface albedo as well as solar-zenith angle for shortwave RF and cloudiness for RF in both longwave and shortwave spectral regions.

In the case of ozone RF, normalized shortwave forcing can be written as:

$$\frac{\Delta F_{\text{sw}}}{\Delta O_3} = S_0 f_{\text{d}} \mu_0 \left(a + \frac{b}{\mu_0} \right) \left(R_{\text{s}} + \frac{c}{\mu_0} \right) \tag{6.1}$$

where normalized forcing is defined as forcing per DU of total ozone; S_0 is solar constant; f_{d} is the day fraction; $\mu_0 = \cos \Theta_0$ (Θ_0 is solar zenith angle); R_{s} is surface albedo. The best fit was obtained for $a = 7.3 \times 10^{-5}\,\text{DU}^{-1}$, $b = 0$ and $c = 0.075$. The constant $b = 0$ means that tropospheric RF, caused by ozone changes, is predominantly determined by diffuse radiation (c/μ_0) and not by direct radiation.

The sensitivity for normalized mean RF values in the Northern Hemisphere between 1850 and 1990 is highest in July (about 0.53 W/m^2 per 10 DU) and minimal in January (0.37 W/m^2 per 10 DU). Annual variation is caused mainly by the seasonal cycle of surface temperature.

Table 6.12 (from van Dorland *et al.* (1997)) illustrates sensitivities for ozone (longwave and shortwave) and for sulphate aerosols (shortwave) between 1850 and 1990 and consequent total radiative forcing (clear sky and cloudy sky) for January and July at three locations in the Northern Hemisphere. The locations selected are in Europe (55°N, 15°E), in the United States (35°N, 275°E) and in South-east Asia (25°N, 115°E), where RF is considerable and reaches a maximum in some cases. These data clearly demonstrate impacts of various factors on radiative forcing, including cloud contribution.

Calculations of the current combined zonal mean RF, because of well-mixed greenhouse gases, tropospheric ozone and sulphate, made by van Dorland *et al.* (1997), indicate that RF values are positive with the lowest values less than 1 W/m^2 in March and April northward of about 50°N. Maximum RF values of about 2.5 W/m^2 are located in the northern subtropics (mainly in summer). There are, however, various areas with negative RF values. Such an area is South-east Asia in January, characterized by total RF up to –2 W/m^2. Four areas of negative RF exist in July over Western Europe, the United States, South-east Asia and Central Africa, all surrounded by large areas of relatively strong positive radiative forcing.

According to the MOGUNTIA transport/chemistry model, maximum TrOZ changes by 2050 are expected to occur around 25°N, showing a southward shift when compared with 1850–1990 changes. Ozone increases in the southern United States/Mexico, India/Pakistan and China up to 20, 14 and 10 DU, respectively have been obtained (of course, these results depend on choosing ozone-producing and depleting-gas dynamics).

Table 6.12. January and July clear sky and average cloudiness sensitivities of ozone and sulphate aerosol and consequent radiative forcing for the industrial period 1850–1990.

	Month	Location			Northern Hemisphere		
		Europe	US	SE Asia	Mean	Ocean	Land
Cloud cover (%)	January	62	49	59	43	45	40
	July	42	37	53	45	50	41
Ozone							
$\tilde{a}_{LW,clr}$ ($W\,m^{-2}$/10 DU)	January	0.27	0.27	0.34	0.39	0.46	0.37
	July	0.42	0.45	0.51	0.51	0.54	0.48
$\tilde{a}_{SW,clr}$ ($W\,m^{-2}$/10 DU)	January	0.03	0.06	0.06	0.05	0.04	0.05
	July	0.09	0.10	0.08	0.09	0.07	0.10
$\tilde{a}_{TOT,clr}$ ($W\,m^{-2}$/10 DU)	January	0.30	0.33	0.40	0.43	0.51	0.37
	July	0.51	0.55	0.59	0.59	0.61	0.58
$\tilde{a}_{TOT,cld}$ ($W\,m^{-2}$/10 DU)	January	0.27	0.30	0.33	0.37	0.42	0.33
	July	0.45	0.46	0.47	0.53	0.53	0.52
$\ddot{A}O_3$ (DU)	January	10.9	11.7	12.4	10.1	9.5	10.7
	July	19.2	18.3	11.8	11.6	9.7	13.5
Sulphate							
$\tilde{a}_{SW,clr}$ ($W\,m^{-2}$/0.01 OD)	January	−0.20	−0.24	−0.26	−0.22	−0.25	−0.20
	July	−0.38	−0.24	−0.21	−0.23	−0.27	−0.23
$\tilde{a}_{SW,cld}$ ($W\,m^{-2}$/0.01 OD)	January	−0.10	−0.12	−0.15	−0.16	−0.17	−0.15
	July	−0.32	−0.22	−0.17	−0.19	−0.18	−0.19
$\ddot{A}SO_4$ (0.01 OD)	January	10.8	9.7	27.4	4.2	2.3	6.1
	July	17.6	25.7	9.2	3.0	0.8	5.0
Ozone + sulphate							
$\ddot{A}F_{clr}$ ($W\,m^{-2}$)	January	−1.82	−1.91	−6.64	−0.45	−0.07	−0.82
	July	−5.60	−5.15	−1.17	0.03	0.39	−0.31
$\ddot{A}F_{cld}$ ($W\,m^{-2}$)	January	−0.77	−0.78	−3.74	−0.26	0.02	−0.54
	July	−4.59	−4.61	−0.96	0.07	0.38	−0.22

$\tilde{a}_{LW,clr}$, $\tilde{a}_{LW,cld}$, $\tilde{a}_{TOT,clr}$ are longwave, shortwave and total clear-sky sensitivities, respectively. $\tilde{a}_{TOT,cld}$ is total sensitivity under average cloudiness conditions (for ozone per 10 DU and for sulphate per 0.01 optical depth). Also shown are calculated O_3 and SO_4 changes and total radiative forcing ($\ddot{A}F_{clr}$ is clear-sky forcing and $\ddot{A}F_{cld}$ is forcing for average cloudiness). OD denotes optical depth.

Assessment of future RF dynamics indicates that by 2050 (expected ozone increase for the next half century is about 70% of the change since pre-industrial times) mean annual mean global RF = 0.28 W/m^2 (with RF = 0.34 W/m^2 in the Northern Hemisphere). Because of enhanced industrial activities in Central America and southern Asia, the regions of maximum ozone-induced change will shift southward (with expected peak values of 0.9 W/m^2) when compared with the 1850–1990 period.

Maximum ozone RF is expected to occur in August, instead of July, when conditions are optimal for the photochemical processes of tropospheric ozone formation as well as for the longwave forcing component. In contrast with the reduced O_3 trend between 1990 and 2050, compared with 1850–1990, sulphate changes are expected to increase in the future, resulting in global average RF = 0.59 W/m^2 (India and South-east Asia will be principal contributors).

Van Dorland *et al.* (1997) pointed out that, compared with current RF because of well-mixed greenhouse gases, tropospheric ozone contribution to the greenhouse effect is about 15%. By 2050 the contribution of ozone to total positive RF since pre-industrial times is expected to be somewhat smaller, but still significant. Increases in sulphate should (over)compensate for positive RF on a regional scale.

Similar RF calculations were obtained by Haywood *et al.* (1998). According to their results the next global annual-mean-adjusted radiative forcing, including clouds, ranges from +0.25 to +0.32 W/m^2 with about 80% contribution from the longwave component.

The problem of numerical modelling of global 3-D distribution of tropospheric ozone changes and respective radiative forcing was considered by Berntsen *et al.* (1997). The ozone-concentration field was calculated using a chemical tracer model (CTM) developed at the University of Oslo, whereas radiative forcing caused by the increase in ozone was computed by two independent radiative transfer models: the University of Reading model (Reading) and the University of Oslo/Norwegian Institute for Air Research model (OsloRad).

The main purpose was to study the impact of increased emissions of NO_x, CO and CH_4 on radiative forcing of climate, by introducing ozone perturbation calculated by the CTM in two different radiative transfer models. The CTM was run with emissions according to 1850 and 1990 conditions, respectively. The CTM had a resolution of 8° latitude by 10° longitude and nine vertical levels below 10 hPa. A comprehensive photochemical scheme with about 50 chemical components and 110 gas-phase reactions was incorporated to simulate the photochemistry of the troposphere. In the stratosphere, concentrations of ozone, NO_x and HNO_3 are fixed according to the observed distribution of these species. The model uses pre-calculated meteorological input data, with 8-hour resolution, from a 1-year climate simulation with the NASA GIS general circulation model, which allows synoptic scale variability to be resolved.

Ozone production in the troposphere is caused by oxidation of NO to NO_2 by HO_2 or RO_2 radicals and the subsequent photolysis of NO_2 by the reactions:

$$NO + HO_2/RO_2 \longrightarrow NO_2 + OH/RO \tag{6.2}$$

$$NO_2 + h\nu \longrightarrow NO + O \tag{6.3}$$

$$O + O_2 + M \longrightarrow O_3 + M \tag{6.4}$$

Since ozone production in the free troposphere is generally limited by the abundance of NO_x, Berntsen *et al.* (1997) emphasized the necessity to prescribe realistic emission inventories for NO_x and CO (Table 6.13).

The numerical modelling, carried out by Berntsen *et al.* (1997), indicated significant increases in upper-tropospheric ozone concentrations at northern mid-

Table 6.13. Emission rates applied in the model study.

Species	1850	1990
NO_x (Tg (N) yr^{-1})		
Surface	11.6	37.8
Ligtning	5.2	5.2
Aircraft	0.0	0.6
CO (Tg yr^{-1})		
Total	290	1400
Anthropogenic	0	810
Biomass burning	100	400
Natural sources[a]	190	190
	267	205
Isoprene (Tg yr^{-1}) / *Other NMHCs*[c] (Tg yr^{-1})	175.7 (64.8[b])	175.7

[a] Natural sources include emissions from vegetation, soils, wildfires and ocean.
[b] Sensitivity experiment with no anthropogenic NMCH emissions.
[c] Other NMCHs includes ethane, butane, hexane, ethene, propene and *m*-xylene.

latitudes (15–45 ppbv, depending on season) at about 10-km altitude. Berntsen *et al.* (1997) pointed out that tropospheric-ozone increase is caused mainly by enhanced *in situ* production, because of transport of ozone precursors from the atmosphere boundary layer (ABL), with a similar contribution by increased transport of ozone produced in the ABL. Because of enhanced global mean concentrations of HO_2 by 40%, as well as OH concentration increase by 6%, the lifetime of ozone in the troposphere decreased by about 35%. The calculated increase in surface ozone concentration in Europe is found to be in good agreement with observations.

The calculated pre-industrial global fields of monthly mean ozone concentration and TOZ for June 1850 are presented in Figures 6.1–6.3 (Berntsen *et al.* (1997)). Under background conditions, two maxima in surface ozone concentrations of about 15 ppb are located on the west coast of continents in the tropics because of *in-situ* production over continents and the easterly flow in the trade winds. The third maximum over northern India and the Himalayas was caused by combination of downward mixing of stratospheric air (probably because of the influence of mountain ranges on the dynamic process in the GCM) and *in situ* production. In the upper troposphere, ozone concentrations are generally low (5–30 ppb) in the tropics, in particular over the Pacific. Although at higher latitudes the net chemical ozone production is negative, ozone concentrations increase towards the poles in both hemispheres because of stratosphere–troposphere exchange.

The increase in CH_4, CO and NO_x concentrations since pre-industrial times led to increased ozone production in the CTM. Table 6.14 illustrates the ozone budget in the 1850 and 1990 cases. The calculations indicate positive net chemical ozone production in the atmospheric boundary layer (ABL) and in the upper troposphere (UT), but net chemical loss in the mid-troposphere (MT, 1.4–7.4 km) because of the low NO_x concentration at this altitude. Berntsen *et al.* (1997) underlined, however,

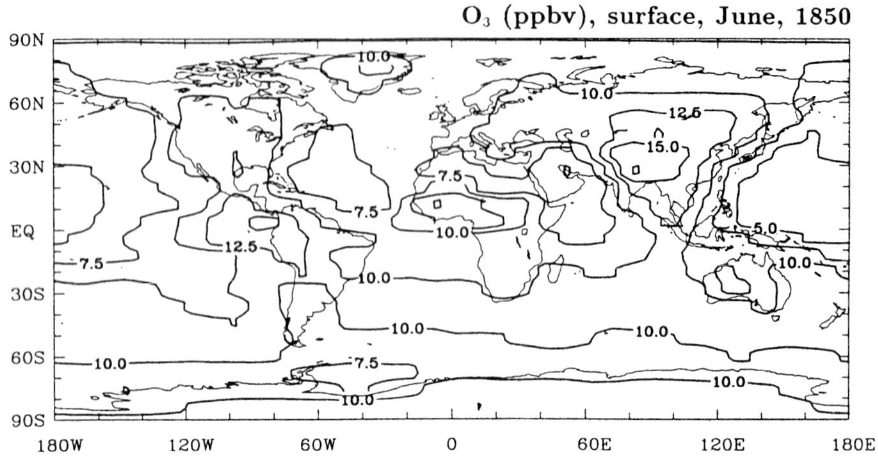

Figure 6.1. Calculated monthly averaged mixing ratio of ozone in the surface layer (i.e. surface to ≈ 500 m) for June for pre-industrial conditions. Isolines are at 5, 7.5, 10, 12.5, 15 and 20 ppbv.

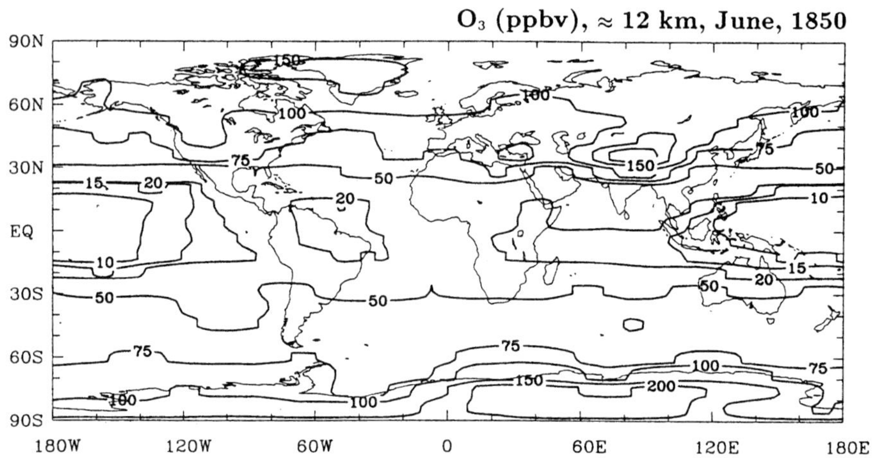

Figure 6.2. Calculated monthly averaged mixing ratio of ozone in the upper troposphere (L $\approx$ 7, $\approx$ 10.3–13.7 km) for June for pre-industrial conditions. Isolines are at 10, 15, 20, 50, 75, 100, 150 and 200 ppbv.

the preliminary nature of these assessments, since some processes have not been taken into account (e.g. heterogeneous processes on cirrus-cloud ice particles and carbonaceous aerosols). Nevertheless, comparison with ozonesonde data indicates in general sufficiently good results. TOZ changes between 1850 and 1990 show clear increase because of ABL ozone over industrialized regions in the Northern Hemisphere.

Radiative transfer calculations were made by Berntsen *et al.* (1997), taking account of the effects of clouds and thermal adjustment in the stratosphere. The calculations indicate that global and annual averaged RF at the tropopause are equal to 0.28 W/m^2 (Reading model) and 0.31 W/m^2 (OsloRad), values in the lower

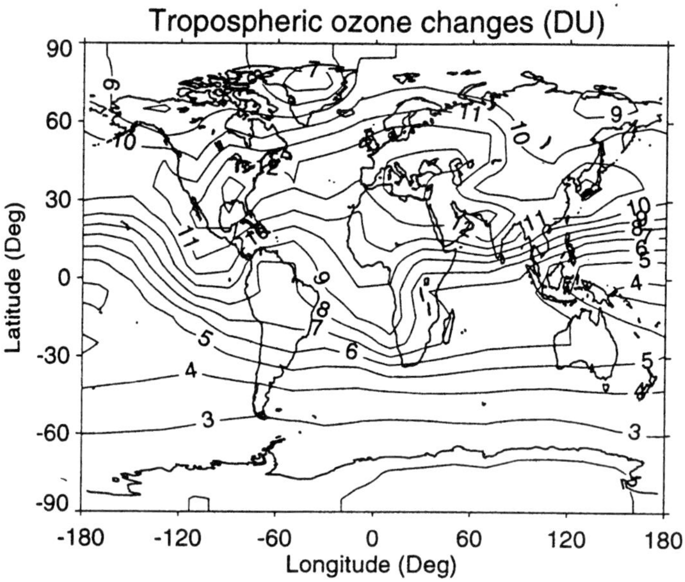

Figure 6.3. Calculated change in annual mean tropospheric ozone in Dobson units between the 1850 and 1990 cases.

Table 6.14. Net chemical production of ozone (PO_3) in the planetary boundary layer (PBL) and the upper troposphere, and net upward flux of ozone from the PBL.

Latitude	Net PO_3 in PBL[a]		Net PO_3 in UT[b]		Net upward flux from PBL[c]	
	Summer	Winter	Summer	Winter	Summer	Winter
1850 Case						
90–32°N	49	9.9	4.9	−0.17	−56	−62
32–0°N	45	32	82	53	−8.7	−42
0–32°S	41	44	65	48	−27	−60
32–90°S	0.12	1.9	0.74	0.74	−30	−44
1990 Case						
90–32°N	176	25	83	21	−95	−114
32–0°N	149	154	189	152	31	−26
0–32°N	75	106	126	110	−32	−75
32–90°S	6.9	5.5	7.2	3.8	−37	−55

Northern Hemisphere summer and Southern Hemisphere winter are defined as April–September, and NH winter and SH summer are October–March. Values are in teragrams per 6 months.

[a] Model layers 1 and 2; < 1.5-km altitude.

[b] Model layers 6 and above; > 7-km altitude.

[c] Through top of layer 2.

part of the IPCC (Houghton *et al.*, 1996) estimated range and coincident with the assessment made by van Dorland *et al.* (1997). The latter conclusion is also true for comparison with negative RF by sulphate aerosols (similar in magnitude to ozone RF, but displaced southward in source regions at northern mid-latitudes). Tropospheric ozone increase has resulted in lower-stratosphere cooling by up to 0.9 K, with possibly half of this cooling occurring in the past two to three decades.

Figures 6.4 and 6.5 show the global distribution of net radiation forcing and its components according to Reading and OsloRad model results. The important conclusion, derived from these data is that regional and seasonal, as well as hemispheric, differences in radiative forcing, because of changes in tropospheric ozone, are higher than the corresponding differences in the RF, because of changes in the concentration of well-mixed greenhouse gases (see e.g. Myhre and Stordal (1997)). This hints at a difference between climatic impacts of ozone and well-mixed greenhouse gases (Wang *et al.*, 1995). Berntsen *et al.* (1997) emphasized, in this context, that detailed experiments with general circulation models are required to establish the extent of this difference.

Besides RF spatial variability, it is very important to take into account its temporal changes, such as seasonal variation (Bernsten *et al.*, 1997). The impact of tropospheric ozone increase on stratospheric temperature is also an important question (Ramaswamy *et al.*, 1996; Forster and Shine, 1997).

6.3 OZONE-INDUCED CLIMATIC IMPACTS

Ozone–climate interaction studies are still at an early stage of development. Only a limited number of attempts have been made to examine the sensitivity to climate dynamics to ozone changes.

Interest in the ozone – climate impact was greatly stimulated by the discovery of the springtime ozone hole over the Antarctic in the mid-1980s (Chapter 2). As mentioned above, according to the initial results of numerical modelling, TOZ depletion leads to climate cooling, while TrOZ increase results in climate warming. Bintanja *et al.* (1997) investigator of climate response to changes in radiative fluxes caused by prescribed changes in stratospheric as well as tropospheric ozone. They used, for this purpose, a simplified climate model which is a combination of the zonal energy-balance atmosphere model and the advection–diffusion zonal ocean model. Radiative fluxes are calculated at 33 vertical levels. Such a coupled model allows simulation of latitudinal and seasonal variations in zonal mean surface air temperature and average lower (12–22 km) and higher (22–100 km) stratospheric temperatures. Two types of numerical experiments were performed: (1) experiments in which the model is perturbed with (uniform) ozone changes after which the model is allowed to reach a (quasi) steady state and (2) transient climate experiments in which the model is forced with temporally varying ozone concentrations (in the form of observed trends). These experiments include altitudinally, latitudinally and seasonally varying ozone changes.

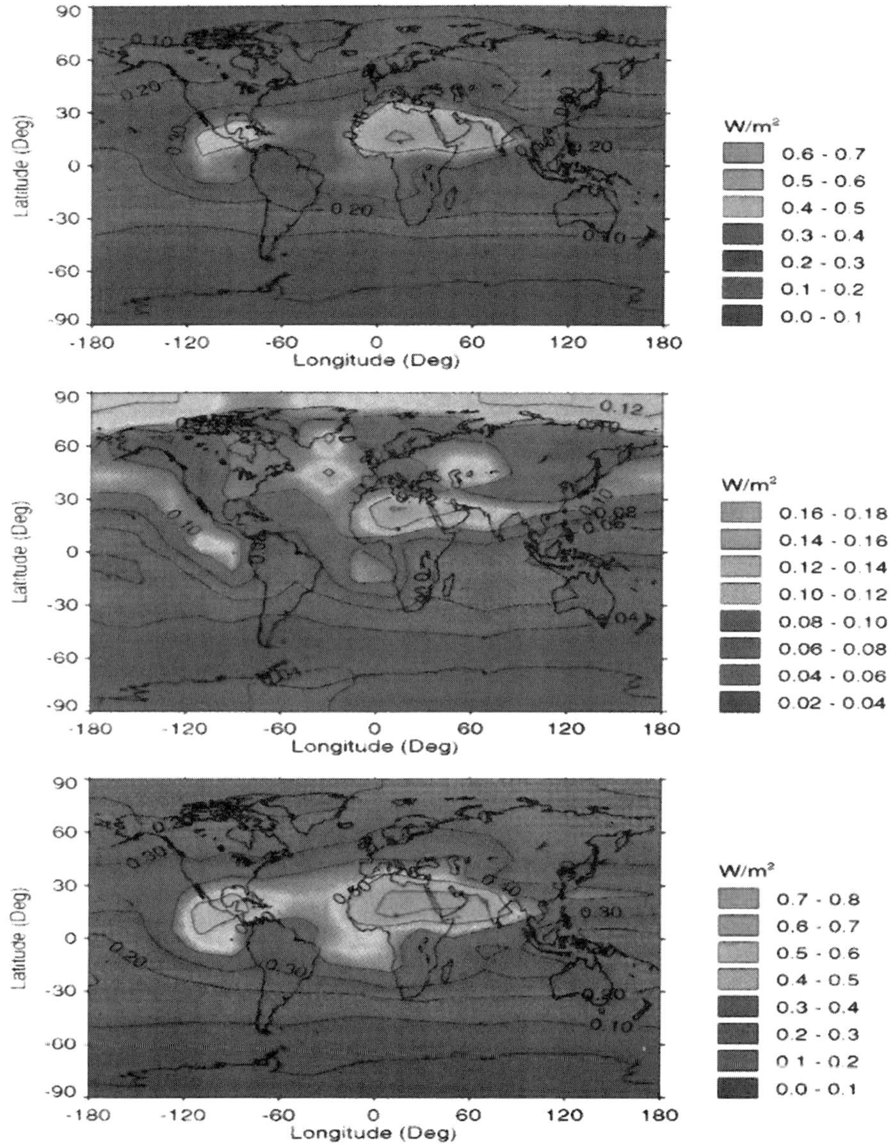

Figure 6.4. Radiative forcing, calculated using the Reading model, because of change in tropospheric ozone since pre-industrial times from the chemical transport model (CTM). Annual mean cloudy-sky-adjusted forcing is shown: (top) thermal infrared; (middle shortwave; (bottom) net.

Bintanja *et al.* (1997) demonstrated that the model is able to successfully simulate the present-day climate, which is an important prerequisite for perturbed climate simulations. Intercomparison between calculated and observed annual mean SAT values as well as annual SAT variations generally indicates that the difference

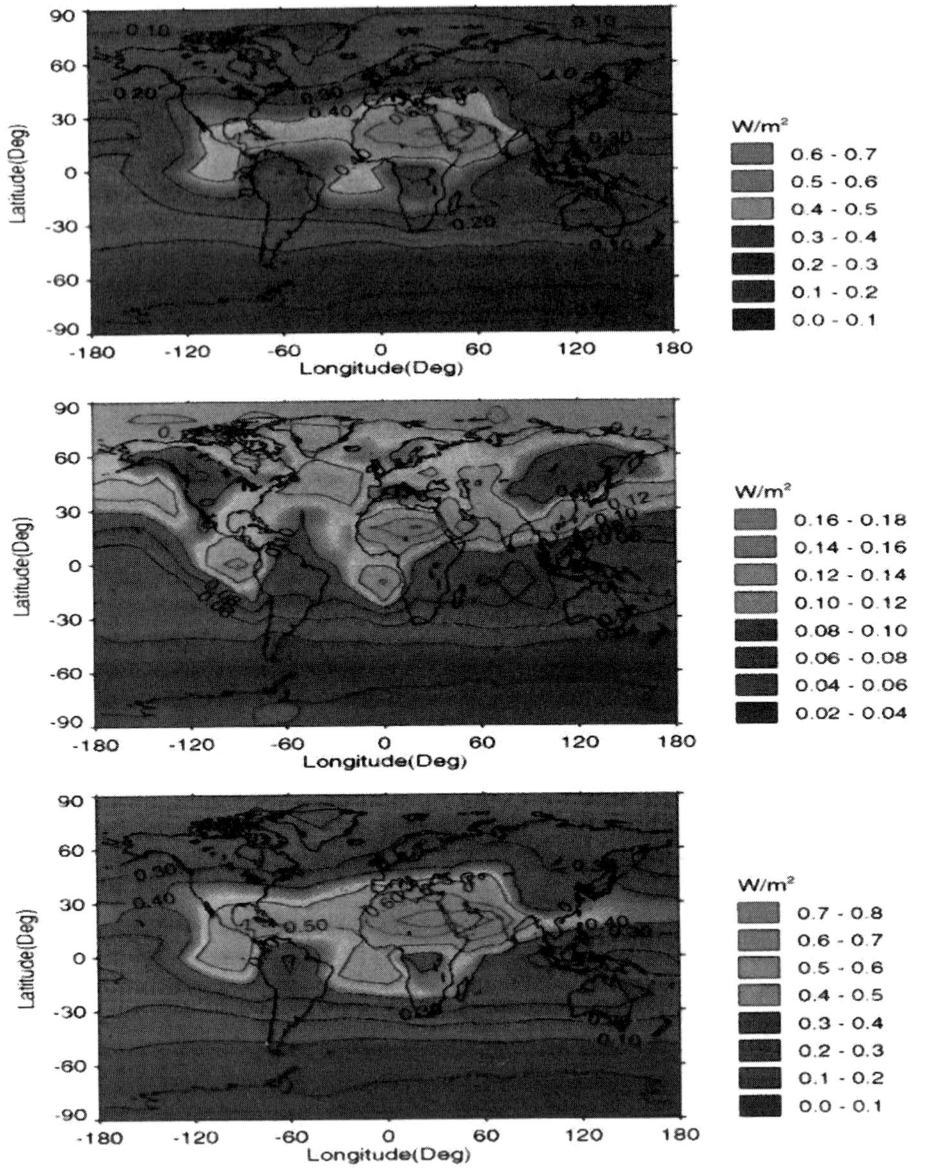

Figure 6.5. As Figure 6.4, but for the OsloRad model.

between model results and observations is small. Simulated temperatures over Antarctica are a few degrees too low, while the annual SAT range is somewhat underestimated in Northern Hemisphere mid-latitudes. There is also agreement with the main essential features of mean meridional energy transport in the atmosphere and ocean. Seasonal and latitudinal distributions of snow cover, surface and planetary albedo, surface radiation fluxes, turbulent fluxes, and ocean temperatures are simulated with reasonable accuracy.

Calculations of quasi-equilibrium climate sensitivity, based on uniform changes in ozone, were made using three versions of the model: (1) ALL, in which all feedback mechanisms are operative; (2) AT, in which the albedo-temperature (α-T) feedback is switched off by keeping the amount of snow and sea ice and the snow and ice albedo at their present-day values; and (3) NTF (no tropospheric feedback), in which the troposphere, earth surface and ocean are kept fixed at their present-day state, only the stratosphere is allowed to change. Radiative forcing (RF) at the tropopause in the NTF model is equivalent to the concept of 'adjusted forcing' (AF), which is the RF at the tropopause with stratospheric temperature adjustment included. In addition to AF, Bintanja *et al.* (1997) also considered 'initial forcing', which is the initial change in net radiation at the tropopause without temperature adjustments.

Figure 6.6 depicts latitudinal and seasonal distribution of the differences in SAT, mean lower stratosphere (STL, 198.6–41.4-hPa layer) and mean higher stratosphere

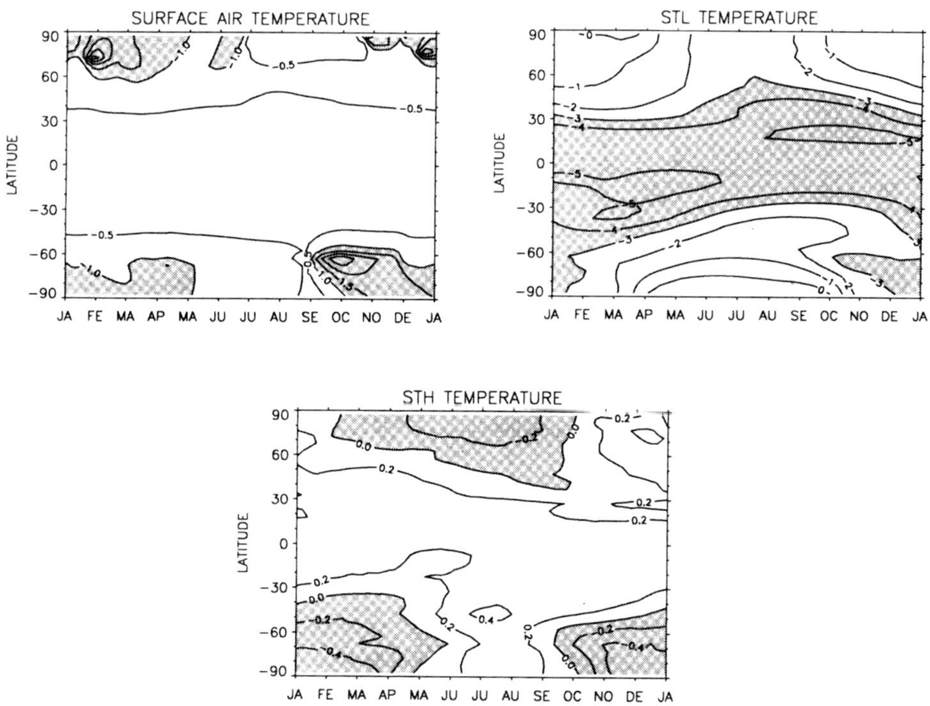

Figure 6.6. Meridional and seasonal distribution of the change in surface air temperature (top left), STL temperature (top right), and STH temperature (bottom) resulting from a reduction in STL ozone by 50%.

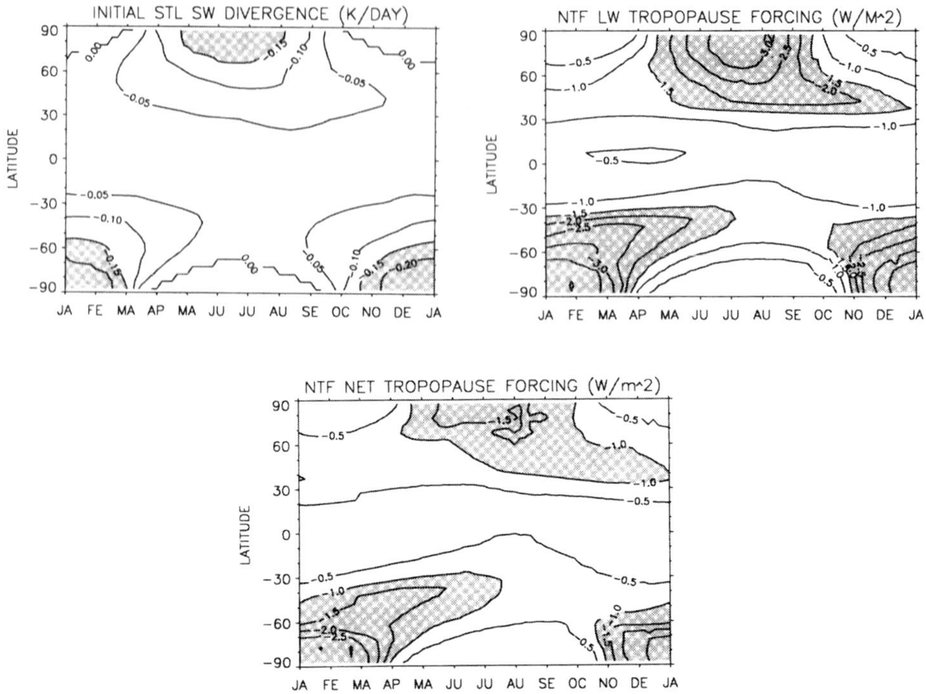

Figure 6.7. Meridional and seasonal distribution of the initial change in shortwave radiative divergence in the STL (top left), 'adjusted' change in longwave radiation at the tropopause (top right) and 'adjusted' change in net radiation at the tropopause (bottom), resulting from a reduction in STL ozone by 50%.

temperature between the perturbed state and the present-day state in the case of a 50% reduction in STL ozone concentration. Figure 6.7 demonstrates the differences in some key radiation quantities. As emphasized by Bintanja *et al.* (1997), the main result of STL ozone reduction is decrease in shortwave absorption within the STL, in particular in polar summer (see Ramanathan *et al.* 1976, 1979, 1985). This is due to high insolation in combination with high surface reflectivity in polar areas and to the fact that absolute ozone changes in polar regions (~66 DU) are larger than those in the tropics. Negative tropopause RF (Figure 6.7c) results in global decrease in SAT (Figure 6a) with strong poleward enhancement in winter because of α-T feedback. Global average cooling is about 3.5 °C in the lower stratosphere with maximum values in the tropics and is 0.46 °C at the surface with maximum cooling in the polar winter (because of α-T feedback).

Bintanja *et al.* (1997) emphasized that the albedo-temperature feedback is consistently stronger in the case of tropospheric and lower stratospheric ozone perturbations than in the case of, for instance, CO_2 perturbations. This can be attributed mainly to differences in the meridional gradient in tropopause radiative forcing.

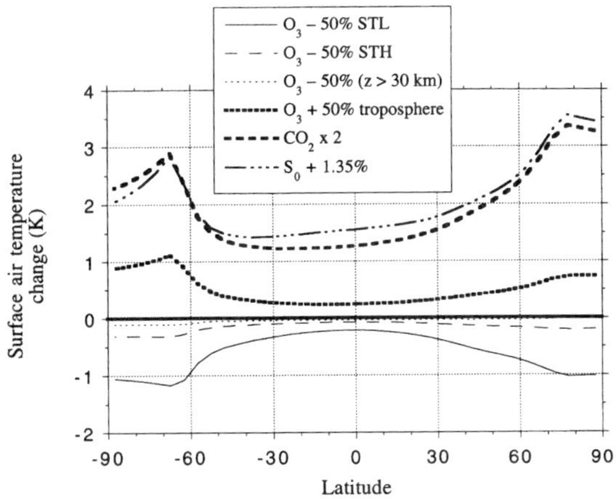

Figure 6.8. Meridional distribution of the change in annual mean surface air temperature for various scenarios.

Similar (50%) ozone reduction in the higher stratosphere (STH, 41.4–0.003-hPa layer) results in a much smaller effect on tropopause RF because of the smaller longwave adjusted forcing and therefore SAT decrease is much smaller.

Since tropospheric ozone increase enhances the atmospheric greenhouse effect, this leads to increased SAT (in particular in the polar regions) and to lower stratospheric temperatures. Figure 6.8 illustrates meridional distribution of the mean annual SAT change for various scenarios, whereas Figure 6.9. shows meridional distribution of the change in mean annual shortwave (top), longwave (middle) and net (bottom) radiation of the tropopause.

Data on meridional and seasonal distribution of the change in STL temperature (K/year) for the TOZ (z) scenario (top) and TOT (φ, z, t) scenario (bottom) are presented in Figure 6.10.

Assessments made by Bintanja *et al.* (1997), in the case of the transient response of the climate system for various more or less realistic ozone perturbation scenarios (based on recent ozone-trend observations), indicate that the combined result of TOZ decrease and TrOZ increase is a small global average cooling from −0.001 to −0.003 K/year, which compensates for about 10% of greenhouse warming caused by greenhouse gases. Relative changes in ozone-induced temperature variations are strongest in the case of tropospheric ozone (0.011–0.015 K/Du) with the principal contribution from the upper mean global troposphere and the lowest from the lower stratosphere (0.008–0.010 K/DU) and the higher stratosphere (0.001 K/DU). Combined mean global sensitivity (taking both stratospheric cooling and tropospheric warming into consideration) is only about 0.002–0.003 K/DU.

In view of the simplified climate model used, Bintanja *et al.* (1997) emphasized that their model can provide qualitative insights into the fundamental processes that determine the sensitivity of climate for ozone changes.

Roelofs *et al.* (1997) discussed results about tropospheric chemistry/climate, using the European Center Hamburg model (ECHAM), by comparing two simulations of

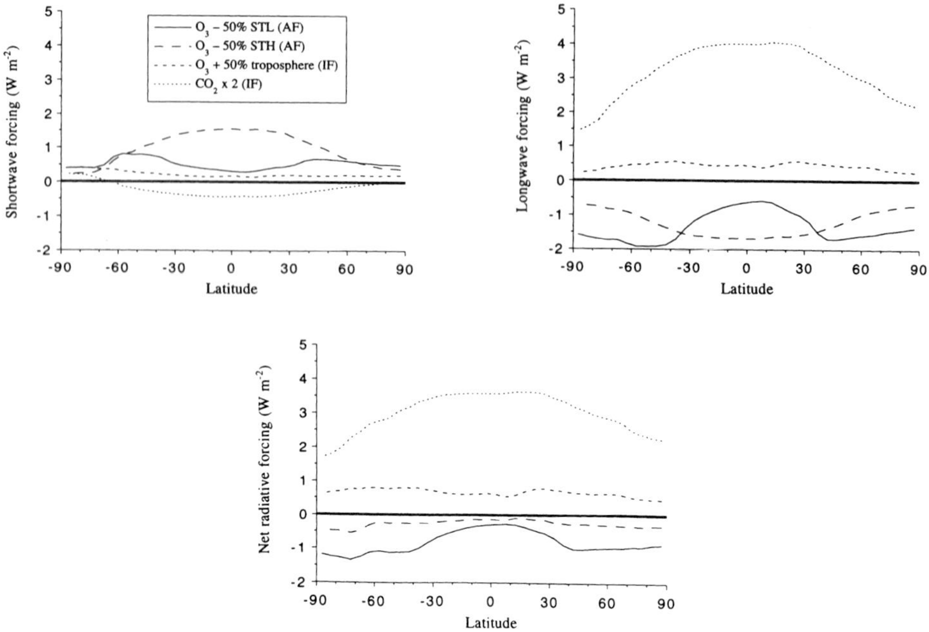

Figure 6.9. Meridional distribution of the change in annual mean shortwave (top left), longwave (top right) and net (bottom) radiation at the tropopause for various scenarios. For ozone changes in the stratosphere, tropopause forcing includes stratospheric temperature adjustment (adjusted forcing), whereas for tropospheric ozone changes and for CO_2 doubling the initial tropopause forcing is shown.

pre-industrial and contemporary emission scenarios. According to these results, photochemical O_3 production from anthropogenically emitted precursors contributes about 30% to the present-day tropospheric O_3 content, which is roughly equal to natural photochemical production. Transport of stratospheric ozone to the troposphere contribute about 40%. Table 6.15 shows more detailed information about tropospheric source and sink terms for ozone and tropospheric O_3 burdens of stratospheric origin (O_3s), O_3 from photochemical formation in the troposphere (O_3t) and total O_3 as well as dry deposition (TgO_3 per year) and net transport.

Photochemical ozone production predominantly takes place through $NO + HO_2$ and $NO + CH_3O_2$ reactions, whereas principal destruction takes place through $O_3 + OH$, $O_3 + HO_2$, and $O(^1D) + H_2O$ reactions. Additional reactions through which relatively small amounts of O_3 are produced/destroyed are HNO_3+OH, N_2O_5 decomposition and NO_3 photolysis. Increase of ozone precursors (NO_x, CH_4 and CO) results in photochemical production increase from 1,732 to 3,415 TgO_3/yr, but increase in photochemical destruction is somewhat smaller. So, at the present time net production of O_3 in the troposphere is calculated, while in the pre-industrial case net destruction of ozone prevails. Net O_3 transport to the troposphere (459 TgO_3/yr) is the result of downward transport of O_3s (930 TgO_3/yr) and upward transport of O_3t (471 TgO_3/yr) (Roelofs and Lilieveld, 1997). In the pre-industrial simulation,

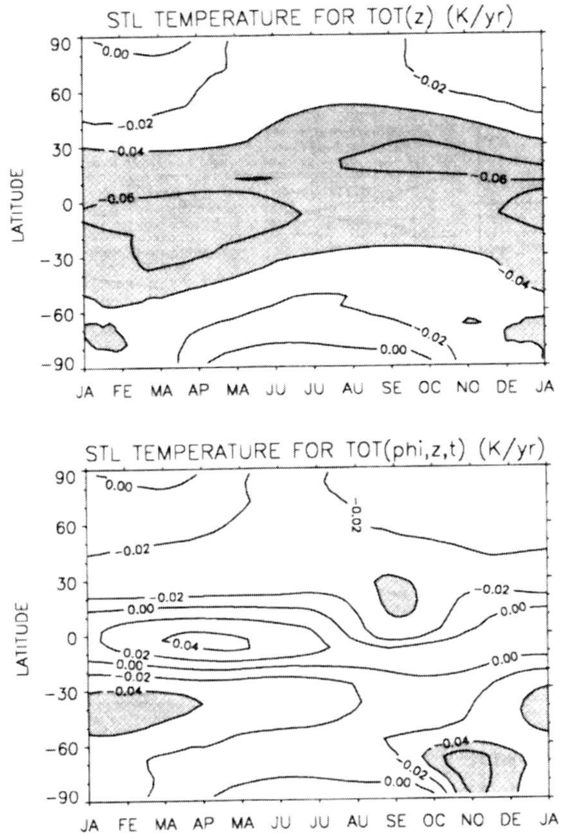

Figure 6.10. Meridional and seasonal distribution of the change in STL temperature (K yr^{-1}) for the TOT (z) scenario (top) and TOT (φ, z, t) scenario (bottom).

downward transport of O_3s is 889 TgO_3/yr and the upward transport of O_3t is 267 TgO_3/yr.

Calculations of total radiative forcing, based on these results, led Roelofs *et al.* (1997) to global and annual average RF, by tropospheric ozone changes, of 0.42 W/m^2 (0.51 W/m^2 in the Northern Hemisphere and 0.33 W/m^2 in the Southern Hemisphere). Respective longwave and shortwave RF components are 0.35 and 0.16 (NH); 0.24 and 0.09 (SH); 0.29 and 0.13 W/m^2 (global). Zonally averaged values indicate the presence of RF maximum near 25°N, especially in the NH summer which coincides with the peak of TOZ change. Geographical distributions of total RF at the tropopause, because of tropospheric O_3 increases during summer and winter seasons, indicate the existence of maxima over northern Africa and the Middle East.

Fuglestveld *et al.* (1999) pointed out that, through production of tropospheric O_3, NO_x emissions lead to positive radiative forcing and warming. But by affecting the concentration of OH radicals, NO_x also reduces the levels of CH_4, thereby giving

Table 6.15. Simulated global budget of tropospheric ozone (Tg/yr).

Components	Present day	Pre-industrial	Anthropogenic
Net STE O_3	459	622	−163
Chemical production	3,415	1,732	1,683
Chemical destruction	−3,340	−2,055	−1,285
Net chemical production/destruction	75	−323	398
Dry deposition	−534	−296	−238
O_3s	110	108	2
O_3t	161	82	79
Total ozone	271	190	81

Table 6.16. Changes in NO_x emissions on the basis of a 20% reduction in emissions in each region and the calculated changes in lifetime ($\Delta\tau_0$) and steady-state concentration of methane. These changes are also normalized to the magnitude of change in NO_x emissions.

Region	E_{NO_x} (TgN yr^{-1})	ΔE_{NO_x} (TgN yr^{-1})	Δ_{τ_0} (yr)	$\Delta_{\tau_0}/\Delta E_{NO_x}$	ΔCH_4 (ppbv)	$\Delta CH_4/\Delta E_{NO_x}$
1. Australia	0.84	−0.169	0.023	−0.136	7.27	−43.1
2. Southern Europe	1.97	−0.393	0.012	−0.031	3.85	−9.8
3. Middle Europe	2.22	−0.444	0.01	−0.023	3.22	−7.3
4. Scandinavia	0.21	−0.043	0.00106	−0.025	0.342	−8.0
5. South east Asia	1.18	−0.236	0.035	−0.148	11.5	−48.6
6. USA	7.53	−1.506	0.048	−0.032	15.7	−10.4
All regions	13.96	−2.79	0.136	−0.049	43.8	−15.7

negative forcing and cooling. The important fact here is that the lifetime of NO_x varies from hours to days, giving large spatial variations in the levels of NO_x.

Keeping all this in view, Fuglestveld *et al.* (1999) used a 3-D global chemical tracer model and a radiation transfer model to study the role of NO_x emissions in radiative forcing of climate. For this purpose, geographical regions were selected representing different chemical and physical conditions, and chemical and radiative effects of reducing NO_x emissions by 20% in each region were studied. The results indicate large geographical differences in the effects of NO_x on O_3 as well as variations in the annual profile of the changes because of non-linearities in O_3 chemistry as well as differences in convective activity. The effect of NO_x emissions on methane depends on the localization of the emissions.

Calculated ozone and the methane RFs are of similar magnitude but of opposite sign. The methane effect is on a global scale with a decay of approximately a decade, while the ozone effect is of regional character and lasts for weeks. Table 6.16 illustrates changes in NO_x emissions on the basis of a 20% reduction in emission in each region, whereas Table 6.17 contains values of total RF.

Table 6.17. Total radiative forcing (10^{-2} W m^{-2}) based on ozone changes for different regions.

		NH	SH	Global
South-east Asia	January	−0.439	−0.360	−0.400
	July	−0.432	−0.764	−0.598
	Annual	−0.522	−0.600	−0.561
USA	January	−0.324	−0.020	−0.171
	July	−1.963	−0.037	−1.000
	Annual	−1.021	−0.021	−0.521
Scandinavia	January	−0.001	0.000	−0.000
	July	−0.057	0.000	−0.028
	Annual	−0.025	0.000	−0.012
Combined	January	−0.911	−1.100	−1.001
	July	−3.513	−1.172	−2.342
	Annual	−2.162	−1.161	−1.662

Sexton *et al.* (1997) conducted numerical climate modelling, using the Hadley Centre Atmospheric GCM (HADAM2a), from late 1948 to the end of 1994 to analyse impacts of different specified atmospheric forcing distributions. Two of the ensemble calculations were made for varying amounts of ozone; one of them (SUL + O_3) took into account both increasing sulphate aerosols and the linearly decreasing trend in ozone concentrations in the lower stratosphere from 1975 based on estimated and observed TOZ decrease. Ozone loss was assumed to occur within about 12 km on a fixed tropopause, with twice the rate of loss in the lower 7 km than the upper part. Another ensemble (SST + O_3) is only forced with decreasing stratospheric ozone since 1975. During this time period, observations show a decrease in mean annual temperature at the 50-hPa level of about 1.3°C, with three short periods of substantial warming because of volcanic eruptions. The observed cooling is very well simulated only when CO_2 increase and stratospheric ozone reduction are taken into account.

Similar results were obtained by Sexton *et al.* (1997) when they modelled the zonal mean temperature distribution between 1961–1975 and 1987–1994. Observed changes are dominated by stratospheric and uppermost tropospheric cooling, and warming lower in the troposphere. The cooling effects of the stratospheric ozone hole in the Antarctic are very evident between 300 and 100 hPa. As far as global land–air surface temperature is concerned, the observed warming trend of 0.5°C is underestimated in all simulations. This numerical modelling clearly shows that stratospheric ozone depletion is an important anthropogenic forcing when considering changes in the vertical temperature structure of the troposphere and stratosphere, and there may be an effect at the surface as well. Sexton et al (1997) emphasized, however, that there is considerable uncertainty in the magnitude and distribution of forcing owing to stratospheric ozone, so there is an urgent need for the ozone dataset to be improved.

Austin and Butchart (1998) discussed the results from the first long integration of the UKMO (UK Meteorological Office) 49-level 'Unified Model' with fully interactive stratospheric chemistry. Photochemistry is active in the 400–1-hPa layer and outside this layer constituents are transported passively (evolution of 32 species is considered, using 15 advected tracers). The photochemistry scheme includes all the processes known to affect stratospheric constituents, except for reactions on sulphate aerosols (this work is currently in progress). The results of this 19-month integration, starting from observations on 15 January 1992, indicate that the simulated chemical behaviour of the atmosphere was reasonably consistent with observations, particularly in the Southern Hemisphere. However, the model TOZ was significantly underpredicted in the Northern Hemisphere, especially during the spring. The principal cause of the discrepancy was probably due to poor meridional transport.

Rind *et al.* (1997) analysed the influence of ozone changes on global warming simulation (in the case of CO_2 doubling) using the GISS Global Climate /Middle Atmosphere Model (GCMAM). When ozone is allowed to respond photochemically, stratospheric cooling is reduced by 20%, with little effect on the troposphere. Shindell *et al.* (1997) discussed application of GCMAM to investigate the response of ozone to a doubled CO_2 climate, and the resulting changes on atmospheric radiation and dynamics. (interactive ozone is considered in the 200–0.1-hPa layer). The results show that the ozone response is opposite to temperature changes, so that cooling in the upper stratosphere results in increased ozone abundances, while lower stratospheric warming reduces ozone in that region. The increased overhead column also reduces the amount of UV reaching the lower stratosphere. The net result is an ozone decrease of up to 7% in the lower stratosphere. Shindell *et al.* (1997) emphasized large differences between their results and those of other models in extratropical latitudes (probably because of stronger tropospheric warming in the GCMAM). When parametrized heterogeneous chemistry on PSCs was included, it was found that the Antarctic ozone hole becomes significantly larger and of longer duration in the doubled CO_2 atmosphere. An ozone hole of variable strength, roughly one-quarter to two-thirds that of the current Antarctic hole, forms in the Arctic as a result of overall cooling in the polar lower stratosphere combined with reduction of sudden warmings in the doubled CO_2 atmosphere.

Perlwitz *et al.* (1997) pointed out the tremendous increase in the number of winter months with a strong circumpolar vortex (CPV) in the stratosphere in comparison with the number of months with mean CPVs. To examine potential causes of this trend, numerical climate modelling was undertaken using the general circulation model ECHAM 3/LSG/T21 developed at the DKRZ (Hamburg). The principal conclusion is that the GHG scenario (enhancement of the greenhouse effect) is in accord with the observed winter mean trend of recent decades both in the stratospheric and tropospheric circulation. At the same time, this is also a naturally occurring feature in the unforced control run. However, this undirectional trend in the control run lasted for at most 40 years. After this time the trend reversed, which had not happened in the case of the GHG experiment (such a persistent trend never occurred in the 700-yr control run). Thus, the increased greenhouse effect enhances and stabilizes the natural baroclinic mode towards a stronger polar night vortex.

Perlwitz *et al.* (1997) emphasized that TOZ decrease since 1979 is another potential candidate for stratospheric change. The results of calculations indicate, however , that the observed changes in stratospheric circulation in winter are not consistent with forcing by reduced ozone. In spring, GHG forcing and ozone act in the same direction: both keep the CPV persistent. The results suggest that ozone anomalies do not actually change the circulation except in spring. As far as winter circulation is concerned, it is circulation changes that determine ozone distribution, not vice versa.

Krivolutsky (1997) suggested, on the basis of 1-D photochemical modelling, that a possible cause of the Antarctic ozone hole may be decrease in air density and, therefore, formation of an 'oxygen hole', which leads to transformed photochemistry. Bazilevskaya *et al.* (1997) substantiated the empirical interrelationship between the long-term trend of ozone depletion and solar and galactic cosmic rays (GCR). Linear regression analysis indicates the influence of solar proton events (SPE) on TOZ of the order of several DU during years of strong SPEs and periodic forcing by GCR with amplitude of several DU.

Christiansen *et al.* (1997) applied the Arpège climate model (3D-level spectral model (T42) extending from the surface up to about 80 km), in perpetual January mode, to assess the sensitivity of the atmospheric general circulation to large perturbations in stratospheric ozone concentration. Three numerical experiments were carried out: uniform reduction in mixing ratio of 50 and 70% and an ozone-hole-type reduction with removal of ozone from 500 hPa and upward until TOZ is reduced to 50% of the unperturbed value. Values of sea ice-cover extent, sea-surface temperatures, surface albedo and solar radiation insolation were prescribed. The model was integrated for 300 days, and averages were taken over the last 100 days (sufficient to reach a new statistical equilibrium). Since a fixed temperature profile was used as the upper boundary condition, the full dynamical model response could only be evaluated below 1 hPa ($\sim$50 km). In accordance with Fels *et al.* (1980), to break down the response of the model into its radiative, thermal and dynamical parts both radiative drive and temperature change were considered, under both fixed dynamical heating approximation (FDH) and full dynamics. In all three experiments, both FDH and full dynamical feedback versions were compared with the control run. An important feature of the model is that it includes a prognostic equation for the ozone mixing ratio, thus ensuring that the sources and sinks of the ozone will be in balance with the dynamics in the control run.

Christiansen *et al.* (1997) pointed out that in all experiments the main effect of dynamics is weakening of the diabatic meridional circulation and associated latitudinal smoothing of the temperature response. Weakening of the diabatic meridional circulation is in agreement with reduction in wave forcing, associated with an extended tongue of low refractive index close to the tropopause. The FDH approximation, in general, represents the pattern of the temperature response well, but the strength of the response is altered when allowing for dynamical effects. In the lower stratosphere, this adjustment is 25–50% of the fixed dynamical heating response, but at the 10-hPa level it reaches 100%. In both uniform reduction experiments, the FDH temperature response is negative everywhere except in the polar night region, where small heating is observed. In the lower stratosphere, the cooling is strongest in the tropical region at 20 hPa. Here cooling is −12 and −25 K with 50 and 75% reduction, respectively. In the upper stratosphere and in the mesosphere the response is large with values up to −35 K and −75 K, respectively. The responses in the ozone-hole-type experiment differ in several aspects from the uniform reduction experiment (Christiansen *et al.*, 1997).

The mean zonal wind is very stable under uniform reductions, but the strength of the winter stratospheric jet stream increases drastically in the Antarctic ozone-hole-type experiment. In the control experiment, variability between winter hemispheres is of an oscillatory nature, and there is no sign in the 300-day experiment that oscillation is merely a transient phenomenon. For a definate conclusion about the persistence the periodicity of the oscillation, a simulation of longer duration is needed. Since the timescale of oscillation is close to 1 year, phase locking with the annual cycle is possible.

The results obtained by Christiansen *et al.* (1997) do not confirm the hypothesis, suggested by Kiehl and Boville (1988), that climate response to ozone depletion is non-linear over the range 0–75% ozone depletion in the sense that only ozone reduction about a certain threshold value would lead to significant climate response. For instance, dynamical response to the 75% reduction experiment is almost identical to dynamical response to the 50% reduction experiment. If the threshold exists, then it must be higher than 75%.

Christiansen *et al.* (1997) emphasized that they did not consider long-term interactions in the climate system (e.g. atmosphere–ocean interaction). This is why the troposphere, as a whole, is relatively insensitive to imposed changes in ozone concentrations. The surface–troposphere temperature response is generally of the order of 1–2 K or smaller, whereas in the stratosphere similar changes are of the order of 5–10 K or more.

Sinha and Toumi (1997) made assessments which indicate that surface temperature–lightning–O_3 feedback, currently absent in general circulation models, could significantly affect anthropogenic climate change. They pointed out that lightning is a major source of NO_x in the troposphere, and thus of tropospheric ozone. On the other hand, Toumi *et al.* (1996) suggested that tropospheric ozone, especially in the tropics, may increase substantially as result of more frequent lightning flashes under warmer climate conditions, thus acting as positive radiation feedback on climate change. Sinha and Tuomi (1997) undertook a study to quantify the extent of this forcing and the associated positive climate feedback.

In this context, they carried out numerical modelling incorporating tropospheric ozone–surface temperature parametrization of the last glacial maximum and of a climate in which greenhouse gas concentrations were doubled. A 1-D radiative–convective model was applied, in which CO_2, CH_4, tropospheric O_3 and, in the case of paleoclimatic simulation, surface albedo were varied both independently and in combination.

Table 6.18. Results of perturbing the climate according to paleoclimatic data.

Case	Variable(s) changed	Change in variable	ΔT_s (K)	E_i (K/W m^{-2})
Single parameter perturbations				
1	CO_2	355 to 175 ppmv	−2.22	0.796
2	CH_4	1,720 to 300 ppbv	−0.66	0.572
3	R_s	0.139(0.331°) to 0.208(0.358[a])	−5.25	0.534
4	O_3 (no LFB[b])	−49%	−0.31	0.554
5	O_3 (with LFB)	−72%[c]	−0.48	0.559
Effect of ozone on paleoclimatic simulations				
6	CO_2, CH_4, R_s	As cases 1, 2 and 3	−8.13	0.579
7	CO_2, CH_4, R_s, O_3 (no LFB)	As cases 1, 2, 3 and 4	−8.46	0.602
8	CO_2, CH_4, R_s, O_3 (with LFB)	As cases 1, 2, 3 and 5	−8.69	0.618

[a] Planetary albedo.
[b] LFB denotes lightning-ozone feedback.
[c] Taken from the equilibrium solution of case 8.

Table 6.18 shows the results of model runs driven by changes in climatic variables since the last glacial maximum. The initial state to which perturbations were applied is that of the present-day climate. The parameter $\varepsilon = \Delta T_s/\Delta F$, where ΔT_s is surface temperature change, ΔF is applied radiative forcing (W/m^2) at the tropopause, called (in accordance with IPCC terminology) 'instantaneous radiative forcing' and ε is 'instantaneous climate-sensitivity parameter'. The upper part of Table 6.18 characterizes the relative importance of individual parameters (cases 1 to 5). As can be seen, largest surface cooling is due to surface-albedo perturbations, while the highest T_s sensitivity to given climate forcing (ε value) takes place in the case of CO_2. The results for the full simulations (cases 6 to 8) show that removing industrial tropospheric O_3 (−50%) increases the temperature drop by 0.33 K (compare cases 6 and 7), whereas the decrease in O_3 because of LFB (a further −22%) adds an additional cooling of 0.23 K (case 7 versus case 8). So, according to Sinha and Toumi (1997), simulations of ozone feedback show an effect that is roughly two-thirds that of industrialization alone. Overall, the effect of changes in tropospheric ozone from both sources increases temperature drop by around 4–7%, that is, it plays only a minor role in determining ΔT_s between the last glacial maximum and the contemporary climate. Dominant contributions belong to surface albedo and CO_2 changes.

Table 6.19 illustrates the situation in the case of global warming because of GHG increases. Being much smaller than warming caused by CO_2 doubling (case 9, $\Delta T_s = 2.20$ K), $\Delta T_s = 0.23$ in case 11 of tropospheric O_3 change still illustrates the significance of taking LFB into account. This is in contrast to the ice-age simulation, indicating that tropospheric O_3 can significantly affect the magnitude of any warming episode. Lightning–ozone feedback augments surface temperature change by around 10%, when CO_2 and CH_4 concentrations are doubled.

Sinha and Toumi (1997) emphasized a number of uncertainties about the simulation results. In particular, the lack of explicit dynamics in the model,

Table 6.19. Results from perturbing the climate according to projected greenhouse-gas-emission data.

Case	Variable(s) changed	Change in variable[a]	ΔT_s (K)	ε (K/W m^{-2})
9	CO_2	355 to 175 ppmv	2.20	0.804
10	CH_4	1,720 to 300 ppbv	0.40	0.556
11	O_3	+34%[b]	0.23	0.869
12	CO_2, CH_4	As cases 9 and 10	2.71	0.785
13	CO_2, CH_4, O_3	As cases 9, 10 and 11	2.95	0.779

[a] In all cases, water-vapour feedback is included.
[b] Taken from the equilibrium results of case 13.

together with restriction to a single dimension, excludes treatment of some potentially non-linear processes. There are also some other sources of significant uncertainties.

A number of new results in the field of ozone–climate simulation were discussed at the First SPARC General Assembly (*Stratospheric* ..., 1997).

Rind *et al.* (*Stratospheric* ..., 1997) undertook simulation of the doubled CO_2 climate using the GISS (Goddard Space Flight Center) Global Climate/Middle Atmosphere Model (GCMAM) in a series of experiments, in which sea-surface temperatures are allowed to adjust to radiation influences. Integration over 58 years indicates global warming by 5.1°C at the surface, while the stratosphere cools by up to 10°C. When ozone is allowed to respond photochemically, stratospheric cooling is reduced by 20%, with little effect in the troposphere. The effect of planetary-wave energy increases in the stratosphere, producing dynamical warming at high latitudes, weakens in the lower stratosphere when O_3 chemistry on polar stratospheric cloud impacts are included.

Discussing specific consequences of including ozone chemistry in coupled chemistry–climate models Shindell *et al.* (*Stratospheric* ..., 1997) analysed the response of ozone to a doubled CO_2 climate. The results indicate that the ozone response is opposite to temperature change, so that cooling in the upper stratosphere results in increased ozone abundance, while lower stratospheric warming reduces ozone in that region. In contrast to other models, net TOZ decrease (0–2%) at mid-latitudes was obtained caused by large tropospheric warming. Taking the impact of parametrized heterogeneous chemistry on PSCs into account results in a simulation of a significantly larger Antarctic ozone hole of longer duration in the doubled CO_2 atmosphere. In addition, an ozone hole of variable strength, roughly one-quarter to two-thirds that of the current Antarctic hole, forms in the Arctic in 4 out of 5 years as a result of overall cooling in the polar stratosphere combined with a reduction of sudden warmings in the doubled CO_2 atmosphere.

Graf (1995) concluded that the observed winter trends in atmospheric circulation since 1957 are characterized in the stratosphere by intensification of the polar night vortex and in the troposphere by intensification of westerlies over the North Atlantic.

Moreover, not only has the mean winter strength of the stratospheric polar vortex increased, but there was also a tremendous increase in the number of winter months with a strong polar vortex in the stratosphere. The ratio of the number of winter months with strong versus weak polar night vortex has increased from 1.1 (1957/1958–1971/1972) to 6.8 (1981/1982–1995/1996). Perlowitz *et al.* (*Stratospheric . . .*, 1997) explored two possibilities to explain this phenomenon: (1) the effect of increasing greenhouse-gas concentration; (2) the effect of decreasing stratospheric-ozone concentration. Since observational timeseries are too short for this kind of analysis, simulation modelling has been undertaken using the coupled climate model ECHAM3/LSG/T21, developed at the Climate Center of the University of Hamburg. The principal results obtained by Perlowitz *et al.* indicate that:

- the observed changes in stratospheric circulation in winter are consistent with greenhouse-gas forcing. In winter, they are not consistent with forcing by reduced ozone. However, ozone distribution depends on changes in stratospheric circulation.
- In spring, greenhouse-gas forcing and ozone act in the same direction. Both keep the polar night vortex persistent.

Thus, the results suggest that ozone anomalies do not actively change the circulation except in spring.

A new important step in assessing the present-day understanding of ozone climate impact was taken at the Fifth Session of the Scientific Steering Group on Stratospheric Processes and their Role in Climate, which took place at Port Jefferson, NY, 17–20 November 1997 (Chanin, 1998). The principal purpose of the session was discussion of the results and perspectives of making quantitative estimates about the impact of stratospheric processes on climate; in particular, impacts of optically active components of the stratosphere such as ozone and aerosols, which are present in both the upper troposphere (UT) and lower stratosphere (LS). A number of scientific problems are of interest in this context, especially stratosphere–troposphere exchange, influence of ozone and aerosols on climate change, and many others. Despite the great significance of the upper troposphere and lower stratosphere (UT/LS) for climate formation, relevant observations are not as yet sufficient and numerical modelling requires further development.

The dynamics of the upper troposphere and chemistry of the lower stratosphere must take high priority in future studies. It was expected that ongoing processing and analysis of the global net of 180 radiosonde stations would result, in particular, in obtaining adequate global data on climatology of gravitational waves (GW) by the beginning of 1999. One of the concrete tasks of observational data analysis is to study the dependence of GW characteristics, at the time of entering the stratosphere, on intensity, temporal change and small-scale structure of convection in the troposphere as well as to verify satellite information on convection induced by gravity waves. Special attention should be paid to studying the interaction between microphysics of cirrus clouds and chemical, radiation and dynamic processes.

From the viewpoint of the stratosphere–troposphere exchange (STE), the following priorities have been defined: STE quantitative characteristics and their annual variations for various minor gaseous components (MGC), including ozone, water vapour and others; mesoscale structure of MGC spatial distributions; processes in the tropical tropopause layer and water vapour; causes of model failures in the UT/LS layer. To solve these and other problems it is necessary, in particular, to obtain more complete and reliable data for: tropopause height; MGC concentration in the UT/LS layer; wind field in the tropics (in view of its significance for MGC transport); water vapour and its condensates concentrations within the tropopause layer. Of exceptional importance are intercomparison between observational data and numerical modelling results.

There is still not enough information for the lower stratosphere on the destruction of ozone due to Cl and Br compounds. Further analysis is necessary of the impact of volcanic eruptions (such as El Chichón and Pinatubo) with regard to stratospheric aerosol influence on temperature and ozone concentration in the LS layer. The problem of impacts of subsonic and supersonic aviation on the UT/LS (including climate change) has not as yet been solved.

The UT/LS is that part of the atmosphere which is characterized by low temperatures and high ozone concentration. The processes responsible for ozone changes here are far from being clear (especially with regard to relationship between *in situ* formation and stratosphere–troposphere exchange). Special attention should be paid to specific features of processes in UT and LS (rapid mixing in UT, but slow mixing in LS; different NO_x concentrations in these layers, etc.). An important aspect of the problems under discussion is the influence of quasi-biennial oscillations (QBO) in the stratosphere on the troposphere and their role in STE. From the viewpoint of stratospheric indicators of climate change, trends in stratospheric temperatures and vertical ozone concentration profiles are very interesting. In this context, the question of the impact of ozone depletion on cooling trends in the lower stratosphere in mid and high latitudes deserves serious attention.

A rather controversial situation exists in the upper stratosphere where not only amplitude but even the sign of temperature variation changes in dependence on latitude. A number of problems are relevant to water-vapour dynamics in the lower stratosphere: further accumulation of observational data on water-vapour climatology in the lower stratosphere; studies of those processes responsible for water vapour change; filtering out long-term trends in water-vapour content; development of models able to adequately simulate water-vapour and impacts on water-vapour changes.

Intercomparison of 13 simulation models (GRIPS Project) indicates that: (1) all models realistically simulate latitude and altitude structure of the mean zonal state of the atmosphere, although certain biases take place with regard to temperature calculations, location and intensity of basic jetstreams, etc., the causes of which have not as yet been understood; (2) there is sometimes significant disagreement between calculated and observed planetary waves in the troposphere and stratosphere of the Northern Hemisphere; (3) the quality of simulations is fairly low in cases of low-frequency oscillations in the tropical stratosphere (this is especially true for QBO simulations).

A number of field programmes are being directed toward solving these and other problems. Satellite remote-sensing data, such as that from UARS/HALOE, SAGE-II and III, is contributing sub-stantially to solution of these problems. The International Workshop on Stratospheric Change and its Role in Climate which was held in Tsukuba (Japan), 20–22 October 1997, was devoted to discussion of SPARC problems in the context of the ATMOS-C1 satellite observation programme (Uchino, 1998). At the first stage of this programme (1995–1997) the following aims were reached: (1) monitoring stratospheric variability; (2) investigations of chemical processes in the stratosphere; (3) observations and numerical modelling of stratospheric changes and their impact on tropospheric climate.

The principal aims of the ATMOS-C1 mission have been remote sensing of the optically active minor components of the stratosphere as well as monitoring the stratospheric ozone and climate parameters. The study of stratospheric temperature decrease because of ozone depletion at various levels is another high-priority task (adequate information on vertical ozone concentration profile changes is very important in this respect, including stratospheric warmings dynamics). Other priority subjects are interaction between gravity waves and convection; climatic consequences of water-vapour changes in the stratosphere, interrelationships between frequency of hurricanes in the Atlantic and El Niño, consideration of cloud and ozone contributions in predictions of anthropogenic climate change, polar ozone depletion and climate change, interactions between the stratosphere and troposphere, between processes in high and mid-latitude stratosphere, small-scale processes and ozone change near the UT/LS boundary.

An important step in the study of ozone as a climate gas was marked by papers presented at the European Geophysical Society Assembly in Vienna (*Annales* ..., 1998). For instance, Christiansen presented a series of general-circulation-model experiments with prescribed ozone perturbations at different vertical intervals. The relevant results were discussed by Christiansen *et al.* (1997) who emphasized that climate is particularly sensitive to ozone changes occurring in the vicinity of the tropopause. The large value for climate sensitivity and the corresponding large positive feedback were found for perturbations in the upper stratosphere, which should be compared with the relatively weak feedbacks found both for ozone perturbations in the lower stratosphere and in the troposphere and in a doubled CO_2 experiment (*Annales* ..., 1998).

Fortuin *et al.* (*Annales* ..., 1998) demonstrated the importance of using adequate observational ozone data to simulate climatic influence of ozone changes. It was found that systematically lower ozone values for the new climatology around the tropopause leads here to lower temperature, accompanied by increased convection and high cloudiness, especially in the tropics. The influence of the ozone hole over Antarctica – in contrast with the ozone spring maximum of the previous climatology – is mainly a cooling and enhancement of the polar vortex in October, but also in November when the ozone hole has almost recovered.

Ponater *et al.* (*Annales* ..., 1998) pointed out that aircraft-induced NO_x emissions lead to significant modifications in atmospheric ozone concentrations and the resulting climate impact is also important, even for present-day subsonic air traffic (a comprehensive 3-D model of the atmosphere–ocean system was applied for these assessments). It has been emphasized, however, that the relation between radiative forcing and climate change (in terms of surface temperature change) is not as straightforward as in greenhouse-gas experiments. Even in the case of present-day aircraft-emission levels, considerable non-linear dependence on global mean temperature response from radiative forcing can be found. RF seems to be a bad predictor for future aircraft scenarios.

The special case of ozone impact on climate is connected with the consequences of subsonic and supersonic aircraft flights. Fortuin *et al.* (1995) assessed radiative forcing in northern mid-latitudes caused by aircraft emissions up to 1990. The results indicate local RF at the tropopause, which should be negative in summer (−0.5 to 0.0 W/m^2) and either negative or positive in winter (−0.3 to 0.2 W/m^2). Moreover, the indirect effect of contrails has to be added, which for the North Atlantic Flight Corridor cover the range from −0.2 to 0.3 W/m^2 in summer and 0.0 to 0.3 W/m^2 in winter. Fortuin *et al.* (1995) emphasized that major positive contributions come from contrails, stratospheric water vapour in winter and ozone in summer. The direct effect of NO_2 is negligible and the contribution of CO_2 is relatively small. Sausen *et al.* (1997) pointed out that NO_x emissions from the fleet of conventional subsonic aircraft significantly enhance (up to a factor of 2) the atmospheric NO_x concentration at levels near the tropopause of the northern extra-tropics. A number of model calculations indicate that enhanced atmospheric NO_x concentrations cause an increase in ozone concentration, in particular at altitudes between 400 and 50 hPa; that is, in the vicinity of the tropopause, where radiative forcing (RF) induced by ozone increment has a maximum. Factors influencing ozone chemistry in aircraft plumes need urgent consideration (see Moulik and Milford, 1999).

Sausen *et al.* (1997) conducted several sensitivity experiments, using the general circulation model ECHAM 4, to calculate the impact of RF changes on monthly averaged mean zonal air temperature. ECHAM 4 is a global spectral model at T21 horizontal resolution. Diabatic processses are calculated on a Gaussian grid of approximately 5.6 × 5.6°. The model has 19 vertical levels from the surface up to 10 hPa. Land-surface temperature and snow cover are calculated interactively, whereas sea-surface temperature is prescribed according to the climatological annual cycle.

A set of five model simulations in annual cycle, each spanning 30 model years mode were carried out: (1) control calculations (for climatological ozone distribution) and four versions of ozone perturbations (ΔO_3); (2) $\Delta O_3'$ perturbations (MOGUNTIA); (3) $5 \times \Delta O_3'$; (4) $\Delta O_3''$ (CTMK); (5) $5 \times \Delta O_3''$. The results indicate that in July (when both ozone increase and solar radiation in the Northern Hemisphere are largest) the temperature response to NO_x emissions reaches its maximum. Although, at first glance, the signal looks rather chaotic (with warming and cooling extremes of similar magnitude), statistical tests (Student *t*-test) revealed certain changes on a 95% significance level. Consistent features were discovered, such as: significant warming in the Northern Hemisphere troposphere, a warming in the upper tropical troposphere and cooling in the northern extra-tropical stratosphere. Sausen *et al.* (1997) emphasized, however, that since atmospheric temperature changes are usually spatially coherent and because the area with significant changes is small, the total response still might be nothing more than noise.

To separate signal from noise, the fingerprint technique was used to demonstrate that ozone changes because of aircraft emissions could lead to a statistically significant change in meridional temperature profile. The magnitude of the signal appears to depend non-linearly on the magnitude of ozone increase. Zonal mean temperature changes are in the range of ± 0.2 K, which is about 5–10% of the response that the same model simulates for doubling CO_2 in the upper troposphere. However, the signal cause by ozone change is less coherent. Sausen *et al.* (1997) pointed out some limitations of these results: (1) prescribing SST, which can be assumed to dampen the temperature signal; (2) application of MOGUNTIA and CTMK models, which might lead to enhancing the temperature signal.

Wyser and Ström (1998) emphasized that aircraft exhaust may reduce crystal size in natural cirrus clouds and assessed the change in cloud RF from such a size reduction by assuming a constant ice–water content. Application of a 1-D model to compute the radiative transfer for an atmospheric column showed that the negative shortwave cloud forcing is enhanced with smaller crystals as they mainly increase the reflectivity of clouds. The change in longwave cloud RF is always positive although its magnitude depends strongly on the ice–water path. The net RF depends on mean crystal size, surface albedo and ice–water content. It appears that there is a range of diameters, between 15 and 25 μm, where the response to reduction in crystal size is fairly insensitive. Below and above this range the change is negative or positive, respectively. In regions of dense air traffic, the magnitude of the change in cloud forcing could be of the order of 0.3 W/m^2 assuming a 20% decrease in mean crystal size from about 30 μm to 20 μm. Wyser and Ström (1998) stressed that, even though aircraft exhaust has the potential to affect climate, the results should be treated with caution as they are based on parametrized optical properties for cirrus clouds.

The substantial impact of solar activity on the ozone layer could result in climate change (Kondratyev, 1998). Important related studies were recently carried out by Haigh (1994, 1996, 1998). Haigh (1998) applied the 3-D spectral (T42) UGCM model with 19 vertical levels (up to 10 hPa), developed within the UK Universities Global Atmospheric Modelling Programme (UGAMP), to simulate the impact of solar variability on climate. Two versions of this modelling were considered: (1) control run I for conditions of minimum activity during the solar 11-year cycle; (2) runs IIa and IIc for solar maximum. In both solar maximum runs, solar irradiance is increased by 0.08%. The difference between the two solar maximum runs is measured by the increase (from run I) in stratospheric ozone amounts.

The results confirm the conclusion (Haigh, 1996) that changes in stratospheric ozone may provide a mechanism whereby small changes in solar irradiance can cause a measurable impact on climate. It was earlier proposed (Haigh, 1996) that changes in the thermal structure of the lower stratosphere could cause strengthening of low-latitude upper-tropospheric easterly winds. Haigh (1998) calculated that this easterly acceleration can came about as a result of a lowering of the tropopause and

Table 6.20. Climatic impacts of 20% reduction in NO_x.

Region	Ä(CH_4) (ppb)	Ä*F*(CH_4) (10^{-2} W/m^2)	Ä*F*(CH_4)/ΔE_{NO_x} (10^{-2} W/m^2/TgN yr^{-1})	Ä*F*(O_3)/ΔE_{NO_x} (10^{-2}W/m^2/TgN yr^{-1})
Australia	7.27	0.317	−1.88	–
South Europe	3.85	0.168	−0.43	–
East Europe	3.22	0.141	−0.32	–
Scandinavia	0.34	0.015	−0.35	0.29
South-east Asia	11.5	0.501	−2.12	2.37
USA	15.7	0.684	−0.45	0.35

deceleration of poleward meridional winds in the upper portions of Hadley cells. One model experiment, in which a very different change in ozone structure was assumed, produced results which are consistent with this hypothesis.

Labitzke and van Loon (*Stratospheric* ..., 1997, p. 107) studied annual mean and interannual variability in TOZ (TOMS data for 1979–1993) to assess spatial correlations of ozone with the 11-year sunspot cycle. They found that correlations reach their highest and most statistically significant values between 5 and 30° latitude on either hemisphere, but especially in the summer hemisphere. The equatorial regions where ozone is produced show comparatively small statistically insignificant correlations, and subpolar regions where the ozone is most plentiful show little or no correlation with the solar cycle.

An important contribution to studying climatic implications of ozone and ozone-related processes was made by papers presented at the International Ozone Symposium in 1996 (*Atmospheric* ..., 1998). As a continuation of earlier investigations by Wang and Isaksen (1995) into tropospheric ozone as a climate gas, Fuglestvedt *et al.* (1998) assessed impacts of reduced NO_x emissions on radiative forcing through changes in tropospheric O_3 and CH_4. For this purpose, they selected six geographical regions with different climatic and physical conditions and prescribed 20% reduction in NO_x emissions for these regions. Table 6.20 illustrates some of the results. Here $\Delta(CH_4)$ and $\Delta F(CH_4)$ are changes (because of NO_x reduction) in CH_4 concentration and radiance forcing, respectively. The sensitivity factor is defined as:

$$\alpha = \frac{\Delta \bar{O}_3}{\Delta E_{NO_x}} \text{ (ppbv/TgN/yr)}$$

where $\Delta\bar{O}_3$ is global mean ocean ozone change; ΔE_{NO_x} is the change in NO_x emissions. Fuglestvedt *et al.* (1998) emphasized that changes in NO_x emissions will affect not only O_3 but also OH and thereby also the level of CH_4. The following sequency of events will take place:

$$\Delta E_{NO_x} \longrightarrow \Delta[OH]_0 \longrightarrow \Delta[CH_4]_0 \longrightarrow \Delta[OH] \longrightarrow \Delta[CH_4]$$
$$\longrightarrow \Delta[OH] \longrightarrow \text{etc.}$$

Table 6.21. Longwave, shortwave and net radiative forcing because of stratospheric ozone; global, Northern Hemispheric and Southern Hemisphere averages for the four seasons and the yearly average (YA).

	DJF	MAM	JJA	SON	YA
Global					
LW	−0.148	−0.177	−0.173	−0.167	−0.167
SW	0.188	0.217	0.235	0.212	0.213
Net	0.039	0.040	0.061	0.044	0.046
Northern Hemisphere					
LW	−0.127	−0.252	−0.246	−0.151	−0.194
SW	0.135	0.272	0.329	0.192	0.232
Net	0.08	0.020	0.082	0.041	0.038
Southern Hemisphere					
LW	−0.170	−0.103	−0.100	−0.184	−0.139
SW	0.240	0.163	0.141	0.232	0.194
Net	0.071	0.060	0.040	0.048	0.055

The results obtained by Fugelstvedt *et al.* (1998) indicate large variation in sensitivity of chemical responses and radiative forcings because of changes in O_3 and CH_4 between the regions. An important conclusion is that O_3 and CH_4 forcings are of opposite sign and generally of similar magnitude. Although the calculated O_3 forcing, caused by 20% reduction in NO_x emissions, is noticeable on a regional basis, in terms of global annual averages the effects are negligible. The general conclusion is that emissions of NO_x have complex and potentially important impacts on climate through atmospheric chemistry interactions.

Myhre *et al.* (1998) calculated radiative forcing and stratospheric temperature change because of changes in stratospheric ozone between 1960 and 2004 using the 2-D stratospheric chemistry–transport model and taking into account both shortwave and longwave components of RF.

Table 6.21 shows some of the results. More detailed results reveal (in accord with earlier results) that RF at the tropopause is very sensitive to the altitude of stratospheric ozone changes.

While ozone reductions in the upper stratosphere give positive RF (especially in low latitudes), reductions in ozone in the lower stratosphere give negative RF. As is typical for atmospheric processes in general, final results are small differences in much larger values (Table 6.19). Net RF is positive for all seasons and for the yearly average. RF in 1996, with 1969 as a reference year, was calculated to be $0.05\,W/m^2$. Previous RF calculations, based on observed changes in ozone from the SAGE satellites, found negative forcing of $-0.1\,W/m^2$ (SAGE data probably overestimate ozone reduction, which is much smaller according to ozonesonde data). Myhre *et al.* (1998) emphasized that, according to their global-scale calculations, heating of the troposphere–surface system because of ozone reductions in the upper and mid-stratosphere nearly balances the cooling caused by reductions in lower stratospheric ozone. This is why RF because of stratospheric ozone is probably very small on the global scale.

Wauben *et al.* (1998) investigated the sensitivity of radiative forcing (net downward radiation at the tropopause level) because of changes in the global distribution of ozone from aviation. The global distribution of radiative forcing was computed for a horizontal grid 10° longitude × about 8° latitude for January and July. The results of calculations indicate that RF is generally most sensitive to ozone changes just below the tropopause (increasing from the surface up to the tropopause and rapidly decreasing in the stratosphere), with the highest RF sensitivity in the tropics and strong seasonal dependence in the polar regions. The sensitivity is characterized by large latitudinal, but also longitudinal variations. Assessment of ozone and relevant RF changes because of aircraft NO_x emissions show that RF attains maximum values of 0.13 W/m^2 in the Mediterranean in July, whereas maximum values over the northern subtropical region of the Atlantic (0.07 W/m^2) take place in January. These maxima are located a little more equatorward than regions of maximum ozone perturbation.

Wauben *et al.* (1998) emphasized that aircraft emissions do not only affect atmospheric composition by changing ozone concentration, but other components are changed as well (see Chapter 3). Of special interest in this context are impacts of aerosols and contrails on radiative forcing. There are serious uncertainties, however, in assessments of aircraft-induced RF changes.

Using TOMS total ozone data from Nimbus-7 and Meteor-3 satellites between 1979 and 1994, Vargin (1998) investigated TOZ non-zonal global structure (especially for the Southern Hemisphere) and calculated non-zonal atmospheric heating rates caused by ozone absorption of solar UV radiation in Hartley (200–300 nm), Huggins (300–350 nm) and Chappuis (450–750 nm) bands (similar calculations were made by Bojkov (1976)). The principal conclusion reached by Vargin (1998) was that large-scale ozone disturbances have regularly been observed in September, during the whole time period considered, within the 55°S–70°N latitude belt. Maximum amplitudes of the first TOZ harmonics in the Southern Hemisphere reached about 100 DU (September 1988). Strong QBO contribution to long-term TOZ variations at high latitudes (55°S–70°S) has been observed. The non-zonal part of ozone-induced solar radiative heating in September 1988 was about 2 K/day in the 40–45-km layer and 0.2 K/day at the 20 km altitude.

Christiansen and Gulberg (1998) carried out a general-circulation-model study of the dynamical and radiative effects of ozone perturbations, including uniform 50 and 75% ozone reductions, and a hole-type reduction concentrated in the lower part of the stratosphere and the upper part of the troposphere (see also Christiansen *et al.*, 1997). Ramaswamy *et al.* (1996) revealed the possibility of lower stratospheric cooling caused by dynamical feedback at latitudes where no ozone perturbations are imposed. This has strengthened the belief that ozone depletion may be responsible for the main part of observed cooling at all latitudes in the lower stratosphere. Christiansen and Gulberg (1998) emphasized, in this context, that changes in the dynamical structure of the stratosphere may alter stratospheric temperature response and thereby radiative forcing on the troposphere. It is therefore very important to study the role of dynamical feedback in radiative forcing on the troposphere–surface system.

To study the contribution of various aspects of ozone impacts, comparison has been made between pure radiative effects (no thermal or dynamical adjustment), the effect of thermal adjustment alone (fixed dynamical heating (FDH) approximation) and the effect of the full dynamical system (Arpège model). The principal conclusions reached by Christiansen and Gulberg were (1) the major effect of the dynamical adjustment is a weakening in meridional circulation and a smoothing of the temperature response in the meridional direction (these results question the validity of the FDH approach); (2) dynamical temperature adjustment in the lower stratosphere plays an important role in latitudinal distribution of radiative forcing; (3) in agreement with previous studies, thermal adjustment in the lower stratosphere changes the sign of the forcing.

6.4 CONCLUSION

The presence of ozone in the atmosphere results in a number of problems which should be considered in the context of global environmental change. Of course, the most important problem is conservation of the ozone layer in the stratosphere for the protection of life on Earth against harmful ultraviolet solar radiation. The cessation of chlorofluorocarbon (CFC) emissions into the atmosphere (in accordance with the Montreal and London Protocols) is a prerequisite to solution of this problem. A number of important scientific questions remain unanswered, which indicates the necessity for further study of those photochemical processes in the stratosphere that are responsible for ozone-layer dynamics. A special role belongs to studies of solar activity impacts on the ozone layer, especially in connection with the potential role of stratospheric processes in solar climatic interrelations. An important aspect of stratospheric ozone studies is heterogeneous chemical reactions on surfaces of aerosol particles of various (natural and anthropogenic) origin, including the impact of stratospheric aviation.

The significance of tropospheric-ozone studies, from the viewpoint of its impact on human health and on ecosystems as well as in connection with global climate changes (tropospheric ozone as a greenhouse gas), is undoubtedly growing.

The fact that all these problems are interrelated (*Atmospheric* ..., 1998; Kondratyev, 1989, 1998; Schneider *et al.*, 1993; *Statospheric* ..., 1998) results in the necessity to apply a systematical approach for solving the relevant problems. Since ozone problems are located in the boundary layer between the World Climate Research Programme (WCRP) and the International Geosphere–Biosphere Programme (IGBP), it would be worthwhile to work out a special interdisciplinary ozone programme, embracing all specific ozone problems to be solved, on the basis of development of a more adequate global ozone observing system and further progress in numerical simulation modelling.

A recent fundamental survey of atmospheric ozone research – WMO/UNEP Ozone Assessment 1998 (*Scientific* ..., 1999) as well as EOS Science Plan (King,

1999) contains an almost exhaustive analysis of present-day results and future perspectives. So, we shall limit ourselves to just a few brief comments about some unsolved problems.

6.4.1 Stratospheric ozone dynamics and its determining factors

This complicated problem connected with the need to explain the numerous dynamic and chemical processes that determine changes in total ozone (TOZ), in conditions of anthropogenic forcing caused by emissions to the atmosphere of ozone-destroying substances (ODS) such as CFCs, requires, apart from chemical processes (primarily heterogeneous chemical reactions on atmospheric aerosol particles), an understanding about the stratosphere–troposphere interaction as well as the air exchange between high and middle, middle and tropical latitudes.

From the point of view of obtaining observational data, this signifies the need for: (1) monitoring ozone with high spatial and temporal resolutions and of most important ODS (ClO, NO_x), aerosol, water vapour, temperature and other meteorological parameters; (2) obtaining adequate information to meet the needs of numerical modelling of stratospheric variability. The problem of global dynamics of the ozone layer should be solved in the context of varying concentrations of CFCs and their substitutes (HCFCs), possible intensification of stratospheric aviation, as well as probable changes in temperature and circulation in the stratosphere (stratospheric climate) under the influence of increasing GHG concentrations (of course, interactivity between climatic and ozone-layer dynamics should be borne in mind).

Solution of these and other problems should be supported by the EOS CHEM programme, accomplishment of which will provide global databases on concentrations of O_3, ClO, NO_3, $ClONO_2$, HCl, HNO_3, N_2O_5, CFCs, H_2O, CH_4, as well as on aerosol and temperature with vertical resolutions 2–3 km (MLS) and 1 km (HIRDLS). Stratosphere–troposphere exchange will be studied using the combined data of aircraft measurements, with vertical resolution $\sim$100 m along horizontal routes from 100 km to 500 km, and satellite remote sounding. The launch on 19 December 1999 of the Terra satellite is an important step forward.

In connection with accomplishment of the EOS programme CHEM, two additional developmental aspects appear: (1) observations of UV radiation flux and substantiation of requirements for these observations; (2) importance of including in the programme measurements of concentrations of OH or other HO_x radicals by some modification to MLS instruments.

The urgency of the problem of hydroxyl is determined by its important role in atmospheric chemistry for several reasons: (a) OH controls conversion of CH_4 to H_2O; (b) reactions in presence of HO_x are most important as mechanisms for ozone losses in the upper and lower stratosphere; (c) reactions in presence of OH determine the rate of oxidation of gaseous sulphur compounds (SO_2, OCS) to give sulphate aerosol; (d) OH competes with heterogeneous chemical reactions over control of the transformation of radicals and 'reserve' components in the NO_y and Cl_x systems of families. So, for example, reaction $OH + NO_2 \longrightarrow HNO_3$ competes both with the inverse reaction ($OH + HNO_3$) and with reactions:

$$ClO + NO_2 \longrightarrow ClNO_2 \qquad \text{and} \qquad OH + HCl \longrightarrow Cl + H_2O$$

Since OH content is expressed in models of chemical process, as a rule, by concentrations of other gaseous components, the data are needed which characterize the dependence of OH on concentration of such components in a wide range of real conditions.

Although ozone in the mid and upper stratosphere contributes relatively little to TOZ and therefore weakly affects the near-surface UV radiation flux, it is very interesting as one of the factors responsible for formation of the temperature field and atmospheric circulation in the stratosphere, which may affect the structure and composition of the lower stratosphere. In this connection, two unsolved problems are most urgent: (1) the remaining differences between observational data and the results of numerical modelling in the zone of dominating photochemical processes at altitudes above 40 km; (2) poorly studied heterogeneous chemical processes in the lower stratosphere (23–35 km).

6.4.2 Tropospheric processes

The combined use of data of direct measurements and remote sounding in association with numerical modelling will open up prospects for solving the most urgent problems. In this section, we outline some of the most important unsolved problems.

Levels of concentration of most substantial GHGs (H_2O, CO_2, CH_4, N_2O, HCFCs, O_3) and their determining factors. While for long-lived gaseous components, such as CO_2, CH_4, N_2O and HCFCs that are rather uniformly mixed up in the atmosphere, the representative information can be obtained from direct surface observational data, in the case of water vapour and tropospheric ozone, characterized by strong spatial and temporal variabilities, remote sounding data are important.

Levels of concentration of oxidants in the troposphere, ozone and hydroxyl in particular. Oxidizers such as ozone, hydroxyl and others (H_2O_2 in particular) form in the troposphere as a result of a complicated set of photochemical reactions in the presence of nitrogen oxides, hydrocarbons and water vapour (the relative role of which remains poorly studied). The primary importance of the heterogeneous chemistry of the atmosphere needs not only observational data about the numerous trace gases and aerosols but also needs problem-oriented field experiments to study, in particular, the gas exchange between the atmosphere and biosphere, the atmosphere and ocean (in this context, the correlation approaches of direct measurements of TG fluxes in the atmosphere can play a special role). Satellite remote sounding should become an important source of information about global fields of precursors (and indicators) of oxidants in the troposphere, such as NO_x, CO, hydrocarbons, water vapour, which are short-lived and strongly variable (King, 1999).

The substantial source of ozone and NO_x for the troposphere is the stratosphere, quantitative data on which are disputable. Since gas exchange between the stratosphere and troposphere is realized apparently by means of mesoscale processes, direct aircraft measurements with high spatial resolution are very important. Satellite remote sensing and *in-situ* data will supplement these measurements with respect to spatial representativity and the role of large-scale processes.

Sources and properties of tropospheric aerosol. Understanding (only recently achieved) of the urgency of the problem greenhouse warming and aerosol cooling on climate determines the need for complex studies of atmospheric aerosols of the impact of, including wide application of surface (lidar) and satellite remote-sensing techniques.

6.5 REFERENCES

Annales Geophysicae (1998) *Part III. Space and Planetary Science*, Supplement III to Volume 16, European Geophysical Society, Vienna, C941–C952.

Atmospheric Ozone, Proceedings XVIII Quadrennial Ozone Symposium, 12–21 September 1996, L'Aquila, Italy. (1998) R. D. Bojkov, G. Visconti (eds.), Vols. 1, 2, Parco Scientifico e Tecnologico, Abruzzo, Italy.

Austin, J. Butchart, N. (1998) Simulations of stratospheric constituents in a climate model with interactive ozone. *Proceedings First SPARC General Assembly*, Vol. 2, WCRP-99 (WMO/TD-N814), WMO, Geneva, Switzerland, pp. 449–452.

Baker, M. B. (1997) Cloud microphysics and climate. *Science*, **276**, N5,315, 1,072–1,078.

Bazilevskaya, G., Krivolutsky, A., Ivanova, I., Nazarova, M., Perejaslova, N., Petrenko, I. (1997) Cosmic rays and ozone variability. *Proceedings First SPARC General Assembly*, Vol. 2, WCRP-99 (WMO/TD-N814), WMO, Geneva, Switzerland, pp. 469–472.

Berntsen, T., Isaksen, I. S. A. (1997) A global three-dimensional chemical transport model for troposphere. 1. Model description and CO and ozone results. *J. Geophys. Res.*, **102**, 21,239–21,280.

Berntsen, T., Isaksen, I. S. A., Myhre, G., Fuglestvedt, J. S., Stordal, F., Larsen, D. A., Freckleton, R. S., Shine, K. P. (1997) Effects of anthropogenic emissions on tropospheric ozone and its radiative forcing. *J. Geophys. Res.*, **102**, ND23, 28,101–28,126.

Bintanja, R., Fortuin, J. P. F., Kelder, H. (1997) Simulation of the meridionally and seasonally varying climate response caused by changes in ozone concentrations. *J. Climate*, **10**, N6, 1,288–1,311.

Bojkov, R. D. (1976) Global net heating rates due to absorption by ozone. *Proceedings IAMAP/WMO Symposium on Atmospheric Ozone, Dresden*, Vol. 2, pp. 375–381, World Meteorological Organisation, Geneva.

Bojkov, R. (1997) Global ozone change and possible climate implications. In: *Solar Ultraviolet Radiation*, C. S. Zerefos, A. F. Bais (eds.), NATO ASI Series Vol. 52, Springer-Verlag, Berlin, pp. 303–324.

Brasseur, G. (ed.) (1997) *The Stratosphere and Its Role in the Climate System*, Springer, Berlin, 368 pp.

Brauers, T., Dom H.-P., Koch, H., Kraus, A. B., Plass-Duelmer, C. (1998) Meteorological aspects, ozone and solar radiation measurements during POPCORN 1994. *J. Atmos. Chem.*, **31**, N1/2, 33–52.

Chanin, M.-L. (ed.) (1993) *The Role of the Stratosphere in Global Change*, NATO ASU Series, Springer-Verlag, Heidelberg.

Chanin, M.-L. (1998) Fifth session of the Scientific Steering Group on Stratospheric Processes and their Role in Climate, 17–20 November 1997, Port Jefferson, NY. *SPARC Newsletter*, N10, 1–5.

Chanin, M.-L., Geller, M. (1998) Report on the WCRP IJSC Meeting. *SPARC Newsletter*, N11, 1–2.

Christiansen, B., Guldberg, A., Hansen, A. W., Riishøjgaard, L. P. (1997) On the response of a three-dimensional general circulation model to imposed changes in the ozone distribution. *J. Geophys. Res.*, **102**, ND11, 13,051–13,077.

Christiansen, B., Guldberg, A. (1998) Dynamical and radiative effects of idealized ozone perturbations. In: *Atmospheric Ozone*, R. D. Bojkov, G. Visconti (eds.), Parco Scientifico e Tecnologico, Abruzzo, Italy, Vol. 2, pp. 805–808.

Daniel, J. S., Solomon, S. (1998) On the climate forcing of carbon monoxide. *J. Geophys. Res.*, **103**, ND11, 13,245–13,260.

Dingmin, L., Shine, K. P., Gray, L. S. (1995) The role of ozone-induced diabatic heating anomalies in the quasi-biennial oscillation. *Quart. J. Roy. Meteorol. Soc.*, **121**, Part B, N524, 937–944.

van Dorland, R., Fortuin, J. P. F. (1994) Simulation of the observed stratospheric temperature trends 1967–1987 over Antarctica due to ozone hole deepening. In: *Non-*CO_2 *Greenhouse Gases*, J. Van Ham and Co-editors, Kluwer Acad. Publ., pp. 237–245.

van Dorland, R., Dentener, F. J., Lelieveld, J. (1997) Radiative forcing due to tropospheric ozone and sulfate aerosols. *J. Geophys. Res.*, **102**, ND23, 28,079–28,100.

Engelen, R. J., Stephens, G. L. (1997) Infrared radiative transfer in the 9.6-μm band: Application to TIROS operational vertical sounder ozone retrieval. *J. Geophys. Res.*, **102**, ND6, 6,925–6,939.

Fels, S. B., Mahlman, J. D., Schwarzkopf, M. D., Sinclair, R.W. (1980) Stratospheric sensitivity to perturbations in ozone and carbon dioxide: Radiative and dynamical response. *J. Atmos. Sci.*, **37**, 2,265–2,297.

Forster, de F. P. M., Shine, K. P. (1997) Radiative forcing and temperature trends from stratospheric ozone changes. *J. Geophys. Res.*, **102**, ND9, 10,841–10,856.

Forster, de F. P. M., Shine, K. P. (1998) Radiative forcing and temperature trends from stratospheric ozone changes. *Proceeding First SPARC General Assembly*, WCRP-99 (WMO/TD-No. 814), WMO, Geneva, Switzerland, pp. 169–172.

Fortuin, J. P. F., van Dorland, R., Wauben, W. M. F., Kelder, H. (1995) Greenhouse effects of aircraft emissions as calculated by a radiative transfer model. *Ann. Geophys.*, **13**, 413–418.

Fuglestvedt, J. S., Berntsen, T. K., Isaksen, I. S. A., Mao, H., Liang, X.-Z., Wang, W.-C. (1998) Impacts of reduced NO_x emissions on radiative forcing through changes in tropospheric O_3 and CH_4. In: *Atmospheric Ozone*, R. D. Bojkov, G. Visconti (eds.), Parco Scientifico e Tecnologico, Abruzzo, Italy, Vol. 2, 809–812.

Fuglestvedt, J. S., Berntsen, T. K., Isaksen, I. S. A., Mao, H., Liang, X.-Z., Wang, W.-C. (1999) Climatic forcing of nitrogen oxides through changes in tropospheric ozone and methane. Global 3D model studies. *Atmos. Environ.*, **33**, N6, 961–977.

Geller, M. A. (1998) Presentation of SPARC at the WCRP Conference. SPARC future prospects. *SPARC Newsletter*, N10, 6–10.

Graf, H.-F. (1995) Recent winter climate trends, ozone changes and increased greenhouse gas forcing. *Contr. Atmos. Phys.*, **68**, 233–248.

Haigh, J. D. (1994) The role of stratospheric ozone in modulating the solar radiative forcing on climate. *Nature*, **370**, 544–546.

Haigh, J. D. (1996) The impact of solar variability on climate. *Science*, **272**, 981–984.

Haigh, J. D. (1998) Stratospheric ozone and the impact of solar variability on climate. *Proceeding First SPARC General Assembly*, Vol. 2, WCRP-99 (WMO/TD-N814), WMO, Geneva, Switzerland, pp. 12–15.

Hansen, J., Sato, M., Ruedy, R. (1997) Radiative forcing and climate response. *J. Geophys. Res.*, **102**, 6,831–6,864.

Haywood, J. M., Schwarzkopf, M. D., Ramaswamy, V. (1998) Estimates of radiative forcing due to modeled increases in tropospheric ozone. *J. Geophys. Res.*, **103**, ND14, 16,999–17,007.

Highwood, E. J., Shine, K. P., Hurley, M. D., Wallington, T. J. (1999) Estimation of direct radiative forcing due to non-methane hydrocarbons. *Atmos. Environ.*, **33**, N5, 759–767.

Houghton, J. T., Meiro Filho, L. G., Callander, B. A., Harris, N., Kattenberg, A., Maskell, K., (eds.) (1996) *Climate change 1995: The Science of Climate Change*, Cambridge University Press, Cambridge, 584 pp.

Joiner, J., Lee, H.-T, Strow, L. L., Bhartia, P. K., Hannon, S., Miller, A. S., Rokke, L. (1998) Radiative transfer in the 9.6 μm HIRS ozone channel using collocated SBUV-determined ozone abundances. *J. Geophys. Res.*, **103**, ND15, 19,213–19,230.

Kiehl, J. T., Boville, B. A. (1988) The radiative-dynamical response of a stratospheric-tropospheric general circulation model to changes in ozone. *J. Atmos. Sci.*, **45**, 1,798–1,817.

King, M. D. (1999) EOS Science Plan. The state of science in the EOS Program. NASA Publ., Washington, D.C., 397 pp.

Kondratyev, K. Ya. (1981) *Stratosphere and Climate*, VINITI, Moscow, 223 pp. (in Russian).

Kondratyev, K. Ya. (1989) Ozone Dynamics. Progress in Science and Technology. *Geomagnetism and Upper Atmosphere*, **11**, VINITI, Moscow, 212 pp. (in Russian).

Kondratyev, K. Ya. (1998) Aerosol and climate: Some results and perspectives of remote sensing. 1. Complex nature of climate change and variability of aerosol properties. *Ecol. Chemistry*, **7**, N2, 73–85 (in Russian).

Kondratyev, K. Ya. (1999) *Ecodynamics and Geopolicy. Global Problems*, Russian *Acad. Sci. Publ.*, St. Petersburg, pp. 678.

Krivolutsky, A. A. (1997) Possible nature of spring ozone decreation and its sensitivity to long-term variability. *Proceedings First SPARC General Assembly*, Vol. 2, WCRP-99 (WMO/TD-N814), WMO, Geneva, Switzerland, pp. 465–468.

Lacis, K., Wuebbles, D.J., Logan, J. A. (1990) Radiative forcing of climate by changes in the vertical distribution of ozone. *J. Geophys. Res.*, **95**, 9,971–9,981.

Lastovicka, J. (1998) Second IAGA/IAMAS Workshop: Solar activity forcing of the middle atmosphere. *SPARC Newsletter*, N10, 26–27.

Lelieveld, J., van Dorland, R. (1995) Ozone chemistry changes in the troposphere and consequent radiative forcing of climate. In: *Atmospheric Ozone as a Climate Gas*, NATO ASI Series I, Vol. 32, W.-C. Wang, I. S. A. Isaksen (eds.), Springer-Verlag, New York, pp. 227–258.

Levy, H., II, Kasibhatla, P. S., Moxim, W. J., Klonecki, A. A., Hirsch, A. I., Oltmans, S. J., Chameides, W. L. (1997) The global impact of human activity on tropospheric ozone. *Geophys. Res. Lett.*, **24**, N7, 791–794.

Mahlman, J. D., Pinto, J. P., Umscheid, L. J. (1994) Transport, radiative, and dynamical effects of the Antarctic ozone hole: A GFDL 'SKYHI' model experiment. *J. Atmos. Sci.*, **51**, 489–508.

Moulik, M. D., Milford, J. B. (1999) Factors influencing ozone chemistry in subsonic aircraft plumes. *Atmos. Environ.*, **33**, 869–880.

Myhre, G., Stordal, F. (1997) Role of spatial and temporal variations in the computation of radiative forcing and GWP. *J. Geophys. Res.*, **102**, ND11, 11,181–11,200.

Myhre, G., Stordal, F., Rognerud, B., Isaksen, I. S. A. (1998) Radiative forcing due to stratospheric ozone. In: *Atmospheric Ozone*, R. D. Bojkov, G. Visconti (eds.), Parco Scientifico e Tecnologico, Abruzzo, Italy, Vol. 2, pp. 813–816.

Nicoli, M. P., Visconti, G. (1982) Impact of coupled perturbations of atmospheric trace gases on world's climate and ozone. *Pure and Appl. Geophys.*, **120**, N4, 626–641.

Ozone Depletion, Greenhouse Gases, And Climate Change (1989) National Academy Press, Washington, DC, 122 pp.

Perlwitz, J., Graf, H.-F., Kirchner, I. (1997) Increasing greenhouse effect and ozone trends. Proceedings First SPARC General Assembly, Vol. 2, WCRP-99 (WMO/TD-N814), WMO, Geneva, Switzerland, pp. 461–464.

Phillips, C. (ed.) (1997) *Stratospheric Processes and Their Role in Climate*, WCRP project, SPARC Office, Verrières-le-Buisson, 8 pp.

Pitari, G., Palermi, S., Visconti, G., Prinn, R. G. (1992) Ozone response to a CO2 doubling: Results from a stratospheric circulation model with heterogeneous chemistry. *J. Geophys. Res.*, **97**, ND5, 5,953–5,962.

Portmann, R. W., Solomon, S., Fishman, J., Olson, J. R., Kiehl, J. T., Briegleb, B. (1997) Radiative forcing of the Earth's climate system due to tropical tropospheric ozone production. *J. Geophys. Res.*, **102**, ND8, 9,409–9,417.

Ramanathan, V., Callis, L. B., Boughner, R. E. (1976) Sensitivity of surface temperature and atmospheric temperature to perturbations in the stratospheric concentration of ozone and nitrogen dioxide. *J. Atmos. Sci.*, **33**, 1,092–1,112.

Ramanathan, V., Dickinson, R. E. (1979) The role of stratospheric ozone in the zonal and seasonal radiative energy balance of the earth-troposphere system. *J. Atmos. Sci.*, **36**, 1,084–1,104.

Ramanathan, V., Cicerone, R. J., Singh, H. B., Kiehl, J. T. (1985) Trace gas trends and their potential role in climate change. *J. Geophys. Res.*, **90**, 5,547–5,566.

Ramaswamy, V., Bower, M. M. (1994) Effects of changes in radiatively active species upon the lower stratospheric temperatures. *J. Geophys. Res.*, **99**, 18,909–18,921.

Ramaswamy, V., Schwarzkopf, M. D., Randel, W. J. (1996) Fingerprints of ozone depletion in the spatial and temporal pattern of recent lower-stratospheric cooling. *Nature*, **382**, 616–618.

Rind, D., Lonergan, D. (1995) Modeled impacts of stratospheric ozone and water vapor perturbations with implications for high-speed civil transport aircraft. *J. Geophys. Res.*, **100**, 7,381–7,396.

Rind, D., Shindell, D., Lonergan, D., Balachandran, N. K. (1997) Climate change and the middle atmosphere; The doubled CO_2 climate revised. *Proceedings of First SPARC General Assembly*, Vol. 2, WCRP-99 (WMO/TD-N814), WMO, Geneva, Switzerland, pp. 453–456.

Roelofs, G. J., Lelieveld, J. (1997) Model study of the influence of cross-tropopause O_3 transports on tropospheric O_3 levels. *Tellus*, Ser. B., **49**, 38–55.

Roelofs, G.-J., Lelieveld, J., van Dorland, R. (1997) A three-dimensional chemistry/general circulation model simulation of anthropogenically derived ozone in the troposphere and its radiative climate forcing. *J. Geophys. Res.*, **102**, ND19, 23,389–23,401.

Rosenfield, J. E., Schoeberl, M. R. (1986) A computation of stratospheric heating rates and the diabatic circulation for the antarctic spring. *Geophys. Res. Lett.*, **13**, N12, 1,339–1,342.

Sausen, R., Feneberg, B., Ponater, M. (1997) Climatic impact of aircraft-induced ozone changes. *Geophys. Res. Lett.*, **24**, N10, 1,203–1,206.

Schneider, H. R., Ko, M. R. W., Shia, R.-L., Sze, N.-D. (1993) A two-dimensional model with coupled dynamics, radiative transfer, and photochemistry. 2. Assessment of the response of stratospheric ozone to increased levels of CO_2, N_2O, CH_4, and CFC. *J. Geophys. Res.*, **98**, 20,444–20,449.

Schwartzkopf, M. D., Ramaswamy, V. (1993) Radiative forcing due to ozone in the 1980's: Dependence on altitude of ozone change. *Geophys. Res. Lett.*, **20**, 205–208.

Scientific Assessment of Ozone Depletion: 1998: WMO Global Ozone Research and Monitoring Project Report No. 44 (1999) WMO, Geneva, Switzerland.

Sexton, D. M. H., Karoly, D. J., Folland, C. K, Rowell, D. P., Parker, D. E. (1997) GCM simulation of the climate response to observed ozone decreases. *Proceedings the First SPARC General Assembly*, Vol. 2, WCRP-99 (WMO/TD-N814),WMO Geneva, Switzerland, pp. 177–180.

Shi, G.-Y., Wang, W.-C., Ko, M. R. W. (1986) Radiative heating due to stratospheric aerosols over Antarctica. *Geophys. Res. Lett.*, **13**, N12, 1,335–1,338.

Shindell, D., Rind, D., Lonergan, P. (1997) The doubled CO_2 climate inclunding ozone chemistry. *Proceedings First SPARC General Assembly*, Vol. 2, WCRP-99 (WMO/TD-N814), WMO, Geneva, Switzerland, pp. 457–460.

Shindell D., Rind D., Balachandran N., Lean J., Lonergan P. (1999). Solar cycle variability, ozone, and climate. *Science*, **284**, N5412, 305–308.

Shine, K. D. (1986) On the modeled thermal response of the Antarctic stratosphere to a depletion of ozone. *Geophys. Res. Lett.*, **13**, 1,331–1,334.

Shine, K. P., Briegleb, A. S., Grossman, A. S., Hauglestaine, D., Mao, H., Ramaswamy, V., Schwartzkopf, M. D., van Dorland, R., Wang, W.-C. (1995) Radiative forcing due to changes in ozone: A comparison of different codes. In: *Atmospheric Ozone as a Climate Gas*, NATO ASI Series Vol. 32, W.-C. Wang, I. S. A. Isaksen (eds.), Springer-Verlag, New York, pp. 373–396.

Shine, K. P., Freckleton, R. S., Forster, P. M. de F. (1998) Comment on 'Climate forcing by stratospheric ozone depletion calculated from observed temperature trends' by Zhong *et al.* Author's reply. *Geophys. Res. Lett.*, **25**, N5, 663–665.

Sinha, A., Toumi, R. (1997) Tropospheric ozone, lightning, and climate change. *J. Geophys. Res.*, **102**, ND9, 10,667–10,671.

Solomon, S., Daniel, J. S. (1996) Impact of the Montreal Protocol and its amendments on the rate of change of global radiative forcing. *Clim. Change*, **32**, 7,017.

Stenchikov, G. L., Kirchner, I., Roback, A., Graf, H.-F., Antuña, J. C., Grainger, R. G., Lambert, A., Thomason, L. (1998) Radiative forcing from the 1991 Mount Pinatubo volcanic eruption. *J. Geophys. Res.*, **103**, ND12, 13,837–13,857.

Stevenson, D. C., Johnson, C. E., Collins, W. J., Derwent, R. G., Shine, K. P., Edwards, J. M. (1998). Evolution of tropospheric ozone radiative forcing. *Geophys. Res. Lett.*, **25**, N20, 3,819–3,828

Stratospheric Processes and Their Role in Climate (SPARC). (1997) *Proceedings SPARC General Assembly*, Vols. I, II, (1997) WCRP-99 (WMO/TD-No. 814), WMO, Geneva, Switzerland, 672 pp.

Stratospheric Processes and Their Role in Climate. Implementation Plan (1998) WCRP-1059 (WMO/TD-No. 914), WMO, Geneva, Switzerland, 131 pp., Appendices.

Toumi, R., Haigh, J. D., Law, K. S. (1996) A tropospheric ozone-lightning climate feedback. *Geophys. Res. Lett.*, **23**, 1,037–1,040.

Uchino, O. (1998) Tsukuba International Workshop on stratospheric change and its role in climate and on the ATMOS-C1 satellite mission. 20–22 October 1997, Tsukuba, Japan. *SPARC Newsletter*, N10, 28–32.

UNEP Scientific Assessment of Ozone depletion, WMO Global Ozone Research and Monitoring Project. Report No. 37 (1994 WMO, Geneva, Switzerland.

Vargin, P. N. (1998) Climatological structure of nonzonal heating rates in the stratosphere caused by large-scale inhomogeneity of total ozone. In: *Atmospheric Ozone*, R. D. Bojkov, G. Visconti (eds.) Parco Scientifico e Tecnologico, Abruzzo, Italy, Vol. 2, pp. 817–820.

Varotsos, C. (ed.) (1997) *Atmospheric Ozone Dynamics, Proceedings NATO Advanced Research Workshop*, Springer, Berlin, Vol. 53, 336 pp.

Wang, W.-C., Sze, N. D. (1980) Coupled effects of atmospheric N_2O and O_3 on the Earth's climate. *Nature*, **286**, 589–590.

Wang, W.-C., Zhuang, Y. C., Bojkov, R. D. (1993) Climate implications of observed changes in ozone vertical distributions in middle and high latitudes of the Northern Hemisphere. *Geophys. Res. Lett.*, **20**, 1,567–1,570.

Wang, W.-C., Isaksen, I. S. A. (eds.) (1995) *Atmospheric Ozone as a Climate Gas: General Circulation Model Simulations*, Springer-Verlag, Berlin, 459 pp.

Wang, W. C., Liang, X.-Z, Dudek, M. P., Pollard, D., Thompson, S. L. (1995) Atmospheric ozone as a climate gas. *Atmos. Res.*, **37**, 247–256.

Wang, C., Prinn, R. G., Sokolov, A. (1998a) A global interactive chemistry and climate model: Formulation and testing. *J. Geophys. Res.*, **103**, ND3, 3,399–3,418.

Wang, W.-C., Mao, H., Isaksen, I. S. A., Fuglestvelt, J. S., Karlsdottir, S. (1998b) Indirect effects of increasing atmospheric methane on the radiative forcing through climate-chemistry interactions. In: *Atmospheric Ozone*, R. D. Bojkov, G. Visconti (eds.), Parco Scientifico e Tecnologico, Abruzzo, Italy, Vol. 2, pp. 821–826.

Wauben, W. M. F., Fortuin, J. P. F., Kelder, H. M. (1998) Sensitivity of radiative forcing due to changes in the global distribution of ozone with application to ozone changes from aviation. In: *Atmospheric Ozone*, R. D. Bojkov, G. Visconti (eds.), Parco Scientifico e Tecnologico, Abruzzo, Italy, Vol. 2, pp. 827–830.

Wuebbles, D. J. (1996) Three-dimensional chemistry in the greenhouse. An editorial comment. *Clim. Change*, **34**, N3–4, 397–404.

Wuebbles, D. J., Wei, C.-F., Patten, K. O. (1998) Effects of stratospheric ozone and temperature during the Maunder Minimum. *Geophys. Res. Lett.*, **25**, N4, 523–526.

Wyser, K., Ström, J. (1998) A possible change in global radiative forcing due to aircraft exhaust. *Geophys. Res. Lett.*, **25**, N10, 1,673–1,676.

Zhong, W. R., Toumi, R., Haigh, J. D. (1996) Climate forcing by stratospheric ozone depletion calculated from observed temperature trends. *Geophys. Res. Lett.*, **23**, 3,183–3,186.

Zillman, J. W. (1997) Atmospheric science and public policy. *Science*, **276**, N5,315, 1,084–1,086.

Zwick A. (1998). *Climate Change Research and Policy: Update No. 11*, Joint Research Centre, EUR 18088 EN, Seville, Spain 78 pp.

Additional references

Alados, I., Alados-Arboledas, L. (199) Validation of an empirical model for photosynthetically active radiation. *Int. J. Climatol.*, **19**, N10, 1,145–1,152.

Alexandris, D., Varotsos, C., Kondratyev, K. Ya., Chronopoulos, G. (1999) On the altitude dependence of solar effective UV. *Phys. Chem. Earth (C)*, **24**, N5, 515–517.

Alvarez, R. J. II, Senff, C. J., Hardesty, R. M., Parrish, D. D., Luke, W. T., Watson, T. B., Daum, P. H., Gillani, N. (1998) Comparisons of airborne lidar measurements of ozone with airborne in situ measurements during the 1995 Southern Oxidant Study. *J. Geophys. Res.*, **103**, ND23, 31,155–31,171.

American Geophysical Union (1999) U.S. Interagency UV Monitoring Program established and operating, *EOS Trans.*, **80**, N10, 113–116.

Atherton, C.S. (1997) Chemistry implications of climate change. *Third Conf. on Atmos. Chemistry, 2–7 February 1997, Preprints. Long Beach, CA*, American Meteorological Society, Boston, MA, pp. 70–71.

Bacmeister, J. T., Kuell, V., Offermann D., Riese, M., Elkins, J. W. (1999) Intercomparison of satellite and aircraft observations of ozone, CFC-11, and NO_y using trajectory mapping. *J. Geophys. Res.*, **104**, ND13, 16,379–16,390.

Bais, A. F., Blumthaler, M., Webb, A. R., Groebner, J., Kirsch, P. J., Gardiner, B. G., Zerefos, C. S., Svenoe, T., Martin, T. J. (1997) Spectral UV measurements over Europe within the Second European Stratospheric Arctic and Midlatitude Experiment acitivites. *J. Geophys. Res.*, **102**, ND7, 8,731–8,736.

Baranka, G. (1999) Near surface ozone concentration evaluation and prediction in Budapest. *Idojaras*, **103**, N2, 107–121.

Barrie, L. A., Platt, U., Shepson, P. (1998) Surface ozone depletion at polar sunrise fueled by sea-salt halogens. *IGACtivities Newsl.*, N14, 4–7.

Bates, T. S., Huebert, B. J., Gras, J. L., Griffiths, F. B., Durkee, P. A. (1998) International Global Atmospheric Chemistry (IGAC) Project's First Aerosol Characterization Experiment (ACE 1): Overview. *J. Geophys. Res.*, **103**, ND13, 16,297–16,318.

Bell, G. D., Halpert, M. S., Ropelewski, Ch. F., Konsky, V. E., Douglas, A. V., Schnell, R. C., Gelman, M. F. (1999) Climate assessment for 1998. *Bull. Amer. Meteorol. Soc.*, **12**, N5, Part 2, 5.1–5.48.

Bengtsson, L., Roeckner, E., Stendel, M. (1999) Why is the global warming proceeding much slower than expected? *J. Geophys. Res.*, **104**, ND4, 3,865–3,876.

Benjamin, M. T., Winer, A. M. (1999) Response to comments on 'Estimating the ozone forming potential of urban trees and shrubs' by Benjamin and Winer. *Atmos. Environ.*, **33**, N26, 4,341.

Bertaux, J. L., Hauchecorne, A., Margin, A., Cot, C., Talagraad, D., Simon, P., Kyrölä, E., Roscoe, H., Hembise, O., Brasseur, G. P. (1999) The MSDOL Project: Assimilation of Gomos Ozone Data in a 3-D chemistry-transport model. *Phys. Chem. Earth (C)*, **24**, N5, 435–437.

Blom, J. G , Verwer, J. C. (1999) A comparison of integration methods for atmospheric transport chemistry problems. Report Mas/Cent. Wisc. Inf., No. MAS-R 9,910, pp. 1–16.

Bonasoni, P., Evangelisti, F., Bonafe, U., Feldmann, H., Memmesheimer, M., Stohe, A., Tositti, L., Kromp-Kolb, L. H., Colombo, T. (1999) Stratospher--troposphere exchanges: Case studies recorded at Mt. Cimore during VOTALP Project. *Phys. Chem. Earth (C)*, **24**, N5, 443–446.

Borrell, P. M., Borrell, P. (1999) *EUROTRAC 98. Transport and Chemical Transformation in the Troposphere*, WIT Press, Southampton, Vol. I, 856 pp.; Vol. II, 952 pp.

Borrell, P., Builtjes, P., Greenfelt, P., Høv, Ø., von Alst, R., Fowler, D., Megie, G., Moussiopoulos, N., Warneck, P., Volz-Thomas, A., Wayne, R. (1997) EUROTRAC: Applications to photo-oxidants and the future project. *Proc. NATO Adv. Res. Workshop 'Ground Level and Satell. Ozone Observ. Change Mediterranean Reg.', Athens, 31 October–4 November 1995*, Springer, Berlin, pp. 1–7.

Bouchet, V. S., Laprise, R., Torlaschi, E., McConnell, J. C. (1999) Studying ozone climatology with a regional climate model, 1, Model description and evaluation. *J. Geophys. Res.*, **104**, ND23, 30,351.

Bouchet, V. S., Laprise, R., Torlaschi, E., McConnell, J. C., Plummer, D. A. (1999) Studying ozone climatology with a regional climate model, 2, Climatology. *J. Geophys. Res.*, **104**, ND23, 30,373.

Brasseur, G. P. (Ed.) (1997) *The Stratosphere and Its Role in the Climate System*, Springer-Verlag, Heidelberg, 368 pp.

Brasseur, G. P. (1999) Aerosols and photooxidants. *IGACtivities Newsletter*, N18, p. 1.

Brasseur, G. P., Orlando, J. J., Tyndall, G. S. (eds.) (1997) *Atmospheric Chemistry and Global Change*, Oxford University Press, Oxford, 672 pp.

Brebbia, C. A., Uso, J. L. (eds.) (1999) *Ecosystems and Sustainable Development. II*, WIT Press, Southampton, 416 pp.

Brönnimann, S. (1999) Early spring ozone episodes: Occurrence and case study. *Phys. Chem. Earth (C)*, **24**, N5, 531–536.

Browning, K. A., Gurney, R.J. (Eds.) (1999) *Global Energy and Water Cycles*, Cambridge University Press, 304 pp.

Cabrera, S., Fuenzalida, A. (1999) Interannual variations of global UV radiation in Santiago, Chile (33.5°S). *Geophys. Res. Lett.*, **26**, N19, 2,945.

Caldarary, F., Patrascu, S., Caldarary, M., Parashhiv, A., Nucolaesen, D. (1999) The far to equilibrium time evolution of the ozone layer: Steady-state and critical behaviour. *Atmos. Environ.*, **33**, N26, 4,243–4,254

Calm, J. M., Wuebbles, D. J., Jain, A.K. (1999) Impacts on global ozone and climate from use and emission of 2,2-dischloro-1,1,1-trifluoroethane (HCFC-123). *Climatic Change*, **42**, N2, 439–454.

Campolongo, F., Saltelli, A., Jensen, N. R., Wilson, J., Hjorth, J. (1999) The role of multiphase chemistry in the oxidation of dimethylsulphide (DMS). A latitude dependent analysis. *J. Atmos. Chem.*, **32**, N3, 327–356.

Carson, D. J. (1999) Climate Modelling: Achievements and prospects. *Quart. J. Roy. Meteorol. Soc.*, **125**, Part A, N553, 1–27.

Cebula, R. P., DeLand, M. T., Hilsenrath, E. (1998) NOAA 11 Solar Backscattered Ultraviolet, model 2 (SBUV/2) instrument solar spectral irradiance measurements in 1989–1994. 1. Observations and long-term calibration. *J. Geophys. Res.*, **103**, ND13, 16,235–16,250.

Cevolani, G., Hajduk, A., Haidukova, M., Porubcan, V., Trivellone, G. (1999) Ozone concentration at meteor heights determined from forward-scatter radar echoes. *J. Atmos. & Solar-Terrestr. Phys.*, **61**, N7, 539–543.

Chernikov, A. A., Byuro, F. D., Zvyagintsev, A. M., Ivanova, N. S., Kruchenitskii, Potanin, Yu. N. (1999). The ozone over Russia and adjacent territories in 1998. *Meteorol. & Hydrol.*, N2, 118–125 (in Russian).

Chow, J. C., Watson, J. G., Lowenthal, D. H., Egami, R. T., Solomon, P. A., Thuillier, R. H., Magliano, K., Ranzieri, A. (1998) Spatial and temporal variations of particulate precursor gases and photochemical reaction products during SJVAQS/AUSPEX ozone episodes. *Atmos. Environm.*, **32**, N16, 2,835–2,844.

Christiansen, B. (1999) Radiative forcing and climate sensitivity: The ozone experience. *Quart. J. Roy. Meteorol. Soc.*, **125** (in print).

DeLand, M. T., Cebula, P. (1998) NOAA 11 Solar Backscattered Ultraviolet, model 2 (SBUV/2) instrument solar spectral irradiance measurements in 1989–1994. 2. Results, validation and comparisons. *J. Geophys. Res.*, **103**, ND13, 16,251–16,274.

Delmas, R. J., Jaffrozo, J.-L., Dibb, J., Bales, R. (1998) Air–snow transfer function: Where are we? *IGACtivities Newsl.*, N14, 16–18.

Delmas, R. J., Legrand, M. (1998) Trends recorded in Greenland in relation with Northern Hemisphere anthropogenic pollution. *IGACtivities Newsl.*, N14, 19–22.

Dentener, F. J., Carmichael, G. R., Zhang, Y., Lelieveld, J., Crutzen, P. (1996) Role of mineral aerosol as a reactive surface in the global troposphere. *J. Geophys. Res.*, **101**, ND17, 22,869–22,890.

Derwent, R. G., Jenkin, M. E., Saunders, S. M., Pilling, M. J. (1999) Photochemical ozone creation potentials for organic compounds in Northwest Europe calculated with a Master Chemical Mechanism. *Atmos. Environ.,* **32**, N14/15, 2,429–2,441.

Derwent, R. C., Carslaw, N., Simmonds, P. G., Bassford, M., O'Doherty, S., Ryall, D. B., Pilling, M. J., Lewis, A. C., McQuaid, J. B. (1999) Hydroxyl radical concentrations estimated for measurements of trichlorethylene during the EASE/ACSOE campaign of Mace Head, Ireland, during July 1998. *J. Atmos. Chem.,* **34**, N2, 185–205.

De Santis, F. (1999) Will a new European vegetation ozone standard be fair to all European countries? *Atmos. Environ.,* **33**, N23, 3,873–3,874.

Dickerson, R. R., Rhoads, K. P., Carsey, T. P., Oltmans, S. J., Burrows, J. P., Crutzen, P. J. (1999) Ozone in the remote marine boundary layer: A possible role for halogens. *J. Geophys. Res.,* **104**, ND17, 21,385–21,396.

Durient, E., Fiorani, L., Calpini, B., Flamm, M., Saquet, L., van den Bergh, H. (1998) Tropospheric ozone measurements over the Great Athens Area during the MEDCAPHOT-TRACE campaign with a new shot-per-shot dial instrument: Experimental system and results. *Atmos. Environm.,* **32**, N12, 2,141–2,150.

Ebel, A., Memmesheimer, M., Jacobs, H. J. (1997) Regional modelling of tropospheric ozone distribution and budgets. *Proc. NATO Adv. Res. Workshop 'Ground Level and Satell. Ozone Observ. Change Mediterranean Reg.', Athens, 31 October–4 November 1995*, Springer, Berlin, pp. 37–58.

Eftathiou, M. (1999) On the statistical analysis of the total ozone content over Athens, Greece. *Phys. Chem. Earth (C)*, **24**, N5, 503–506.

Egorova, T., Zubou, V., Jagovkina, S., Rosanov, E. (1999) The simulation of atmospheric ozone distribution with a 2-D model and validation against UARS and TOMS data. *Phys. Chem. Earth (C)*, **24**, N5, 465–471.

Egorova, T., Zubou, V., Jagovkina, S., Rosanov, E. (1999) Lightning production of NOx and ozone. *Phys. Chem. Earth (C)*, **24**, N5, 473–479.

Eichmann, K.-U., Bramstedt, K., Weber, M., Rozanov, V. V., Hougen, R., Burrows, J. P. (1999) O_3 profiles from GOME satellite data-II: Observations in the Arctic spring 1997 and 1998. *Phys. Chem. Earth (C)*, **24**, N5, 453–457.

Eisinger, M., Burrows, J. P. (1998) Tropospheric sulfur dioxide observed by the ERS-2 GOME instrument. *Geophys. Res. Lett.,* **22**, N15, 4,177–4,180.

Engel, A., Schmidt, U. (1999) An estimate of the trend of the stratospheric chlorine loading based on in-situ balloon-borne observations. *SPARC Newsl.,* N13, 14–16.

Erickson, D. J. (1999) Nitrogen deposition, terrestrial carbon uptake and changes in the seasonal cycle of atmospheric CO_2, *Geophys. Res. Lett.*, **26**, N21, 3,313.

Eskes, H. J., Piters, A. J. M., Levelt, P. F., Allaart, M. A. P., Kelder, H. M. (1999) Variational assimilation on GOME total-column ozone satellite data in a 2-D latitude-longitude tracer-transport model. *J. Atmos. Sci.*, **56**, N20, 3,560–3,572.

EUROTRAC-2 (1998) *The EURECA Project on the Transport and Chemical Transformation of Environmentally Relevant Trace Constituents in the Tropo-*

sphere over Europe; Second Phase. Project Description and Handbook, Munich, 69 pp.

Evans, W. F. J., Puckrin, E. (1999) Remote sensing measurements of tropospheric ozone by ground-based thermal emission spectroscopy. *J. Atmos. Sci.,* **56**, N2, 311–318.

Fabian, C. P., Singh, O. N. (Eds.) (1999) *Reactive Halogen Compounds in the Atmosphere*, Springer, Berlin, Vol. XIII, 227 pp.

Fairlie, T. D., Pierce, R. B., Al-Saadi, J. A., Grose, W. L., Russell, J. M., III, Proffitt, M. H., Webster, C. R. (1999) The contribution of mixing in Lagrangian photochemical predictions of polar ozone loss over the Arctic in summer 1997. *J. Geophys. Res.*, **104**, ND21, 26,597.

Fehsenfeld, F. C., Meagher, J. F., Cowling, E. (1997) Integrated observations related to regional ozone formation: Results from the 1995 SOS Nashville/Middle Tennessee Study. *Third Conference on Atmospheric Chemistry, 2–7 February 1997, Preprints, Long Beach, CA*, American Meteorological Society, Boston, MA.

Fenneteaux, I., Colin, P., Etienne, A., Boudries, A., Dudot, A. L., Perros, P. E., Toupance, G. (1999) Influence of continental sources on oceanic air composition at the eastern edge of the North Atlantic Ocean, TOR 1992–1995. *J. Atmos. Chem.*, **32**, N2, 233–280.

Ferry, G. V., Pueschel, R. F., Strawa, A. W., Kondo, Y., Howard, S. D., Verma, S., Mahoney, M. J., Bui, T. P., Hannan, J. R., Fuelberg, H. E. (1999) Effects of aircraft on aerosol abundance in the upper troposphere. *Geophys. Res. Lett.*, **26**, N15, 2,399.

Fiorani, L., Calpini, B., Jaquet, L., van den Bergh, H., Duriex, E. (1998) A combined determination of wind velocities and ozone concentrations for a first measurement of ozone fluxes with a DIAL instrument during the MEDCAPHOT–TRACE campaign. *Atmos. Environm.,* **32**, N12, 2,151–2,159.

de F. Forster, P-M., Shine, K. P. (1999) Stratospheric water vapour changes as a possible contributor ot observed stratospheric cooling. *Geophys. Res. Lett.*, **26**, N21, 3,309.

Foyo-Moreno, I., Vida, J., Alados-Arboledas, L. (1998) Ground-based ultraviolet (290-985 nm) and broadband solar radiation measurements in Southern Spain. *Int. J. Climatol.*, **18**, 1,389–1,400.

Franser, P. J., Oram, D. E., Reeves, C. E., Penkett, C. E., McCulloch, A. (1999) Southern hemispheric halon trends (1978–1998) and global halon emissions. *J. Geophys. Res.,* **104**, ND13, 15,985–16,000.

Francey, R. J., Steele, L. P., Langenfelds, R. L., Pak, B. C. (1999) High precision long-term monitoring of radiatively active and related trace gases at surface sites and from aircraft in the Southern Hemisphere atmosphere. *J. Atmos. Sci.,* **56**, N2, 279–285.

Fraser, P. J., Prather, M. J. (1999) Atmospheric chemistry: Uncertain road to ozone recovery. *Nature*, **398**, N6,729, 663–664.

Friedl, R. R. (1999) Unraveling aircraft impacts. *Science*, **286**, N5,437, 57–58

Frost, G. J., Trainer, M., Mauldin, R. L., III, Eisele, F. L., Prevot, A. S. H., Flocke, S. J., Madronich, S., Kok, G., Schillawski, R. D., Baumgardner, D., Bradshaw, J. (1999) Photochemical modeling of OH levels during the First Aerosol Characterization Experiment (ACE 1). *J. Geophys. Res.*, **104**, ND13, 16,041–16,052.

Fusco, A. C., Salby, M. L. (1999) Interannual variations of total ozone and their relationship to variations of planetary wave activity. *J. Climate*, **12**, N6, 1,619–1,629.

Gan, J., Yates, S. R., Ohr, H. D., Sims, J. J. (1998) Production of methyl bromide by terrestrial higher plants. *Geophys. Res. Lett.*, **25**, N19, 3,595–3,598.

Georgantzis, N., Barreda, I. (Eds.) (1999) *Spatial Economics and Ecosystems. The Interaction between Economics and the Natural Environment*, WIT Press, Southampton, 250 pp.

Gerlach, T. M., Doukas, M. P., McGee, K. A., Kessler, R. (1999) Airborne detection of diffuse carbon dioxide emissions at Mammoth Mountain, California. *Geophys. Res. Lett.*, **26**, N24, 3,661.

Gierens, K., Jensen, E. (1998) A numerical study of the contrail-to-cirrus transition. *Geophys. Res. Lett.*, **22**, N15, 4,341–4,344.

Glatthor, N., Blom, C. E., von Clarmann, T., Fischer, H., Gulde, T., Piesch, C., Erde, F., Pfeilsticker, K., Chipperfield, M. P., Lee, A. M., Pyle, J. A. (1998) Airborne remote sensing of NO2 in the Arctic winter of 1994–1995 and comparison with a three-dimensional chemical transport model. *J. Geophys. Res.*, **103**, ND11, 13,315–13,326.

Global Aspects of Atmospheric Chemistry (1999) Springer, Berlin, Vol. X, 334 pp.

Griffin, R. J., Cocker, D. R. III, Seinfeld, J. H., Dabdub, D. (1999) Estimate of global atmospheric organic aerosol from oxidation of biogenic hydrocarbons. *Geophys. Res. Lett.*, **26**, N16, 2,721.

Grosjean, F., Grosjean, D. (1999) The reaction of unsaturated aliphatic oxygenates with ozone. *J. Atmos. Chem.*, **32**, N2, 205–232.

Guild, L. S., Kauffman, J. B., Ellingson, L. J., Cummings, D.,L., Castro, E.,A., Babbitt, R.,E., Ward, D. E. (1998) Dynamics associated with total aboveground biomass, C, nutrient pools, and biomass burning of primary forest and pasture in Rondonia, Brazil during SCAR-B. *J. Geophys. Res.*, **103**, ND24, 72,091–72,100.

Halett, J., Hudson, J. G., Rogers, F., Teets, E., Yum, S. S., Xie, Y. (1997) The influence of natural and aircraft exhaust aerosol on cirrus formation. *Third Conference on Atmospheric Chemistry, 2–7 February 1997, Preprints, Long Beach, CA*, American Meteorological Society, Boston, MA, pp. 45–53.

Hanson, H. M., Hanson, D. M. (1998) The application of the revised algorithm for the prediction of the formation of exhaust condensation trails by jet aircraft. *J. Appl. Meteorol.*, **37**, N4, 436–440.

Hauglustaine, D. A., Brasseur, G. P., Levine, J. S. (1999) A sensitivity simulation of tropospheric ozone changes due to the 1997 Indonesian fire emissions. *Geophys. Res. Lett.*, **26**, N21, 3,305.

Hauglustaine, D. A., Madronich, S., Ridley, B. A., Flocke, S. J., Cantrell, C. A., Eisele, F. L., Shetter, R. E., Tanner, D. J., Ginoux, P. (1999) Photochemistry and budget of ozone during the Mauna Loa Observatory Photochemistry Experiment (MLOPEX 2*), J. Geophys. Res.*, **104**, ND23, 30,275.

Heidam, N. Z., Wahlin, P., Christensen, J. H. (1999) Tropospheric gases and aerosols in northeast Greenland. *J. Atmos. Sci.*, **56**, N2, 261–178.

Herman, J. R., Krotkov, N., Celarier, F., Larko, D., Labow, G. (1999) Distribution of UV radiation at the Earth's surface from TOMS-measured UV-back-scattered radiances. *J. Geophys. Res.*, **104**, ND10, 12,059–12,076.

Herman, J. R., Larko, D., Ziemke, J. (1999) Regional changes in global UV reflectivity from clouds and aerosols. *J. Geophys. Res.*, **104** (in print).

Hess, P. G., Madronich, S. (1997) On tropospheric chemical oscillations. *J. Geophys. Res.*, ND102, 15,949–15,965.

Hewett, C. N. (1999) *Reactive Hydrocarbons in the Atmosphere*, Academic Press, New York, NY, 322 pp.

Holm, E. V., Untch, A., Simmons, A., Saunders, R., Bouttier, F., Andersson, E. (1999) The SODA EU Project 1996–1999: Final report from ECMWF, Abstract, 12 pp

Hood, L. L., Zhou, S. (1999) Stratospheric effects of 27-day solar ultraviolet variations: The column ozone response and comparisons of solar cycles 21 and 22. *J. Geophys. Res.*, 26,473.

Hoogen, R., Rozanov, V. V., Bramstedt, K., Eichmann, K.-U., Weber, M., Burrows, J. P. (1999) O_3 profiles from GOME satellite data – I: Comparison with ozonesonde measurements. *Phys. Chem. Earth (C)*, **24**, N5, 447–452.

Horrech, G. (1995) Quantification of the biological effectivness of environmental UV radiation. *J. Photochem. Photobiol. B: Biology*, **31**, 43–49.

Hubbard, M. C., Cobourn, W. G. (1998) Development of a regression model to forecast ground-level ozone concentration in Louisville, KY, *Atmos. Environm.*, **32**, N14, 2,637–2,647.

Iatrou, T. (1999) On the ozone content of the free troposphere over Athens, Greece as deduced from ozonesoundings. *Phys. Chem. Earth (C)*, **24**, N5, 499–501

Jacob, D. J., Logan, J. A., Murti, P. P. (1999) Effect of rising Asian emissions on surface ozone in the United States. *Geophys. Res. Lett.*, **26**, N14, 2,175.

Jaegle, L., Jacob, D. J., Ridley, B. A., Weinheimer, A. J., Brune, W. H., Talbot, R. W., Dibb, J. E., Sachse, G. W., Anderson, B. E., Collins, J. E. (1997) Upper tropospheric chemistry during SECCESS. *Third Conf. on Atmos. Chemistry, 2–7 February 1997, Preprints, Long Beach, CA*, American Meteorological Society, Boston, MA, pp. 37–39.

Jaegle, L., Jacob, D. J., Brune, W. H., Faloona, I. C., Tan, D., Kondo, Y., Sachse, G. W., Anderson, B., Gregory, G. L., Vay, S., Singh, H. B., Blake, D. R., Shetter, R. (1999) Ozone production in the upper troposphere and the influence of aircraft during SONEX: Approach of NO_x-saturated conditions, *Geophys. Res. Lett.*, **26**, N20, 3,081.

Jaffrezo, J.-L., Davidson, C. I., Kuhns, H. D., Bergin, M. H., Hillamo, R., Maenhaut, W., Kahl, J. W., Harris, J. M. (1998) Biomass burning signatures in the atmosphere of central Greenland. *J. Geophys. Res.,* **103**, ND23, 31,067–31,078.

Jenkins, G. S., Morris, V. R., Washington, R. (1997) Linking Central Africa to elevated tropospheric ozone concentrations in the tropical Atlantic: Transports of anthropogenic sources. *Third Conf. on Atmos. Chemistry, 2–7 February 1997, Preprints, Long Beach, CA*, American Meteorological Society, Boston, MA, pp. 27–28.

Jobson, B. T., McKeen, S. A., Parrish, D. D., Fehsenfeld, F. C., Blake, D. R., Goldstein, A. H., Schaulfler, S. M., Elkins, J.W. (1999) Trace gas mixing ratio variability versus lifetime in the troposphere and stratosphere: Observations. *J. Geophys. Res.,* **104**, ND13, 16,091–16,115.

Jorquera, H., Pérez, R., Cipriano, A., Espejo, A., Letelier, M. V., Acuña, G. (1998) Forecasting ozone daily maximum levels at Santiago, Chile. *Atmos. Environm.*, **32**, N20, 3,415–3,424.

Kalachev, L. V., Field, R. J. (1998) Absence of multiple steady states and a transition from steady state to monotonic growth behaviour of [CO] and [O_3] in a simple, nonlineary, oscillatory model of tropospheric photochemistry. *Geophys. Res. Lett.*, **25**, N24, 4,505–4,508.

Karcher, B., Meilinger, S. K. (1998) Perturbation of the aerosol layer by aviation-produced aerosols: A parameterization of plume processes. *Geophys. Res. Lett.*, **25**, N24, 4,465–4,468.

Kawa, S. R., Anderson, D. F. (1999) Assessing the impact of aircraft emissions on the stratosphere. *SPARC Newsl.*, N13, 21–24.

Kawamura, K., Legrand, M., Cachier, K. (1998) Organic matter in polar aerosol, snow, and ice. *IGACtivities Newsl.*, N14, 12–15.

Kent, G. S., Trepte, C. R., Lucker, P. L. (1998) Long-Term Stratospheric Aerosol and Gas Experiment I and II measurements of upper tropospheric aerosol extinction. *J. Geophys. Res.,* **103**, ND22, 28,863–28,874.

Kesselmeier, J., Staudt, M. (1999) Biogenic volatile organic compounds (VOC): An overview on emission, physiology and ecology. *J. Atmos. Chem.,* **33**, N1, 23–88.

Khalil, M. A. K. (ed.) (1999) *Atmospheric Methane*, Springer, Berlin, 335 pp.

Khosravi, R., Brasseur, G. P., Smith, A. K., Rusch, D. W., Waters, J. W., Russell, J. M. III. (1998) Significant reduction in the stratospheric ozone deficit using a three-dimensional model constrained with UARS data. *J. Geophys. Res.,* **103**, ND13, 16,203–16,220.

Kiehl, J. T., Schneider, T. L., Portmann, R. W., Solomon, S. (1999) Climate forcing due to tropospheric and stratospheric ozone. *J. Geophys. Res.*, **104**, ND24, 31,239.

Kjellstrom, E., Fiechter, J., Sausen, R., Hein, R. (1999) The contribution of aircraft emissions to the atmospheric sulfur budget. *Atmos. Environm.,* **33**, N21, 3,455–3,465.

Klemm, O., Stockwell, W. R., Schlager, H., Ziereis, H. (1998) Measurements of nitrogen oxides from aircraft in the northeast Atlantic flight corridor. *J. Geophys. Res.,* **103**, ND23, 31,127–31,229.

Klimkin, V. M., Sokovikov, V. G., Fedorishchev, V. N., Chikurov, V. A. (1999) Two types of the ozone diurnal variation in boundary atmospheric layer. *Optics of Atmos. & Ocean*, **12**, N7, 628–631.

Knyazikhin, Y., Martonchik, J. V., Diner, D. J., Myneni, R. B., Verstraete, M., Pinty, B., Gobron, N. (1998) Estimation of vegetation canopy leaf area index and fraction of absorbed photosynthetically active radiation from atmosphere-corrected MISR data. *J. Geophys. Res.*, **103**, ND24, 32,239–32,256.

Koelemeijer, R. B. A., Stammes, P. (1999) Validation of Global Ozone Monitoring Experiment cloud fractions relevant for accurate ozone column retrieval. *J. Geophys. Res.*, **104**, ND15, 18,801–18,814

Kondratyev, K. Ya. (1993) Atmospheric pollution monitoring on the European continent: EUROTRAC Programme. *Optics of Atmos. & Ocean*, **6**, N9, 1,159–1,168.

Kondratyev, K. Ya. (1998) Biosphere and Climate Relationship. In: *Approaches to Vulnerable Environment. Hydrosphere-Geosphere Interaction*, Oxford & IBH, New Delhi, Calcutta, pp. 1–21.

Kondratyev, K. Ya. (1999) *Ecodynamics and Geopolicy*. Vol. 1. *Global Problems*. Russian Academy Science, St. Petersburg, 1,040 pp. (in Russian).

Kondratyev, K. Ya., Varotsos, C. A. (1999) Total and tropospheric ozone changes: observations and numerical modelling. *Il Nuovo Cimento*, **22C**, N2, 219–246.

Kondratyev, K. Ya., Lindskog, A., Varotsos, C. (2000) TOR-2 Workshop in Athens, 19-21 September 1999: Tropospheric ozone and its precursor dynamics in Europe. *EUROTRAC Newsl.* (in print)

Krzyscin, J. W., Puchalski, S. (1998) Aerosol impact on the surface UV radiation from the ground-based measurements taken at Belsk. *J. Geophys. Res.*, **103**, ND13, 16,175–16,182.

Kumar, H. D., Häder, D. P. (1999) *Global Aquatic and Atmospheric Environment*, Springer, Berlin, Vol. XII, 393 pp.

Labitzke, K. G. (1999) *The Stratosphere. Phenomena, History, and Relevance*, Springer, Berlin, Vol. XII, 179 pp.

Lambert, J.-C., Roosendael, M., Mazière, M., Simon, P. C., Pommereau, J.-P., Goutail, F., Sarkissian, A., Gleason, J. F. (1999) Investigation of pole-to-pole performances of spaceborne atmospheric chemistry sensors with the NDSC. *J. Atmos. Sci.*, **56**, N2, 176–193.

Langford, A. O. (1999) Stratosphere-troposphere exchange at the subtropical jet: Contribution to the tropospheric ozone budget at midlatitudes, *Geophys. Res. Lett.*, **26**, N16, 2,449.

Lawrence, M. G., Crutzen, P. J., Rasch, P. J. (1999) Analysis of the CEPEX ozone data using a 3D chemistry–meteorology model. *Quart. J. Roy. Meteorol. Soc.*, **125** (in print).

Lee-Taylor, J. M., Doney, S. C., Brasseur, G. P., Mueller, J.-F. (1998) A global three-dimensional atmosphere-ocean model of methyl bromide distributions. *J. Geophys. Res.*, **103**, ND13, 16,039–16,058.

Legrand, M., Leopold, A., Domine, F. (1996) Acidic gases (HCl, HF, HNO_3, HCOOH and CH_3COOH): A review of the core data and some preliminary discussions on their air-snow relationships. *Proceedings NATO Advanced Research Workshop 'Process. Chem. Exchange: Between Atmos. and Polar Snow', 19–23 March 1995, Il Ciocco, Italy*, Springer, Berlin , pp. 19–43.

Le Prestre, P. G., Reid, D. D., Morehouse, E. T. Jr. (eds.) (1998) *Protecting the Ozone Layer. Lessons, Models, and Prospects*, Kluwer, Amsterdam, 256 pp.

Loats, K. V., Rebbeck, J. (1999) Interactive effects of ozone and elevated carbon dioxide on the growth and physiology of black cherry, green ash, and yellow-poplar seedings. *Environm. Pollution*, **106**, N2, 237–348.

Logan, J. A., Megretskaia, I. A., Miller, A. J., Tiao, G. C., Choi, D., Zhang, L., Stolarski, R. S., Labow, G. J., Hollandsworth, S. M., Bodeker, G. E., Claude, H., De Muer, D., Kerr, J. B., Tarasick, D. W., Oltmans, S. J., Johnson, B., Schmidlin, F., Staehelin, J., Viatte, P., Uchino, O. (1999) Trends in the vertical distribution of ozone: A comparison of two analyses of ozonesonde data. *J. Geophys. Res.*, **104**, ND21, 26,373.

Longo, K. M., Thomson, A. M., Kirchhoff, V. W. J. H., Remer, L. A., Freitas, S. R., Silva Dias, M. A. F., Artaxo, P., Hart, W., Spinhirne, J. D., Yamasoe, M. A. (1999) Correlation between smoke and tropospheric ozone concentration in Cuiaba during Smoke, Clouds, and Radiation–Brazil (SCAR–B). *J. Geophys. Res.,* **104**, ND10, 12,113–12,130.

Lucke, R. L., Korwan, D. R., Bevilacqua, R. M., Hornstein, J. S., Shettle, E. P., Chen, D. T., Daehler, M., Lumpe, J. D., Fromm, M. D., Debrestian, D., Neff, B., Squire, M., König-Langlo, G., Davies, J. (1999) The Polar Ozone and Aerosol Measurement (POAM) III instrument and early validation results. *J. Geophys. Res.*, **104**, ND15, 18,785–18,795.

Lyons, W. A., Armstrong, R. A. (1997) NO_x production within and above thunderstorms: the contributions of lightning and sprites. *Third Conference on Atmospheric Chemistry, 2–7 February 1997, Preprints, Long Beach, CA*, American Meteorological Society, Boston, MA, pp. 3–12.

Macalady, D. L. (ed.) (1998) *Perspectives in Environmental Chemistry*, Oxford University Press, 512 pp.

MacKenzie, I. .A., Harwood, R. S., Stott, P. A., Watson, G. C. (1999) Radiative-dynamic effects of the Antarctic ozone hole and chemical feedback. *Quart. J Roy. Meteorol. Soc.,* **125 Part B**., N558, 2,171–2,203.

Matthijsen, J., Suhre, K., Rosset, R., Eisele, F. L., Mauldin, R. L. III, Tanner, D. J. (1998) Photodissociation and UV radiative transfer in a cloudy atmosphere: Modeling and measurements. *J. Geophys. Res.* , **103**, ND13, 16,665–16,676.

Mauzerall, D. L., Jacob, D. J., Fan S.-M., Bradshow, J. D., Gregory, G. L., Sachse, G. W., Blake, D. R. (1996) Origin of tropospheric ozone at remote high northern latitudes in summer. *J. Geophys. Res.,* **101**, ND2, 4,175–4,188.

McDermid, I. S., Bergwerff, J. B., Bodeker, G., Boyd, I. S., Brinksma, E. J., Connor, B., J., Farmer, R., Gross, M. R., Kimvilakani, P., Matthews, W. A., McGee, T. J., Ormel, F. T., Parrish, A., Singh, U., Swart, D. P. J., Tsou, J. J., Wang, P. H., Zawodny, J. (1998) OPAL: Network for the detection of stratospheric change ozone profiler assessment at Lauder, New Zealand. 1. Blind inter-comparison. *J. Geophys. Res.*, **103**, ND22, 28,683–28,692.

McDermid, I. S., Bergwerff, J. B., Bodeker, G., Boyd, I. S., Brinksma, E. J., Connor, B. J., Farmer, R., Gross, M. R., Kimvilakani, P., Matthews, W. A., McGee, T. J., Ormel, F. T., Parrish, A., Singh, U., Swart, D. P. J., Tsou, J. J., Wang, P. H., Zawodny, J. (1998) OPAL: Network for the detection of stratospheric change ozone profiler assessment at Lauder, New Zealand. 2. Intercomparison of revised results. *J. Geophys. Res.,* **103**, ND22, 28,693–28,700.

McDonald, M. K., Turnbull, D. N., Donovan, D. P. (1999) Steller Brewer, ozonesonde, and DIAL measurements of Arctic O_3 column over Eureka, Northwest Territories during 1996 winter/spring. *Geophys. Res. Lett.*, **26**, N15, 2,383.

McKendry, I. G., Steyn, D. G., Banta, R. M., Strapp, W., Anlauf, K., Pottier, J. (1998) Daytime photochemical pollutant transport over a tributory valley lake in southwestern British Columbia. *J. Appl. Meteorol.,* **37**, N4, 393–404.

McKenzie, R., Connor, B., Bodeker, G. (1999) Increased summertime UV radiation in New Zealand in response to ozone loss. *Science,* **285**, N5434, 1,709–1,711.

McLinden, C. A., McConnell, J. C., McElroy, C. T., Griffioen, E. (1999) Observations of stratospheric aerosol using CPFM polarized limb radiances. *J. Atmos. Sci.,* **56**, N2, 233–240.

Mentzka, S. A., Butler, J. H., Elkins, J. W., Thompson, T. M., Clarke, A. D., Lock, L. T. (1999) Present and future trends in the atmospheric burden of ozone-depleting halogens. *Nature*, **398**, N6,729, 690.

Mickley, L. J., Murti, P. P., Jacob, D. J., Logan, J. A., Koch, D. M., Rind, D. (1999) Radiative forcing from tropospheric ozone calculated with a unified chemistry-climate model. *J. Geophys. Res.*, **104**, ND23, 30,153

Möller, D. (Ed.) (1999) *Atmospheric Environmental Research. Critical Decisions Between Technological Progress and Preservation of Nature*, Springer, Berlin, Vol. XIV, 200 pp.

Monod, A., Varlier, P. (1999) Impact of clouds on the tropospheric ozone budget: direct effects of multiphase photochemistry of soluble organic compounds. *Atmos. Environ.*, **33**, N27, 4,431–4,446.

Munro, R., Siddans, R., Reburn, W. J., Kerridge, B. J. (1998) Direct measurements of tropospheric ozone distributions from space. *Nature*, **392**, N6,672, 169–171.

Münzenberg-St. Denis, A., Renner, F. (1999) Numerical investigation of the influence of biogenic emission on ozone over Saxony (Germany). *Phys. Chem. Earth (C)*, **24**, N5, 487–490.

Nagao, I., Matsumoto, K., Tanaka, H. (1999) Sunrise ozone destruction found in the sub-tropical marine boundary layer. *Geophys. Res. Lett.*, **26**, N22, 3,377.

Noël, S., Bovensmann, H., Burrows, J. P., Frerick, J., Chance, K. V., Goede, A. H. P. (1999) Global atmospheric monitoring with SCIAMACHY. *Phys. Chem. Earth (C)*, **24**, N5, 427–434.

O'Dowd, C., McFiggans, G., Creasey, D. J., Pirjola, L., Hoell, C., Smith, M. H., Allan, B. J., Plane, J. M. C., Heard, D. E., Lee, J. D., Pilling, M. J., Kulmälä, M. (1999) On the photochemical production of new particles in the coastal boundary layer. *Geophys. Res. Lett.*, **26**, N12, 1,707–1,710.

Ollerenshaw, J. H., Lyons, T. (1999) Impacts of ozone on the growth and yield of field-grown winter wheat. *Environm. Pollution*, **106**, N1, 67–72.

Olszyna, K. J., Parkhurst, W. J., Meagher, J. F. (1998) Air chemistry during the 1995 SOS/Nashville intensive determined from level 2 network. *J. Geophys. Res.*, **103**, ND23, 31,143–31,153.

Oltmans, S. J. (1997) Distribution and trends of ozone in the troposphere. *Third Conference on Atmospheric Chemistry, 2–7 February 1997, Preprints, Long Beach, CA*, American Meteorological Society, Boston, MA, pp. 56–60.

Ozone-depleting gases in the atmosphere: Will they continue to decline? (1999) *Bull. Amer. Meteorol. Soc.*, **80**, N6, 1,176–1,177.

Pachart, E., Lenoble, J., Brogniez, C., Masserot, D., Bocquet, J. L. (1999) Ultraviolet spectral irradiance in the French Alps: Results of two campaigns. *J. Geophys. Res.*, **104**, ND14, 16,777–16,784

Papayannis, A. D., Porfenerve, J., Balis, D., Zerefos, C., Galani, E. (1999) Design of a new DIAL system for tropospheric and lower stratospheric ozone monitoring in Northern Greece. *Phys. Chem. Earth (C)*, **24**, N5, 439–442

Park, J., Cho, S. Y. (1998) A long range transport of SO2 and sulfate between Korea and East China. *Atmos. Environm.,* **32**, N6, 2,745–2,756.

Penkett, S. A., Plane, J. M. C., Comes, F. J., Clemitshaw, K. C. (1999) The Weybourne Atmospheric Observatory. *J. Atmos. Sci.,* **33**, N2, 107–110.

Pepper, D. W., Brebbia, C. A., Zanetti, P. (Eds.) (1999) *Development and Application of Computer Techniques to Environmental Studies*, WIT Press, Southampton, 400 pp.

Pérez, A., Jaque, F. (1998) On the Antarctic origin of low ozone events at the South American continent during the spring of 1993 and 1994. *Atmos. Environm.,* **32**, N21, 3,665–3,668.

Peterson, M. C., Honrath, R. E. (1999) NO_x and NO_y over the northwestern North Atlantic: Measurements and measurement accuracy. *J. Geophys. Res.,* **104**, ND9, 11,695–11,708.

Pickering, K. E., Wang, Y., Tao, W.-K., Price, K., Mueller, L.-F. (1998) Vertical distributions of lightning Nox for use in regional and global chemical transport models. *J. Geophys. Res.,* **103**, ND23, 31,203–31,216.

Pirjola, L. (1999) Effects of the increased UV radiation and biogenic VOC emissions on ultrafine sulfate aerosol formation. *J. Aerosol Science*, **30**, 355–367

Platt, U., Moortgat, C. K. (1999) Heterogeneous and homogeneous chemistry of reactive halogen compounds in the lower troposphere. *J. Atmos. Chem.,* **34**, N1, 1–8.

Poellot, M. R., Hallett, J., Arnott, W. P. (1999) In situ observations of contrail microphysics and implications for their radiative impact. *J Geophys. Res.,* **104**, ND10, 12,077–12,084.

Poppe, D., Koppmann, R., Rudolph, J. (1998) Ozone formation in biomass burning plumes; Influence of atmospheric dilution. *Geophys. Res. Lett.,* **25**, N20, 3,823–3,826.

Preston, B. L., Snell, T. W., Kneisel, R. (1999) UV-B exposure increases acute toxicity of pentachlorphenol and mercury to the rotifer *Brachionus calyciflorus*. *Environm. Pollution*, **106**, N1, 23–31.

Rabbe, A., Larsen, S. H. H. (1999) Ozone distribution in winter at high latitudes in the Northern hemisphere. *J. Atmos. & Solar–Terrestr. Phys.*, **61**, N7, 531–537.

Randell, W. J., Wu, F. (1999) Cooling of the Arctic and Antarctic polar stratospheres due to ozone depletion. *J. Climate*, **12**, N5, Part 2, 1,467–1,479.

Randel, W. J., Stolarski, R. S., Cunnold, D. M., Logan, J. A., Newchurch, M. J., Zawodny, J. M. (1999) Trends in the vertical distribution of ozone. *Science*, **285**, 1,689–1,692.

Randel, W. J., Wu, F. (1999) A stratospheric ozone trends data set for global modeling studies. *Geophys. Res. Lett.*, **26**, N20, 3,089.

Rex, M., Salawitch, R. J., Toon, G. C., Sen, B., Margitan, J. J., Osterman, G. B., Blavier, J.-F., Gao, R. S., Donnelly, S., Keim, E., Neumann, J., Fahey, D. W., Webster, C. R., Scott, D. C., Herman, R. L., May, R. D., Moyer, E. J., Gunson, M. R., Irion, F. W., Chang, A. Y., Rinsland, C. P., Bui, T. P. (1999) Subsidence, mixing, and denitrification of Arctic polar vortex air measured during POLARIS. *J. Geophys. Res.*, **104**, ND21, 26,611.

Richter, A., Wittrock, F., Eisinger, M., Burrows, J. P. (1998) GOME observations of tropospheric BrO in Northern Hemisphere spring and summer 1997. *Geophys. Res. Lett.*, **25**, N14, 2,683–2,686.

Riese, M., Tie, X., Brasseur, G., Offermann, D. (1999) Three-dimensional simulation of stratospheric trace gas distributions measured by CRISTA. *J. Geophys. Res.*, **104**, ND13, 16,419–16,435.

Roberts, J. M., Bertman, S. B., Parrish, D. D., Fehsenfeld, F. C., Jobson, B. T., Niki, H. (1998) Measurement of alkyl nitrates at Cheboque Point, Nova Scotia during the 1993 North Atlantic Regional Experiment (NARE) intensive. *J. Geophys. Res.*, **103**, ND11, 13,569–13,580.

Rosenfield, J. E., Douglass, A. R. (1998) Doubled CO_2 effects on NO_y in a coupled two-dimensional model. *Geophys. Res. Lett.*, **22**, N15, 4,381–4,384.

Rozanov, E., Zubov, V., Schlesinger, M., Yang, F., Andronova, N. (1999) Three-dimensional simulations of ozone in the stratosphere and comparison with UARS data. *Phys. Chem. Earth (C)*, **24**, N5, 459–463.

Rozanov, E. V., Zubov, V. A., Schlesinger, M. E., Yang, F., Andronova, N. G. (1999) The UIUC three-dimensional stratospheric chemical transport model: Description and evaluation of the simulated sources of gases and ozone. *J. Geophys. Res.*, **104**, ND13, 11,755–11,786.

Sahm, P., Moussiopoulos, N. (1999) The OFIS model: A new approach in urban scale photochemical modelling. *EUROTRAC Newsl.*, N21, 22–28.

Saylor, R. D., Chameides, W. L., Cowling, E. B. (1998) Implications for the new ozone National Ambient Air Quality Standards for compliance in rural areas. *J. Geophys. Res.*, **103**, ND23, 31,137–31,141.

Schulz, A., Rex, M., Steger, J., Harris, N., Braathen, G. O., Reimer, E., Alfier, R., Beck, A., Alpers, M., Cisneros, J., Claude, H., De Backer, H., Dier, H., Dorokhov, V., Fast, H., Godin, S., Hansen, G., Kondo, Y., Kosmidis, E.,

Kyrö, E., Molyneux, M. J., Murphy, G., Nakane, H., Parrondo, C., Ravegnani, F., Varotsos, C., Vialle, C., Yushkov, V., Zerefos, C., von der Gathen, P. (2000) Match observations in the Arctic winter 1996/97: High stratospheric ozone loss rates correlate with low temperatures deep inside the polar vortex, *Geophys. Res. Lett.*, **27**, N2, 205

Semenov, S. M., Kunina, I. M., Kukhta, B. A. (1998) A comparison of surface O_3, SO_2 and CO_2 concentrations anthropogenic changes with ecological criteria. *Dokl. Russian Acad. Sci.*, **361**, N2, 275–279.

Semenov, S. M., Kunina, I. M., Kukhta, B. A. (1999) *Tropospheric Ozone and Plant Growth in Europe*, Meteorology & Hydrology, Moscow, 208 pp.

Senff, C. J., Hardesty, R. M., Alvarez, R. J. II, Mayor, S. D. (1998) Airborne lidar characterization of power plant plumes during the 1995 Southern Oxidants Study. *J. Geophys. Res.*, **103**, ND23, 31,173–31,189.

Simmonds, P. G., Cunnold, D. M., Weiss, R. F., Prinn, R. G., Fraser, P. J., McCulloch, A., Alyea, F. N., O'Doherty, S. (1998) Global trends and emission estimates of Ccl4 from in situ background observations from July 1978 to June 1996. *J. Geophys. Res.*, **103**, ND13, 16,017–16,028.

Siroic, A., Pudykjcwicz, A., Kallaur, A. (1999) A comparison between simulated and observed ozone mixing ratios in eastern North America. *J Geophys. Res.*, **104**, ND17, 21,397–21,424.

Siskind, D. E., Froidevaux, L., Russell, J. M., Leon, J. (1998) Implications of upper stratospheric trace constituent changes observed by HALOE for O_3 and ClO from 1992 to 1995. *Geophys. Res. Lett.*, **25**, N18, 3,513–3,516.

Smith, W. L., Stamnes, K. (eds.) (1997) IRS '96: Current problems in atmospheric radiation. *Proceedings International Radiation Symposium, August 1996, Fairbanks, Alaska*, A. Deepak, Hampton, VA, 1,067 pp.

Soja, G., Soja, A.M. (1999) Ozone indices based on simple meteorological parameters: potentials and limitations of regression and neural network model. *Atmos. Environ.*, **33**, N26, 4,299–4,307

Solomon, S., Portmann, R. W., Sanders, R. W., Daniel, J. S., Madsen, W., Bartram, B., Dutton E.G. (1999) On thc role or nitrogen dioxide in the absorption of solar radiation. *J. Geophys. Res.*, **104**, ND10, 12,047–12,058.

Steinbrecht, W., Neuber, R., von der Gathen, P., Wahl, P., McGee, T. J., Gross, M. R., Klein, U., Langer, J. (1999) Results of the 1998 Ny-Ålesund Ozone Monitoring Intercomparison. *J. Geophys. Res.*, **104**, ND23, 30,515

Stockwell, D. Z., Chipperfield, M. P. (1999) A tropospheric chemical transport model: Development and validation of the model transport schemes. *Quart. J. Roy. Meteorol. Soc.*, **125**, Part A, N557, 1,747–1,783.

Stockwell, D. Z., Giannakopoulos, C., Plantevin, P.-H., Carver, G. D., Chipperfield, M. P., Law, K. S., Pyle, J. A., Shalcross, D. E., Wang, K.-Y. (1999) Modelling NOx from lighning and its impact on global chemical fields. *Atmos. Environ.*, **33**, N27, 4,477–4,493.

Stratospheric Processes and Their Role in Climate. Implementation Plan (1998) WCRP-105, WMO/TD-N914, Geneva, 131 pp.

Stutz, J., Heberstreit, K., Alicke, B., Platt, U. (1999) Chemistry of halogen oxides in the troposphere: Comparison of model calculations with recent field data. *J. Atmos. Chem.*, **34**, N1, 65–85.

Su, X., Mohnen, V. A. (1997) Relationships between synoptic systems and ozone at Whiteface mountain. *Third Conference on Atmospheric Chemistry, 2–7 February 1997, Preprints, Long Beach, CA*, American Meteorological Society, Boston, MA, pp. 33–35.

Sugita, T., Kondo, Y., Nakajima, H., Schmidt, U., Engel, A., Oelhaf, H., Wetzel, G., Koike, M., Newman, P. A. (1998) Denitrification observed inside the Arctic vortex in February 1995. *J. Geophys. Res.*, **103**, ND13, 16,221–16,234.

Sumner, A. L., Shepson, P. B. (1999) Snowpack production of formaldehyde and its effect on the Arctic troposphere. *Nature*, **398**, N6,724, 230–233.

Sun, J., Massman, W. (1999) Ozone transport during the California Ozone Depletion Experiment. .J *Geophys. Res.*, **104**, ND10, 11,939–11,948.

Suppan, P., Fabian, P., Vyras, L., Gryning, S. E. (1998) The behaviour of ozone and peroxyacetylnitrate concentrations for different wind regimes during the MEDCAPHOT-TRACE campaign in the Greater Area of Athens, Greece. *Atmos. Environm.*, **32**, N12, 2,089–2,102.

Svensson, G. (1997) Model study of the air quality in Athens, Greece. *Third Conference on Atmospheric Chemistry, 2–7 February 1997, Preprints, Long Beach, CA*, American Meteorological Society, Boston, MA, pp. 29–32.

Tabazadeh, A., Jacobson, M. Z., Singh, H. B., Toon, O. B., Lin, J. S., Chatfield, R. B., Thakur, A. N., Talbot, R. W., Didd, J. E. (1998) Nitric acid scavenging by mineral and biomass burning aerosols. *Geophys. Res. Lett.*, **22**, N15, 4,185–4,188.

The ozone layer: Burnt by the Sun down under (1999) *Science*, **285**, N5,434, 1,647–1,648.

Thijsse, Th. R., Roemer, M. G. M., van Oss, R. F. (1999) Trends in large-scale VOC concentrations in the Southern Netherlands between 1991 and 1997. *Atmos. Environ.*, **33**, N23, 3,803–3,812

Third Conference on Atmospheric Chemistry, 2–7 February 1997, Preprints, Long Beach, CA (1997) American Meteorological Society, Boston, MA, 151 pp.

Thomas, G. E., Stamnes, K. (1999) *Radiative Transfer in the Atmosphere and Ocean*. Cambridge University Press, Cambridge, 544 pp.

Timmreck, C., Craf, H.-F., Kirchner, I. (1999) A one and half year interactive MA/ECHAM 4 simulation of Mount Pinatubo aerosol. *J. Geophys. Res.*, **104**, ND8, 9,337–9,359.

Toon, G. C., Blavier, J.-F., Sen, B., Salawitch, R. J., Osterman, G. B., Notholt, J., Rex, M., McElroy, C. T., Russell, J. M. III (1999) Ground-based observations of Arctic O3 loss during spring and summer 1997. *J. Geophys. Res.*, **104**, ND21, 26,497.

Turunen, M., Heller, W., Stich, S., Sandermann, H., Sutinen, M. L., Norokorpi, Y. (1999) The effects of UV exclusion on the soluble phenolics of young Scots pine seedlings in the subarctic. *Environm. Pollution*, **106**, N2, 219–228.

Udelhofen, P. M., Gies, P., Roy, C., Randel, W. J. (1999) Surface UV radiation over Australia, 1979–1992: Effects of ozone and cloud cover changes on variations of UV radiation. *J. Geophys. Res.*, **104**, ND16, 19,135–19,159.

Untch, A., Simmons, A. (1999) Increased stratospheric resolution in the ECMWF forecasting system. *SPARC Newsl.*, N13, 18–21.

Varinou, M., Kallos, G., Tsiligiridis, G., Sistla, G. (1999) The role of anthropogenic and biogenic emissions on tropospheric ozone formation over Greece. *Phys. Chem. Earth (C)*, **24**, N5, 507–513.

Varotsos, C., Alexandris, D., Chronopoulos, G. (1999) On the role of the lower-stratospheric circulation to the vertical ozone structure. *Phys. Chem. Earth (C)*, **24**, N5, 481–485.

Varotsos, C. A., Kondratyev, K. Ya., Cracknell, A. P. (2000) New evidence for ozone depletion over Athens. *Int. J. Remote Sensing* (in print).

Varotsos, C., Kondratyev, K. Ya., Efstathiou, M. (2000) On the seasonal variation of the surface ozone in Athens Greece. *Atmos. Environ.* (in print).

Veretenenko, S. V., Pudovkin, M. I. (1999) Variations of solar radiation input to the lower atmosphere associated with different helio/geophysical factors. *J. Atmos. & Solar–Terrestr. Phys.*, **61**, N7, 521–529.

Walcek, C. J., Yuan, H.-H. (1997) Short- and longer-term impacts of heterogeneous HO_x chemistry on ozone formation in the troposphere. *Third Conference on Atmospheric Chemistry, 2–7 February 1997, Preprints, Long Beach, CA*, American Meteorological Society, Boston, MA, pp. 15–18.

Wang, Y., Jacob, D. J. (1998) Anthropogenic forcing on tropospheric ozone and OH since preindustrial times. *J. Geophys. Res.*, **103**, ND23, 31,123–31,135.

Wang, Y., Jacob, D. J., Logan, J. A. (1998a) Global simulation of tropospheric O_3-NO_x-hydrocarbon chemistry. 1. Model formulation. *J. Geophys. Res.*, **103**, ND9, 10,713–10,725.

Wang, Y., Jacob, D. J., Logan, J. A. (1998b) Global simulation of tropospheric O_3-NO_x-hydrocarbon chemistry. 2. Model evaluation and global ozone budget. *J. Geophys. Res.*, **103**, ND9, 10,727–10,755.

Wang, H. J., Cunnold, D. M., Froidevaux, L., Russell, J. M. (1999) A reference model for middle atmospheric ozone in 1992–1993. *J. Geophys. Res.*

Wesely, M. L., Gao, W., Cook, D. R., Martin, T. J., Qiu, J. (1997) Micrometeorological studies of NO_x and O_3 fluxes above agricultural crops. *Third Conference on Atmospheric Chemistry, 2–7 February 1997, Preprints, Long Beach, CA*, American Meteorological Society, Boston, MA, pp. 138–140.

Winter-Sorkina, R. (1997) Impact of ozone layer depletion on tropospheric ozone photodissociation rates. *Third Conference on Atmospheric Chemistry, 2–7 February 1997, Preprints, Long Beach, CA*, American Meteorological Society, Boston, MA, pp. 118–127.

Woyke, T., Müller, R., Stroh, F., McKenna, D. S., Engel, A., Margitan, J. J., Rex, M., Carslaw, K. S. (1999) A test of our understanding of the ozone chemistry in the Arctic polar vortex based on in situ measurements of ClO, BrO, and O_3 in the 1994/1995 winter. *J. Geophys. Res.*, **104**, ND15, 18,755–18,768.

Zhadin, E. A. (1999) Low-frequency variations of ozone and stratospheric circulation. *Meteorol. & Hydrol.,* N2, 68–80 (in Russian).

Ziemke, J. R., Chandra, S. (1999) Seasonal and interannual variabilities in tropical tropospheric ozone. *J. Geophys. Res.*, **104**, ND17, 21,425–21,442.

Ziemke, J. R., Chandra, S., Herman, J., Varotsos, C. (1999) Erythemally weighted UV trends over northern latitudes derived from Nimbus 7 TOMS measurements. *J. Geophys. Res.,* **104** (in print).

Ziomas, I. C., Tzoumaka, P., Balis, D., Melas, D., Zerefos, C. S., Klemm, O. (1998) Ozone episodes in Athens, Greece. A modelling approach using data from the MEDCAPHOT-TRACE. *Atmos. Environm.,* **32**, N12, 2,313–2,321.

Zuev, V. V., Marichev, V. N., Khryapov, P. A. (1999) Peculiarities of stratospheric ozone vertical distribution over Tomsk. *Optics of Atmos. & Ocean*, **12**, N7, 632–634.

Zuev, V. V., Marichev, V. N., Smirnov, S. V., Khryapov, B. A. (1999) Optical monitoring of nondisturbed ozonosphere at Siberian lidar station. *Optics Atmos. Ocean,* **12**, N10, 902–910.

Index